I enjoy sharing
my books as
I do my
friends, asking
only that you treat
them well and see
them safely home
Persinger

Aquatic Humic Substances

ADVANCES IN CHEMISTRY SERIES **219**

Aquatic Humic Substances

Influence on Fate and Treatment of Pollutants

I. H. Suffet, EDITOR
Environmental Studies Institute
Drexel University

Patrick MacCarthy, EDITOR
Colorado School of Mines

Developed from a symposium sponsored
by the Division of Environmental Chemistry
of the American Chemical Society
and the International Humic Substances Society
at the 193rd National Meeting
of the American Chemical Society,
Denver, Colorado,
April 5–10, 1987

American Chemical Society, Washington, DC 1989

Library of Congress Cataloging-in-Publication Data

Aquatic humic substances: influence on fate and treatment of pollutants/I. H. Suffet, editor, Patrick MacCarthy, editor.

p. cm.—(Advances in chemistry series, ISSN 0065-2393; 219)

"Developed from a symposium sponsored by the Division of Environmental Chemistry of the American Chemical Society and the International Humic Substances Society at the 193rd National Meeting of the American Chemical Society, Denver, Colorado, April 5-10, 1987."

Includes bibliographies and indexes.

ISBN 0-8412-1428-X

1. Humic acid—Congresses. 2. Organic water pollutants—Congresses.

I. Suffet, I. H. II. MacCarthy, Patrick. III. American Chemical Society. Division of Environmental Chemistry. IV. International Humic Substances Society. V. American Chemical Society. Meeting (193rd: 1987: Denver, Colo.) VI. Series.

QD1.A355 no. 219
[QD341.A2]
540 s—dc19
[628.1′68]

88-38029
CIP

PRINTED IN THE UNITED STATES OF AMERICA

Advances in Chemistry Series

M. Joan Comstock, *Series Editor*

FOREWORD

The ADVANCES IN CHEMISTRY SERIES was founded in 1949 by the American Chemical Society as an outlet for symposia and collections of data in special areas of topical interest that could not be accommodated in the Society's journals. It provides a medium for symposia that would otherwise be fragmented because their papers would be distributed among several journals or not published at all. Papers are reviewed critically according to ACS editorial standards and receive the careful attention and processing characteristic of ACS publications. Volumes in the ADVANCES IN CHEMISTRY SERIES maintain the integrity of the symposia on which they are based; however, verbatim reproductions of previously published papers are not accepted. Papers may include reports of research as well as reviews, because symposia may embrace both types of presentation.

ABOUT THE EDITORS

I. H. (MEL) SUFFET is P. W. Purdom Professor of Environmental Chemistry at Drexel University. He received his Ph.D. from Rutgers University, his M.S. in chemistry from the University of Maryland, and his B.S. in chemistry from Brooklyn College. Among his numerous awards are the American Chemical Society's Zimmerman Award in Environmental Science, Drexel University Research Achievement Award, Pennsylvania Water Pollution Control Association Service Award, and Instrument Society of America Service Award.

Suffet has coauthored more than 100 research papers and monograph chapters on environmental and analytical chemistry. His research expertise in the field of environmental chemistry focuses on phase equilibria and transfer of hazardous chemicals. This expertise allows him to work on the analysis, fate, and treatment of hazardous chemicals and has led to his current studies on humic materials.

He has organized and chaired numerous technical society meetings. In addition, he has served on the Safe Drinking Water Committee of the National Academy of Sciences, for which he chaired the Subcommittee on Adsorption, and was a consultant to the Water Reuse Panel of the National Academy of Sciences.

Suffet coedited, with M. Malaiyandi, Advances in Chemistry Series 214, *Organic Pollutants in Water: Sampling, Analysis, and Toxicity Testing*; with M. J. McGuire, Advances in Chemistry Series 202, *Treatment of Water by Granular Activated Carbon*; and a two-volume set, *Activated Carbon Adsorption of Organics from the Aqueous Phase*. He also edited a two-volume treatise, *The Fate of Pollutants in the Air and Water Environments*; he was a journal editor for a special issue of the *Journal of Environmental Science and Health, Part A–Environmental Science and Engineering*; and he served on the editorial board of the companion journal, *Journal of Environmental Science and Health, Part B–Pesticides,*

Food Contaminants, and Agricultural Wastes. He serves on the editorial boards of the journals *Chemosphere* and *CHEMTECH.* He has recently completed a four-year term as treasurer of the prospering ACS Division of Environmental Chemistry.

PATRICK MACCARTHY is Professor of Chemistry and Geochemistry at the Colorado School of Mines. He received his Ph.D. in analytical chemistry from the University of Cincinnati, his M.S. from Northwestern University, and M.Sc. and B.Sc. (Hons.) degrees in chemistry from University College, Galway, Ireland. MacCarthy has served on the faculty of the University of Georgia, Athens, and spent a two-year period from 1985 through 1987 on sabbatical leave at the U.S. Geological Survey in Denver.

MacCarthy has been studying the chemistry of humic substances since his introduction to this field by his first academic mentor, Seán O'Cinneide, at University College, Galway. MacCarthy's interests span a broad range of fundamental and applied aspects of humic substances research. In particular, he is interested in the characterization of humic substances and in investigating their interactions with metal ions.

As one of the founding members of the International Humic Substances Society, MacCarthy has served as secretary-treasurer of that organization since its establishment in 1982 and continues in that position today. He has also served as Chairman of the Colorado Section of the American Chemical Society.

MacCarthy has authored or coauthored more than 70 research papers and chapters in the areas of humic substances, soil science, environmental chemistry, and analytical chemistry. He is coeditor of the book *Humic Substances in Soil, Sediment, and Water: Geochemistry, Isolation, and Characterization* (Wiley-Interscience, 1985).

He is also a successful inventor, and several of his toy inventions are currently marketed worldwide.

CONTENTS

INFLUENCES OF COAGULATION PROCESSES
ON WATER TREATMENT

SORPTION ONTO ACTIVATED CARBON:
INFLUENCES ON WATER TREATMENT

INFLUENCES OF OZONATION AND CHLORINATION
PROCESSES ON WATER TREATMENT

PREFACE

HUMIC SUBSTANCES ARE UBIQUITOUS IN NATURE. They represent a unique category of natural products in which the essence of the material appears to be its heterogeneity. Because of their complex, multicomponent nature, humic substances cannot be described in specific molecular terms; rather, they must be treated in a more operational manner. Thus, when dealing with humic substances, researchers are forced by circumstances to express the amount of material in terms of organic matter or as dissolved organic carbon. Despite these difficulties in the study of humic substances, this book presents many compelling, practical reasons for their study. Furthermore, the unique difficulties associated with the study of humic substances make the investigation of these materials a particularly challenging and exciting endeavor.

The influence of aquatic humic substances on the fate and treatment of pollutants has not been explored in a cohesive and comprehensive manner, and we are only beginning to understand the multifarious ways in which humic substances influence the fate of chemicals in the aquatic environment. At present, the literature on the environmental influences of humic substances is scattered. It is frequently hidden as the secondary issue in numerous research publications. Therefore, we saw the need to consolidate the theoretical and experimental approaches to investigation of the environmental impacts of humic substances in order to develop a consistent body of knowledge in this important area. This book serves to partially fulfill that need.

Humic substances directly exert many environmental influences because of their interactions with other species in the environment. Specifically, humic substances can influence the transport of metals and the redox state of metal ions. Humic substances also interact with nonionic compounds and thus immobilize such compounds in sedimentary organic matter; conversely, the association of slightly soluble organic compounds with dissolved or suspended organic matter in the aquatic environment contributes to the mobilization of these compounds. These considerations are certainly relevant to the mobilization of pesticides and other contaminants in the environment. Whereas such association between contaminant species and humic substances can lead to a greater persistence of contaminants in the environment, the binding may actually cause them to have a decreased availability to organisms and consequently a diminished toxicity.

Understanding the environment is the essence of this discourse. It is a theme open to several frames of reference, running the gamut from understanding the natural environment to environmental protection. The concept of the environmental fate and treatment of pollutants attempts to bridge these various approaches. It illustrates how each thought process evolves and becomes interconnected within the confines of "Planet Earth". In pursuit of the fate of pollutants, we observe how one pollutant's fate can become another pollutant's source and how humic substances affect them. Humic substances are like Joseph's coat of many colors: They are different from place to place and may form and react differently, but nevertheless, they are still part of the same environmental fabric. Finding the common threads and defining and scientifically evaluating the differences are the challenges to the authors of this book.

The welcome interaction of fresh viewpoints continually reshapes the traditional disciplinary approach to science. This is especially true in the environmental area, and for the study of humic substances in particular. This treatise emphasizes the value of fruitful interdisciplinary endeavors in an attempt to break the bonds of tradition by focusing on the need to fulfill the goal of understanding our environment and to ensure environmental protection in a technological society. We took advantage of the desire of workers from different disciplines to focus their message on a common theme: humic substances. The result is the first integrated sourcebook that draws together perspectives that overlap in a synergistic manner and relate to the fate and treatment of hazardous chemicals in the environment.

Acknowledgments

This book is based on a symposium entitled "The Influences of Aquatic Humic Substances on the Fate and Treatment of Pollutants" that was held in Denver, CO, April 5–10, 1987, at the National Meeting of the American Chemical Society, in the Division of Environmental Chemistry. Individual chapters are based on invited and contributed papers that were presented at that symposium. A few additional chapters were included to give balance to the book. The symposium was sponsored jointly by the International Humic Substances Society (IHSS) and the Division of Environmental Chemistry of the American Chemical Society.

A project such as this one requires the expertise and help of many people. Unfortunately, there is not sufficient space to thank everyone. We know and appreciate how important their contributions were. The support of Ronald Malcolm, of the U.S. Geological Survey, Past President of the IHSS, is particularly acknowledged. It was his idea; we carried it out. IHSS and the Division of Environmental Chemistry of the American Chemical Society provided financial assistance for travel funds and guest registrations of renowned scientists from other countries. We are grateful for the guidance and support of each organization. Although many of the chapters in this book describe research that was fully or partially funded by U.S. federal agencies, the individual chapters do not reflect the views of the agencies and no endorsement should be inferred.

Many individuals have participated in the preparation of this book. Each chapter has been technically reviewed by at least two outside referees and the editors to ensure that the content meets the scientific rigor of a technical or review journal and that the combination of papers is within the framework of the goal of the book. We commend the reviewers for their candor. We are deeply indebted to the speakers and session chairpersons of the symposium for their contributions and efforts before, during, and after the symposium. The symposium on which this book is based was exceptionally well attended and required double the projected space that was planned to accommodate the overflow. We thank Barbara (Hodsdon) Ullyot of the ACS Meetings and Divisional Activities Department and Bob Jolley of the ACS Division of Environmental Chemistry for helping us accommodate the larger-than-expected audience that was interested in this symposium.

The task of managing the review process, as well as all production phases of this book, was expertly handled by Colleen P. Stamm of the ACS Books Department. We enjoyed working with her, Janet S. Dodd, Paula M. Bérard, and the entire ACS staff. We appreciate their patience, professionalism, and hair-shirt attitude at appropriate times.

We acknowledge several people at Drexel University and at Colorado School of Mines for their editing, typing, and clerical help. Suffet expresses his gratitude for the work of Beverly Henderson and her staff at Drexel University, who handled the correspondence and typing chores with professional dedication. MacCarthy gratefully acknowledges the valuable secretarial help of Anna Papadopoulos and Peggy Ballard at the Colorado School of Mines.

The lilting Irish brogue of the Galway native and the nasal New York accent of the Brooklyn native did harmonize, joining forces many times to accomplish a common goal and, surprisingly, without any squabbles. In fact, the biggest problem we faced, considering the equal effort that had been put forth by both editors, was the choice of whose name would appear first on the cover. In our typical approach to scientific problem solving based on undisputable statistical logic, we agreed to flip a coin transcontinentally between Golden and Philadelphia with the help of the Denver Nuggets basketball team. A game was chosen; the first author would be Suffet if Denver scored an odd number of points, MacCarthy if Denver scored an even number of points. In this prodigious battle, Denver tied the score at the buzzer of regulation time (even point score), but won in overtime by an odd point score, and the rest is folklore. Unfortunately, disputes about the nature of humic substances and their environmental influences are not so readily resolved.

We both gratefully acknowledge, with deepest affection and appreciation, the understanding, support, and encouragement of our wives, Eileen Suffet and Helen MacCarthy. They remain our sources of perspective, understanding, and patience. Without their concern and help, this book would not have been possible.

I. H. (MEL) SUFFET
Environmental Studies Institute
Drexel University
Philadelphia, PA 19104

PATRICK MACCARTHY
Department of Chemistry and Geochemistry
Colorado School of Mines
Golden, CO 80401

August 1988

To Our Wives and Families
Eileen, Alison, and Jeffrey Suffet
and
Helen, Patrick, Michael, Catherine, Mary, Cara, and Erin MacCarthy

Introduction

Aquatic Humic Substances and Their Influence on the Fate and Treatment of Pollutants

Patrick MacCarthy

Department of Chemistry and Geochemistry, Colorado School of Mines, Golden, CO 80401

I. H. (Mel) Suffet

Department of Chemistry and Environmental Studies Institute, Drexel University, Philadelphia, PA 19104

THE TERM *HUMIC SUBSTANCES* refers to organic material in the environment that results from the decomposition of plant and animal residues, but that does not fall into any of the discrete classes of compounds such as proteins, polysaccharides, and polynucleotides. Humic substances are ubiquitous, being found in all soils, sediments, and waters. Although these materials are known to result from the decomposition of biological tissue, the precise biochemical and chemical pathways by which they are formed have not been elucidated.

An endeavor to establish such pathways is a formidable task because humic substances consist of an extraordinarily complex mixture of compounds. As an indication of the molecular heterogeneity and complexity of humic substances, these materials have defied all attempts at separation into discrete components. Virtually every separation technique that has been developed by chemists and biochemists has been applied to humic substances. Many of these attempts at fractionation have succeeded in diminishing the degree of heterogeneity of the samples, but none of them comes close to isolating a material that could be referred to as a *pure* humic substance in the classical meaning of the term *pure* chemical or even a *pure* group of chemicals. In this regard, humic substances represent a unique category of natural products in which the essence of the material appears to be heterogeneity per se.

This inability to define humic substances in specific chemical terms forces us to use a more vague operational-type definition, such as that given in the opening sentence of this chapter. A more detailed discussion of the heterogeneous nature of humic substances and of the concomitant difficulties in their study is given in ref. 1.

Because of the poorly defined nature of humic substances, compared to the more discrete types of materials that most chemists and biochemists are familiar with handling, it might appear that any fundamental study of humic substances would be a rather futile endeavor. Adoption of such a viewpoint might be further provoked by the realization that there is, in fact, no analytical method for uniquely assaying humic substances. How could one develop a quantitative chemical analysis for a material that is so complex and ill-defined that researchers must be satisfied with a rather vague operational definition? Thus, when dealing with humic substances, researchers must resort to such crude measures as expressing the amount of material in terms of organic matter (e.g., as determined by loss on ignition), or expressing the concentration of humic substances as dissolved organic carbon (DOC).

Even the term *dissolved* in this context lacks the precise meaning that it usually has in chemistry. In this environmental context, *dissolved* generally refers to materials in an aqueous system that pass through a 0.45-μm (or other arbitrarily chosen pore size) filter. In reality, some of these materials may be in colloidal suspension rather than in solution.

Constraints in dealing with humic substances are evident throughout the chapters of this book. However, despite these difficulties in the study of humic substances, this book presents many compelling, practical reasons for studying humic substances. Furthermore, the unique difficulties associated with the study of humic substances make the investigation of these materials a particularly challenging and exciting endeavor.

It should not be surprising that a material that occurs in all soils, sediments, and waters exerts significant influences on many agricultural, geochemical, environmental, and pollutant-treatment processes. The agronomic importance of humus was recognized long before the chemistry of humic substances was ever addressed. Humic substances are vital for maintaining the agriculturally important crumb structure of soils, for sustaining the water regime of soils, for holding micronutrients for plant growth in the soil, and for acting as an acid–base buffer in the soil.

In geochemical and environmental contexts, humic substances can influence the transport of metals and the redox state of metal ions and other species. Humic substances interact with metal ions by complexation and ion-exchange mechanisms. Depending upon the circumstances, this interaction may lead to solubilization or immobilization of the metal.

Humic substances can also interact with nonionic compounds and thus cause the immobilization of such compounds in sedimentary organic matter;

conversely, the association of slightly soluble compounds with dissolved or suspended organic matter in the aquatic environment imbues an enhanced solubility on these compounds. Such enhanced solubility can result in an increased mobility of these compounds in rivers and other aquatic environments. These considerations clearly are relevant to the mobilization of pesticides and other contaminants in the environment.

The potential impact of a contaminant in the environment depends not only on the concentration and mobility of the species, but also on its bioavailability. The association of inorganic and organic species with humic substances can lead to a masking of these species by the organic matter, with a concomitant decrease in the toxicity of the contaminants to organisms in the environment. This association may contribute to a greater persistence of the contaminants, but to a diminished toxicity. The multifarious roles of humic substances in influencing the fate of chemicals in the aquatic environment are only beginning to be understood, and much further research remains to be done in this area.

In addition to the more direct environmental effects of humic substances resulting from their interaction with potential pollutants, these materials also have an environmental impact on water- and waste-treatment processes. For example, chlorination of indigenous humic substances during municipal water disinfection can produce carcinogenic compounds; ozonation of the biologically refractory humic substances can produce biodegradable byproducts and thereby promote microbial growth; and humic substances can compete with hazardous chemicals during the adsorption process in activated-carbon treatment, as well as cause fouling of ion-exchange resins and reverse-osmosis membranes. An understanding of the mechanisms by which humic substances interfere with treatment processes, and the development of protocols to eliminate these problems, has considerable economic advantage.

This book is directed to studies of the environmental influences of humic substances in aquatic environments, with an emphasis on how these materials affect the fate of organic and inorganic pollutants and how their presence in natural waters impacts on treatment processes.

Historical Perspective

Humic substances have been studied for more than 200 years (*2*). Most of these studies have dealt with humic substances isolated from soils, although Berzelius in the early 1800s did investigate humic substances isolated from water (*3*). Since the early 1970s, there has been a major upsurge in the number of studies dealing with humic substances isolated from water. A significant factor leading to this increased focus on aquatic humic substances was the report by Rook in 1974 (*4*) that chlorination of natural waters leads to the formation of chloroform and other potentially hazardous chlorinated hydrocarbons. Chlorination is one of the most universal disinfection processes in water and waste treatment, and many chlorinated chemicals are suspected carcinogens.

Development of new technologies for the isolation of reasonable quantities of humic substances from natural waters (5) has helped to promote research on these aquatic materials. However, many workers like to study the interaction of humic substances with pollutants without concentrating the samples, to observe their effect in a more natural state. This approach has become possible because of the ability to analyze nanogram-per-liter to microgram-per-liter concentrations of pollutants routinely by modern analytical chemical techniques. The obvious difference in the approaches is determined by the end objectives of the particular study.

Isolation of Humic Substances

Humic substances do not occur alone in the environment. Rather, they are mixed with, or chemically or physically associated with, other classes of materials. For example, in the dissolved state in natural waters, humic substances are mixed with amino acids, sugars, various aliphatic and aromatic acids, and a host of other organic compounds. In soils and sediments, and in suspension in aquatic systems, humic substances are frequently bound to the mineral components. Of course, the ultimate objective in the present context is to understand the role of humic substances in the real environment, where all of these other substances and interactions occur. Consequently, some researchers choose to work with whole water or with unextracted soil and sediment samples when evaluating the environmental impact or treatment of humic substances. Other investigators believe that all studies should actually be conducted in situ without removing the substrate from its natural location.

Most researchers isolate the humic substances from the nonhumic materials and independently evaluate the characteristics and behavior of the isolated materials. These researchers believe that a part is simpler than the whole, and that the chemistry of these materials can be evaluated more rigorously in the controlled conditions of a laboratory. In this regard, humic-substance researchers generally devote considerable effort to obtaining low-ash samples. These efforts can be contrasted with those of researchers whose primary focus is on the mineral constituents of soils and sediments and who make correspondingly intense efforts to isolate the mineral constituents with as little organic "contamination" as possible. These approaches focus upon two extremes of the true situation, and are largely a manifestation of the research interests of the individual investigators. Ideally, in the long run the data from both approaches can be merged to provide a more meaningful picture of the true environmental situation.

Over the years, many subfractions of humic substances have been isolated and given special names. Of these, only three fractions have stood the test of time as being generally useful, namely, humic acid, fulvic acid, and humin. Because of the difficulty in establishing a uniform set of definitions that is applicable to a complicated mixture that can be isolated from a solid substrate (e.g., soil) and a nominally dissolved substrate (natural waters), these three fractions are operationally defined as follows (*6*):

- **Humic acid:** the fraction of humic substances that is not soluble in water under acidic conditions (pH <2.0) but is soluble at higher pH values
- **Fulvic acid:** the fraction of humic substances that is soluble in water under all pH conditions
- **Humin:** the fraction of humic substances that is not soluble in water at any pH value

These definitions reflect the methods of isolating the various fractions from the original substrate. Although there are literally hundreds of variations of the extraction techniques for humic substances, the essential features embodied in all of these techniques are as follows.

Isolation from Solid Substrates. When dealing with a solid substrate (e.g., soil or sediment), the sample is mixed with a sodium hydroxide solution; after centrifugation, the alkaline supernate contains humic and fulvic acids in the salt form. This solution is decanted and acidified to pH 2.0 with HCl. The humic acid precipitates, and the fulvic acid, which remains in solution, is decanted after centrifugation. The humic acid and fulvic acid are then desalted by any of various techniques. The insoluble residue remaining after the alkaline extraction of the organic substrate contains the humin, generally mixed with insoluble mineral matter or undecomposed fibrous plant material. Any soluble nonhumic material that is mixed with the humic acid is removed during the subsequent washing with water during desalting of the humic acid.

The supernate that remains following acidification of the alkaline extract is more correctly referred to as the fulvic acid fraction, rather than as fulvic acid. The fulvic acid fraction contains discrete compounds such as amino acids and simple sugars, in addition to the humified material (fulvic acid). These discrete compounds can be separated from the fulvic acid by passing the fulvic acid fraction through a hydrophobic resin. The more hydrophilic compounds (e.g., amino acids, sugars) pass through the column, and the less hydrophilic constituents (fulvic acid) sorb to the resin. The fulvic acid can subsequently be eluted from the resin with dilute base. The distinction between fulvic acid and fulvic acid fraction is not always made, even within chapters of this book, but such a distinction needs to be recognized if data are to be properly interpreted and compared. Methods for extracting and fractionating humic substances from soils are discussed in refs. 7 and 8, respectively. These techniques are also applicable to sediment samples.

Isolation from Aqueous Substrates. Isolation of humic substances from natural waters is a much more formidable task, even for those waters that have a relatively high concentration of humic substances. Generally, the water is acidified to approximately pH 2.0 with HCl and is passed through a column containing an essentially hydrophobic resin, such as the methyl methacrylate cross-linked polymer, XAD–8. The humic substances sorb to the column while the more hydrophilic, nonhumic materials pass through the column. The column is then eluted at pH 7.0 to remove the fulvic acid. Subsequent elution with 0.1 M NaOH desorbs the humic acid. The humic and fulvic acids are then converted to the hydrogen form by passage through a strong cation exchanger in the hydrogen form. Any humin in the water is removed by a prior filtration. Methods for the isolation and fractionation of humic substances from waters are reviewed in refs. 5 and 9, respectively.

Because of the low concentration of humic substances in natural waters, the isolation of reasonable quantities of humic and fulvic acids can be a very time-consuming and tedious task. Nevertheless, it is a necessary and critical step in the investigation of aquatic humic substances.

Some workers have chosen to avoid the difficulty of isolating real aquatic humic substances. Instead, they have used surrogate humic acids as substitutes for aquatic humic and fulvic acids in an attempt to assess the environmental influences of humic substances in natural water. For example, commercial humic acids have been used by many researchers as analogues of true water and soil humic substances, as is evident from reading some of the chapters in this book. It has been shown (*10*) that the commercial humic acids are not representative of water or soil humic and fulvic acids; this subject is discussed further in Chapter 4 of this book.

Composition of Humic Substances

Despite the limitations in our knowledge of the chemical nature of humic substances, a great deal is known about the occurrence and composition of these materials. This compositional information affords a basis for understanding many of the environmental and geochemical effects of humic substances. Humic substances vary in composition, depending on their source, method of extraction, and other parameters. Overall, however, the similarities between different humic substances are more pronounced than their differences.

About 50% of the dissolved organic carbon in uncolored surface waters of the United States consists of humic substances. The average concentration of the humic substances in these surface waters is 2.2 mg of C/L or 4.4 mg of humic substances/L. (The humic substances are approximately 50% car-

Table I. Elemental Compositions of Representative River and Soil Humic Substances

Sample	*C*	*H*	*O*	*N*	*S*	*P*	*Total*	*Ash (%)*
Ohio River, dissolved fulvic acid	55.03	5.24	36.08	1.42	2.00	0.34	100.35	0.38
Ohio River, dissolved humic acid	54.99	4.84	33.64	2.24	1.51	0.06	98.76	1.49
Sanhedron Al, soil fulvic acid	48.71	4.36	43.35	2.77	0.81	0.59	100.59	2.25
Sanhedron Al, soil humic acid	58.03	3.64	33.59	3.26	0.47	0.10	99.09	1.19

NOTE: Elemental compositions are expressed on a percent-by-weight, ash-free, and moisture-free basis; adapted from ref. 10.

bon by weight.) Although the DOC of colored surface waters is extremely variable, it typically ranges from about 5 to more than 50 mg of C/L. Also, in these colored waters, the fraction of the total DOC in the form of humic substances varies considerably and can be as large as 80%. Typically 90% of the dissolved humic substances in natural waters consists of fulvic acid, and the remaining 10% consists of humic acid. This composition is in contrast to humic substances from soils, where the humic acid is in very large excess over the fulvic acid. These data are taken from Malcolm (*11*), who presents a detailed discussion of the occurrence and distribution of humic substances in streams.

Many workers use the terms DOC and humic substances interchangeably. Such usage is not correct and fails to make the appropriate distinction between humic substances and a mixture of humic substances plus nonhumic substances. With the growing awareness of the subtleties involved in humic substances research and terminology, important distinctions between terms such as *humic substances* and *DOC* should be more adequately accommodated in the future.

Aquatic humic acids differ from aquatic fulvic acids in elemental and functional group compositions, average molecular weights, and other characteristics. Aquatic humic and fulvic acids also differ from their corresponding soil counterparts. Elemental compositions of representative water and soil humic substances are given in Table I. The major functional groups in humic substances are carboxyl, phenolic hydroxyl, and alcoholic hydroxyl. Typical number average molecular weights of aquatic fulvic acids are 800–1000 daltons, and those of aquatic humic acids are 2000–3000 daltons. In contrast, the molecular weights of soil humic acids are reported to be as large as several hundred thousand daltons (*8*, *12*).

It is not possible to integrate all of the compositional data into neat structural models for humic substances. These data are, nevertheless, very useful in accounting for many of the environmental influences of these materials. For example, materials with a greater oxygen content will have a greater concentration of functional groups. This composition will be likely to cause the material to be more hydrophilic, and consequently less effective in the uptake of nonionic organic compounds. However, the higher concentration of oxygen-containing functional groups will likely render this material more acidic and more effective in complexing metal ions. Refs. 13–15 contain more detailed discussions of the chemistry of humic substances.

Standard Humic Substances

A major problem that has confronted researchers in the area of humic substances has been the lack of standard material through which they could objectively compare their results (*16–18*). This lack of standard material is apparently why commercial humic materials are used in research, but it has now been remedied by the International Humic Substances Society, which has established standard humic substances from water, soil, peat, and leonardite. Information relating to the nature and availability of these standard humic substances is available from the society.

Environmental Effects of Aquatic Humic Substances

Evaluation of the fate of potentially hazardous pollutants in the environment or during water treatment is concerned with how, where, in what form, and in what concentration pollutants are distributed (*19*). This concern for an understanding of pollutant impact or removal generally involves dealing with a trace concentration of a pollutant in solution (microgram-per-liter concentrations or less) within the background matrix of milligram-per-liter concentrations of other natural organic materials. The influence of the background organic matter on the fate and treatment of pollutants is potentially great and is only beginning to be understood.

Influences of aquatic humic substances on the fate and treatment of pollutants have been investigated only recently. The remainder of this introduction will indicate how humic substances behave in the aquatic environment and during treatment processes, and how they affect the fate and treatment of hazardous chemical pollutants.

Both situations are important because humic substances themselves are precursors of hazardous chemicals (e.g., chlorination of humic substances is known to produce haloforms in drinking-water treatment; humic substances also contribute undesirable color to drinking water). Humic substances can neutralize hazardous chemicals by complexation with toxic metals or by association with toxic organic chemicals. The presence of humic substances in the environment may have beneficial or deleterious consequences, and therefore an evaluation of the impacts of humic substances in each situation must be carefully conducted. Humic substances, and indeed DOC as a whole, comprise a complex mixture of chemicals that are constantly subject to change by chemical and biological forces. Therefore, only gross primary effects may remain the same from time to time. The chemistry of complex mixtures usually cannot be described as the sum of the properties of the individual constituents because intermolecular interactions can modify component characteristics (Leenheer et al., Chapter 2 in this book).

The concentration of a pollutant as it changes with time is a function of mixing and transport (i.e., turbulent diffusion and advection), emission rate, sinks, and reactivity (e.g., biological and chemical). The continuity of mass and heat must also be satisfied. The fate of hazardous trace organic chemicals in the environment, as well as during treatment processes, is becoming the main criterion for assessing pollution impact. Elucidation of the chemical, physical, and biological influences acting upon compounds in the environment is of primary and immediate concern. Therefore, the impacts of humic substances in the environment and during treatment processes must be described to gain an understanding of hazardous chemical effects. Chemical influences include physicochemical reactions of the compounds (such as sorption, coagulation, acid–base interactions, and complexation reactions) and chemical changes (oxidation, reduction, hydrolysis, and photochemical reactions). Physical influences include processes that transport or disperse chemicals. Biological influences include uptake and depuration by organisms and metabolic changes associated with receptor organisms and food chains.

Models are used to depict simplified segments of the environment or treatment processes on a scale that can be conceptualized and understood. The primary types of models that have been used to describe the behavior of chemicals in the environment are equilibrium, steady-state, and homeostatic models. Equilibrium models involve closed thermodynamic systems (time-invariant), whereas steady-state models involve open thermodynamic systems with equal input and output changes. Homeostatic models include the concept of feedback and feedforward kinetic control. The importance of humic substances in sediments and aquatic environments has only recently been recognized as a factor in the environmental reactivity and treatment of chemicals. As a result, the effects of humic substances are only now beginning to be incorporated into models of environmental fate (e.g., Caron and Suffet, Chapter 9 in this book) and treatment processes (e.g., Weber and Smith, Chapter 30 in this book).

Influences of Aquatic Humic Substances on Fate of Pollutants

The association or binding of nonpolar highly hydrophobic compounds to DOC has been observed to increase the solubility (*20*) and thus the mobility of chemical pollutants. However, it was only in the 1980s that Carter and Suffet (*21*), Landrum et al. (*22*), and others started to quantify the phenomena. The binding of nonpolar highly hydrophobic compounds to DOC was also observed to decrease sorption of these compounds to sediments and suspended sediments (*23–25*), to decrease their bioavailability to aquatic organisms (*26–30*), to decrease the volatility rate of PCBs (*31, 32*), to decrease the rate of alkaline hydrolysis of certain pesticides (*33*), and to influence the photochemistry of hazardous chemicals in natural waters (Cooper et al.; Hoigné et al., Chapters 22 and 23 in this book).

Energy from sunlight can cause photochemical changes in certain pollutants. Sunlight can also interact with constituents of natural waters to produce changes in the DOC. The DOC acts as a sensitizer or precursor for the production of reactive intermediates (photoreactants) by producing singlet oxygen, peroxy radicals, hydrogen peroxide, etc., when photooxidized. Thus, sunlight can react with hazardous organic chemicals directly by photochemical pathways or indirectly via photoreactants produced from DOC (Cooper et al.; Hoigné et al., Chapters 22 and 23 in this book).

Carter and Suffet (*34*), Kile and Chiou (Chapter 10 in this book), and others have concluded that the variability in solubility of hazardous organic chemicals in natural waters is probably caused by the inconsistent structural configuration of aquatic organic matter in the natural environment. Kile and Chiou (Chapter 10 in this book) observed that polarity, size, configuration of the DOC, and hydrophobicity of the solute were controlling factors in solubility enhancement of trace organic substances in aquatic environments.

In aquatic systems, binding of organic or inorganic contaminants to humic substances can alter the bioavailability of the contaminants (*28*). Organic contaminants associated with humic substances appear to be essentially unavailable for uptake by the biota. This decrease in bioavailable contaminant concentration decreases the toxicity of the hazardous organic chemicals in the environment. The association of nonionic organic solutes to humic substances is more pronounced for the most hydrophobic compounds (i.e., those compounds with octanol–water partition coefficients $>10^4$). In most cases, toxic metals associated with humic substances have reduced uptake and less toxic effects. However, interactions among complexing metal ions, major cations in solution, and the carrier proteins on biological membranes make it difficult to generalize and predict any reduction in accumulation and toxicity of metals.

Influences of Aquatic Humic Substances on Treatment Processes

The presence of humic substances in a water supply is undesirable for a number of reasons (Vik and Eikebrokk, and Baker et al., Chapters 24 and 31 in this book). Major problems result from the fact that humic substances:

1. produce esthetically undesirable problems such as color in the water;
2. serve as precursors of potentially hazardous trihalomethanes during chlorination processes;
3. act as precursors of other low- and high-molecular-weight chlorine-containing organic compounds that are produced during chlorination;

4. serve as precursors of low- and high-molecular-weight organic compounds that are formed by oxidation during ozonation processes;
5. stabilize dispersed and colloidal particles during coagulation processes;
6. lead to the formation of biodegradable organic compounds during ozonation and thereby enhance regrowth of microorganisms within the water-distribution systems;
7. can compete with pollutant compounds for adsorption sites in GAC adsorption; and
8. can precipitate in the distribution system; this precipitation can lead to deterioration of tap water quality and increase the need for interior cleaning of pipes.

The health effects of high concentrations of humic substances per se (mixtures containing high-molecular-weight materials) are unknown.

Water-treatment processes have been designed primarily to remove pathogens and turbidity through the use of coagulation–flocculation–sedimentation–filtration and disinfection with ozone and chlorine processes. As is true in any serial set of processes, the efficiency of the first (coagulation) process will affect subsequent processes. Because humic substances stabilize dispersed and colloidal particles, the removal of humic substances has been a focus of the coagulation process.

During coagulation, stabilization of dispersed and colloidal particles can occur through the adsorption of the higher-molecular-weight humic substances on the surface of the mineral particles. A strong interaction has often been observed between the dissolved humic substances and the flocculants. In most cases, higher humic-substance concentrations mean larger flocculant dosages and therefore higher treatment costs.

An alternative to coagulation for particle removal during water treatment is membrane filtration. The main problem with membrane filtration is irreversible fouling of the membrane by complex mixtures of inorganic materials and humic substances. The characterization and control of membrane fouling is currently under investigation (Mallevialle et al., Chapter 41 in this book).

Chlorine can combine with aquatic humic substances to form chlorinated organic compounds, such as chloroform (*3*, *35*), other low-molecular-weight disinfection byproducts (e.g., haloacetonitriles, haloacids, haloaldehydes, haloketones, chlorophenols, chloropicrin, and cyanogen chloride) (Stevens et al., Chapter 38 in this book), and complex high-molecular-weight chlorinated compounds (*36*). All of these compounds may have adverse effects on human health.

Oxidation processes also can change the chemical structure of humic substances and alter the biodegradability of these compounds. In German waterworks, ozonation is very often used to render humic substances from surface water more biodegradable. However, the biodegradable compounds need to be removed before entering the water-distribution system, where they can enhance regrowth of microorganisms (Kruithof et al., Chapter 37 in this book).

Background DOC can also influence the treatment of pollutants by reducing the capacities and rates of the carbon adsorption unit process for the removal of smaller molecules that present a known health hazard (Weber and Smith, Baker et al., Chapters 30 and 31 in this book). Thus, the lower the concentration of humic substances present in the influent water to a GAC contactor, the more efficient the process will be for pollutant removal.

Conclusions

Although humic substances have been studied for many years, it has only been since the early 1970s that a major research effort has been devoted to the investigation of aquatic humic substances and their environmental influences. As in the case of humic substances from other environments, one cannot assign unique compositions or formulas to aquatic humic substances. However, the compositional and other data that are available allow us to understand many of the environmental effects of these materials.

If evaluation of the fate of potential hazardous pollutants in the environment and during treatment is concerned with how, where, in what form, and in what concentration pollutants are distributed (*19*), then an understanding of the background DOC matrix as it directly influences the fate of pollutants must be carefully developed. At present, the understanding of the influences of humic materials is only sufficient to define site-specific effects on the pollutants. Thus, site-specific investigations must be completed to define the fate of hazardous chemicals. In the future, as information is accumulated, it is hoped that a more general understanding of the effects of humic substances on hazardous chemicals will be developed to help predict effects in unknown situations.

References

1. MacCarthy, P.; Rice, J. A. In *Humic Substances in Soil, Sediment, and Water: Geochemistry, Isolation, and Characterization*; Aiken, G. R.; McKnight, D. M.; Wershaw, R. L.; MacCarthy, P., Eds.; Wiley–Interscience: New York, 1985; pp 527–559.
2. Achard, F. K. *Crell's Chem. Annu.* **1786,** *2*, 391–403.
3. Berzelius, J. J. *Lehrbuch der Chemie,* 3rd ed.; translated by Wohler; Dresden & Leipzig, 1839.
4. Rook, J. J. *Water Treatment Exam.* **1974,** *23*, 234–243.

5. Aiken, G. R. In *Humic Substances in Soil, Sediment, and Water: Geochemistry, Isolation, and Characterization*; Aiken, G. R.; McKnight, D. M.; Wershaw, R. L.; MacCarthy, P., Eds.; Wiley–Interscience: New York, 1985; pp 363–385.
6. Aiken, G. R.; McKnight, D. M.; Wershaw, R. L.; MacCarthy, P. In *Humic Substances in Soil, Sediment, and Water: Geochemistry, Isolation, and Characterization*; Aiken, G. R.; McKnight, D. M., Wershaw, R. L.; MacCarthy, P., Eds.; Wiley–Interscience: New York, 1985; pp 1–9.
7. Hayes, M. H. B. In *Humic Substances in Soil, Sediment, and Water: Geochemistry, Isolation, and Characterization*; Aiken, G. R.; McKnight, D. M.; Wershaw, R. L.; MacCarthy, P., Eds.; Wiley–Interscience: New York, 1985; pp 329–362.
8. Swift, R. S. In *Humic Substances in Soil, Sediment, and Water: Geochemistry, Isolation, and Characterization*; Aiken, G. R.; McKnight, D. M.; Wershaw, R. L.; MacCarthy, P., Eds.; Wiley–Interscience: New York, 1985; pp 387–408.
9. Leenheer, J. A. In *Humic Substances in Soil, Sediment, and Water: Geochemistry, Isolation, and Characterization*; Aiken, G. R.; McKnight, D. M.; Wershaw, R. L.; MacCarthy, P., Eds.; Wiley–Interscience: New York, 1985; pp 409–429.
10. Malcolm, R. L.; MacCarthy, P. *Environ. Sci. Technol.* **1986,** *20*, 904–911.
11. Malcolm, R. L. In *Humic Substances in Soil, Sediment, and Water: Geochemistry, Isolation, and Characterization*; Aiken, G. R.; McKnight, D. M.; Wershaw, R. L.; MacCarthy, P., Eds.; Wiley–Interscience: New York, 1985, pp 181–209.
12. Wershaw, R. L.; Aiken, G. R. In *Humic Substances in Soil, Sediment, and Water: Geochemistry, Isolation, and Characterization*; Aiken, G. R.; McKnight, D. M.; Wershaw, R. L.; MacCarthy, P., Eds.; Wiley–Interscience: New York, 1985; pp 477–492.
13. Hayes, M. H. B.; Swift, R. S. In *The Chemistry of Soil Constituents*; Greenland, D. J.; Hayes, M. H. B., Eds.; Wiley–Interscience: New York, 1978; pp 179–230.
14. Stevenson, F. J. *Humus Chemistry: Genesis, Composition, Reactions*; Wiley–Interscience: New York, 1982; 443 pp.
15. *Humic Substances in Soil, Sediment, and Water: Geochemistry, Isolation and Characterization*; Aiken, G. R.; McKnight, D. M.; Wershaw, R. L.; MacCarthy, P., Eds.; Wiley–Interscience: New York, 1985; 692 pp.
16. MacCarthy, P. *Geoderma* **1976,** *16*, 179–181.
17. Malcolm, R. L.; MacCarthy, P. In *Trace Organic Analysis: A New Frontier in Analytical Chemistry*; Chesler, S. N.; Hertz, H. S., Eds.; Proceedings of 9th Materials Research Symposium, National Bureau of Standards: Gaithersburg, MD, 1979; pp 789–792.
18. MacCarthy, P.; Malcolm, R. L. In *Trace Organic Analysis: A New Frontier in Analytical Chemistry*; Chesler, S. N.; Hertz, H. S., Eds.; Proceedings of 9th Materials Research Symposium, National Bureau of Standards: Gaithersburg, MD, 1979; pp 793–796.
19. Suffet, I. H. In *Fate of Pollutants in the Air and Water Environment*; Suffet, I. H., Ed.; Wiley–Interscience: New York, 1977; Chapter 1.
20. Wershaw, R. L.; Burcar, P. J.; Goldberg, M. C. *Environ. Sci. Technol.* **1969,** *3*, 271–273.
21. Carter, C. W.; Suffet, I. H. *Environ. Sci. Technol.* **1982,** *16*, 735–740.
22. Landrum, P. F.; Nihart, S. R.; Eadie, B. J.; Gardner, W. S. *Environ. Sci. Technol.* **1984,** *18*, 187–192.
23. Hassett, J. P.; Anderson, M. A. *Water Res.* **1982,** *16*, 681–686.
24. Brownawell, B. J.; Farrington, J. W. In *Marine and Estuarine Geochemistry*; Sigleo, A. C.; Hattori, A., Eds.; Lewis: Chelsea, MI, 1985; pp 97–120.

25. Caron, G.; Suffet, I. H.; Belton, T. *Chemosphere* **1985,** *14*, 993–1000.
26. Leversee, G. J.; Landrum, P. F.; Giesy, J. P.; Fannin, T. *Can. J. Fish Aquat. Sci.* **1983,** *40* (*Suppl.* 2), 63–69.
27. Adams, W. J. Presented at U.S. Environmental Protection Agency Sediment Workshop, Florissant, CO, August 1984.
28. McCarthy, J. F.; Jimenez, B. D. *Environ. Toxicol. Chem.* **1985,** *4*, 511–521.
29. Landrum, P. F.; Reinhold, M.; Nihart, S. R.; Eadie, B. J. *Environ. Toxicol. Chem.* **1985,** *4*, 459–467.
30. Henry, L. Ph.D. Thesis, Drexel University, 1988.
31. Griffin, R. A.; Chian, E. S. K. *Attenuation of Water Soluble Polychlorinated Biphenyls by Earth Materials*; U.S. Environmental Protection Agency, 1980; EPA Publication EPA–600/2–80–027.
32. Hassett, J. P.; Milicic, E. *Environ. Sci. Technol.* **1985,** *19*, 638–643.
33. Perdue, E. M.; Wolfe, N. L. *Environ. Sci. Technol.* **1982,** *16*, 847–852.
34. Carter, C. W.; Suffet, I. H. In *Fate of Chemicals in the Environment*; Swann, R. L.; Eschenroeder, A., Eds.; ACS Symposium Series 225; American Chemical Society: Washington, DC, 1983; pp 215–229.
35. Bellar, T. A.; Lichtenberg, J. J.; Kroner, R. C. *J. Am. Water Works Assoc.* **1974,** *66*, 703–706.
36. Stevens, A. A.; Dressman, R. C.; Sorrell, R. K.; Brass, H. J. *J. Am. Water Works Assoc.* **1985,** *77*, 146–154.

RECEIVED September 22, 1988.

CHARACTERIZATION

1

Chemical Degradation of Humic Substances for Structural Characterization

L. B. Sonnenberg, J. D. Johnson, and R. F. Christman

Department of Environmental Sciences and Engineering, School of Public Health, University of North Carolina–Chapel Hill, Chapel Hill, NC 27514

Chemical degradations of humic substances are reviewed, with emphasis on the identity of the products and how these compounds relate to the overall structure of the humic macromolecule. The methods described include (1) reduction using metals, hydrogen, and metal hydrides; (2) oxidation with copper oxide, permanganate, and chlorine; and (3) alkaline hydrolysis. Experimental results are reported for the reductive and solvolytic degradation of isolated aquatic humic acid using metallic sodium dissolved in liquid ammonia. A number of highly oxygenated compounds, many that commonly result from oxidative procedures, are identified in the reduction mixture. The presence of these structures in the original humic macromolecule is suggested.

DEGRADATIVE TECHNIQUES CAN BE EFFECTIVE TOOLS for structural elucidation of complex natural polymers (*1*). Prediction of the types of bonds that are unstable to a particular method of degradation, coupled with identification and quantification of products, can indicate what substructures are included in the macromolecule and how they are linked together. Ideally, these techniques will produce untransformed subunits in a mechanistically predictable manner.

Degradative methods have long been used to examine the subunits that exist in humic substances and their mode of attachment to the parent molecule. Aquatic humic material has been degraded hydrolytically and oxidatively by a variety of methods. However, because of the uncertainty about

0065–2393/89/0219–0003$06.25/0

the reactions that are taking place in these complex reaction mixtures, compounds identified may not represent structures in the parent molecule. The low weight yields generally obtained in oxidative and hydrolytic work (1–25%) also make extrapolation to the macromolecular structure difficult. As yet no single analytical method, degradative or instrumental, can provide data for unequivocal characterization of humic structure. Therefore, the use of several techniques, chemical and instrumental, and the comparison and corroboration of data obtained from each of these methods is a sound approach to the complex issue of humic composition.

Although seldom applied to aquatic humic material, reductive degradation is a potentially important element in a multifaceted approach to its characterization. Reduction of humic substances can be viewed as a complement to oxidation. The two processes are similar, in that structural features affecting their outcome are likely to be similar. Because the fundamental mechanisms of the initial step of cleavage are different, however, the sites of fission and secondary reactions are not likely to be the same.

In this chapter, some important results of oxidative and hydrolytic degradation of aquatic humic material are reviewed. Because reductive degradation has not been commonly applied to aquatic humic substances, a review of methods and results for the reduction of nonaquatic humic materials and humiclike compounds is given. The application of a new reductive degradative method to aquatic humic acid is reported.

Oxidation and Hydrolysis of Aquatic Humic Materials

Some degradative procedures that have produced measurable quantities of subunits from aquatic humic and fulvic acid include alkaline hydrolysis (*2*) and oxidation using copper oxide (*3–6*), potassium permanganate (*2*, *7*), and chlorine (*8–10*). Classes of compounds that represent the majority of the compounds produced by each method are given in Table I.

Liao (*2*, *11*) exposed isolated humic and fulvic acid from two aquatic sources (Black Lake, North Carolina, and Lake Drummond, Virginia) to 0.5 N NaOH for 1.5 h to get under 2% by weight of acidic ether-extractable products. Although fission of certain activated ethers (as in lignin) and of esters might be expected, ester cleavage was postulated (*2*, *12*). The alcohols expected from the cleavage of both of these bond types were not identified, a result suggesting that the alcoholic fragments may be on a very large molecule.

Distributions of hydrolysis and permanganate oxidation products were similar, but permanganate gave 20–25% weight yields of acidic ether-extractable products. An exception to the similarity in product identities was the formation of phenylglyoxylic acids in the permanganate procedure (*2*). Analysis of permanganate degradation products indicated the presence of

Table I. Major Classes of Common Degradation Products from Aquatic Humic Material

Product Classes	*NaOH*	*CuO*	*$KMnO_4$*	*Cl_2*
Aliphatic				
monoacids	+[a]	+	+	+ +[b]
diacids	+ +	+	+ +	+ +
triacids	+		+	
Carboxylic				
benzenes	+ +	+	+ +	+ +
phenols–anisoles	+ +	+ +	+ +	
furans	+		+	+
Phenolic–anisolic		+		
Phenylglyoxylic acids			+	+
Aromatic				
aldehydes		+ +		
ketones		+ +		

NOTE: References from which this table is compiled are given in the text.
[a]Single pluses indicate only that the class of products was found in the reaction products.
[b]Two pluses indicate that the class of compounds was present as major reaction products.

aromatic compounds with three to six substituents (alkyl, ketone, carboxylic acid, or hydroxyl), short-chain aliphatics, polycyclic ring structures, and carbohydrates. These substructures were thought to be joined to the humic macromolecule through carbon–carbon bonds (2), because permanganate cleaves double bonds and alkyl side chains of aromatic compounds.

Reuter et al. (7) obtained about 12% yields of similar products for the permanganate treatment of nonmethylated humic material. Premethylated humic substances gave a higher yield with a different product distribution. Reuter et al. concluded that the Satilla River humus they degraded was more aliphatic in nature than previously believed. They based their conclusion on the prevalence of oxalic acid when premethylated humus was degraded in methylene chloride containing crown ether (for permanganate dissolution). The oxalic acid identified in the reaction mixture was also judged to indicate a high degree of unsaturation and oxygenation of the aliphatic portion of the macromolecule. Phenolic acids were determined to be the dominant form of aromatic compounds on the basis of the same experiments. The production of most compounds (including furanmonocarboxylic acid) was explained in terms of humic structure. However, furandicarboxylic acid, which was also found by Liao et al. (*11*), was considered an artifact of the oxidative treatment of premethylated humic material because its distribution varied with treatment conditions.

Although treatment of humic substances with copper oxide is a milder degradative procedure than permanganate oxidation, it has produced a number of identifiable compounds. Christman and Ghassemi (3) provided conclusive evidence for the presence of aromatic structures in aquatic humic material by using copper oxide degradation. Phenolic products, both lignin- and non-lignin-derived, were considered structural nuclei. In addition, alkyl and ether linkages were postulated on the basis of these oxidative data. Ertel and Hedges (*4*) obtained 0.8–5.2% of the original starting carbon in aquatic humic material in the form of lignin-derived phenols. The suite of phenols that were released from humic matter, including those with carboxylic, alkyl, and aldehydic functional groups, were related to the biogeochemical nature of the humic material investigated (*4, 13*).

Norwood (*5, 6*), who degraded aquatic humic material with copper oxide to carry out structural studies, found both aliphatic (*5*) and aromatic (*5, 6*) material. Although the presence of phenolic structures is confirmed through copper oxide degradation results and solid-state ^{13}C NMR spectra, their relative contribution to the overall structure of fulvic acid is small for the aquatic humic material examined, as shown by solid-state ^{13}C NMR spectra (*5, 6*). In fact, ^{13}C NMR data reported by Norwood et al. (*6*) suggested that aliphatic carbon is 51% of the total humic carbon; only 21% of the total carbon is aromatic. When alkaline copper oxide was applied to soil humic acid and the results were compared to the products of alkaline hydrolysis, it was found that the weight yields were similar, but the oxidation products were smaller and more easily identified (*14*). Degradation by copper oxide appears to be the only oxidation method reviewed to give noncarboxylic phenols. Although it is likely that these compounds exist as such in the parent molecule, decarboxylation of *o*-phenolic acids can occur with copper oxide treatment (*3, 15*).

Chlorination of aquatic material has been conducted to release humic subunits from the parent molecule, as well as to investigate its role in by-product formation during disinfection of drinking water and wastewater. Chlorination studies at high pH (*8, 16*) (using aquatic humic acid) and at neutral pH (*10*) (using aquatic fulvic acid) under high chlorine-to-carbon molar ratios gave an array of chlorinated and nonchlorinated products distributed as indicated in Table I. The weight yield of products at high chlorine-to-carbon ratios is intermediate to those of base hydrolysis and permanganate oxidation (*12*). Structural inferences made from chlorination experiments include the following: (1) a highly cross-linked structure is consistent with the formation of aromatic compounds that are highly substituted with carboxyl groups from oxidative carbon–carbon cleavage (*16*); (2) unsaturated short-chain acids are derived from ring cleavage of anisolic and phenolic compounds (*16*); (3) saturated acids arise from saturated side chains (*12*); and (4) fused ring systems in the humic macromolecule may produce phenylglyoxylic acids (*10*).

As shown in Table I, a high degree of oxygenation is found in the major products of oxidative and hydrolytic degradation of aquatic humic material. The presence of polycarboxyl functional groups is a striking feature of the aliphatic, benzoic, phenolic, and furan products. It is not surprising that oxidation and hydrolysis primarily produce compounds with high oxidation states. Whether these functional groups originate from the degradative processes or are present in the original molecule is unclear. However, data from permanganate and hydrolysis studies have been used to demonstrate method-specific structural relationships of humic material (*12*). Models of structures that could produce the amounts and types of compounds found for each degradation method are shown in Charts I and II.

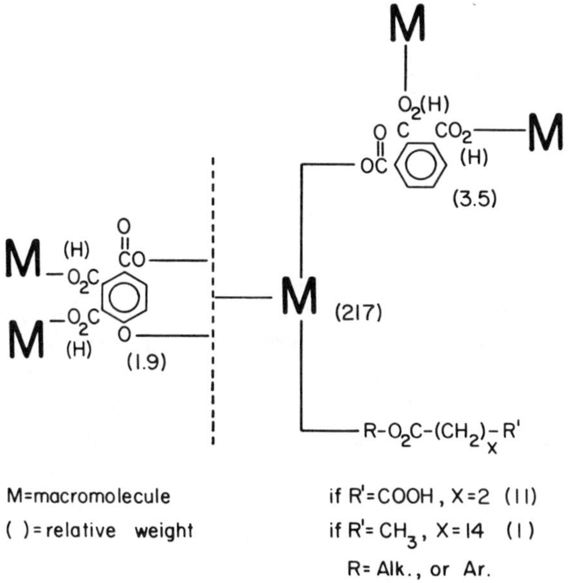

Chart I. Hypothetical structural relationships in aquatic humic material based on degradation products of alkaline hydrolysis. (Reproduced with permission from ref. 12. Copyright 1989 John Wiley & Sons.)

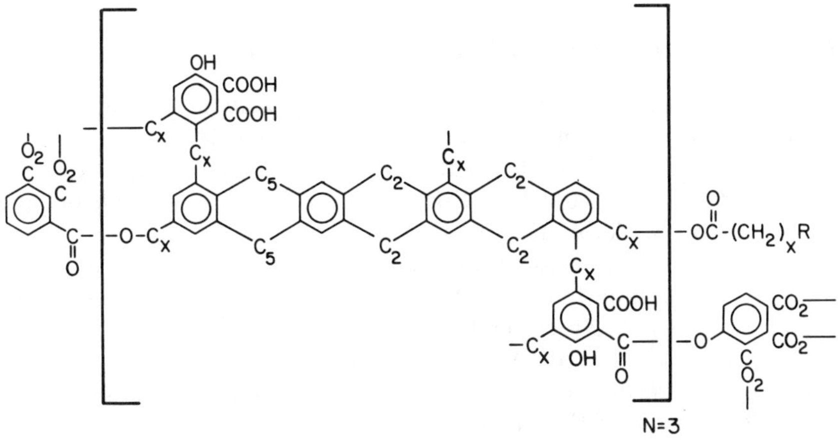

Chart II. Hypothetical structural relationships in aquatic humic material based on degradation products of permanganate. (Reproduced with permission from ref. 12. Copyright 1989 John Wiley & Sons.)

Reductive Degradation

A class of reactions used in lignin and soil humic studies but seldom applied to aquatic humic work is reductive degradation. Reduction can be achieved by a number of methods, including catalytic hydrogenation and the use of metals, metallic hydrides, and nonmetallic reducing compounds such as those containing phosphorus and sulfur. Several of these procedures have been employed to evaluate the properties of humic and humiclike material.

Zinc dust distillation of peat humic acid (*17, 18*) and soil humic acid (*19*) and zinc dust fusion of the latter humic material (*19*) yielded polyaromatic hydrocarbons. Interestingly, Chesire and various co-workers (*17, 18*) saw evidence of a polyaromatic hydrocarbon core in peat humic acid with permanganate oxidation followed by decarboxylation. These same investigators found complex aromatic compounds when peat humic acid was treated with hydrogen iodide and red phosphorus (*17, 18*).

Hydrogenation of lignin with copper chromite and Raney nickel catalysts has given evidence of guaiacyl and syringyl nuclei linked by β- and γ-ethers, as well as by propyl groups (*20–23*). Condensed aromatic ring systems with side chains linked through oxygen bridges were postulated by Kukharenko and Savelev (*24, 25*) to be constituents of the neutral fraction of humic acids hydrogenated with a nickel catalyst. Volatile and nonvolatile carboxylic acids and phenols were other products of the hydrogenation. Felbeck (*26*) degraded organic matter from a muck soil by using high-pressure hydrogenolysis followed by hydrogenation with Raney nickel. Results suggested that at least one carbon–carbon double bond exists for every four carbon atoms in the nonhydrolyzable fraction. An n–C_{25} or n–C_{26} hydrocarbon, detected in one vacuum distillate, may have arisen from chains of unsaturated heterocyclic compounds connected by carbon–carbon bonds (*26*).

Alkali Metal Reduction of Humic Materials

The reductive procedure most used for the degradation of humic material entails the use of sodium amalgam, usually in alkaline solution. Mostly soil humic acid has been degraded, although one application to aquatic organic matter is discussed. The different procedures followed and results obtained are summarized in Table II and Chart III.

Many of the sodium amalgam investigations were undertaken to differentiate sources of humic material, both on a geographic basis and with respect to their biogenesis. To these ends, a great deal of emphasis has been placed on the array of phenols produced, as indicated in Chart III. One of the first groups of investigators of sodium amalgam degradation, Burges et al. (*27, 28*), assigned most of the phenolic products to two groups: those derived from lignin and those derived from either flavonoids or soil microbial processes. The differences in thin-layer chromatograms of reaction products were deemed to be of value as a “fingerprint technique” for humic origin

Table II. Sodium Amalgam Reaction Conditions and Yields

	Humic Acid		*Na–Hg*		*Reductant–*		
Reference	*Amount*	*Source*	*Amount (g)*	*Strength (%)*	*Substrate (wt ratio)*	*Time (h)*	*Yields*[a] *(% wt)*
28	0.5 g	AH soil[b]	50	3	3.0	3	1 × 10–14
							>1 × 30–35
29	0.7–2.0 g	AH soil	50	3	2.1–0.8	4	1 × 2–5
							4 × 10–15
39	25–50 mg	UH soil[b]	30	5	0.15	3	1 × 10–12
45	250 mg	AH soil	20	5	4.0	3	2 × 5–8
46	2 g	AH soil	50	3	0.8	4	3 × 32
33	25–40 mg	soil, peat	25	5	30.0–50.0	3–4	1 × 8–20
42	0.5 g	AH soil	50	3	3.0	1–2	2 × 8–20
43	1 g	AH soil	100	5	5.0	3	1 × 12–16 (methylated extracts)

[a]Yields are based on weights of ether-extractable products. The number of sequential degradations required to produce the yields is given.
[b]AH, acid hydrolyzed; UH, unhydrolyzed.

(28,39,46) (28,33) (46) (39) (28,39,42,45,46)

(29) (28,33,39) (28,33,45,46) (28,33,39,45,46) (46) (39)

(28) (28,46) (33,39,46) (33,39) (29)

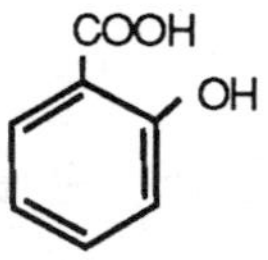

(39) 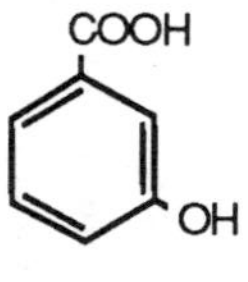 (33,39,42,46) 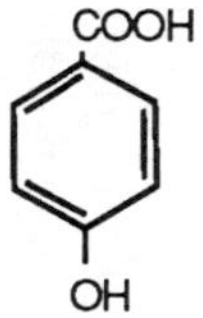 (28,33,39,46) 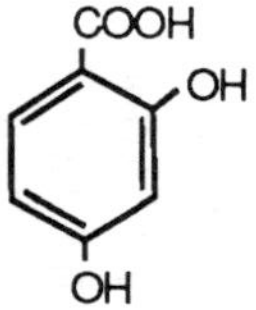 (33,39) 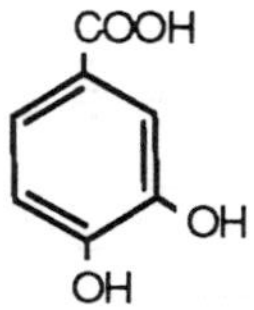 (28,33,39,42,45,46)

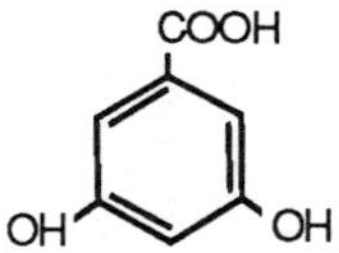

(28,33,39,45) 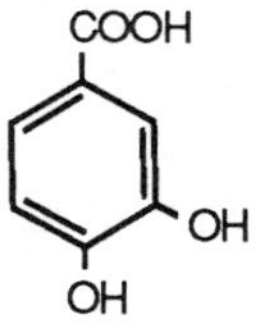 (28,29,33,39,42,45,46) 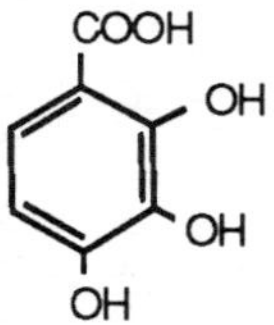 (33) 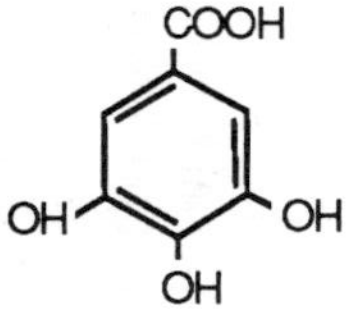 (39) 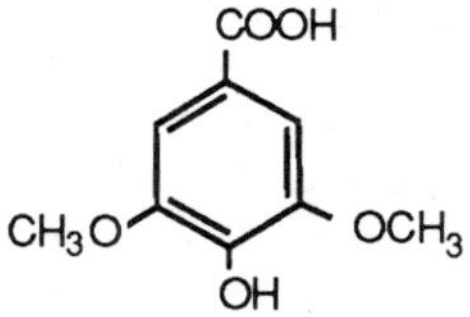 (28,29,33,39,42,45,46)

Chart III. Sodium amalgam degradation products from soil humic acid. Numbers in parentheses represent references.

classification. Degradation of cyanidin as a model flavonoid produced several fully aromatic polyhydroxylated compounds.

Evidence for a large fraction of aliphatic material was found by Mendez and Stevenson (*29, 30*) in mixtures of sodium amalgam degradation products of soil humic acid. None of the nonaromatic components were identified, and the five lignin-type phenols that were identified contributed little to the total ether-extractable yield. An optimum reaction time was 3–5 h for highest yield. However, 8-h treatments of model phenols produced significant structural alteration. Mendez and Stevenson also found that the IR spectrum of products obtained after 10 h of degradation was very similar to that of products obtained from 4-h treatment of humic acid that had previously been degraded and reisolated. In general, they were unable to match the results of Burges et al. (*28*) on the basis of yields or compounds identified.

Martin and co-workers (*31, 32*) used the sodium amalgam technique to investigate the structure of microbially synthesized humiclike polymers. They intended to evaluate the importance of the microbial process to the structure of soil humic acid. In a method investigation by Martin et al. (*33*), phenolic acids, phenols, and lignin-type phenols were identified as reaction products, as indicated in Chart III. However, they also discovered in model polymer degradation studies that a large number of these types of compounds were poorly recovered in their original form. Martin et al. referred simply to the "destruction" of the compounds, but saturation of the aromatic systems may have occurred without detection of the dihydro and tetrahydro derivatives. The saturation of aromatic compounds is well-documented in metallic reductions with the Birch method (*34, 35*) and other alkali metal methods (*36*). Helbig et al. (*37*) found hydrocaffeic acid as a product of the sodium amalgam reduction of a caffeic acid polymer; this finding indicates that aromatic saturation occurred in addition to fission.

Piper and Posner (*38*) first applied results of a sodium amalgam reduction to the question of amino acid contents of soil humic acid. Later they extended the method to study several different soils with different origins and degrees of humification (*39*). First they investigated potential cleavage patterns and tried to optimize the method to enhance phenol production from soil humic acids (*40*). Degradation of five phenolic polymers indicated to the authors that aryl–aryl ether bonds, as well as phenyl bonds of a *p*-hydroxybiphenyl polymer, were broken. Although the quantitative methods were crude, Piper and Posner claimed a lower yield of phenol from unsubstituted diaryl ether than from the *p*-hydroxy derivative. This finding is not consistent with postulated mechanisms of metallic reductive cleavage, and the authors described a possible mechanism where nascent hydrogen electrophilically attacks the oxygen in ether. They applied their optimized method (*see* Table II) to 26 humic acids of different origins. Products that were found are shown in Chart III. An average of 6% (2–10%) of the original material was recovered as nonacidic phenols (i.e., phenols without any carboxylic groups), and the average recovery of acidic phenols (i.e., phenols with carboxylic groups) was 11% (2–24%). Their final assessment of the method as a source differentiator was that, although it could detect differences in widely disparate types of humic acid, it could not distinguish materials with smaller variations. They noted that less decayed humic acid released more phenols and phenolic acids, a result suggesting an increase in resistance to degradation with increasing humification.

Matsui and Kumada (*41*) examined the action of sodium amalgam on soil humic acids with respect to production of ether-soluble compounds, loss of color, and elemental and functional group transformation of residual humic material. They also found that ether-extractable products decreased with an increase in the degree of soil humification. The production of ether-soluble compounds was so low as to be ineffective in structural elucidation, except for the least humified acids. A later work by these same researchers (*42*), in which phenolic products were examined by gas chromatography with flame ionization detection, showed that the identified phenols made up 0.2–0.6% of the original material; total ether-extractable material was 10–20%. In addition to releasing an ether-soluble fraction, the reduction unsurprisingly increased the hydrogen-to-carbon ratio while decreasing the oxygen-to-carbon ratio of the residual humic acid, except for the acid from the most humified soil. Although the authors suggested that side-chain saturation occurred along with carbonyl reduction in the macromolecule, some saturation of aromatic rings probably occurs.

Schnitzer and de Serra (*43*) reduced two soil humic acids and a fungal "humic acid" with sodium amalgam. Weight yields of methylated ether extracts were similar to those of Burges et al. (*28*) and Piper and Posner (*39*). Identified products accounted for 3–4% of the original material. They looked for but detected no flavonoid phenols and concluded that the reductive method is inefficient for structure determination, compared to permanganate degradation of premethylated humic material.

The sodium amalgam method was also judged by Dormaar (*44*) to be ineffective for complete differentiation of soil zones and horizons. He detected no flavonoid phenols either, but noted that flavone is susceptible to cleavage by alkali and found phenol after treating flavone with reaction conditions described in Table II.

Conversely to Schnitzer and de Serra (*43*) and Dormaar (*44*), Tate and Goh (*45*) found reductive degradation to be an effective tool for differentiating three soils on the basis of phenolic products formed, even though overall yields were low. Thin-layer chromatographic patterns of the reductive products for the soils were distinctly different. They also found that reisolation of the remaining humic residue after reduction and subjecting it to subsequent reductions gave product yields and distributions similar to the initial reduction. This result suggests that the original material is homogeneous, at least with respect to its degradability. An expected increase in aliphatic

IR absorption of the residue after reduction was accompanied by an inexplicable increase in a C–O stretch of ketonic and carboxylic C–O.

Peat humic material was also reduced by Zhorobekova (*46*). Using the Piper and Posner (*39*) method of sodium amalgam reduction, he released phenols and phenolic acids that were considered indicative of different biogenetic series.

Aquatic organic matter was degraded with sodium amalgam in alkaline solution by Hall and Lee (*47*). They were unable to identify any of the compounds isolated, but they obtained a 15–19% ether-extractable yield. By comparing IR and UV spectra of degradation products from copper oxide and sodium amalgam procedures, they were able to make the following conclusions: the reductive process is less destructive to saturated carbon chains than the oxidative procedure, ether linkages are present in the macromolecule, and ester and hydrocarbon chains are possible interaromatic structures. They noted that the aliphatic components produced from reduction could be derived from cleavage of flavonoid structures rather than exist as direct bridges to the aromatic nuclei.

The majority of structural analyses of complex natural polymers using sodium-in-ammonia reductive cleavage have been applied to lignan. Shorygina and various co-workers (*48, 49*) have systematically used this method to ascertain the presence of various lignin subunits. Usually the lignin was treated with sodium in ammonia for up to several days. The residue remaining after solvent evaporation was covered with wet ether for a day, after which time water was added. A great deal of information about the phenylpropane subunits of lignin has been obtained by using the sodium-in-ammonia method of degradation. One reduction released 2.6% ether-soluble products from the alkaline solution, and nine treatments gave an 8.4% yield (*48*).

More recent lignin work with the sodium-in-ammonia method entailed the use of gas chromatography with flame ionization to isolate and identify simple phenolic, syringyl, and guaiacyl compounds, some of which had unsaturated propyl side chains. Some of these aromatic compounds had hydroxyls on the α or γ position of the propyl side chain (*50, 51*). Significant ether extract yields (about 10%) were obtained at both pH 8 and 2. Chloroform (at pH 2) extracted an additional 5% of product weight from hydrochloric acid lignin of cotton plant stems (*51*).

An almost direct application of the lignin method to soil and coal humic acids was conducted by Maximov and Kraskovskaya (*52*). For each treatment conducted, 9–12% of the total starting soil humic acid was obtained as water-soluble products. After five reductions only 43–45% of the undegraded humic residue remained. The ether extracts of the reduced soil humic acid contained hydroxybenzoic and dihydroxybenzoic acids (1.4%) and polycarboxylic benzene (0.7%), with no detection of nonacidic phenols. About half of the ether-soluble products were "acid oligomeric degradation products" that were not gas chromatographable. Cleavage of aryl ether linkages in the original macromolecule to release the phenolic acids was hypothesized. Maximov and Krasovskaya (*52*) concluded that the carboxylic groups on these aromatic compounds were present in the original molecule.

Preliminary Study of the Application of Alkali Metal Reduction to Aquatic Humic Acid

Background. Alkali metal reduction of aquatic humic acid was selected for investigation for several reasons. The chemical behavior of alkali metal is conducive to humic acid structural characterization. It reductively cleaves bonds that are likely to be prevalent in humic material (*53, 54*). Selectivity options available through alteration of reaction conditions also make it attractive. In addition, alkali metal reductions have been applied to nonaquatic humic material by other investigators with some success. There appear to be many areas where the process can be enhanced and optimized by careful method development and product characterization.

Metallic reductions probably initially involve the transfer of "free" or solvated electrons to the reactive molecule to form radical anion or dianion intermediates, with subsequent additions of protons (*55–57*). The electrons are provided by the conversion of atomic metals to their cationic form. This conversion can be achieved either in solution or from a metallic surface. Although it has been suggested that "nascent" hydrogen may be a second actively reducing species in dissolving metal reactions (*58*), Shaede and Walker (*59*), using kinetic studies of sodium amalgam in water, showed that solvated electrons predominate over the hydrogen species. The mechanism of reductive fission implicates several parameters as important to the outcome of the reaction: the solvent(s) (for dissolution of reductant and substrate and for stabilization of the intermediate), the proton donor, the reductant, and the substrate.

Structure elucidation applications have developed chiefly through the fission capabilities of the method (*60*). Linkages relevant to humic substance structure that tend to be susceptible to cleavage by metallic reduction are given in List I and were compiled chiefly from ref. 61. The major category of labile bonds are ethers activated by unsaturation. Examples of structural investigations include the fission of benzyl alkyl ethers to determine the structure of the lignin gmelinol (*53*) and benzyl phenyl ether cleavage of the methylated flavonols, catechin and epicatechin (*54*). These metallic reductions were instrumental in final assignments of these fairly complex natural compounds.

Experimental Methods. The reaction apparatus consisted of a three-necked round-bottom flask fitted with a cold-finger condenser in the center neck. Anhydrous, gaseous, ultrapure ammonia was admitted into the flask through the

List I. Linkages Cleaved by Metallic Reduction

Ethers	*Alcohols*
Alkyl–vinyl	Allyl
Allyl–aryl	Benzyl
Alkyl–benzyl	
Aryl–alkyl	*Other*
Aryl–benzyl	Activated aliphatic
Diaryl	Esters
Dibenzyl	
Diallyl	
Catechol acetates	

condenser and was allowed to escape the vessel, in gaseous form, through tubing fitted to a second neck of the flask. The discharge tubing contained soda lime and calcium chloride traps to minimize atmospheric water and carbon dioxide in the reaction mixture. The reaction vessel was charged with the ammonia for about 1 min before dry ice–acetone was introduced into the well of the condenser to collect about 50 mL of ammonia into the round-bottom flask containing the humic acid. The sodium was then added in approximately 0.5-g portions through the third neck of the flask. An excess of sodium was considered desirable so that the reaction would not become reductant-limited. This excess was achieved with 2 g of sodium in 50–75 mL of ammonia and up to 0.5 g of humic acid.

After about 4 h, the ammonia was allowed to evaporate at room temperature until a damp sludge consisting of products, solid humic acid, unreacted metal, and a small amount of ammonia remained. Acidified water (50 mL of distilled deionized water with 15 mL of HCl) was then very slowly and carefully added to the mixture under a nitrogen flow. The acidic water served three purposes: (1) to provide a strong proton donor after electron addition; (2) to precipitate unreacted, acid-insoluble humic acid; and (3) to quench any active sodium. A thick slurry that resulted (pH <1) was centrifuged. Diethyl ether was the primary solvent used to extract (three times with 25 mL) the clear yellow, aqueous solution obtained after centrifugation. Gas chromatography–flame ionization detection demonstrated that a second extraction with ethyl acetate yielded little additional product. The extracts were dried by freezing out the water with dry ice–acetone and placing over magnesium sulfate overnight. Diazomethane was used to methylate the extracts, which were concentrated either by nitrogen blow-down or rotoevaporation.

Gas chromatography was carried out on a 15-m capillary column (DB–1701; J&W Scientific, Inc.); helium served as a carrier gas, with a flow of about 2 mL/min. Temperature programming for ether extracts consisted of an initial temperature of 35 °C, a 1-min hold, a 4 °C/min ramp to 290 °C, and a final hold of 10 min. Flame ionization screening was conducted on a gas chromatograph (Carlo Erba HRGC 5160) equipped with a flame ionization detector. Initial mass spectrometric analyses were conducted on a quadrupole (Hewlett Packard 5985) and data system in 70-eV electron-impact mode. Tentative identifications were based on a greater than 80% match of sample mass spectra with library mass spectra. Later analyses were conducted on a quadrupole mass spectrometer (VG 12–250, with a PDP11/25 Plus data system). Identifications were made on the basis of electron-impact and chemical-ionization mass spectra, manual match of literature spectra with sample spectra, and fragmentation analyses.

SOLVOLYSIS EXPERIMENTS. Humic acid is likely to be susceptible to degradation by the highly basic ammonia medium in which the reduction occurs; therefore, solvolytic experiments were conducted in addition to the reduction procedures. The procedure just described was followed, except that the reductant was never added for the solvolysis experiments. Although reagent blanks (no humic acid) were clean, the solvolysis samples were not "blank". Exposing humic acid to ammonia causes a significant production of chromatographable material. Preliminary experiments have repeatedly shown that it is necessary to conduct solvolysis with reduction experiments if the reduction products are to be differentiated from the ammonolysis products. Indeed, although early experiments suggest a significant degree of reduction over solvolysis (Figure 1), recent work shows a striking similarity between reduction and solvolysis chromatograms (Figure 2). Because humic acid is known to be susceptible to alkaline hydrolysis, which must occur during oxidations, it was postulated that an analogous situation may exist with the ammonia–reductant system. The strongly basic amide anion, generated during the course of the reduction, may be an actively cleaving species similar to the hydroxide anion in hydrolyses. Exposing humic acid

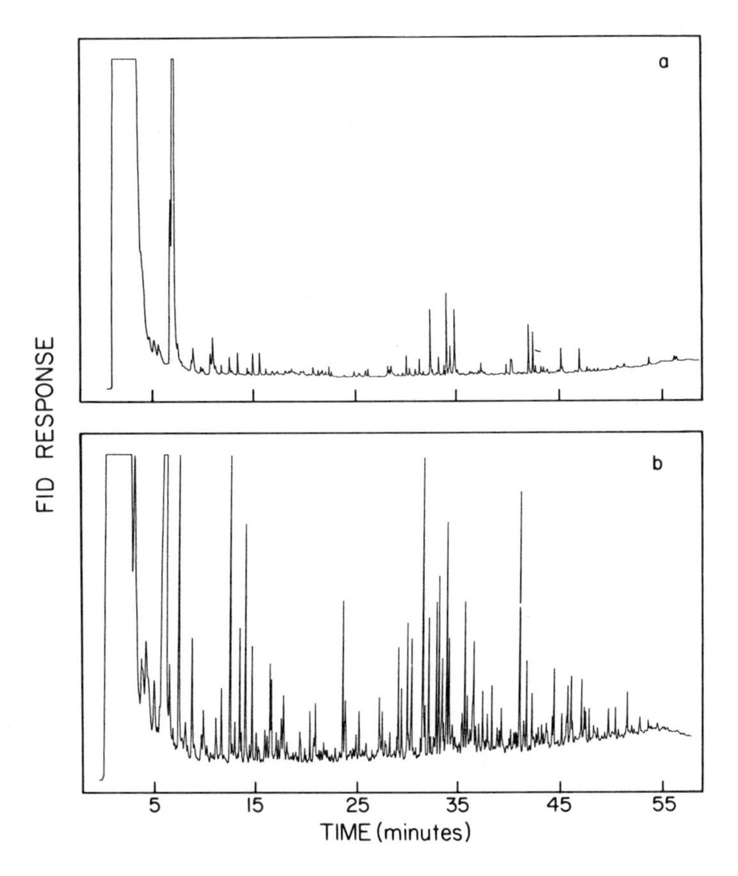

Figure 1. Flame ionization chromatograms of degradation products of aquatic humic acid: (a) treatment with ammonia only and (b) treatment with ammonia and sodium.

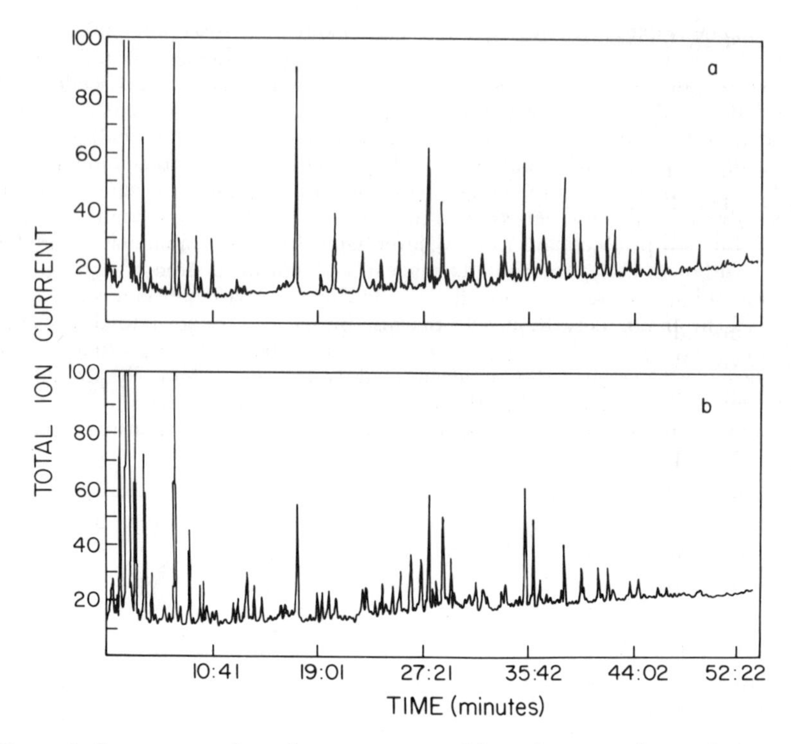

Figure 2. Reconstructed gas chromatograms of degradation products of aquatic humic acid: (a) treatment with ammonia only and (b) treatment with ammonia and sodium.

in ammonia to excess sodium amide produced many of the same compounds but not in the same relative concentrations. Therefore, the amide anion does not appear to be the sole species responsible for product generation in the solvolysis reaction.

PROTON DONOR EXPERIMENTS. An experiment was conducted in which ammonia was the only proton donor available throughout the sample work-up. After the reduction was performed, the residual humic material, together with the products, and the unreacted sodium were washed three times with anhydrous ether. The ether washes were combined, concentrated, derivatized, and analyzed by using GC/FID. Few products were found. The preliminary interpretation of these results is that few anionic intermediates are produced that are basic enough to abstract a proton from the ammonia (pK_a = 23).

Results and Discussion. At this preliminary stage of experimentation, the extent to which reduction occurs over ammonolysis has not been completely addressed. However, the sodium in ammonia can produce a large number of compounds, as indicated in Figures 1 and 2. The chemical classes of major products from ammonolysis–reduction are shown in List II, and the specific compounds identified are shown in Table III. Most of

List II. Classes of Major Products of Ammonolysis–Reduction of Aquatic Humic Acid

- *Dibasic aliphatic acids*: C_2–C_9
- *Benzoic acids*: mono-, di-, tri-, tetraacids
- *Phenolic–anisolic acids*: hydroxylated mono-, diacids; methoxylated mono- to tetraacids
- *Acetates*: propyl, others

the compounds identified have been found in nonreductive degradation procedures. Regardless of the mechanism of cleavage in the ammonolysis–reduction reactions, it is interesting that so many of these highly oxidized compounds (similar to those identified in oxidative and hydrolytic degradations) have been produced in a reductive atmosphere. The similarities imply that these types of subunits are present as such in the humic macromolecule and that their highly oxidized functional groups are not introduced simply by oxidative and hydrolytic processes.

Some of the products, which need further confirmation, promise to yield new information about the aliphatic nature of aquatic humic material. These products include the aliphatic alcohols and the acetates. Interaromatic and terminal structures may be implicated. The very-long-chain aliphatic mono- and diacids that have been identified in other procedures have not been identified in ammonolysis–reduction procedures yet. Simple saturated aliphatic esters are subject to base hydrolysis, although they tend to be resistant to cleavage by ammonia (*61*).

The possibility of the formation of nonacidic phenols tends to corroborate findings of Norwood (*5, 6*) and supports the conclusion that some of the aromatic compounds are not acidic or are not joined to the macromolecule through phenyl ester bonds. Of course, bonding of the form where the aromatic is on the alcoholic portion of an ester linkage is not precluded by the presence of nonacidic phenols. Indeed, perhaps phenols are the "missing" alcohols from postulated ester cleavages. The source of the small number of nonaromatic hydroxylic compounds identified may be esters or ethers. Their presence in the ammonolysis reaction would suggest that the former type of bond is cleaved. However, an apparent increase in concentration occurs in the monoalcohol when going from solvolysis to the reduction, so perhaps this net production represents fission of ethers.

Finally, the polycarboxylated aromatic compounds found show that these functional groups may not necessarily represent a high level of carbon–carbon cross-linking in the humic macromolecule (*16*). They may be present as free carboxyls, as concluded by Maximov and Krasovskaya (*52*), or as esters in the original structure. Quantitative investigations will address this issue further.

Table III. Ammonolysis–Reduction Products of Aquatic Humic Acid

	Also Found After			
Methylated Ammonolysis Reduction Products	*NaOH*	*CuO*	*MnO_4*	*Cl_2*
Aliphatic compounds				
2-butoxyethanol[a]				
2,3-butanediol[b]				
propyl acetate[a]				
propyl diacetate[a]				
other acetates[a]				
propanedioic acid dimethyl ester[a]	×		×	×
butanedioic acid dimethyl ester[c]	×	×	×	×
pentanedioic acid dimethyl ester (two isomers)[c]	×		×	×
hexanedioic acid dimethyl ester[a]	×		×	×
octanedioic acid dimethyl ester[a]	×		×	×
nonanedioic acid dimethyl ester[a]	×		×	×
4-oxopentanoic acid methyl ester[c]		×	×	
Aromatic compounds				
phenol[b]				
dimethylbenzene[b]				
dimethoxybenzaldehyde[b]				
dimethoxybenzenediacetic acid dimethyl ester[a]				
benzeneacetic acid methyl ester[b]				
Phenolic–anisolic acids				
hydroxybenzoic acid methyl ester (three isomers)[b]		×	×	
hydroxymethoxybenzoic acid methyl ester[c]		×		
methoxybenzoic acid methyl ester (two isomers)[c]	×		×	
dimethoxybenzoic acid dimethyl ester (two isomers)[c]	×		×	
hydroxybenzenedicarboxylic acid dimethyl ester (three isomers)[a]	×		×	
methoxybenzenedicarboxylic acid dimethyl ester[a]	×		×	
methoxybenzenetricarboxylic acid trimethyl ester (four isomers)[a]	×		×	
methoxybenzenetetracarboxylic acid tetramethyl ester (two isomers)[a]	×		×	
Benzoic acids				
benzoic acid methyl ester[c]	×	×	×	×
benzenedicarboxylic acid dimethyl ester (three isomers)[c]	×		×	×
benzenetricarboxylic acid trimethyl ester (two isomers)[c]	×		×	×
benzenetetracarboxylic acid tetramethyl ester (two isomers)[a]	×		×	×
benzenedicarboxylic acid butylphenyl ester[b]				

[a]Identified by >80% match of low-resolution EI spectra with library spectra.
[b]Identified by low-resolution EI, CI, fragmentation analysis, and manual match of literature spectra.
[c]Identified by both methods in footnotes *a* and *b*.

Conclusions

Oxidative, hydrolytic, and reductive degradations of humic substances release a number of identifiable compounds. Because the products from these disparate processes are so similar, these compounds probably represent subunits of the humic macromolecule. The presence of highly oxidized compounds, such as polycarboxylic phenols, in a reduction mixture corroborates information obtained from oxidative and hydrolytic studies. The supposition that these types of compounds exist as such in the original macromolecule is supported. The modes of attachment of these subunits to the parent molecule are not yet completely understood. However, a mechanistic interpretation of the products formed from various degradation methods suggests that the major covalent linkages involved are ether and alkyl bonds activated by unsaturation, and esters.

Much of the information gleaned from individual chemical degradation studies is qualitative, because of the low and variable product yields often obtained. More comprehensive product isolation and the performance of repetitive degradation experiments may be useful. These techniques could maximize yields and thereby minimize constraints on structural interpretations.

Comparison of the results obtained from a variety of methods, both instrumental and chemical, can provide new and corroborative structural information. Reductive degradation should be an important element in a comprehensive effort to characterize humic structure.

References

1. Thurmond, E. M.; Malcolm, R. L. In *Aquatic and Terrestrial Humic Materials*; Christman, R. F.; Gjessing, E. T., Eds.; Ann Arbor Science: Ann Arbor, 1983; pp 1–23.
2. Liao, W. T. Ph.D. Thesis, University of North Carolina–Chapel Hill, 1981.
3. Christman, R. F.; Ghassemi, M. *J. Am. Water Works Assoc.* **1966,** *58*, 723–741.
4. Ertel, J. R.; Hedges, J. I.; Perdue, E. M. *Science (Washington, D.C.)* **1984,** *223*, 48–55.
5. Norwood, D. L. Ph.D. Thesis, University of North Carolina–Chapel Hill, 1985.
6. Norwood, D. L.; Christman, R. F.; Hatcher, P. G. *Environ. Sci. Technol.* **1987,** *21*, 791–798.
7. Reuter, J. H.; Ghossal, M.; Chian, E. S. K.; Giabbai, M. In *Aquatic and Terrestrial Humic Materials*; Christman, R. F.; Gjessing, E. T., Eds.; Ann Arbor Science: Ann Arbor, 1983; pp 107–126.
8. Johnson, J. D.; Christman, R. F.; Norwood. D. L.; Millington, D. S. *EHP, Environ. Health Perspect.* **1982,** *46*, 63–71.
9. Christman, R. F.; Norwood, D. L.; Millington, D. S.; Johnson, J. D. *Environ. Sci. Technol.* **1983,** *17*, 625–628.
10. Norwood, D. L.; Johnson, J. D.; Christman, R. F.; Millington, D. S. In *Water Chlorination: Environmental Impact and Health Effects*; Jolley, R. L.; Brungs, W. A.; Cumming, R. B.; Jacobs, V. A., Eds.; Ann Arbor Science: Ann Arbor, 1983; Vol. 4, pp 191–200.

11. Liao, W. T.; Christman, R. F.; Johnson, J. D.; Millington, D. S. *Environ. Sci. Technol.* **1982,** *16*, 403–410.
12. Christman, R. F.; Norwood, D. L.; Seo, Y.; Frimmel, F. H. In *Humic Substances II: In Search of Structure*; Hayes, M. H. B.; MacCarthy, P.; Malcolm, R. L.; Swift, R. S., Eds.; Wiley: Chichester, England, in press.
13. Ertel, J. R.; Hedges, J. T. *Geochim. Cosmochim. Acta* **1984,** *48*, 2065–2074.
14. Hayes, M. B.; Swift, R. S. In *The Chemistry of Soil Constituents*; Greenland, D. J.; Hayes, M. B., Eds.; Wiley–Interscience: New York, 1978; pp 179–319.
15. March, J. *Advanced Organic Chemistry*; Wiley and Sons: New York, 1985; pp 600–601.
16. Christman, R. F.; Johnson, J. D.; Pfaender, F. K.; Norwood, D. L.; Webb, M. K.; Hass, J. R.; Bobenreith, M. In *Water Chlorination: Environmental Impact and Health Effects*; Jolley, R. L.; Brungs, W. A.; Cumming, R. B.; Jacobs, V. A., Eds.; Ann Arbor Science: Ann Arbor, 1980; Vol. 3, p 75.
17. Chesire, M. V.; Cranwell, P. A.; Falshaw, C. P.; Floyd, A. J.; Haworth, R. D. *Tetrahedron* **1967,** *23*, 1669–1682.
18. Chesire, M. V.; Cranwell, P. A.; Haworth, R. D. *Tetrahedron* **1968,** *24*, 5155–5167.
19. Hansen, E. H.; Schnitzer, M. *Soil Sci. Soc. Am. Proc.* **1969,** *33*, 29–36.
20. Brewer, C. P.; Cooke, L. M.; Hibbert, H. *J. Am. Chem. Soc.* **1948,** *70*, 57–59.
21. Pepper, J. M.; Hibbert, H. *J. Am. Chem. Soc.* **1948,** *70*, 67–71.
22. Cooke, L. M.; McCarthy, J. L.; Hibbert, H. *J. Am. Chem. Soc.* **1941,** *63*, 3052–3056.
23. Cooke, L. M.; McCarthy, J. L.; Hibbert, H. *J. Am. Chem. Soc.* **1941,** *63*, 3056–3061.
24. Kukharenko, T. A.; Savelev, A. S. *Dokl. Akad. Nauk SSSR* **1951,** *76*, 77–78; *Chem. Abstr. 45*, 8451h.
25. Kukharenko, T. A.; Savelev, A. S. *Dokl. Akad. Nauk SSSR* **1951,** *86*, 729–732; *Chem. Abstr. 47*, 8037c.
26. Felbeck, G. T. *Soil Sci. Soc. Am. Proc.* **1965,** *29*, 48–55.
27. Burges, N. A.; Hurst, H. M.; Walkden, B.; Dean, F. M.; Hirst, M. *Nature (London)* **1963,** *199*, 696–697.
28. Burges, N. A.; Hurst, H. M.; Walkden, B. *Geochim. Cosmochim. Acta* **1964,** *28*, 1547–1554.
29. Mendez, J.; Stevenson, F. J. *Soil Sci.* **1966,** *102*, 85–93.
30. Stevenson, F. J.; Mendez, J. *Soil Sci.* **1967,** *103*, 383–388.
31. Martin, J. P.; Richards, S. J.; Haider, K. *Soil Sci. Soc. Am. Proc.* **1967,** *21*, 657–661.
32. Martin, J. P.; Haider, K.; Wolf, D. *Soil Sci. Soc. Am. Proc.* **1972,** *36*, 311–315.
33. Martin, J. P.; Haider, K.; Saiz-Jiminez, C. *Soil Sci. Soc. Am. Proc.* **1974,** *38*, 760–765.
34. Birch, A. J. *J. Chem. Soc.* **1947,** *Part 1*, 102–105.
35. Subba Rao, G. S. R.; Shyama Sundar, N. *J. Chem. Soc. Perkin Trans. 1* **1982,** No. 3, 875–880.
36. Paul, D. E.; Lipkin, D.; Weissman, S. I. *J. Am. Chem. Soc.* **1956,** *78*, 116–120.
37. Helbig, B.; Hartung, J.; Klocking, R.; Sprossing, K.; Graser, H. *Geoderma* **1985,** *35*, 225–261.
38. Piper, T. J.; Posner. A. M. *Soil Sci.* **1968,** *106*, 188–192.
39. Piper, T. J.; Posner, A. M. *Soil Biol. Biochem.* **1972,** *4*, 513–523.
40. Piper, T. J.; Posner, A. M. *Soil Biol. Biochem.* **1972,** *4*, 525–531.
41. Matsui, Y.; Kumado, K. *Soil Sci. Plant Nutr. (Tokyo)* **1977,** *23*, 341–354.
42. Matsui, Y.; Kumado, K. *Soil Sci. Plant Nutr. (Tokyo)* **1977,** *23*, 491–501.

43. Schnitzer, M.; Oritz de Serra, M. I.; Ivarson, K. *Soil Sci. Soc. Am. Proc.* **1973,** *37*, 229–236.
44. Dormaar, J. F. *Plant Soil* **1969,** *31*, 182–184.
45. Tate, K. R.; Goh, K. M. *N. Z. J. Sci.* **1973,** *16*, 59–69.
46. Zhorobekova, Sh. Zh. *Khim. Tverd. Topl.* (*Moscow*) **1977,** *11*, 81–82 (abstract).
47. Hall, K. J.; Lee, G. F. *Water Res.* **1974,** *8*, 239–251.
48. Shorygina, N. N.; Kefeli, T. Ya. Semechkina, A. F. *J. Gen. Chem. USSR* (*Engl. Transl.*) **1949,** *19*, 1569–1577.
49. Shorygina, N. N.; Kefeli, T. Ya. *J. Gen. Chem. USSR* (*Engl. Transl.*) **1950,** *20*, 1243–1252.
50. Saidalimov, S. A.; Smirnova, L. S.; Abduazimov, Kh. A. *Chem. Nat. Compd.* (*Engl. Transl.*) **1976,** *12*, 574–576; *Khim. Prir. Soedin.* **1976,** *5*, 643–646.
51. Geronikaki, A. A.; Abduazimov, Kh. A. *Chem. Nat. Compd.* (*Engl. Transl.*) **1976,** *12*, 577–579; *Khim. Prir. Soedin.* **1976,** *5*, 646–648.
52. Maximov, O. B.; Krasovskaya, N. P. *Geoderma* **1977,** *18*, 227–228.
53. Birch, A. J.; Hughes, G. K.; Smith, E. *Aust. J. Chem.* **1955,** *7*, 83–86.
54. Birch, A. J.; Clark-Lewis, J. W.; Robertson, A. V. *J. Chem. Soc.* **1957,** 3586–3594.
55. Birch, A. J. *Q. Rev. Chem. Soc.* **1950,** *4*, 69–93.
56. House, H. O. *Modern Synthesis Reactions*; W. A. Benjamin: New York, 1965; pp 51–77.
57. Smith, M. In *Reduction: Techniques and Applications*; Augustine, R. L., Ed.; Dekker: New York, 1968; pp 95–171.
58. Watt, G. W. *Chem. Rev.* **1950,** *46*, 317–379.
59. Shaede, E. A.; Walker, D. C. *The Alkali Metals: An International Symposium*; Special Publication No. 22; Chemical Society: London, 1967; pp 277–283.
60. Birch, A. J.; Smith, H. *Q. Rev. Chem. Soc.* **1958,** *12*, 17–33.
61. Smith, H. *Organic Reactions in Liquid Ammonia*; Interscience: New York, 1963.

RECEIVED for review August 19, 1987. ACCEPTED for publication December 4, 1987.

2

Implications of Mixture Characteristics on Humic-Substance Chemistry

J. A. Leenheer, P. A. Brown, and T. I. Noyes

U.S. Geological Survey, Denver Federal Center, Denver, CO 80225

The chemistry of complex mixtures usually cannot be described as the sum of the characteristics of the individual components because of intermolecular interactions that modify component characteristics. A study of both inter- and intramolecular interactions in the Suwannee River fulvic acid, by measurement of the dependence of density on concentration in various solvents, indicated that polar functional groups (carboxyl, phenolic, hydroxyl, and ketone) were responsible for most of the molecular interactions in water. A two-stage fractionation of Suwannee River fulvic acid on silica gel resulted in 31 fractions that had variable molecular weight distributions and acid-group contents. Intramolecular interactions were determined to predominate over intermolecular interactions for the Suwannee River fulvic acid in all solvents at weight–volume concentrations less than 1%.

HUMIC SUBSTANCES ARE HETEROGENEOUS with respect to the diversity of molecules that compose the mixture, and with respect to the diverse functional groups, structural units, and configurations that compose individual molecules. Intermolecular interactions between molecular components in the mixture, or intramolecular interactions between organic functional groups within a molecule, may change chemical and physical properties of humic substances among various environments. Molecular interactions of humic substances are dependent on temperature, pH, ionic strength, type of solvation, degree of hydration (on drying), type of countercations, and concentration of the humic substance in solution. Environmentally significant phenomena that may be affected by molecular interactions of humic substances include organic-contaminant partitioning, trace-element inter-

0065–2393/89/0219–0025$06.00/0

actions, and rates of decomposition of humic substances in various environments.

Humin and humic acid are reactive substances that form amorphous organic phases in soil, sediment, and water; types of interactions and ordering of these interactions into a membrane model are discussed by Wershaw and Marinsky (*1*). Fulvic acid exists as a solute in natural waters more commonly than does humic acid, and molecular interactions are controlled by the chemistry of water. A question frequently asked concerning fulvic acid is whether fulvic acid exists in water as a solute dissolved in an ideal solution, or does it interact to form nonideal molecular aggregates? If fulvic acid exists as a molecular aggregate in water, the chemistry of fulvic acid cannot be described as the sum of the characteristics of the individual components because of intermolecular interactions that modify component characteristics.

This chapter will investigate the mechanisms of molecular interactions in a stream fulvic acid isolated from the origin of the Suwannee River at the Okefenokee Swamp in southern Georgia. Variations in solution densities and molecular-size distributions with solvents of differing polarity and concentrations were used to distinguish intermolecular from intramolecular interactions. A two-stage fractionation of the fulvic acid on silica gel was designed to disrupt heterogeneous molecular aggregates (if they exist) and obtain smaller-sized particles in solutions of the fractions that are more homogeneous than fulvic acid.

Both polar and nonpolar interactions undoubtedly are responsible for the often unique chemical and physical properties of extremely complex mixtures that are called humic substances. Polar interactions such as hydrogen bonding will be emphasized in this chapter over nonpolar interactions because of the predominantly polar character of the Suwannee River fulvic acid. Molecular aggregates of fulvic acid caused by nonpolar interactions are not likely to form in natural waters, because concentrations of fulvic acid at natural concentrations are less than critical micelle concentrations of known detergents and because of the lack of significant nonpolar moieties in fulvic acid structure. However, polar interactions, both intramolecular and intermolecular, will increase the nonpolar character of fulvic acid, and they may increase partitioning of nonpolar contaminants into molecular structures of fulvic acid. The findings of polar-interaction mechanisms in fulvic acid also can be related to interactions between polar moieties in humin and humic acid structures. However, nonpolar interactions also should be recognized as relevant to humin and humic acid.

Experimental Methods and Materials

The Suwannee River was sampled at its origin at the outlet of the Okefenokee Swamp near Fargo, Georgia, during November 1983. Onsite, 8104 L of water were processed through a filtration and column-adsorption system that fractionated and isolated organic solutes into hydrophobic–neutral, strong hydrophobic fulvic acid, weak hydrophobic fulvic acid, hydrophilic acid, and hydrophilic–neutral fractions (2). The theory and chemical significance underlying these operational fractionation procedures are given in previous reports (3–7). Most (94.6%) of the fulvic acid was contained in the strong hydrophobic fulvic acid fraction; 382.4 g of this fraction was isolated from water and used in the study described in this chapter. The strong hydrophobic fulvic acid fraction was 66% of the dissolved organic-carbon concentration, 38.4 mg/L of the Suwannee River water.

Molecular-weight distributions of the Suwannee River fulvic acid were estimated by the degree of polydispersity determined by the ratio of M_w (weight-average molecular weight) to M_n (number-average molecular weight). M_w and M_n were determined by equilibrium centrifugation. Equilibrium–ultracentrifugation experiments were conducted on an ultracentrifuge (Beckman L8–70 M, equipped with a Beckman Prep UV scanner) that assayed solute concentrations at 280 nm throughout the ultracentrifuge cells during the run. A four-phase titanium rotor (AnF, 5.7-cm radius) was used at speeds of 20,000 to 40,000 rpm.

Equilibrium between sedimentation and molecular diffusion of the solutes was virtually complete between 24 and 48 h of centrifugation. Analytical ultracentrifugation cells were constructed of aluminum bodies with centerpieces of Kel–F (for aqueous 0.2 M KCl solvent) or aluminum (for organic solvents). Each cell had a sample and reference compartment with quartz windows in the centerpiece that contained 300 μL of solution. The immiscible fluorocarbon oil (FC–43), used to determine the bottom of the solution in the cell, was used only for aqueous solvents and not for organic solvents that partially solubilized the oil. The M_w and M_n values were determined from the equilibrium–ultracentrifugation curves by the method of Lansing and Kraemer (8).

Solution densities of the Suwannee River fulvic acid at 20 °C were determined by pycnometry (9) in the solvent used for equilibrium ultracentrifugation. Pycnometers of 2-mL capacity gave results accurate to 0.02 g/mL for the samples at 1–2% w/v solute concentrations; 10-mL pycometers were used at the 0.4–1% solute concentrations. Partial molar volume or apparent molar volumes (for nonideal solutions) were calculated as the reciprocal of the solution density. All concentrations of humic substances in this chapter are expressed on a weight-per-solution-volume basis.

Solvents used for density determinations included distilled water, aqueous 0.2 M KCl, tetrahydrofuran (freshly distilled), dioxane (freshly distilled), *N,N*-dimethylformamide (dried over calcium hydride and distilled), acetonitrile, and glacial acetic acid. Equilibrium ultracentrifugation was conducted with aqueous 0.2 M KCl, tetrahydrofuran, and glacial acetic acid. Both tetrahydrofuran and glacial acetic acid have small dielectric constants, so that the carboxyl group does not ionize to produce negative charges on fulvic acid solutes. These negative charges interfere with solute sedimentation during centrifugation. Tetrahydrofuran was preferred for use as a solvent for centrifugation because of its minimal density and viscosity; however, glacial acetic acid was used when samples were not soluble in tetrahydrofuran. Fulvic acid ionization was suppressed by 0.2 M KCl for centrifugation of samples dissolved in water.

Fractionation of fulvic acid on silica gel was conducted with 100–200-mesh activated silica (ICN Biochemicals). The first-stage silica fractionation used a 2-L bed volume of silica contained in a 50-mm-i.d. × 950-mm glass column. The column was conditioned by passing 0.1 M tetrabutylammonium acetate in chloroform through until solution equilibrium of the tetrabutylammonium acetate with the silica was obtained. Six grams of the Suwannee River fulvic acid was titrated in water with tetrabutylammonium hydroxide to pH 8; then the sample was freeze-dried.

The tetrabutylammonium salt of fulvic acid was dissolved in chloroform, and a total volume of 22 mL was added to the silica column. Fraction 1 eluted with

3100 mL of chloroform; fraction 2 eluted with 970 mL of methyl ethyl ketone; fraction 3 eluted as a distinct band at the interface of methyl ethyl ketone and acetonitrile in 400 mL of solvent; fraction 4 eluted in 2000 mL of acetonitrile; fraction 5 eluted in 1425 mL of 75% acetonitrile and 25% 2-propanol; fraction 6 eluted in 1310 mL of 75% acetonitrile and 25% water; and fraction 7 eluted in 1350 mL of 75% acetonitrile and 25% water, with 0.25 M oxalic acid dissolved in the mixed solvent. The volumes of solvent used during the elution sequence were determined by observing the cessation of color eluted from the column by each solvent.

After elution, the solvents were removed by vacuum-rotary evaporation; the fractions were redissolved in 50:50 acetonitrile–water; then the samples were converted back to free acids by passing them through columns of cation-exchange resin in the hydrogen form (MSC–1H, Dow Chemical). The fractions were (1) vacuum-evaporated to about 20% of their original volume to remove acetonitrile, (2) adjusted to pH 2 with HCl, (3) applied to a 500-mL bed volume resin column (XAD–8), (4) washed with 1 L of 0.01 M HCl to remove acetic acid and oxalic acid, and (5) eluted with 300 mL of acetonitrile. The acetonitrile was removed by vacuum evaporation, and the water was removed by freeze-drying. The dried fractions were weighed to calculate yield.

The second-stage fractionation separated free-acid fulvic acid fractions 2, 4, 5, and 6 on silica gel. Silica gel was packed in glass columns, with bed volumes corresponding to a loading of 2.5 mg of sample per 1 mL of bed volume. Silica was packed with chloroform as the initial mobile phase and the fractions were applied and then dissolved in tetrahydrofuran at a volume of 5% of the bed volume. The elution sequence was (1) chloroform, (2) diethyl ether, (3) 75% diethyl ether and 25% methyl ethyl ketone, (4) methyl ethyl ketone, (5) acetonitrile, (6) 75% acetonitrile and 25% isopropyl alcohol, (7) 75% acetonitrile and 25% water, and (8) 75% acetonitrile and 25% acetonitrile with 0.25 M oxalic acid. Varying quantities of solvents were used for each fraction; cessation of eluted color was the sign for the beginning of the next solvent. Each solvent was removed by vacuum-rotary evaporation; the oxalic acid was removed from fraction 8 through readsorption on resin (XAD–8) by the procedure discussed in the previous paragraph.

Unfractionated fulvic acid was methylated with diazomethane. Nitrogen gas was slowly bubbled in a polytetrafluoroethylene (Teflon) tube through two 25-mm i.d. × 150-mm test tubes, which were sealed with rubber stoppers. The first test tube was two-thirds filled with diethyl ether; the second test tube, kept in an ice bath, contained 8 mL of diethyl ether, 8 mL of 2-(2-ethoxyethoxy)ethanol, 8 mL of 30% KOH solution; and 3 g of 99% *N*-methyl-*N*-nitroso-*p*-toluenesulfonamide. Diazomethane gas (CH_2N_2), generated by the reagents, was bubbled with nitrogen through 100 mg of fulvic acid sample and then suspended in methylene chloride in an ice bath while being dispersed by ultrasonic vibration. The fulvic acid sample was dissolved in methylene chloride during methylation; the reaction was stopped after complete dissolution, and methylene chloride and excess diazomethane were removed by vacuum evaporation.

Results and Discussion

Density–concentration curves for Suwannee River fulvic acid in water, tetrahydrofuran, dioxane, and *N*,*N*-dimethylformamide are shown in Figure 1. Pimentel and McClellan (*10*) state that intramolecular interactions are concentration-independent in their effects on density, whereas intermolecular interactions result in increases in density with concentration. Che-

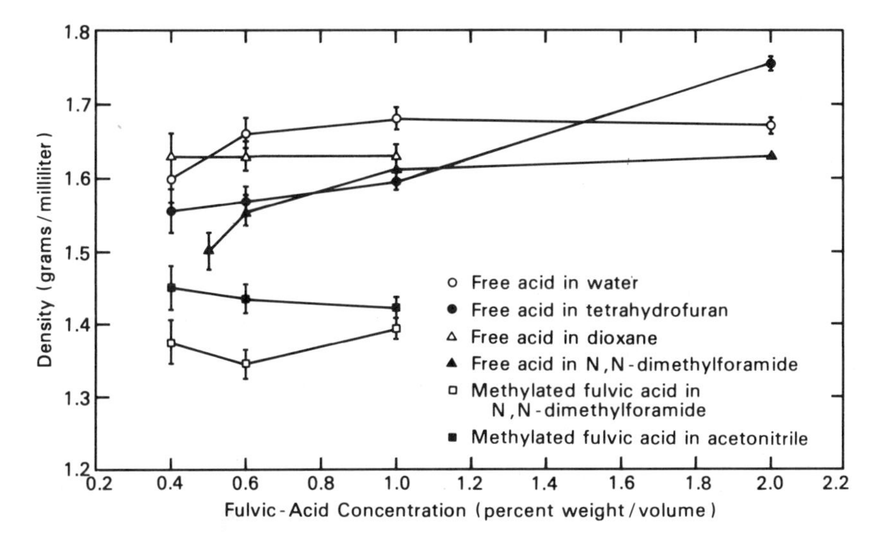

Figure 1. Density–concentration curves for Suwannee River fulvic acid in various solvents.

baevskii et al. (*11*) reported that soil fulvic acid aggregates in water near 1% concentrations, as determined by density measurements. Both the water and *N,N*-dimethylformamide density–concentration curves in Figure 1 indicate similar density–concentration dependence, with little or no density increase beyond 1% concentration. However, the density indicated by the *N,N*-dimethylformamide curve is 0.1–0.2 g/mL less than that indicated by the water curve. The tetrahydrofuran and dioxane curves are concentration-independent from 0.4–1.0% concentration (within limits of error); the larger value for density in tetrahydrofuran at 2.0% may be due to lack of complete solubility, because a slight precipitate was observed at this concentration. The range of density values in different solvents of different polarity indicates that some type of molecular interaction is occurring. The concentration independence of the density data for tetrahydrofuran and dioxane indicates that one type of interaction is intramolecular.

A conclusion for intramolecular interactions compared with intermolecular interactions cannot be made from the data of Figure 1 alone because of the lack of data from 0–0.4% concentration, where the analytical error for pycnometric measurements of density becomes too large to detect significant differences. However, a minimum density for the sample can be calculated from Traube's rule (*12*), which states that the molar volume of an organic liquid with no interactions is equal to the summation of atomic volumes, plus a correction term for double bonds and aromatic rings, plus a correction term for end groups. Brown and Leenheer (*13*) applied Traube's rule and used elemental analysis, molecular weight data, ^{13}C nuclear magnetic resonance (NMR) data, ^{1}H NMR data, and titrimetric data for the

Suwannee River fulvic acid. They calculated a minimum solution-state density of 1.45 g/mL.

Methylation is known to decrease hydrogen-bonding interactions (*9*), and is likely to produce a decrease in density compared to free-acid samples. However, a methyl ester group is less dense (d = 1.26 g/mL) than a free carboxyl group (d = 1.90 g/mL) (*12*), and methylation will decrease the minimum calculated density of the sample. The degree of methylation of the sample was measured by ^{1}H NMR spectroscopy (*14*). The calculated minimum density of the sample, using Traube's rule (*12*), decreased to 1.37 g/mL. The solution-state density values for the methylated sample in *N,N*-dimethylformamide (Figure 1) are identical (within limits of error) to the calculated minimum density; the density values for the methylated samples in acetonitrile are about 0.1 g/mL larger. *N,N*-Dimethylformamide is a good hydrogen-bonding solvent, whereas acetonitrile cannot hydrogen-bond. Perhaps unmethylated hydroxyl groups in the fulvic acid dissolved in acetonitrile are interacting with carbonyl and ether groups of fulvic acid, but not in *N,N*-dimethylformamide, which hydrogen-bonds to these groups.

Conversion of a free carboxyl group to a carboxylate anion by an increase in pH in water increases hydrogen-bonding interactions with un-ionized weak-acid groups such as phenols (*10*, *15*) because of the greater dipole of the carboxylate anion. However, the creation of negative charges with the pH increase tends to disrupt molecular interactions because of electrostatic repulsion (*16*). Density–concentration curves of the potassium salt of fulvic acid at pH 8 in water and in 0.2 M KCl are shown in Figure 2. Densities of potassium salts of fulvic acid in water are much greater than density of

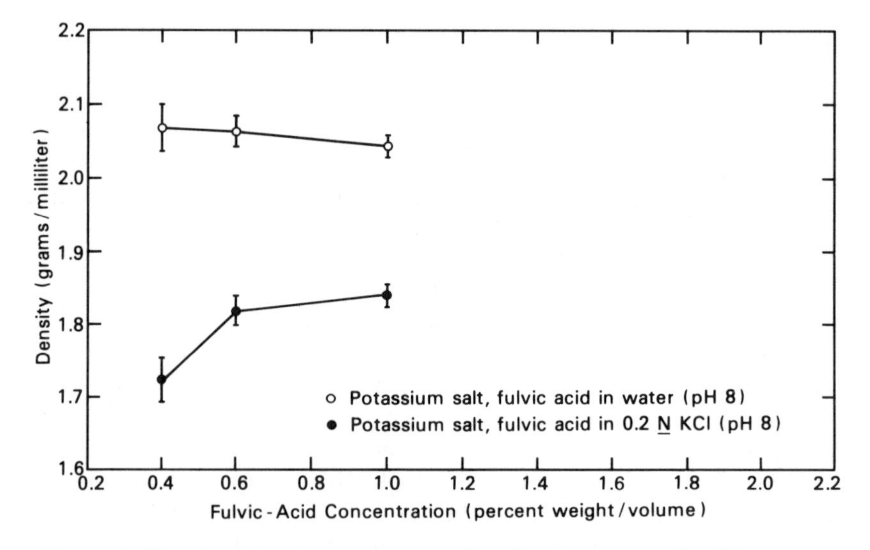

Figure 2. Density–concentration curves for the potassium salt of Suwannee River fulvic acid at pH 8 in water and in 0.2 M KCl.

hydrogen-saturated fulvic acid in water (Figure 1). The initial interpretation of these densities is that molecular interactions are enhanced markedly in the salt–anion forms of fulvic acid. However, electrolyte salts are known to increase the density of water by closer packing of the water molecules around the ions in a process known as electrostriction (*17*).

To determine the density and molecular weight effects of electrostriction on a potassium–carboxylate ion pair, solution-density determinations in distilled water were performed on a series of 16 acids in free-acid and potassium-salt forms. The resulting data are presented in Table I. Carboxyl-group contribution to compound density, calculated from Traube's rule (*12, 13*), generally increased by more than a factor of 2 between the free-acid and potassium-salt forms. The magnitude of the increase from acetic through benzoic acids in Table I can be accounted for by adding four to six water molecules per potassium–carboxylate ion pair. This water is the bound water in the electrical double layer. Larger densities of potassium–carboxylate groups in phthalic through salicylic acids in Table I are probably caused by inter- and intramolecular interactions in these acids. The carboxylate group seems to form an intramolecular chelate with the free carboxyl group in monopotassium phthalate, with aromatic π electrons in phenylacetic acid, and with the phenolic hydroxyl group in salicylic acid. Carboxylate–phenol interactions apparently cause a dimer with *p*-hydroxybenzoic acid.

The large density values of the potassium salts of the Suwannee River fulvic acid presented in Figure 2 can be explained best by electro-

Table I. Solution Densities for Carboxylic Acids and Their Salts

	Acid Form[a]		*Potassium Salt Forms*[b]	
Carboxylic Acid	*Compound Density* (g/mL)	*Carboxyl Group Density*[c] (g/mL)	*Compound Density* (g/mL)	*Carboxylate Group Density*[c] (g/mL)
Acetic	1.12	1.92	2.02	3.73
Propionic	1.04	1.99	1.91	4.62
Succinic	1.43	2.04	2.54	4.25
Adipic	1.28	2.21	2.07	4.59
Pimelic	1.22	2.17	1.88	4.51
Azelaic	1.14	2.23	1.70	4.74
Sebacic	1.11	2.25	1.63	4.84
Tartaric	1.85	1.93	2.90	4.05
Citric	1.85	2.37	3.10	4.88
Benzoic	1.20	2.08	1.61	4.81
Phthalic[d]	1.47	2.10	1.83	5.12
p-Hydroxybenzoic	1.39	2.16	1.89	5.52
Phenylacetic	1.19	2.85	1.59	6.07
Salicylic	—	—	2.25	7.99

[a]Densities determined in *N,N*-dimethylformamide at 20 °C at 1% (w/v).
[b]Densities determined in water at 20 °C at 1% (w/v).
[c]Calculated as carboxyl group contribution to compound density using Traube's rule.
[d]Monopotassium salt.

lyte–electrostriction effects on water, rather than by molecular interactions within and among fulvic acid solutes, although the density–concentration dependence of the 0.2 M KCl curve may indicate intermolecular interactions. The smaller solution-density values of 0.2 M KCl curve in Figure 2 compared to the water curve illustrates the loss of the bound water layer at greater ionic strength. Not as many water molecules are bound to potassium carboxylate ion pairs in fulvic acid as are bound with dissociated potassium and carboxylate ions at lesser ionic strength.

Equilibrium ultracentrifugation was used to determine molecular weight distribution data in different solvents at small concentrations (0.002–0.004%). Changes in molecular weight distribution of fulvic acid dissolved in solvents of differing polarity may indicate changes in molecular conformation and aggregate states at concentrations approaching infinite dilution, where accurate density measurements were not available. Data for M_n, M_w, and degree of polydispersity of the Suwannee River fulvic acid in water, tetahydrofuran, *N,N*-dimethylformamide, and acetonitrile are given in Table II.

The data of Table II has only semiquantitative significance because a UV detector measures only compounds with aromatic nuclei that vary in their molar absorptivity. Furthermore, computation of M_n by the method of Lansing and Kraemer (*8*) provides only a semiquantitative estimate. Table II clearly indicates that the molecular weights of the potassium salt of the fulvic acid are much greater. However, the degree of polydispersity of the potassium salt data is much less than the data for fulvic acid in organic solvents.

Larger molecular weights were expected for the potassium salt of the fulvic acid because of the additional weight added by the potassium ions and bound water; however, the reason for the large decrease in the degree of polydispersity was unknown. The lesser degree of polydispersity for potassium salt in water, coupled with the large density, indicated that the presence of stable aggregates at a small concentration should be considered a possi-

Table II. Molecular-Weight Distribution Data for Suwannee River Fulvic Acid in Various Solvents

Sample Form[a]	*Solvent*	*Number Average* (M_n)	*Weight Average* (M_w)	*Degree of Polydispersity*
Potassium salt (density = 1.8 g/mL)	0.2 N KCl, pH 8 in water	1060	1340	1.27
Free acid (density = 1.55 g/mL)	tetrahydrofuran	470	1110	2.36
Methylated (density = 1.37 g/mL)	*N,N*-dimethylformamide	530	1250	2.36
Methylated (density = 1.43 g/mL)	acetonitrile	530	1190	2.25

[a]Density values were derived from 0.4% weight–volume values of fulvic acid concentrations in Figures 1 and 2.

bility. Intermolecular aggregation is not likely to increase the degree of polydispersity in organic solvents because molecular weights were lower.

Two-stage silica-gel fractionation was designed as a chemical fractionation on the basis of polarity differences within the fulvic acid sample. If stable aggregates of fulvic acid were held together by polar hydrogen-bonding forces, two-stage silica-gel fractionation would have a large probability of disaggregating the sample because of competitive polar-adsorption effects of the silica gel. The fractionation and the yields of the silica-gel fractionation are shown in Figure 3, and the yields are shown in Table III. Losses (10.9% for the first stage, 12.2% for the second stage) for the two-stage fractionation were within expectations for normal handling and processing losses, and the yield for each fraction can be regarded as representative of the whole sample.

Molecular-weight distributions of 15 of the major fractions from silica-gel fractionation are presented in Table IV. Molecular-weight distributions were determined in tetrahydrofuran, rather than in water, to avoid electrostriction effects, which result in variable densities and large uncertainties in molecular weight determinations. The 15 fractions in Table IV accounted for nearly 88% of the weight of the sample. Large variations in molecular weight distributions were determined among the 15 fractions, but the summations of M_n and M_w adjusted for the yield of each fraction (Table IV) gave virtually the same M_n and M_w as the whole sample. Correspondence of the M_n and M_w values of the fractions with the M_n and M_w values of the whole sample provide definitive evidence that Suwannee River fulvic acid is not aggregated in tetrahydrofuran.

The 15 fractions also were titrated with 0.1 N NaOH to pH 8.5 to determine acid-group content. Variable acid-group contents occurred, as shown in Table IV, and will produce variable electrostriction effects on density for the potassium-salt form of the fulvic acid. These effects are probably the reason that the potassium salt of fulvic acid is less polydisperse in water than in tetrahydrofuran (Table II). Some indication exists that fractions of large acid-group content (such as fraction 6–2 in Table IV) have lesser M_w and degrees of polydispersity than fractions with small acid-group contents. Fractions with large acid-group contents would indicate the largest increase in density and apparent molecular weight on conversion to potassium salt. Fulvic acid molecular weight distribution, skewed to low-molecular-weight solutes in tetrahydrofuran, would appear to be less polydisperse in aqueous 0.2 M KCl.

The silica-gel fractionation data coupled with molecular weight distribution data do not support the existence of stable aggregate structures for fulvic acid at small concentrations (0.003%) in either tetrahydrofuran or aqueous 0.2 M KCl. Therefore, density increases of 0.1–0.2 g/mL (Figure 2) from the minimum calculated values for fulvic acid can be ascribed to intramolecular interactions for the underivatized sample in tetrahydrofuran and dioxane, and for the methylated sample in acetonitrile. The more polar solvents with good hydrogen-bonding characteristics (water and *N*,*N*-di-

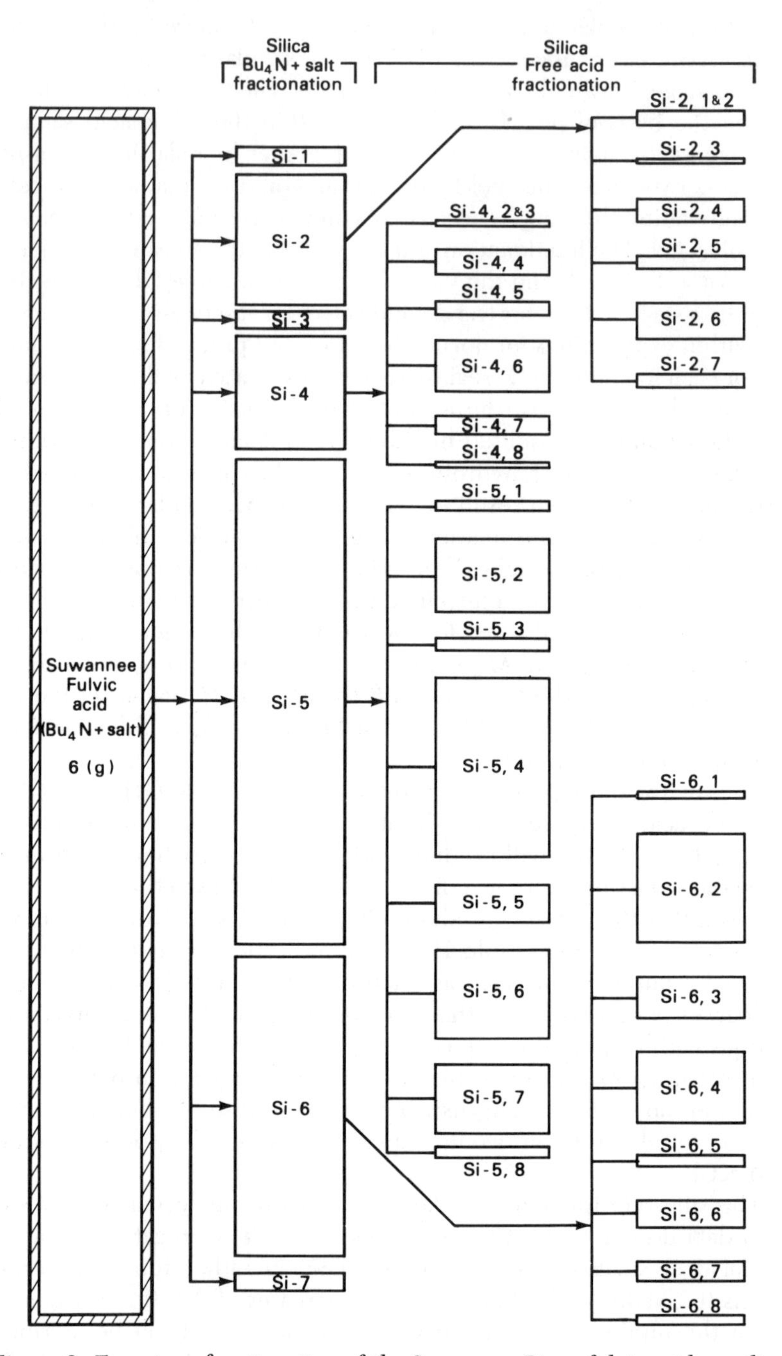

Figure 3. Two-stage fractionation of the Suwannee River fulvic acid on silica gel. Bar sizes are proportional to yields.

Table III. Yields of Silica-Gel Fractionation of Suwannee River Fulvic Acid

Fraction Number (from Figure 3)	*Elution Solvent*	*Weight (mg)*	*Percent Yield*
Si–1	chloroform	72	1.35
Si–2	methyl ethyl ketone	658	12.32
Si–3	50% methyl ethyl ketone, 50% acetonitrile	66	1.23
Si–4	acetonitrile	559	10.45
Si–5	75% acetonitrile, 25% 2-propanol	2420	45.21
Si–6	75% acetonitrile, 25% water	1500	28.15
Si–7	75% acetonitrile, 25% water, 0.25 M oxalic acid	70	1.30
Si–2,1 and 2	chloroform and diethyl ether	57	1.21
Si–2,3	75% diethyl ether, 25% methyl ethyl ketone	23	0.49
Si–2,4	methyl ethyl ketone	105	2.23
Si–2,5	acetonitrile	49	1.04
Si–2,6	75% acetonitrile, 25% 2-propanol	187	4.00
Si–2,7	75% acetonitrile, 25% water	58	1.23
Si–4,2 and 3	diethyl ether and 75% diethylether, 25% methyl ethyl ketone	6.8	0.14
Si–4,4	methyl ethyl ketone	127	2.72
Si–4,5	acetonitrile	54	1.15
Si–4,6	75% acetonitrile, 25% 2-propanol	244	5.21
Si–4,7	75% acetonitrile, 25% water	62	1.31
Si–4,8	75% acetonitrile, 25% water, 0.25 M oxalic acid	3.6	0.08
Si–5,1	chloroform	25	0.54
Si–5,2	diethyl ether	360	7.68
Si–5,3	75% diethyl ether, 25% methyl ethyl ketone	42	0.89
Si–5,4	methyl ethyl ketone	913	19.47
Si–5,5	acetonitrile	162	3.45
Si–5,6	75% acetonitrile, 25% 2-propanol	391	8.33
Si–5,7	75% acetonitrile, 25% water	362	7.71
Si–5,8	75% acetonitrile, 25% water, 0.25 M oxalic acid	67	1.26
Si–6,1	chloroform	15	0.33
Si–6,2	diethyl ether	548	11.69
Si–6,3	75% diethyl ether, 25% methyl ethyl ketone	200	4.27
Si–6,4	methyl ethyl ketone	336	7.18
Si–6,5	acetonitrile	37	0.80
Si–6,6	75% acetonitrile, 25% 2-propanol	136	2.89
Si–6,7	75% acetonitrile, 25% water	88	1.88
Si–6,8	75% acetonitrile, 25% water, 0.25 M oxalic acid	31	0.65

methylformamide) are more effective at disrupting intramolecular hydrogen bonds in fulvic acid. Extrapolation of the density–concentration curves for these solvents in Figure 1 to zero concentration would result in values reasonably similar to the minimum calculated density of 1.45 g/mL. The occurrence of intramolecular interactions in the potassium-salt form of the Suwannee River fulvic acid cannot be deduced from the density–

Table IV. Molecular-Weight Distributions of Suwannee River Fulvic Acid Fractions from Silica-Gel Fractionation

Fraction[a]	*Percent Weight*	*Number Average*[b] (M_n)	*Weight Average*[b] (M_w)	*Degree of Polydispersity* $\left(\frac{M_n}{M_w}\right)$	*Acid-Group Content*[c] *(meq/g)*
2–4	2.23	200	1060	5.27	3.85
2–5	1.04	400	1700	4.26	3.65
2–6	4.00	570	2920	5.11	3.46
2–7	1.23	410	1270	3.07	2.96
4–4	2.72	460	1170	2.56	5.29
4–6	5.21	750	820	1.09	4.30
5–2	7.68	730	1060	1.45	6.97
5–4	19.47	380	1160	3.06	5.63
5–5	3.45	280	1050	3.73	5.70
5–6	6.50	290	1500	5.17	4.59
5–7	8.34	510	1200	2.30	4.74
6–2	11.69	440	520	1.17	7.16
6–3	4.27	640	730	1.14	6.57
6–4	7.18	880	1110	1.26	5.29
6–6	2.89	1540	1670	1.09	4.40
Weight yield	87.9				

$$\sum M_n \times \%W = 476 \qquad \sum M_w \times \%W = 1008$$

$$\overline{M}_n = \frac{476}{0.879} = 542 \qquad M_w = \frac{1008}{0.879} = 1147$$

$$\frac{\overline{M}_n \text{ fractions}}{M_n \text{ whole sample}} = \frac{542}{470} = 1.15 \qquad \frac{M_w \text{ fractions}}{M_w \text{ whole sample}} = \frac{1147}{1110} = 1.03$$

[a]First stage–second stage.
[b]M_n and M_w are determined by equilibrium ultracentrifugation in tetrahydrofuran.
[c]Measured by the base titer to pH 8.5.

concentration data of Figure 2 because of the large and variable effects of electrostriction of water on density.

Implications of Findings

Intermolecular association is indicated by the density–concentration dependence of Suwannee River fulvic acid at concentrations of about 1% in water and *N*,*N*-dimethylformamide. This indication, together with findings of intramolecular association in organic solvents with lesser polarity than water, has important implications regarding partitioning of nonpolar organic contaminants into humic substances. Both intermolecular and intramolecular hydrogen bonding (illustrated in reactions 1 and 2) dehydrate a polar portion of a polar molecule. Dehydration renders it more nonpolar, such that nonpolar contaminants can more readily partition into the hydrogen-bonded structure if this structure is sufficiently large to accommodate the contaminant.

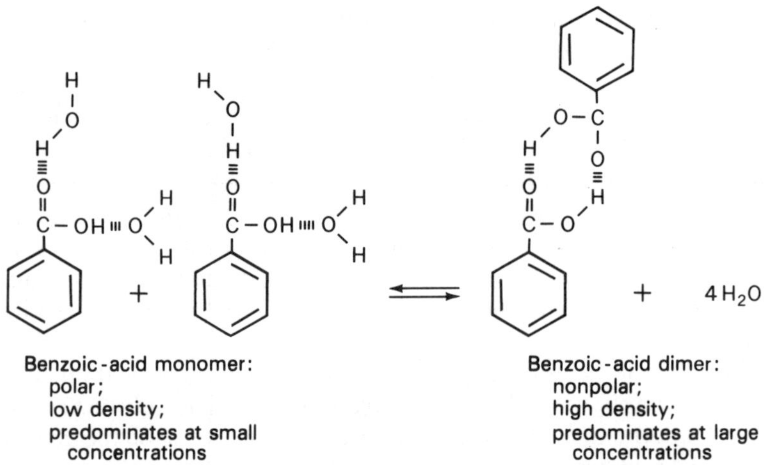

+ $4\,H_2O$

Benzoic-acid monomer:
polar;
low density;
predominates at small concentrations

Benzoic-acid dimer:
nonpolar;
high density;
predominates at large concentrations

Reaction 1. Intermolecular hydrogen bonding.

+ $2\,H_2O$

Salicyclic acid:
low density;
polar

H-bonded chelate:
high density;
less polar

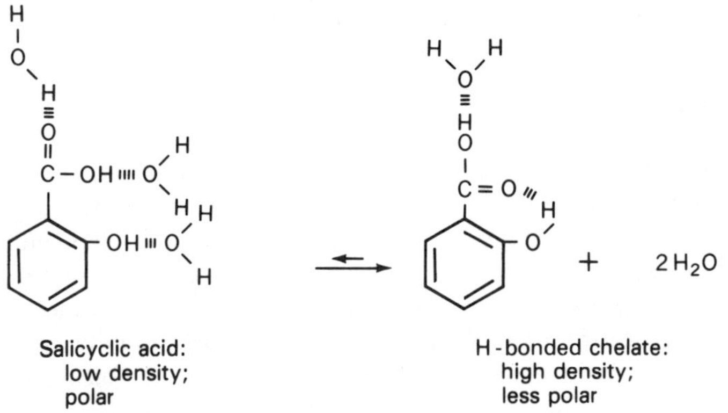

Reaction 2. Intramolecular hydrogen bonding.

If polar humic substances like Suwannee River fulvic acid exist at large concentrations, such as on moist soil or sediment surfaces, intermolecular interactions can decrease the polarity and enhance contaminant partitioning. If polar humic substances are dried in soils or sediments, increased intermolecular and intramolecular interactions will decrease polarity and enhance partitioning of nonpolar contaminants. The lack of molecular interactions for the Suwannee River fulvic acid in water at small concentrations is significant. This lack of interactions substantiates the findings by Chiou et al. (*18*) that the Suwannee River fulvic acid at small concentrations in water has little affinity for various nonpolar contaminants. However, Chiou et al. (*18*) and Carter and Suffet (*19*) determined that other aquatic fulvic acids and humic acids from different environments had significant affinities for nonpolar contaminants.

Polar interactions of inorganic substances with humic substances are likely to occur in many environments. Hydrogen bonding with silica gel was used to fractionate the Suwannee River fulvic acid. The last fraction desorbed interacted so strongly with silica that oxalic acid had to be added to the water to break up the complex. In addition to weakly acidic silica, boric acid, acids of sulfides and polysulfides, acids of sulfites and polythionates, phosphoric acid, selenious acid, arsenous acid, and a number of other weak inorganic acids may hydrogen-bond to electron-donating groups in humic substances. Inorganic electron-donating groups that may hydrogen-bond to weak acid sites in humic substances include the oxides of iron, manganese, uranium, and other metals.

The successful chromatographic fractionation of fulvic acid on silica gel has relevant implications for chromatography of humic substances to gain structural information. The variable acid-group content of the Suwannee River fulvic acid was unknown before silica-gel fractionation. Additional chromatography using basic adsorbents, such as magnesium oxide or alumina, may resolve humic substances into more homogeneous fractions that will provide more useful structural information.

Summary

This study illustrates some of the difficulties encountered because of the complex mixture characteristics of humic substances. When chemical properties of constituent fractions were additive, such as the molecular weight distributions and acid-group contents of Table IV, the skewed nature of the molecular weight distribution and the differential effects of electrostriction of water on acid-group density required the use of silica-gel fractionation to interpret the density–concentration curves of Figure 2. Because molecular interactions within fulvic acid change the properties of polar functional groups, the concept of minimum density calculation by Traube's rule (*12*) had to be introduced to detect the effect. Intermolecular interactions could not be distinguished from intramolecular interactions without silica-gel fractionation data. This study, as well as a number of other reviews (*20*, *21*), continues to identify the need to separate the complex mixture of humic substances into more homogeneous fractions, in order to extend our knowledge of their molecular nature and properties.

References

1. Wershaw, R. L.; Marinsky, J. A. Presented at the 193rd National Meeting of the American Chemical Society, Denver, CO, April 1987.
2. Leenheer, J. A.; Noyes, T. I. U.S. Geological Survey Water Supply Paper 2230 1984, 16 p.
3. Leenheer, J. A. *Environ. Sci. Technol.* **1981,** *15,* 578–587.

4. Leenheer, J. A.; Huffman, E. W. D., Jr. *J. Res. U.S. Geol. Surv.* **1976,** *4,* 737–751.
5. Leenheer, J. A.; Huffman, E. W. D., Jr. *Water Resour. Invest. (U.S. Geol. Surv.)* **1979,** *79–4,* 16 pp.
6. Thurman, E. M.; Malcolm, R. L.; Aiken, G. R. *Anal. Chem.* **1978,** *50,* 775–779.
7. Aiken, G. R. In *Organic Pollutants in Water: Sampling, Analysis, and Toxicity Testing*; Suffet, I. H.; Malaiyandi, M., Eds.; Advances in Chemistry 214; American Chemical Society: Washington, DC, 1987, p 295–307.
8. Lansing, W. D.; Kraemer, E. O. *J. Am. Chem. Soc.* **1935,** *57,* 1369–1377.
9. American Society for Testing and Materials. *Petroleum Products and Lubricants: Method D–941–55*; American Society for Testing and Materials: Philadelphia, 1966; pp 17, 310–315.
10. Pimentel, G.; McClellan, A. *The Hydrogen Bond*; W. H. Freeman: San Francisco, 1960; Chapter 5, pp 167–192.
11. Chebaevskii, A. I.; Tuev, N. A.; Stepanova, N. P. *Pochvovedenie* **1971,** *7,* 31–37.
12. Traube, J. *Ber. Dtsch. Chem. Ges.* **1895,** *28,* 2722.
13. Brown, P. A.; Leenheer, J. A. In *Humic Substances in the Suwannee River, Florida and Georgia: Interactions, Properties, and Proposed Structures*; Averett, R. C., Ed.; U.S. Geological Survey Water Supply Paper, in press.
14. Noyes, T. I.; Leenheer, J. A. In *Humic Substances in the Suwannee River, Florida and Georgia: Interactions, Properties, and Proposed Structures*; Averett, R. C., Ed.; U.S. Geological Survey Water Supply Paper, in press.
15. Davis, M. M. *Acid–Base Behavior in Aprotic Organic Solvents*; National Bureau of Standards Monograph 105; U.S. Government Printing Office: Washington, DC, 1968; 151 pp.
16. Hayes, M. H. B. In *Humic Substances in Soil, Sediment, and Water: Geochemistry, Isolation, and Characterization*; Aiken, G. R.; McKnight, D. M.; Wershaw, R. L.; MacCarthy, P., Eds.; John Wiley and Sons: New York, 1985; pp 329–362.
17. Gurney, R. W. *Ionic Processes in Solution*; Constable and Company: London, 1953; pp 190–195.
18. Chiou, C. T.; Kile, D. E.; Brinton, T. I.; Malcolm, R. L.; Leenheer, J. A. *Environ. Sci. Technol.* **1987,** *21*(12), 1231–1234.
19. Carter, C. W.; Suffet, I. H. *Environ. Sci. Technol.* **1982,** *16,* 735–740.
20. Swift, R. S. In *Humic Substances in Soil, Sediment, and Water: Geochemistry, Isolation, and Characterization*; Aiken, G. R.; McKnight, D. M.; Wershaw, R. L.; MacCarthy, P., Eds.; John Wiley and Sons: New York, 1985; 387–408.
21. Leenheer, J. A. In *Humic Substances in Soil, Sediment, and Water: Geochemistry, Isolation, and Characterization*; Aiken, G. R.; McKnight, D. M.; Wershaw, R. L.; MacCarthy, P., Eds.; John Wiley and Sons: New York, 1985; 409–429.

RECEIVED for review October 21, 1987. ACCEPTED for publication December 21, 1987.

3

Characterization of a Stream Sediment Humin

James A. Rice[1] and Patrick MacCarthy

Department of Chemistry and Geochemistry, Colorado School of Mines, Golden, CO 80401

The isolation of a stream sediment humin by a methyl isobutyl ketone (MIBK) liquid–liquid partitioning procedure is described. An extension of the MIBK method is used to fractionate the humin into a bound humic acid fraction, a lipid fraction, and an inorganic fine-grained residue. The humin, bound humic acid, humic acid, and fulvic acid, all isolated by the MIBK method, were characterized by IR spectroscopy, acidic functional group analysis, and elemental analysis. The lipids were characterized by gas chromatography–mass spectrometry (GC–MS). The humin is similar to the humic acid, although the bound humic acid characteristics are intermediate between those of humic acid and fulvic acid. The lipid associated with the humin was separated into extractable (bitumen) and nonextractable (bound lipid) fractions. Each lipid fraction exhibited distinct n-*alkane and* n-*fatty acid profiles by GC–MS. These observations are discussed with respect to the nature of humin and the fate of hydrophobic organic chemicals in the environment.*

HUMIN IS THE FRACTION OF HUMIC MATERIALS that is insoluble in an aqueous system at any pH value. Thus, including humin in a book devoted to the influence of *aquatic* humic materials on the fate and treatment of pollutants might appear to be a mistake. However, in well-mixed waters such as shallow lakes, streams, or rivers, water is continually brought into contact with bottom sediments that contain humin.

[1]Current address: Department of Chemistry, South Dakota State University, Brookings, SD 57007

0065–2393/89/0219–0041$06.00/0

The organic carbon of humin constitutes the largest fraction of the organic solid phase in modern sediments, representing as much as 80% of the total organic carbon in the sediment (*1–3*). Thus, humin could exert considerable influence on the mobility and fate of pollutants in these aquatic systems. The nature of humin as established in this study supports such a hypothesis. Despite the environmental implications, this material has not been studied to any appreciable extent. It is the least understood of the three main humic-material fractions (humic acid, fulvic acid, and humin).

The lack of research interest in humin is probably due to its insolubility, which makes it difficult to characterize by most conventional analytical methods. The apparent lack of interest in humin may also be attributed to the manner in which humin has traditionally been isolated. Humin is commonly obtained as the solid residue that remains after a soil, sediment, or peat sample has been stirred with an alkaline aqueous solution and centrifuged. MacCarthy et al. (*4*), however, describe a technique whereby the humin in a sample can be isolated directly, as opposed to the traditional "residual debris" method. The humin isolated by the method of MacCarthy et al. (*4*) is obtained as the result of an active isolation step and is not simply the residue that remains after extracting humic acid and fulvic acid.

The ability to isolate humin from other insoluble but nonhumic residues facilitates the study of humin. However, most humin samples represent a substance in which relatively little organic material is dispersed in a much more abundant inorganic matrix. To concentrate this organic component to a level at which most analytical characterization methods can be applied, most workers have digested the sample in a mixture of HF and HCl to dissolve silicate minerals (*5, 6*).

The most serious limitations of the HF–HCl treatment are the loss of organic matter to the acid solution and alteration of the organic residue during silicate dissolution (*2, 7–10*). In general, the effect of acid digestion (to remove silicates) on the organic constituents of humin is greater on geologically young samples than on older samples (*9*). Thus, it could be expected that humins from unlithified sediments would be altered by HF–HCl digestion.

The actual effect of HF–HCl digestion on the organic constituents of humin has not been clearly demonstrated through direct comparison. Even the use of HCl alone can alter the components present in a particular humic sample (*11*). The hydrolysis of a sample with 6 M HCl to remove carbohydrate and proteinaceous material associated with humic materials (*12, 13*) will almost certainly alter the organic concentrate's original composition.

An extension of the method of MacCarthy et al. (*4*) provides an alternative, and much milder, method for separating the organic components of humin from its inorganic constituents. This procedure, unlike the HF–HCl method, is no more chemically aggressive than the method by which the humin was originally isolated. In addition, it accomplishes a separation of the organic components of humin into recognized fractions.

The objective of this chapter is to describe (1) a new method for the isolation of humin, (2) the characterization of a stream sediment humin, (3) the separation of humin into three primary fractions, (4) the characterization of the organic components of humin, and (5) a potential role for humin in influencing the fate of contaminants in an aquatic environment.

Experimental Methods

The Methyl Isobutyl Ketone Method. Because this isolation technique and the concomitant observations are inherent to the development of the ideas on which this chapter is based, they will be described in some detail. This technique consists of a series of steps that partition the organic substances in a sample (sediment, soil, etc.) between an aqueous phase and methyl isobutyl ketone (MIBK) according to their varying solubilities as a function of pH. The isolation of humin from a sample is summarized in Figure 1 and is performed as follows:

1. The dried sample (1 g of sediment, soil, etc.) is added to 100 mL of 0.5 M NaOH solution and stirred for 24 h.
2. The entire mixture from step 1 is transferred to a separatory funnel along with 75 mL of MIBK and acidified to pH 1 with concentrated HCl.
3. The mixture is shaken vigorously and the organic matter is allowed to partition between the organic and aqueous phases. Humin and humic acid enter the MIBK phase as a *suspension*, leaving most of the fulvic acid in the aqueous phase. The aqueous phase containing most of the fulvic acid is discharged.
4. A fresh 100-mL aliquot of 0.5 M NaOH is added to the separatory funnel, the contents are shaken vigorously, and the organic matter is allowed to partition between the two phases. The humic acid is extracted from the MIBK phase into the aqueous alkaline phase. This separation leaves behind, *suspended* in the MIBK phase, a material that conforms to the definition of humin (i.e., the dried material, in the acid or salt form, does not dissolve in an aqueous system, regardless of the system's pH).

A humic material that conforms to the operational definition of humin exists only after the completion of step 4. Insoluble residues may settle out at various points in the procedure. These inorganic residues (≤1.0% organic) consist of mineral matter.

Step 5A is performed to obtain humin in the hydrogen form, and step 5B is performed to fractionate the humin.

5A. The alkaline aqueous phase from step 4 is discharged and the humin is shaken vigorously with 0.1 M HCl. The phases are allowed to separate, the aqueous phase is discharged, and this process is re-

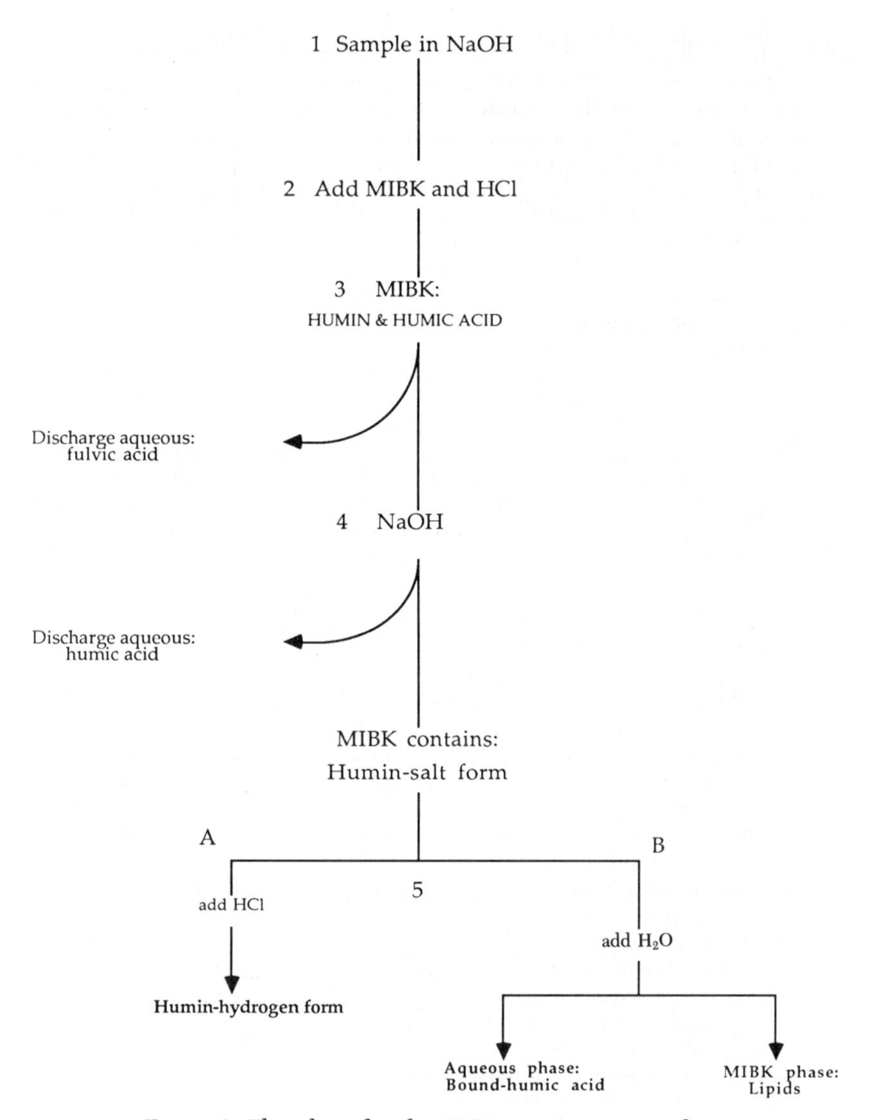

Figure 1. Flowchart for the MIBK extraction procedure.

peated until the aqueous phase is colorless. The humin remains *suspended* in the organic phase and can be collected by evaporation of the MIBK.

5B. The alkaline aqueous phase from step 4 is discharged and 100 mL of deionized water is added to the MIBK phase. After vigorous shaking the phases are allowed to separate.

Step 5B causes the humin to separate into an insoluble nonhumic residue, a fraction that resides in the MIBK phase, and a fraction that is dissolved in the alkaline aqueous phase. The nonhumic residue is typically 1% organic matter or less (demonstrated by combustion). This fraction is referred to here as the "insoluble residue". The fraction dissolved in the MIBK, in the hydrogen form, conforms to the geochemical definition of lipids as stated by Bergman (*14*): ". . . organism-produced substances that are soluble in water with difficulty but are extractable by one or another of the so-called fat solvents such as chloroform, carbon tetrachloride, ethers, aliphatic and aromatic hydrocarbons and acetone". A lipid fraction is obtained, although its abundance is diminished, even when the dried humin sample is exhaustively extracted with organic solvents (e.g., 72 h in a Soxhlet apparatus) prior to being subjected to steps 1–5B of the MIBK method.

The fraction dissolved in the alkaline aqueous phase present after step 5B is brown and precipitates when the solution is acidified; it thus conforms to the operational definition of humic acid. Humin, a humic fraction that is insoluble in an aqueous solution regardless of that solution's pH, no longer exists. Evidently step 5B causes a disaggregation of the humin into three distinct components.

Humin was fractionated into an insoluble residue, a humic-acid-like fraction, and lipids by means of the MIBK method from every soil, sediment, peat, or lignite sample that we studied. The humic-acid-like fraction will be referred to as bound humic acid. In addition, humin obtained by the traditional alkali extraction procedure will also fractionate into a bound humic acid fraction, lipids, and an insoluble residue when subjected to steps 1–5B of the MIBK method.

Characterization of a Stream Sediment Humin. SOURCE MATERIAL. All humic and lipid materials were isolated from a stream sediment collected from South Clear Creek, Clear Creek County, Colorado. Random grabs were collected from the top 5 cm of the undisturbed stream bed. The bulk stream sediment contained, on a dry basis, 7.0% organic carbon. Humin organic carbon constituted 58.4% of the total organic carbon (TOC) in the sediment, and humic acid represented 35.9% of the TOC (*1*).

ISOLATION OF HUMIN, HUMIC ACID, AND FULVIC ACID. Humin, humic acid, and fulvic acid were isolated from the stream sediment by the MIBK method (Figure 1). The humic acid from step 4 (Figure 1) was taken through an additional two cycles (steps 1–4). It was then repartitioned into a fresh MIBK phase and collected by evaporation of the MIBK. The fulvic acid obtained from step 2 (Figure 1) was desalted by the method of Thurman and Malcolm (*15*), which uses acrylic–ester resin (XAD–8) and a strong-acid cation-exchange resin in the hydrogen form. It was necessary to remove the MIBK from the solution containing fulvic acid prior to passing the solution through the XAD–8 resin because MIBK saturates the resin and greatly diminishes its capacity for fulvic acid. This removal was accomplished by adjusting the fulvic acid solution to pH 7 and rotoevaporating the solution. The solution was then treated as described by Thurman and Malcolm (*15*).

QUANTIFICATION OF THE FRACTIONS OF HUMIN. The humin (1 g, hydrogen form) isolated by the MIBK procedure was extracted for 72 h with a benzene–methanol azeotrope (3:1, v:v) in a Soxhlet apparatus. The solvent was evaporated and the residue weighed. This lipid fraction (the residue) is referred to as the bitumen. The solvent-extracted humin samples were then subjected to steps 1–5B of the MIBK procedure (Figure 1). After step 5B the bound humic acid was obtained by acidifying the alkaline aqueous phase, shaking it with a fresh MIBK phase, washing it once with deionized water, and evaporating the MIBK. The residue (bound humic acid) was then weighed. The lipids remaining in the humin after Soxhlet extraction of the sample are referred to here as bound lipids. The bound lipids from step 5B were converted to the hydrogen form by shaking the MIBK phase with 0.1 M HCl, the solvent was evaporated, and the residue (bound lipids) was weighed. The insoluble residue was acidified with 50 mL of 0.1 M HCl, centrifuged, shaken with deionized water, centrifuged, dried in a 105 °C oven, and weighed. Triplicate samples of the stream sediment humin were treated in this fashion.

CHARACTERIZATION OF THE HUMIC MATERIALS. Elemental analysis consisted of direct determination of C, H, N, and S contents; O content was determined by difference. The results are presented on a moisture-free, ash-free basis. Limitations associated with this approach to elemental analysis, and elemental analysis in general, have been discussed by Rice and MacCarthy (*16*). Acidic functional group determinations consisted of barium hydroxide titration for measurement of total acidity (*17*) and a modified acetate method for measurement of carboxyl content (*1*). Following convention, the weakly acidic hydroxyl content was taken as the difference between total acidity and carboxyl content (*13*). Infrared spectra were obtained by the KBr pellet method over the range 4000–600 cm^{-1}.

CHARACTERIZATION OF THE LIPID FRACTION. The isolated bitumen and bound lipids were fractionated into hexane, benzene, and methanol eluates by column chromatography. Activated silica was employed as the stationary phase. The hexane eluate was analyzed directly by gas chromatography–mass spectrometry (GC–MS). A portion of the methanol eluate was methylated with 14% BF_3–methanol. The derivatized samples were then analyzed by GC–MS. The gas chromatograph was programmed from 70 to 300 °C at 4 °C/min, with an initial hold of 1.5 min and a final hold of 10 min. Bitumen and bound lipid samples were injected in the split mode, with a split ratio of 5 to 1. The carrier gas was He at a linear flow rate of 34 cm/s through a 30-m DB–5 (nonpolar) bonded–stationary-phase capillary column. The mass spectrometer was operated in the electron impact mode at an acceleration voltage of 4 kV, electron energy of 70 eV, a source temperature of 200 °C, and low resolution conditions.

Results and Discussion

Quantification of the Humin Fractions. The weight percents of the stream sediment humin fractions on a dry basis are bitumen, 26.3 ± 2.1; bound lipids, 1.7 ± 0.7; bound humic acid, 11.0 ± 1.9; and insoluble residue, 67.8 ± 2.7. (Errors are absolute standard deviations.) The total (106.8%) may reflect inadequate removal of salt from the samples prior to drying.

The bitumen component represents a larger fraction of the humin than the bound humic acid. The total lipids in the humin (bitumen and bound lipids) are the largest component of humin organic matter in this stream sediment sample. The bound lipids represent a small fraction compared to the bitumen in the unextracted humin. Once the bitumen has been extracted, however, the bound lipids represent 13.4% of the remaining organic matter in the stream sediment humin.

Infrared Spectra. The IR spectra of the stream sediment humin, bound humic acid, humic acid, and fulvic acid in the hydrogen form are given in Figure 2. The humin spectrum exhibits strong bands at 3720 and

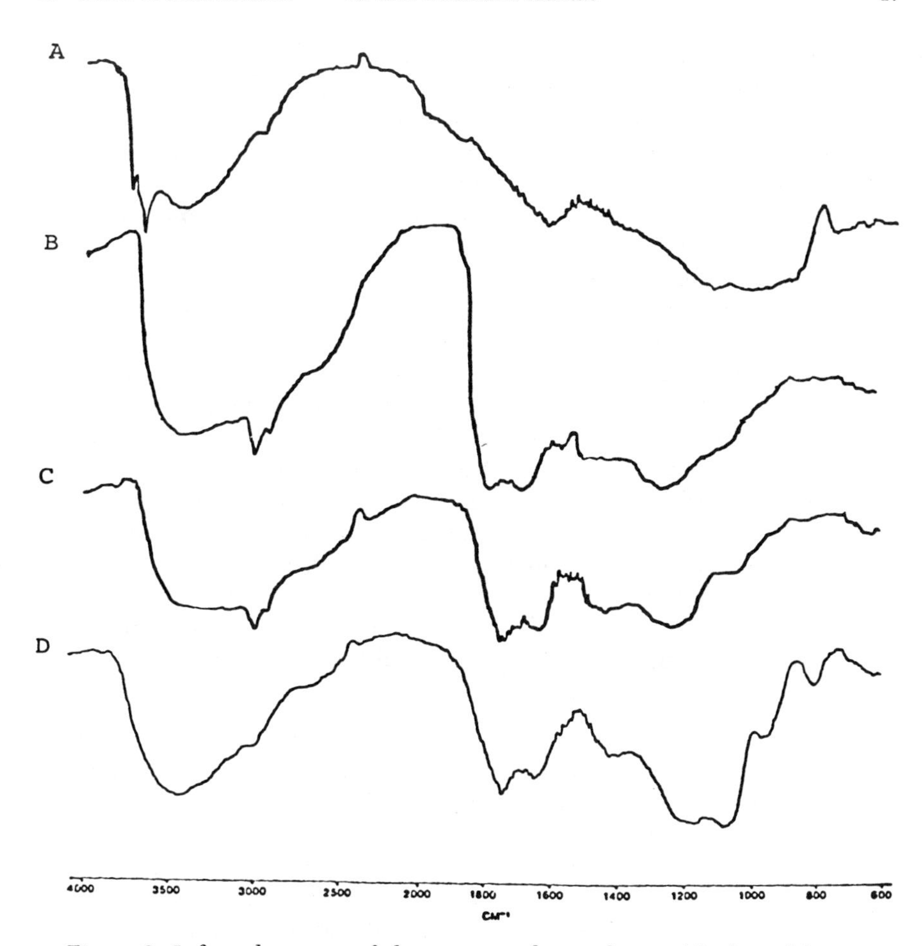

Figure 2. Infrared spectra of the stream sediment humin (A), bound humic acid (B), humic acid (C), and fulvic acid (D), in the hydrogen form, isolated by the MIBK method.

3640 cm^{-1} that can be attributed to the stretching vibration of the SiO–H bond (*18*). When the stream sediment humin is treated with HF–HCl to reduce the mineral content of the sample, these bands completely disappear. Another band that is usually attributed to an Si–O stretching vibration of silicate minerals (*18, 19*) is centered at about 1050 cm^{-1}. This band also disappears after the stream sediment humin has been treated with HF–HCl. The IR spectra of the stream sediment humin, humic acid, and fulvic acid exhibit absorption bands at 2920 and 2860 cm^{-1}. These absorptions are attributed to the asymmetric (2920 cm^{-1}) and symmetric (2860 cm^{-1}) stretching vibrations of aliphatic C–H bonds in methyl and methylene groups. These absorbances in the humin IR spectrum are seen only as an ill-defined shoulder on the OH absorption (3400 cm^{-1}).

An absorbance at 1720 cm^{-1} is usually attributed mainly to the C=O vibration of un-ionized carboxyl groups. Both humic acid and fulvic acid, in the hydrogen form, exhibit more pronounced bands at this wave number than humin in the hydrogen form. The mineral signature displayed by most humin samples that have not been subjected to HF–HCl digestion is overwhelming, and humin characterization is difficult. It is more informative to use the IR spectrum of the bound humic acid as representative of the humic component of humin and compare it to the spectra of humic and fulvic acids.

The IR spectra of the bound humic acid and humic and fulvic acids (Figure 2) differ from each other in several ways. The bound humic acid spectrum exhibits more intense aliphatic absorptions (2920 and 2860 cm^{-1}), relative to the OH absorbance (3400 cm^{-1}), than either humic or fulvic acid. The bound humic acid spectrum exhibits a broad absorption at 2600 cm^{-1} that appears as a shoulder on the OH band. In this respect, it is similar to both humic and fulvic acids. Both humic and fulvic acid exhibit absorbances at 1720 cm^{-1} that are more intense (relative to the band at 1650 cm^{-1}) than that observed in the bound humic acid spectra. Below 1650 cm^{-1} the bound humic acid and humic acid spectra are generally similar. Thus on the basis of a comparison of their IR spectra, the bound humic acid and humic acid from this stream sediment may be similar, although the bound humic acid is more aliphatic.

Elemental Analysis. The elemental compositions and atomic H/C and O/C ratios of the humin, humic acid, fulvic acid, and bound humic acid isolated by the MIBK method are given in Table I. The results are presented on a moisture-free, ash-free basis.

The stream sediment humin exhibits carbon and oxygen weight percents intermediate between those of the stream sediment humic acid and fulvic acid. The humin's hydrogen content is somewhat higher than that of the humic or fulvic acids. The H/C atomic ratio of the humin, a measure of a material's degree of aliphaticity (*20*), is intermediate between the values displayed by the humic acid and fulvic acid. These values indicate that, for these samples, humin is more aliphatic than humic acid. This trend is consistent with other work (*2*, *6*, *10*).

Table I. Elemental Compositions of the Stream Sediment Humic Materials

Material	*C*	*H*	*N*	*S*	*O*	*H/C*	*O/C*
Humin	53.85	4.27	<0.30	0.21	41.67	0.95	0.73
Humic acid	57.06	3.94	3.37	0.55	35.08	0.83	0.46
Fulvic acid	38.19	3.48	1.77	0.58	55.98	1.09	1.10
Bound humic acid	52.07	4.83	3.02	1.01	39.07	1.11	0.56

NOTES: C, H, N, S, and O contents are expressed as weight percents. H/C and O/C values are atomic ratios. All results are expressed on a moisture-free, ash-free basis. Oxygen was determined by difference.

The elemental composition of the stream sediment bound humic acid is typical of that observed for humic acids (*12, 13*). The H/C ratio of the bound humic acid is similar to that of the fulvic acid. This value indicates that bound humic acid is more aliphatic than either humin or humic acid.

Acidic Functional Group Analysis. Table II gives the acidic functional group contents of the stream sediment humin, bound humic acid, humic acid, and fulvic acid.

Humin exhibits a total acidity similar to that of humic acid. Despite being isolated by a nontraditional method, the acidic functional group contents of both humic acid and humin are typical of values measured from more traditionally isolated samples (*12, 13*). The fulvic acid exhibits higher values for each parameter than either humin or humic acid. This relationship is usually observed when the total acidities of humic acid, fulvic acid, and humin are compared (*12, 13*). The acidic functional group content of bound humic acid is intermediate between those of fulvic acid and the humin and humic acid (Table II).

The Lipid Fractions. Figure 3 summarizes the saturated *n*-alkanes identified in the hexane eluate of the stream sediment bitumen. The alkanes ranged from C_{14} to C_{33}, with the C_{27} *n*-alkane predominating. The chromatogram of the bound lipid hexane eluate showed a single peak that was identified as dibutyl phthalate. This compound was traced to an impurity in the hexane. The only evidence of branched alkanes in the hexane eluate of the bitumen was attributed to compounds that were identified as 2,6-dimethyldodecane and 2,6-dimethylhexadecane.

The distribution of saturated monobasic fatty acid methyl esters (FAMEs) formed by methylation (with BF_3–methanol) of the stream sediment bitumen and bound lipids is given in Figure 4. The distribution of the bitumen FAME is dominated by the methyl ester of palmitic acid (C_{16}) and secondarily by the methyl ester of lignoceric acid (C_{24}). Also identified in the bitumen were the isobranched monobasic acids methyl 14-methyl-

Table II. Acidic Functional Group Contents of the Stream Sediment Humic Materials

Material	*Total Acidity*	*Carboxyl Content*	*Weakly Acidic Hydroxyl Content*
Humin	3.26 ± 0.11	1.95 ± 0.04	1.31
Humic acid	3.61 ± 1.39	1.47 ± 0.11	2.14
Fulvic acid	9.13	4.85	4.28
Bound humic acid	4.84 ± 0.69	2.61 ± 0.30	2.23

NOTES: Contents are reported in milliequivalents per gram. Errors are absolute standard deviations calculated from at least three measurements; fulvic acid values represent one measurement.

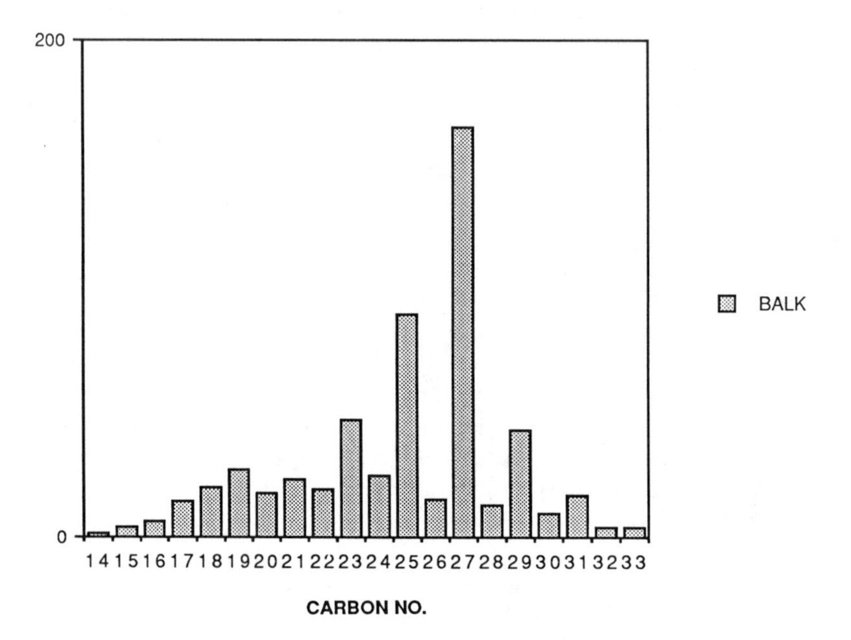

Figure 3. Distribution of saturated n-*alkanes in bitumen of stream sediment humin.*

pentadecanoate, methyl 15-methylhexadecanoate, and methyl 16-methylheptadecanoate. Unsaturated FAMEs identified were n-$C_{16:1}$ (methyl hexadecenoate) (16:1 indicates a 16-carbon chain with one double bond, 16:2 indicates two double bonds, etc.), n-$C_{18:1}$ (methyl octadecenoate), a branched $C_{18:1}$ (methyl 11-methyloctadecenoate), n-$C_{18:2}$ (methyl octadecenoate), and n-$C_{19:1}$ (methyl nonadecenoate).

The distribution of the FAME in the stream sediment bound lipids in Figure 4 is different from that of the bitumen FAME. The bound lipid distribution is dominated by C_{16} and C_{18} FAMEs. Though the observed FAMEs range from C_{14} to C_{30} and their distribution appears to be bimodal, the higher-carbon-number FAMEs are of considerably less abundance in the bound lipids than in the bitumen. The unsaturated FAME $C_{18:1}$ was also observed in the stream sediment bound lipids.

Conclusions

Because of the variability inherent in humic materials, it is difficult to make generalizations about their nature or behavior, especially on the basis of only one set of samples. Within this limitation, some comparisons will be made among humin, bound humic acid, humic acid, and fulvic acid.

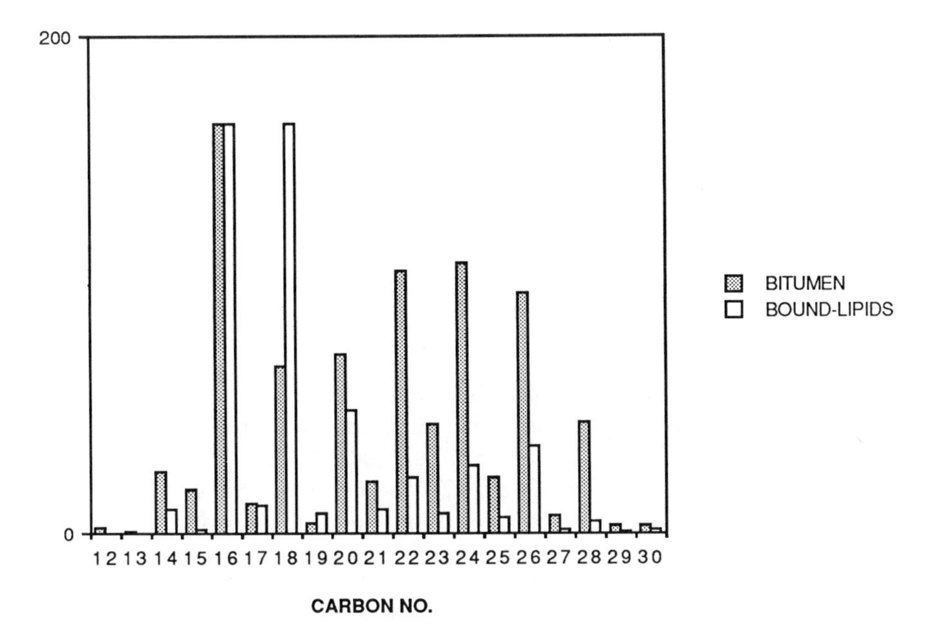

Figure 4. Distribution of saturated n-*FAME in bitumen and bound lipids of stream sediment humin.*

On the basis of IR spectra, the humin isolated from this stream sediment appears to be more similar to humic acid than to fulvic acid. Despite this similarity, there are differences between the humic acid and the bound humic acid. The IR spectra suggest that bound humic acid may be more aliphatic (based on aliphatic C–H absorbances at 2920 and 2860 cm^{-1}) than humic acid. This possibility is consistent with IR spectroscopic comparisons of HF–HCl-treated humin samples and humic acids from soils (*5, 21*) and recent sediments (*2, 19*). Elemental analysis (Table I) supports this conclusion; bound humic acid exhibits a larger H/C ratio (which indicates a more aliphatic nature) than humic acid.

Solid-state ^{13}C-NMR studies of humins isolated from recent sediments (*22, 23*) have also ascribed a more aliphatic nature to humin than to the corresponding humic acid. The total acidity and carboxyl content of the bound humic acid (Table II) are intermediate between those of the humic acid and fulvic acid. The result of this characterization is that the bound humic acid displays some similarities to, and some differences from, humic acid. These differences indicate that bound humic acid is a distinct component of humin, not merely unextracted humic acid.

Both the *n*-alkane and FAME profiles show differences between the bitumen and bound lipid fractions. Both compound classes are observed in the bitumen, but *n*-alkanes are not observed in the bound lipids. Fatty acids are observed in the bound lipids, but the higher-molecular-weight acids seem to have been preferentially extracted. The dominant FAMEs in the bound lipid fraction correspond to the C_{16} and C_{18} acids. The differences in the distribution profiles of the bitumen and bound lipid FAME indicate that the bound lipids are not simply unextracted bitumen. These results suggest that fatty acids, dominated by the n-C_{16} and n-C_{18} saturated acids, are more intimately associated with humin than are other high-molecular-weight fatty acids.

Implications for the Environmental Fate of Pollutants

The organic solid phase of a soil or sediment (i.e., humin or humic acid) is an important control on the fate of an anthropogenic hydrophobic organic compound introduced into the environment (*24, 25*). In a manner analogous to interactions in wet soils (*24, 26, 27*), the contributions of a sediment's mineral component (i.e., the stream sediment insoluble residue) to the overall interaction of the sediment with a hydrophobic pollutant could be expected to be negligible. The interaction of the hydrophobic organic chemical with the sediment's organic components has been described as a solute partitioning between the aqueous phase and organic solid phase (*24, 26–29*). From this perspective, the sediment organic matter may be viewed as the organic phase in a two-phase (i.e., organic solvent–water) system. Unlike a simple two-phase system in a solvent extraction, an aqueous solute would not be interacting with a homogeneous organic phase in a natural aquatic system. As Isaacson and Frink (*25*) have pointed out, ". . . the apparent uniformity of the sorbent character of organic matter may be partially the result of 'averaging' which is inherent in observations of such complex material." As the previous results have shown, humin apparently consists of three distinct organic components: bitumen, bound humic acid, and bound lipids.

The present study raises several questions about the interaction of a hydrophobic organic chemical with sedimentary organic matter. First, because the bulk of a sediment's organic content can be expected to conform to the definition of humin, the relative and absolute contribution of humin as a sorbent for a hydrophobic organic chemical in a water–sediment system needs to be assessed. For example, size fractionation of a sediment is frequently employed to obtain organic-matter fractions (*25*). Though humic materials are associated with the clay-sized fractions (*30*), they are not readily extractable (*31*) and thus conform to the definition of humin. The smaller (clay-sized) fractions frequently exhibit the greatest affinity for hydrophobic organic chemicals (*25*). These observations suggest that humin may exhibit an affinity for hydrophobic organic chemicals. Nearpass (*32*) found that a peat humin was a better sorbent for 4-amino-3,5,6-trichloropicolinic acid (picloram) than either the bulk peat or the peat humic acid.

Second, this study has shown that the organic components of humin can easily be separated into solvent-extractable organic matter (bitumen) and non-solvent-extractable organic matter (bound humic acid and bound lipids). The ease with which the bitumen can be extracted implies that it is much less intimately associated with humin than are the bound lipids. Thus, the bound lipids may represent a distinct class of organic matter within the humin. The absolute and relative affinities of the bitumen and the solvent-extracted humin for hydrophobic organic compounds need to be assessed.

Third, the MIBK partitioning method has provided a means for separating the non-solvent-extractable organic components of humin from the inorganic components by a relatively mild procedure. Humin consists of a humic-acid-like substance, a bound lipid fraction, and, in the stream sediment studied here, a mineral fraction consisting of fine-grained material. The presence of the bound lipid fraction also suggests that a mechanism exists that causes some compounds to be more intimately associated with the humin. Isaacson and Frink (25) observed that up to 90% of some chlorophenols bound to a sediment cannot be extracted. Thus, an understanding of how the bound lipid fraction is associated with the bound humic acid may provide insights into the mechanisms of the interaction of hydrophobic organic chemicals with the organic solid phase in an aquatic sedimentary environment.

References

1. Rice, J. A. Ph.D. Thesis, Colorado School of Mines, 1987.
2. Vandenbroucke, M.; Pelet, R.; Debyser, Y. In *Humic Substances in Soil, Sediment, and Water: Geochemistry, Isolation, and Characterization*; Aiken, G. R.; McKnight, D. M.; Wershaw, R. L.; MacCarthy, P., Eds.; Wiley–Interscience: New York, 1985; pp 249–273.
3. Stuermer, D. H.; Peters, K. E.; Kaplan, I. R. *Geochim. Cosmochim. Acta* **1978,** *42,* 989–997.
4. MacCarthy, P.; Allen, R. S., III; Rice, J. A. *Geoderma* **1988,** in review.
5. Russell, J. D.; Vaughan, D.; Jones, D.; Fraser, A. R. *Geoderma* **1983,** *29,* 1–12.
6. Hatcher, P. G.; Breger, I. A.; Maciel, G. E.; Szeverenyi, N. In *Humic Substances in Soil, Sediment, and Water: Geochemistry, Isolation, and Characterization*; Aiken, G. R.; McKnight, D. M.; Wershaw, R. L.; MacCarthy, P., Eds.; Wiley–Interscience: New York, 1985; pp 275–302.
7. Huc, A. Y.; Durand, B. *Fuel* **1977,** *56,* 73–80.
8. Povoledo, D.; Pitze, M. *Soil Sci.* **1979,** *128,* 1–8.
9. Durand, B.; Nicaise, G. In *Kerogen*; Durand, B., Ed.; Editions Technip: Paris, 1980; pp 35–53.
10. Ishiwatari, R. In *Humic Substances in Soil, Sediment, and Water: Geochemistry, Isolation, and Characterization*; Aiken, G. R.; McKnight, D. M.; Wershaw, R. L; MacCarthy, P., Eds.; Wiley–Interscience: New York, 1985; pp 147–180.

11. Steelink, C. In *Humic Substances in Soil, Sediment and Water: Geochemistry, Isolation, and Characterization*; Aiken, G. R.; McKnight, D. M.; Wershaw, R. L.; MacCarthy, P., Eds.; Wiley–Interscience: New York, 1985; pp 457–475.
12. Schnitzer, M. In *Soil Organic Matter*; Schnitzer, M.; Khan, S. U., Eds.; Elsevier: New York, 1978; pp 1–64.
13. Stevenson, F. J. *Humus Chemistry: Genesis, Composition and Reactions*; Wiley: New York, 1982; 433 pp.
14. Bergman, W. In *Organic Geochemistry*; Breger, I. A., Ed.; MacMillan: New York, 1963; pp 503–542.
15. Thurman, E. M.; Malcolm, R. L. *Environ. Sci. Technol.* **1981,** *15,* 463–466.
16. Rice, J. A.; MacCarthy, P. *Geochim. Cosmochim. Acta* **1988,** in review.
17. Schnitzer, M.; Gupta, U. C. *Soil Sci. Soc. Am. Proc.* **1965,** *29,* 274–277.
18. Farmer, V. C. In *The Infrared Spectra of Minerals*; Farmer, V. C., Ed.; Mineralogical Society: London, 1974; pp 331–363.
19. Bellamy, L. J. *The Infrared Spectra of Complex Minerals*; Methuen: London, 1958; 423 pp.
20. van Krevelen, D. W. *Fuel* **1961,** *29,* 269–284.
21. Skjemstad, J. O.; Dalal, R. C.; Baron, P. F. *Soil Sci. Soc. Am. J.* **1986,** *50,* 354–359.
22. Hatcher, P. G.; Rowan, R.; Mattingly, M. A. *Org. Geochem.* **1980,** *2,* 77–85.
23. Hatcher, P. G.; Breger, I. A.; Dennis, L. W.; Maciel, G. E. In *Aquatic and Terrestrial Humic Materials*; Christman, R. F.; Gjessing, E. T., Eds.; Ann Arbor Science: Ann Arbor, 1983; pp 37–81.
24. Chiou, C. T.; Porter, P. E.; Schmedding, D. W. *Environ. Sci. Technol.* **1983,** *17,* 227–231.
25. Isaacson, P. J.; Frink, C. R. *Environ. Sci. Technol.* **1984,** *18,* 43–48.
26. Chiou, C. T.; Peters, L. J.; Freed, V. H. *Science (Washington, DC)* **1979,** *206,* 831–832.
27. Chiou, C. T.; Peters, L. J.; Freed, V. H. *Science (Washington, DC)* **1981,** *213,* 684.
28. Carter, C. W.; Suffet, I. H. *Environ. Sci. Technol.* **1982,** *16,* 735–740.
29. Choi, W.-W.; Chen, K. Y. *Environ. Sci. Technol.* **1976,** *10,* 782–786.
30. Wershaw, R. L.; Pinckney, D. J. In *Contaminants and Sediments*, Vol. 2; Baker, R. A., Ed.; Ann Arbor Science: Ann Arbor, 1980; pp 207–219.
31. Young, J. L.; Spycher, G. *Soil Sci. Soc. Am. J.* **1979,** *43,* 324–328.
32. Nearpass, D. C. *Soil Sci.* **1976,** *121,* 272–277.

RECEIVED for review March 31, 1988. ACCEPTED for publication June 29, 1988.

4

The Nature of Commercial Humic Acids

Patrick MacCarthy

Department of Chemistry and Geochemistry, Colorado School of Mines, Golden, CO 80401

Ronald L. Malcolm

U.S. Geological Survey, Box 25046, Mail Stop 408, Denver Federal Center, Denver, CO 80225

In a follow-up to a study based primarily on solid-state ^{13}C-NMR spectroscopy, commercial humic acids were investigated by infrared spectroscopy. Pronounced similarities were observed between the infrared spectra of all the commercial materials, and these "humic acids" appear to be partly in the salt form or complexed with metals. The commercial materials differ somewhat from each other in terms of their mineral compositions. The infrared spectra of the commercial materials are not remarkably different from humic acids isolated from peat, soils, and leonardite, but pronounced differences between these two classes of materials are revealed by ^{13}C-NMR spectroscopy. The use of commercial humic acids as analogues of soil and water humic substances is criticized in this chapter. The adverse impact of high ash content on elemental analyses is illustrated, and the desirability of using direct methods for elemental determinations is discussed.

USE OF COMMERCIAL HUMIC ACIDS AS ANALOGUES of soil and water humic substances has been escalating over the past decade. The literature on this subject has been reviewed (*1*). The same article contains a detailed characterization of various commercial humic acids by elemental analysis and solid-state ^{13}C-NMR spectroscopy, and compares the commercial products with humic and fulvic acids that were isolated from soils and streams. Although the commercial products bear some similarities to the soil and stream humic substances, there are, nevertheless, significant differences between

0065–2393/89/0219–0055$06.00/0

the two classes of materials. As a result, the use of commercial humic acids for simulating the properties and behavior of true soil and water humic substances needs to be reevaluated. Problems with the use of the commercial humic acids are further compounded by the lack of information on the source and methods of isolation and treatment, if any, of these materials.

The objectives of this chapter are to provide further information on the commercial humic acids from five different suppliers, to discuss the nature of the commercial humic acids, and to illustrate the adverse impact of high ash content on the elemental analysis of humic substances.

Experimental Materials and Methods

The following commercial humic acids were used: No. 1: K&K humic acid (lot No. 215316, 1974) obtained from K&K Laboratories, Plainview, NY; No. 2: CPL humic acid (1974) obtained from Chemicals Procurement Laboratories, College Point, NY; No. 3: Fluka–Tridom humic acid (lot No. 159128115, 1974) from Fluka Chemical Corp., Hauppauge, NY; No. 4: Pfaltz and Bauer humic acid (batch 1, 1974) and No. 5: Pfaltz and Bauer humic acid (batch 2, 1981) obtained from Pfaltz and Bauer, Stamford, CT; No. 7: Aldrich humic acid (lot No. 113027, 1974); and No. 8: Aldrich humic acid, sodium salt (lot No. LE3601KE, 1981) from Aldrich Chemical Company, Milwaukee, WI.

Dopplerite (No. 10) was obtained from an outcrop at Stinking Springs, Fremont County, near Jackson, WY; this sample was used as dug from the outcrop and was not treated in any way. Peat for this study was obtained at a depth of about 1 m at the Peatland Experimental Station, Glenamoy, County Mayo, Ireland. Leonardite was taken from a homogenized 1000-kg sample obtained at the Biscoyne Mine, Bowman County, SD. This large sample of leonardite represents one of the bulk reference materials of the International Humic Substances Society, from which one of the standard humic acids (No. 12) has been extracted. The Irish peat humic acid (No. 11) (*2*) and the leonardite humic acid were obtained from the bulk materials by extraction with sodium hydroxide. The ash content of two of the commercial materials (Nos. 5 and 8) was diminished by subjecting the materials to liquid–liquid extraction with methyl isobutyl ketone (*3*), and the resulting materials are designated as sample Nos. 6 and 9, respectively.

Soil humic acid (No. 13) was extracted from the Sanhedron A1 horizon near Petrolia, Mendocino County, in the Mattole River Valley of Northern California, by using a modified sodium hydroxide extraction procedure for soils (*4*). Dissolved stream humic acid (No. 14) was isolated from Ohio River water with a methyl methacrylate resin (XAD–8) extraction procedure (*5*).

Elemental and ash contents were determined by Huffman Laboratories, Golden, CO. All elemental contents were determined by direct methods. Ash content was determined by combustion to constant weight at 750 °C in an oxygen atmosphere. The elemental and ash contents are given in Table I. Infrared spectra were recorded using KBr disks (1% sample) on an IR spectrophotometer (Perkin Elmer Model 580). Solid-state ^{13}C-NMR spectra were measured by G. Maciel, Colorado State University, Fort Collins, CO. All NMR spectra were run by the cross-polarization magic-angle spinning (CPMAS) technique.

Table I. Elemental Contents of Commercial Humic Acids and Selected Humic Acids from Dopplerite, Peat, and Leonardite

Sample	*C*	*H*	*O*	*N*	*S*	*P*	*Total*	*Ash (%)*
1. K&K humic acid	63.25	5.17	32.22	0.68	4.49	0.15	106.0	22.48
2. CPL humic acid	59.89	5.07	34.81	0.81	0.72	<0.05	101.3	9.28
3. Fluka–Tridom humic acid	65.79	5.51	37.79	0.71	3.16	<0.05	113.0	32.82
4. P&B humic acid (batch 1)	63.88	5.69	35.23	0.80	3.04	<0.05	108.6	24.77
5. P&B humic acid (batch 2)	62.84	5.38	34.99	0.80	3.68	<0.05	104.9	15.01
6. P&B humic acid (purified batch 2)	61.87	4.45	28.92	0.89	3.35	<0.05	99.5	4.44
7. Aldrich humic acid	63.15	5.60	34.98	0.80	4.58	<0.05	108.9	22.56
8. Aldrich humic acid, sodium salt	68.98	5.26	43.45	0.74	4.24	<0.05	122.7	31.21
9. Aldrich humic salt (purified to H^+ form)	65.31	5.94	25.05	0.51	3.36	<0.05	100.2	4.46
10. Wyoming dopplerite	61.16	4.39	43.59	1.61	2.36	<0.05	113.1	31.02
11. Irish peat humic acid	60.70	5.22	32.59	1.29	0.90	<0.05	100.7	1.05
12. Leonardite humic acid	63.25	3.64	31.05	1.17	0.84	<0.05	100.0	2.47

NOTE: All results are expressed on a percent-by-weight, ash-free, and moisture-free basis.

Results and Discussion

Infrared Spectroscopy. The infrared spectra of all samples are shown in Figure 1. Overall, the infrared spectra of the commercial materials (No. 1–9) are generally similar to those of dopplerite (No. 10), peat (No. 11), and leonardite (No. 12) humic acids; however, some differences are discernible. Spectra of all the untreated commercial humic *acids* (samples

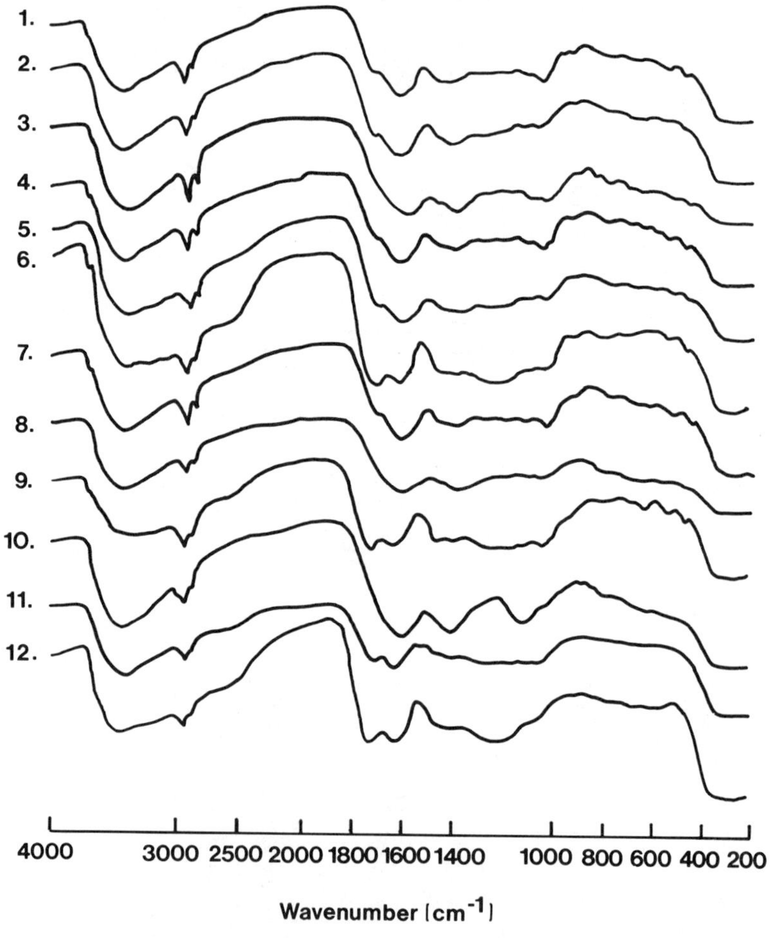

Figure 1. Infrared spectra of commercial humic acids (Nos. 1–9) and other humic materials (Nos. 10–12): (1) K&K humic acid; (2) CPL humic acid; (3) Fluka–Tridom humic acid; (4) Pfaltz and Bauer humic acid (batch 1); (5) Pfaltz and Bauer humic acid (batch 2); (6) "purified" Pfaltz and Bauer humic acid (batch 2); (7) Aldrich humic acid; (8) Aldrich humic acid, sodium salt; (9) "purified" Aldrich humic acid from No. 8; (10) Wyoming dopplerite; (11) Irish peat humic acid; and (12) leonardite humic acid.

No. 1–5 and 7) are very similar to each other, the major differences being in the ~1750-cm^{-1} region of sample No. 3.

Because of the absence of a reasonably well-defined absorption band at ~1750 cm^{-1} in sample No. 3, the spectrum resembles that of the salt of humic acid more than humic acid itself. The resemblance becomes evident on comparison of this spectrum with that of sample No. 8, which is known to be in the salt form. The major difference between the spectrum of untreated (No. 5) and partially de-ashed (No. 6) Pfaltz and Bauer humic acid resides in the increased absorbance of the 1750-cm^{-1} band in the latter. A similar, but more pronounced, change was observed on treating sample No. 8, which was known to be in the salt form, to produce sample No. 9.

The 1750-cm^{-1} band is due to the C=O stretching vibration of the un-ionized and uncomplexed carboxyl group. Evidently, the carboxyl group in commercial humic acids Nos. 1, 2, 4, 5, and 7 is partly in a salt or complexed form. The nominal "humic acid" No. 3 is evidently largely in the salt form, as is the Aldrich humic acid salt, sample No. 8. In the de-ashing procedure that was applied to samples No. 6 and 9, the humic acids are contacted with dilute HCl solution in the last step of the treatment, which ensures that all carboxyl groups are in the undissociated state. Wyoming dopplerite (No. 10) occurs naturally as a salt (*6*). This state is consistent with its infrared spectrum, which resembles that of the commercial humic acid salt (No. 8). The spectra of samples No. 11 and 12 show a pronounced absorption at ~1750 cm^{-1} because of the un-ionized carboxyl group. This characteristic is consistent with the method of preparing these samples.

The predominant ash components in many humic substances are sesquioxides of iron, aluminum, and manganese, as well as amorphous silica and exchangeable metal ions such as sodium, calcium, magnesium, iron, and aluminum. Crystalline aluminosilicate clay minerals are also frequently associated with humic substances.

Sample No. 5 (Pfaltz and Bauer humic acid, purchased in 1981) had about 10% less ash than sample No. 4 (purchased in 1974) (Table I). The difference suggests either that two different source materials were used or that the 1981 sample had been subjected to some type of purification. This decrease in ash content apparently reflects a combination of fewer metal ions associated with carboxyl groups (increase in absorption because of un-ionized and uncomplexed carboxyl at ~1750 cm^{-1} in the spectrum of sample No. 5 as compared to that of No. 4; *see* Figure 1) and the presence of less silica and aluminosilicate clay minerals (decrease in absorption in the 950-cm^{-1} region of sample No. 5).

The decrease in ash content from 15% in sample No. 5 to 4.4% in sample No. 6, caused by liquid–liquid extraction, was largely caused by replacement of exchangeable metal ions with hydrogen ions at carboxylate sites. This replacement is shown by the increase in the intensity of absorption of the ~1750-cm^{-1} band because of the un-ionized carboxyl group and by the corresponding decrease in the absorption intensity in the 1650–1575-cm^{-1} region, caused by the ionized carboxyl group, in the spectrum of sample No. 6.

The decrease in ash content after liquid–liquid extraction is apparently not caused by a reduction in amorphous silica or aluminosilicate clay minerals, because the contribution to the Si–O stretching region near 1100 cm^{-1} is essentially constant in the spectra of sample Nos. 5 and 6. Decrease in ash content because of a reduction in sesquioxides cannot be discerned from the infrared spectra.

The band at ~1750 cm^{-1} that indicates the un-ionized carboxyl group in sample No. 7 is absent from the spectrum of sample No. 8, and the band centered at ~1650 cm^{-1} has been broadened and intensified in sample No. 8. Sample No. 8 contains less amorphous silica and aluminosilicate clay minerals than sample No. 7, as indicated by the smaller absorption due to Si–O (~1100 cm^{-1}) in the former. The large decrease in ash content from 31.21% in sample No. 8 to 4.46% in sample No. 9 is again caused, at least in part, by replacement of metal ions from carboxylate groups by hydrogen ions, as shown by the large increase in intensity of the ~1750-cm^{-1} band and decrease in intensity of the carboxylate band at ~1650 cm^{-1} from sample No. 8 to sample No. 9.

Influence of Ash Content on Elemental Analysis. The elemental and ash contents for all samples are given in Table I. Ash interferes with elemental analyses (*4*), and some of the problems caused by high ash content in the study of soil organic matter have been discussed by Goh (*7*). As substantiated by the data in Table I, the elemental analysis (on an ash-free basis) of humic and fulvic acids with greater than 10% ash seldom yields a summation of elemental constituents (carbon, hydrogen, oxygen, nitrogen, sulfur, and phosphorus) within 100 ± 5%.

The considerable improvement in the quality of elemental analyses with diminished ash content is apparent from a comparison of the values in the "total" column for the "purified" samples (Nos. 6 and 9) with the corresponding Table I values for the original samples (Nos. 5 and 8, respectively). The data also indicate that high ash content leads to anomalously high values for the organic oxygen content. Of the noncommercial samples, dopplerite (No. 10), with an ash content of 31.02%, gives a poor value for the summation of elemental content; the peat humic acid (No. 11) and leonardite humic acid (No. 12), both of which have low ash content, show good values for the summation of elemental content.

All six elements listed in Table I were determined by direct methods, and the discrepancies in the elemental analyses would not be evident if one of the elements had been determined by difference, as is frequently the case in humic-substances research.

^{13}C-NMR Spectroscopy. The spectrum of sample No. 7 shown in Figure 2 is typical of the solid-state ^{13}C-NMR spectra of the commercial humic acids investigated in this research, all of which were found to be very similar to each other (*1*). The ^{13}C-NMR spectra of the commercial products (e.g., sample No. 7) were quite different from the spectra of humic acids isolated from peat (No. 11), soil (No. 13), and streams (No. 14), as illustrated in Figure 2. The remarkable similarity between the NMR spectra of the commercial humic acids and that of raw untreated dopplerite (sample No. 10) suggests a close similarity among these materials. A more detailed discussion of the ^{13}C-NMR spectra of the commercial humic acids and of humic acids isolated from peat, soils, and waters is provided in ref. 1.

Conclusions

Infrared spectra of commercial humic acids are similar to those of other humic acids. The commercial humic materials that are nominally in the acid form are, in fact, partly or largely in the salt or complexed form. However, these materials can be converted readily to the acid form. Infrared spectroscopy is not a sensitive technique for discriminating between commercial humic acids and humic acids isolated from soils and waters. However, distinct and significant differences between these two classes of materials are evident in the ^{13}C-NMR (*1*). There are also differences between these two classes of materials in elemental composition (*1*), but these differences are not as pronounced as those observed in the ^{13}C-NMR spectra.

Commercial humic acids have been used extensively by many researchers as analogues of soil and water humic substances. Many deductions concerning the environmental influence of humic substances have been based on studies involving such commercial samples. The inability of the more classical techniques, such as IR spectroscopy and elemental analysis, to clearly distinguish between the commercial materials and "true" humic substances isolated from soils and waters may have contributed to the hesitation of many environmental scientists to abandon the use of the commercial materials as analogues of soil and water humic substances, despite prior criticism (*8*) of the use of the commercial materials.

However, in view of the distinct differences between these classes of materials, as clearly revealed by ^{13}C-NMR spectroscopy, the commercial materials should not be used as analogues of soil and water humic substances in environmental studies. The problem involved in the use of the commercial materials is exacerbated by the almost complete lack of available information on the origin and pretreatment of these materials. A more recent quantitative study showed significant differences between the water solubility enhancement characteristics of commercial humic acids and those of stream-derived humic substances (*9*).

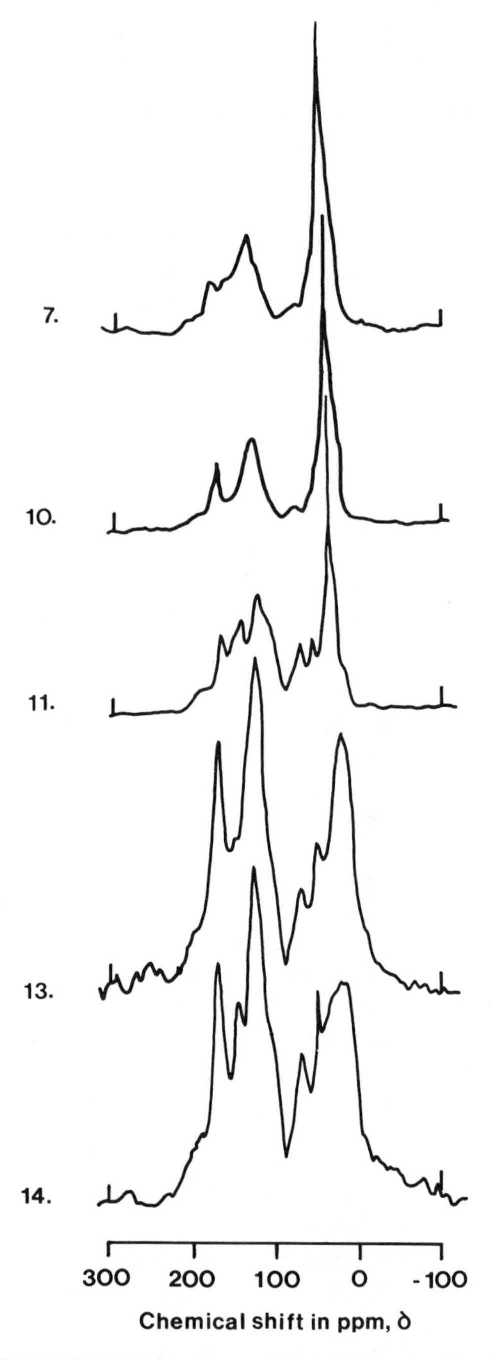

Figure 2. CPMAS ^{13}C-NMR spectra of Sample No. 7, Aldrich humic acid; No. 10, Wyoming dopplerite; No. 11, Irish peat humic acid; No. 13, Sanhedron soil humic acid; and No. 14, Ohio River humic acid.

In addition, high ash content leads to poor elemental analyses of humic substances. This results primarily from the anomalously high oxygen values and leads to a summation of elemental contents in excess of 100%. Reduction of the ash content results in considerably improved analyses. This points to a potentially serious problem in the common practice of determining the oxygen content of humic substances by difference.

References

1. Malcolm, R. L.; MacCarthy, P. *Environ. Sci. Technol.* **1986,** *20,* 904–911.
2. MacCarthy, P.; Mark, H. B., Jr. *Soil Sci. Soc. Am. Proc.* **1975,** *39,* 663–668.
3. Allen, R. S., III; MacCarthy, P. In *Agronomy Abstracts*; American Society of Agronomy: Madison, 1982; p 180.
4. Malcolm, R. L. *J. Res. U.S. Geol. Surv.* **1976,** *4,* 27–40.
5. Thurman, E. M.; Malcolm, R. L. *Environ. Sci. Technol.* **1981,** *15,* 463–466.
6. Adam, Z. M.S. Thesis, Colorado School of Mines, 1985.
7. Goh, K. M. *N. Z. J. Sci.* **1979,** *13,* 669–686.
8. MacCarthy, P.; Malcolm, R. L. In *Trace Organic Analysis: A New Frontier in Analytical Chemistry*; Chesler, S. N.; Hertz, H. S., Eds.; Proceedings of 9th Materials Research Symposium, National Bureau of Standards: Gaithersburg, MD, 1979; pp 793–796.
9. Chiou, C. T.; Kile, D. E.; Brinton, T. I.; Malcolm, R. L.; Leenheer, J. A.; MacCarthy, P. *Environ. Sci. Technol.* **1987,** *21,* 1231–1234.

Received for review March 28, 1988. Accepted for publication June 27, 1988.

5

Analysis of Humic Substances Using Flow Field-Flow Fractionation

Ronald Beckett

Water Studies Centre, Chisholm Institute of Technology, 900 Dandenong Road, Caulfield East, Victoria 3145, Australia

James C. Bigelow, Zhang Jue, and J. Calvin Giddings

Department of Chemistry, University of Utah, Salt Lake City, UT 84112

A new method for the determination of molecular-weight distributions of humic substances using flow field-flow fractionation (flow FFF) is outlined. Fairly good agreement between the results obtained by flow FFF and other methods for some humic reference substances was achieved using poly(styrene sulfonate) molecular-weight calibration standards. The molecular-weight distributions obtained for a variety of humic samples were all fairly broad and, in contrast to the data sometimes reported for gel permeation chromatography, did not show any indication of multiple peaks. The number- and weight-average molecular weights can be estimated, and these were shown to vary considerably ($\overline{M}_w$ from about 4400 to 19,000) for humic acids extracted from different environments. The method is also capable of fractionating a humic sample. However, because it is a very small-scale separation (<1 mg of sample), quite sensitive analytical detection methods would be required to make use of this feature.

INTEREST IN HUMIC SUBSTANCES GOES BACK at least to the 19th century when the dark organic matter from soil was extracted with alkali and studied (*1*, *2*). In recent years environmental scientists have become increasingly aware of the importance of humic substances in many geochemical processes that affect the fate of pollutants in natural aquatic systems and in water-treatment processes. This interest is evidenced by the growing number of

0065–2393/89/0219–0065$06.00/0

studies being reported in the past decade, either investigating the interaction of pollutants with humic substances or attempting to elucidate the structures of these complex organic compounds.

Not only do humic substances form metal complexes of varying stability with most trace metals (*3–5*) but they can also bind and help solubilize nonpolar organic compounds (*6–8*). In addition, they are adsorbed to many mineral surfaces, and thus they affect the properties of soils, sediments, and suspended particulate matter. In particular, they control the surface charge of aquatic particles (*9*, *10*) and substantially influence their colloid stability (*11–13*) and adsorption behavior (*14–16*). Despite much effort and considerable advances, our knowledge of the structure of humic substances is still inadequate. No doubt this lack stems largely from the fact that this very heterogeneous class of materials is defined largely by the method of extraction used to isolate its components. Certainly the introduction of better fractionation schemes is beginning to help simplify the problem of structure determination (*17*).

A good illustration of the dearth of reliable information on humic substances is the diversity of views held by scientists in the field on the molecular weight of these compounds. They are very commonly described as having high molecular weight, yet others dispute this claim (*18*). The molecular weights reported in the literature vary enormously, with values in the range of 500–200,000 being recorded (*19*). There appear to be several plausible reasons for this observation. Certainly, different workers have used samples extracted from various sources and by various methods. However, the question remains of whether the huge variations reported for different samples are real or artifacts of the methods and conditions used. The introduction of reference samples by the International Humic Substances Society (*20*) should eventually help to answer this question, and researchers should be encouraged to use such standards as reference materials in their work (*21*).

A number of different techniques have been used to determine the molecular weight of humic substances. Each technique is based on a different physical principle and is invariably subject to its own set of limitations and approximations. This variation may be one of the reasons for analytical discrepancies. Another factor that may account for the very high molecular weight sometimes reported is the suggestion that humic substances may aggregate under some solution conditions. Aggregation may be promoted, for example, by high solution concentration, increased ionic strength (particularly in the presence of multivalent cations), or low pH. Wershaw and co-workers (*22*, *23*) have even suggested that humic substances may be capable of forming micellelike agglomerates that can be used to account for properties such as the ability of humic substances to solubilize nonpolar substances (e.g., pesticides, polychlorinated biphenyls [PCBs], polycyclic aromatic hydrocarbons [PAHs]). On the other hand, perhaps degradation reactions may occur under some conditions and lead to lower apparent molecular weights.

In this chapter we discuss the development of flow field-flow fractionation (flow FFF) as a new method for determining the molecular-weight distributions of humic substances. The aim is to describe the experimental methodology, to outline current limitations that need to be overcome, and to contrast this technique with other methods that are currently in use. We also speculate on how the separation capabilities of flow FFF may be used in environmental studies.

Field-Flow Fractionation

Field-flow fractionation refers to a series of separation techniques that are similar in operation to chromatography, but in which the separation of molecules or particles occurs in thin (0.1–0.5-mm) unpacked rectangular channels (*24*). In normal FFF operation the sample is introduced at one end of the channel; a driving force is then applied across the thin dimension of the channel and causes the sample to migrate toward one wall, the accumulation wall. There the sample particles form a cloud whose mean thickness (l) depends on the velocity induced by the field (U) and their diffusion coefficient (D) as follows (*24*):

$$l = \frac{D}{U} \tag{1}$$

Thus, the value of l generally depends on the particle or molecular size (or mass). When carrier flow begins, laminar flow is maintained as a consequence of the thin channel cross-section. The carrier flow is greatest in the center of the channel and decreases to zero at the two boundary walls. Indeed, the distribution of flow velocities across the channel is parabolic, as illustrated in Figure 1. As the sample clouds interact with the carrier flow distribution, the particles migrate down the channel with a velocity that depends on the value of l. Sample components will thus elute at characteristic retention volumes (V_r). It can be shown that V_r and l are related to the key retention parameter, the retention ratio (R), by the relationship (*25*):

$$R = \frac{V^0}{V_r} = 6\lambda \left[\coth \frac{1}{2\lambda} - 2\lambda\right] \tag{2}$$

where V^0 is the channel void volume, λ is the dimensionless cloud thickness l/w, and w is the channel thickness.

Different FFF subtechniques make use of different types of applied fields. The most widely used subtechniques are gravitational or centrifugal sedimentation, thermal gradient, and crossflow methods. Electrical and magnetic fields are less commonly used. Other possibilities exist, including fluid shear induced forces (*26*). The general strategy of FFF is to measure V_r for the various sample components. V_r is used to calculate R and then λ or l,

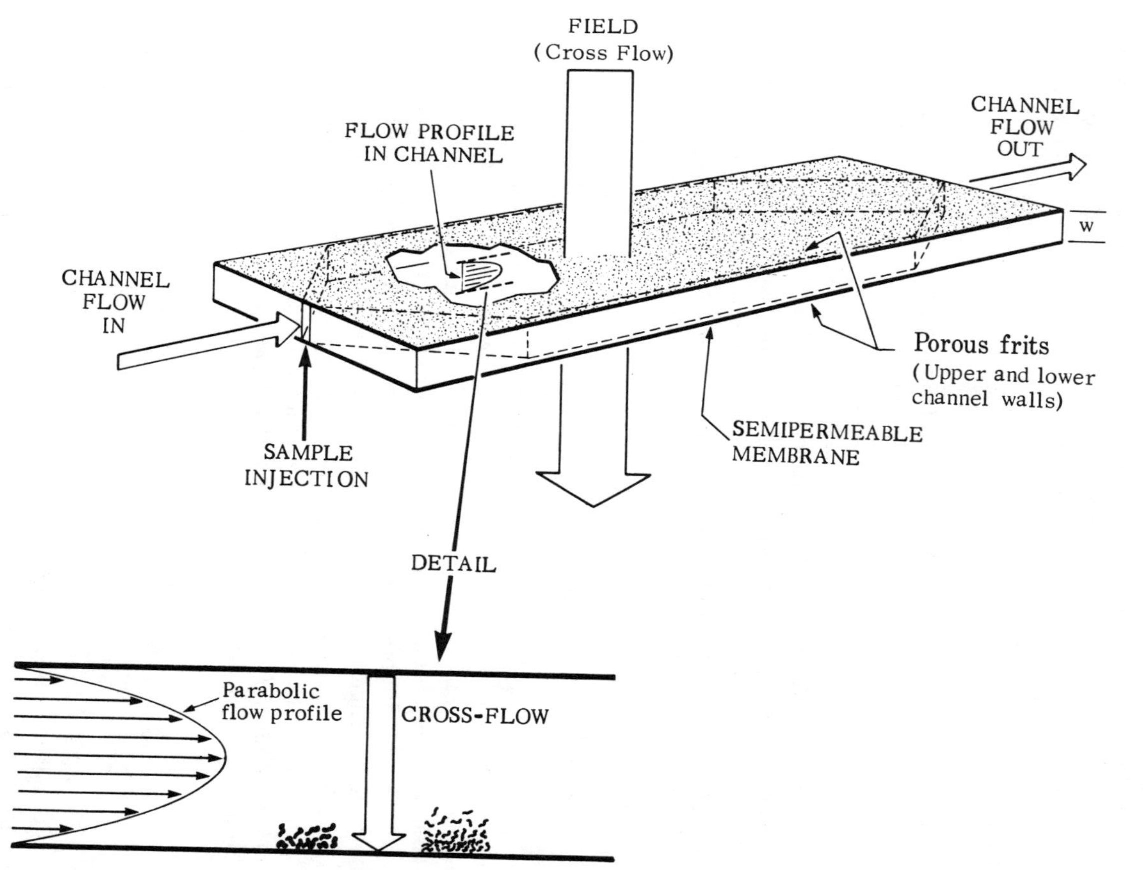

Figure 1. Schematic diagram of the flow FFF channel showing the semipermeable bounding walls, consisting of frits and a membrane. Inset is a cross section of the channel illustrating the separation mechanism.

which in turn can be related to certain parameters of the particles or molecules (e.g., mass, diffusion coefficient, mobility) by equations derived for the particular field (subtechnique) being used.

Flow FFF is the subtechnique that uses a crossflow of fluid to drive the molecules to the accumulation wall (Figure 1). This "field" flow is maintained at right angles to and superimposed on the normal channel flow by constructing the channel walls from a porous frit material with semipermeable membranes at the accumulation wall to retain the sample in the channel. In this system the field-induced sample velocity (U) is simply the linear crossflow velocity. Therefore it follows from Equation 1 that (*27*):

$$\lambda = \frac{D}{Uw} = \frac{DV^0}{\dot{V}_c w^2} \tag{3}$$

where $\dot{V}_c$ is the volumetric crossflow rate.

Thus, in flow FFF the molecular diffusion coefficient can be estimated from the measured elution volume.

In addition to its usefulness for characterizing colloidal particles (*28*, *29*), flow FFF has been shown to be effective for separating proteins (*30*) and water-soluble polymers (*31*) with good resolution, as illustrated in Figure 2. The theoretical diameter-based selectivity is unity, a value higher than that for gel permeation chromatography (*32*).

Experimental Details

The equipment, samples, and method of operation have been described in detail elsewhere (*33*). In addition to the polysulfone membrane (Millipore PTGC) with a nominal molecular-weight cutoff of 10,000 daltons for globular proteins, which was used as the accumulation wall in the previous study, some experiments were run

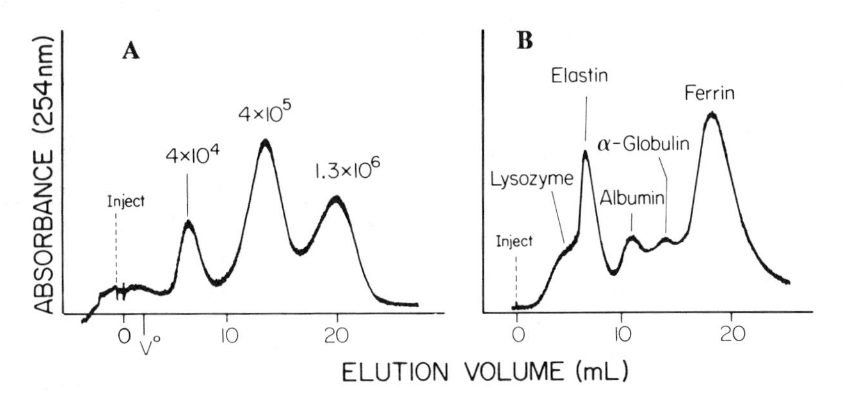

Figure 2. Flow FFF fractograms showing good resolution in the separation of (A) poly(styrene sulfonate) molecular-weight standards and (B) proteins.

here with a cellulose membrane (Amicon, YC05) with a specified 500-dalton pore size. Most fractograms were collected with conventional UV–visible detectors and chart recorders. Several were recorded with a photodiode array detector (Varian Polychrom 9060) capable of collecting data between 190 and 367 nm. The Suwannee River fulvic and humic acids used were the International Humic Substances Society reference stream samples, usually with 40-μL injections of 0.25-mg/mL solution buffered at pH 7.9. The carrier solvent in all cases was 0.05 M tris(hydroxymethyl)aminomethane (TRISMA, Aldrich), 0.0268 M HNO_3, and 0.00308 M NaN_3, and the pH was 7.9.

Poly(styrene sulfonate) molecular-weight standards (Polysciences, Inc.) covering the range 4000–100,000 daltons were made up in water (1 g/L), as were some biological test samples of bovine serum albumin, ovalbumin, transfer RNA, and RNase.

Results and Discussion

Fractograms of Humic Substances. Fractograms of the reference Suwannee River fulvic and humic acids are shown in Figure 3A. In these plots the ordinate is the absorbance (arbitrary units) of the sample at 254 nm; it should represent the sample concentration if the molar extinction coefficient remains constant across the molecular-weight distribution. The abscissa is the elution volume, which is related to the diffusion coefficient through equations 2 and 3. Thus, the abscissa is not linear with respect to molecular weight.

The fractogram is a smooth single-peaked curve, unlike many of the complex chromatographs based on gel permeation chromatography (GPC). This finding was true of all the samples we have run so far, which have included aquatic, soil, peat, and coal humic substances. The consistency leads us to suspect that the multiple peaks generated by the GPC method are due to anomalous effects, such as adsorption and charge repulsion, involving the sample and gel.

To investigate whether different molecular-weight components show different UV signals, which may indicate a distinct change in chromophore, some runs were made with a photodiode array detector (Varian) capable of scanning between 190 and 367 nm. An example of the detailed information obtainable from such a fractogram is shown in the three-dimensional plot of Figure 4. However, because of the featureless spectrum of humic substances and the fact that no shift in peak maximum occurs as the molecular weight changes, a rather simple picture is obtained in this case. The diagram does indicate rather clearly the fact that 254 nm (highlighted curve) is a good choice of wavelength for recording fractograms of humic substances.

Several inherent problems could arise with the flow FFF technique, two of which are associated with the membrane used on the accumulation wall to contain the sample molecules within the channel. These problems are the permeability of the membrane, which may allow some of the sample to leak from the channel, and the possibility of sample adsorption onto the

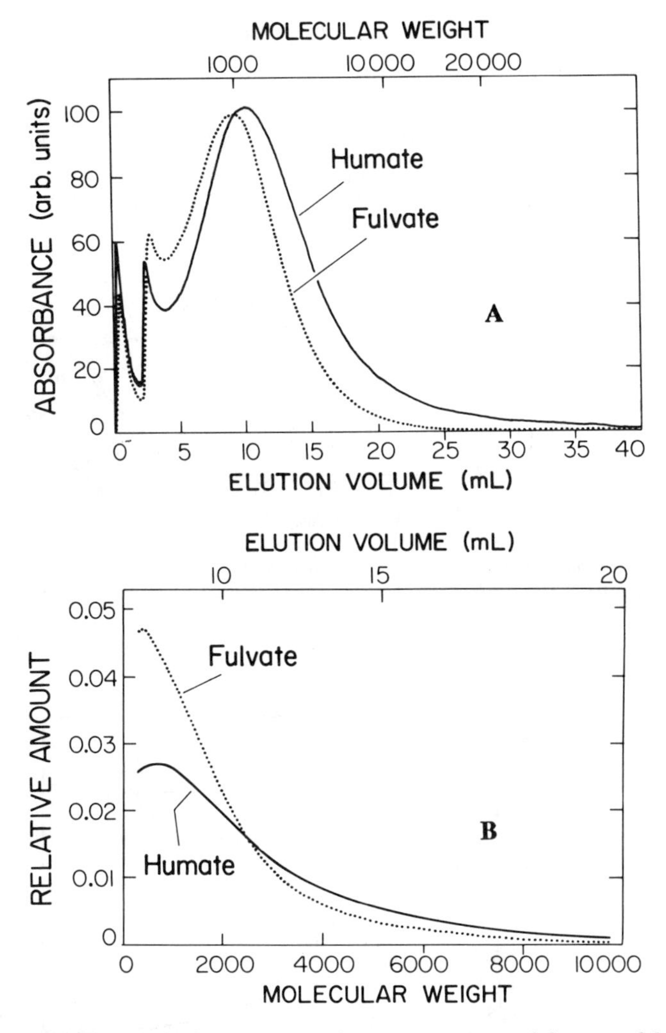

Figure 3. (A) Flow FFF fractograms of Suwannee River fulvate and humate samples at pH 7.9 with Millipore membrane. (B) Calculated molecular-weight distributions obtained with molecular-weight calibration line for PSS standards given in Figure 6.

membrane. An indication that one or both of these may be occurring to some extent is given in Figure 5, which shows that the peak area of the eluted sample decreases as the stop flow time is increased. (The stop flow time is the time the sample is held without channel flow at the head of the channel, while being forced to the accumulation wall by crossflow.) Presumably a longer stop flow time gives the sample a greater opportunity to be lost by these two mechanisms. In all other fractograms reported here, a stop flow time of 1.5 min was used to minimize these effects.

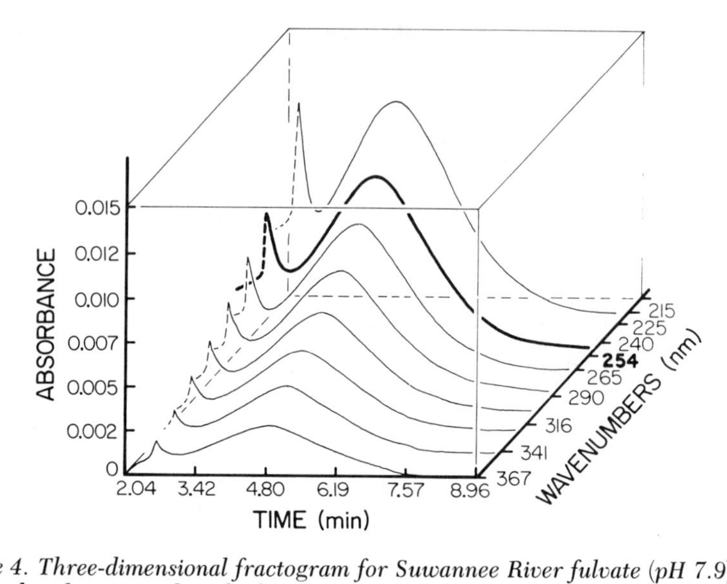

Figure 4. Three-dimensional fractogram for Suwannee River fulvate (pH 7.9) obtained with Varian photodiode array multiwavelength detector and Amicon membrane.

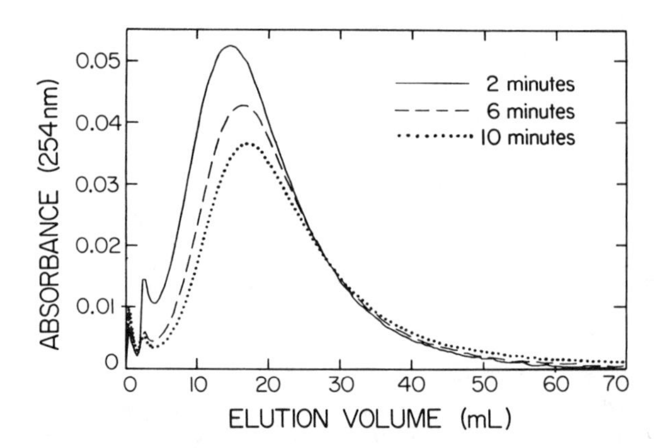

Figure 5. Fractograms of Suwannee River fulvate (pH 7.9) obtained with Millipore membrane and different stop flow times.

An attempt was made to produce a mass balance on the sample injected by using two detectors to monitor the peak areas. One detector was connected to the normal channel outlet and the other to the crossflow outlet, where it would measure sample escaping through the membrane. These peak areas were compared with the area of an equivalent sample aliquot injected directly into the detector. Although only semiquantitative, the results indicate that as much as 15–20% of the sample may be lost by each of the processes (adsorption and leakage).

Another potential interference is sample overloading effects. The standard FFF equations used require the sample to form an equilibrium cloud whose distribution is not affected by intermolecular interactions. This effect was not apparent for the concentrations of humic substances tested in this study (0.25–1.0 mg/mL), as the retention times obtained for runs with the same experimental conditions were not dependent on sample concentration. However, overloading effects were indicated with the PSS standards, as discussed later.

Molecular-Weight Calibration. Separation of components in flow FFF is achieved because of differences in their diffusion coefficients (D). Because the relationship between D and the molecular weight (M) of a macromolecule depends on several parameters such as its molecular conformation (shape) and flexibility, the conversion from elution volume to molecular weight generally will require the use of calibration standards. For many polymers the relationship is of the form:

$$D = \frac{A}{M^b} \tag{4}$$

where A and b are constants for a particular sample type and carrier solution (*34*). The exponent b is usually in the range 0.33–1.0, depending on whether the molecules are globular, random coil, rodlike, etc.

In this study two sets of molecular-weight standards were tested: poly(styrene sulfonate) (PSS) samples, which are linear random-coil molecules subject to charge repulsion effects, and some more rigid proteins and RNase. For the PSS samples, the peaks shifted to slightly lower relative elution volumes as the applied field increased. This trend resulted in the D_{max} values decreasing by up to 15% in the range tested (crossflow = 1–4 mL/min). This result could have been caused by either intermolecular interactions or membrane adsorption effects, both of which would be expected to be more pronounced as the field increased and the sample became more compressed against the accumulation wall.

To minimize the influence of this effect on the molecular-weight determination, the diffusion coefficients plotted on the calibration graph were those corresponding to about the same λ value (0.08) as attained in the humic-substance runs. This correspondence was achieved by plotting D_{max} versus λ for the data obtained at different crossflows and reading off the diffusion coefficient ($D_{0.08}$) at $\lambda = 0.08$ for each PSS standard. The resulting calibration curves (log D versus log M) are shown in Figure 6. Both curves are linear, but with different slopes; the random-coil PSS molecules are displaced to lower diffusion coefficient, as would be expected.

To test which of these curves may be best suited for the estimation of molecular weight of the humic substances, a few reference fulvic and humic substances were obtained whose molecular weights had been determined

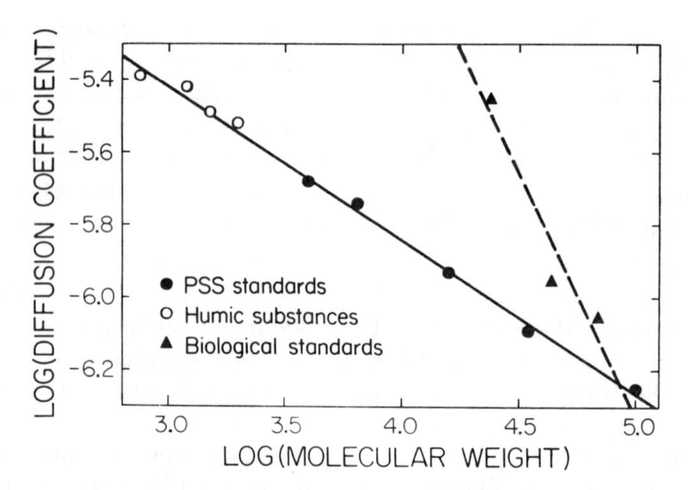

Figure 6. Calibration lines obtained with either poly(styrene sulfonate) or protein molecular-weight standards (solid symbols) with Millipore membrane. Also plotted are points (open circles) corresponding to the reference Suwannee River and Mattole soil fulvic and humic acid samples.

by a number of different methods (R. Malcolm, U.S. Geological Survey, Denver, CO, personal communication). The points corresponding to the diffusion coefficients at the FFF peak maximum and the molecular weight determined independently for these humic substances are also plotted on the calibration graph. Because these points fall satisfactorily on the PSS calibration line, it was chosen for use in all the work involving the Millipore 10,000-dalton polysulfone membrane, yielding $A = 7.05 \times 10^{-5}$ and $b = 0.422$ (with D expressed in square centimeters per second) for the constants in equation 4. Somewhat different calibration constants ($A = 2.58 \times 10^{-5}$, $b = 0.320$) were obtained subsequently when the Amicon 500-dalton cellulose membrane was installed.

Molecular-Weight Distributions. The fractograms were digitized (150 points) and the abscissa transformed from elution volume (V_r) to molecular weight by using a calibration equation obtained as already described. Because the conversion between V and M is nonlinear, the ordinate must also be modified in order that the area under the curve is proportional to the mass of sample (m) between given molecular-weight limits (*35*). The ordinate of the molecular-weight distribution should represent dm/dM, whereas in the case of the original fractogram, the ordinate represents the detector signal, which is proportional to dm/dV. Because $dm/dM = dm/dV \cdot dV/dM$, the required transformation is achieved by multiplying the amplitude values of the digitized fractogram by $\Delta V/\Delta M$, where ΔV is the difference between the elution volume for consecutive points and ΔM is the difference in the molecular weight for the same points.

The molecular-weight distributions were also normalized so that the total area under the curve was 100. Thus, the area between any two molecular-weight limits is the percentage by weight of the sample in that range. The minimum measurable molecular weight on these distributions is determined by interference from the void volume peak and the deteriorating resolution as R increases toward unity. For the conditions used here, the lower limit is about 300 daltons.

The effect of this molecular-weight transformation is illustrated in Figure 3B. The procedure heavily weights the points occurring at lower elution volumes; thus, the maximum in the fractogram is almost eliminated in the molecular-weight distribution. The result is a fairly broad molecular-weight distribution that decreases smoothly with increasing molecular weight, without any indication of multiple peaks, and tails off toward 10,000 daltons or beyond.

From these distributions it is possible to calculate both number (M_n) and weight (M_w) average molecular weights of the sample. However, a number of factors will restrict the accuracy of these parameters. The lower molecular-weight limit of 300 will particularly affect M_n, whereas uncertainties in the higher molecular-weight values will mainly affect M_w. Contributions to errors in the latter will be the difficulty in establishing the exact baseline in the long tail of the distribution and uncertainties in the calibration.

Comparison with Other Methods

Various methods have been used to measure molecular size or weight of humic substances, primarily ultrafiltration, gel permeation chromatography (GPC), low-angle X-ray scattering (LAXRS), colligative properties, ultracentrifugation, and viscosity. All techniques have limitations, and the results must always be treated cautiously. Excellent critical appraisals of these methods can be found in the recent reviews by Wershaw and Aiken (*18*) and Thurman (*19*).

One common problem with many of the available methods, including flow FFF, is the need to find an appropriate set of standards on which the molecular-weight determination can be based. Only colligative properties are free of this problem; unfortunately, they yield only the number average molecular weight, with no indication of the polydispersity of the sample. The work reported here showed that a random-coiled polyelectrolyte (PSS) could be used to give molecular-weight results for some reference fulvic and humic acids that were more consistent with data from other methods (ultracentrifuge, LAXRS, vapor pressure osmometry) than when proteins are used as standards. This finding is not surprising, considering the likely structure of humic substances, which would be expected to behave more like the PSS molecules than the more rigid proteins. Nonetheless, proteins have been commonly used as molecular-weight standards in the past, particularly for ultrafilters and in GPC. Despite the apparent success obtained in using PSS standards, it is likely that even better reference materials could be found, as the humic molecules are probably somewhat more branched than the linear PSS molecules.

The most comprehensive molecular-weight studies have been done using the International Humic Substances Society Suwannee River fulvic acid reference material. The results obtained with several methods are shown in Table I. The satisfactory agreement between the flow FFF results and the other methods (36) is evidence for the validity of the technique despite some of the potential difficulties, which still need further development to overcome.

Commercially available humic acids (e.g., Aldrich, Fluka) are commonly used in laboratory experiments on humic substances, although it is doubtful that they are a particularly good model, especially for aquatic humic substances (37). We can compare the results of flow FFF and GPC for Aldrich humic acid by using data of El-Rehaili and Weber (38). The raw fractionation output is given for each method in Figure 7A. The GPC elution volume was converted to molecular weight by using protein standards, whereas in the FFF determination, PSS standards were used. Of course, with GPC the molecular weight decreases with increasing elution volume. We have plotted it from right to left to be more comparable with the FFF fractogram. The most noticeable difference is the occurrence of two peaks in the GPC trace.

The GPC curve was transposed to a molecular-weight distribution by using the same procedure as described for the FFF fractogram. This transformation has rarely been attempted with the GPC results reported in the literature. The two molecular-weight distributions are compared in Figure 7B. The major difference is the occurrence of a quite sharp peak in the GPC curve. This peak was close to the column volume (77 mL) and is outside the range covered by the calibration standards. It probably does not accurately represent the true molecular-weight distribution in this region, and for this reason we show it as a dotted line in Figure 7B. FFF is generally characterized by higher selectivity and resolution than GPC (39), so it is likely that this peak would also be resolved by flow FFF unless some nonideal phenomenon (e.g., intermolecular interaction) was occurring with these humic samples.

The GPC molecular-weight distribution still contains, in addition, a high-molecular-weight peak that is absent in the FFF distribution. This peak

Table I. Molecular-Weight Studies for the Suwannee River Fulvic Acid Sample

Method	M_n	M_w	*Reference*
Flow FFF	1150	1910	33
Ultracentrifuge	655	1335	36
VP osmometry	823	—	36

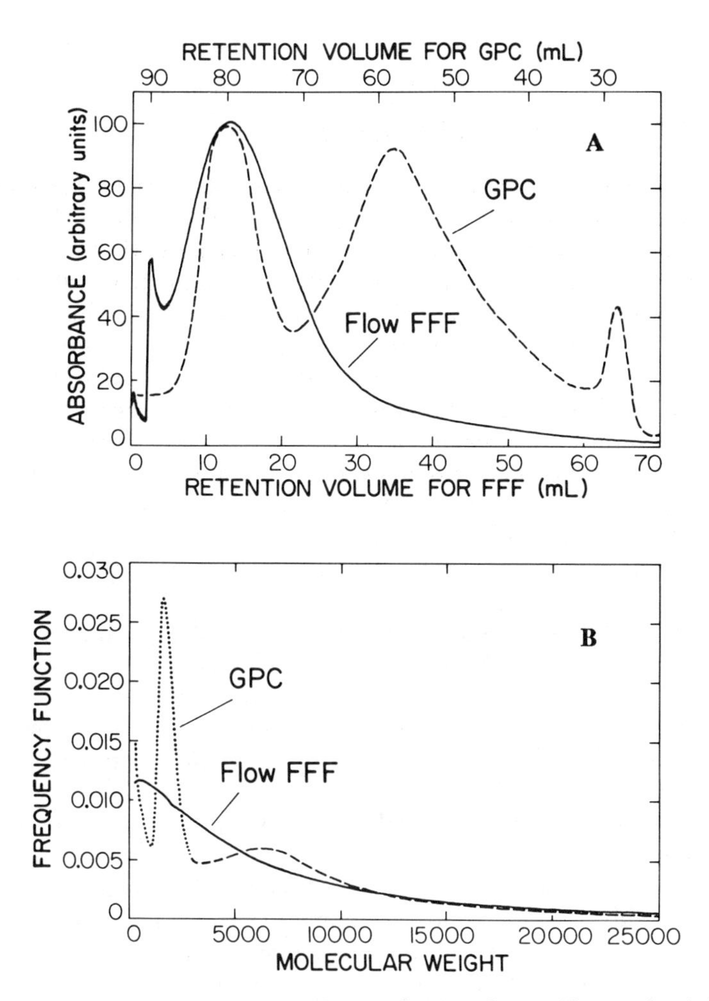

Figure 7. Comparison of flow FFF (from ref. 33) and GPC (from ref. 38) data for Aldrich humic acid sample. (A) Flow FFF fractogram and GPC chromatogram. Volume scales are different. (B) Calculated molecular-weight distributions. The dotted portion of the GPC distribution is subject to considerable uncertainty, as discussed in text.

is much less pronounced than the first. Possibly this peak, which would correspond to a fairly high apparent molecular weight, is due to low-molecular-weight components that have been excluded from the gel by charge repulsion. Alternatively, this peak may be missing in the FFF fractogram due to adsorption or aggregation phenomena. The presence of multiple peaks in GPC analysis of humic substances is a common occurrence that we believe may be an artifact of the method. However, further work is required before this matter can be resolved completely. We also stress that the plots of the raw output from fractionation techniques such as these can be rather misleading and may be a poor representation of the true molecular-weight distribution.

Potential Applications of Flow FFF

Flow FFF offers the potential to provide molecular-weight distributions of humic substances. It could thus become a valuable additional tool in the quest to understand the origin, nature, and behavior of this important class of naturally occurring organic compounds. In combination with other analytical methods, it should be of help in unraveling the biogeochemical changes that occur in these materials, both in the short term in aquatic systems and in the longer term in ground water, soils, and sediments.

We previously (*33*) reported the molecular-weight distributions of humic substances from a number of aquatic and terrestrial sources. Some of the data reproduced in Table II demonstrate a distinct trend toward increased molecular weights as we progress from stream, soil, and peat to coal humate types. Of course, this information alone will not answer the many questions that remain, and a multidisciplinary approach to these problems is most desirable.

Preliminary experiments have indicated a shift to higher molecular weight as the solution pH is decreased or if calcium salts are added. The relatively high pH of 7.9 was used in this study in an attempt to repress any tendency of the humic substances to aggregate and thus to obtain estimates of the primary particle molecular weight. This possible aggregation of humic substances would have important consequences in their behavior in natural systems and their role in affecting the fate of pollutants. Certainly more work is warranted in this area of environmental research.

The fact that flow FFF is also a fractionation technique should make it useful to study the interactions of humic substances with environmental pollutants. The small scale of the separation, which usually involves much less than 1 mg of sample, would be a limitation here. However, with the use of ultratrace analytical methods, a very powerful method for investigating the importance of different size fractions of humic substances responsible for binding pollutants could be developed. Radiotracer techniques and mass spectrometry are examples of methods that would provide detection systems sensitive enough for such experiments.

Table II. Molecular Weights of Humate Samples as Determined by Flow FFF Data

Sample	M_n	M_w	M_w/M_n
Suwannee stream humate	1,580	4,390	2.78
Mattole soil humate	1,940	6,140	3.16
Washington peat humate	3,020	17,800	5.89
Leonardite coal humate	3,730	18,700	5.01

NOTE: Data are from ref. 33.

Summary

Flow FFF is capable of yielding the molecular-weight distributions of humic substances, thus providing more detailed molecular-weight information than any other method, with the possible exception of GPC. Flow FFF should be less prone to anomalous sample interaction effects than GPC, because with this method the separation is carried out in unpacked channels of low surface area. However, indications are that some sample interaction with the membrane on the accumulation wall of the FFF channel does still occur. We hope that further work will find membrane materials and solution conditions that eliminate these undesirable effects.

The availability of a quick and reliable method for molecular-weight determination should provide a useful addition to the tools available to assist with humic-substances research. Furthermore, the fractionating ability of flow FFF could be used to provide valuable information on the interaction of pollutants with humic substances.

Acknowledgments

This work was supported by the Australian Research Grant Scheme and the Department of Science in Australia and Department of Energy Grant DE–FG02–86ER60431 in the United States.

References

1. de Saussure, T. *Paris Annu.* **1804,** *12,* 162.
2. Sprengel, C. *Arch. Gesammte Naturlehre.* **1826,** *8,* 145.
3. Schnitzer, M.; Khan, S. U. *Soil Organic Matter*; Elsevier: New York, 1978.
4. Mantoura, R. F. C.; Dickson, A.; Riley, J. P. *Estuarine Coastal Mar. Sci.* **1978,** *6,* 387.
5. Hart, B. T. *Environ. Technol. Lett.* **1981,** *2,* 96.
6. Wershaw, R. L.; Burcar, P. J.; Goldberg, M. C. *Environ. Sci. Technol.* **1969,** *3,* 271.
7. Chiou, C. T.; Porter, P. E.; Schmeddling, D. W. *Environ. Sci. Technol.* **1983,** *17,* 227.
8. Chiou, C. T.; Kile, D. E.; Malcolm, R. L.; Brinton, T. I. *Environ. Sci. Technol.* **1986,** *20,* 502.
9. Hunter, K. A.; Liss, P. S. *Nature (London)* **1979,** *282,* 823.
10. Beckett, R. In *The Role of Particulate Matter in the Transport and Fate of Pollutants*; Hart, B. T., Ed.; Chisholm Institute of Technology: Melbourne, 1986; p 113.
11. Gibbs, R. J. *Environ. Sci. Technol.* **1983,** *17,* 237.
12. Tipping, E. *Mar. Chem.* **1986,** *18,* 161.
13. Jekel, M. R. *Water Res.* **1986,** *20,* 1543.
14. Karickhoff, S. W. *Chemosphere* **1981,** *10,* 833.
15. Tipping, E.; Griffith, J. R.; Hilton, J. *Croat. Chim. Acta* **1983,** *56,* 613.
16. Davis, J. A. *Geochim. Cosmochim. Acta* **1984,** *48,* 679.

17. Aiken, G. R.; McKnight, D. M.; Wershaw, R. L.; MacCarthy, P., Eds. *Humic Substances in Soil, Sediment and Water*; Wiley–Interscience: New York, 1985.
18. Wershaw, R. L.; Aiken, G. R. In *Humic Substances in Soil, Sediment, and Water*; Aiken, G. R.; McKnight, D. M.; Wershaw, R. L.; MacCarthy, P., Eds.; Wiley–Interscience: New York, 1985; p 477.
19. Thurman, E. M. *Organic Geochemistry of Natural Waters*; Martinus Nijhoff/Junk: The Hague, Netherlands, 1985; p 304.
20. Malcolm, R. L.; MacCarthy, P. In *Trace Organic Analysis: A New Frontier in Analytical Chemistry*; Chester, S. N.; Hertz, H. S., Eds.; U.S. National Bureau of Standards: Gaithersburg, MD; U.S. National Bureau of Standards Special Publication No. 519, p 789.
21. MacCarthy, P. *Geoderma* **1976,** *16*, 179.
22. Wershaw, R. L. *J. Contam. Hydrol.* **1986,** *1*, 29.
23. Wershaw, R. L.; Thorn, K. A.; Pinckney, D. J.; MacCarthy, P.; Rice, J. A.; Hemond, H. F. In *Peat and Water*; Fuchsman, C. H., Ed.; Elsevier: New York, 1986; p 133.
24. Giddings, J. C.; Myers, M. N.; Caldwell, K. D.; Fisher, S. R. In *Methods of Biochemical Analysis*; Glick, D., Ed.; Wiley: New York, 1980; Volume 26, p 79.
25. Hovingh, M. E.; Thompson, G. H.; Giddings, J. C. *Anal. Chem.* **1970,** *42*, 195.
26. Giddings, J. C. *Sep. Sci. Technol.* **1984,** *19*, 831.
27. Giddings, J. C.; Yang, F. J.; Myers, M. N. *Anal. Chem.* **1976,** *48*, 1126.
28. Giddings, J. C.; Lin, G. C.; Myers, M. N. *J. Colloid Interface Sci.* **1978,** *65*, 67.
29. Giddings, J. C.; Yang, F. J.; Myers, M. N. *J. Virol.* **1977,** *21*, 131.
30. Giddings, J. C.; Yang, F. J.; Myers, M. N. *Science (Washington, D.C.)* **1976,** *193*, 1244.
31. Wahlund, K.-G.; Winegarner, H. S.; Caldwell, K. D.; Giddings, J. C. *Anal. Chem.* **1986,** *58*, 573.
32. Myers, M. N.; Giddings, J. C. *Anal. Chem.* **1982,** *54*, 2284.
33. Beckett, R.; Jue, Z.; Giddings, J. C. *Environ. Sci. Technol.* **1987,** *21*, 289.
34. Tanford, C. *Physical Chemistry of Macromolecules*; Wiley: New York, 1961; Chapter 6.
35. Giddings, J. C.; Myers, M. N.; Yang, F. J. F.; Smith, L.K. In *Colloid and Interface Science*; Kerker, M., Ed.; Academic: New York, 1976; Volume IV, p 381.
36. Averett, R. C., Ed. *Humic Substances in the Suwannee River, Florida and Georgia: Interaction, Properties, and Proposed Structures*; Water Supply Paper; U.S. Geological Survey, in press.
37. Malcolm, R. L.; MacCarthy, P. *Environ. Sci. Technol.* **1986,** *20*, 904.
38. El-Rehaili, A. M.; Weber, W. J. *Water Res.* **1987,** *21*, 573.
39. Gunderson, J. J.; Giddings, J. C. *Anal. Chim. Acta* **1986,** *189*, 1.

RECEIVED for review November 3, 1987. ACCEPTED for publication February 29, 1988.

ENVIRONMENTAL IMPACTS

6

Interactions of Hazardous-Waste Chemicals with Humic Substances

Stanley E. Manahan

Department of Chemistry, University of Missouri, Columbia, MO 65211

Hazardous-waste chemicals may interact in a variety of ways with humic substances. These interactions include solubilization and desolubilization of metal compounds and organic compounds by humic substances, precipitation reactions, and oxidation–reduction phenomena. All three major types of humic substances—fulvic acid, humic acid, and humin—may interact with hazardous-waste chemicals. Among the more important classes of hazardous waste chemicals that may be influenced by the presence of humic substances are sparingly soluble organic compounds, particularly organohalides; polycyclic aromatic hydrocarbons; heavy metal ions; soluble oxidants; iron compounds; aluminum compounds; and strong acids and bases. This chapter discusses the interaction of humic substances with hazardous-waste chemical species and the possible uses of humic substances for the treatment of such chemicals.

In assessing the fates and effects of hazardous-waste chemicals, it is important to consider the potential environmental chemical interactions of these wastes with humic substances from water, soil, codisposed municipal refuse, or the wastes themselves. Among the types of interactions possible between humic substances and hazardous wastes, and the effects that occur from such interactions, are the following: (1) precipitation of humic acid by reaction with waste acid, such as steel pickling liquor acid waste; (2) solubilization of humic acid by reaction with waste alkali; (3) complexation of heavy metal ions by humic substances; (4) reduction of metal species by humic substances [e.g., reduction of iron(III) to iron(II) or chromium(VI) to

0065–2393/89/0219–0083$06.00/0

chromium(III)]; (5) effects of humic substances on the solubilities of organic wastes; and (6) effects of humic substances on the sorption of hazardous-waste species (e.g., complexation of metal cations).

Humic substances sometimes can be used in the treatment and detoxification of hazardous-waste chemicals. Examples include the application of peat as a matrix for the growth of white rot fungus used to degrade organic wastes and the reduction–sorption of chromate wastes by surface-oxidized low-rank coal.

Hazardous Wastes

A chemical waste may be deemed hazardous because of toxicity, ignitability, reactivity, corrosivity, radioactivity, or a combination of these characteristics. Relatively large numbers of people may be exposed to low levels of toxicants through drinking-water sources contaminated by hazardous-waste constituents leached from disposal sites, a process that can be affected by interaction with humic substances.

Numerous methods are used to reduce the hazards of waste chemicals. These methods can be broadly categorized as destruction, deactivation, and stabilization.

The most common means of destruction is the incineration of organic chemicals to CO_2, H_2O, and other inorganic species (such as HCl in the case of chlorinated hydrocarbons). Wet oxidation of chemicals in water can be accomplished with air at high temperatures (175–345 °C) and high pressures (20.4–204 atm) (*1*).

The objective of deactivation is to convert the waste to a nonhazardous material disposable in a conventional landfill or, for aqueous wastes, in a publicly owned treatment works. Dilute hazardous-waste solutes may be removed from water by processes such as activated carbon adsorption of organic compounds, sulfide precipitation of heavy metals, or reverse osmosis.

Some wastes, such as radionuclides, cannot be destroyed or deactivated at a reasonable cost with available technology, and stabilization is required prior to disposal (*2*). Stabilization prevents contamination of air or water by hazardous substances and allows safer handling and transport. In favorable cases, stabilized wastes can be incorporated with landfill at construction sites or even added to building materials (*3*). Stabilization technologies include mixing wastes with cement, conversion of inorganic wastes to pozzolanic concrete materials produced by mixing lime with fly ash and other siliceous materials, and vitrification to produce a glassy solid.

Hazardous-waste chemicals include a wide variety of species. Insofar as interaction with humic substances is concerned, the most important of these species include sparingly soluble organic compounds, particularly organohalides; polycyclic aromatic hydrocarbons; heavy metal ions; soluble oxidants; iron compounds; aluminum compounds; and strong acids and bases.

Humic Substances

Humic substances are colored organic substances found in natural waters, soil, peat, and other sources where plant material has undergone partial degradation. The exact nature of humic substances defies a concise definition because of their variable compositions and properties. They are normally defined in terms of three somewhat overlapping kinds of substances—fulvic acid, humic acid, and humin (*4*). Humic acid and fulvic acid are soluble in strong basic solution, from which humic acid precipitates upon acidification, leaving fulvic acid in solution. Humin is not extractable in aqueous solution of even strong base; it can be isolated as a suspension in methyl isobutyl ketone (*5*).

Despite the widespread occurrence of humic material in natural waters and terrestrial systems and detailed study by a host of investigators, knowledge of the exact nature and formation mechanism of humic material is still incomplete. Its formation probably involves a combination of degradation and condensation of plant residual material. Soil scientists long ago recognized the importance of "humus" in maintaining soil texture, moisture content, and fertility (*6*). More recently, environmental scientists have become interested in humic substances because of their interactions with soil and water pollutants and their influence on treatment processes. Prominent among the latter are problems encountered in trying to remove humic-bound metals (e.g., iron) in water treatment and the role of humic substances in forming trihalomethane contaminants in water chlorination (*7*).

Beliefs about the properties of humic substances have undergone some modification in recent years in light of more sophisticated studies. For example, although they are somewhat aromatic, aquatic humic substances may have less aromaticity than was once widely believed (*8*). The literature abounds with diverse estimates of the molecular weights of humic substances, ranging from a few hundred (fulvic acid) to several hundred thousand (humin materials). More recent estimates have tended to be lower (*9*, *10*), for example, about 800 for fulvic acid and 1500–3000 for humic acid (*4*). With the use of newly developed flow field-flow fractionation, molecular weight distributions have been determined for fulvates and humates from various sources (*4*). The molecular weight values obtained ranged from 860 for a Suwannee stream fulvate to 4050 for a Leonardite coal humate. Intermediate values obtained were 1010 for a Mattole soil fulvate, 1490 for a Suwannee stream humate, and 2430 for a Washington peat humate. Each of these numbers represents a distribution of values for each source of humic substance.

Regardless of uncertainties such as molecular weight, humic substances have some well established functionalities and properties. They are polyelectrolytes with various functional groups attached to the hydrocarbon skeleton. Prominent among these groups are carboxyl and phenolic functionalities responsible for the acid–base, complexation, and salt formation capabilities of humic substances. Methoxyl groups are also present. Humic molecules are subject to strong intermolecular association (*11*).

Acid–Base Solubility Behavior of Humic Substances with Hazardous Wastes

Interaction between humic substances and hazardous-waste chemicals is most likely to occur when such chemicals have been disposed of underground. In older sites containing chemical wastes mixed with municipal refuse, the decay of the refuse can generate humic materials. Humic substances are present in soil lining and covering underground disposal sites and are mixed with waste chemicals placed in lagoons, trenches, or pits. Infiltration of surface water or groundwater containing humic substances can result in their contacting disposed chemical wastes. It is even plausible that humic materials such as peat or weathered, surface-oxidized coal residues may be codisposed with waste chemicals. There are several potential sources of humic substances, a condition that should be considered in assessing environmental chemical processes involving landfilled chemicals.

A major factor in considering interaction of humic substances with waste chemicals is the solubility of the humic materials, including the influence of chemical species on the solubility, which depends predominantly on the acid–base precipitation behavior of humic substances. The most likely route for the mobilization of large quantities of humic material is through contact with strong base from waste caustics, such as that discarded after use for removal of sulfur compounds from petroleum products. Humic acid solubilized by this route could move some distance from a waste disposal site and precipitate upon neutralization of the base or dilution. However, lower molecular weight fractions could stay in solution, acting as water pollutants per se or in the form of chelates with metal ions.

Humic substances in solution can be precipitated by contact with waste chemicals in two major ways. Acidification, such as by waste steel pickling liquor, causes humic acid to precipitate. This phenomenon is unlikely to cause any problems; humic-chelated heavy metal ions could conceivably be released, perhaps in a more mobile form.

Soluble humic substances can be precipitated and colloidal humic materials aggregated by contact with multiply charged metal ions. The most likely way for this to happen is through contact with Ca^{2+} ion, which abounds in water at many waste-disposal sites because of the widespread use of lime in waste treatment. Fly ash, sometimes used to cover disposal lagoons for closure, is another possible source of calcium. The precipitation of calcium humates should be beneficial in removing humic organic matter from waste-site leachate. Coprecipitation of heavy metal ions with calcium humate would be beneficial in removing these pollutant species. Even some organic species might be removed from aqueous solution or suspension by the humate precipitate. It is difficult to imagine any harmful effects that might arise from formation of insoluble humates.

Interaction of Humic Substances with Waste Heavy Metal Ions

The ability of humic substances to act as chelating agents for metal ions, including trace metals in water (*12*), is well known. The knowledge base in this area continues to expand steadily, and extensive coverage of this topic is found in the scientific literature. The particular effects that humic substances have on chelatable metals in hazardous wastes depend upon the types of metal species, the nature of the humic substances, and the chemical environment with respect to acidity–alkalinity, oxidation–reduction, and the presence of competing species. (Examples of competing species are calcium in competition with other metals for humic ligands or complexing species such as cyanide that compete with humic ligands for metal ions). Under some conditions, fulvate species act to keep metal ions in solution as predominantly anionic chelates that are relatively hard to remove from water (hazardous-waste leachate) by natural or treatment processes. However, interactions of fulvates with metal ions are involved in the removal of fulvic acid from water by coagulation with alum (*13*) or iron(III) salts. Spent steel pickling liquor is a common source of iron(III) salts at waste chemical disposal sites. In some cases the chelation of metal ions as soluble humic species may prevent precipitation of the metals by precipitate-forming anions such as CO_3^{2-}, OH^-, and S^{2-}. Precipitation by S^{2-} could be particularly significant under anoxic conditions, in which the presence of sulfide ion is often assumed to keep heavy metal ions in an insoluble form.

Most research on humic substances, including their metal-binding characteristics, has concentrated on the fulvic acid and humic acid fractions, and comparatively little attention has been given to the humin fraction (*5*). The lower priority given humin is understandable in light of its lack of solubility and widely held reputation for intractability. A recently published method for the extraction of at least some fractions of humin material (*5*) may enable more complete investigations of this class of humic substance. Clearly, humin deserves careful consideration in assessing the fates of heavy metal ions from hazardous-waste disposal sites. Because humin is a water-insoluble material with ion-exchange and metal-binding capabilities, it can act as an ion exchanger and sink for heavy metals from waste chemical leachate. This effect is generally beneficial because soil humin acts to immobilize heavy metal ions in leachate passing through it. More research is needed in assessing the influence of humic substances on the fates of heavy metal wastes because of binding between the two types of materials.

Humic substances influence speciation of metals by chelation. The other major influence of humic substances on metal speciation is through oxidation–reduction reactions, which can occur in two major ways. Humic material is an active oxidation–reduction system with an E^0 value for humic acid estimated to be 0.70 V (*see* Table I) (*14*); it normally functions as a reducing agent. The second major influence of humic substances on oxidation–reduction of metal species is stabilization of the reduced cationic form by chelation. For example, as shown in Table I, the unchelatable oxoanion $Cr_2O_7^{2-}$ is reduced by humic acid to chelatable cationic Cr^{3+}, which tends to shift the half-reaction for the reduction of $Cr_2O_7^{2-}$ to the right in the presence of humic substance.

In addition to the reductions of metal species shown in Table I, humic acid has been reported to reduce ionic mercury to elemental mercury(0) (*15*). The accumulation of vanadium and molybdenum in peats (which are humic materials) and coals (which undergo a humic stage as they develop) has been attributed to the reduction by humic substance of soluble VO_3^- and MoO_4^{2-} to chelatable cationic species (*16*).

The reduction of acidic iron(III) to iron(II) and subsequent retention of the iron(II) product has been demonstrated for surface-oxidized bituminous coal (*17*). This granular solid material produced by contacting the coal with 6 M nitric acid and bubbling air through the suspension, followed by washing with base to remove base-soluble organic products, can be regarded as a humin material.

Interaction of Humic Substances with Waste Organic Compounds

A large fraction of waste chemicals consigned to landfill disposal consists of water-insoluble, hydrophobic organic compounds. Typical of these compounds and prominent among them are numerous chlorinated hydrocarbons. Another example is bis(2-ethylhexyl) phthalate, widely used as a plasticizer and found ubiquitously in environmental samples. Most of the hydrophobic organic waste compounds are poorly biodegradable (i.e., they are biorefractory). Those compounds more dense than water (predominantly halogenated hydrocarbons) belong to the class of insoluble sinking pollutants,

Table I. Oxidation–Reduction Reactions Involving Humic Acid

Reaction	E^0 (V)
$MnO_4^- + 8H^+ + 5e^- \leftrightharpoons Mn^{2+} + 4H_2O$	1.69
$Cr_2O_7^{2-} + 14H^+ + 6e^- \leftrightharpoons 2Cr^{3+} + 7H_2O$	1.33
$MnO_2 + 4H^+ + 2e^- \leftrightharpoons Mn^{2+} + 2H_2O$	1.23
$V(OH)_4^+ + 2H^+ + e^- \leftrightharpoons VO^{2+} + 3H_2O$	1.00
$Fe^{3+} + e^- \leftrightharpoons Fe^{2+}$	0.77
Humic acid(ox) + $xe^- \leftrightharpoons$ Humic acid(red)	0.70

which tend to settle to the bottom of lagoons or other disposal sites and to be transported as "globs" of liquid along the bottoms of natural water systems, such as aquifers (*18*).

The role that humic substances may play in the transport, reactivity, and fate of hydrophobic organic compounds has been succinctly summarized by Caron and Suffet (*19*). The transport, reactivity, and fate of nonpolar organic compounds in the aquatic and terrestrial environments are influenced by their association with dissolved, colloidal, and undissolved humic substances. The degree of association is affected by the nature of the compound and humic material, the concentrations of both, pH, calcium ion concentration, and the presence of other organic and ionic solutes. The octanol–water partition coefficient (K_{ow}) is useful for predicting the degree of association, which is usually significant when K exceeds 10^5. Attempts to correlate the degree of association with measurable characteristics of humic substances have not been notably successful, and Caron and Suffet point out the shortcomings of the widespread practice of using base-extracted humic substances and commercial humic acids for studies of association. The development of reliable methods for predicting the association between hydrophobic organic waste compounds and the humic substances that they are likely to contact in waste-disposal sites would be very useful in predicting the fates and effects of the wastes.

The enhancement of the solubility of nonpolar organic compounds by dissolved humic substances appears to increase with organic compound molecular weight and with lower polarity of dissolved humic material (*20*) when tested with the compounds *p,p'*-DDT, {1,1'-(2,2,2-trichloroethylidene)bis[4-chlorobenzene]}, 2,4,5,2',5'-PCB, 2,4,4'-PCB, 1,2,3-trichlorobenzene, and lindane. Findings such as these suggest that humin, the highest-molecular-weight, least-polar fraction of humic substance, should have an especially high affinity for nonpolar organic compounds. However, because humin is insoluble, it should have the effect of decreasing the solubilities of nonpolar organic compounds and immobilizing them. This phenomenon could be a factor in restricting the movement of nonpolar organic compounds at hazardous-waste sites.

Effects of Humic Substances on Hazardous Waste Leachate Treatment

One of the most common remedial actions taken at a hazardous-waste disposal site is the treatment of leachate water from the site to remove heavy metals, toxic inorganic species (e.g., cyanide), and organic contaminants. The presence of humic substances can have significant effects upon the processes for removing leachate solutes. One of the more obvious examples is the solubilization of metal ions as humic complexes. Solubilization may prevent the removal from leachates of dissolved heavy metals by standard techniques, such as precipitation.

In some cases the presence of humic substances can lead to undesirable byproducts as the result of side reactions from processes used to remove other contaminants from leachate. For example, the destruction of dissolved cyanide by chlorination would lead to the formation of trihalomethane byproducts in the presence of humic substances.

Biological treatment of biodegradable organic contaminants, usually by an activated sludge process, is an operation commonly applied to hazardous-waste leachates. The presence of humic substances may well affect biological waste treatment. In some cases the effects could be beneficial; for example, toxic heavy metal ions may be bound to humic materials, and the bound metals are less toxic to the microorganisms carrying out the biodegradation processes. In general, association of humic substances with nonionic organic compounds reduces the toxicity and bioaccumulation of the organic compounds, apparently by reducing their ability to traverse biological membranes (*21*). This phenomenon could reduce the biodegradation of organic compounds.

Activated carbon sorption is often employed to remove refractory organic compounds and other solutes from hazardous chemical waste leachate. By competing with other organic matter, humic materials in the leachate should generally reduce the capacity and effectiveness of activated carbon to remove organic contaminants from leachate.

Humic material in hazardous-waste leachate can affect membrane processes (reverse osmosis) and resin processes (ion exchange) used to treat the leachate. Humic substances are known to foul reverse osmosis membranes, a phenomenon that can be prevented by precipitation and coagulation of the humic material with iron(III) or aluminum(III) (*22*). Membrane fouling by humic substances normally leads to a lower flux and reduces separation efficiency. In some cases the presence of a layer of organophilic humic substance on a membrane can have the beneficial effect of making it a more effective filter for the retention of organic solutes, but the overall effect is generally too unpredictable for deliberate use in leachate treatment.

Uses of Humic Substances for Hazardous Waste Chemical Treatment

The preceding discussion has pointed out some of the possible, largely detrimental, effects of the interactions of humic substances with hazardous waste chemicals, particularly heavy metals and nonpolar organic compounds. However, humic substances have several important properties that suggest their beneficial uses for the treatment and detoxification of hazardous chemicals. Prominent among these properties are the ability to chelate heavy metals, reduce oxidized metal species, and bind to nonpolar organic compounds.

For the treatment and immobilization of hazardous chemical species, the insoluble humin fraction is the most useful. It is readily available in inexpensive raw materials, such as peat treated to remove soluble constituents or surface-oxidized low-rank coal. Some waste chemical byproducts can be used to prepare and pretreat humic materials to be employed for waste treatment. Examples include waste caustic for the removal of soluble humic fractions in the preparation of humin; waste nitric acid for the surface oxidation of low-rank coal to give it humic properties; and lime, aluminum salts, and iron salts for the coagulation and precipitation of humic materials to be used in waste treatment.

As an example of humic materials applied to wastewater treatment, the use of humic acid–fly ash mixtures has been described (*23*). Applications of humic acids or humates for the purification of highly polluted water, such as hazardous-waste leachate, include the following:

1. Neutralization of acids by exchange of calcium, magnesium, or sodium on humates for H^+ ion in acidic water, accompanied by the formation and settling of insoluble humic acid.
2. Removal of heavy metals by chelation.
3. Removal of anions, such as phosphates, cyanide, and organic anions by mixed ligand complexation.
4. Sorption of organics from water.
5. Clarification of suspended matter by precipitation and flocculation of humic acid and insoluble metal humates.

As an example of the use of humic substances for the removal of potential air pollutants, the sorption of sulfur dioxide by humic acid–fly ash mixtures (*24*) and by sodium humate (*25*) has been described. The major mechanism for sorption of SO_2 was found to be formation of hydrogen sulfite (HSO_3^-) in the presence of basic humates. However, under conditions of low pH and high temperature, evidence was found for the formation of a humate complex of SO_2(aq).

The concentration and immobilization of hazardous-waste chemicals, such as heavy metals and refractory organic compounds, on relatively small masses of humic material provide several possibilities for hazardous-waste chemical treatment. These include incineration of sorbed organics and humic material under conditions that retain heavy metals with the ash and biological treatment and degradation of organic compounds on the humic substance matrix.

References

1. Kiang, Y.-H.; Metry, A. A. *Hazardous Waste Treatment Technology*; Ann Arbor Science: Ann Arbor, 1982.

2. Tucker, S. P.; Carson, G. A. *Environ. Sci. Technol.* **1985,** *19,* 215–220.
3. Malone, P.; Jones, L. *Guide to the Disposal of Chemically Stabilized and Solidified Waste*; U.S. Environmental Protection Agency. U.S. Government Printing Office: Washington, DC, 1980; SW–872.
4. Beckett, R.; Jue, Z.; Giddings, J. C. *Environ. Sci. Technol.* **1987,** *21,* 289–295.
5. Rice, J. A.; MacCarthy, P. *Abstracts of Papers,* 193rd National Meeting of the American Chemical Society, Denver, CO; American Chemical Society: Washington, DC, 1987; ENVR 199.
6. Bohn, H. L.; McNeal, B. L.; O'Connor, G. A. In *Soil Chemistry,* 2nd ed.; Wiley–Interscience: New York, 1985; Chapter 5, pp 135–152.
7. Christman, R. F.; Norwood, D. L.; Millington, D. S.; Johnson, J. D.; Stevens, A. A. *Environ. Sci. Technol.* **1983,** *17,* 625–628.
8. Thurman, E. M.; Malcolm, R. L. In *Aquatic and Terrestrial Humic Materials*; Christman, R. F.; Gjessing, E. T., Eds.; Ann Arbor Science: Ann Arbor, 1983; pp 1–23.
9. Malcolm, R. L. In *Humic Substances in Soil Sediment and Water*; Aiken, C. R.; McKnight, D. M.; Wershaw, R. L.; MacCarthy, P., Eds.; Wiley–Interscience: New York, 1985; Chapter 7, pp 181–209.
10. Thurman, E. M. *Organic Geochemistry of Natural Waters*; Martinus Nijhoff/ Junk: The Hague, 1985; pp 304–312.
11. Wershaw, R. L.; Pinckney, D. J. *J. Res. U.S. Geol. Surv.* **1973,** *1,* 701–707.
12. Hart, B. T. *Environ. Technol. Lett.* **1981,** *2,* 95–110.
13. Dempsey, B. A.; Ganho, R. M.; O'Melia, C. R. *J. Am. Water Works Assoc.* **1984,** *76,* 141–150.
14. Szilagyi, M. *Fuel* **1974,** *53,* 26.
15. Alberts, J. J. *Science (Washington, D.C.)* **1974,** *184,* 895.
16. Szalsy, A.; Szilagyi, M. *Advances in Organic Geochemistry 1965,* Proceedings of the 4th International Meeting on Organic Geochemistry; Pergamon Press: New York, 1968.
17. Harlan, Sandra J. M.A. Thesis, University of Missouri, 1974.
18. Meyer, R. A.; Kirsch, M.; Marx, L. F. *Detection and Mapping of Insoluble Sinking Pollutants*; U.S. Environmental Protection Agency, Municipal Environmental Research Laboratory: Cincinnati, OH, 1981; EPA–600/S2–81–198.
19. Caron, G.; Suffet, I. H. *Abstracts of Papers,* 193rd National Meeting of the American Chemical Society, Denver, CO; American Chemical Society: Washington, DC, 1987; ENVR 4.
20. Kile, D. E.; Brinton, T. I.; Chiou, C. T. *Abstracts of Papers,* 193rd National Meeting of the American Chemical Society, Denver, CO; American Chemical Society: Washington, DC, 1987; ENVR 5.
21. McCarthy, J. F. Preprint of an extended abstract, Division of Environmental Chemistry, 193rd National Meeting of the American Chemical Society, Denver, CO; American Chemical Society: Washington, DC, 1987, Vol. 27, No. 1, pp 286–288.
22. Schippers, J. C.; Verdous, J.; Hofman, J. M. *Desalination* **1980,** *32,* 103–112.
23. Green, J. B.; Manahan, S. E. In *Chemistry of Wastewater Technology*; Rubin, A. J., Ed.; Ann Arbor Science: Ann Arbor, 1978; pp 373–401.
24. Green, J. B.; Manahan, S. E. *Fuel* **1981,** *60,* 330–334.
25. Green, J. B.; Manahan, S. E. *Fuel* **1981,** *60,* 488–494.

RECEIVED for review August 4, 1987. ACCEPTED for publication December 1, 1987.

7

Effect of Humic Substances on the Treatment of Drinking Water

A. Bruchet, C. Anselme, J. P. Duguet, and J. Mallevialle

Centre de Recherche Lyonnaise des Eaux, 38 rue du Président Wilson, 78230 Le Pecq, France

Nonspecific parameters such as dissolved organic carbon, UV absorbance, and fluorescence are no longer sufficient to optimize new water-treatment processes. Specific determinations with gas or liquid chromatographic techniques identify only 5–15% of the dissolved organic carbon present in raw waters. The remaining high-molecular-weight compounds are still poorly characterized. This chapter presents a pyrolysis–gas chromatographic–mass spectrometric technique used to study the background organic matrix of natural waters. Application to various soil and fresh-water samples indicates wide variations in the relative contributions of carbohydrates, polyhydroxyaromatic substances, amino sugars, and proteinaceous materials to the high-molecular-weight fraction. Possible consequences for water-treatment processes are examined. The change of concentration of these biopolymers during a clarification process is also reported.

THE INFLUENT OF A DRINKING-WATER-TREATMENT PLANT usually contains between 2 and 6 mg of dissolved organic carbon (DOC) per liter. Specific organic chemicals detected by gas or liquid chromatographic techniques associated with mass spectrometry (GC–MS or LC–MS) typically represent only 5–15% of the DOC (*1*). The remaining high-molecular-weight fraction (85–95%) is usually referred to as "humic substances". These so-called humic substances, which have been operationally defined on the basis of their water solubility at acidic or basic pH (*2*), are not amenable to analysis with current analytical techniques and hence are poorly characterized.

0065–2393/89/0219–0093$06.00/0

The various models of humic substances advanced in the literature are derived from the study of their oxidation or hydrolysis byproducts or based on examination of their elemental composition, spectroscopic behavior, and other physicochemical properties. The Schnitzer–Kahn model (*3*), which involves a polyhydroxyaromatic core made up of phenolic and aromatic acids, has contributed greatly to the consensus that humic substances are highly aromatic and phenolic in nature. Though the predominantly aliphatic character of humic substances has been demonstrated in more recent studies (*2*), the aromatic–phenolic concept will probably prevail for a long time in the water-treatment industry.

Early studies on the behavior of humic substances during water-treatment processes often involved a molecular-weight separation followed by the evaluation of nonspecific parameters such as UV absorbance, fluorescence, or DOC for each fraction generated (*4–6*). More recently, gel permeation chromatography combined with pyrolysis–gas chromatography–mass spectrometry (Py–GC–MS) has provided better insight into the structure of the organic matrix, as well as the relationship of structures to molecular size and the treatment process (*7*).

This chapter reports on careful evaluation of the use of Py–GC–MS to characterize humic substances from various sources. Two soil fulvic fractions and various natural waters are investigated to determine the structural variability of humic substances and its possible consequences on water-treatment processes. An example of the behavior of aquatic humic substances during a clarification process is also described.

Experimental Materials and Methods

Samples. Water samples were collected from the River Houlle near Dunkirk (France). This natural water usually contains a high background organic concentration, with typical DOC values ranging from 4 to 12 ppm. Seasonal algal blooms are followed by increases in DOC and trihalomethane formation potential (THMFP). The Houlle raw water is treated at the Moulle treatment plant with breakpoint prechlorination followed by coagulation with $FeSO_4Cl$, flotation, and activated-carbon filtration. Various surface or ground-water samples collected from different locations throughout France were characterized.

Two different fulvic acids were used. The so-called "INRA" (Institut National Recherche Agronomique) fulvic acid was extracted from a podzol soil according to the method described by Holtzclaw and Sposito (*8*). The commercially available sample (Contech, ETC Limited–Ottawa, Canada) has been studied and some of its properties can be found in the literature (*9, 10*).

Sample Preparation, Py–GC–MS. The methods used for water-sample concentration, molecular-weight separation, and Py–GC–MS have been reported elsewhere (*7*). Briefly, water samples are concentrated by vacuum rotary evaporation at 40 °C in order to obtain a DOC between 100 and 200 ppm. The concentrates (10-

mL aliquots) are injected into a column (Sephadex, Sephadex Pharmacia, Uppsala, Sweden) and eluted with ultrapure water at a rate of 100 mL/h. Detection of the fractions is by UV absorbance at 260 nm and total organic carbon (TOC) measurement on a TOC meter (Dohrman DC 80, Envirotech Corp., Santa Clara, CA). The Sephadex fractions are concentrated by rotary evaporation down to a few milliliters and further reduced under a nitrogen stream to 100 μL.

Replicates (50-μL) are then deposited into quartz tubes and allowed to dry slowly at room temperature. The quartz sample holders are inserted into a filament pyrolyzer (Pyroprobe 100, Chemical Data Systems) and heated to 750 °C at a rate of 20 °C/ms. The resulting final temperature inside the quartz tube is controlled with a type K thermocouple minithermometer (Cole Parmer Instrument Company) at 625 ± 5 °C. After pyrolysis, the fragments are separated on a 30-m fused silica capillary column (DB WAX) programmed from 25 to 220 °C at a rate of 3 °C/min and identified by mass spectrometer (R 10–10 C, Ribermag, Rueil–Malmaison, France) operated at 70 eV and scanned from 20 to 400 m/z.

Data Interpretation

To simplify comparison of pyrolysis data, pyrochromatograms were coded and presented as pie charts. To draw the pie charts, a wide set of standards* representative of the various types of biological macromolecules (bovine serum albumin, chitin, starch, cellulose, cellulose acetate, cellobiose, *n*-acetylmuramyl L-alanyl D-isoglutamine, purified lignin) was submitted to flash pyrolysis. Pyrolysis fragments characteristic of each type of biopolymer (*11*) (e.g., acetamide for *N*-acetylamino sugars; pyrrole, 4-methylphenol, and phenol for proteins; furfural, methylfurfural, levoglucosenone, and carbonyl compounds for carbohydrates; phenols and methoxyphenols for polyhydroxyaromatic compounds) were selected. Intensities of these selected fragments weighed against the amount of material pyrolyzed (usually between 300 and 500 μg) were then used to determine average relative response factors and to correct for differences in sensitivity between the four types of biopolymers. The percentages indicated in the pie charts are not absolute, but only relative to the set of standards used.

To discriminate between the phenolic peaks arising from polyhydroxyaromatic substances and tyrosine-containing proteins, the following correction was performed. Because tyrosine leads to similar quantities of phenol and 4-methylphenol (in contrast to polyhydroxyaromatic compounds, which yield predominantly phenol), the response for 4-methylphenol was subtracted from the area obtained for phenol and the difference was used to calculate the proportion of polyhydroxyaromatic compounds.

Contact us for the list of selected standards, along with the individual and average relative response factors of their characteristic fragments used for these calculations. The percentages indicated in the pie charts are not absolute, but only relative to the set of standards used.

*The list of selected standards, along with the individual and relative response factors of their characteristic fragments used for these calculations, can be obtained from us.

Results and Discussion

Comparison of Fulvic Fractions. The pyrochromatograms obtained for the INRA and Contech fulvic samples are shown in Figures 1 and 2. Their elemental analysis and THMFP are reported in Table I. The main difference between the two fractions lies in the much higher N/C, O/C, and H/C ratios observed for the INRA sample. The INRA sample shows a higher nitrogen content (N = 3.7%). The low H/C ratio of the Contech sample indicates a more aromatic nature. In good agreement with the unusually high O/C ratio observed (0.95), most of the fragments detected in the pyrochromatogram of INRA fulvic acid (furfural derivatives, anhydro sugars, and carbonyl compounds such as hydroxypropanone, cyclopenten-1-one, acetic acid, and a lactone) are typical decomposition products of carbohydrates (*12*). The lactone detected (3-hydroxy-2-penteno-1,5-lactone) is characteristic of nonreducing xylopyranosyl residues (*13*). The nitrogen content of this fulvic fraction (3.7%) is found under the form of both *N*-acetylamino sugars (indicated by the acetamide peak) and peptidic material (presence of pyrrole). The low intensity of the phenolic peaks indicates that the polyhydroxyaromatic content is extremely low.

By contrast, the dominant phenolic peak observed on the Contech pyrochromatogram points to a highly purified structure from which the soil

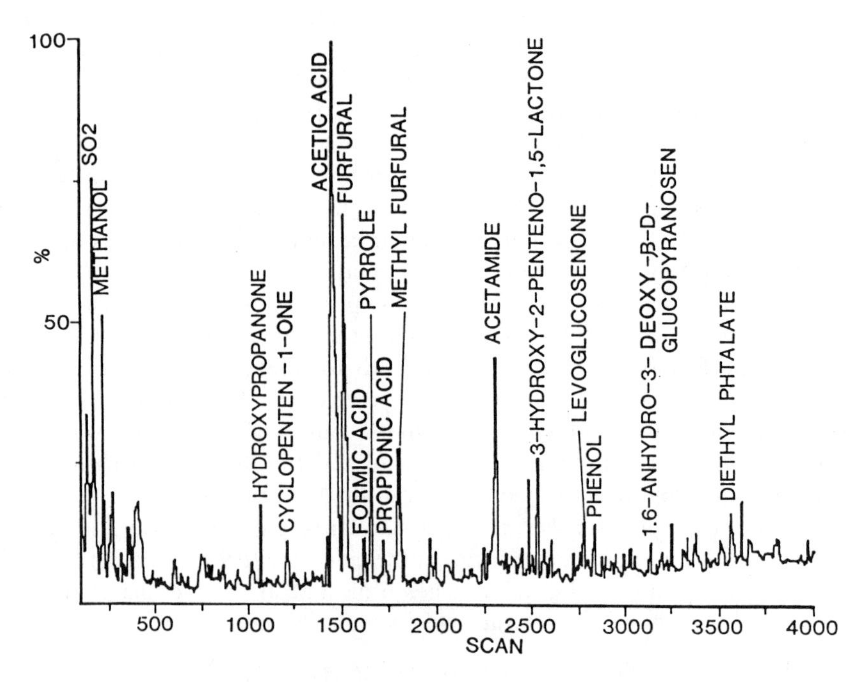

Figure 1. Pyrochromatogram of INRA fulvic acid.

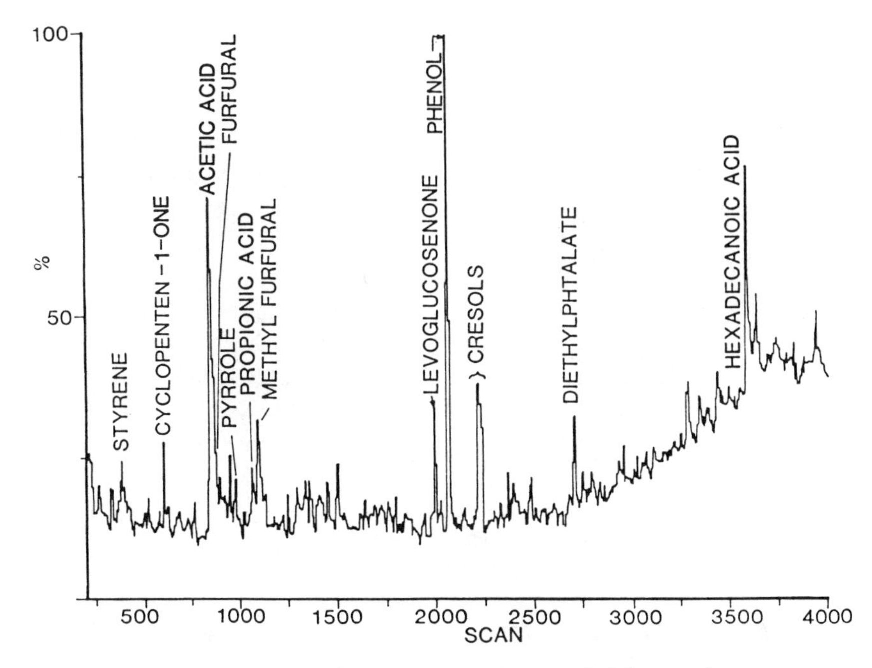

Figure 2. Pyrochromatogram of Contech fulvic acid.

Table I. Elemental Analysis and THMFP of the Investigated Fulvic Fractions

Element	*INRA*	*Contech*
C	36	50.9
H	4.8	3.35
O	46	44.75
N	3.7	0.75
H/C	1.60	0.79
N/C	0.088	0.013
O/C	0.95	0.66
THMFP[a]	42	163

NOTE: Results are expressed in micrograms per liter.
[a]The 10 mg/L of each fraction reacted with 10 mg/L of Cl during 72 h.

polysaccharides have been almost entirely removed. In agreement with the lower O/C ratio (0.66 as compared to 0.95), typical carbohydrate pyrolysis fragments show only minor intensities. The Contech fulvic acid appears to be almost devoid of nitrogenous compounds. Compounds such as the phthalates, which are more likely to be volatilization products than pyrolysis fragments, have been classified as miscellaneous on the pie charts.

The drastic differences observed between the two soil fulvic fractions investigated are illustrated by the corresponding pie charts in Figure 3. As

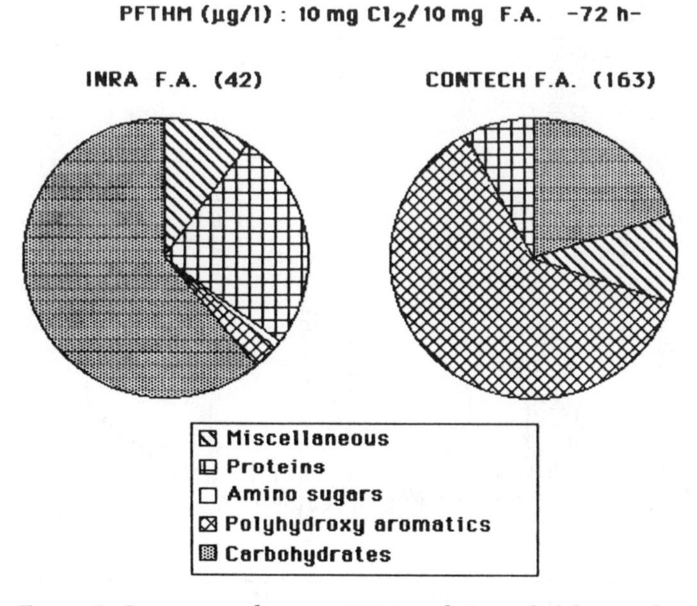

Figure 3. Comparison between INRA and Contech fulvic acids.

far as water-treatment processes are concerned, the Contech fulvic acid shows a much higher THMFP (163 µg/L as opposed to 42 µg/L). This THMFP value is assumed to be directly related to its high polyhydroxyaromatic content. Additional examples of the variability of humic substances and the resulting effect on the THMFP have been reported elsewhere (*14*). Humic and fulvic fractions are defined only in an operational way; therefore, any substance soluble in water above or below pH 2 can be classified as part of a humic or fulvic fraction, respectively. As a result of this definition, compositional variations in space and time are the rule, rather than the exception. This example also raises the question of the use of standard or commercial humic substances in batch or pilot experiments intended to study specific water-treatment processes. It appears better to use the actual water without disturbing the matrix in any attempt to conduct a realistic pilot plant for design purposes.

Comparison of Aquatic Humic Substances. Pie charts obtained for three fresh waters from different locations in France are presented in Figure 4. These pie charts take into account the distribution observed on the three fractions usually collected after Sephadex gel permeation chromatography and therefore represent the global composition of the water. However, the elution of the Sephadex columns is carried out with ultrapure water instead of a buffer solution to avoid salt-induced catalytic effects that

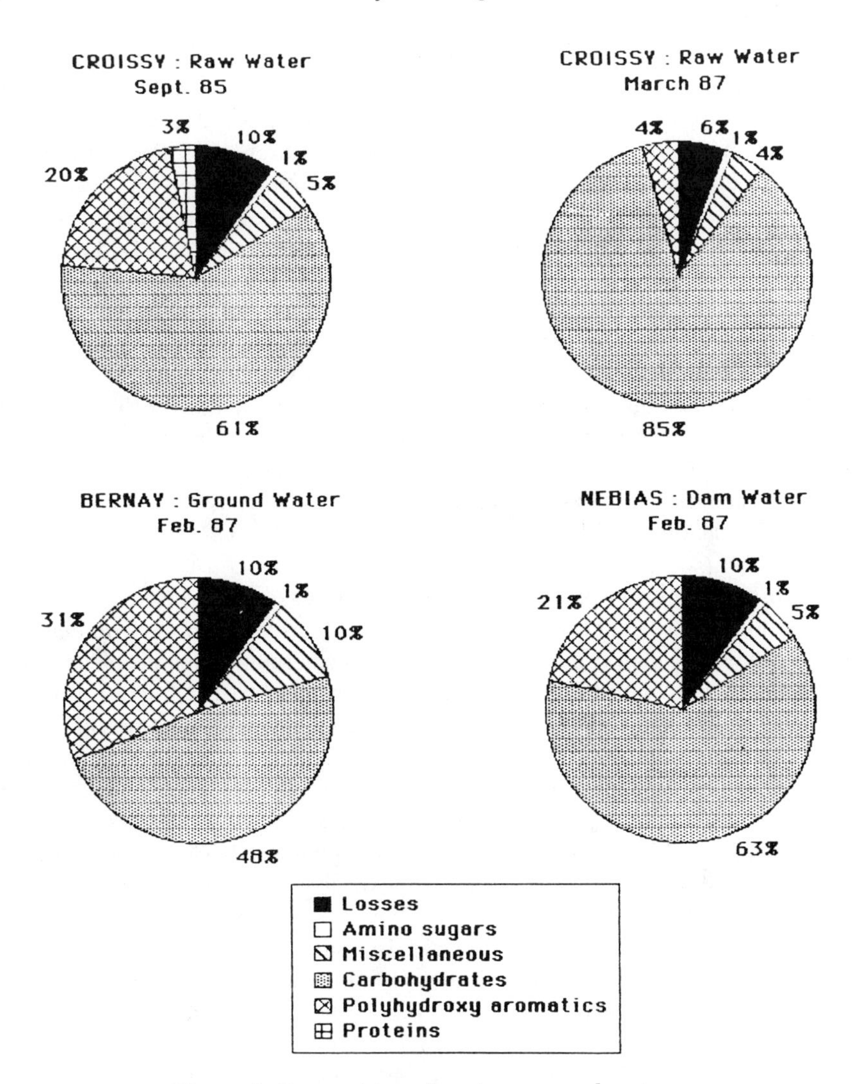

Figure 4. Composition of various natural waters.

may occur during the pyrolysis of the concentrated fractions (*11*). Because of this substitution, care should be exercised in relating the elution position with molecular weight. DOC losses during the concentration step by rotary evaporation (usually less than 20%) are also indicated. Percentages reported for each class of biopolymers should not be considered absolute, but relative to the set of standards used (*see* Data Interpretation). Because the exact structure of the compounds present is not known, true quantitative work is impossible to achieve. However, the conventions used to draw the pie charts are the same, and so the results are directly comparable.

Nebias dam water exhibits a high DOC (6.1 mg/L); losses during concentration are small (only 10%) and the composition is dominated by carbohydrates (63%) and polyhydroxyaromatic substances (21%). Two samples collected from the River Seine at Croissy in March 1987 and September 1985 give an example of probable seasonal fluctuations; polyhydroxyaromatic compounds represent 20% in September, but only 4% in March. In both cases, carbohydrates are predominant (85 and 61%, respectively, for March and September). The ground water investigated (Bernay), which shows a quick turnover during the rainy season, contains 48% carbohydrates and 31% polyhydroxyaromatic substances.

The examples included in Figure 4 clearly illustrate the wide variations encountered in high-molecular-weight aquatic organic matter. The proportions of the four types of biological macromolecules most frequently detected in natural waters (carbohydrates, polyhydroxyaromatic substances, *N*-acetylamino sugars, and peptidic material) may change drastically as a function of sample location and history.

Behavior During a Clarification Process. Gel permeation chromatograms obtained at the Moulle treatment plant for both raw and clarified water are reported in Figure 5. The Sephadex fractions collected are referred to as G1, G2, G3, and G4; these labels correspond to the fraction number by order of elution. Calibration of the Sephadex gels with dextrans of known molecular weights indicates the following apparent molecular weights: >5000 daltons for G1, between 1500 and 5000 daltons for G2, and <1500 daltons for G3 and G4. The Houlle raw water exhibits an unusually high exclusion peak (G1), which accounts for as much as 31.5% of the DOC.

Table II shows the DOC values (both absolute and relative) and the ratio of optical density at 260 nm to DOC (OD_{260}/DOC) for each fraction before and after clarification. Because the same concentration factor was applied to both water samples, Table II values are directly comparable to the gel permeation chromatograms (GPC). The clarification process mainly concerns DOC removal from the higher-molecular-weight fractions (decreased by 75% (G1), 63% (G2), and 40% (G3), respectively); the lower-molecular-weight fraction (G4) is practically unchanged. Also, the different molecular-weight fractions show a modification of their UV/DOC ratios that indicates selective removal of certain types of molecules during the clarification process.

The corresponding Py–GC–MS results are indicated in Figure 6. The proportion of proteins and carbohydrates is kept constant (13 and 31%) while the DOC is decreased by 50%, which means that both types of biopolymers were removed with approximately 50% efficiency. The corresponding efficiency for the polyhydroxyaromatic compounds is around 75%; the efficiency for miscellaneous compounds is about 25%. To summarize, the efficiency of the clarification process is favored in this case by the presence of a significant

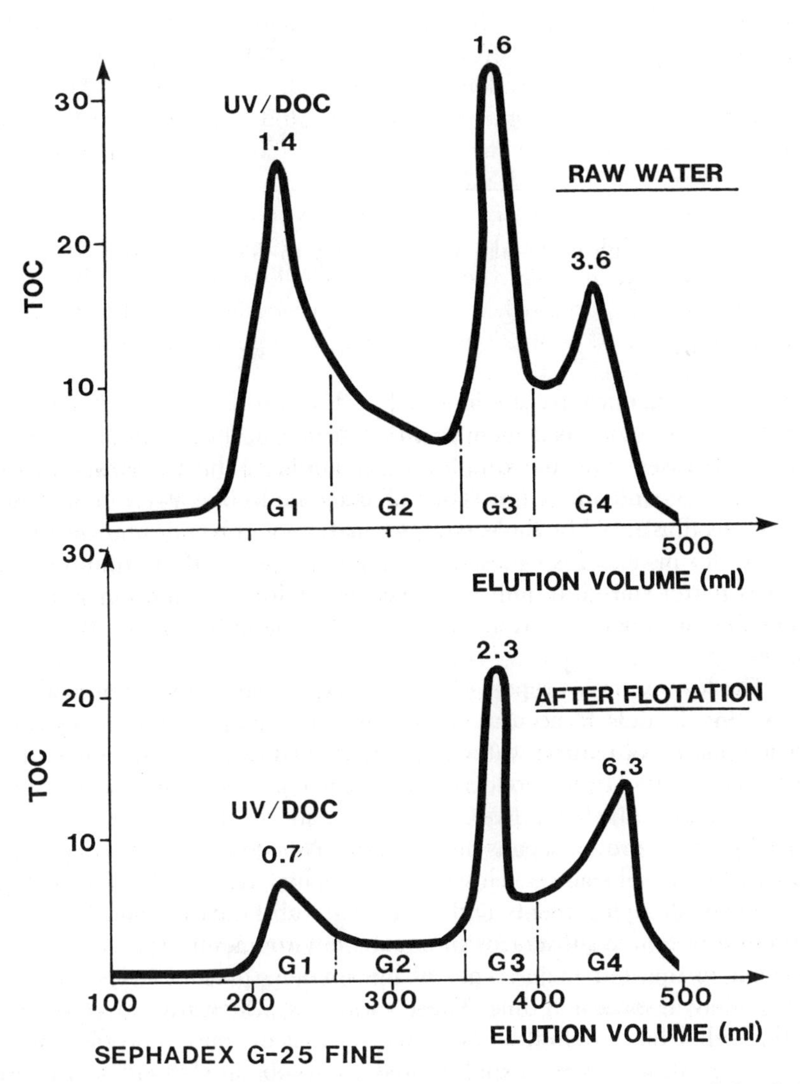

Figure 5. Sephadex gel permeation chromatograms of Houlle raw and clarified water.

Table II. Evolution of Houlle Sephadex Fractions During Clarification at the Moulle Plant

Sample	G1	G2	G3	G4
Raw water				
DOC (μg)	1323	802	1171	903
DOC (%)	31.5	19.1	27.9	21.5
OD_{260}/DOC	1.4	—	1.6	3.6
Clarified water				
DOC (μg)	325	292	695	886
DOC (%)	14.8	13.3	31.6	40.3
OD_{260}/DOC	0.7	—	2.3	6.3

proportion of high-molecular-weight, easy-to-flocculate, polyhydroxyaromatic compounds.

Conclusions

Nonspecific chemical parameters such as UV absorbance, fluorescence, and DOC have become insufficient to allow further optimization of water-treatment processes. Specific problems encountered during these treatment processes depend upon the types of macromolecules present in the raw waters, as illustrated in Table III. Elimination of polyhydroxyaromatic compounds, for instance, is necessary to decrease the THMFP; removal of proteins is better suited to limit bacterial regrowth in distribution systems. To meet the challenge, a better knowledge of aquatic humic substances is required.

The few examples reported in this chapter have shown the inadequacy of existing models to accurately describe the complex and ever-changing organic matrix of natural waters. The highly condensed aromatic cores described by most of these models often appear as a minor constituent of aquatic organic matter. Instead, Py–GC–MS reveals the ubiquitous presence of carbohydrates, proteinaceous material, and *N*-acetylamino sugars in soil and aquatic humic substances. This result is in good agreeement with the composition of decaying plants and algae that substantially contribute to the structure of humic substances in aquatic environments (*15*). Also, the proportions of the four major types of biopolymers present in natural waters vary widely in space and time. These variations, which are expected because of the rather fuzzy definition of humic substances, raise a problem in comparison or extrapolation of experimental results obtained thoughout the world with such highly inconstant mixtures.

Finally, the methods developed in this chapter require further refinement to improve their quantitative aspect. However, they probably represent an invaluable tool for obtaining an overall view of the composition of aquatic humic substances and their behavior in various water-treatment processes.

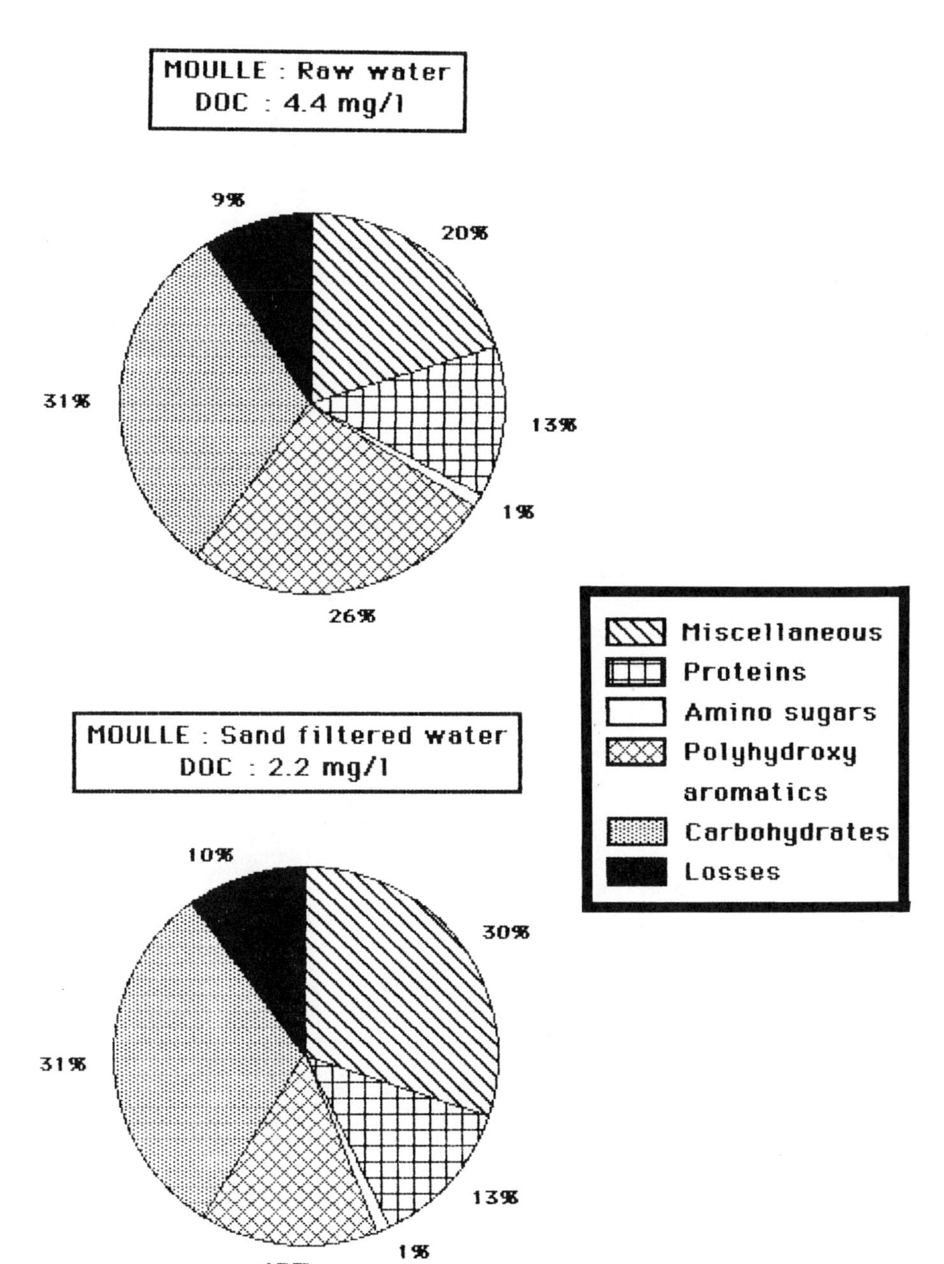

Figure 6. Comparison between raw and clarified water at the Moulle treatment plant.

Table III. Problems Caused by the Presence of Organic Matter in Water

Macromolecules	*Sequestration (Metals or Pesticides)*	*Bacteria Regrowth*	*Adsorption Processes*[a]	*Oxidation–Disinfection Processes*	*Chlorination*	*Ozonation*
Proteins	+	+ + +	+	+	Problems of tastes and toxicity due to formation of THM +, formation of TOX and organic chloramines + + +, and formation of aldehydes	Problems of fruity tastes and odors due to formation of aldehydes and acids +
Polyhydroxyaromatic compounds	+ + +	+	+ +	+ +	Problems of tastes and toxicity due to formation of THM and TOX + + + and formation of aldehydes and acids +	Problems of fruity tastes and odors, bacteria regrowth, and possibly toxicity due to formation of aldehydes and acids +
Polysaccharides	0	+	+ ?	0	0	0
Amino sugars	0	?	?	?	Problems of tastes and possibly toxicity due to formation of THM, TOX, and organic chloramines	0

NOTE: Direct problems are sequestration, bacteria regrowth, and competition in the removal of micropollutants or microorganisms by adsorption or oxidation–disinfection. Chlorination and ozonation are problems caused by the formation of oxidation byproducts.
[a]Dependent on molar weight.

References

1. Watts, C. D. *Mass Spectrometric Identification of Non-volatile Organic Compounds*; WRC Technical Report TR110; Water Research Centre: Oxfordshire, England, April 1979.
2. Aiken, R. G.; McKnight, D. M.; Wershaw, R. L.; MacCarthy, P. *Humic Substances in Soil, Sediment, and Water*; Wiley–Interscience: New York, 1985.
3. Schnitzer, M.; Kahn, S. U. *Humic Substances in the Environment*; Dekker: New York, 1972.
4. Gjessing, E. T. *Schweiz. A. Hydrol.* **1973,** *35*, 286–294.
5. Gloor, G.; Hans, L. *Anal. Chem.* **1979,** *51*(6), 645.
6. Glaze, W. H.; Jones, P. C.; Saleh, F. W. In *Advances in the Identification and Analysis of Organic Pollutants in Water*; Keith, L. H., Ed.; Ann Arbor Science: Ann Arbor, 1981; pp 371–382.
7. Bruchet, A.; Tsutsumi, Y.; Duguet, J. P.; Mallevialle, J. In *Organic Pollutants in Water*; Suffet, I. H.; Malaiyandi, M., Eds.; Advances in Chemistry Series 214; American Chemical Society: Washington, DC, 1987; pp 381–399.
8. Holtzclaw, K. M.; Sposito, G. *Soil Sci. Soc. Am. J.* **1976,** *40*, 254–258.
9. Gamble, D. S.; Schnitzer, M. *Trace Metals and Metal Organic Interactions in Natural Waters*; Ann Arbor Science: Ann Arbor, 1973; pp 265–302.
10. Sposito, G.; Holtzclaw, K. M.; Levesque, C. S.; Johnston, C. T. *Soil Sci. Soc. Am. J.* **1982,** *46*, 265–270.
11. Bruchet, A. Thèse de 3ème cycle, Université de Poitiers, 1985.
12. Irwin, W. J. *Analytical Pyrolysis: A Comprehensive Guide*; Chromatographic Science Series, Vol. 22; Dekker: New York, 1982.
13. Ohnishi, A.; Kato, K.; Takagi, E. *Carbohydrate Res.* **1977,** *58*, 387–395.
14. Bruchet, A.; Anselme, C.; Duguet, J. P.; Mallevialle, J. Presented at the Sixth Conference on Water Chlorination, Environmental Impact and Health Effects, Oak Ridge, TN, May 3–8, 1987.
15. Gadel, F.; Bruchet, A. *Water Res.* **1987,** *21*(10), 1195–1206.

RECEIVED for review October 15, 1987. ACCEPTED for publication May 18, 1988.

8

Separation of Humic Substances and Anionic Surfactants from Ground Water by Selective Adsorption

E. M. Thurman

U.S. Geological Survey, Campus West, Lawrence, KS 66046

Jennifer Field

Geochemistry Department, Colorado School of Mines, Golden, CO 80401

Surface-active organic compounds, surfactants, may be isolated from ground water effectively on polymeric resins with a procedure nearly identical to the isolation procedure for humic substances. Surfactants, such as alkylbenzenesulfonates and alkyl sulfates, are sorbed from both distilled water and ground water onto XAD–4 (styrene–divinylbenzene) resin at neutral pH (6–7). At this pH the majority of the humic substances are not sorbed, but pass through the XAD–4 resin. The effluent from the column is collected and the pH adjusted to 2.0 with concentrated hydrochloric acid. The solution is then passed through a second column of XAD–8 (methyl methacrylate) resin, and humic substances are sorbed. The surfactants and some colored substances are eluted from the XAD–4 column with methanol, and the humic substances are eluted from the XAD–8 column with an aqueous solution of 0.1 N sodium hydroxide. The humic substances from the XAD–8 column are then ready for characterization without interference from anionic surfactants. Unfortunately, the surfactant fraction from the XAD–4 column does contain some colored humic material.

THE STUDY OF AQUATIC HUMIC SUBSTANCES in surface and ground water is an important area of environmental chemistry because they are so wide-

0065–2393/89/0219–0107$06.00/0

spread (*1–3*) and they play an important role in chemical reactions such as trihalomethane production (*4*), trace-metal complexation (*5*), cosolubilization of pollutants (*6, 7*), and other environmental chemical reactions (*3*). Furthermore, studies of the structure of humic substances (*1*) require humic fractions that are free of synthetic organic contaminants. Isolation procedures, therefore, are the important first step in the study of aquatic humic substances.

The two general isolation approaches for aquatic humic substances are ion-exchange (*8–10*) and adsorption chromatography (*11*). Both approaches isolate 50–90% of the dissolved organic carbon (*3*) and most of the organic color from natural waters (*3, 11, 12*). However, both ion-exchange resins and polymeric resins also isolate nonionic contaminants from wastewater (*13*), as well as some anionic surfactants (*14, 15*) such as linear alkylbenzenesulfonate surfactants (LAS) and branched-chain alkylbenzenesulfonate surfactants (ABS). Thus, there is the potential for co-isolation of surfactants and humic substances in ground and surface waters contaminated with wastewater; sewage is a common contaminant source (*15, 16*). Because we were studying ground water contaminated by sewage at Otis Air Base, Falmouth, Massachusetts (*16*), we became interested in selective isolation of both anionic surfactants (*15, 17*) and humic substances.

Past studies have found that the sodium salts of ABS and LAS surfactants were efficiently sorbed onto XAD–8 (a methyl methacrylate resin) as ions. The hydrophobic alkylbenzene chain interacts with the resin by the hydrophobic effect (*15*), and the sulfonic acid salt is oriented into the water phase. This mechanism has been named the dynamic ion-exchange model in ion-pair chromatography (*18–20*).

XAD–2 and XAD–4 (styrene–divinylbenzene resins) have been used to isolate both nonionic (*21–22*) and anionic surfactants from water. They are effective resins for isolation of small molecules (less than 500 molecular weight), but are less effective for humic substances because their small pore size excludes humic substances (*12*). On the other hand, XAD–8 effectively isolates humic substances from water (*11–12*).

We used both XAD–4 for surfactants and XAD–8 for humic substances to take advantage of the best characteristics of both for the selective isolation of humic substances and anionic surfactants. Results discussed here include laboratory and field experiments.

Experimental Procedures

Reagents and Supplies. XAD–4 and XAD–8 resins (Rohm and Haas, Philadelphia), styrene–divinylbenzene and methyl methacrylate copolymers, were extracted with 0.1 N NaOH, followed by methanol Soxhlet extraction. The resins, which had been stored in methanol, were rinsed with 50 bed volumes of distilled water prior to use to remove the methanol. Surfactant standards were reagent grade, including the sodium salt of linear alkylbenzenesulfonate (Fluka, Ronkonkoma, New

York), and sodium dodecyl sulfate (SDS) (Fluka, Ronkonkoma, New York). The ABS surfactant was technical grade (Association of American Soap and Glycerine Producers, New York). The humic material was Suwannee River fulvic acid, which was isolated by the method of Thurman and Malcolm (*11*) and has been discussed in previous work (*23*).

Procedures. The XAD–4 and XAD–8 resins were packed into 20-mL columns in water and rinsed with 50 bed volumes of distilled water. The individual solutions of surfactants in the form of sodium salts and humic substances, at 2 mg/L each in distilled water, were passed through the XAD resins at 15 bed volumes per hour. The anionic surfactant breakthrough curve was measured by determining the methylene blue active substances (MBAS) in the column effluent (*24*). The fulvic acid concentration was measured by absorbance at 320 nm. The fulvic acid did not interfere in the MBAS test.

Field Analysis. The ground water on Cape Cod, Massachusetts (*17*), was sampled by stainless steel submersible pump at well F347–67 for surfactant and humic substances analysis. The water samples were collected in 4-L glass bottles and shipped to the laboratory for analysis.

The 8-L aliquots were pumped through a 20-mL column of XAD–4 at neutral pH (15 bed volumes per hour). The water sample was adjusted to pH 2 with concentrated hydrochloric acid and passed through a 20-mL XAD–8 column.

The surfactants were measured on the XAD–8 and XAD–4 columns by the MBAS colorimetric test (*24*). This method uses the formation of a complex between the anionic surfactant and methylene blue (a cationic dye), which is extracted into chloroform and quantified by colorimeter. The humic substances were measured as color (absorbance at 320 nm).

Results and Discussion

Surfactant Recovery on XAD. LAS and SDS are efficiently sorbed on XAD–4, with 100% sorption of each (Table I). Although distribution coefficients were not measured on the resins, the minimum distribution coefficient can be calculated from the experimental conditions. This calculation gives a distribution coefficient of greater than 166 for LAS and SDS

Table I. Percent Recoveries of Anionic Surfactants and Aquatic Humic Substances at Neutral pH (7.0)

	Percent Sorbed[a]		*Percent Eluted with CH_3OH*[b]	
Compound	*XAD–4*	*XAD–8*	*XAD–4*	*XAD–8*
Linear alkylbenzenesulfonate	100 ± 5	100 ± 5	77 ± 5	88 ± 5
Branched-chain alkylbenzenesulfonate	ND[c]	100 ± 5	ND[c]	88 ± 5
Sodium dodecyl sulfate	100 ± 2	41 ± 2	72 ± 2	86 ± 2
Suwannee River fulvic acid	0	0	—	—

[a]Percent sorbed is the percent of the total surfactant passed through the column.
[b]Percent eluted is the percent of the amount sorbed, which is then eluted.
[c]ND, experiment not done.

on the XAD–4. ABS was not determined on XAD–4, but because of its structure (nearly identical to LAS) and the complete sorption of SDS, it is reasonable to assume that ABS would also have large sorption efficiency on XAD–4.

A minimum value for the distribution coefficient, K_D, can be estimated from the size of the column (6 g of resin dry weight) and the interstitial volume of water in the column (8.0 mL):

$$K_D = \frac{\text{concentration on resin/g of resin}}{\text{concentration in water/mL of water}}$$

Thus, for complete removal of 2 mg of surfactant from 1 L of water at a concentration of 2 mg/L, the calculation gives:

$$K_D = \frac{2 \text{ mg/6 g}}{0.002 \text{ mg/mL}}$$

$$K_D > 166$$

We realize that this calculation is not a true K_D, but actually a point on the isotherm. Because the column was not saturated with surfactant, the actual capacity of the resin is much greater than this estimated value.

The sorption of ABS, LAS, and SDS on XAD–8 was different. All of the ABS and LAS was sorbed, but only 41% of the SDS was removed. This result indicates a K_D of less than 66. Because the SDS contains only 12 carbon atoms and ABS contains 18 carbon atoms, there is much less of a hydrophobic effect upon sorption for SDS. This conclusion is based upon previous studies of XAD sorption of hydrophobic organic compounds (*25*). On the other hand, the XAD–4 has considerably more surface area than XAD–8 (650 versus 140 m^2/g), and the SDS was completely sorbed to XAD–4. Although XAD–4 and XAD–8 have different functional groups (styrene–divinylbenzene and methyl methacrylate), past studies have shown that surface area is of key importance on sorption (*12*). These data suggest that XAD–4 is more efficient for the removal of the more water-soluble anionic surfactants.

During elution the XAD–4 released 72–77% of the anionic surfactants (Table I), and XAD–8 released 86–88% of the sorbed surfactants. Larger pore size is credited with giving XAD–8 more efficient elution properties (*12*) than XAD–4. Our data support that hypothesis. On the basis of these data, XAD–4 seems to be a better sorbent for a wide range of water-soluble surfactants. On the other hand, if only the alkylbenzenesulfonates are of interest, then the XAD–8 is more efficient.

The Suwannee River fulvic acid (*11, 23*) did not adsorb at neutral pH on either the XAD–4 or the XAD–8 resin. This result was expected because of previous studies (*11, 12*) and the initial isolation of the Suwannee River fulvic acid on XAD–8 resin at pH 2. Previous studies (*11, 12, 23*) have shown that, without pH adjustment, aquatic humic substances do not concentrate on the XAD resin because the many ionic carboxyl groups present on the humic material make the compounds water-soluble. When the pH is adjusted to 2.0, the carboxyl groups are protonated and the fulvic acid is much less soluble. Thus, a considerable enhancement in sorption occurs at pH 2.0.

However, some natural surface-active compounds in water (for instance, long-chain fatty acids and some pigments) may sorb onto XAD resins at neutral pH. Thus, many natural compounds may still be present in the surfactant fraction. The question of whether the pigments and other colored substances are humic substances is the subject of a different paper (*26*). The isolation of natural surfactants at neutral pH would be an interesting research project for future work.

Humic Isolation and Surfactant Co-isolation. Although humic material does not co-isolate with the surfactants at neutral pH, there may be a major problem with the co-isolation of humic substances and surfactants at pH 2.0. Humic substances are efficiently sorbed on XAD–8 at pH 2.0 (*11, 12*), but anionic surfactants are also efficiently sorbed at this pH. The amount sorbed in both cases is 100 ± 5%, and the amount eluted is 2 ± 5%. Co-isolation, which would be most pronounced in surface waters receiving sewage effluent, is a minor problem if the eluent used is 0.1 N NaOH. We found that only a trace amount (2%) of the surfactant sorbed with the humic material co-eluted with the humic substance at alkaline pH. However, if a mixture of methanol and ammonium hydroxide is used as eluent, then the surfactants would co-elute. Methanol–ammonium hydroxide is a commonly used eluent for the removal of humic substances from sea water (*27, 28*). Harvey (*29*) reported that alkylbenzenes found in marine fulvic acid may originate from sewage (*30*). An alternative hypothesis, based on our results about surfactant removal, is that ABS and LAS surfactants would co-isolate with the marine fulvic acid and then be co-eluted in methanol–ammonium hydroxide. These substances would then interfere in the structural studies of the marine fulvic acid. Thus, we have designed the following simple two-column isolation scheme for aquatic fulvic acid.

Selective Humic Isolation. For selective isolation, the water sample is passed through an XAD–4 column at neutral pH; the anionic surfactants, long-chain fatty acids, and some pigments are removed. However, the majority of the humic substances pass through the column. The amount sorbed of linear alkylbenzenesulfonates is 100 ± 5%, and 0 ± 5% of Suwannee River fulvic acid. The effluent is recovered, then the pH is lowered to 2.0 with HCl. The sample is processed a second time through XAD–8; during this process the humic substances are sorbed. The amount sorbed of linear alkylbenzenesulfones is 0 ± 5%, and 100 ± 5% of Suwannee River fulvic acid. The humic substances are then back-eluted with 0.1 N NaOH and treated with the procedure described by Thurman and Malcolm (*11*).

The anionic surfactants are efficiently separated and recovered in the first step (>90% recovery of anionic surfactants from a 2-mg/L solution of surfactants), and the humic substances are efficiently recovered in the second column (>90% recovery at 2 mg/L). Thus, the modification of the isolation procedure is an effective method to remove surfactant substances from the humic material. An added benefit of this procedure is that nonionic contaminants are also effectively removed in this first step (*13*). Biodegradation products of surfactants (carboxylated alkylbenzenesulfonates) may also be present. These substances would most likely not isolate in either step 1 or step 2 in the humic fraction. This topic is the subject for future research on surfactants in surface and ground waters.

Selective Surfactant Isolation from Ground Water. This result, that surfactants may be isolated separately from the humic fraction, was tested on ground water contaminated by sewage on Cape Cod, near Falmouth, Massachusetts. The ground water contained anionic surfactants that had been mapped by MBAS in previous studies (*16*, *17*).

Table II shows the recovery of anionic surfactant (MBAS) and of aquatic humic substances (color at 320 nm) for well F347–67, which is 300 m down gradient of the sewage disposal beds and known to contain a mixture of surfactants and humic substances (*16*, *17*). Table II shows that 64% of the MBAS substances were isolated onto the XAD–4 resin (0.105 mg/L). This fraction is currently being examined by linked-scan fast-atom bombardment to determine if LAS is present.

Approximately 18% of the color was isolated on the XAD–8 resin and 11% on the XAD–4 resin. The remainder of the color was not removed by either column. The small XAD–8 column (20 mL) was overloaded with sample (8.0 L), which forced much of the colored organic matter through XAD–8. Only 2% of the MBAS co-isolated on XAD–8, with most sorbing to the XAD–4 (62%). For future work a larger XAD–8 column will be used, approximately 20 mL of resin per liter of water. The fact that color sorbed without pH adjustment indicates that a colored surfactantlike material is present in the wastewater. Thus, the method seems to be partially effective for the separation of anionic surfactants and humic substances during the

Table II. Recovery of Surfactants and Humic Substances from Well F347–67

	Color Removed		*MBAS[a] Removed*	
Resin	*mg/L*	*%*	*mg/L*	*%*
XAD–4	0.316	11	0.105	62
XAD–8	0.515	18	0.004	2

[a]Methylene blue active substances.

sorption phase. However, the humic substances isolated on XAD–8 are relatively free of surfactants.

Summary

When water samples are passed through XAD–8 resin at pH 2.0, anionic surfactants and humic substances are effectively isolated. Because of this success, we have modified a previously published isolation scheme for humic substances. The new procedure includes a precolumn of XAD–4 that isolates the surfactants during sorption at neutral pH. The aquatic humic substances are isolated in a second step by passing the water through a XAD–8 column at pH 2.0. The major improvement of this new method over previous humic-substances research methods is the finding that surfactants will isolate with humic substances, even in an ionic form. Thus, when large volumes of water are passed through XAD columns and eluted with methanol–ammonium hydroxide (commonly done in marine humic studies), there is the possibility of eluting surfactants into a humic fraction. Unfortunately, colored hydrophobic organic acids are isolated on XAD–4 without pH adjustment. Thus, it is not possible to separate humic substances and surfactants totally in a single adsorption–elution procedure.

References

1. Aiken, G. R.; McKnight, D. M.; Wershaw, R. L.; MacCarthy, P. *Humic Substances in Soil, Sediment, and Water: Geochemistry, Isolation, and Characterization*; John Wiley and Sons: New York, 1985.
2. Stevenson, F. J. *Humic Chemistry*; John Wiley and Sons: New York, 1982.
3. Thurman, E. M. *Organic Geochemistry of Natural Waters*; Martinus Nijhoff: Dordrecht, 1985.
4. Rook, J. J. *Environ. Sci. Technol.* **1977,** *11*, 478.
5. Mantoura, R. F. C. In *Marine Organic Chemistry*; Duursma, E. K.; Dawson, R., Eds.; Elsevier: Amsterdam, 1981; 179–224.
6. Chiou, C. T.; Malcolm, R. L.; Brinton, T. I.; Kile, D. E. *Environ. Sci. Technol.* **1986,** *20*, 502.
7. Carter, C. W.; Suffet, I. H. *Environ. Sci. Technol.* **1982,** *16*, 735.
8. Sirotkima, I. S.; Varshal, G. M.; Lu're, Y. Y.; Stepanovas, N. P. *Zh. Anal. Khim.* **1974,** *29*, 1626.
9. Abrams, I. M. *Ind. Eng. Chem. Prod. Res. Dev.* **1975,** *14*, 108.
10. Leenheer, J. A. *Environ. Sci. Technol.* **1981,** *15*, 578.
11. Thurman, E. M.; Malcolm, R. L. *Environ. Sci. Technol.* **1981,** *15*, 463.
12. Aiken, G. R.; Thurman, E. M.; Malcolm, R. L.; Walton, H. F. *Anal. Chem.* **1979,** *51*, 1799.
13. Junk, G. A.; Richard, J. J.; Grieser, M. D.; Witiak, D.; Witiak, J. L.; Arguello, M. D.; Vick, R.; Svec, J. J.; Fritz, J. S.; Calder,G. V. *J. Chromatogr.* **1974,** *99*, 745.
14. Gabriel, D. M. *J. Soc. Cosmet. Chem.* **1974,** *25*, 33.
15. Thurman, E. M.; Willoughby, T.; Barber, L. B., Jr.; Thorn, K. A. *Anal. Chem.* **1987,** *59*, 1798.

16. LeBlanc, D. R. U.S. Geological Water-Supply Paper, 1984, No. 2218.
17. Thurman, E. M.; Barber, L. B., Jr.; LeBlanc, D. *Contam. Hydrol.* **1986,** *1,* 143.
18. Bidlingmeyer, B. A.; Deming, S. N.; Price, W. P.; Sachok, B.; Petrusek, M. J. *J. Chromatogr.* **1979,** *186,* 419.
19. Bidlingmeyer, B. A. *J. Chromatogr. Sci.* **1980,** *18,* 525.
20. Bidlingmeyer, B. A. *LC Mag.* **1983,** *1,* 344.
21. Jones, P.; Nickless, G. *J. Chromatogr.* **1978,** *156,* 87.
22. Osburn, W. W. *J. Am. Oil Chem. Soc.* **1986,** *25,* 33.
23. Thurman, E. M.; Malcolm, R. L. In *Aquatic and Terrestrial Humic Materials*; Christman, R. F.; Gjessing, E. T., Eds.; Ann Arbor Science: Ann Arbor, 1983; 1–23.
24. American Public Health Association. *Standard Methods for the Examination of Water and Wastewater,* 16th ed.; American Public Health Association: Washington, DC, 1985.
25. Thurman, E. M.; Malcolm, R. L.; Aiken, G. R. *Anal. Chem.* **1978,** *50,* 775.
26. Thurman, E. M.; Aiken, G. R.; Ewald, M. J.; Fischer, W. R.; Forstner, U.; Hack, A. H.; Mantoura, R. F. C.; Parsons, J. W.; Pocklington, R.; Stevenson, F. J.; Swift, R. S.; Szpakowska, B. *Humic Substances and Their Role in the Environment, 1988 Proceedings of Dahlem Konferenzen*; Frimmel, F. H.; Christman, R. F., Eds.; John Wiley and Sons: New York, 1988; pp 31–43.
27. Mantoura, R. F. C.; Riley, J. P. *Anal. Chim. Acta* **1975,** *76,* 97.
28. Stuermer, D. H.; Harvey, G. R. *Mar. Chem.* **1978,** *6,* 55.
29. Harvey, G. R. *Mar. Chem.* **1985,** *16,* 187.
30. Eganhouse, R. P.; Blumfield, D. L.; and Kaplan, I. R. *Environ. Sci. Technol.* **1983,** *17,* 523.

RECEIVED for review July 24, 1987. ACCEPTED for publication March 7, 1988.

INTERACTIONS IN NATURAL WATERS WITH ORGANIC CONTAMINANTS

9

Binding of Nonpolar Pollutants to Dissolved Organic Carbon

Environmental Fate Modeling

Gail Caron[1] and I. H. Suffet

Environmental Studies Institute, Drexel University, Philadelphia, PA 19104

Nonpolar compounds associate with organic carbon in the environment. The interaction between pollutants and dissolved organic carbon in natural waters is not as well defined as that between pollutants and sedimentary organic matter. The limitations of experimental techniques and extraction and concentration procedures are partially responsible for the incomplete description of pollutant–DOC (dissolved organic carbon) interactions. Despite the lack of complete understanding of the phenomenon, the association of nonpolar compounds with natural DOC can exert a significant influence on their environmental partitioning. Mathematical models of environmental behavior should include dissolved organic carbon in both overlying and sedimentary interstitial waters as compartments for equilibrium partitioning.

NUMEROUS PHYSICAL, CHEMICAL, AND BIOLOGICAL PROCESSES act upon organic chemicals that are released into the environment. The interaction of these factors determines the ultimate environmental fate of pollutant compounds, as well as the hazard they pose to living organisms. To assess the risk associated with a released chemical, it is necessary to understand how the compound will behave in the environment. In view of the large

[1]**Current address: U.S. Environmental Protection Agency, Region 3, 841 Chestnut Street, Philadelphia, PA 19107**

0065–2393/89/0219–0117$06.00/0

and ever-increasing number of organic chemicals being produced, experimental study of individual compounds is an impossible task.

Considerable effort is currently being directed toward developing mathematical models to accurately predict the environmental distribution of organic chemicals. Simple compartmental models such as the quantitative water, air, and sediment interactive (QWASI) fugacity model of Mackay et al. (*1*) and the chemical equilibrium partitioning and compartmentalization (CEPAC) model of McCall et al. (*2*) predict the environmental distribution of pollutants from physical–chemical properties of the compound that determine its affinity for various media. More complex models add the consideration of transformation reactions and transport processes.

The various environmental transport processes are poorly understood, especially for compounds associated with dissolved humic materials in the environment. We have a new approach to the modeling of hydrophobic organic pollutant behavior in the aquatic environment, in which dissolved humic materials play an important role.

Binding of Nonpolar Organic Compounds to Sedimentary Organic Carbon

The association of nonpolar organic pollutants with soils and sediments has been studied extensively and identified as a major process affecting the environmental fate and distribution of these compounds. The binding of nonpolar organic compounds to sedimentary organic carbon is important background information related to the association of these compounds to dissolved humic materials.

The distribution of hydrophobic organic compounds between aquatic sediments and the overlying water column has typically been viewed as a surface adsorption phenomenon and, as such, has been studied with batch sorption isotherm techniques. Adsorption isotherms of nonpolar organic compounds on a number of soils and sediments are linear over a wide range of equilibrium solute concentrations (*3–5*). This behavior can be expressed as

$$C_{\text{sed}} = K_{\text{p}} \times C_{\text{aq}} \qquad (1)$$

where C_{sed} and C_{aq} are sorbed and dissolved concentrations of a compound, respectively; and K_{p} is the distribution, or partition, coefficient describing the ratio of the equilibrium concentration of a compound in the sediment to its equilibrium concentration in the water.

A number of studies have shown that the binding of nonpolar organic compounds to natural sediments is highly correlated with the organic carbon content of the solid material. Because of the important influence of organic carbon, sediment–water distribution coefficients are often normalized to the organic-carbon fraction of the sediment (f_{oc}) by the expression

$$K_{oc} = \frac{K_p}{f_{oc}} \tag{2}$$

For a given compound, the magnitude of K_{oc} is relatively constant among sediments (*6, 7*). K_{oc} values, therefore, provide good predictions of sorptive behavior. The value of K_{oc} constants for describing the distribution of organic compounds between sedimentary organic carbon and water is further enhanced by the fact that K_{oc} values can be closely correlated with a chemical's octanol–water partition coefficient (K_{ow}) and water solubility (*3, 6, 8*). K_{oc} values that have not been experimentally determined thus may be estimated from measured K_{ow} values for the same compound.

Lambert (*9*) and Chiou et al. (*3, 4*) have proposed that the association between nonpolar compounds and the organic carbon fraction of sediments, soils, and natural waters is better described as a liquid–liquid partitioning phenomenon than as a surface adsorption process. An organic-matter partitioning process is supported by a number of observations, including

1. linear sorption isotherms to near aqueous saturation concentrations of nonpolar organic substances, with no evidence of isotherm curvature at the higher concentration range; isotherm curvature at higher concentrations is predicted by adsorption theories;
2. small temperature effects on solute sorption;
3. absence of competition in experiments using binary solute systems; and
4. data covering seven orders of magnitude in which sediment–water partition coefficients were inversely proportional to aqueous solubility and well correlated to octanol–water partition coefficients.

The actual physical mechanism of the reaction between nonpolar organic compounds and natural organic matter is still a matter of controversy. The terms *sorption* and *partitioning* will, therefore, be used loosely in this chapter.

A number of workers have attempted to describe the association between nonpolar organic compounds and humic material on a molecular level. Schnitzer and Khan (*10*) proposed that the humic polymer consists of an aromatic core to which peptides, carbohydrates, metals, and phenols are attached. This proposed structure is an open network, and it has been suggested that organic molecules are trapped inside the spaces of the humic structure.

Freeman and Cheung (*11*) picture humic material as highly branched polymer chains that form a three-dimensional randomly oriented network. Interconnections between the chains prevent the network from dissolving in liquids. Instead, liquids may be absorbed, and absorption is typically accompanied by swelling of the network to form a gel. Freeman and Cheung suggested that humic substances bind organic chemicals by a process of incorporation into the humic gel structure, and that the binding of hydrophobic compounds is controlled by the relative affinity of the compound for the aqueous and gel phases.

At present, it is not known which of the proposed structures best describes the molecular configuration of naturally occurring humic material. Further research is necessary in this area.

Relatively recent evidence indicates that dissolved organic matter in natural waters can, like sedimentary organic carbon, "sorb" or bind nonpolar organic compounds. Dissolved organic carbon is composed largely of dissolved humic material. The binding of nonpolar organic chemicals with dissolved organic carbon (DOC) can be described by an equilibrium distribution coefficient, K_{doc}, where

$$C_{doc} = K_{doc} \times C_{aq} \quad (3)$$

where C_{doc} is the concentration of the chemical associated with the DOC at equilibrium.

Dissolved organic carbon in natural waters must be considered as a separate environmental compartment in a model of pollutant behavior. Mathematical models developed to date have not included the nonpolar-organic-pollutant–DOC interaction. Where DOC concentrations are high, this interaction can exert an important influence on the environmental behavior of nonpolar organic materials, especially those with a strong tendency to bind to dissolved humic substances.

Systems that contain naturally high levels of DOC include bogs, swamps, and interstitial waters of soils and sediments. Interstitial water (porewater) is formed by the entrapment of water during sedimentation, which isolates it from the overlying water. Porewater is considered to be in equilibrium with the sedimentary solid phase and separate from the overlying water column, or bulk water (*12, 13*). Dissolved organic carbon concentrations in sedimentary porewater can exceed 100 mg/L, whereas overlying surface waters typically contain less than 5 mg/L of DOC (*14*).

For modeling purposes, a kinetic boundary can be hypothesized at the sediment–water interface, as illustrated in Figure 1. The hypothesized boundary would describe conditions in lakes, reservoirs, and slow-moving streams, where the rates of dispersion and diffusion between the sediment and water column are orders of magnitude slower than those within the

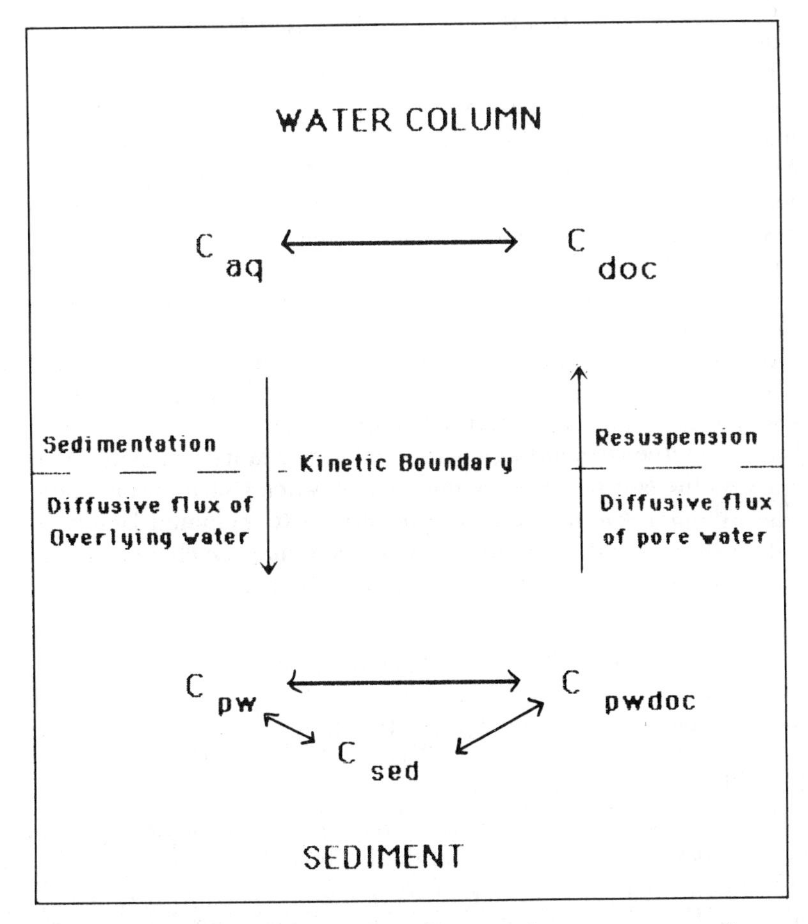

Figure 1. Hypothetical kinetic boundary at sediment–water interface.

separate phases. The sediment and its associated porewater can then be treated as separate environmental compartments in an equilibrium model. The equilibrium reactions occurring in such a model are presented in List I.

Measurements of Binding of Nonpolar Organic Pollutants to DOC

The interaction between nonpolar organic pollutants and DOC is not as well understood as that between nonpolar organic pollutants and sedimentary organic carbon. Several experimental difficulties are responsible for the limited understanding of pollutant–DOC binding. The development of adequate methods for studying the binding of nonpolar organic compounds has been

List I. Equilibrium Relationships for Proposed Model

$K_1 = C_{sed}/C_{pw}$
$K_2 = C_{sed}/C_{pwdoc}$
$K_3 = C_{pwdoc}/C_{pw}$
$K_4 = C_{doc}/C_{pw}$
$K_5 = C_{pwdoc}/C_{aq}$
$K_6 = C_{doc}/C_{aq}$
$K_7 = C_{sed}/C_{aq}$
$K_8 = C_{sed}/C_{doc}$
$K_9 = C_{sed}/(C_{pwdoc} + C_{pw})$
$K_{10} = C_{sed}/(C_{aq} + C_{doc}) = C_{sed}/C_w$

where C_{sed} is the concentration bound to sedimentary organic matter
C_{pw} is the free concentration dissolved in porewater
C_{pwdoc} is the concentration bound to porewater DOC
C_{aq} is the free concentration dissolved in water column
C_{doc} is the concentration bound to water column DOC
C_w is the $C_{aq} + C_{doc}$

difficult. A number of methods have been tried, but none has proven suitable for studying all nonpolar organic compounds (*15, 16*). These methods include gel permeation chromatography (*17*), ultrafiltration (*18*), reverse-phase liquid chromatography (*19*), equilibrium dialysis (*20*), solubility methods (*21, 22*), and gas-phase partitioning (*23–25*).

Gel permeation chromatography and reverse-phase separation techniques are based on the theory that the fraction of a nonpolar organic compound bound to DOC will not be retained by the gel or reverse-phase column. DOC-sorbed compounds will be excluded from the pore spaces of gel permeation columns and will not bind to reverse-phase column material at pH levels above 5. Free compound will be retained in either column type. The amount of bound compound measured increases with increasing flow rate. This relationship suggests that measurements of equilibrium distributions may not be accurate because of rapid desorption of bound material (*16, 19*).

Dialysis and ultrafiltration methods rely on physical separation of free and bound forms with semipermeable membranes. Ultrafiltration techniques are limited to nonpolar organic compounds, which do not interact with the ultrafiltration membrane (*15, 16, 19*). Similarly, equilibrium dialysis is limited to nonpolar organic compounds that will readily pass through the membrane. Dialysis membranes strongly adsorb some nonpolar compounds (*19*). Carter and Suffet (*21*) reported some discrepancies between K_{doc} measured by dialysis and by other methods. However, for compounds that do not significantly interact with the membrane, equilibrium dialysis is a promising method for measuring the binding of nonpolar organic compounds to DOC (*15, 16, 20*).

Solubility methods measure the effect of dissolved organic carbon on the apparent solubility of nonpolar organic compounds (*21, 22*). The difference in measured solubility in the presence and absence of DOC is attributed to binding of the compound to the dissolved organic matter. The advantage of solubility determination is that the technique is applicable to virtually all nonpolar organic compounds. However, there are disadvantages to the method. Binding constants are measured at saturation, which is normally much greater than concentrations found in the environment. The activity of the compound in the aqueous and organic phases may change as saturation is approached (*15, 16, 20*). Furthermore, the accuracy and precision of solubility measurements of very hydrophobic compounds are often adversely affected by dispersion of the compound rather than true dissolution and by the presence of suspended microcrystals in solution.

The dynamic coupled column liquid chromatographic technique of May et al. (*26*) is designed to eliminate these problems. The method is based on pumping water through a column containing glass beads coated with the compound of interest. Whitehouse (*22*) successfully applied the technique to study the effect of dissolved humic substances on the aqueous solubility of polynuclear aromatic hydrocarbons.

Gas-phase partitioning methods have been used to study the interaction of nonpolar organic compounds with DOC (*16, 23–25, 27*). The technique is based on the fact that, according to Henry's law, the vapor concentration of the compound under study is directly proportional to the freely dissolved concentration. The concentration of the compound in the vapor phase is measured and used to calculate the aqueous concentration. The procedure avoids the problem of incomplete phase separations, which is inherent in other methods. However, the procedure is limited to compounds with significant vapor pressures and is complicated by sorption of the compound to container walls (*16*).

In summary, a number of methods have been applied to the study of DOC–nonpolar-organic-compound interactions. Each method has its advantages and its drawbacks; no one method has proven applicable to all nonpolar organic compounds.

Uncertainties in Binding of Nonpolar Organic Compounds to DOC

Current information indicates that the mechanism of pollutant binding to DOC is similar to that of binding to the organic carbon fraction of soils and sediments (*15, 16, 20*). The evidence for this mechanistic similarity includes the observation of linear binding "isotherms" for nonpolar organic compounds and DOC; the absence of competition in multisolute systems; and the close agreement between K_{oc} values measured for a given compound with sediments and DOC.

K_{doc} values for nonpolar organic compounds associated with DOC are, like sedimentary K_{oc} values, highly correlated with K_{ow} values of the compounds. However, some studies indicate that a large variability in K_{doc} values for a given compound depends on the source of the dissolved organic carbon (*19–21, 27, 28*). Carter and Suffet (*20, 21*) found a range of K_{doc} values for DDT (1,1-dichloro-2,2-bis(*p*-chlorophenyl)ethane) with dissolved humic and fulvic acids from different sources. The largest differences were observed between humic and fulvic acids. One fulvic acid (Suwannee River fulvic acid) did not bind DDT at all. Attempts to correlate the extent of DDT binding to measurable characteristics of the dissolved humic materials were unsuccessful.

A possible explanation for this seemingly anomolous behavior may be found in the common practice of using base-extracted humic materials in the study of the binding of nonpolar organic compounds to DOC. At present, it is not known how the humic material is affected by the extraction and cleanup procedures used in laboratory studies. The chemical structure of humic material is not definitively known; it may be rather fragile and easily disrupted by the extractions.

Perhaps some humic materials are more fragile than others. In the work of Carter and Suffet (*20, 21*) Suwanee River fulvic acid showed little tendency to bind DDT, in contrast to other humic materials used. Suwannee River fulvic acid was the only humic material in that study that was subjected to extensive cleanup procedures. This treatment may have further changed the molecular structure and binding ability of the material.

Humic substances in natural waters are mixtures rather than pure substances, and they vary in composition from one environment to another. If natural organic matter is a mixture, can absolute consistency be expected? More likely, there will be variability in the behavior of organic materials from different environments, as reported by Garbarini and Lion (*27*). In the face of such variability, caution should be used in analyzing data from studies using soil-derived humic material as a model for DOC in natural waters. Future research should be directed at determining whether the differences in reported K_{doc} values are caused by differences in the organic matter itself, or are the result of isolation and extraction procedures or techniques used to study the DOC–organic-pollutant interaction.

Importance of DOC in Interstitial Waters

Dissolved organic carbon is found in high concentrations in marine sedimentary interstitial waters (*29*). The study of freshwater sedimentary interstitial waters is relatively new, but available data indicate elevated DOC concentrations in these porewaters as well (*13, 14*). DOC concentrations in the porewater of the Brandywine River, which was used in the work reported here, ranged from 15 to 30 ppm. Overlying-water DOC concentrations were 3 ppm or less. Interstitial-water DOC thus constitutes a significant environmental compartment to be considered in environmental models.

Some studies report that interstitial-water DOC will bind nonpolar organic compounds. Eadie et al. (*30*) found that polycyclic aromatic hydrocarbons (PAH) in porewaters of Lake Michigan sediments were associated with very fine particles, colloids, and humic substances. Brownawell and Farrington (*31, 32*) studied the partitioning of polychlorinated biphenyls (PCB) in sediments and interstitial waters of New Bedford Harbor, Massachusetts. PCB concentrations were highly elevated in the porewaters. The observed partitioning of PCBs between sediments and interstitial waters could not be explained if sediment–solution partitioning were the only process involved. A three-phase equilibrium model that includes PCBs associated with colloidal organic material in the porewater, in the dissolved phase, and sorbed to sedimentary particulate material adequately explained the observed PCB distribution.

"Five-Phase" Model of the Aquatic Environment

Our equilibrium model views the abiotic aquatic environment as consisting of five "phases" or compartments:

1. the overlying water column,
2. DOC in the water column,
3. interstitial water,
4. DOC in the interstitial water, and
5. sediment particles with their "coating" of organic carbon.

Nonpolar organic compounds released into the environment are distributed among these phases according to the equilibrium relationships summarized in List I. Several of the partition coefficients are combined equilibria, yielding apparent partition coefficients that appear anomalous if the individual compartments are not considered. For example, the distribution of a compound between interstitial water with its associated DOC and sedimentary particulate matter yields the apparent K_{oc} designated K_9 in the proposed model. Because of a considerable level of DOC in the porewater, this partition coefficient indicates decreased binding of a given compound to the sediment in the presence of porewater with respect to overlying water. Consideration of the interaction of the compound with DOC accounts for the apparently low partition coefficient.

The analytical scheme developed to test the interaction hypothesis is diagramed in Figure 2. A modification of the generator column method of May et al. (*26*) was used to measure the solubility of the test compound in the overlying water and interstitial water compartments to determine chemical partitioning between the aqueous compartments and their associated dissolved organic carbon in these phases. Partitioning of a compound between the aqueous compartments and sedimentary solids was determined by using a column technique in which a spiked solution was passed through a column containing a measured amount of sediment. The experimental design includes passing the solution through a granular activated carbon (GAC) column to remove the DOC without fractionation. The sediment column apparatus is diagramed in Figure 3.

Initial results using 2,2′,4,4′-tetrachlorobiphenyl as the test compound indicated considerable binding to interstitial-water DOC. Water, sediment, and interstitial-water grab samples taken from the Brandywine River in southeastern Pennsylvania were used. DOC concentrations in the overlying water were below 3 mg/L, although porewater DOC concentrations ranged from 15 to 30 mg/L. The sedimentary organic carbon content was 1.2%.

Binding to DOC in the overlying water had no effect on the binding of the test compound to sediments in the Brandywine River. From the model, K_6 was undeterminable, and as a result, K_{10} is equivalent to K_7. Binding to DOC in the interstitial water, on the other hand, showed a significant effect. From the model, K_3 and K_1 are combined to calculate K_9, the apparent K_{oc} for the interstitial water. The mean value of the sediment–water partition coefficient, K_{oc}, for 2,2′,4,4′-tetrachlorobiphenyl in the Brandywine River was 3.61×10^4. The apparent K_{oc} for tetrachlorobiphenyl in the presence of Brandywine River porewater was 6.98×10^3. This value indicates a considerable effect of porewater DOC on sediment–water partitioning.

The decreased sorption of tetrachlorobiphenyl to sedimentary material was attributed to binding of the compound to interstitial-water DOC. The distribution coefficient, K_{pwdoc} (K_3 of the model), represents the ratio of equilibrium concentrations bound to porewater DOC and freely dissolved in the porewater. For 2,2′,4,4′-tetrachlorobiphenyl in the Brandywine River, K_{pwdoc} was determined by solubility enhancement to be 1.63×10^5.

The distribution coefficient K_{pwdoc} for 2,2′,4,4′-tetrachlorobiphenyl bound to DOC in the interstitial water of the Brandywine River was approximately an order of magnitude greater than the binding constant of that compound to the organic carbon fraction of the sedimentary solid material. This observation confirms that DOC in the porewater of natural sediments can influence the environmental fate of nonpolar organic compounds to a considerable extent. Further research is needed to determine the applicability of the sediment-column experimental design to other compounds and sediment types.

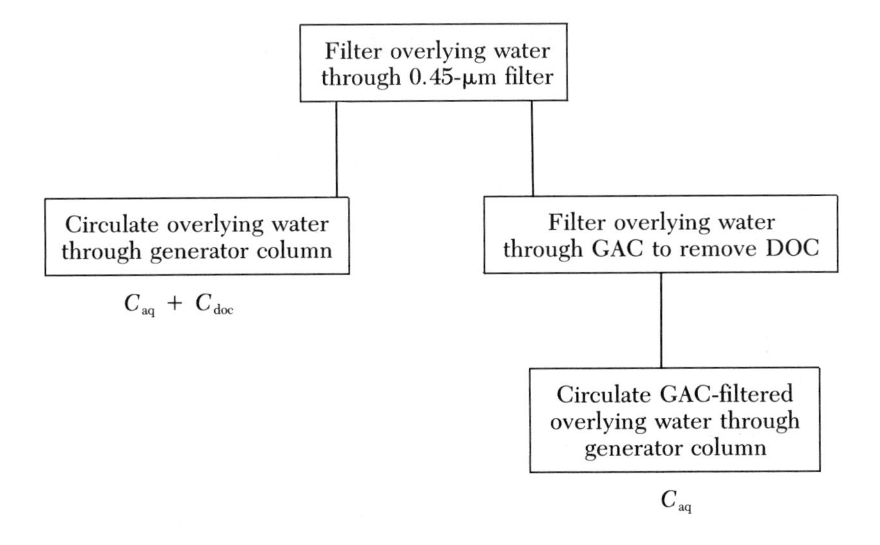

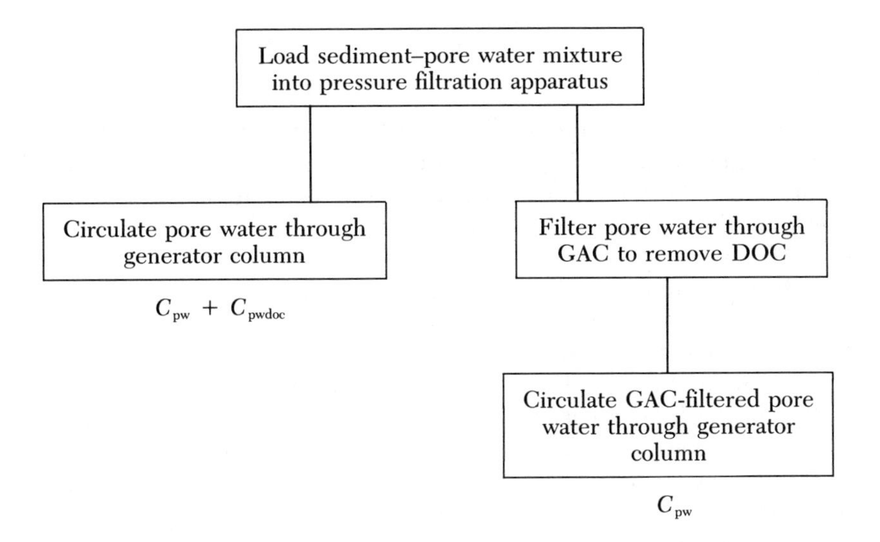

Figure 2. Analytical scheme for determining the concentration of compound partitioned in each environmental compartment.

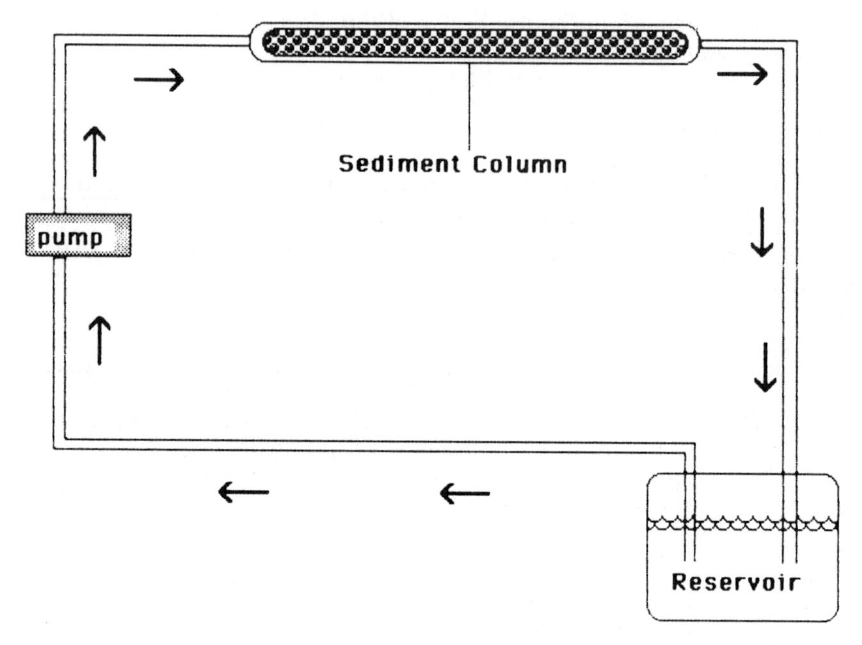

Figure 3. Sediment column apparatus used for equilibration of aqueous phases with sedimentary particulate material.

Summary

Dissolved organic carbon in natural waters interacts with and influences the environmental behavior of nonpolar organic compounds. A number of methods have been developed to study and quantify this interaction. At present, no universally applicable technique has been defined. Further research is necessary to develop new methods that will overcome the experimental difficulties encountered with existing procedures.

The extent of binding of nonpolar organic compounds to DOC is a function of the octanol–water partition coefficient and aqueous solubility of the compound. Present data indicate that the magnitude of binding is a function of the humic material as well. It is not known whether there are indeed differences in the binding ability of DOC from different sources or whether the discrepancies result from alteration of the structure of natural organic matter during sampling, isolation, and extraction.

An experimental design using DOC as found in the environment (i.e., with no fractionation, extraction, or chemical alteration) was used to investigate the importance of DOC–pollutant interactions in the aquatic environment. Results of this research indicate that DOC in the interstitial water of natural sediments can significantly affect the behavior of hydrophobic compounds exhibiting high K_{ow} values. Environmental models should include this interaction in the prediction of the ultimate fate and transport of nonpolar pollutants.

Acknowledgment

We thank R. Lee Lippincott for help in the development of the equilibrium model presented in this chapter and for providing computer graphics.

References

1. Mackay, D.; Paterson, S.; Joy, M. In *Fate of Chemicals in the Environment*; Swann, R. L.; Eschenroeder, A., Eds.; ACS Symposium Series 225; American Chemical Society: Washington, DC, 1983; pp 175–196.
2. McCall, P. J.; Swann, R. L.; Laskowski, D. A. In *Fate of Chemicals in the Environment*; Swann, R. L.; Eschenroeder, A., Eds.; ACS Symposium Series 225; American Chemical Society: Washington, DC, 1983; pp 105–123.
3. Chiou, C. T.; Peters, L. J.; Freed, V. H. *Science (Washington, DC)* **1979,** *206,* 831.
4. Chiou, C. T.; Porter, P. E.; Schmedding, D. W. *Environ. Sci. Technol.* **1983,** *17,* 227–231.
5. Karickhoff, S. W. *Chemosphere* **1981,** *10,* 833–846.
6. Karickhoff, S. W.; Brown, D. S.; Scott, T. A. *Water Res.* **1979,** *13,* 241–248.
7. Means, J. C.; Wood, S. G.; Hassett, J. J.; Banwart, W. L. *Environ. Sci. Technol.* **1980,** *14,* 1524–1528.
8. Perdue, E. M. In *Aquatic and Terrestrial Humic Materials*; Christman, R. F.; Gjessing, E. T., Eds.; Ann Arbor Science: Ann Arbor, 1983; pp 441–460.
9. Lambert, S. M. *J. Agric. Food Chem.* **1967,** *15,* 572–576.
10. Schnitzer, M.; Khan, S. U. *Humic Substances in the Environment*; Marcel Dekker: New York, 1972.
11. Freeman, D. H.; Cheung, L. S. *Science (Washington, DC)* **1981,** *214,* 790–792.
12. Glass, G. E.; Poldoski, J. E. *Verh. Int. Ver. Limnol.* **1975,** *19,* 405–420.
13. Batley, G. E.; Giles, M. S. *Water Res.* **1979,** *13,* 879–886.
14. Thurman, E. M. *Organic Geochemistry of Natural Waters*; Kluwer Academic: Hingham, MA, 1985.
15. Carter, C. W. Ph.D. Thesis, Drexel University, 1982.
16. Brownawell, B. J. Ph.D. Thesis, Massachusetts Institute of Technology and Woods Hole Oceanographic Institute, 1986.
17. Hassett, J. P.; Anderson, M. A. *Environ. Sci. Technol.* **1979,** *13,* 1526–1529.
18. Means, J. C.; Wijayaratne, R. *Science (Washington, DC)* **1982,** *215,* 968–970.
19. Landrum, P. F.; Nihart, S. R.; Eadie, B. J.; Gardner, W. S. *Environ. Sci. Technol.* **1984,** *18,* 187–192.
20. Carter, C. W.; Suffet, I. H. *Environ. Sci. Technol.* **1982,** *16,* 735–740.
21. Carter, C. W.; Suffet, I. H. In *Fate of Chemicals in the Environment*; Swann, R. L.; Eschenroeder, A., Eds.; ACS Symposium Series 225; American Chemical Society: Washington, DC, 1983; pp 215–229.
22. Whitehouse, B. G. *Estuarine Coastal Shelf Sci.* **1985,** *20,* 393–402.
23. Diachenko, G. W. Ph.D. Thesis, University of Maryland, 1981.
24. Hassett, J. P.; Milicic, E. *Environ. Sci. Technol.* **1985,** *19,* 638–643.
25. Yin, C.; Hassett, J. P. *Environ. Sci. Technol.* **1986,** *20,* 1213–1217.
26. May, W. E.; Wasik, S. E.; Freeman, D. H. *Anal. Chem.* **1978,** *50,* 175–179.
27. Garbarini, D. R.; Lion, L. W. *Environ. Sci. Technol.* **1985,** *19,* 1122–1128.

28. Chiou, C. T.; Malcolm, R. L.; Brinton, T. I.; Kile, D. E. *Environ. Sci. Technol.* **1986,** *20,* 502–508.
29. Krom, M. D.; Scholkovitz, E. R. *Geochim. Cosmochim. Acta* **1977**, *41*, 1565–1573.
30. Eadie, B. J.; Landrum, P. F.; Faust, W. *Chemosphere* **1982,** *11*, 847–858.
31. Brownawell, B. J.; Farrington, J. W. In *Marine and Estuarine Geochemistry*; Sigleo, A. C.; Hattori, A., Eds.; Lewis Publishers: Chelsea, MI, 1985; pp 97–120.
32. Brownawell, B. J.; Farrington, J. W. *Geochim. Cosmochim. Acta* **1986,** *50*, 157–169.

RECEIVED for review November 5, 1987. ACCEPTED for publication July 19, 1988.

10

Water-Solubility Enhancement of Nonionic Organic Contaminants

Daniel E. Kile and Cary T. Chiou

U.S. Geological Survey, Denver Federal Center, Denver, CO 80225

Water-solubility enhancement of selected organic solutes by dissolved organic matter was studied by using a variety of dissolved soil and aquatic humic materials, commercial humic acids, and synthetic organic compounds. The solubility enhancement of relatively water-insoluble solutes (e.g., polychlorinated biphenyls and dichlorodiphenyltrichloroethane) by natural and commercial humic materials is accounted for by a partitionlike interaction of the solutes with the nonpolar organic environment of the dissolved organic matter. Solute-solubility enhancement is minimal when poly(acrylic acid) and phenylethanoic acid are used as dissolved organic matter species. The polarity, size, and configuration of the dissolved organic matter and the hydrophilicity of the solute are controlling factors for solubility enhancement. Solubility enhancement is much greater with commercial humic acids than with the aquatic humic materials.

INTERACTIONS OF CERTAIN ORGANIC CONTAMINANTS with dissolved natural humic substances can often significantly influence the transport and fate of the compounds in aquatic systems. A fundamental understanding of the mechanism(s) of interaction in relation to the structure and composition of dissolved humic substances is essential for assessing the behavior of organic contaminants in natural waters. An evaluation of this effect in relation to the properties of solute and dissolved organic matter is presented in this chapter. Data on the solubility enhancement of a variety of organic solutes by different types of dissolved humic substances (of soil and aquatic origins) are used as a basis for elucidating the mechanism involved.

0065–2393/89/0219–0131$07.75/0

In general, solutes that are relatively insoluble in water are most susceptible to enhanced water solubility by dissolved organic matter. Wershaw et al. (*1*) found that the apparent water solubility of DDT (1,1′-(2,2,2-trichloroethylidene) bis[4-chlorobenzene]) increased more than 200 times in an aqueous solution of 0.5% (soil-derived) sodium humate, and Poirrier et al. (*2*) showed that an organic–mineral colloid present in a natural surface water concentrated DDT by a factor of 15,800. Boehm and Quinn (*3*) reported significant increases in the apparent water solubility of some highly water-insoluble alkanes when in contact with fulvic acid derived from a marine sediment. Later, Carter and Suffet (*4*) and Landrum et al. (*5*) found significant enhancements of the concentrations of DDT and other sparingly water-soluble organic compounds in water caused by aquatic humic substances and by Aldrich humic acid. The magnitudes of enhancement were found to be dependent on the source of the organic matter and on the solution pH.

Because concentrations (or apparent solubilities) in water of many sparingly soluble organic solutes can be influenced by dissolved humic substances, equilibrium constants and process rates related to the solute concentrations in water may exhibit certain anomalies when the effect on solute concentration is not taken into account. For example, Hassett and Anderson (*6*) observed a decreased recovery of cholesterol from water due to a substantial binding of the compound to dissolved humic substances; Griffin and Chian (*7*) and Hassett and Milicic (*8*) showed lower rates of volatilization for certain PCBs (polychlorinated biphenyls). Similarly, Perdue and Wolfe (*9*) reported an apparent decrease in the hydrolysis rate of the 1-octyl ester of 2,4-D (2,4-dichlorophenoxyethanoate) at a relatively low concentration of dissolved Aldrich humic acid in water as a result of the sorption of the solute by dissolved humic acid.

A more commonly recognized consequence of the interaction between solute and dissolved humic substance is the modification of the distribution of the organic solute to other aquatic compartments. Hassett and Anderson (*10*) showed that the sorption coefficients of cholesterol and 2,5,2′,5′-tetrachlorobiphenyl between river sediment and water are decreased substantially in the presence of dissolved humic substances, compared to those in the presence of pure water. Similarly, Caron et al. (*11*) observed that the apparent sorption coefficient of DDT with aquatic sediments decreased in the presence of dissolved humic substances; they attributed this decrease to a strong association of DDT with the dissolved humic substances. Gschwend and Wu (*12*) found a similar decrease of the sediment–water sorption coefficient for some polychlorinated biphenyls with an increasing concentration of particulate material in water, and they attributed this effect to a simultaneous partitioning of the solutes to the suspended particulate-bound organic matter. Dissolved humic substances have also been shown to reduce the intrinsic bioconcentration factors for some polynuclear aromatic hydrocarbons (*13–15*).

These studies demonstrate that interactions of many organic contaminants with naturally occurring dissolved humic substances can lead to significant modifications of the apparent solubility (and thus the mobility and fate) of the compound in aquatic systems. Because enhancement of solute water solubility by dissolved humic substances is presumably the primary cause for many related modifications of solute behavior in aquatic systems, evaluation of the significance of this effect for various solutes with different types of humic substances is warranted for environmental-impact assessments.

Theoretical Considerations

An acceptable model for the interaction between solutes and dissolved humic substances must take into account properties of the solutes and of humic substances that are consistent with the observed data. The magnitude of the solubility enhancement of a nonionic organic solute increases with a decrease in the intrinsic solubility of the solute in pure water. For example, Boehm and Quinn (*3*) showed that in the presence of dissolved humic substances (at dissolved organic carbon concentrations ranging from 1.8 to 17.5 mg L^{-1}) there was a significant enhancement of the apparent solubility of relatively water-insoluble higher alkanes, but not of more water-soluble aromatic compounds. Caron et al. (*11*) found that a dissolved humic acid isolated from a sediment (at a dissolved organic carbon concentration of 6.95 mg L^{-1}) strongly influenced the sorption coefficient of DDT with sediments, but had little effect on the corresponding coefficient of lindane. Similarly, Haas and Kaplan (*16*) demonstrated that Aldrich humic acid (at a dissolved organic carbon concentration of 70 mg L^{-1}) has a minimal effect on the apparent solubility of toluene, a compound with a relatively high water solubility compared to DDT. Thus, from the standpoint of the solute, one would expect that only nonionic organic compounds with very low water solubilities would be significantly affected by dissolved humic substances.

In principle, dissolved organic materials (such as humic substances at sufficient concentrations) can promote the apparent solubility of a nonionic organic solute in aqueous solution. This increased solubility can be accomplished either by changing the overall solvency of the solution (analogous to a conventional mixed-solvent effect) or through a direct solute interaction by either adsorption or partitioning. Humic substances, at the relatively low concentrations typically found in natural aquatic systems, are not likely to have a significant impact on water solvency. This point is substantiated by data showing that octanol-saturated water (containing about 600 ppm octanol) increases the water solubility of DDT by a factor of approximately 2.5 (*17*). A solubility enhancement of this magnitude is too low to account for the observed enhancement by dissolved humic substances at concentrations considerably less than that of the octanol in water. Also, the pronounced solubility enhancement of a solute such as DDT is unlikely to be a result of complexation or specific interaction because the hydrophilic functional groups of a humic molecule would be preferentially associated with the highly polar water molecules. Considering this condition, and the fact that dissolved humic substances are relatively high-molecular-weight species, Chiou et al. (*18*) hypothesized that the observed solute-solubility enhancement results from a partitionlike interaction between solute and dissolved organic matter, in which the dissolved high-molecular-weight organic matter is regarded as a "microscopic" organic phase; that is, interactions between solute and dissolved macromolecules are governed primarily by van der Waals forces.

The kind of partition interaction envisioned here is similar mechanistically to that for the solubilization of relatively water-insoluble organic solutes in micelles, where a microscopic organic phase is formed through aggregation of surfactant monomers (*19–23*). However, the effectiveness of this partition effect with a dissolved humic material is a function of the size, polarity, configuration, and conformation of the humic molecule. On this basis, dissolved organic matter with a sufficiently large molecular size and nonpolar molecular environment would effectively promote the partitioning of a sparingly water-soluble nonionic solute, while a dissolved low-molecular-weight organic substance or a highly polar organic macromolecule would not exhibit a strong effect.

To facilitate a better understanding of the presumed partitionlike interaction of organic solutes with dissolved humic substances, some characteristics that differentiate a partition effect from an adsorption effect need to be considered. In general, the term *adsorption* refers to the condensation or specific association of a solute (or a vapor) onto surfaces or interior pores of a solid (adsorbent) through physical or chemical bonding forces. Adsorption is necessarily competitive between solutes (or vapors) because of the constraints of the available surfaces or specific sites. Such condensation or specific association leads to a decrease in entropy for the solute, and therefore adsorption of the solute is accompanied by a relatively high exothermic heat. The adsorption isotherm (i.e., the plot of the amount adsorbed per unit mass of adsorbent versus the equilibrium concentration in solution) is rarely linear over a wide range of solute concentrations because surfaces or active sites in the adsorbent are seldom energetically homogeneous.

In contrast, the term *partition* or *partitioning* is used to describe a model in which the solute is distributed between two immiscible or partially miscible phases (e.g., an organic solvent and water) by forces common to solution. Some distinctive characteristics are associated with equilibrium partitioning of organic solutes: the partition coefficients for different compounds between an organic solvent and water are closely related to the reciprocals of their water solubilities; the partition isotherms are highly linear over a wide range of solute concentrations (*24–28*); the system shows an absence of competitive effect in the concurrent partitioning of binary solutes

(*25*, *26*); and the equilibrium heat for solute partitioning is relatively small and constant because of the partial cancellation of heats of solution for the transfer of solute from water to the organic phase (*24*).

Assuming that the water-solubility enhancement of a nonionic organic solute by dissolved organic matter results from a partitionlike interaction between the solute and the macromolecule, the magnitude of this effect for a solute with respect to a specific dissolved organic matter sample can be defined as

$$S_w^* = S_w + XC_o \quad (1)$$

where S_w^* is the apparent solubility of the solute in water containing dissolved organic matter (at concentration X), S_w is the solubility of the solute in pure water, and C_o is the mass of solute partitioned into a unit mass of dissolved organic matter. In accordance with this model, the solute partition coefficient (K_{dom}) between dissolved organic matter and pure water can be defined as

$$K_{dom} = \frac{C_o}{S_w} \quad (2)$$

Substituting equation 2 into equation 1 gives

$$S_w^* = S_w(1 + XK_{dom}) \quad (3)$$

Alternatively, if C_o and X are defined in terms of the organic carbon content of the dissolved organic matter, equation 3 can be written as

$$S_w^* = S_w(1 + XK_{doc}) \quad (4)$$

where K_{doc} is the corresponding organic-carbon-based partition coefficient. According to equation 3 or 4, the magnitude of solute solubility enhancement is controlled by the XK_{dom} (or XK_{doc}) term, and a substantial enhancement therefore takes place when the magnitude of this term is significant relative to 1. The magnitude of the partition coefficient (K_{dom} or K_{doc}) is a function of the physical properties of the solute and the nature of the dissolved organic matter. At a given concentration of dissolved organic matter (X), the product term XK_{dom} (or XK_{doc}) should increase with decreasing water solubility for a series of solutes (or with increasing octanol–water partition coefficients) (*17*). The K_{dom} can be experimentally determined by measuring the apparent solute solubility over a range of dissolved organic matter concentrations. A plot of S_w^* versus X should yield a straight line with a slope equal to S_wK_{dom} and an intercept equal to S_w (equation 3).

The assumed partitionlike interaction of nonionic solutes with dissolved organic matter can be independently evaluated by comparing the observed enthalpy of interaction with that found in a typical solvent–water mixture (*24*). The enthalpy change ($\Delta\overline{H}$), for the transfer of a solute from water to the solvent phase is described as

$$\Delta\overline{H} = \Delta\overline{H}_o - \Delta\overline{H}_w \tag{5}$$

where $\Delta\overline{H}_o$ is the molar heat of solution in the organic phase and $\Delta\overline{H}_w$ is the molar heat of solution in water (i.e., $-\Delta\overline{H}_w$ is the molar heat of condensation from water). A nonionic organic solute with a low water solubility usually gives a high (positive) $\Delta\overline{H}_w$ because of its high incompatibility with water. Conversely, the corresponding $\Delta\overline{H}_o$ should be much smaller because of the greater compatibility of an organic solute with the organic solvent. Both $\Delta\overline{H}_o$ and $\Delta\overline{H}_w$ are relatively independent of solute concentrations. Thus, the molar heat of partitioning ($\Delta\overline{H}$) would be relatively small and independent of solute concentrations, and less exothermic (negative) than the heat of condensation from water ($-\Delta\overline{H}_w$). This characteristic is in contrast to the heat of adsorption, which should be much more exothermic (deriving from solute condensation from solution and additional exothermic interactions with adsorbent) in order to compensate for the decrease in entropy. The difference in the equilibrium heat between adsorption and partition processes becomes more pronounced for a high-melting-point solid solute because of its large heat of fusion ($\Delta\overline{H}_f$), which leads to a high heat of solution. The heat of solution of a solid compound in an organic phase can be expressed as

$$\Delta\overline{H}_o = \Delta\overline{H}_f + \Delta\overline{H}_o^{ex} \tag{6}$$

and the heat of solution of a solid compound in water can be expressed as

$$\Delta\overline{H}_w = \Delta\overline{H}_f + \Delta\overline{H}_w^{ex} \tag{7}$$

where $\Delta\overline{H}^{ex}$ represents the excess (molar) heat of solution of the supercooled liquid (*29*). The molar heat of fusion cancels in a partition equilibrium (equation 5) because it is the same in both the $\Delta\overline{H}_o$ and $\Delta\overline{H}_w$ terms, whereas in adsorption equilibria it adds to the overall heat of adsorption.

The foregoing account of the mechanistic differences of partition and adsorption equilibria serves as a useful basis for evaluating the cause of the water-solubility enhancement of organic solutes by dissolved humic substances, in terms of the properties of solutes and the structural features of humic molecules.

Influences of Molecular Properties

Humic substances isolated from different sources have significantly different effects on solubility enhancement; this fact has been well documented (*4, 18*). These results have been effectively explained by Chiou et al. (*18, 30*) in terms of a partitionlike interaction between solute and dissolved humic substances. The magnitude of the solubility enhancement of a nonionic solute is related to its intrinsic solute solubility in water and to the molecular size, structure, and polarity of the humic macromolecule. In the study of Chiou et al. (*18, 30*), highly purified soil and aquatic humic samples were used to minimize complications resulting from mineral constituents in the sample. Soil-derived humic and fulvic acids were isolated from the surface (A1) horizon of Sanhedron soil obtained from the Mattole River Valley in northern California, and stream-derived humic and fulvic acids were isolated from the Suwannee River, Georgia. These samples had ash contents less than 4%. Elemental analyses for these humic samples are given in Table I. Intrinsic water solubilities of a group of investigated organic compounds are given in Table II.

Partition coefficients (K_{dom}) for the pairs of dissolved humic substances and solutes were determined (according to equation 3) by plotting the apparent solubility (S_w^*) against the concentration of the dissolved humic substance. The slope gives $S_w K_{dom}$, and the intercept gives S_w. The results are shown in Figures 1 through 4. Except as noted, all the solubility-enhancement experiments were conducted at pH 6.5 or below, and at a temperature of 24 ± 1 °C. The observed partition coefficients (i.e., K_{dom} or K_{doc} values) for various solute-dissolved humic-substance systems are given in Table III. These data show that: (1) a high degree of linearity exists between apparent water solubility and concentration of dissolved organic matter for all solutes tested; (2) the observed solubility enhancements of solutes with a given dissolved organic matter (i.e., the K_{dom} values) are inversely related to the solubilities of the solutes in water (S_w), in the order of DDT > 2,4,5,2′,5′-PCB > 2,4,4′-PCB, or in a linear relation to their log K_{ow} values (Table III), indicating similarity in the nature of the equilibria between the two systems; and (3) no competitive interference is found for individual solutes in the binary-solute system (Figure 1). Conversely, no discernible enhancement is noted for the relatively water-soluble lindane and 1,2,3-trichlorobenzene throughout the concentration range of Sanhedron soil humic acid (Figure 5). These findings are consistent with a partitionlike interaction of solutes with dissolved organic matter.

A comparison of the data in Figures 1 through 4 shows that the Sanhedron soil humic acid was most effective in enhancing solute solubility. It was about 4 times as effective as the soil-derived fulvic acid, and 5 to 7 times as effective as the stream-derived humic and fulvic acids. A partition interaction would be more effective with a large-sized macromolecule that contains regions of large nonpolar volumes. The greater enhancement for the

Table I. Ash-Free Elemental Contents of Various Humic and Fulvic Acids

Sample	*C*	*H*	*O*	*N*	*S*	*P*	*Total*	*Ash*
Sanhedron soil humic acid[a]	58.03	3.64	33.59	3.26	0.47	0.10	99.09	1.19
Sanhedron soil fulvic acid[a]	48.71	4.36	43.35	2.77	0.81	0.59	100.59	2.25
Suwannee River humic acid[a]	54.22	4.14	39.00	1.21	0.82	0.01	99.40	3.18
Suwannee River fulvic acid[a]	53.78	4.24	40.28	0.65	0.60	0.01	99.56	0.68
Aldrich humic acid, sodium salt (lot no. 1204 PE, 1984)[b]	69.42	5.04	39.29	0.75	4.25	0.15	118.9	31.0
Fluka–Tridom humic acid (lot no. 159128115, 1974)[b]	65.79	5.51	37.79	0.71	3.16	<0.05	113.0	32.8
Calcasieu River humic extract[b]	56.68	4.69	35.72	1.14	0.64	—	98.87	3.63

NOTE: Values are percents on moisture-free basis.
[a]Elemental data are from ref. 18.
[b]Elemental data are from ref. 30.

Table II. Intrinsic Water Solubilities of Selected Organic Solutes

Compound	*Water Solubility* (mg/L^{-1})	*Temperature* (°C)
p,p'-DDT	5.5×10^{-3}	25
2,4,5,2',5'-PCB	1.0×10^{-2}	24
2,4,4'-PCB	0.115	20
1,2,3-Trichlorobenzene	16.3	23
Lindane	7.80	25

NOTE: Cited in ref. 18.

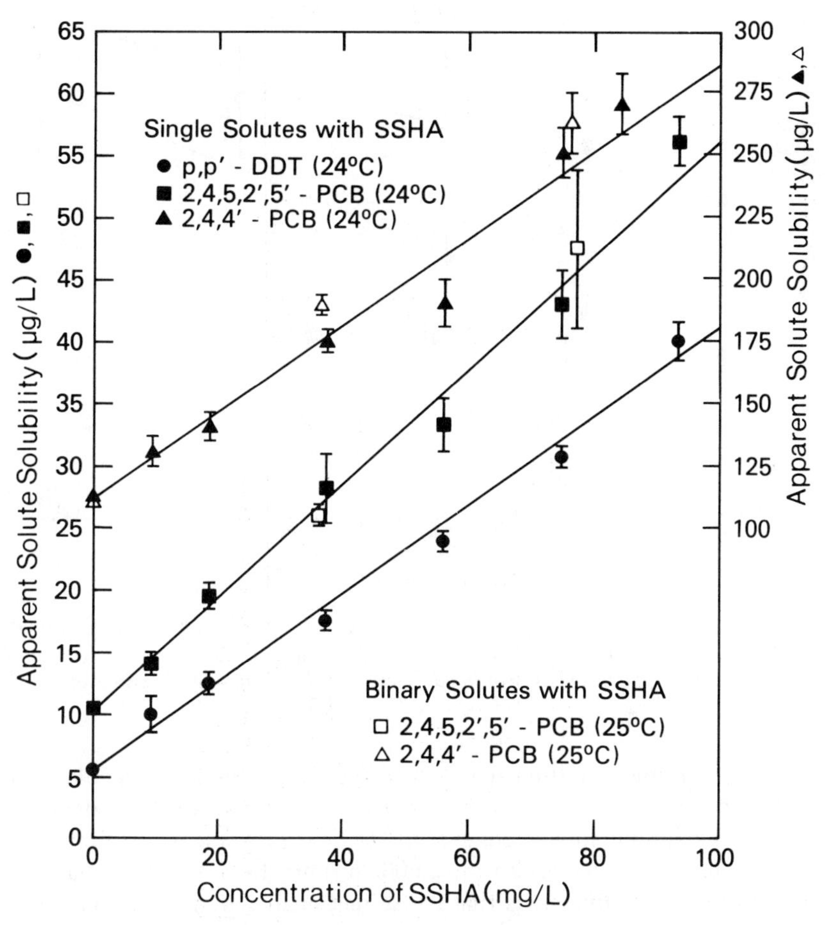

Figure 1. Dependence of the apparent water solubilities of p,p'-*DDT, 2,4,5,2',5'-PCB, and 2,4,4'-PCB on the concentration of Sanhedron soil humic acid at pH 6.5. (Reproduced from ref. 18. Copyright 1986 American Chemical Society.)*

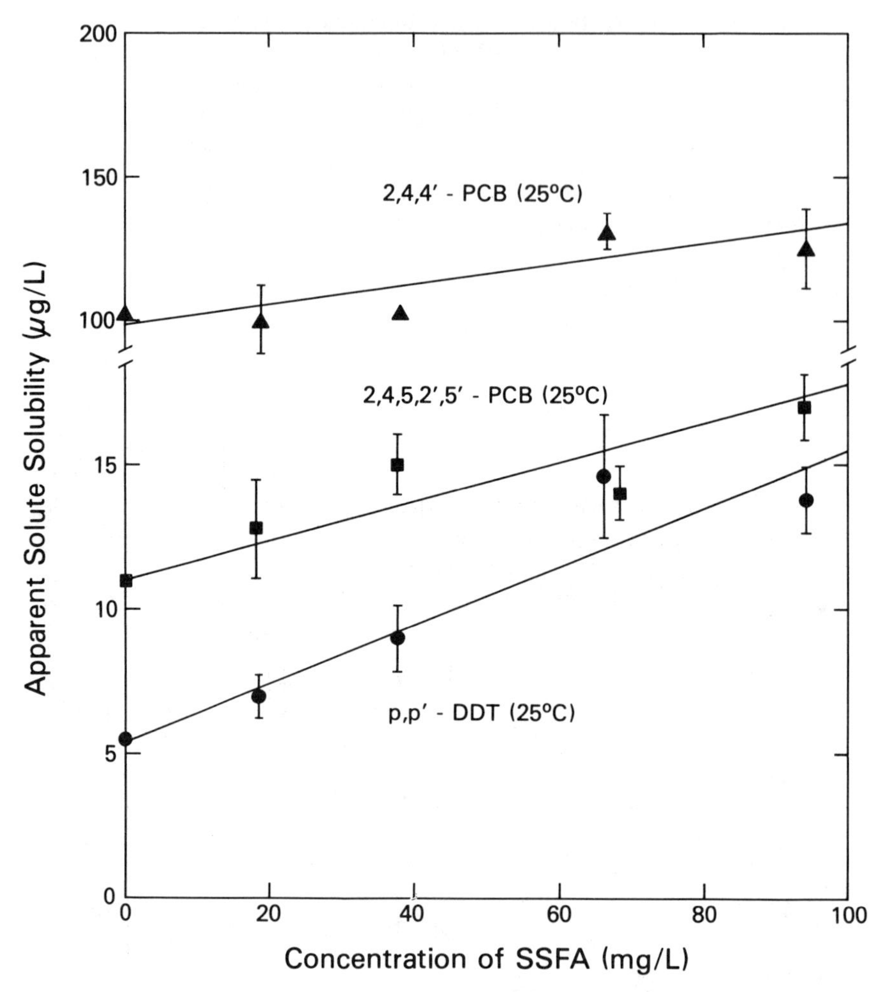

Figure 2. Dependence of the apparent water solubilities of p,p'*-DDT, 2,4,5,2',5'-PCB, and 2,4,4'-PCB on the concentration of Sanhedron soil fulvic acid at pH 6.5. (Reproduced from ref. 18. Copyright 1986 American Chemical Society.)*

soil-derived humic acid thus may be attributed to its higher molecular weight and lower polar-group content (based on its lower oxygen content) relative to the other humic substances tested. Humic substances are highly polydisperse; molecular weights from 2600 to more than 1 million have been reported for soil humic acids that were fractionated by gel chromatography and filtration through graded porosity membranes (*31*). Molecular weights for aquatic humic acids have been quoted in the range of 1000–10,000, while molecular weights for aquatic fulvic acids have been quoted in the range of 500–2000 (*32*). The proposed molecular weight of Suwannee River fulvic acid is approximately 800 (*33*).

Thus, from the standpoint of molecular size and polarity, soil-derived humic acids should provide larger regions of nonpolar intramolecular environment that can more effectively promote the partition interactions with nonionic organic solutes. The enhancement effect caused by the low-molecular-weight phenylethanoic acid as dissolved organic matter is quite small on a unit weight basis of the dissolved material (Figure 6). This fact illustrates the inability of a small dissolved organic molecule to effectively promote a partitionlike interaction, regardless of its polarity. These results also indicate that such solubility enhancements are not caused by specific interactions of

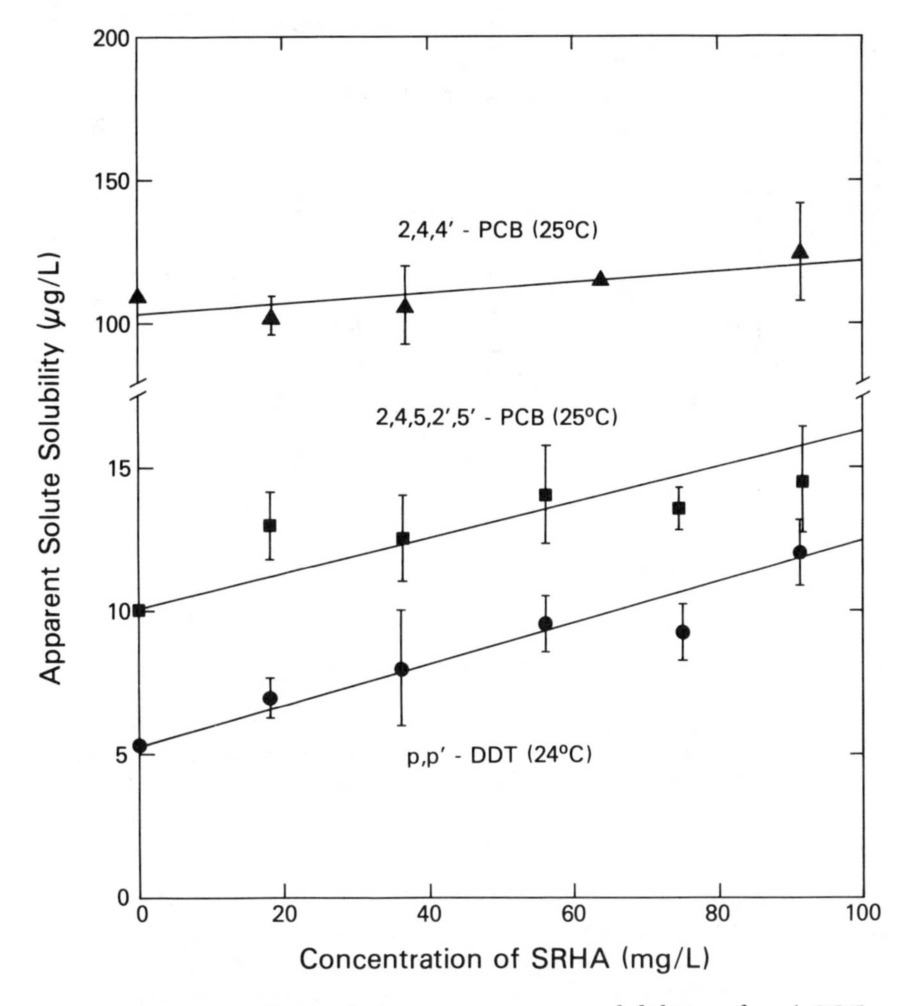

Figure 3. Dependence of the apparent water solubilities of p,p'*-DDT, 2,4,5,2',5'-PCB, and 2,4,4'-PCB on the concentration of Suwannee River humic acid at pH 4.1 to 5.5. (Reproduced from ref. 18. Copyright 1986 American Chemical Society.)*

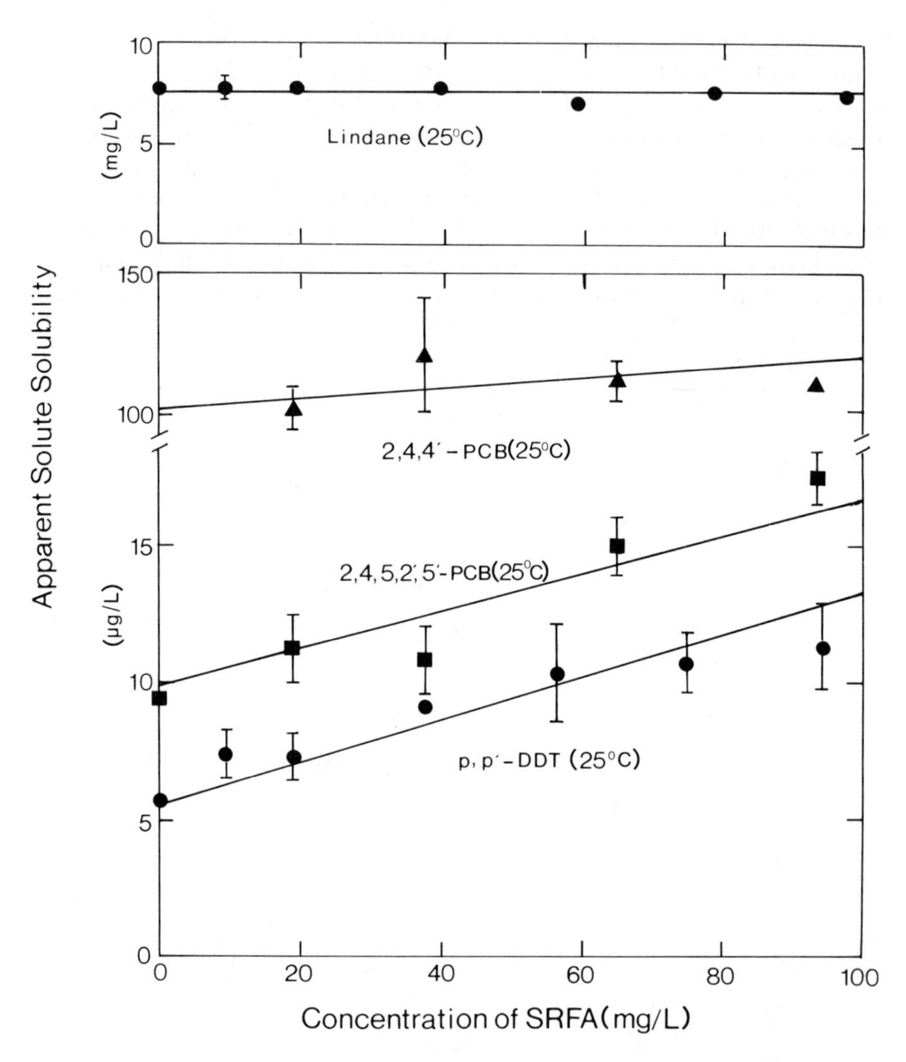

Figure 4. Dependence of the apparent water solubilities of p,p′-*DDT, 2,4,5,2′,5′-PCB, 2,4,4′-PCB, and lindane on the concentration of Suwannee River fulvic acid at pH 3.9 to 5.7. (Adapted from ref. 18.)*

solutes with polar groups of the dissolved humic molecules. If they were caused by specific interactions, then phenylethanoic acid would most likely be as effective as humic substances for promoting the solubility of DDT.

Given that the molecular sizes of dissolved organic substances are sufficiently large (compared to the solute molecules), their efficiencies in enhancing solute solubility by partition effect would be a function of their polarities, configurations, and conformations. This point is illustrated by the

Table III. Solute Octanol–Water Partition Coefficients (log K_{ow}) and log K_{dom}/log K_{doc} Values for Selected Organic Solutes

		log K_{dom}/log K_{doc} [b]			
Compound	log K_{ow} [a]	SSHA[c]	SSFA[d]	SRHA[e]	SRFA[f]
p,p'-DDT	6.36	4.82/5.06	4.27/4.58	4.12/4.39	4.13/4.40
2,4,5,2′,5′-PCB	6.11	4.63/4.87	3.81/4.12	3.80/4.07	3.83/4.10
2,4,4′-PCB	5.62	4.16/4.40	3.58/3.89	3.27/3.54	3.30/3.57
1,2,3-Trichlorobenzene[g]	4.14	~2.8/3.0	~2.0/2.3	~1.7/2.0	~1.7/2.0
Lindane[g]	3.70	~2.5/2.7	~1.5/1.8	~1.2/1.5	~1.2/1.5

[a]Values from ref. 18.
[b]Data from ref. 18, pH ≤6.5.
[c]SSHA, Sanhedron soil humic acid.
[d]SSFA, Sanhedron soil fulvic acid.
[e]SRHA, Suwannee River humic acid.
[f]SRFA, Suwannee River fulvic acid.
[g]Approximate log K_{dom}/log K_{doc} values for 1,2,3-trichlorobenzene and lindane are obtained by linear extrapolations of the plot of log K_{dom}/log K_{doc} versus the log K_{ow} of *p,p'*-DDT, 2,4,5,2′,5′-PCB, and 2,4,4′-PCB.

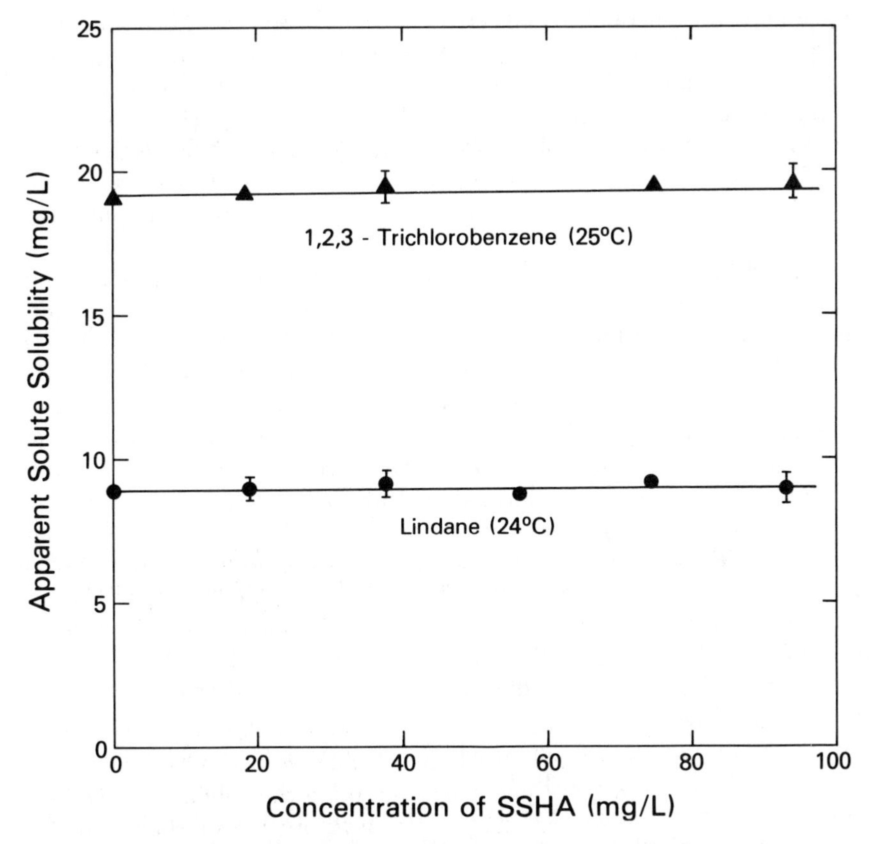

Figure 5. Dependence of the apparent water solubilities of lindane and 1,2,3-trichlorobenzene on the concentration of Sanhedron soil humic acid at pH 6.5. (Reproduced from ref. 18. Copyright 1986 American Chemical Society.)

comparable enhancements caused by the Suwannee River fulvic acid and the Suwannee River humic acid (Figures 3 and 4), even though differences in their molecular weights are probably large. The correspondence of the enhancement effects by these two acids is consistent with their comparable carbon–oxygen ratios, as indicated by elemental analyses. Thus, although molecular size is important, polarity, configuration, and conformation appear to be the controlling factors for high-molecular-weight dissolved humic substances. The much larger enhancement effect observed for the Sanhedron soil humic acid relative to the enhancement for aquatic humic substances probably results more from its fewer polar groups per unit weight and certain molecular configurations than from its larger molecular size.

The importance of molecular configuration is illustrated by the observation that poly(acrylic acid) is unable to promote enhancement of solute solubility (Figure 7), although its molecular weight (ranging from 2000 to 90,000; data for MW 90,000 are not shown) is sufficiently large, and that its composition (C = 50%, O = 44.4%, H = 5.6%) is comparable to that of some humic substances. The ineffectiveness of poly(acrylic acid) in promoting the solubility enhancement of a solute appears to be related to its nonbranched structure and regular attachment of the polar carboxyl groups to the carbon chain, which do not permit the formation of a sizable nonpolar environment required for solute partitioning. Similarly, poly(ethylene glycol), with molecular weights ranging from 200 to 3400, exhibited no noticeable solubility enhancement of DDT at ambient pH values (our unpublished results).

The effect of pH on solute-solubility enhancement, while probably not very significant in most natural aquatic environments, nevertheless provides useful information on the mechanism of the interaction between solute and dissolved humic substance. This effect can be explained in terms of the influence of pH on the structure and conformation of the humic macromolecule. At an alkaline pH, the hydration of the dissociated acidic groups and the concomitant repulsion between ionized functional groups would lead to more extended molecular conformations and thus decrease the intensity of the nonpolar environment. The data presented in Figure 8 for DDT and two PCBs with Suwannee River fulvic acid show a substantial decrease of solute-solubility enhancement at a pH of 8.5, relative to that reported earlier by Chiou et al. (*18*) at pH values of 4.0–6.5. The calculated K_{dom} values at pH 8.5 for DDT, 2,4,5,2′,5′-PCB, and 2,4,4′-PCB are 31%, 61%, and 35%, respectively, of the corresponding K_{dom} values obtained at the lower pH. A similar decline in the K_{dom} of solutes with the Sanhedron soil humic acid was noted at pH 8.5, in which the observed K_{dom} values for DDT, 2,4,5,2′,5′-PCB, and 2,4,4′-PCB were decreased by 54%, 61%, and 42%, respectively (Figure 9 and Table IV). Carter and Suffet (*4*) reported a similar decrease in the binding constant of DDT with dissolved humic acid when the pH of the solution was increased from 6.0 to 9.2. Thus, the effect of pH on nonionic organic solute-solubility enhancement appears to be more significant in the alkaline pH range.

The effectiveness of certain dissolved humic macromolecules as a microscopic organic medium for the partitioning of relatively nonpolar organic solutes leads one to speculate that humic molecules contain sizable regions that are predominantly nonpolar. Stream fulvic acids with relatively high oxygen contents would have a more limited distribution of such nonpolar regions. This point is substantiated by nuclear magnetic resonance (NMR) studies that show a higher degree of aromaticity for soil humic and fulvic acids than for stream humic substances (*34*). Molecules such as stream fulvic acids would be likely to assume more extended (linear) conformations because of the hydrophilicity of the polar functional groups and the mutual repulsion of the ionized groups. Presumably, the presence of large regions

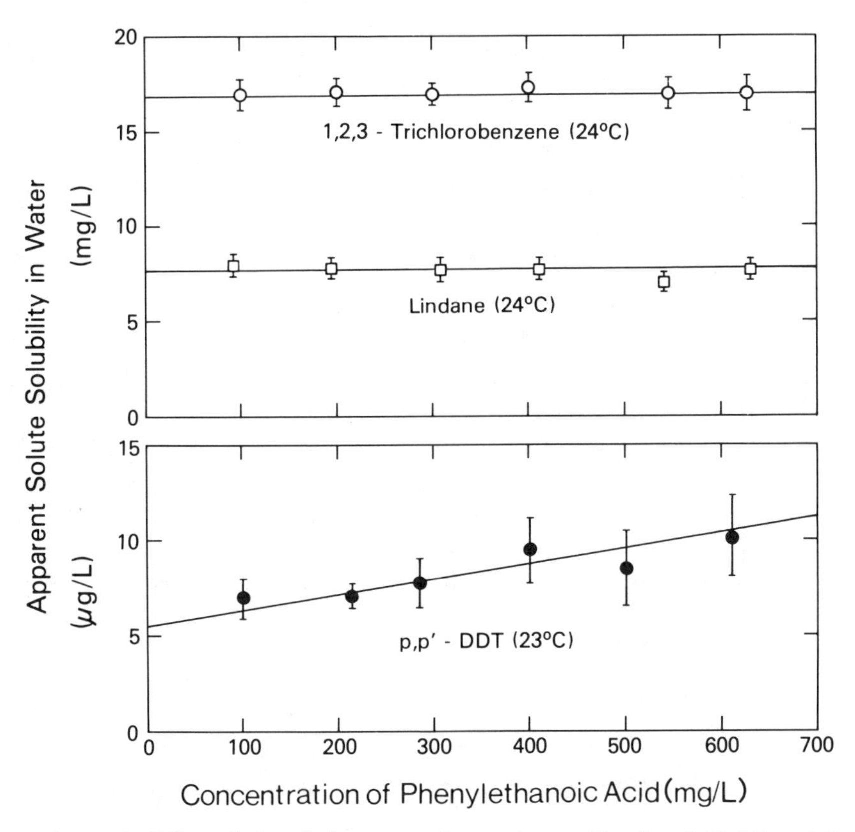

Figure 6. Effect of phenylethanoic acid at ambient pH values (pH 3.7 to 4.1) on the water solubilities of p,p'-*DDT, lindane, and 1,2,3-trichlorobenzene. (Reproduced from ref. 18. Copyright 1986 American Chemical Society.)*

of relatively nonpolar moieties (such as branched aliphatic and aromatic structures) in humic molecules would more effectively promote partition interactions of humic substances with (nonpolar) organic solutes by London forces.

Solubility enhancement of various nonionic organic solutes by dissolved organic matter is significant only for solutes whose intrinsic water solubilities are at least 2 orders of magnitude lower than concentrations of dissolved organic matter (35). This dependence of the enhancement effect on solute solubility is characteristic of a partition equilibrium. Therefore, at a given

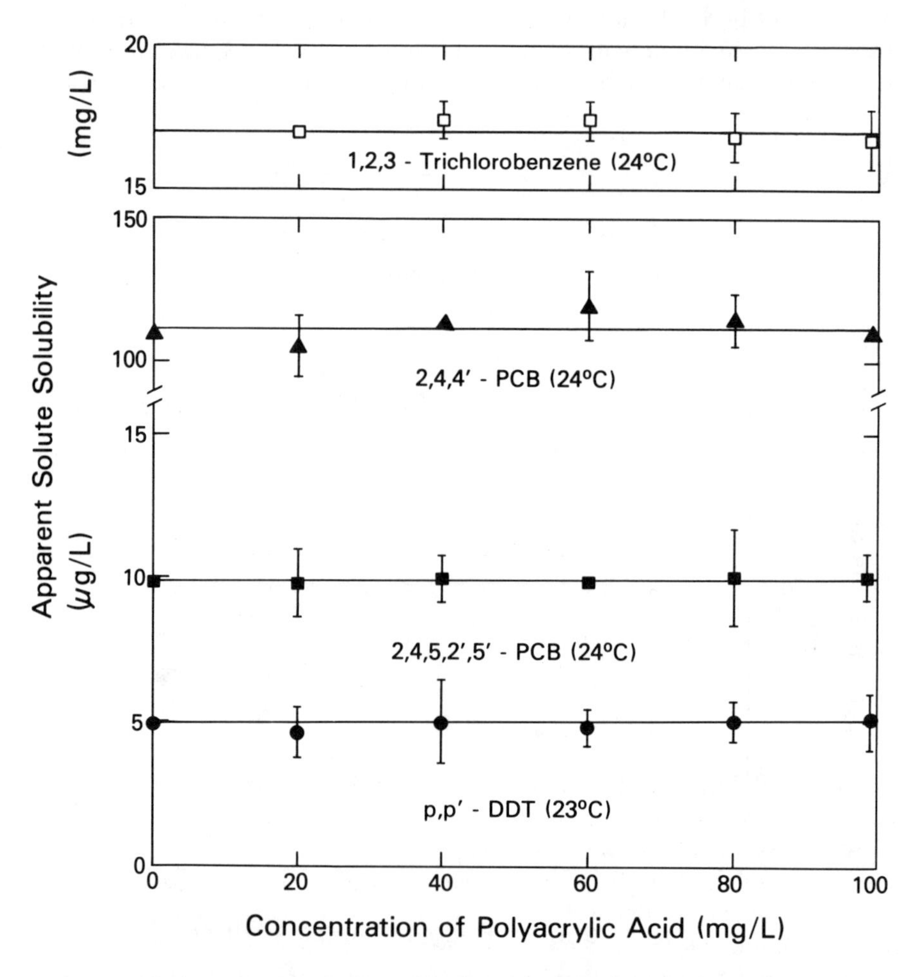

Figure 7. Effect of poly(acrylic acid) (MW 2000) at ambient pH values (pH 4.1 to 4.7) on the water solubilities of p,p'*-DDT, 2,4,5,2',5'-PCB, 2,4,4'-PCB, and 1,2,3-trichlorobenzene. (Reproduced from ref. 18. Copyright 1986 American Chemical Society.)*

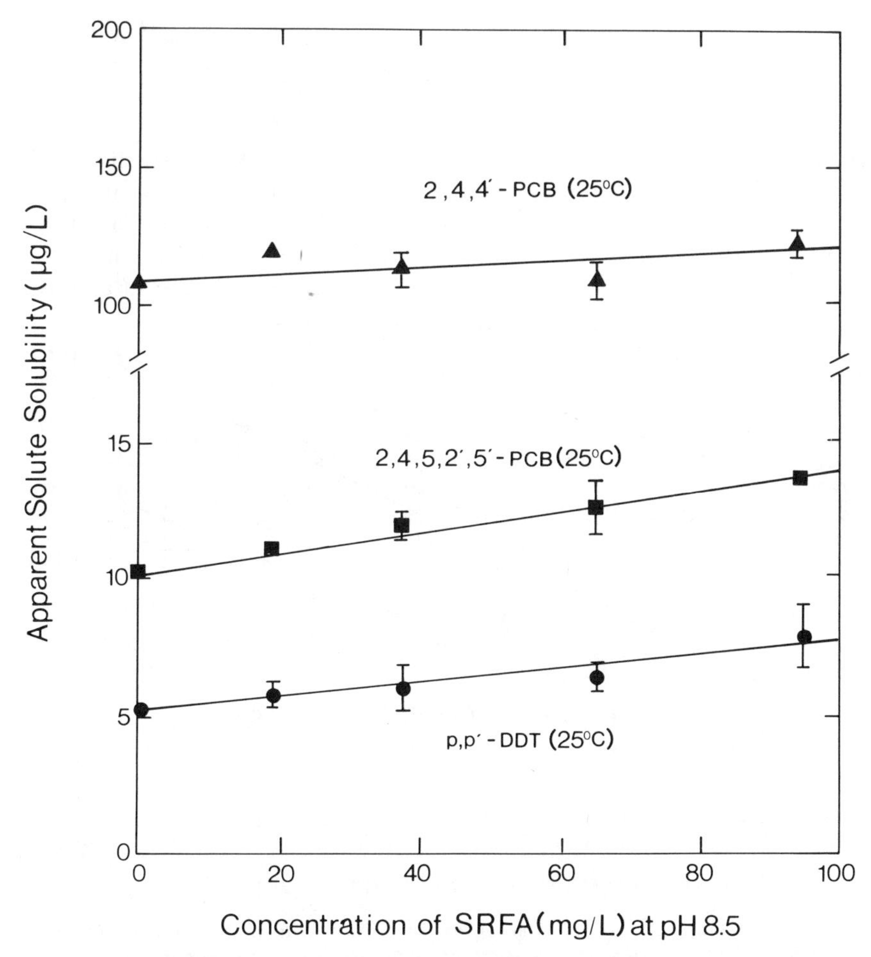

Figure 8. Dependence of the apparent water solubilities of p,p′*-DDT, 2,4,5,2′,5′-PCB, and 2,4,4′-PCB on the concentration of Suwannee River fulvic acid at pH 8.5.*

concentration of dissolved organic matter, the effect on apparent water solubility of highly insoluble solutes may be rather pronounced, but the effects would be less significant for highly water-soluble solutes. As a result, apparent water solubilities of compounds such as DDT, some PCBs, and higher alkanes are sensitive to low levels of dissolved humic substances. Their related partition constants (such as soil sorption coefficients and bioconcentration factors) in aqueous environments would be concomitantly decreased when the solubility enhancement effect is not taken into account.

Similarly, the decreased volatility and hydrolysis rates, as noted for some sparingly soluble solutes in water containing dissolved humic sub-

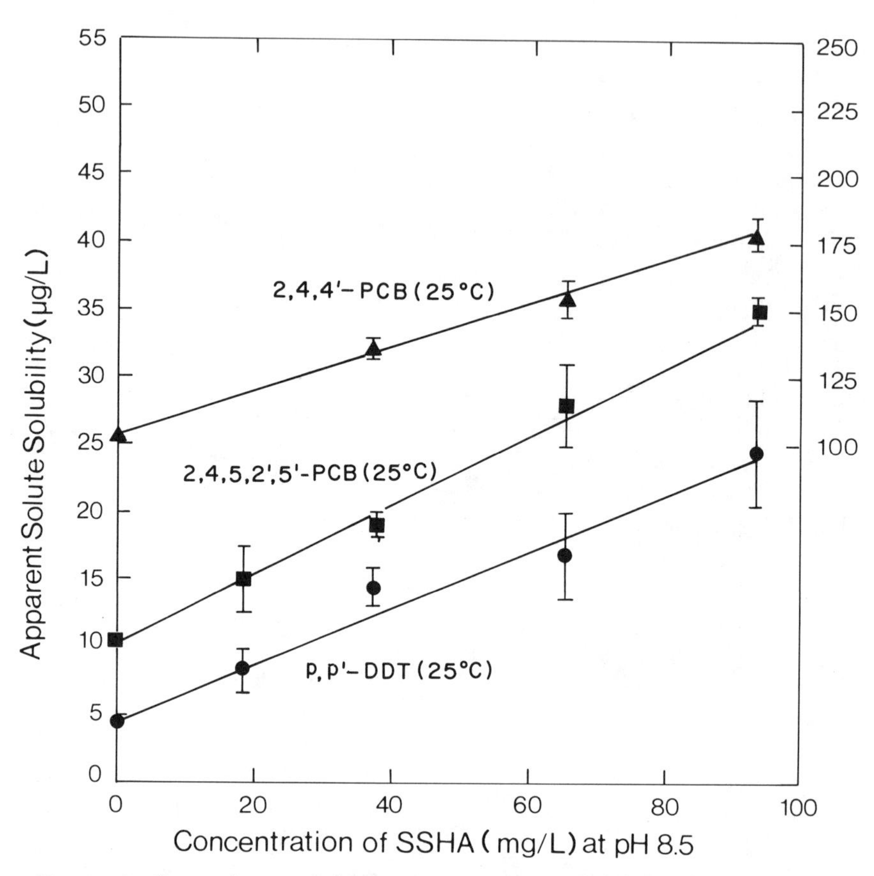

Figure 9. Dependence of the apparent water solubilities of p,p'*-DDT, 2,4,5,2',5'-PCB, and 2,4,4'-PCB (right ordinate) on the concentration of Sanhedron soil humic acid at pH 8.5.*

Table IV. Comparison of log K_{dom} Values for Selected Organic Solutes

	log K_{dom}, pH ≤ 6.5		log K_{dom}, pH = 8.5	
Compound	*SRFA*[a]	*SSHA*[b]	*SRFA*	*SSHA*
p,p'-DDT	4.13	4.63	3.62	4.55
2,4,5,2',5'-PCB	3.83	4.63	3.61	4.41
2,4,4'-PCB	3.30	4.16	2.84	3.78

[a]SRFA, Suwannee River fulvic acid.
[b]SSHA, Sanhedron soil humic acid.

stances, may be considered a result of the use of the (higher) apparent solute concentration in manipulating the data. Compounds that are more water soluble, such as toluene, 1,2,3-trichlorobenzene, and lindane, would not be sensitive to low levels of dissolved humic substances.

Effect of Temperature

It is instructive to substantiate the type of interactions between dissolved humic substances and organic solutes in terms of the equilibrium enthalpy for the solutes. The enthalpy associated with the transfer of an organic solute from water to an organic phase in a partition equilibrium ($\Delta\overline{H}$) is usually less exothermic than the heat of condensation from water, $-\Delta\overline{H}_w$ (i.e., the reverse heat of solution). Consequently, the equilibrium partition coefficient of the solute would be less sensitive to temperature than the equilibrium adsorption coefficient. Up to this time very little data has been available on the effect of temperature on the interaction of organic solutes with dissolved organic matter, from which the enthalpy value can be determined. Carter and Suffet (*36*) provided the only published data describing the temperature effect on the binding of DDT to dissolved aquatic- and sediment-derived humic acid. These studies showed a nearly twofold increase in the binding constant of DDT when the temperature was decreased from 25 to 9 °C. The molar enthalpy value, $\Delta\overline{H}$, for the binding of DDT with dissolved humic acid can be calculated by the van't Hoff equation

$$\Delta\overline{H} = -R\frac{\ln(K_2/K_1)}{1/T_2 - 1/T_1} \tag{8}$$

where $\Delta\overline{H}$ expresses the molar heat of equilibrium, R is the gas constant, K_2 is the binding constant (i.e., K_{doc}) at temperature T_2 (K), and K_1 is the binding constant at T_1 (K). The observed K_1 (at 282 K) and K_2 (at 298 K) values for DDT bound to a freshwater humic acid are about 222,000 and 145,000, respectively, and the corresponding K_1 and K_2 values of DDT with the sediment-derived humic acid are 718,000 and 444,000, respectively. The calculated $\Delta\overline{H}$ value is approximately –20.9 kJ mol^{-1} for DDT with both humic acids used in this study.

By comparison, the heat of solution of DDT in water ($\Delta\overline{H}_w$), as determined from its water solubility at 10 and 25 °C, is approximately 39 kJ mol^{-1} (our unpublished results). As expected, this value is greater than the heat of fusion for DDT, which is about 25 kJ mol^{-1} as reported by Plato and Glasgow (*37*). The observed binding of DDT to dissolved humic acids in water is therefore much less exothermic than the $-\Delta\overline{H}_w$ value. Apparently, the interaction between DDT and dissolved humic substances is not mechanistically consistent with surface adsorption or other specific interaction (such as complexation) because the molar heat for these processes should be greater than the $-\Delta\overline{H}_w$, unless the effects are extremely weak. The observed heat effect is largely consistent with a partition interaction of DDT with dissolved humic acids according to equations 5–7.

Source of Dissolved Humic Substances

Substantial variations can occur in the solubility enhancement of a nonionic solute by dissolved humic substances from different sources because of the effects of polarity, molecular size, and configuration of the organic matter on partition interactions. Probably, the structure and composition of dissolved humic substances are largely controlled by local environmental conditions (such as pH and climate) and by biological processes specific to the aquatic system in which they occur. Thus, humic substances from different surface water systems may display some distinct molecular characteristics. For example, humic substances derived from more acidic surface water sources appear to contain a higher percentage of oxygen (i.e., more polar groups) compared to samples from neutral waters. This point is illustrated by a comparison of the humic and fulvic acids derived from the Suwannee River, Georgia (pH 4.8), with the corresponding fractions from the Calcasieu River, Louisiana (pH 6.5). The humic and fulvic acids from the Suwannee River have oxygen contents of about 40% by weight, and the combined humic and fulvic acids from the Calcasieu River have oxygen contents of only 36%. Similarly, elemental data for aquatic humic samples from other near-neutral-pH rivers, such as the Ohio River (*34*) and the Missouri River (R. L. Malcolm, U.S. Geological Survey, unpublished data) give oxygen contents of about 34–36%, which is comparable to that for the Calcasieu River sample.

Elemental data for various humic substances (Table I) and partition coefficients (K_{doc}), obtained for DDT and other solutes with humic extracts from the Suwannee and Calcasieu Rivers, substantiate the difference in composition between these stream humic substances. Chiou et al. (*30*) presented data elucidating the relation between the enhancement by these various humic fractions and their sources; results of these studies are shown in Figures 10–12 and in Table V. A plot of the apparent water solubility of DDT versus the dissolved organic carbon concentration of the Calcasieu River humic extract (a near-neutral river), the Suwannee River humic and fulvic acids (an acidic source), and whole-water samples from the Suwannee River, Georgia and the Sopchoppy River, Florida, all at pH 6.5, is presented in Figure 10. Similar plots for 2,4,5,2′,5′-PCB and 2,4,4′-PCB are shown in Figures 11 and 12, respectively. The apparent solute solubility is again linear with respect to the dissolved organic carbon content, as predicted by equation 4, and the order of solute enhancements follows the same sequence as noted before (DDT > 2,4,5,2′,5′-PCB > 2,4,4′-PCB). Solubility en-

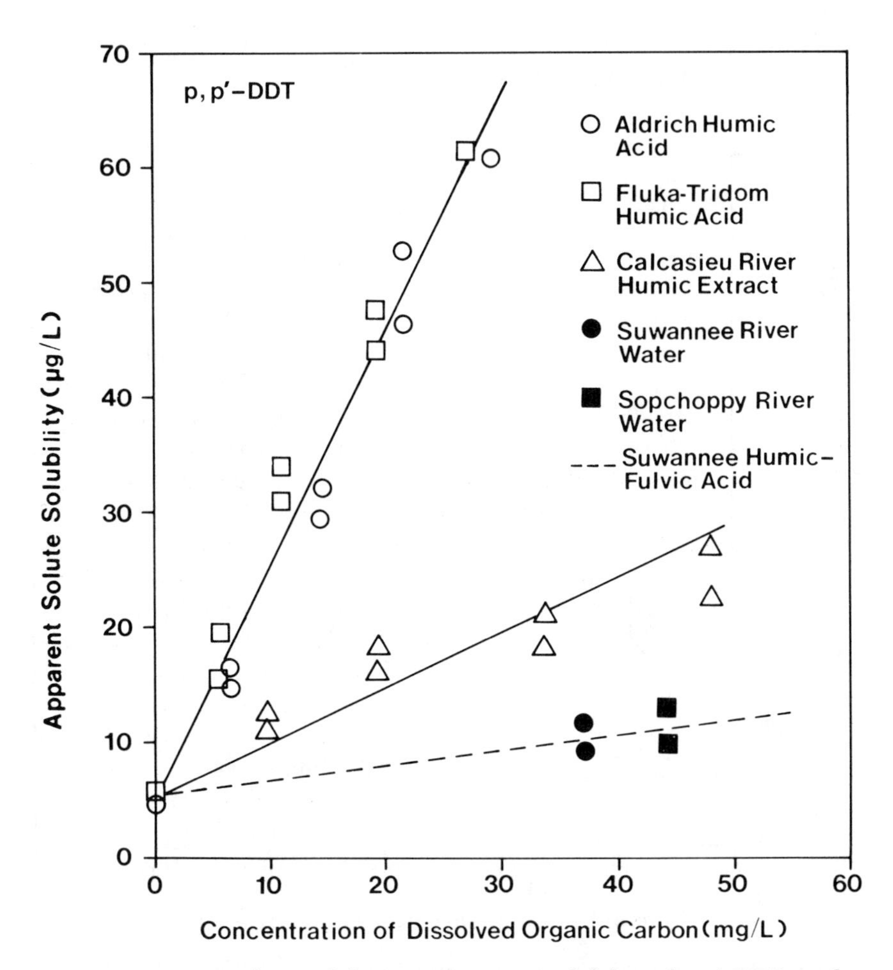

Figure 10. Dependence of the apparent water solubility of p,p′*-DDT on the dissolved-organic-carbon concentration of selected humic materials and natural waters. (Reproduced from ref. 30. Copyright 1987 American Chemical Society.)*

hancement data for two commercial humic acid samples are also shown in these figures. The organic carbon-based partition coefficients (K_{doc}) are again determined from the slope (which gives $S_w K_{doc}$) and the intercept (which gives S_w) of the plot. K_{doc} is used in these studies (rather than K_{dom}) because of the high ash content of the commercial humic materials. The calculated K_{doc} values for various solute–humic-substance pairs are summarized in Table V.

The substantially higher K_{doc} of the Calcasieu humic–fulvic extract relative to the K_{doc} for the Suwannee River humic and fulvic acids is in accord with the higher polarity (inferred by the carbon–oxygen ratio) of the humic

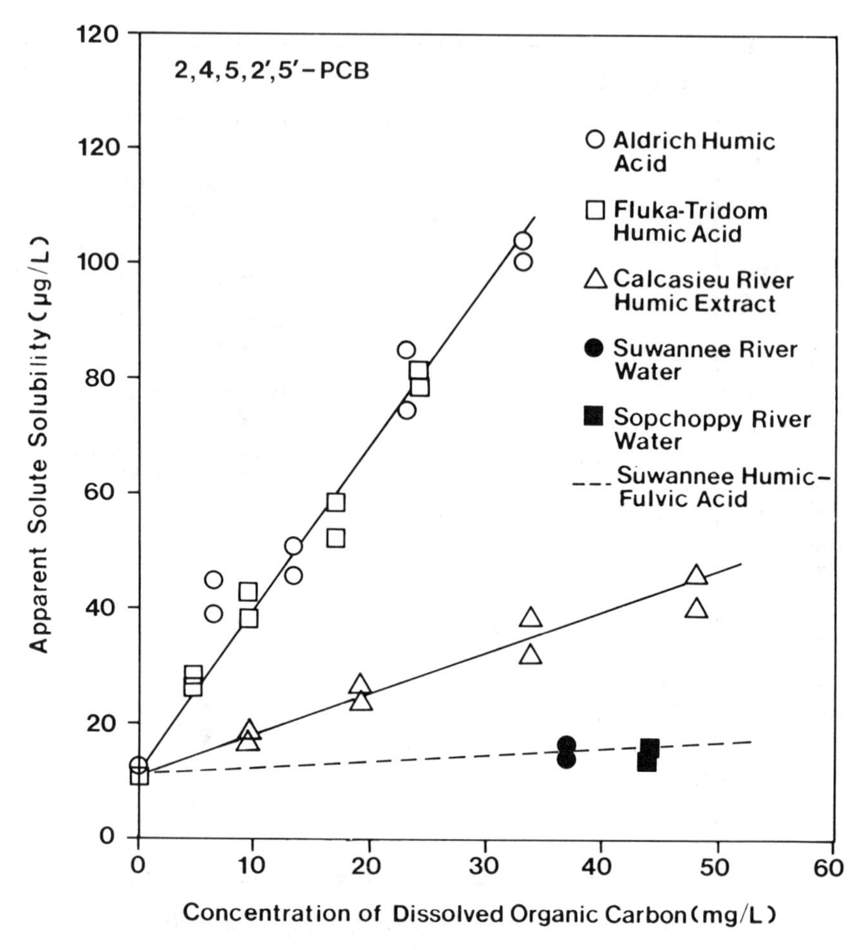

Figure 11. Dependence of the apparent water solubility of 2,4,5,2′,5′-PCB on the dissolved-organic-carbon concentration of selected humic materials and natural waters. (Reproduced from ref. 30. Copyright 1987 American Chemical Society.)

substances derived from the more acidic water of the Suwannee River. Similarly, the slightly smaller solubility-enhancement effect with the Calcasieu humic extract, in comparison to the soil-derived humic acid, is in agreement with the greater oxygen and smaller carbon contents of the sample. The K_{doc} values of solutes with natural whole-water samples (Sopchoppy River and Suwannee River) are identical to those with the humic and fulvic acids from the Suwannee water, a result suggesting that the humic- and fulvic-acid fractions, which are usually the predominant dissolved organic matter components in natural waters, are primarily responsible for the observed solubility enhancement. On the basis of these results, it may be concluded that differences in water-solubility enhancements by aquatic

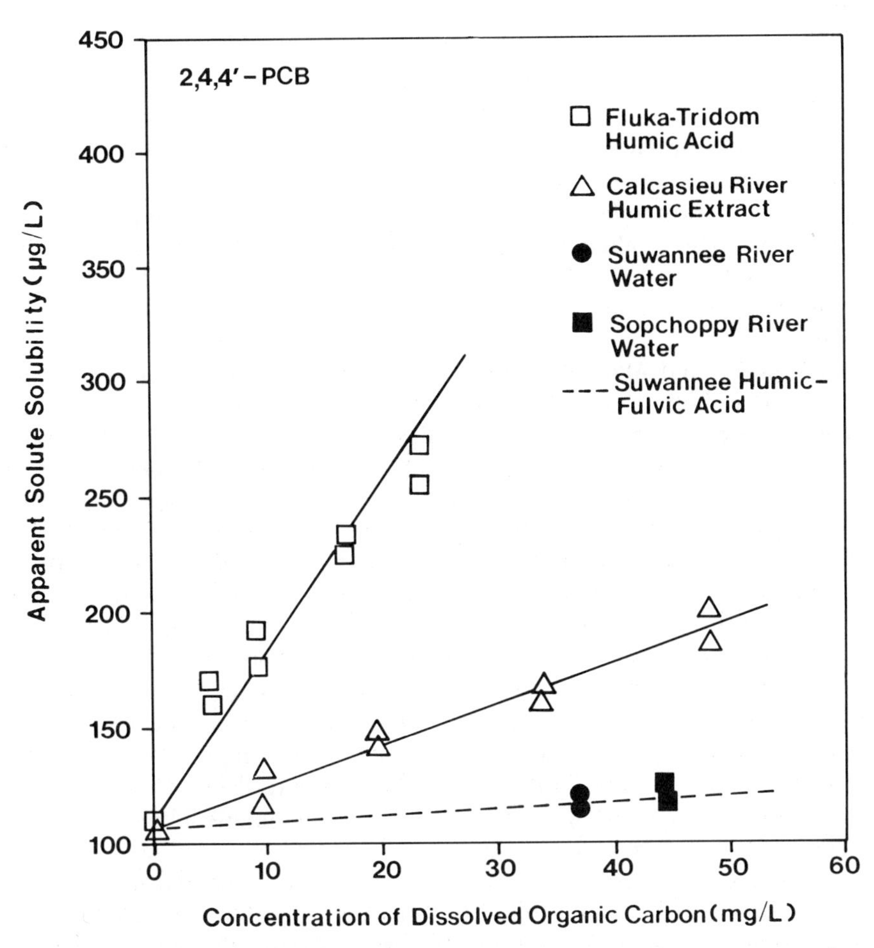

Figure 12. Dependence of the apparent water solubility of 2,4,4'-PCB on the dissolved-organic-carbon concentration of selected humic materials and natural waters. (Reproduced from ref. 30. Copyright 1987 American Chemical Society.)

Table V. Comparison of log K_{doc} Values

	log K_{doc}		
Dissolved Organic Carbon	p,p'-*DDT*	*2,4,5,2',5'-PCB*	*2,4,4'-PCB*
Aldrich humic acid sodium salt	5.56	5.41	—
Fluka–Tridom humic acid	5.56	5.41	4.84
Calcasieu River humic extract	4.93	4.81	4.24
Suwannee River water sample	4.39	4.09	3.53
Sopchoppy River water sample	4.39	4.01	3.57

SOURCE: Data from ref. 30.

humic substances are largely related to the composition of the humic molecules.

Commercial Humic Acids as Models

Because the solubility enhancement of a solute depends on the type and composition of the dissolved organic matter, it is pertinent to evaluate the appropriateness of some commercial humic acids as models for aquatic humic substances. This evaluation is particularly important because certain conclusions reached in previous studies (*4, 5, 8, 9, 14–16*) were based on experiments that used such commercial materials as models for dissolved humic substances.

Malcolm and MacCarthy (*34*) gave a critical evaluation of the use of these materials as analogs of naturally occurring humic substances, based on analyses of elemental composition, IR spectra, and ^{13}C-NMR data. Some compositional and structural features of the commercial humic acids were shown to be significantly different from those of the humic and fulvic acids isolated from natural soil and freshwater sources. Such analyses suggest that the commercial humic samples are not representative of normal aquatic or soil humic substances. One of the important findings in that study was the low carboxyl and carbohydrate contents of the commercial (Aldrich and Fluka–Tridom) humic samples, which would presumably make the molecule more effective in enhancing the water solubility of sparingly soluble organic solutes. Solubility enhancement data obtained with these two commercial humic acids and several humic extracts of stream origin (Figures 10–12) confirm this expectation. Solutions of the commercial humic acids in salt form were filtered through a 0.45-μm silver membrane filter to remove suspended particulate matter. Results of the solubility studies were expressed in terms of dissolved organic carbon because of the high ash contents of the commercial humic acids. The calculated partition coefficients (K_{doc}) are presented in Table V. These results show a markedly greater enhancement effect with both Aldrich and Fluka–Tridom samples, compared to both soil-derived and stream-derived humic and fulvic acids.

The fact that these two commercial samples exhibit virtually identical enhancement effects supports the earlier contention by Malcolm and MacCarthy (*34*) that the samples are essentially identical in nature (and may have come from the same source). As shown in Table V, the K_{doc} values obtained with the commercial humic acids are about 3 times as large as those for the soil-derived humic acid (Table III), approximately 4 times as large as those of the Calcasieu River humic extract, and about 20 times larger than the values for two acidic water samples and the Suwannee River humic and fulvic acids. Similar results with Aldrich humic acid have been reported by others. For example, Carter and Suffet (*4*) gave a log K_{doc} of 5.61–5.74 for DDT, and Hassett and Milicic (*8*) gave a log K_{doc} of 4.86 for 2,5,2′,5′-PCB.

The stronger enhancement effect shown by these commercial materials may be attributed to their lower polarity and possibly to their higher molecular weights, as well as to some unique molecular structures. The reported high oxygen contents of commercial humic acids appear to contradict the observed large solubility-enhancement effects. However, it has been suggested that the oxygen content for high-ash commercial samples is subject to substantial error, presumably because the oxygen associated with ash is also counted into the elemental oxygen content (*34*). In support of this speculation, a purified (H^+ form) low-ash Aldrich humic acid shows an elemental content of C = 63.31%, H = 5.94%, O = 25.05%, N = 0.51%, S = 3.36%, and P < 0.05% on an ash-free and moisture-free basis, when the ash content is decreased to 4.46% (*34*).

On the basis of the preceding solubility-enhancement data, the commercial samples cannot be treated as analogs of normal water and soil humic substances. Their use may lead to an improper estimation of the behavior of organic solutes in natural waters.

Summary

An evaluation of the interaction mechanism between dissolved humic substances and nonionic organic contaminants is based on the intrinsic properties of humic macromolecules and solutes. This interaction, as manifested in apparent water-solubility enhancements of relatively water-insoluble organic solutes, can best be described by a partition interaction. The assumed partition mechanism effectively explains the solubility-enhancement data and other modifications of solute behavior, which influence the transport and fate of pollutants in natural aquatic systems. The interaction characteristics for nonionic organic solutes with dissolved humic substances are similar to those found with bulk soil organic matter, a result indicating that a common partition mechanism is involved for both systems. However, the effectiveness of the partition interaction with dissolved humic substances appears to be more limited in magnitude than that for the bulk soil organic matter because of the relatively smaller organic environment and the normally more polar nature of the dissolved organic matter.

At a given concentration of dissolved humic substances, the observed solubility enhancement is significantly greater for highly water-insoluble compounds, such as DDT and 2,4,5,2′,5′-PCB. The effect decreases with increasing water solubility of the solute. In general, for a solute to show a significant solubility enhancement, the concentration of the dissolved organic matter has to be more than 2 orders of magnitude greater than the intrinsic water solubility of the solute. Critical properties of the dissolved organic matter for enhancing the solubility of the solute are polarity, molecular size, configuration, and conformation. The composition of the humic molecule (and therefore the enhancement effect) is related to the source and environment from which the humic substance originated. The soil-derived humic acid is generally more effective in enhancing solute solubility than stream-derived humic substances.

Commercial humic acids enhance the solubility of nonionic organic compounds to a significantly greater extent than soil-derived and freshwater humic substances. These commercial materials are on the order of 3 times more effective than a soil-derived humic acid, and from 4 to 20 times more effective than stream-derived humic samples. The greater enhancement effects of the commercial humic acids appear to be related to their low polar-group contents, and possibly to other molecular properties such as molecular sizes and structures. Use of some commercial humic samples as substitutes for naturally occurring humic substances should be regarded with caution, as their use can lead to gross overestimation of the impact of dissolved organic matter on the solubility of nonionic compounds.

Many observed modifications of the behavior of sparingly soluble organic solutes in the presence of dissolved organic matter (such as hydrolysis, bioaccumulation factor, and soil–water sorption coefficient) appear to be mainly consequences of the increase in the solute's apparent water solubility.

References

1. Wershaw, R. L.; Burcar, P. J.; Goldberg, M. C. *Environ. Sci. Technol.* **1969,** *3,* 271–273.
2. Poirrier, M. A.; Bordelon, B. R.; Laseter, J. L. *Environ. Sci. Technol.* **1972,** *6,* 1033–1035.
3. Boehm, P. D.; Quinn, J. G. *Geochim. Cosmochim. Acta* **1973,** *33,* 2459–2477.
4. Carter, C. W.; Suffet, I. H. *Environ. Sci. Technol.* **1982,** *16,* 735–740.
5. Landrum, P. F.; Nihart, S. R.; Eadie, B. J.; Gardner, W. S. *Environ. Sci. Technol.* **1984,** *18,* 187–192.
6. Hassett, J. P.; Anderson, M. A. *Environ. Sci. Technol.* **1979,** *13,* 1526–1629.
7. Griffin, R. A.; Chian, E. S. K. *Attenuation of water-soluble polychlorinated biphenyl by earth materials*; U.S. Environmental Protection Agency: Cincinnati, 1980; EPA–600/2–80–028.
8. Hassett, J. P.; Milicic, E. *Environ. Sci. Technol.* **1985,** *19,* 638–643.
9. Perdue, E. M.; Wolfe, N. L. *Environ. Sci. Technol.* **1982,** *16,* 847–852.
10. Hassett, J. P.; Anderson, M. A. *Water Res.* **1982,** *16,* 681–686.
11. Caron, G.; Suffet, I. H.; Belton, T. *Chemosphere* **1985,** *14,* 993–1000.
12. Gschwend, P. M.; Wu, S.-C. *Environ. Sci. Technol.* **1985,** *19,* 90–96.
13. Leversee, G. J.; Landrum, P. F.; Giesy, J. P.; Fannin, T. *Can. J. Fish. Aquat. Sci.* **1983,** *40*(Suppl. 2), 63–69.
14. McCarthy, J. F. *Environ. Contam. Toxicol.* **1983,** *12,* 559–568.
15. McCarthy, J. F.; Jimenez, B. D. *Environ. Sci. Technol.* **1985,** *19,* 1072–1076.
16. Haas, C. M.; Kaplan, B. M. *Environ. Sci. Technol.* **1985,** *19,* 643–645.
17. Chiou, C. T.; Schmedding, D. W.; Manes, M. *Environ. Sci. Technol.* **1982,** *17,* 227–231.
18. Chiou, C. T.; Malcolm, R. L.; Brinton, T. I.; Kile, D. E. *Environ. Sci. Technol.* **1986,** *20,* 502–508.

19. McBain, M. E. L.; Hutchinson, E. *Solubilization and Related Phenomena*; Academic Press: New York, 1955.
20. Elworthy, P. E.; Florence, A. T.; McFarlane, C. B. *Solubilization by Surface Active Agents*; Chapman and Hall: London, 1968.
21. Tanford, C. *The Hydrophobic Effect*; Wiley: New York, 1973.
22. Nagarajan, R.; Ruckenstein, E. In *Surfactants in Solution*; Mittal, K. L.; Lindman, B., Eds.; Plenum: New York, 1984; Vol. 2, pp 923–947.
23. Moroi, Y.; Sata, K.; Noma, H.; Matuura, R. In *Surfactants in Solution*; Mittal, K. L.; Lindman, B., Eds.; Plenum: New York, 1984; Vol. 2, pp 963–979.
24. Chiou, C. T.; Peters, L. J.; Freed, V. H. *Science (Washington, DC)* **1979,** *206*, 831–832.
25. Chiou, C. T.; Porter, P. E.; Schmedding, D. W. *Environ. Sci. Technol.* **1983,** *17*, 227–231.
26. Chiou, C. T.; Shoup, T. D.; Porter, P. E. *Org. Geochem.* **1985,** *8*, 9–14.
27. Karickhoff, S. W.; Brown, D. S.; Scott, T. A. *Water Res.* **1979,** *13*, 241–248.
28. Means, J. C.; Wood, S. G.; Hassett, J. J.; Banwart, W. L. *Environ. Sci. Technol.* **1980,** *14*, 1524–1528.
29. Chiou, C. T. In *Hazard Assessment of Chemicals: Current Developments*; Saxena, J.; Fisher, F., Eds.; Academic Press: New York, 1981; pp 117–153.
30. Chiou, C. T.; Kile, D. E.; Brinton, T. I.; Malcolm, R. L.; Leenheer, J. A.; MacCarthy, P. *Environ. Sci. Technol.* **1987,** *21*, 1231–1234.
31. Cameron, R. S.; Thornton, B. K.; Swift, R. S.; Posner, A. M. *J. Soil Sci.* **1972,** *23*, 394–408.
32. Thurman, E. M.; Wershaw, R. L.; Malcolm, R. L.; Pinckney, D. *Org. Geochem.* **1982,** *4*, 27–35.
33. Aiken, G. R.; Brown, P.; Noyes, T.; Pinckney, D. In *Humic Substances in the Suwannee River, Florida and Georgia: Interactions, Properties, and Proposed Structures*; Averett, R. C., Ed.; U.S. Geological Survey: Reston, VA, in review; Water Supply Paper 1507–B.
34. Malcolm, R. L.; MacCarthy, P. *Environ. Sci. Technol.* **1986,** *20*, 904–911.
35. Chiou, C. T.; Porter, P. E.; Shoup, T. D. *Environ. Sci. Technol.* **1984,** *18*, 295–297.
36. Carter, C. W.; Suffet, I. H. In *Fate of Chemicals in the Environment*; Swann, R. L.; Eschenroeder, A., Eds.; ACS Symposium Series 225; American Chemical Society: Washington, DC, 1983; pp 215–229.
37. Plato, C.; Glasgow, R. *Anal. Chem.* **1969,** *41*, 330–336.

RECEIVED for review July 24, 1987. ACCEPTED for publication December 21, 1987.

11

Sorption of Chlorinated Hydrocarbons in the Water Column by Dissolved and Particulate Organic Material

Linda L. Henry[1] and I. H. Suffet

Environmental Studies Institute, Drexel University, Philadelphia, PA 19104

Steven L. Friant

Academy of Natural Science, Philadelphia, PA 19103

Partition coefficients for dissolved and particulate organic material (K_{dom} and K_{pom}, respectively) and five organic chemicals were measured in surface waters and leachates from soils and sediments. K_{dom} was measured by using changes in Daphnia magna *dry-weight bioconcentration factors, and K_{pom} was measured by centrifugation. The results showed that sorption by dissolved organic matter can dominate the distribution of organic chemicals in the water column and affects calculation of K_{pom}. The linear relationship of log K_{dom} and log K_{pom} values to log K_{ow} (octanol–water partition coefficient) was similar to the relationship found by Kenaga and Goring for soils and sediments. This resemblance indicates that organic materials in the water column, soils, and sedimentary material sorb chemicals by a similar mechanism.*

SORPTION OF HYDROPHOBIC ORGANIC CHEMICALS by naturally occurring organic matter in soils and aquatic systems is an important factor in determining their distribution in the environment. Incorporation of sorption by the organic matter in soils and sedimentary material into fate and transport

[1]**Current address: BCM Engineers, 1 Plymouth Meeting, Plymouth Meeting, PA 19462**

0065–2393/89/0219–0159$06.00/0

models has been facilitated by the commonality of the behavior of hydrophobic organic chemicals (*1*). Research has led to predictive relationships that correlate the partition coefficient (K_{oc}) for organic chemicals in soils and sedimentary material to the octanol–water partition coefficient (K_{ow}) and water solubility (*2, 3*).

Sorption by dissolved organic material (DOM, less than 0.45 μm) and suspended particulate organic material (POM) in the water column, particularly for naturally occurring concentrations, has received less attention than sorption by soils and sedimentary material. The importance of sorption interactions in the water column is illustrated by the changes in fate and transport processes for organic chemicals bound to DOM. DOM sorption interactions have been shown to retard volatilization, increase solubility, and change chemical reactivity and the rate of bioaccumulation (*4–8*). Studies indicate that sorption or association of contaminants by DOM may be a key component in controlling their distribution in the water column (*9–12*). For example, in Delaware River water, an average of 34% ($n = 12$) of benzo[*a*]pyrene was bound to less than 1 mg/L of DOM (*13*).

Research with naturally occurring DOM has shown that sorption of certain chemicals can be correlated with DOM concentrations, but not necessarily between different sources of DOM (*9, 14, 15*). The variability among sources suggests that K_{dom} may not be as amenable for use in predictive models as K_{pom}. However, further research on naturally occurring DOM is needed. Most research on K_{dom} has used humic material either from a commercial source (e.g., Aldrich humic acid) or extracted from aquatic systems (*9, 11, 16*), which may not truly represent DOM in its natural state.

The relative importance of sorption by DOM and POM in the water column and the effect of the DOM partition coefficient (K_{dom}) on the calculation of the partition coefficient for POM (K_{pom}) have yet to be clarified. To date, K_{pom} has been calculated by assuming that the fraction of chemicals remaining in the water after removal of the particulates represents the freely dissolved fraction of the chemicals (two-compartment model). However, in natural waters, DOM is present and should be included as a third equilibrium compartment (Figure 1).

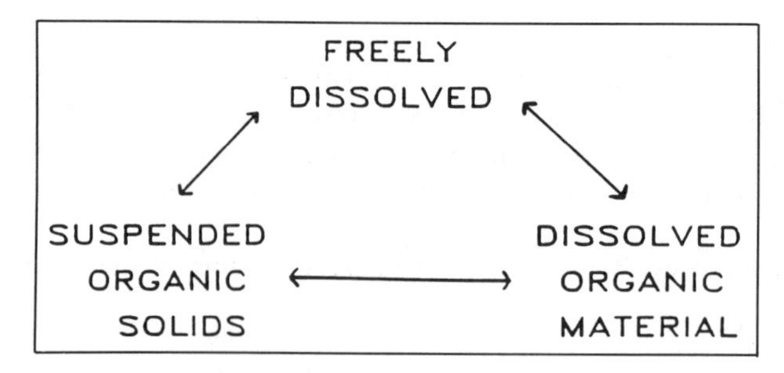

Figure 1. Three-compartment model for the distribution of organic chemicals in the water column.

The chemicals remaining in the water after removal of particulate material include a fraction bound to the DOM and a freely dissolved fraction. A more correct method of estimating K_{pom} would be to subtract the fraction of chemicals bound to DOM from the total chemicals remaining in the water (three-compartment model) (*17, 18*). Depending on the significance of DOM–pollutant interactions, current methods may be underestimating K_{pom}.

To date, experimental measurements of K_{dom} have not been included in K_{pom} studies, although their importance is recognized. The presence of DOM has been used to explain decreases in K_{pom} as the concentration of particulate matter increases (*17, 18*). The effect of organic matter concentration has also been observed for K_{dom} (*9*). The phenomena are controversial for both K_{pom} and K_{dom}. The observed decrease in K_{pom} has been attributed, in some cases, to the presence of DOM in the test water. When an estimated value for K_{dom} was included in the calculation of K_{pom}, the decrease in K_{pom} was not seen (*17, 18*). In both of these studies, K_{dom} was incorporated with literature values rather than experimentally determined values on the same samples (synoptic).

The importance of K_{dom} in the determination of K_{pom} has yet to be demonstrated experimentally, and synoptic measures of K_{dom} and K_{pom} are considered an important component of the study design for the research presented here.

The objective of this research was to evaluate the predictability of K_{dom} and K_{pom} for five chlorinated hydrocarbons by using different components of the organic matter and the effect of K_{dom} on K_{pom} measurements in natural waters. The tests, designed to simulate conditions in the water column, included measures of sorption by DOM and suspended POM from a surface water and leachates of soils and sedimentary material.

Experimental Methods

Study Compounds. Five chemicals—lindane, dieldrin, *p,p'*-DDT [1,1-dichloro-2,2-bis(*p*-chlorophenyl)ethane], and two PCB isomers (2,4,4'-trichlorobiphenyl and 2,3,4,5,6-pentachorobiphenyl) (Ultra Chemical Co.)—were chosen to represent a range of sorption potentials (log K_{ow} 3.35–6.7). Greater than 99% purity was verified with gas chromatography. Actual concentrations of the chemicals in the test solutions varied, but all were less than half of the solubility limits. Lindane (log K_{ow} 3.35) was not expected to sorb to either DOM or POM at the experimental levels and was included to monitor for experimental error.

Experimental Design. Each experiment measured K_{pom} and K_{dom} in surface waters and leachates from sediment or soil. Natural waters were collected from six sources: a large river (the Delaware River, which runs between Pennsylvania and New Jersey), two major tributaries (Schuylkill and Cooper Rivers), and three small ponds in southern New Jersey (mesotrophic Lincoln Lake, eutrophic Harris Lake, and highly dystrophic Pakim Pond).

The waters were collected in 5-gal polyethylene containers and used within 48 h. Surface soils and sedimentary materials were also collected from each of the rivers or ponds. Leachates were prepared by mixing the soil or sediment (1:20 wet weight) with the surface water for 18 h (*19*). After allowing at least 5–6 h settling time, the supernatant liquid was siphoned off and used as the experimental water. K_{pom} was measured in the surface water or leachate with all three compartments present: freely dissolved, DOM, and POM. K_{dom} was measured in the same water after the particulate material had been removed by centrifugation.

Characterization of DOM and POM. The effect of source and the relative ability of certain factors to predict K_{dom} (total DOM, humic acids, and DOM >1000 daltons) and K_{pom} (total suspended solids and POM) was evaluated. Total suspended solids (TSS) was determined as mg/L dry weight (evaporated at 103 °C for 24 h) removed by centrifugation at 5000 rpm for 1 h; the value for POM was estimated as half the ash-free dry weight of the TSS as milligrams of carbon per liter (*20*).

Total DOM, humic acid content, and DOM >1000 daltons were measured on 0.45-μm filtered water as dissolved organic carbon by a carbon analyzer (Dohrmann DC-80). Humic acid content was defined as the amount of dissolved carbon removed by a second 0.45-μm filtration 24 h after acidifying the sample to pH 2. DOM >1000 daltons was measured as the amount of carbon that did not pass through a prewashed dialysis membrane (Spectrapor 6) after dialyzing natural water against distilled water.

Sorption to DOM. K_{dom} was measured by using bioconcentration factors (BCF) for *Daphnia magna*, with and without DOM present in the test water. Chemicals sorbed to DOM are not accumulated by amphipods and daphnids, and changes in BCF have been used to measure the fraction of chemical bound to the DOM (*16, 21, 22*).

K_{dom} was measured in all water types after the particulate matter was removed. The fraction bound to the DOM was measured as the decrease in BCF found in *Daphnia magna* juveniles exposed to the study chemicals in the natural water with DOM (POM removed) relative to BCF in control water without DOM. The study chemicals were added in methanol stock solutions (<0.1% methanol). The daphnids were <48 h old and came from well-fed, low-density cultures. Details on culture technique are given elsewhere (*23*).

In the first three experiments the control water was the surface water, with 90% of the DOM removed by exposure to a 1200-W UV source for 24 h in a flow-through system. In one of the three experiments, acute toxicity and surface film entrapment were observed in the control water. As a result, another control water was added: a well water with low DOM (<0.2 mg/L of DOC), diluted with distilled–deionized water to match the surface water in pH and conductivity. Additional tests showed that BCF was not affected by conductivity within the range of the test waters (0–400 μS). Data from both controls was combined, unless toxicity was apparent in the UV-treated water.

To obtain BCF values, 10 juvenile daphnids were added to 200 mL of test water and chemicals in a 250-mL Erlenmeyer flask and kept in dim light at 20 °C. Preliminary experiments had established that adding the daphnids 24 h after the test chemicals were added had no effect on BCF. This stability indicated that sorption of the chemicals to DOM occurred rapidly. The daphnids accumulated <10% of the total amount of chemical and did not increase DOM. After 24 h, water samples were taken for hexane extraction. The daphnids were rinsed in clean water, dried over a desiccant for 30 min, and weighed on a microbalance (Cahn). The dried material was kept in acetone–hexane (1:1) for 24 h before gas chromatographic analysis. Each BCF value was based on four to six tissue samples from one flask. The samples were quantitated on a gas chromatograph (Tracor 560) with an electron-capture detector after separation on a 2-m packed column (1.5% SP-2250:1.95% SP-2401 on 100–120-mesh Supelcoport, Supelco, Bellefonte, PA) run isothermally at 230 °C.

Sorption to POM. A mixture of the five study chemicals in methanol was added to 100-mL flasks of surface water or soil and sediment leachates. The flasks were covered with foil and rotated on a shaker table for 24 h. At the end of the test, the water was sampled before and after centrifugation at 5000 rpm for 30 min. The fraction bound to the solids was the amount of the chemical removed by centrifugation after correction for losses in control flasks without particulate material. This way of calculating K_{pom} gave the same results as a mass balance approach that included the amount remaining in the centrifuge tube.

Calculation of K_{dom} and K_{pom}. The calculation of both partition coefficients was based on the fraction bound to the DOM or POM and the fraction freely dissolved in the water. For K_{dom}, the fraction of chemical bound to DOM was calculated using the difference in *Daphnia magna* BCFs with and without DOM present. The freely dissolved fraction of chemical (f_f) is calculated as ratio of BCFs with and without DOM (equation 1). The fraction of chemical bound to DOM is 1 minus the fraction bound (equation 2).

$$f_f = \frac{\mathrm{BCF_d}}{\mathrm{BCF_p}} \quad (1)$$

where f_f is free fraction, $\mathrm{BCF_d}$ is BCF with DOM present, and $\mathrm{BCF_p}$ is BCF in control water.

$$f_b = 1 - f_f \quad (2)$$

where f_b is the fraction bound to DOM.

For K_{pom}, the fraction bound to the POM was calculated as the fraction removed from the water during centrifugation. The amount remaining in the water was corrected for losses in centrifuged samples without particulate material present.

K_{dom} and K_{pom} were obtained by rearranging the standard equation (equation 3) into a linear form (equation 4). The partition coefficient (K) was the slope of the line when the fraction bound divided by the free fraction was regressed against the concentration of the organic material as milligrams of carbon per liter.

$$K = \frac{f_b \times 10^6}{f_f \times \text{mg of C/L}} \quad (3)$$

where K is K_{pom} or K_{dom}, f_b is the fraction bound to DOM or POM, and f_f is the fraction freely dissolved.

$$\frac{f_b}{f_f} = K \times 10^{-6} \times \text{mg of C/L} \quad (4)$$

The use of linear regression allows for comparison of the treatment effects (e.g., total DOM or humic acid material) on the partition coefficient via substitution into the regression equation. The ability of a component to predict K_{dom} or K_{pom} was measured by using the coefficient of determination (r^2) from a least squares linear regression. The r^2 is an estimate of the proportion of variation in the fraction bound/fraction free that is accounted for by the fitted regression (*24*).

Results

Table I lists the values for components of DOM and POM measured for the surface waters and leachates. Total DOM ranged from 2.46 mg of C/L for Delaware River sediment leachate to 18.72 mg of C/L for Pakim Pond. The percent of total DOM measured as humic acids was variable and ranged from 1.7 to 64%. In 12 of 15 samples, the DOM >1000 daltons was found in larger amounts than the humic material. This proportion suggested that large-molecular-weight nonhumic material was present.

The POM ranges from 0.81 to 60.5 mg of C/L, with the higher values in soil leachates. The highest value for water or sediment leachate was 6.9 mg of C/L for Delaware River sediment leachate.

Mixing surface water and sediment does not necessarily increase DOM. In fact, DOM was lower for leachates of Delaware and Lincoln sedimentary material than in the surface water. This difference indicates sorption of DOM by sedimentary material. DOM remained unchanged in Pakim Pond leachate and increased only slightly in Cooper River leachate. The difference in DOM and POM between the two Lincoln soil samples (Table I) is attributed to inadequate premixing of soils.

Sorption to DOM. To calculate K_{dom}, linear regressions were run with total DOM, humic acid content, or DOM >1000 daltons. Table II lists

Table I. Components of DOM and POM in Surface Waters and Soil and Sediment Leachates

	DOM[a]			*POM*	
Source	*Total DOM*	*Humic Acids*	*DOM >1000 Daltons*	*TSS*[b]	*POM*[a]
Harris Lake[c,d]	5.84	0.47	2.51	10.9	0.81
Harris Lake[c,d]	6.48	0.39	1.43	14.9	1.44
Delaware River[d]	3.38	0.38	0.57	—	—
Delaware River sediment[e]	2.46	0.54	1.62	74.4	6.9
Delaware River soil[e]	9.33	6.00	5.08	282.0	43.7
Lincoln Lake[d]	6.58	3.00	1.07	—	—
Lincoln Lake sediment[e]	4.69	0.80	2.91	303.0	27.0
Lincoln Lake soil[e,f]	7.75	1.93	5.27	198.2	32.9
Lincoln Lake soil[e,f]	5.81	4.00	2.56	378.0	60.5
Schuylkill River[d]	3.24	0.70	0.90	19.8	5.3
Schuylkill River soil[e]	4.25	0.95	1.25	208.3	24.3
Cooper River[d]	5.68	0.10	0.25	17.7	2.4
Cooper River sediment[e]	6.88	1.30	1.37	25.7	2.5
Pakim Pond[d]	18.67	2.17	11.1	3.3	1.4
Pakim Pond sediment[e]	18.72	2.22	1.1	5.5	2.7

[a]All values are milligrams of carbon per liter.
[b]TSS values are milligrams dry weight per liter.
[c]Two different collections.
[d]Surface water.
[e]Leachate.
[f]Replicates from same collection.

Table II. Coefficients of Determination and Partition Coefficients for DOM Components

Component	*Total DOM*	*Humic Acids*	*DOM >1000 Daltons*
3-Cl PCB			
r^2	0.51	0.68	0.54
$K \times 10^{-6}$	0.097	0.238	0.229
5-Cl PCB			
r^2	0.77	0.57	0.68
$K \times 10^{-6}$	0.144	0.272	0.301
Dieldrin			
r^2	0.46	0.39	0.48
$K \times 10^{-6}$	0.034	0.067	0.076
DDT			
r^2	0.75	0.53	0.68
$K \times 10^{-6}$	0.115	0.211	0.242

NOTE: All results were calculated by using equation 4.

the variables, r^2 and K_{dom}. Based on the coefficient of determination (r^2), total DOM accounted for 77 and 75% of the variation in sorption for 5-Cl PCB and DDT, respectively. Humic acids gave the highest r^2 (0.68) for 3-Cl PCB. K_{dom} for dieldrin gave the worst fit, r^2 <0.5. In most cases, the difference in r^2 is small for different components of the DOM. Lindane did not sorb to DOM or POM in any of the experiments. The mean coefficient of variation between replicate samples was less than 10% for all chemicals. The consistency suggests that the error is not in measurement technique. Other sources of experimental error include collection techniques, sample storage, and treatment.

There was no statistically significant difference nor an obvious trend in K_{dom} attributable to source for DDT, dieldrin, and the two PCBs. The absence of any consistent effect on K_{dom} was observed regardless of which component of the organic material was used in the K_{dom} calculation. The data for the fraction of bound DDT/fraction of free DDT versus total DOM are given as a representative example (Figure 2). Pakim Pond samples (four points to the far right) gave anomalous results, possibly because of its unusual nature—acidic (pH 3.8) and high DOM (17.8 mg of C/L). These data were omitted from the analyses.

The precision of using BCF to measure K_{dom} was evaluated by the coefficient of variation between duplicate sets of BCF values for all treatments (control and DOM). The mean coefficients of variation for BCF values for were 8.3, 6.8, 17.4, and 9.4% for 3-Cl PCB, 5-Cl PCB, dieldrin, and DDT, respectively, for 15 replicate data sets.

Sorption to POM. K_{pom} values were calculated in the two- or three-compartment model by using POM or TSS (Table III). For all the study chemicals, K_{pom} was highest for the three-compartment model with POM;

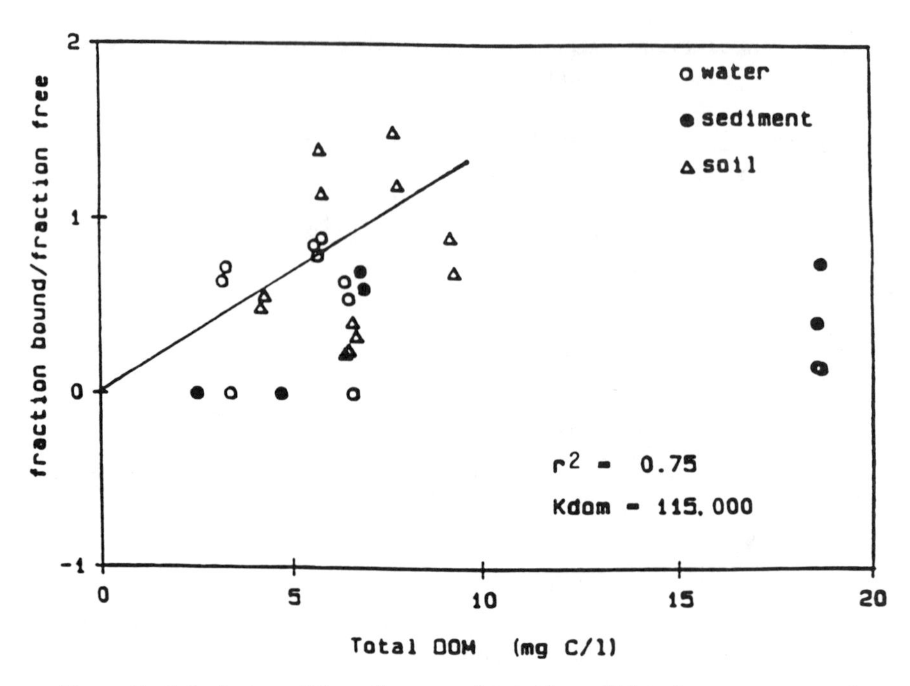

Figure 2. Calculation of K_{dom}: *fraction of DDT bound/free fraction regressed against total DOM.*

the lowest values were found for the two-compartment model with TSS. K_{pom} calculated with POM increased between the two- and three-compartment models by more than 50% for all the chemicals except dieldrin. The three-compartment model with POM accounted for 67, 79, 76, and 75% of the variability in K_{pom} for 3-Cl PCB, 5-Cl PCB, dieldrin, and DDT, respectively. Although this model accounted for the greatest amount of variability,

Table III. Coefficients of Determination and Partition Coefficients for POM Components

	Two-Phase		*Three-Phase*	
Component	*POM*	*TSS*	*POM*	*TSS*
3-Cl PCB				
r^2	0.66	0.61	0.67	0.57
$K \times 10^{-6}$	0.046	0.005	0.066	0.007
5-Cl PCB				
r^2	0.68	0.73	0.79	0.68
$K \times 10^{-6}$	0.026	0.004	0.053	0.008
Dieldrin				
r^2	0.77	0.72	0.76	0.70
$K \times 10^{-6}$	0.012	0.002	0.014	0.002
DDT				
r^2	0.67	0.70	0.75	0.71
$K \times 10^{-6}$	0.019	0.003	0.036	0.005

the differences between the coefficients of determination for the other models were not significant.

K_{pom} values based on different sources of the particulate material showed partition coefficients for surface water POM 2–3 times higher than for solids leached from soils and sedimentary material. However, the difference was not significant (i.e., the slopes of the individual lines were not statistically different). The data for DDT are presented in Figure 3.

Discussion

To develop a model for the distribution of hydrophobic organic chemicals sorbed to dissolved and particulate organic material in the water column, one needs to know if sorption of a chemical by the organic material can be predicted from some measurable component of the organic material and with what degree of confidence the prediction can be made. This information is particularly important if DOM and POM sorption is to be useful in a range of surface waters. Also, if the model is to have a broad applicability, the partition coefficient for chemicals with DOM and POM should be related to K_{ow} or a similar property.

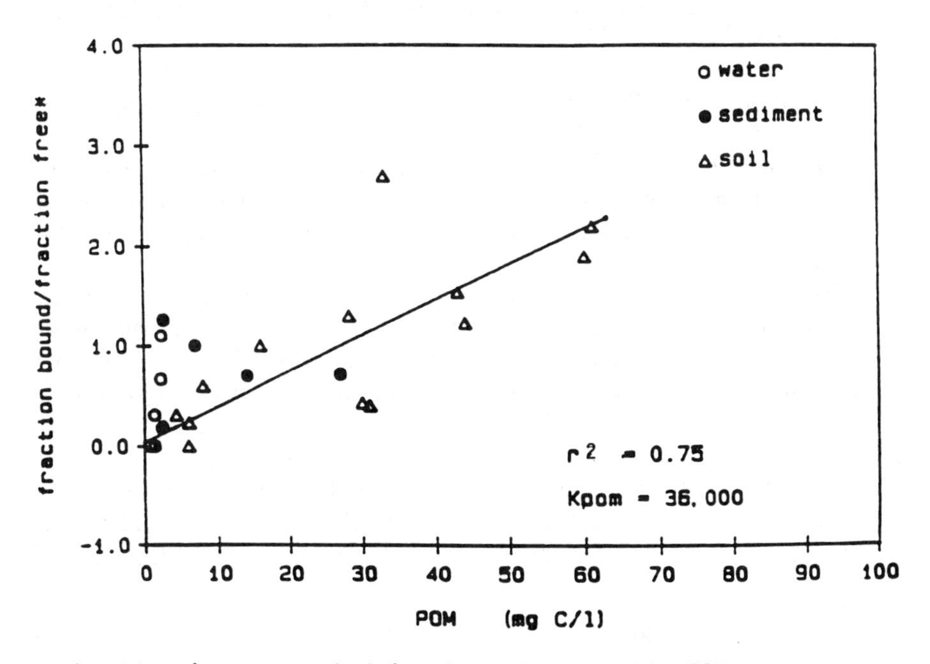

Figure 3. Calculation of K_{pom}*: fraction of DDT bound/free fraction regressed against POM.*

In most cases, 70% of the variation in sorption of the study chemicals by DOM and POM can be accounted for by using total DOM and POM, respectively. However, the variation in the partition coefficient accounted for by total DOM and POM, compared with the other variables tested, was not large.

The selection of the organic-material component was based on "goodness of fit" by using the coefficient of variation (r^2) from the partition-coefficient equation (equation 4). Emphasis was placed on the "average" ability of a component to give the best fit across compounds as a basis for a generalized model. In addition, total DOM and POM are commonly and easily measured parameters.

The values for K_{dom} and K_{pom} are summarized in Table IV, along with K_{ow} and sediment and soil partition coefficients (K_{oc}) (3). Total DOM and POM in a three-compartment model were used, respectively, to calculate K_{dom} and K_{pom}. As shown, the values for K_{dom} and K_{pom} are in close agreement with the values calculated from K_{ow} by Kenaga and Goring (3, equation 5), where

$$\log K_{oc} = 0.54 \times \log K_{ow} + 1.38 \qquad (r^2 = 0.74, n = 45) \qquad (5)$$

Carter and Suffet (25) reported values for K_{dom} for a variety of chemicals. Including their data with ours results in equation 6, which is similar to that of Kenaga and Goring (3):

$$\log K_{dom} = 0.59 \times \log K_{ow} + 1.37 \qquad (r^2 = 0.80, n = 15) \qquad (6)$$

The similarity between the partition coefficients for dissolved, particulate, and soil and sedimentary material suggests that organic matter sorbs hydrophobic chemicals by a similar mechanism, regardless of source or form in this study.

The variability of K_{dom} not accounted for by the model remains unexplained, and there is no clear indication as to the reason. The consistency of replicate K_{dom} measurements indicates that the BCF method is not a signficant source of error. The low variability found between replicates is

Table IV. Comparison of Partition Coefficients from This Study with Literature Values

Coefficients	*Lindane*	*3-Cl PCB*	*5-Cl PCB*	*Dieldrin*	*DDT*
$\log K_{ow}$	3.35	5.70	6.45	5.15	6.36
$\log K_{dom}$ [a]	—	4.98	5.15	4.35	5.06
$\log K_{pom}$ [a]	<3.50	4.80	4.72	4.15	4.56
$\log K_{oc}$ [b]	3.19	4.48	4.88	4.17	4.83

[a] Current study.
[b] Ref. 3; K_{oc} is sediment organic carbon.

attributed to the standardized culture conditions and age of the test animals (<48 h old), which result in a consistent lipid content. The toxicity encountered in the control water should not present a serious limitation to further use of this method to measure K_{dom}.

There was some indication that aeration of the UV-treated water for 24 h prior to adding the daphnids would remove the toxicity. The formation of products toxic to *Daphnia magna* has also been observed for ozonation of humic acids (*26*). A promising approach for the control water is to use well water that is naturally low in DOM, or perhaps reconstituted water. However, continued study of the effect of inorganic factors on accumulation is recommended.

The three-compartment model did not account for all the variability in K_{pom}. Other possible factors are an effect of POM concentration unrelated to DOM or to the source of POM. Theories to explain the effect of the solids concentration on K_{pom} other than the presence of DOM have been suggested (*27, 28*). The results from tests done in conjunction with this research that consider the effect of the concentration of the organic matter will be presented in a separate publication.

There was some indication that POM from the water column may have a higher sorptive potential than particulates leached from soils. However, a larger data set is needed to evaluate this tendency. It is difficult to separate possible effects of solids concentrations because POM levels in surface waters were much lower than in the leachates from sedimentary material and soils.

Including contaminant sorption by DOM in the calculation of K_{pom} increased K_{pom} values, but did not result in a significant increase in r^2. A key aspect for incorporation of K_{dom} into K_{pom} measurements is that the difference in K_{pom} between the two- and three-compartment models is controlled by the level of DOM. For this research the level of DOM was generally less than 5 mg/L and resulted in a difference of approximately 50% between K_{pom} values for the two models. As the level of DOM increases, the difference in K_{pom} values will increase. Whenever high levels of DOM are encountered, as in interstitial water, sorption by DOM will have a greater impact on calculation of K_{pom}.

The results show that sorption by DOM dominates the distribution of organic chemicals in the water column in most streams, lakes, and rivers where POM is generally less than 10% of the DOM (*29, 30*). Three possible scenarios for the distribution of 5-Cl PCB in typical water columns are given in Table V. For DOM and POM concentrations of 5 and 0.5 mg of C/L, respectively, the bulk of the 5-Cl PCB is distributed between the DOM (40%) and freely dissolved (56%) compartments. For higher concentrations of DOM and POM (10 and 5 mg of C/L, respectively), the percent of the chemical drops to 32% in the freely dissolved compartment, increases slightly to 46% in DOM, and rises to 23% in the POM. The POM compartment becomes dominant (80% of the 5-Cl PCB) when the level reaches 50 mg of

Table V. Percent Distribution of Pentachlorobiphenyl in the Water Column of Three Typical Aquatic Systems

	Organic Matter (mg/CL)		*Percent of 5-Cl PCB in Each Compartment*		
System	*DOM*	*POM*	*Freely Dissolved*	*DOM*	*POM*
1	5.0	0.5	56	40	4
2	10.0	5.0	32	46	23
3	5.0	50.0	12	8	80

C/L, as in the third example. There is 8% in the DOM at 5 mg of C/L and 11% in the freely dissolved state.

Conclusions

The similarity between K_{pom} and K_{dom} values and their relationship with log K_{ow} and with published equations for log K_{oc} support the hypothesis that organic matter behaves by a similar mechanism, regardless of particle size and source. Indeed, the similarity between the different forms of carbon is expected because the distinction between DOM and POM is arbitrary.

Our findings indicate the importance of including the three-compartment model in predictive relationships for sorption in fate models. Although not all of the variability in the sorption can be accounted for, continued research on sorption by DOM and POM will allow refinement of the model.

Our results indicate that DOM, which represents most of the natural organic material in sea, ground, and surface waters (*30*), is the dominant factor in controlling the distribution of hydrophobic organic chemicals while in the water column.

References

1. McCall, P. J.; Swann, R. L.; Laskowski, D. A. In *Fate of Chemicals in the Environment*; Swann, R. L.; Eschenroeder, A., Eds.; ACS Symposium Series 225; American Chemical Society: Washington, DC, 1983; pp 105–123.
2. Karickhoff, S. W.; Brown, D. S.; Scott, T. A. *Water Res.* **1979,** *13,* 241–248.
3. Kenaga, E. E.; Goring, C. A. I. In *Aquatic Toxicology*; Eaton, J. G.; Parrish, P. R.; Hendricks, A. C., Eds.; American Society for Testing Materials: Philadelphia, 1980; ASTM STP 707, pp 78–115.
4. Boehm, P. D.; Quinn, J. G. *Geochim. Cosomochim. Acta* **1973,** *37,* 2459–2477.
5. Perdue, E. M.; Wolfe, N. L. *Environ. Sci. Technol.* **1982,** *16,* 1033–1035.
6. Spacie, A.; Hamelink, J. L. *Environ. Toxicol. Chem.* **1982,** *1,* 304–320.
7. Wershaw, R. L.; Burcar, P. J.; Goldberg, M. C. *Environ. Sci. Technol.* **1969,** *3,* 271–273.
8. Zepp, R. G.; Baughman, G. L.; Scholtzhauer, P. F. *Chemosphere* **1981,** *10,* 109–117.
9. Carter, C. W.; Suffet, I. H. *Environ. Sci. Technol.* **1982,** *16,* 735–740.
10. Friant, S. L.; Henry, L. L. *Chemosphere* **1985,** *14,* 1897–1907.
11. Landrum, P. F.; Nihart, S. R.; Eadie, B. J.; Gardner, W. S. *Environ. Sci. Technol.* **1984,** *18,* 187–191.

12. Caron, G.; Suffet, I. H.; Belton, T. *Chemosphere* **1985,** *14,* 735–740.
13. Friant, S. L.; Henry, L. L.; Suffet, I. H. Presented at Fifth Annual Symposium for the Society of Environmental Toxicology and Chemistry, Arlington, VA, November 4–7, 1984.
14. Moorehead, N. R.; Eadie, B. J.; Lake, B.; Landrum, P. F.; Berner, D. *Chemosphere* **1986,** *15,* 403–412.
15. Whitehouse, B. G. *Estuarine Coast. Shelf Sci.* **1985,** *20,* 393–402.
16. McCarthy, J. F.; Jimenez, B. D.; Barbee, T. *Aquat. Toxicol.* **1985,** *7,* 15–24.
17. Gschwend, P.; Wu, S. *Environ. Sci. Technol.* **1985,** *19,* 90–96.
18. Voice, T. C.; Rice, C. P.; Weber, W. J. *Environ. Sci. Technol.* **1983,** *17,* 513–518.
19. *Standard Test Method for Shake Extraction of Solid Waste with Water*; American Society for Testing Materials: Philadelphia, 1986; ASTM D 3987–85.
20. Clark, J. R.; Dickson, K. L.; Cairns, J. In *Methods and Measurements of Periphyton Communities: A Review*; Weitzel, R. L., Ed.; American Society for Testing Materials: Philadelphia, 1979; pp 116–141; ASTM STP 690.
21. Landrum, P. F.; Reinhold, M. D.; Nihart, S. R.; Eadie, B. J. *Environ. Toxicol. Chem.* **1985,** *4,* 459–467.
22. Henry, L. L. Ph.D. Thesis, Drexel University, 1987.
23. Goulden, C. E.; Comotto, R. M.; Hendrickson, J. A., Jr.; Hornig, L. L.; Johnson, K. L. *Aquatic Toxicology and Hazard Assessment*; Pearson, J. G.; Foster, R. D.; Bishop, W. E., Eds.; American Society for Testing Materials: Philadelphia, 1984; pp 139–160; Fifth Conference, ASTM STP 766.
24. Zar, J. H. *Biostatistical Analysis*; Prentice–Hall: Englewood Cliffs, 1974.
25. Carter, C. W.; Suffet, I. H. *Org. Geochem.* **1985,** *8,* 145–146.
26. Imbenotte, M.; Pommery, J.; Pommery N. *J. Int. Ozone Assoc.* **1986,** *8,* 37–48.
27. DiToro, D. M.; Mahony, J. D.; Kirchgraber, P. R.; O'Byrne, A. L.; Pasquale, L. R.; Piccirilli, D. C. *Environ. Sci. Technol.* **1986,** *20,* 55–61.
28. Mackay, D.; Powers, B. *Chemosphere* **1987,** *16,* 745–757.
29. Wetzel, R. G. *Limnology*; Saunders: Philadelphia, 1983.
30. Thurman, E. M. In *Organic Carcinogens in Drinking Water*; Ram, N. M.; Calabrese, E. J.; Christman, R. F., Eds.; Wiley, New York, 1986.

RECEIVED for review October 29, 1987. ACCEPTED for publication July 12, 1988.

12

Charge-Transfer Interaction Between Dissolved Humic Materials and Chloranil

Michael E. Melcer, Margaret S. Zalewski, and John P. Hassett

Department of Chemistry, State University of New York, College of Environmental Science and Forestry, Syracuse, NY 13210

Marion A. Brisk

City University of New York Medical School, New York, NY 10038

Using UV-difference spectroscopy, we have established the electron-donating ability of humic materials and introduced a simple, direct method for measuring binding constants. Chloranil (2,3,5,6-tetra-chloro-p-benzoquinone) was combined with Aldrich humic materials and two dissolved organic matter samples. These mixtures gave rise to shifts in the chloranil 290-nm absorption band. Spectra of such solutions versus a chloranil reference contained a negative peak at this wavelength due to the binding interaction. Binding with the Aldrich humic material continued to occur over a period of several weeks. Binding with the dissolved organic materials occurred on the order of a few hours. A charge-transfer mechanism can be postulated for the chloranil–humic system and for similar systems.

DISSOLVED HUMIC MATERIALS BIND ORGANIC CONTAMINANTS that are released into the environment (*1–8*). Bound pollutants behave differently from freely dissolved molecules, with changes in chemical processes such as hydrolysis and photolysis rates (*9, 10*). Physical parameters of bound organic compounds in the water column are also affected. These parameters include an increase in their apparent aqueous solubility (*11*), a decrease in their partitioning onto particulate matter, and a reduction in their rates of

0065–2393/89/0219–0173$06.00/0

vaporization (*8, 12*). The bioavailability of organic pollutants to aquatic organisms is also altered (*7, 13*). Because the complexation of organic contaminants may markedly change their chemical and physical pathways in the environment, an understanding of the binding interaction is crucial in order to predict the fate of chemical pollutants. Binding constants have been determined for several organic molecules by measuring the effect of binding on some physical property of the molecule (e.g., solubility enhancement, dialysis, vaporization depression, and sorption). However, the nature of the interaction has not been elucidated.

Lindquist (*14, 15*) demonstrated electron-donating and -accepting properties of humic materials by analyzing the effect of *p*-benzoquinone, a known electron acceptor, and hydroquinone, an electron donor, on the visible spectrum of humic materials. Mathus (*16*) also reported electron donation by triazenes to dissolved organic material, detected by electron spin resonance (ESR) and IR shifts. Mudambi (*10*) found that the presence of humic material altered the photochemical action spectrum of mirex, a result implying an electronic interaction between the two materials.

Humic materials contain both electron-rich and electron-deficient sites; this structure would account for their electron-donating and -accepting properties (*17*). Electron-donating structures (such as hydroquinones, ethers, alcohols, nitrogen-containing compounds, and phenolic moieties) can form electron donor–acceptor complexes (charge-transfer complexes) with suitable electron acceptors such as benzoquinones and many chlorinated compounds in aqueous solution. Our earlier results using UV-difference spectroscopy also support a charge-transfer interaction mechanism between humic materials and organic molecules (*18*). Here we will use this technique to compare the binding constants and kinetics of interaction of chloranil with humic materials from three sources.

Experimental Methods

Aqueous solutions of chloranil (2,3,5,6-tetrachloro-*p*-benzoquinone) were mixed with Aldrich humic acid and water samples from Labrador Hollow (a marshy pond in Tully, NY) and Limestone Creek (a stream in Fayetteville, NY). A humic acid stock solution was prepared by dissolving 2 g of Aldrich humic acid in 1 L of distilled water. This solution was stirred for 24 h and filtered through a glass fiber filter (Whatman, 934–AH) to remove suspended material. A solution containing 6.0×10^{-6} M chloranil with 2.7 mg of C/L of humic acid was prepared from the stock humic acid solution. Water from the two field sources was collected, filtered, diluted with distilled water, and mixed with chloranil such that the final solutions contained 3.0 mg of DOC/L and 5×10^{-6} M chloranil. All solutions were stored in the dark at room temperature.

The chloranil–humic acid mixtures were each analyzed by UV-difference spectroscopy. Difference spectra of the Aldrich humic acid mixtures were determined upon mixing and were repeated at various intervals for up to 32 days. Dissolved organic matter samples were analyzed at shorter intervals for 2 or 4 h from the time of mixing. Two sets of spectra were obtained for each solution: spectra of sample mixtures versus a chloranil reference of equal concentration and spectra of sample mixtures versus a humic material reference of equal concentration. All spectra were obtained with a spectrophotometer (Varian DMS 100) with sample and reference solutions in matched 1-cm cuvettes. The baseline was corrected with an internal baseline correction program referenced to a water blank. All spectra were scanned from 190 to 350 nm at a rate of 100 nm/min.

Analysis of Difference Spectra

Chloranil is a well-known electron acceptor in charge-transfer complexes (*19–24*). The 290-nm absorption band of free chloranil is sensitive to complexation; a red shift is typically observed with binding to an electron donor, presumably due to an increase in electron density that destabilizes the molecular orbitals involved in the transition. UV spectra of an aqueous chloranil (6.0×10^{-6} M) and Aldrich humic acid (2.7 mg of C/L) solution referenced to a 2.7 mg of C/L of humic acid solution indicate that over a period of 32 days the 290-nm band broadens and shifts to lower energy (Figure 1). The

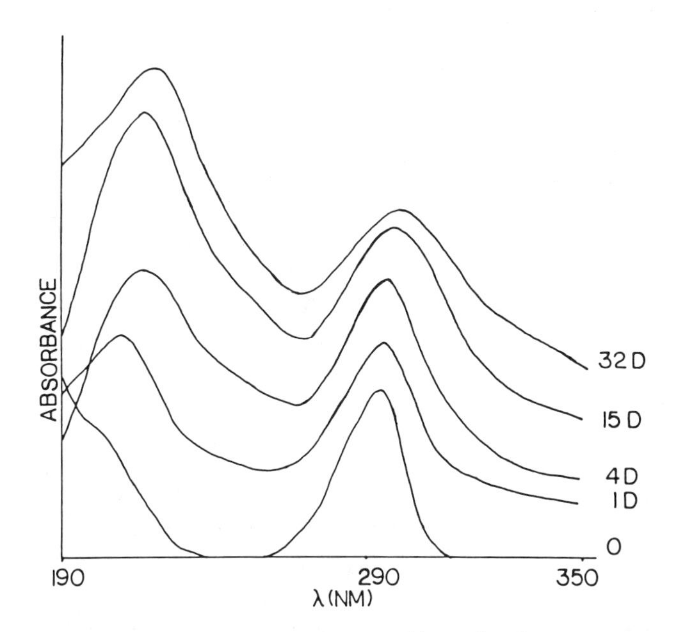

Figure 1. Difference spectra of a humic–chloranil solution measured over 32 days after mixing. Sample: Aldrich humic acid (2.7 mg of C/L) and chloranil (6.0×10^{-6} M). Reference: Aldrich humic acid (2.7 mg of C/L). (Reprinted with permission from ref. 18. Copyright 1987 Pergamon Journals Ltd.)

shift is consistent with a charge-transfer model in which an increase in the chloranil electron density results from complexation with humic molecules. The broadening and shifting of the 290-nm band indicates that the fraction of bound–unbound chloranil increases with time.

The UV-difference spectra of the same chloranil–Aldrich humic solution compared to a 6.0×10^{-6} M chloranil solution (Figure 2) shows the same trend. Because of complexation of the chloranil to humic materials, the concentration of the free dissolved chloranil in the sample is less than the concentration of chloranil in the reference; a negative peak at 290 nm results. This negative peak is therefore proportional to the concentration of the bound chloranil. Thus, the increasing negative peak intensity indicates a growing bound–unbound ($[Ch]_b/[Ch]_u$) ratio with time (Figure 2).

The negative peak intensity at 290 nm is also a function of the total chloranil concentration (*18*). As the total chloranil concentration is increased, the negative peak intensity also increases, a result indicating a rise in the

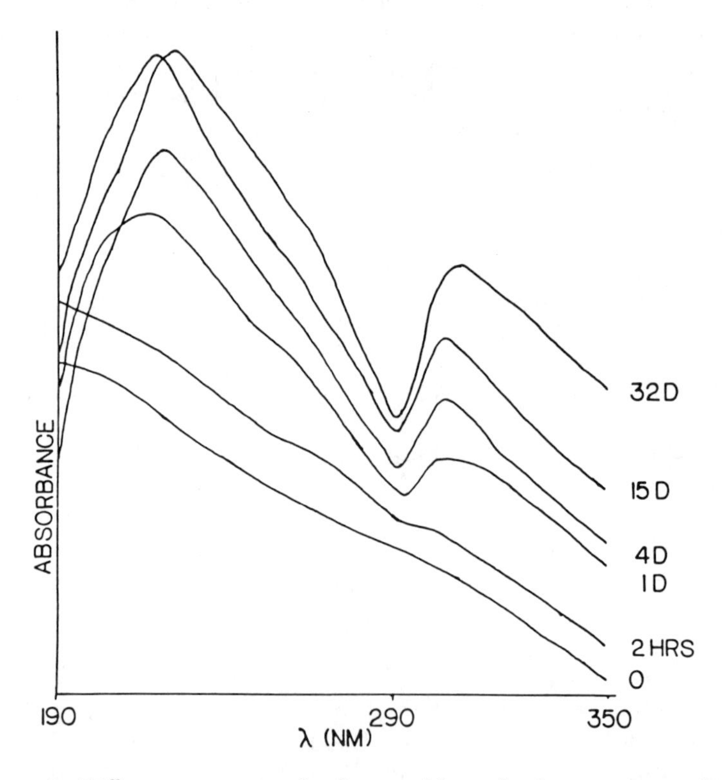

Figure 2. Difference spectra of a humic–chloranil solution measured over 32 days after mixing. Sample: Aldrich humic acid (2.7 mg of C/L) and chloranil (6.0×10^{-6} M). Reference: chloranil solution (6.0×10^{-6} M). (Reprinted with permission from ref. 18. Copyright 1987 Pergamon Journals Ltd.)

concentration of complexed chloranil. The concentration of complexed chloranil can be determined directly by measuring the negative peak absorbance and relating it to a Beer's law plot for chloranil in the concentration region of interest. The bound–unbound ratio can then be calculated by using equation 1:

$$\frac{[\mathrm{Ch}]_b}{[\mathrm{Ch}]_u} = \frac{A_b}{A_0 - A_b} \tag{1}$$

where A_0 = absorbance associated with initial concentration before any complexation and A_b = absorbance of complexed chloranil after 4 days (intensity of negative peak).

The ratio of $[\mathrm{Ch}]_b/[\mathrm{Ch}]_u$ for this system was determined to be 3.02. A unitless partition coefficient K_{doc} can be calculated from this ratio:

$$K_{doc} = \frac{[\mathrm{Ch}]_b}{[\mathrm{Ch}]_u} \cdot \frac{1}{\mathrm{DOC}} \tag{2}$$

where DOC is the concentration of dissolved organic carbon in milligrams of C per milligram of H_2O.

After 4 days, K_{doc} was 1.2×10^5 in a 2.7 mg of C/L of humic acid solution; this result indicates that 30% of the total chloranil was bound. A K_{doc} of 5×10^5 (60% bound) was calculated for the 6×10^{-6} M total chloranil–humic acid solution after 32 days. The K_{doc} for the chloranil–humic acid complex appeared to remain constant 4–7 days after mixing. However, after approximately 7 days, negative peak intensity of the 290-nm valley began to increase slowly, leveling off again after approximately 15–20 days from the time of initial mixing. This two-step increase in binding may be due to conformational changes of the humic macromolecule as sites become occupied by chloranil. These conformational changes apparently allow for an increase in availability of binding sites. Zepp et al. (25) also noted a two-step binding process with organic molecules and sediment.

Difference spectra of the Labrador Hollow (Figure 3) and Limestone Creek (Figure 4) chloranil mixtures versus a chloranil reference also show a negative peak at 290 nm. The K_{doc} for the Labrador Hollow dissolved organic matter (DOM) at 2 h and Limestone Creek DOM at 4 h were 9.7×10^4 and 1.1×10^5, respectively. A plot of ln $[\mathrm{Ch}]_{total}/[\mathrm{Ch}]_u$ versus time yields a pseudo-first-order rate constant (initial slope) for the complex formation (Figure 5). Values of this constant for the samples studied are: 0.26 h^{-1} (r = 0.99) for the Labrador Hollow sample; 0.17 h^{-1} (r = 0.99) for the Limestone Creek sample; and 0.0033 h^{-1} (r = 0.98) for the Aldrich humic acid sample. As the DOM–chloranil interactions proceed the plots show curvature that indicates that the first-order approximation is no longer valid. The first-order plot of the Aldrich humic acid–chloranil interaction

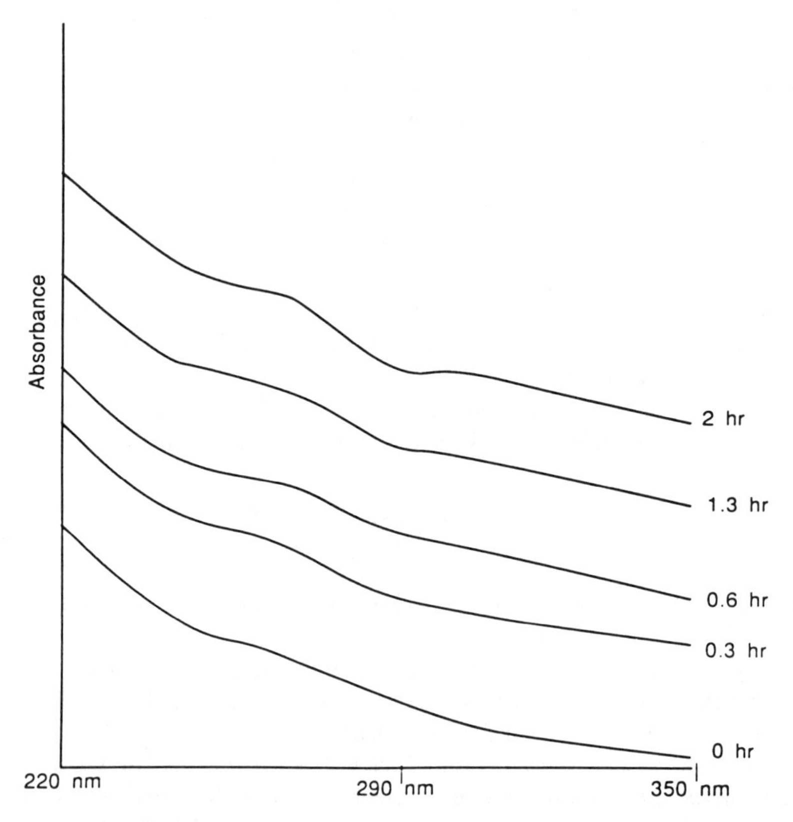

Figure 3. Difference spectra of Labrador Hollow DOC mixed with chloranil over 2 h after mixing. Sample: Labrador Hollow DOC (3.0 mg of C/L) and chloranil (5.0×10^{-6} M). Reference: chloranil (5.0×10^{-6} M).

clearly demonstrates the two-step process (Figure 6). Thus, aside from the differences observed for K_{doc}, the rate of complex formation is also dependent on the humic material source.

Humic substances from different sources are structurally different (*26*). The concept of site-specific binding interaction may explain the different affinities of humic materials from different sources for DDT, as noted by Carter and Suffet (*1*). It would also explain the differences we found in K_{doc} and kinetics of interaction. Both the availability and nature of the binding sites would determine the extent and rate of interaction between a charge acceptor and humic acids.

Conclusions

Difference spectroscopy of humic acids mixed with chloranil in aqueous solutions suggest the presence of binding sites for these organic molecules

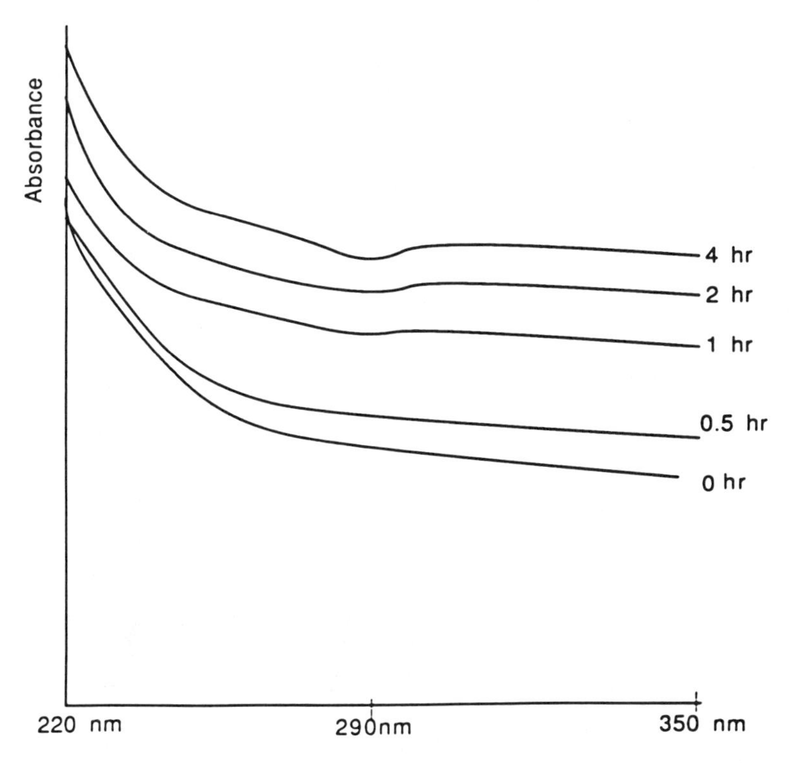

Figure 4. Difference spectra of Limestone Creek DOC mixed with chloranil over 4 h after mixing. Sample: Limestone Creek DOC (3.0 mg of C/L) and chloranil (5.0×10^{-6} M). Reference: chloranil (5.0×10^{-6} M).

on the dissolved humic materials. A charge-transfer or electron donor–acceptor mechanism appears to be responsible for the interaction between sites on the humic acids and the organic molecules. This model is supported by:

- previous studies of chloranil interaction with electron donors, with similar shifts of the 290-nm peak;
- a constant K_{doc} for different concentrations of hydrophobe, which is consistent with complex formation;
- known presence of electron-donating groups on the humic acid molecule; these groups can act as binding sites for electron acceptors;
- slow formation of the complex (over a period of hours to days), which is characteristic of many charge-transfer complexes.

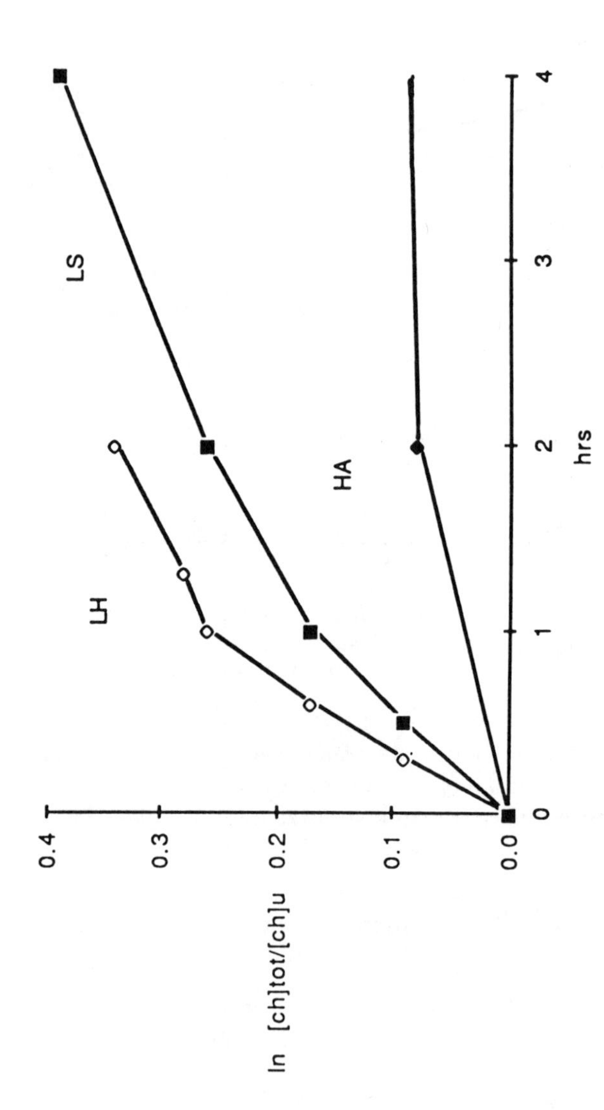

Figure 5. Plot of the natural log of (the total chloranil concentration divided by the unbound chloranil concentration) over 4 h. (LH = Labrador Hollow; LS = Limestone Creek; HA = Aldrich humic acid.)

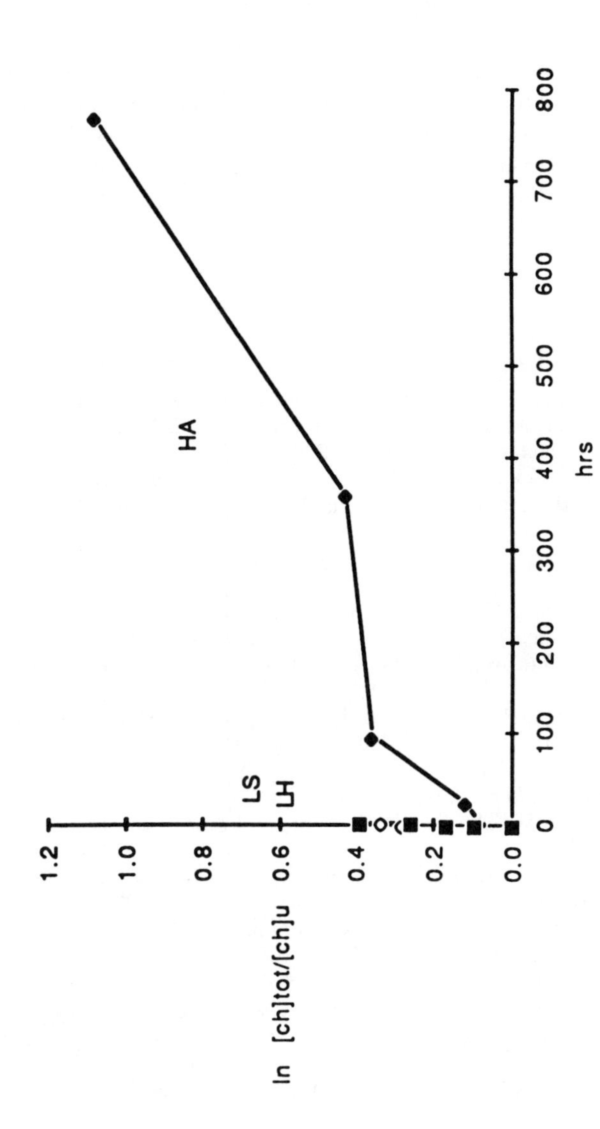

Figure 6. Plot of the natural log of (the total chloranil concentration divided by the unbound chloranil concentration) over 800 h. (LH = Labrador Hollow; LS = Limestone Creek; HA = Aldrich humic acid.)

It is recognized that humic materials bind organic molecules and affect their physical and chemical properties. This work suggests a site-specific electron donor–acceptor mechanism for the chloranil–humic acid complex. Other molecules of environmental importance (i.e., DDT, dioxins, and PCBs) possess electron-accepting properties. In fact, a charge-transfer mechanism has been postulated to be responsible for the photolysis of DDT (27) in the presence of humic materials. The importance of this type of interaction in other systems is unknown, but it is presently under investigation in our laboratories. If in fact the charge-transfer mechanism is significant, then knowledge of the electron-donating ability of humic material may permit a prediction of the K_{doc} for the binding of chlorinated organic compounds.

References

1. Carter, C. W., Suffet, I. H. In *Fate of Chemicals in the Environment*; Swann, R. L.; Eschenroeder, A., Eds.; ACS Symposium Series 225; American Chemical Society: Washington, DC, 1983; pp 201–220.
2. Hassett, J. P.; Anderson, M. A. *Environ. Sci. Technol.* **1979,** *13*, 1526.
3. Wershaw, R. L.; Burcar, P. J.; Goldberg, M. C. *Environ. Sci. Technol.* **1969,** *3*, 271.
4. Poirrier, M. A.; Bordelon, B. R.; Laseter, S. L. *Environ. Sci. Technol.* **1972,** *6*, 1033.
5. Hassett, H.; Anderson, M. A. *Water Res.* **1982,** *16*, 681.
6. Perdue, E. M.; Wolfe, N. L. *Environ. Sci. Technol.* **1982,** *16*, 847.
7. Leversee, G. J.; Landrum, P. F.; Giesy, J. P.; Fannin, T. *Can. J. Fish Aquat. Sci.* **1983,** *40*, 63.
8. Yin, C. Q.; Hassett, J. P. Presented at the 186th National Meeting of the American Chemical Society, Washington, DC, August–September 1983; paper ENVR 135.
9. Zepp, R. G. *Chemosphere* **1981,** *10*, 109.
10. Mudambi, A. R. Ph.D. Thesis, SUNY College of Environmental Science and Forestry, 1987.
11. Matsuda, K.; Schnitzer, M. *Bull. Environ. Contam. Toxicol.* **1971,** *6*, 200.
12. Hassett, J. P.; Milicic, E. *Environ. Sci. Technol.* **1985,** *19*, 638.
13. Boehm, P. D.; Quinn, J. G. *Estuarine Coastal Mar. Sci.* **1976,** *4*, 93.
14. Lindquist, I. *Swed. J. Agric. Res.* **1983,** *13*, 201.
15. Lindquist, I. *Swed. J. Agric. Res.* **1982,** *12*, 105.
16. Mathus, S. P.; Marley, H. V. *Bull. Environ. Contam. Toxicol.* **1978,** *20*, 268.
17. Thurman, E. M. *Organic Geochemistry of Natural Waters*; Martins, Nijhoff, Dr. W. Junk: Boston, 1986.
18. Melcer M. E.; Zalewski M. S.; Brisk M. A.; Hassett, J. P. *Chemosphere* **1987,** *16*, 1115.
19. Mulliken, R. S. *J. Am. Chem. Soc.* **1952,** *56*, 801.
20. Slifkin, M. A.; Walmsley, R. H. *Experientia* **1969,** *25*, 930.
21. Douglass, D. D. *J. Chem. Phys.* **1960,** *32*, 1882.
22. Davies, K. M.; Snart, R. S. *Nature (London)* **1960,** *188*, 724.

23. Slifkin, M. A. *Spectrochem. Acta* **1964,** *20*, 1543.
24. Fulton, A.; Lyons, L. E. *Aust. J. Chem.* **1968,** *21*, 873.
25. Zepp, R. G.; Schlotzhauer, P. F. *Chemosphere* **1981,** *10*, 453.
26. Melcer, M. E.; Hassett, J. P. *Toxicol. Environ. Chem.* **1986,** *11*, 147.
27. Miller, L. L.; Narang, R. S. *Science (Washington, DC)* **1970,** *169*, 368.

RECEIVED for review July 24, 1987. ACCEPTED for publication February 19, 1988.

13

Sorption of Organochlorines by Lake Sediment Porewater Colloids

Paul D. Capel[1] **and Steven J. Eisenreich**

Environmental Engineering Sciences, Department of Civil and Mineral Engineering, University of Minnesota, Minneapolis, MN 55455

Laboratory and field experiments have shown that distribution coefficients (K_D) *for hydrophobic organic compounds (HOC) in sediment–porewater environments are usually in the range of* 10^3 *L/kg. These values are* 10^2–10^3 *times lower than those normally measured in surface waters or predicted by structure–activity relationships. The reduced KD values are attributed to the sorptive abilities of the organic colloids in the porewater. The sorption of HOC in sediments–porewaters was indirectly studied by examining their diffusion in sediments. The processes of sorption and diffusion are closely related for HOC in sediments. By using an experimental sediment purge setup, effective sediment diffusion coefficients* (D_{eff}) *of* 10^{-8}–10^{-12} cm^2/s *were obtained for a series of chlorinated pesticides and polychlorinated biphenyl (PCB) congeners differing in their hydrophobicities. Molecular diffusion of dissolved HOC dominates fluxes of compounds with low-to-moderate octanol–water partition coefficients* (K_{ow}), *whereas colloid-associated HOC may dominate total diffusional fluxes out of sediments for chemicals with a high* K_{ow}.

INTACT LAKE SEDIMENTS PRESENT A UNIQUE ENVIRONMENT to study sorption of hydrophobic organic chemicals (HOC). The sediments, which remain essentially undisturbed for decades once they pass out of the surficial mixed zone, provide an environment conducive to the establishment of sorptive equilibrium. Lake sediments consist of minerals covered by natural organic

[1]**Current address: Swiss Federal Institute for Water Resources and Water Pollution Control (EAWAG), CH–8600 Dübendorf, Switzerland**

0065–2393/89/0219–0185$06.75/0

matter (typically 1–5% organic carbon (OC) by weight) and surrounded by porewater containing relatively high concentrations of organic matter (typically 10–100 mg of OC/L) (*1, 2*). These two pools of natural organic matter are in sorptive competition for the HOC. The sorption of HOC in sediments is identical to their sorption in the water column and laboratory, except that the solid and colloidal sorbent concentrations are orders of magnitude higher.

HOC in the sediment are distributed between the porewater and sediment compartments. The relative distribution between the two phases is a function of the sediment and porewater OC contents and the hydrophobicity of the sorbing organic compound. Elevated concentrations of HOC in the porewater and measured distribution coefficients (K_D), 10^2–10^3 times less than those measured in the water column or predicted by structure–activity relationships, suggest that HOC may be bound to porewater organic colloids (*1, 2*). Here we present the results of a study in which the effective diffusion coefficients (D_{eff}) of HOC in sediments are used to infer the competitive distribution between sediment OC, porewater OC (colloids), and the dissolved phase.

Because of the importance of porewater colloids in the sorption of HOC, care must be taken in defining sorption expressions. An overall theoretical equilibrium expression can be written

$$C_p \underset{}{\overset{p}{\rightleftharpoons}} C_d \underset{}{\overset{K_c}{\rightleftharpoons}} C_c \qquad (1)$$

where C_p, C_d, and C_c are the particulate, dissolved, and colloidal concentrations of an HOC, respectively. Simple equilibrium expressions are

$$K_p = \frac{\mathrm{P}/M_p}{\mathrm{D}/M_w} \quad \text{and} \quad K_c = \frac{\mathrm{COLL}/M_c}{\mathrm{D}/M_w} \qquad (2)$$

where P, D, and COLL are the masses of an HOC in the particulate, dissolved, and colloidal phases; and M_p, M_w, and M_c are the masses of the particles, water, and porewater colloids, respectively. Unfortunately, these idealized fractions (P, D, and COLL) are not differentiated by presently available separation techniques for natural sediments. Separation techniques are limited to an operational distribution between the particulate phases and the sum of the dissolved and colloidal fractions. A distribution coefficient (K_D) obtained by filtration or centrifugation can be defined on this basis as

$$K_D = \frac{\mathrm{P}/M_p}{(\mathrm{D} + \mathrm{COLL})/M_w} \qquad (3)$$

which is related to the equilibrium expression

$$K_D = \frac{K_p}{1 + (K_c\ \mu[\mathrm{PWOC}]\ \mu\ 10^{-6})} \qquad (4)$$

where [PWOC] is the concentration of the porewater organic colloids (mg of OC/L) and K_D, K_p, and K_c have units of liters per kilogram. Throughout this discussion, K_D will be used as the field-measured distribution coefficient, and K_p and K_c will be used as the theoretical equilibrium coefficients.

Diffusion Coefficients

Coefficients (K_D) that describe the distributions of HOC between particulate and aqueous fractions have been determined for several anthropogenic compounds in the water column (*3–5*) and in the laboratory (*6–10*). These studies have elucidated the important parameters that control the distribution process. For HOC, the best predictor of the magnitude of sorption is the organic carbon fraction of the solid (foc) (*6, 9*). Other important factors include the hydrophobicity of the chemical (as quantified by solubility or the octanol–water partition coefficient, K_{ow}), the dissolved–colloidal organic carbon content of the aqueous phase (*10, 11*), the kinetics of sorption–desorption (*12–14*), and the particle concentration (*1, 4, 15–18*).

Because intact sediments have the highest particle concentration of all natural aquatic environments (10^5–10^6 mg of solids/L), the effect of the particle concentration on sorption in sediments must be addressed. From thermodynamic principles, K_p should be independent of solid concentrations at solute concentration far below solubility. However, in both laboratory and field systems, K_D (the closest quantity to K_p that can be measured) is observed to decrease with increasing solids concentration (*15–18*). The reasons proposed for this observation include experimental artifacts, nonlinear sorption isotherms at low concentrations, slow desorption kinetics, and solute complex formation with organic colloids. Presently, there is no comprehensive explanation for this phenomenon, but the effect of the HOC–colloidal complexes will undoubtedly be important, especially in sediments.

Gschwend and Wu (*10*), Means and co-workers (*19–21*), and Caron et al. (*22*) reported that colloidal organic matter binds HOC in lacustrine and marine environments. Baker et al. (*18*) provided field data to support the hypothesis that a sizable fraction of the "dissolved" polychlorinated biphenyls (PCBs) in Lake Superior is bound to colloids and particles <0.6 µm in diameter. This binding is due to the high specific surface area and organic carbon content of the colloids. The measured sorption coefficients of HOC to organic colloids are of the same order of magnitude as particulate organic-carbon-based sorption coefficients (K_{oc}, $K_{oc} = K_D \times$ foc, where foc is fraction of organic carbon), a condition suggesting that both are governed by the same nonspecific partitioning.

HOC in Sediment Porewaters. There have been only a few investigations of HOC in sediment–porewater systems (Table I). Early work by Duinker and Hildebrand (*23*) in a North Sea estuary demonstrated measurable quantities of selected chlorinated hydrocarbons in porewater. Re-

Table I. Hydrophobic Organic Compounds in Lacustrine and Marine Porewaters (ng/L)

Organic Chemical	*North Sea*[a]	*Great Lakes*[b] *Erie*	*Huron*	*Superior*
Total PCB	30 (10–40)	35–72	26–307	12–26
γ-HCH	2 (2–30)	3–4	0.6–4	0.5–2
HCB	7 (4–20)	0.2–0.8	0.5–1	0.02–1.2
p,p'-DDE[c]		2–7	4–72	0.4–13

NOTE: The total PCB in New Bedford Harbor, MA, was 8,000–20,000 ng/L (2), and that for Lake Michigan was 60–340 ng/L (*26*).
[a]Data are from ref. 23.
[b]Data are from ref. 28.
[c]DDE is 2,2-bis(*p*-chlorophenyl)-1,1-dichloroethylene.

cently, Brownawell and Farrington (*2, 24*) measured both sediment solid and porewater PCB at a site in New Bedford Harbor, Massachusetts. They examined the role of organic carbon (particulate and dissolved–colloidal) on the diagenesis of these compounds. Work on freshwater systems has been restricted to investigations in Lake Michigan (PCB and polyaromatic hydrocarbons; *25–26*).

Concentrations of HOC in the porewater are higher than in the overlying water, and the distribution coefficients (K_D) are lower than those observed in the water column (*1, 4, 18, 27*) and laboratory studies (*6, 9*). The elevated concentrations of HOC (including PCB in the sediment porewaters of Lakes Erie, Huron, and Superior) agree with these findings (*28*). This agreement suggests that elevated porewater concentrations of HOC and correspondingly low K_Ds are universal phenomena.

The observations of high concentration of HOC in the porewater, and thus low K_Ds, can be attributed to direct competition between the pools of sediment solid and dissolved–colloidal organic carbon. In sediments a high concentration of porewater organic carbon effectively competes with the particulate organic carbon for the HOC (*28, 29*). The ability of the porewater organic colloids to bind HOC will affect their diagenesis, including mobility, in sediments.

Using the Diffusion of HOCs in Sediments To Study Sorption. Available methodologies to study sorption of HOC in intact sediments are limited if the integrity of the natural sediment conditions is to be preserved. The approach used by Karickhoff and Morris (*30*) and in this study is to measure the effective diffusion coefficient of an HOC in intact sediments and then to calculate the sorption coefficients. The process of sorption is independent of diffusion, but it influences the rate of diffusion significantly. Those molecules associated with the solid phase in the sediments are not free to diffuse through the porewater. The overall effect of sorption will be a retardation of the rate of effective diffusion.

Background on Diffusion in Sediments. Molecular diffusion is the process by which molecules are transported by random movement in a stagnant fluid. In dilute solution, each molecule behaves independently, constantly undergoing random collisions with solvent molecules. The net movement is from high to low concentration. Molecular diffusion in an isotropic medium can be expressed in flux terms by Fick's first law:

$$\text{flux} = -D_o \frac{dC}{dx} \tag{5}$$

where D_o is the diffusion coefficient, a proportionality constant relating the flux to the concentration gradient over a distance normal to that flux; C is concentration; and x is distance. Molecular diffusion in a liquid is a function of temperature, solute size and shape, solvent viscosity, and solvent–solute interactions (*31*).

Diffusion in porous solids (sediments) is slower than in bulk solution because of geometric effects (tortuosity and porosity) and surface interactions. Diffusion can occur in the interstitial water, on the solid "surface", and within the solid. For macroscopic movements of HOC in sediments, diffusion in interstitial water is the only important process (*32*).

The presence of interstitial and intrastitial spaces in sediments hinders diffusive fluxes relative to bulk water by forcing the chemical to deviate from a straight-line diffusional path. Geometric effects are specific to the sediment and are influenced by particle size, solids packing, and pore diameter. Tortuosity, the ratio of the mean free path length to the straight-line distance, has been determined through measurements of electrical resistivity (*33*). Porosity, volumetric water content, and bulk density also contribute to the geometric hindrance on diffusion in sediments. Various investigators have tried to model the effect of the solids on diffusion in various porous solids (Table II). Many of these relationships do not agree under the conditions normally existing in unconsolidated sediments (porosity: 0.6–0.95). Four different predictions from the models in Table II are illustrated in Figure 1. These vary by up to a factor of 2 in the predicted effects in this range of sediment porosities.

The relationship between the independent processes of sorption, molecular diffusion, and overall flux can be described mathematically. Several models (such as Freundlich, Langmuir, and Brunauer–Emmett–Teller (BET) adsorption) describe the distribution of a chemical between the solid and liquid phases (*34*). Crank (*35*), Duursma and Hoede (*36*), and Corwin and Farmer (*37*) have solved a variety of diffusion equations with these sorption models incorporated into the boundary conditions. At low environmental concentrations of HOC, the solid–water

Table II. Theoretical and Empirical Equations for Modeling the Diffusion of a Chemical Species in a Water-Saturated Porous Media

Equation	*Solid*	*Species*	*Porosity Range*	*Ref.*
	No interactions with solids			
$D_{geo} = D_w\phi/\tau^2$	theoretical		0–1	48
$D_{geo} = D_w\phi^{4/3}$	theoretical		—	59
$D_{geo} = D_w\theta/\tau$	theoretical		—	60
$D_{geo} = D_w\phi^2$	sediment	ions	0.2–0.7	49
$D_{geo} = D_w/\tau^2$	sediment	ions	—	32
$D_{geo} = D_w f_L$	soils	Cl^-	0.2–0.4	51
$D_{geo} = D_w\theta\tau^{1/2}$	glass beads	Rb^+	0.2–0.4	42
$D_{geo} = D_w a\tau^2$	soils	$^3H-H_2O$	0.8	46
	Including interactions with solids			
$D_{eff} = D_w\phi/(K + 1)\tau$	theoretical		—	61
$D_{eff} = D_w\tau/K\rho/\phi + 1$	sediments	radionuclides	—	62
$D_{eff} = D_w\gamma\theta\tau^2$	soils	pesticides	—	63
$D_{eff} = D_w B\ (1 - \gamma)\tau^2$	soils (theoretical)	pesticides	—	64
$D_{eff} = D_w\phi\tau/bB + \phi$	soils	pesticides	0.1–0.4	65
$D_{eff} = D_w\phi\tau/KB + \phi$	soils	pesticides	0.2–0.4	66

NOTE: D_{geo}, diffusion coefficient in porous solids; includes the geometrical effects due to the solid matrix; D_w, diffusion coefficient in water; D_{eff}, effective diffusion coefficient in porous solids; includes the effects of geometry of the solids and interactions with the solids; ϕ, porosity; τ, tortuosity; θ, volumetric water content; f_L, impedance factor (2 for Cl^-); K, distribution coefficient between solid and water; ρ, sediment mass per unit volume of sediment; γ, interaction between solid and diffusing species; B, bulk density; and a, relative mobility or fluidity of water.

distribution is adequately expressed by the linear portion of a sorption isotherm (*6, 34*):

$$C_p = KC_{aq} \tag{6}$$

where C_p is the HOC concentration on the solid and C_{aq} is the aqueous HOC concentration. Kinetics must also be considered. In their solutions to diffusion equations, Crank (*35*) and Duursma and Hoede (*36*) considered fast, slow, and irreversible sorption kinetics. The current understanding of HOC sorption in sediments is relatively fast sorption and slow desorption kinetics (minutes to years) (*38*). Overall, this is fast on the time scale of macroscopic diffusion in sediments. Crank (*38*) suggests that the modified diffusion equation for these assumptions (linear sorption isotherms, constant diffusion coefficient, and fast kinetics) should be

$$\frac{dC}{dt} = \frac{D_o}{K + 1}\frac{d^2C}{dx^2} \tag{7}$$

The literature contains numerous expressions that combine the geometric and sorption effects impeding diffusion in porous solids and define

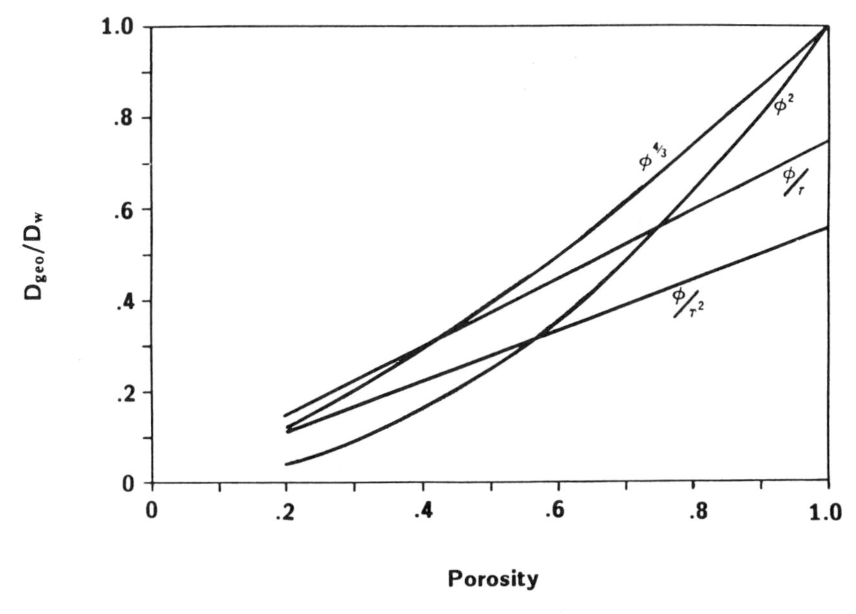

Figure 1. Predicted relationships between D_{geo}/D_w *and porosity* (ϕ) *from Table II.* D_{geo}/D_w *is the ratio of the diffusion coefficient in a porous media that includes the geometrical effects to the diffusion coefficient in water. A ratio of 1 means that there is no effect of the solid on* D_w. *Calculated with a constant tortuosity* (γ) *of 1.8.*

an effective diffusion coefficient, D_{eff} (Table II). Most agree on the manner in which sorption is incorporated in the expression ($K + 1$ in the denominator). In a strict sense, D_{eff} is not a true diffusion coefficient because it incorporates sorption. However, it is useful in describing the relationship between diffusion and sorption.

A few investigators, in both laboratory and field experiments, have measured D_{eff} in sediments for anions, cations, silica, water, and various chemicals. A representative list is presented in Table III, together with the molecular diffusion coefficients of the species in water (D_w). Examination of the effective diffusion of HOC in sediments can help elucidate their sorption behavior. This examination will supplement the data obtained from field measurements of the same HOC in sediment–porewaters.

Estimation of Molecular Diffusion Coefficients in Water. The diffusion coefficients of most HOC have not been measured in water. Semi-empirical equations have been established to estimate these coefficients in dilute liquid solutions. The equation proposed by Wilke and Chang (39) is

$$D_o = (7.4 \times 10^{-8})\,(XM_B)^{0.5}\left(\frac{T}{\mu_B V_a^{0.6}}\right) \tag{8}$$

Table III. Diffusion Coefficients of Inorganic and Organic Chemical Species in Water and Sediment

Species	D_w (*× 10^{-6} cm²/s*)	D_{eff} (*cm²/s*)	*Organic Carbon (%)*	*Temperature (°C)*	*Porosity*	*Ref.*
Water	13.0[a]	3×10^{-6}	2.5	4	—	67
Na^+	13.3	5.8×10^{-6}	—	24	0.7	68
Ca^{2+}	7.9	15.5×10^{-6}				
SO_4^{2-}	10.7	5.3×10^{-6}				
Cl^-	20.3	10.2×10^{-6}				
Hexachlorobenzene		7×10^{-9}	0.8	15–20	0.8	30
Pentachlorobenzene		8×10^{-9}				
Trifluralin		5×10^{-9}				
2,3′,5-PCB		1.4×10^{-14}	2.9	16	—	69
2,2′,4,5-PCB		1.5×10^{-13}				
2,2′,4,5,5′-PCB		9.8×10^{-12}				
2,2′,3,4,5-PCB		2.7×10^{-12}				
Aroclor 1254		1.2×10^{-12}				
2,2′,4,4′,5,5′-PCB		4.6×10^{-11}	2.8	—	0.75	70
Lindane (saturated soil)		1.2×10^{-7}	—	30	—	64

[a]Data are from ref. 43; at 4 °C.

where T is the temperature (K), X is the solvent "association" parameter (2.6 for water), μ_B is the solvent viscosity (cP), M_B is the molecular weight of the solvent, and V_a is the molar volume of the solute. All of the HOC's diffusion coefficients in water (D_w) have been calculated at 23 °C.

Quantification of the Geometrical Effects on Diffusion in Sediments. The effect on diffusion of the geometry of the inter- and intrastitial spaces of the particles has been evaluated by the half-cell technique (*40–42*). $^3H-H_2O$ (1 mCi/mL, New England Nuclear) was used for this determination. The value of D_w in bulk solution is well known. Wang et al. (*43*) showed that the presence of a tritium atom in the water molecule has little effect on its diffusion coefficient (2.44×10^{-5} cm^2/s at 25 °C). A variety of clay-to-sandy sediments, encompassing the range of porosities naturally found in lakes (0.7–0.96), were used in this study. All tests were conducted at room temperature (23 ± 2 °C).

Experimental Methods

The sediments were collected from Lakes Superior and Erie (with a box corer) and from Little Rock Lake and Island Lake, two smaller lakes in northern Wisconsin (with an Eckmann dredge). The sediments were used as originally obtained, centrifuged, or mixed with water, depending on the desired porosity. By these manipulations, the same sediment could be tested at different porosities. Correlations were made between the observed diffusion coefficients and porosity.

The experimental setup consisted of two glass half-cells (cylinders: 2.2-cm i.d. by 1.6-cm length) closed at one end and ground smooth on the open end. The sediment was made up to the desired porosity and homogenized by stirring. One half-cell was filled with the sediment and the end smoothed over so that the sediment was flush with the glass surface. The unused sediment was spiked with about 3 µL of the $^3H-H_2O$ and stirred vigorously to distribute the labeled water. A portion was taken for dry weight analysis (105 °C for >24 h). Porosity was calculated on the basis of dry weight analysis (*44*).

$$\text{porosity} = \frac{\text{mass}_{\text{water lost}}}{(\text{mass}_{\text{dry sed}}/2.65) + \text{mass}_{\text{water lost}}} \qquad (9)$$

where 2.65 is the density ($g\ cm^{-3}$) of the dry sediment. The other half-cell was filled with the labeled sediment and the surface was smoothed. The two half-cells were joined, sealed with tape, and stored with the labeled side down. The presence of labeled water in the upper half-cell can only be due to diffusive flux. The cell was equilibrated 3–4 h (optimum time for the diffusion of H_2O in these cells) and then separated. The sediment was removed and placed in a 50-mL centrifuge tube with 9 mL of water. This mixture was stirred on a vortex mixer for 30 s and centrifuged at 10,000 × *g* for 5 min. The water was decanted and collected in a 50-mL volumetric flask. Another 9 mL of water was added, and the sediment resuspended. The procedure, repeated four times, removed >99.5% of the labeled water. The total volume was made up to 50 mL. Three 1-mL aliquots were each combined with 2.5 mL of scintillation cocktail (Beckman MP) and counted to an accuracy of ±1% of the mean at the 95% confidence level (Beckman LS 1801). Corrections for quench were performed, based on internal Compton electron-quenching measurements from the instrument (*45*).

The diffusion coefficient incorporating the geometric effects (D_{geo}) was calculated with the equation suggested by Schaik et al. (*46*):

$$D_{geo} = \left(\frac{\pi}{t}\right)\left(\frac{Q}{Q_o}\right)^2 L^2 \tag{10}$$

where t is time (s), Q is the activity of the $^3H-H_2O$ that has diffused into the unspiked half-cell, Q_o is the total activity present in the cell, and L is the length (cm) of one half-cell. This equation, derived for the boundary conditions of the half-cell experimental technique, is an approximate solution to the Fick's second law for these boundary conditions. If Q/Q_o is <0.3, then the error in this approximation is <1%. In all of these experiments, this ratio was about 0.2.

Half-Cell Method. The half-cell technique was also used to quantify the D_{eff} of a PCB congener in various sediments. A ^{14}C-labeled PCB congener (2-PCB, IUPAC number 1, 11.2 mCi mmol^{-1}, Pathfinder Laboratories, Inc.) was spiked into eight different sediments and allowed to stand for >1 month. The half-cells were filled, joined, and stored in a water-saturated atmosphere for 30–120 days. After separation, each side was extracted by sonication (Brinkmann Instruments, Polytron) once with CH_3CN and twice with toluene. The water was removed from the combined solvent extracts and the final volume made up to 50 mL. To each of three 10-mL aliquots, 3 mL of Beckman MP scintillation cocktail was added. The samples were counted to an accuracy of <±2% of the mean at the 95% confidence level (Beckman LS 1801).

Purge Method. This technique closely follows the method used by Karickhoff and Morris (*30*). The wet sediment (Duluth Harbor, Lake Superior; 5% sand, 60% silt, 35% clay, 3.3% organic carbon) was spiked with mixtures of chlorinated hydrocarbons (unlabeled PCBs, β- and γ-hexachlorocyclohexane and hexachlorobenzene, or ^{14}C-labeled 2-PCB, Table IV) and allowed to equilibrate for 2 weeks with occasional stirring. The spiked solids were slowly introduced into 1-L small-neck Pyrex cells, avoiding contamination of the side of the cells (Figure 2). The depth of the solids ranged from 1.5 to 2.1 cm in the various jars. The total of nine jars included duplicates and blanks.

Ultrapure water (Milli-Q, Millipore) was introduced slowly to minimize the disturbance of the surficial solids. Once the water was added, each jar was fitted with a two-hole rubber stopper. One hole had a Teflon tube extending down into the water. The other had a glass rod open above the water surface and attached to a column of styrene–divinylbenzene (XAD–2) resin (1.1-cm i.d. by 9-cm length, Rohm and Haas), which served as a sorbent trap for the chlorinated hydrocarbons. Clean wet air (passed through activated carbon, XAD–2, and water) was bubbled into the jar at a rate sufficient to strip the chlorinated hydrocarbons from the water, but not to disturb the surficial sediments (110–170 mL/min). Resin columns were changed every 2 weeks for 8 weeks and then once a month for 2 months. The chlorinated hydrocarbons were removed from the resin by successive batch extractions (once with acetone and six times with hexane). This procedure removed >98% of the compounds from the XAD–2 resin.

The seven extracts were combined and concentrated to about 1 mL under a gentle stream of purified N_2 and analyzed by capillary gas chromatography with ^{63}Ni–electron-capture detection (HP 5840A). A 25-m Hewlett Packard cross-linked

Table IV. Effective Diffusion Coefficients (D_{eff}, cm^2/s) and Distribution Coefficients (K_D, L/kg) from the Sediment Purge Studies

				Purge Studies					
				Duplicates					
Compound	*MW*	*log* K_{ow}[a]	*log* D_w[b]	*log* D_{eff}	*log* D_{eff}	*Average log* D_{eff}	*Estimated Error log* D_{eff}	*Calculated log* $K_{D\text{-}D_{eff}}$	*Measured log* K_D
β-Hexachlorocyclohexane	291	3.80	−5.21	−9.68	−9.98	−9.83	1.48	4.27	3.01
γ-Hexachlorocyclohexane	291	3.72	−5.21	−9.62	−9.51	−9.57	0.88	3.98	2.00
Hexachlorobenzene	249	5.47	−5.18	−10.19	−9.89	−10.04	1.57	4.51	3.25
2-PCB	223	4.60	−5.18	−10.39		−10.39	0.10	4.89	2.55
2,2′,5-PCB	258	5.48	−5.22	−10.26	−9.96	−10.11	0.72	4.60	3.08
2,2′,4-PCB	258	5.76	−5.22	−10.17	−10.83	−10.50	0.85	5.07	3.04
2,2′,3-PCB	258	5.31	−5.22	−11.29	−11.28	−11.29	0.80	5.75	3.10
2,4,4′-PCB + 2,4′,5-PCB	258	5.69	−5.22	−10.84	−11.10	−10.97	1.46	5.45	3.02
2′,3,4-PCB	258	5.57	−5.22	−11.74	−11.49	−11.62	0.60	6.09	3.30
3,4,4′-PCB + 2,2′,3,4′-PCB	292	4.94	−5.24	−10.58	−10.41	−10.50	0.31	4.94	3.43

[a] Data are from ref. 71.
[b] Estimated from the equation of Wilke and Chang (39).

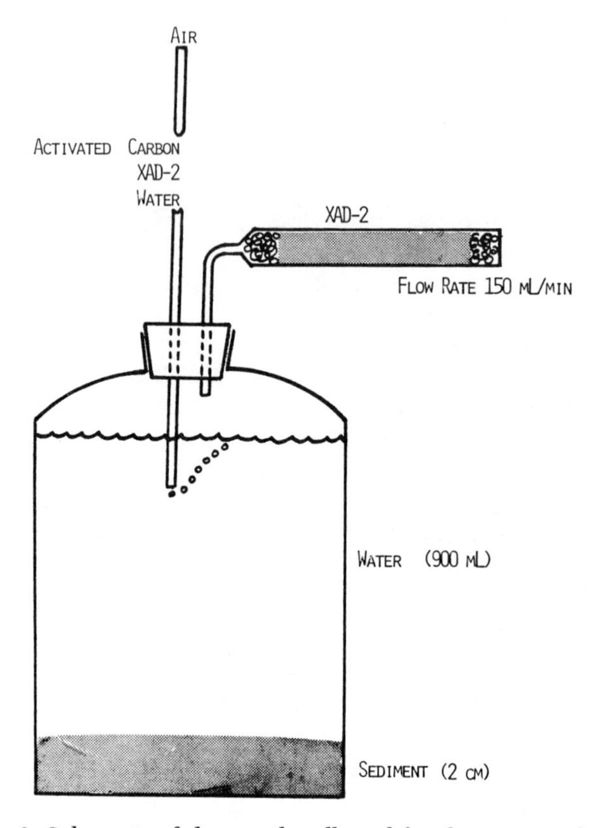

Figure 2. Schematic of the sample cell used for the purge technique.

5% phenyl–methylsilicone capillary column was used with nitrogen as the carrier and makeup gas. Oven conditions consisted of a 20-min hold at 100 °C and a 10 °C/min ramp for 5 min, followed by a 1.3 °C/min ramp of 65 min. The temperature was then raised to 275 °C and held there to the end of a 120-min run. The ^{14}C-2-PCB was extracted and quantified as described previously.

The conditions created by this experimental setup yielded a maximum diffusional flux because the water concentration was maintained near zero by the purging. Karickhoff and Morris (*30*) present the equation to calculate the diffusion coefficient with this set of boundary conditions (i.e., an initially uniform concentration for all distances below the interface; at time ≤ 0, the concentration above the interface is equal to zero). Solving for the mass lost from the sediments, the expression is

$$\text{mass}_{\text{HOC lost}} = \left(\frac{\text{mass}_{\text{HOC in sediment}}}{\text{depth of sediment}}\right)\left(\frac{D_{\text{eff}}}{\text{time}}\right)^{0.5} \tag{11}$$

which is solved for D_{eff} by iteration. The estimate of error in D_{eff} is on the order of $<\pm 1$ log unit.

Results

Diffusion of HOC in Water. The diffusion coefficients in water of the test compounds have been calculated with the equation suggested by Wilke and Chang (*39*). This equation is able to predict D_w within a mean deviation of 10–15% (*47*). The predicted coefficients for the chlorinated hydrocarbons included in this study are all similar (6.0–6.6 $\times$ 10^{-6} cm^2/s, Table III). The values from this calculation are in agreement with values predicted for the same compounds by the method suggested by Othmer and Thaker (*48*). The error in the estimation of D_w will be unimportant in the calculations using the measured D_{eff}.

Effect of Geometry on the Diffusion of HOC in Sediments. The diffusion coefficients of various chemical species have been measured in different types of porous media. A survey of the literature has resulted in a large data set for the diffusion of noninteracting species in porous media. These species include chloride in sediments (*40*, *49*), clay (*46*), and soils (*50*, *51*); bicarbonate and sulfate in sediments (*52*, *53*); and ^3H–H_2O in clays (*46*, *54*) and soils (*55*). These data have been plotted in Figure 3 (open squares) as D_{geo}/D_w versus porosity (ϕ). D_{geo}/D_w is the ratio of the diffusion coefficient, including the geometrical effects of the porous media, to the diffusion coefficient in water. A ratio of 1 means that there is no effect of the solids on

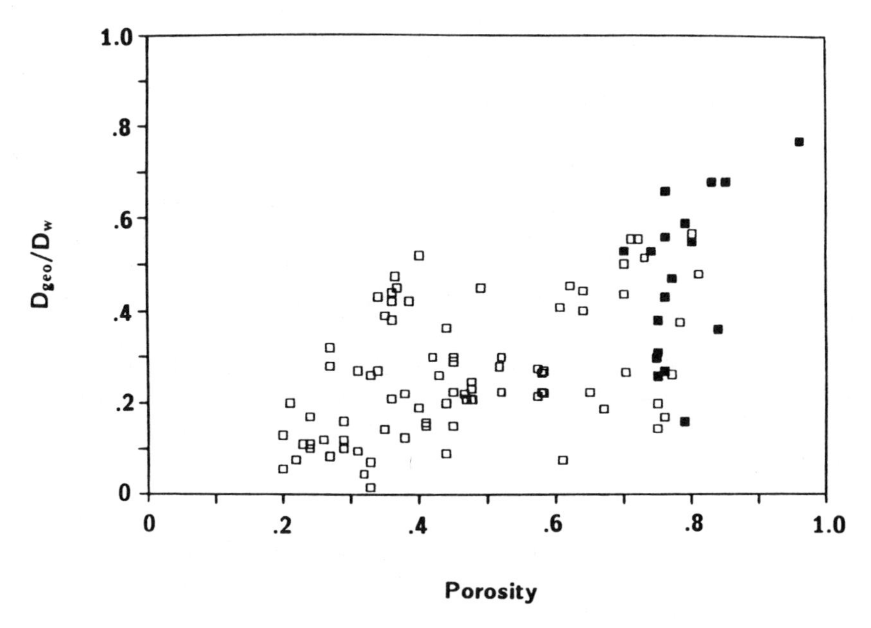

Figure 3. Experimental data illustrating the relationship between D_{geo}/D_w *and porosity.* □, *literature data; and* ■, *this study.*

D_w. The results of the $^3H–H_2O$ diffusion experiments conducted in this study (dark squares) are also plotted in the same figure. These values fall in the same range as those found in the literature. A regression of the complete data set has the equation

$$\frac{D_{geo}}{D_w} = 0.6\ \phi \qquad r^2 = 0.37 \tag{12}$$

This relationship can be used to approximate the effect of geometry, expressed in terms of porosity, on the diffusion coefficient.

The data show considerable scatter because of the inclusion of clays, soils, and various sediments. Each porous solid is unique in its packing, inter- and intrastitial spaces, and tortuosity. The results of this study also demonstrate the scatter and weak relationship between D_{geo}/D_w and porosity found in the other studies. Porosity is obviously not the best predictor of this effect, but it is the only parameter that can be measured easily. Most of the data points fall within the predictions of two theoretical models:

$$\frac{D_{geo}}{D_w} = \frac{\phi}{\gamma} \quad \text{and} \quad \frac{D_{geo}}{D_w} = \frac{\phi}{\gamma^2} \tag{13}$$

if a constant value of 1.8 is used for the tortuosity (γ) (32). Tortuosity varies with porosity, but the exact relationship is unique to each porous solid. For the present discussion of diffusion of HOC in sediments, the most important observation is that the geometrical effects decrease D_w only by a factor of 2–3 over the porosity range normally found in surficial sediments.

Measurement of D_{eff} of HOC in Sediments. The results of the purge and half-cell diffusion experiments are presented in Tables IV and V, respectively. The two techniques overlapped with the measurement of D_{eff} of 2-PCB in Duluth Harbor (Lake Superior) sediment at 23 ± 2 °C. The value of D_{eff} from the half-cell technique (6.6×10^{-11} cm^2/s) is not significantly different than the value derived from the purge method (4.1×10^{-11} cm^2/s). The influence of sampling time for the half-cell experiments was examined by letting identical cells remain intact for 4 and 20 weeks. The resulting D_{eff} values are <0.1 log units different (Table V).

Discussion

The dual experiments using the half-cell and purge techniques provide two different sets of information to evaluate the effect of sorption on the D_{eff} in sediments. The half-cell technique allows an examination of a single compound in a variety of sediments, and the purge technique permits the study of a variety of HOC in a single sediment.

Table V. Results of Half-Cell Experiments

Sediment Identification	*Sediment Type*	*log* D_{eff} *(cm^2/s)*	*Temperature (°C)*	*OC (%)*	*PWOC (mg/L)*	*Porosity*
Effect of temperature (32 days before separation)						
Duluth Harbor (V–84)	silt/clay	–8.70	4	3.34	74.4	0.76
Duluth Harbor (V–84)	silt/clay	–8.29	14	3.34	66.9	0.76
Duluth Harbor (V–84)	silt/clay	–8.20	22	3.34	90.6	0.76
Duluth Harbor (V–84)	silt/clay	–8.12	22	3.34	—	0.76
Effect of sampling time (5 months before separation)						
Duluth Harbor (V–84)	silt/clay	–8.39	14	3.34	—	0.76
Duluth Harbor (V–84)	silt/clay	–8.12	22	3.34	—	0.76
Effect of sediment type						
Lake Superior (2Bx–A)	clay	–8.02	22	2.03	95	0.79
Lake Superior (2Bx–B)	clay	–7.28	22	1.23	109	0.79
Lake Erie (LS/81)	clay	–7.75	22	2.84	160	0.81
Duluth Harbor(V–84/2)	silt/clay	–7.42	22	3.94	169	0.75
V–84 + SDS	silt/clay	–7.34	22	4.60	229	0.75
Little Rock Lake	mucky sand	–7.74	22	23.77	138	0.96
Island Lake	sand/silt	–7.60	22	3.33	160	0.77

In the half-cell experiments, a ^{14}C-labeled PCB congener was used in seven different sediments (Table V). These sediments included fine clays, silts, sandy clays, and a mucky sand. Organic carbon ranged from 1.2 to 23.8%, and porewater organic carbon ranged from 91 to 169 mg/L. A sample of the Duluth Harbor sediment (described previously) was centrifuged to remove a fraction of the porewater. This volume was replaced with a 0.06 M solution of sodium dodecylsulfate (SDS) to artificially increase the porewater organic carbon concentration. SDS has a molecular weight of 288, is negatively charged, and forms micelles at the applied solution concentration. It may behave similarly to natural organic colloids, but probably has stronger surfactant properties (*56*).

There was no relationship observed between the measured D_{eff} values of the PCB congener in the seven sediments and their percent OC. In contrast, a relationship was found between the measured D_{eff} values and the concentration of porewater organic carbon (PWOC) in these sediments (log D_{eff} = 0.42 log PWOC (mg/L) + 5.4; $r^2 = 0.90$; Figure 4). The D_{eff} for the sediment–SDS falls along the extrapolated line. These results indicate an influence of the PWOC on the diffusion of HOC in sediments. The porewater colloids can effectively compete with the sediment solids for the HOC. Therefore as the PWOC concentration increases, more of the HOC will be present in the porewater and able to diffuse through and presumably out of the sediment. Capel and Eisenreich (*29*) estimate from field measurements that >90% of the PCB in the sediment porewaters are associated with the organic colloids. HOC associated with the porewater colloids may be visualized as "piggybacking" on the diffusing colloids. The small HOC (MW = 200–400) are bound to the larger colloids (MW = 10^3–10^5) (*57*). The overall

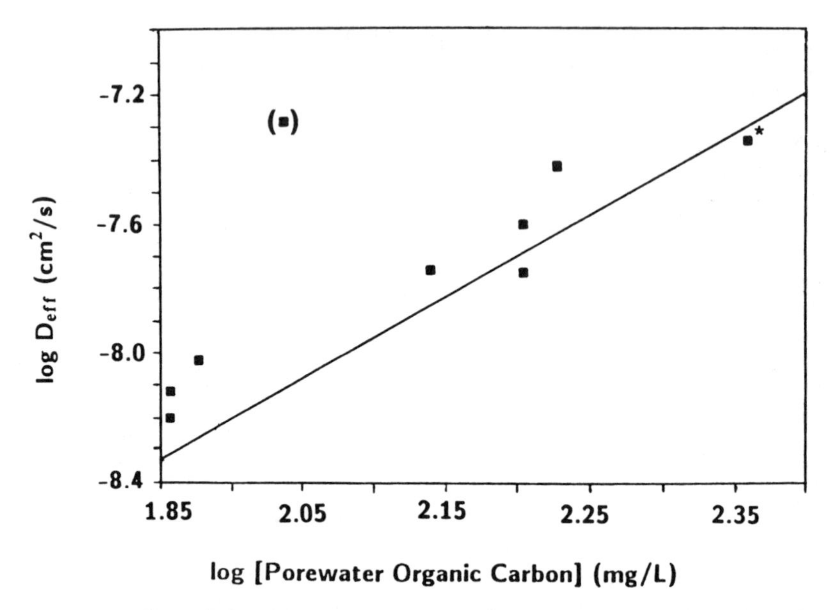

Figure 4. Effect of the porewater organic carbon concentration (mg/L) on the D_{eff} *(cm^2/s) for 2-PCB (PCB IUPAC No. 1) in seven different sediments (Table V). The regression line is log* D_{eff} *= 0.42 log PWOC (mg/L) + 5.4;* r^2 *= 0.90. The two values to the far left are duplicates. The data point marked * is the sediment spiked with SDS. The data point enclosed in parentheses was not used in the regression calculations.*

diffusional flux is then the sum of the fluxes of the HOC–colloid complex plus the truly dissolved HOC. For compounds $>10^3$ daltons, the D_w is insensitive to the molecular weight (MW = 10^3–10^5; D_w = 0.5–2.5 × 10^{-6} cm^2/s) (58).

The purge technique is used to determine the D_{eff} values of a number of chlorinated hydrocarbons in a natural sediment. The results of the sediment purge studies are presented in Table IV and Figure 5. The log D_{eff} has been plotted versus log K_{ow} to illustrate the effect of chemical hydrophobicity and, indirectly, the effect of sorption on the sediment diffusion coefficient. Numerous investigators have shown that sediment–water partition coefficients (K_p, or probably more accurately K_D) are strongly correlated with K_{ow} for HOC (6, 8, 9). Given the equations in Table II (which can be summarized as $D_{eff} \alpha D_w/K_p$), an inverse relationship between D_{eff} and K_{ow} might be expected. This relationship is observed, albeit weak (log D_{eff} = –0.56 log K_{ow} – 7.7; r^2 = 0.38). The lack of a stronger relationship can be attributed to the influence of the porewater colloids. The decreasing D_{eff} with increasing K_{ow} can be attributed to the influence of the physical–chemical properties of the HOC, which determines their diffusional behavior. The more hydrophobic compounds are bound to the solids to a greater extent and, therefore, diffuse at a slower rate.

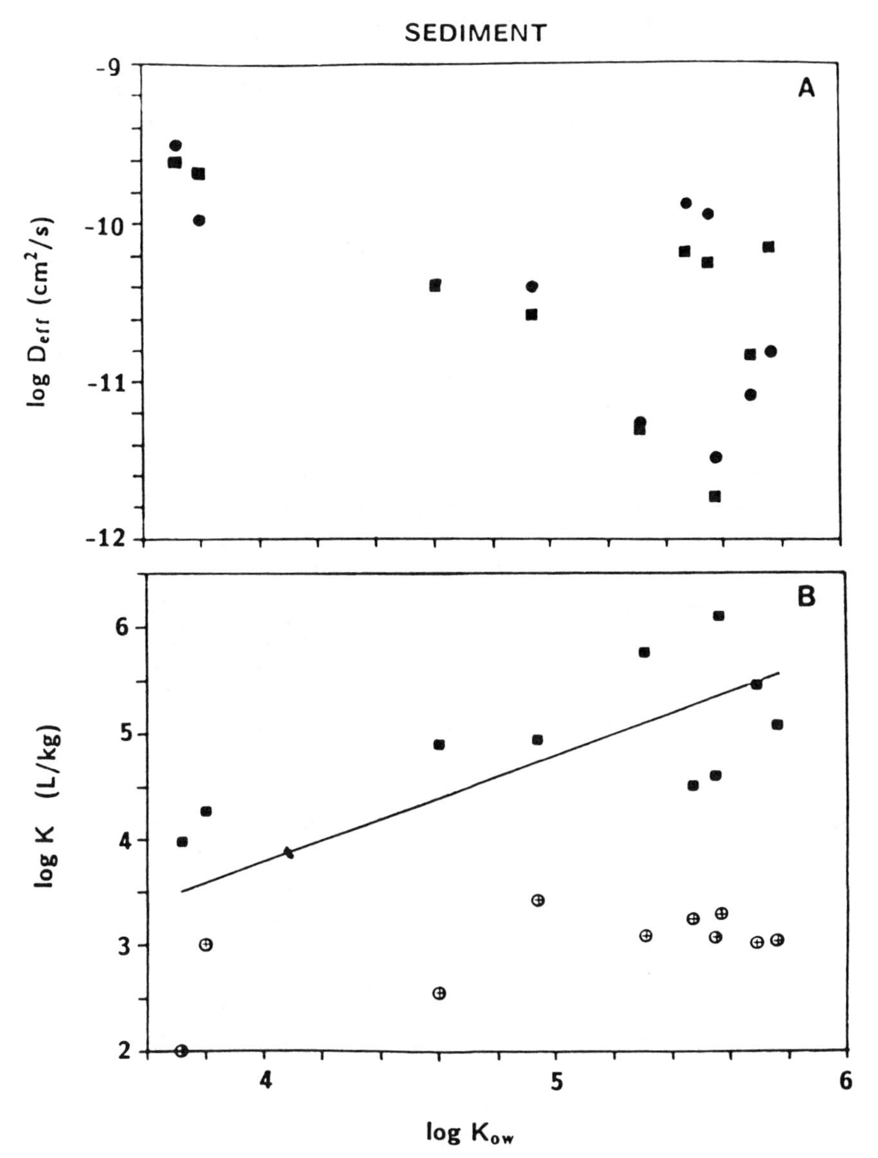

Figure 5. (A) log D_{eff} versus log K_{ow} for PCB congeners and chlorinated hydrocarbons (Table IV) in Duluth Harbor sediment. The two points for each K_{ow} value (squares and circles) represent duplicate samples. (B) log distribution coefficient versus log K_{ow} for the same compounds. Field-measured K_D: circle with crosses; calculated $K_{D\text{-}D_{eff}}$: dark squares; line predicted by log K_{oc}/log K_{ow} relationship of Karickhoff et al. (6): log K_{oc} = 1.0 log K_{ow} − 0.21; log D_{eff} = −0.56 log K_{ow} − 7.7, r^2 = 0.40; log $K_{D\text{-}D_{eff}}$ = 0.59 log K_{ow} + 2.0, r^2 = 0.46; log K_D = 0.18 log K_{ow} + 2.1, r^2 = 0.20.

A distribution coefficient can be calculated on the basis of the measured D_{eff}. In the previous sections, D_w and the effect of the geometry of the sediment solids was quantified and determined to be relatively unimportant (factor of 2–3) compared to the influence of the distribution coefficients of HOC (factor of 10^2–10^6). Following the forms of the equations in Table II, a mathematical description of D_{eff} can be approximated as

$$D_{eff} = \frac{D_w(0.6\ \phi)}{K + 1} \tag{14}$$

This equation can be rearranged and solved for K. The distribution coefficients calculated in this manner will be referred as $K_{D\text{-}D_{eff}}$. These calculated values are shown as solid squares in Figure 5B versus log K_{ow}. The line passing through the points is not the regression line, but rather the K_{oc}/K_{ow} relationship predicted by Karickhoff et al. (*6*). Independent measurements of K_D have been made for this sediment in the field (*29*) and in the laboratory (Table IV; Figure 5B, circles with crosses). The measured values of K_D show no relationship to K_{ow}, but rather demonstrate a constant K_D (about 10^3 L/kg) over a wide range of hydrophobicity. The constant K_D values for HOC between sediments and porewaters, observed in both marine (*2*) and freshwater sediments (*2, 28, 29*), are attributed to the influence of the PWOC competing with the sediment solids for the HOC. K_D can be interpreted as the distribution coefficient between the HOC associated with the solids and the water plus porewater colloids (equation 3), whereas $K_{D\text{-}D_{eff}}$ is the distribution coefficient between the HOC associated with the solids and those in the truly dissolved phase. Probably, $K_{D\text{-}D_{eff}}$ is a measure of desorption coefficient of the HOC from the solid. The reasonable match between the K_{oc}/K_{ow} prediction and the $K_{D\text{-}D_{eff}}$ values supports the use of K_{ow} as an independent variable in the range of log K_{ow} = 3.5–6.0.

Diffusional Flux of HOC Within and From Sediments. The diffusional flux of HOC in and from porewater can be expressed as

$$\text{flux}_{\text{total}} = \text{flux}_{\text{surface}} + \text{flux}_{\text{dissolved}} + \text{flux}_{\text{colloidal-bound}} \tag{15}$$

Berner (*32*) showed that the diffusional flux of a species (i.e., HOC) sorbed to the surface ($\text{flux}_{\text{surface}}$) is quantitatively unimportant. The flux of HOC in the porewater is governed by interactions with the bulk sediment solids and the porewater organic carbon (colloids). The HOC in the porewater exist either in the truly dissolved form or in association with the porewater colloids. Both forms yield diffusional fluxes, $\text{flux}_{\text{dissolved}}$ and $\text{flux}_{\text{colloidal-bound}}$, respectively. The flux of the colloidal–bound HOC is actually governed by the flux of the colloid.

From the results of the purge studies, D_{eff} appears to be dependent on K_{ow}. This finding suggests that the flux of the dissolved HOC is primary and that the HOC–colloid flux is of secondary importance. For extremely hydrophobic compounds, where the amount truly dissolved is extremely low, this may not be the case. The importance of the HOC–colloid flux can be estimated from the purge experiments. The experimental setup presents conditions that approximate the natural sediment–water interface. The flux of the colloids out of the sediment will be close to the maximum rate due to the concentration gradient between the sediment porewater and the overlying water. In this experiment, the flux of the organic colloids is approximated by the increase in the concentration of the dissolved organic carbon (DOC) in the overlying water [DOC (mg/L) × volume of overlying water (L) × duration of experiment]. Typically 90% of the HOC in the porewater are associated with the colloids in this sediment (29). The flux of the HOC–colloid component is calculated as the flux of the porewater organic carbon × porewater HOC–colloid concentration ($mass_{HOC}/mass_{PWOC}$). From the results of the purge experiments, it is estimated that less than 20% of the total amount of the chlorinated hydrocarbons diffusing out of the sediment is bound to the porewater colloids.

Summary and Conclusions

The D_{eff} values in sediments of a series of PCB congeners and three chlorinated pesticides were measured by two different methods. Both methods gave comparable results for the one overlapping compound and sediment. The two methods were able to highlight different aspects of the diffusion–sorption relationship. From the results of the half-cell technique, it appears that the porewater organic carbon concentration is able to influence D_{eff}. As the PWOC concentration increases, it is better able to compete with the sediment solids for the HOC, and thus yield a larger fraction of these compounds in the porewater. This larger porewater fraction will yield a higher diffusional flux (thus a faster D_{eff}) via both molecular diffusion of the dissolved HOC and diffusion of the colloid–HOC pair (piggybacking). The results of the purge technique showed that, in a given sediment–porewater system, the D_{eff} is related to the hydrophobicity of the HOC. By approximating the quantities of D_w and the effect of the geometrical effects on D_w, distribution coefficients ($K_{D\text{–}D_{eff}}$) were calculated from the measured D_{eff} values. These values matched the K_{oc}/K_{ow} structure–activity relationship of Karickhoff et al. (6) reasonably well and suggested that the calculated $K_{D\text{–}D_{eff}}$ is a close approximation of the truly dissolved partition coefficient (K_p). These $K_{D\text{–}D_{eff}}$ values are dissimilar to the field-measured K_D values in both magnitude and relationship to chemical hydrophobicity (i.e., K_{ow}).

Overall, these results are complementary to the previous field measurements of HOC K_D values. The field measurements showed relatively high porewater HOC concentrations and low K_D values when compared to water column studies. Also, the measured K_D values in sediments were far different than would be predicted from structure–activity relationships based on chemical hydrophobicity. The measured sediment–porewater K_D values showed no relationship to K_{ow} and were approximately constant at about 10^3 L/kg. All of these observations are attributed to the sorption of the HOC by the PWOC. The field measurements compared the HOC associated with the solid phase to those in the porewater (truly dissolved plus colloid-associated). The purge diffusion experiments allowed an estimation of the true K_p values of the HOC. Apparently, the laboratory structure–activity relationships developed for sorption of HOC are valid in sediment–porewater regimes, if the influence of the porewater colloids can be factored out of the measurements and calculations.

Acknowledgments

This research was funded in part by the Minnesota Sea Grant Program and the Great Lakes Environmental Research Laboratory of the National Oceanic and Atmospheric Administration in Ann Arbor, Michigan. The assistance of Rob Holtznecht in the laboratory and the stimulating conversations with Joel Baker, Noel Urban, Rob Rapaport, John Robbins, and Brian Eadie are appreciated.

Abbreviations

C	concentration
C_{aq}	concentration of HOC in the water (mass HOC divided by volume water)
C_c	concentration of HOC associated with organic colloids (mass HOC divided by mass of colloids)
C_d	concentration of HOC in the dissolved phase (mass HOC divided by volume water)
C_p	concentration of HOC in the solid phase (mass HOC divided by mass solid)
COLL	mass of HOC associated with colloidal phase
D	mass of HOC that is truly dissolved
D_{eff}	effective diffusion coefficient of HOC in sediments
D_o	diffusion coefficient
D_w	diffusion coefficient in water
foc	fraction of organic carbon of a solid
HCB	hexachlorobenzene
HCH	hexachlorocyclohexane
HOC	hydrophobic organic chemical
K	adsorption coefficient

K_c	theoretical colloidal–water partition coefficient for HOC
K_D	operationally defined particle–(dissolved + colloidal) distribution coefficient for HOC
$K_{D\text{-}D_{eff}}$	distribution coefficients, calculation based on measured D_{eff}
K_{oc}	fraction organic carbon of the solid–water distribution coefficient for HOC
K_{ow}	octanol–water partition coefficient
K_p	theoretical particle–water partition coefficient for HOC
M_B	molecular weight of the solvent (Equation 8)
M_c	mass of colloids
M_p	mass of particles
M_w	mass of water
OC	organic carbon
P	mass of HOC in the particulate phase
PCB	polychlorinated biphenyls
PWOC	porewater organic carbon
SDS	sodium dodecylsulfate
t	time
T	temperature
V_a	molar volume of the solute
x	distance
X	solvent "association" parameter (Equation 8)
γ	tortuosity
μ_B	solvent viscosity
ϕ	porosity

References

1. Eisenreich, S. J. In *Sources and Fates of Aquatic Pollutants*; Advances in Chemistry Series 216, American Chemical Society: Washington, DC, 1987; 393–496.
2. Brownawell, B. J.; Farrington, J. W. *Geochim. Cosmochim. Acta* **1986,** *50*, 157–169.
3. Baker, J. E.; Eisenreich, S. J.; Johnson, T. C.; Halfman, B. M. *Environ. Sci. Technol.* **1986,** *19*, 854–861.
4. Capel, P. D.; Eisenreich, S. J. *J. Great Lakes Res.* **1985,** *11*, 447–461.
5. Swackhamer, D. Ph.D. Thesis, University of Wisconsin, 1985.
6. Karickhoff, S. W.; Brown, D. S.; Scott, T. A. *Water Res.* **1979,** *13*, 241–248.
7. DiTori, D. M.; Horzempa, L. M. *J. Great Lakes Res.* **1982,** *8*, 336–349.
8. Chiou, C. T.; Porter, P. E.; Schmedding, D. W. *Environ. Sci. Technol.* **1983,** *17*, 227–231.
9. Schwarzenbach, R. P.; Westall, J. *Environ. Sci. Technol.* **1981,** *15*, 1360–1367.
10. Gschwend, P. M.; Wu, S.-C. *Environ. Sci. Technol.* **1985,** *19*, 90–96.
11. Voice, T. C.; Weber, W. J. *Environ. Sci. Technol.* **1983,** *17*, 513–518.
12. Karickhoff, S. W. In *Contaminants and Sediments*, Vol. 2; Baker, R. A., Ed.; Ann Arbor Science: Ann Arbor, 1980; pp 193–205.
13. Coates, J. T.; Elzerman, A. W. *J. Contam. Hydrol.* **1986,** *1*, 191–210.
14. Wu, S.-C.; Gschwend, P. M. *Environ. Sci. Technol.* **1986,** *20*, 717–725.

15. O'Connor, D. J.; Connolly, J. P. *Water Res.* **1980,** *14*, 1517–1523.
16. DiToro, D. M.; Horzempa, L. M.; Casey, M. M.; Richardson, W. *J. Great Lakes Res.* **1982,** *8*, 336–349.
17. DiToro, D. M.; Mahoney, J. D.; Kirchgraber, P. R.; O'Byrne, A. L.; Pasquale, L. R.; Piccirilli, D. C. *Environ. Sci. Technol.* **1986,** *20*, 55–61.
18. Baker, J. E.; Capel, P. D.; Eisenreich, S. J. *Environ. Sci. Technol.* **1986,** *20*, 1136–1143.
19. Means, J. C.; Wijayaratne, R. D. *Bull. Mar. Sci.* **1984,** *35*, 449–461.
20. Means, J. C.; Wijayartne, R. D.; Boynton, W. R. *Can. J. Fish Aquat. Sci.* **1983,** *40*, 337–345.
21. Means, J. C.; Wijayaratne, R. D. *Science (Washington, DC)* **1982,** *215*, 968–970.
22. Caron, G.; Suffet, I. H.; Belton, T. *Chemosphere* **1985,** *14*, 993–1000.
23. Duinker, J. C.; Hillebrand, M. T. *Neth. J. Sea Res.* **1979,** *2*, 256–281.
24. Brownawell, B. J.; Farrington, J. W. In *Marine and Estuarine Geochemistry*; Sigleo, A. C.; Hattori, A., Eds.; Lewis Publishers: Ann Arbor, 1985; pp 447–461.
25. Eadie, B. J.; Landrum, P. F.; Faust, W. *Chemosphere* **1982,** *11*, 847–858.
26. Eadie, B. J.; Rice, C. P.; Frez, W. A. In *Physical Behavior of PCBs in the Great Lakes*; Mackay, D.; Paterson, S.; Eisenreich, S. J.; Simmons, M., Eds.; Ann Arbor Science: Ann Arbor, 1983; pp 213–228.
27. Rice, C. P.; Meyers, P. A.; Brown, G. S. In *Physical Behavior of PCBs in the Great Lakes*; Mackay, D.; Paterson, S.; Eisenreich, S. J.; Simmons, M., Eds.; Ann Arbor Science: Ann Arbor, 1983; pp 157–180.
28. Capel, P. D.; Baker, J. E.; Eisenreich, S. J., in review.
29. Capel, P. D.; Eisenreich, S. J. In *COST-641 Workshop on Transfer Processes Between Water and Organic Phases*; Water Pollution Research Report 2, Commission of the European Communities, Dübendorf, Switzerland, EUR 11071; pp 15–22.
30. Karickhoff, S. W.; Morris, K. R. *Environ. Sci. Technol.* **1985,** *19*, 51–56.
31. Van Holde, K. E. *Physical Biochemistry*; Prentice Hall: New York, 1971; pp 85–90.
32. Berner, R. A. *Early Diagenesis: A Theoretical Approach*; Princeton University Press: Princeton, 1980; pp 31–42.
33. Manheim, F. T.; Waterman, L. S. *Initial Rep. Deep Sea Drill. Proj.* **1974,** *22*, 663–670.
34. Adamson, A. W. *Physical Chemistry of Surfaces*, 4th ed.; John Wiley and Sons: New York, 1982.
35. Crank, J. *The Mathematics of Diffusion*; Clarendon Press: New York, 1956.
36. Duursma, E. K.; Hoede, C. *Neth. J. Sea Res.* **1967,** *3*, 423–457.
37. Corwin, D. L.; Farmer, W. J. *Chemosphere* **1984,** *13*, 1295–1317.
38. Karickhoff, S. W. *J. Hydraul. Div. Am. Soc. Civ. Eng.* **1984,** *110*, 707–735.
39. Wilke, C. R.; Chang, P. C. *AIChE J.* **1955,** *1*, 264–270.
40. Li, Y. H.; Gregory, S. *Geochim. Cosmochim. Acta* **1974,** *38*, 703–714.
41. Bloksma, A. H. *J. Colloid Sci.* **1957,** *12*, 40–52.
42. Klute, A.; Letey, *Soil Sci. Soc. Am. Proc.* **1958,** *22*, 213–215.
43. Wang, J. H.; Robinson, C. V.; Edelman, I. S. *J. Am. Chem. Soc.* **1953,** *75*, 466–470.
44. Johnson, T. C.; Evans, J. E.; Eisenreich, S. J. *Limnol. Oceanogr.* **1982,** *27*, 481–491.
45. Long, E. C. Technical Report No. 1096–NUC–77–27, 1977; Beckman Instruments; 27 pp.
46. Van Schaik, J. C.; Kemper, W. D.; Olsen, S. R. *Soil Sci. Soc. Am. Proc.* **1966,** *30*, 17–25.

47. Geankoplis, C. J. *Mass Transport Phenomena*; Edward Brothers: Columbus, OH, 1972; pp 182–184.
48. Othmer, D. F.; Thaker, M. S. *Ind. Eng. Chem.* **1953,** *45*, 589–593.
49. Lerman, A. *Geochemical Processes in Water and Sediment Environments*; Wiley–Interscience: New York, 1979; p 92.
50. Porter, L. K.; Kemper, W. D.; Jackson, J. D.; Stewart, B. A. *Soil Sci. Soc. Am. Proc.* **1960,** *24*, 460–463.
51. Rowell, D. L.; Martin, M. W.; Nye, P. H. *J. Soil Sci.* **1967,** *18*, 204–222.
52. Kepay, P. E.; Cooke, R. C.; Bowen, A. J. *Geochim. Cosmochim. Acta* **1981,** *45*, 1401–1409.
53. Krom, M. D.; Berner, R. A. *Limnol. Oceanogr.* **1980,** *25*, 327–337.
54. Phillips, R. E.; Brown, R. D. *Soil Sci. Soc. Am. Proc.* **1968,** *32*, 302–306.
55. Nakayama, F. S.; Jackson, R. D. *Soil Sci. Soc. Am. Proc.* **1963,** *27*, 255–258.
56. Wershaw, R. L.; Burcar, P. J.; Goldberg, M. C. *Environ. Sci. Technol.* **1969,** *16*, 271–273.
57. Ishiwatari, R. In *Humic Substances in Soil, Sediment and Water*; Aiken, G. R.; McKnight, D. M.; Wershaw, R. C.; McCarthy, P., Eds.; John Wiley and Sons: New York, 1985; p 147.
58. Perry, P. H.; Chilton, C. H. *Chemical Engineer's Handbook*, 5th ed.; McGraw-Hill: New York, 1973; pp 17–78.
59. Millington, R. I.; Quirk, J. P. *J. Chem. Soc. Faraday Trans. 1* **1961,** *57*, 1200–1207.
60. Satterfield, C. N.; Sherwood, T. K. *The Role of Diffusion in Catalysis*; Addison–Wesley: New York, 1963; pp 12–28.
61. Karger, B. L.; Snyder, L. R.; Horvath, C. *An Introduction to Separation Sciences*; Wiley–Interscience: New York, 1973; p 80.
62. Krezoski, J. R.; Robbins, J. A. *J. Geophys. Res. C* **1985,** *90*, 11999–12006.
63. Bode, C. E.; Day, C. L.; Gebhart, M. R. *Weed Sci.* **1973,** *21*, 480–484.
64. Ehlers, W.; Letey, J.; Spencer, W. F.; Farmer, W. J. *Soil Sci. Soc. Am. Proc.* **1969,** *33*, 501–504.
65. Graham-Bryce, I. J. *J. Sci. Food Agric.* **1969,** *20*, 489–494.
66. Walker, A.; Crawford, D. V. *Weed Res.* **1970,** *10*, 126–132.
67. Hesslein, R. H. *Can. J. Fish Aquat. Sci.* **1980,** *37*, 544–551.
68. Li, Y. H.; Gregory, S. *Geochim. Cosmochim. Acta* **1974,** *38*, 703–714.
69. Fischer, J. B.; Petty, R. L.; Lick, W. *Environ. Pollut., Ser. B* **1983,** *25*, 121–132.
70. DiToro, D. M.; Jeris, J. S.; Ciarcia, D. *Environ. Sci. Technol.* **1985,** *19*, 1169–1176.
71. Rapaport, R. A.; Eisenreich, S. J. *Environ. Sci. Technol.* **1984,** *18*, 163–170.

RECEIVED for review July 24, 1987. ACCEPTED for publication February 26, 1988.

14

Sorption of Benzidine, Toluidine, and Azobenzene on Colloidal Organic Matter

J. C. Means

Institute for Environmental Studies, Louisiana State University, Baton Rouge, LA 70803

R. D. Wijayaratne

Chesapeake Biological Laboratory, University of Maryland, Solomons, MD 20688–0038

The sorptive behavior of natural estuarine colloids was investigated with benzidine, toluidine, and azobenzene. In general, the curvilinear equilibrium sorption isotherms were well represented by the Freundlich equation, with 1/n *values of 0.72–1.00. Freundlich sorption constants for each aromatic amine, normalized to the organic carbon content of the colloids* (K_{oc})*, were 3420, 2010, and 1390 at pH 7.9. Sorption was largely controlled by the aqueous-phase pH, with sorption constants increasing as pH decreased to 5. We attempted to establish to what extent benzidine, its derivatives, and azobenzene as related to their molecular structure and thermodynamic properties influence the sorptive process. Interpretation of the sorption data allowed certain mechanistic hypotheses to be formulated. An experimental value for* o*-toluidine water solubility of 75 ppm at pH 7.9 was determined.*

AROMATIC AMINES ARE IMPORTANT IN MANY AREAS of industry and research. They are classically important in dye chemistry (*1*), and the physiological activity of these compounds makes them of interest in the biomedical field. Some of the member compounds are on the U.S. Envi-

0065–2393/89/0219–0209$06.00/0

ronmental Protection Agency's (EPA) list of toxic substances; 8 of the 14 chemicals controlled by the second emergency standard issued by the Occupational Safety and Health Administration (OSHA) are aromatic amines (*2*). Results of limited studies on human health and environmental effects have raised concerns regarding their potential for toxicity, mutagenicity, and carcinogenicity (*3*).

Because organic contaminants eventually find their way into water, largely as a result of industrial discharge, some attention has been focused on the fate of industrial amines in wastewaters and in the environment as a whole. Malaney et al. (*4*) and Baird et al. (*5*) have investigated the removal and fate of several aromatic amines in activated sludge reactors. They concluded, on the basis of oxygen-uptake data, that these carcinogens were toxic and probably refractory to bacterial degradation at 500-mg/L doses. In addition, they stated that many of the monoaromatic substances could be metabolized to some extent by acclimated systems. It has also been reported (*6*) that a strong correlation exists between the mutagenic activity of sediment extracts from the Buffalo River (New York) and the proximity of the sampling sites to a dye-manufacturing plant.

Bond-stability calculations and model reactions of aromatic amines with monomeric constituents of humic substances suggest that covalent binding of the amine residue may occur by at least two distinct mechanisms, in a hydrolyzable (probably anil and anilinoquinone) and in a nonhydrolyzable (probably heterocyclic rings and ether bonds) manner (*7*). More recently, Parris (*8*) reported the results of binding experiments with aromatic amines and compounds that serve as models of humate functional groups (e.g., carbonyls and quinones). He postulated that primary ring-substituted anilines bind covalently to soil organic matter (e.g., humates) via carbonyl and quinone moities. The proposed mechanism of binding involves two phases. A reversible rapid equilibrium is initially established with the formation of an imine linkage with the humate carbonyls. Subsequently, a slow reaction involves 1,4-addition to a quinone ring, followed by tautomerization and oxidation to give an amino-substituted quinone.

Although the sorption behavior of aniline in soils is known to a limited extent, the cycling of the broader class of compounds in aquatic environments is very poorly understood. Because aromatic amines can be introduced into waters via industrial discharge and surface run-off from land, assessing the fate of these compounds in aquatic environments is important. We recently investigated the sorptive properties of natural colloids with polynuclear aromatic hydrocarbons (PAHs) and the herbicides linuron and atrazine. Colloidal organic material was found to be a factor of 10 times better as a sorptive substrate for these compounds than soil or sediment organic matter (*9–11*). These results indicate that natural colloids are potentially important substrates that can significantly influence the movement and persistence of organic contaminants in aquatic environments.

We have investigated the sorptive behavior of estuarine colloidal organic matter with azobenzene, *o*-toluidine, and benzidine. The dependence of sorption on aqueous-phase pH will be discussed. We attempted to establish to what extent the molecular structure and thermodynamic properties of benzidine and its derivatives influence the sorptive process. Interpretation of the sorption isotherm data allows formulation of certain mechanistic hypotheses that need to be investigated further. The water solubility of *o*-toluidine was also determined experimentally.

Experimental Methods

Materials. Benzidine was purchased from Columbia Organic Chemicals Co. (Columbia, SC); azobenzene and *o*-toluidine were obtained from Eastman (Rochester, NY). All chemicals were 99+% pure. These amines were further purified by recrystallization from a methanolic solution. The minimum purity was 99.5%, confirmed by high-performance liquid chromatographic (HPLC) and mass spectral analyses.

Distilled-in-glass-grade acetonitrile (Burdick and Jackson, Muskegon, MI), monobasic potassium phosphate (Fisher Scientific, Fairlawn, NJ), and glass-distilled nanopure water were used in preparing buffers.

Apparatus. All separations were performed on an HPLC system equipped with an M–45 solvent delivery pump (Waters Associates), a Rheodyne model 7125 rotary valve injector with a 20-μL sample loop, a fixed-wavelength (254-nm) model 440 UV absorbance detector (Waters Associates), and a Perkin–Elmer 650–10S fluorescence spectrophotometer having a 10-μL flow cell. Reverse-phase HPLC was performed under ambient conditions on a radial compression module (Waters Associates) with an A_{10} cartridge (C_{18} packing, 10 cm × 5 mm i.d.). Chromatograms were recorded on a Hewlett Packard model 3390A electronic integrator.

Sample Collection and Sorption Methods. Ten 20-L natural-water samples were collected in the Chesapeake Bay estuary from the mouth of the Patuxent River in Solomons, Maryland. These samples were collected at 0.5 m below the surface with acid-washed, distilled-water-rinsed Nalgene containers. Samples were filtered through 0.45-μm filters (Millipore) to remove suspended particulate matter. The filtrates containing the dissolved organic carbon and colloidal carbon fractions were then subjected to ultrafiltration with an Amicon H_1P_5 hollow-fiber filtration system, having a nominal molecular weight cutoff of 5000. Thus, the original 200 L of estuarine water was divided into two fractions in this step, an ultrafiltrate (approximately 195 L) that contained the truly dissolved organic carbon and a colloidal fraction (approximately 5 L) that was enriched by a factor of 50. The colloidal fraction is never dried or concentrated on an ultrafilter membrane in this procedure, but remains in an enriched suspension.

Aliquots (1 L) of the ultrafiltrate were spiked with known amounts of the amines (below the solubility limit of the test compound), and 50–70 mL of the enriched colloid fraction was then added back to each ultrafiltrate solution contained in 2-L glass Eyrlenmeyer flasks. Triplicate spiked samples for each amine concentration were equilibrated in a table shaker for 18 h at 20 °C in the dark. Kinetic experiments indicated that this equilibration time was sufficient to reach >95% of true equilibrium. After equilibration, each sample (1.05 L) was passed through the hollow-fiber filtra-

Table I. Chromatographic Conditions for the Separation of Aromatic Amines

Condition	*Benzidine*	o-*Toluidine*	*Azobenzene*
Mobile-phase $CH_3{\equiv}N{:}KH_2PO_4$	35:65	35:65	35:65
Flow rate (mL/min)	2	2	2
Retention time (min)	4.27	7.73	8.51
UV (λ) (nm)	—	—	254
Fluorescence, λ_{ex}, λ_{em} (nm)	340, 480	280, 480	—

[a]All separations were performed on a radial compression module (Waters Associates) with an A_{10} cartridge (C_{18} packing, 10 cm × 5 mm i.d.).

tion system again to yield an ultrafiltrate fraction (~1 L) that contained dissolved organic carbon and truly dissolved amine, and a colloid fraction (~50 mL) that contained the colloidal-bound amine and dissolved amine. This step took approximately 20 min, a period of time shown in control experiments to be short enough to prevent any significant redistribution of the solutes. The hollow fibers were rinsed with 20 mL of distilled water to ensure complete recovery of the colloidal fraction from the device.

Mass balances for all compounds in all experiments were greater than 97%, which is within the range of analytical error for the amine quantification. We then determined the amine concentration contained in both fractions by HPLC on a reverse-phase C_{18} (μBondapak) column. The amount sorbed to the colloids was determined by determining the excess amount of amine contained in the colloid fraction after correcting for the amount in solution. Each fraction was analyzed directly by HPLC in quadruplicate. Each sorption isotherm was determined in duplicate independent experiments.

Separations were accomplished by using an isocratic mixture of acetonitrile and water at a flow rate of 2 mL/min. The column effluent was passed through an UV detector set at a wavelength of 254 nm or a fluorescence spectrophotometer equipped with a 10-μL flow cell. The aromatic amines were identified on the basis of retention time by comparison with standards. These compounds were well resolved on the column; thus identification was unambiguous. The chromatographic conditions used for the separation of aromatic amines are listed in Table I. A mass balance for each aromatic amine was performed to verify that no significant losses or degradation of material occurred during the sorption experiments. The experiments determining the effects of pH on sorption were performed by adjusting the pH of the solution with either NaOH or HCl. The resultant sorption constant was measured after equilibration by the techniques already discussed in this section.

The organic carbon content of the bulk water, ultrafiltrate, and enriched colloidal fractions was measured on an analyzer (Oceanography International) by using the persulfate method of Menzel and Vaccaro (*12*). Results on blanks indicated that less than 2% of the carbon in the samples originated in the filters. The water solubility of *o*-toluidine was determined by using the procedures outlined by Means et al. (*13*).

Results and Discussion

The sorption of aromatic amines by estuarine colloids produced isotherms that conformed to the Freundlich equation:

$$C_s = K_d C_w \frac{1}{n} \quad (1)$$

where C_s is the equilibrium concentration of the compound on the sorbent, C_w is the equilibrium concentration of compound in solution, and K_d and n are constants related to the sorptive capacity of the sorbent. Because the sorption data were normalized to the organic carbon content of the colloids, the partition coefficient of the Freundlich expression corresponds to the sorption constant (K_{oc}), which is derived from the relation:

$$K_{oc} = \frac{K_d \times 100}{\text{percent organic carbon}} \tag{2}$$

Each sorption isotherm was plotted with the averages of duplicate experiments determined at four or five concentration points run in triplicate. These concentrations were considerably below the aqueous solubility limit of the test compounds.

The thermodynamic equilibrium constant for a sorption reaction can be defined as

$$K_0 = \frac{a_s}{a_e \cdot a_c} = \frac{\gamma_s C_s}{\gamma_e C_e} \tag{3}$$

where a refers to the activity, γ to the activity coefficient, and C to the concentration of the organic compound (μg/mL). The subscripts s and e denote the compound–colloid complex and equilibrium solution, respectively. Because the colloid may be considered as a solid phase in this system, the activity (a_c) of this component of the system is by convention equal to unity. The concentration C_s (μg sorbed/g of C) was normalized to the organic carbon content of natural colloids. In ideal solutions, $\gamma = a/m$. As the solute becomes more dilute, the activity a_e approaches the modality m. Therefore, by the infinite dilution convention, as the concentration (molal units) of the solute approaches 0, γ approaches 1. Also, under these conditions the value of $1/n$ approaches unity. Thus equation 3 may be rewritten as

$$K_0 = \frac{a_s}{a_e} = \lim_{C_s \to 0} \frac{C_s}{C_e} \tag{4}$$

Because concentrations of organic pollutants in environmental samples are often low, equation 4 is fundamentally significant in interpreting sorption results. The standard free energy ($\Delta G°$) of the sorption reaction at equilibrium and constant T is related to the thermodynamic equilibrium constant by the relationship:

$$\ln K_0 = \frac{-\Delta G}{RT} \tag{5}$$

Values of $-\Delta G$ were obtained by extrapolating to $C_s = 0$ from plotted values of $-RT \ln C_s/C_e$ vs. C_s. This extrapolation also allows calculation of the thermodynamic equilibrium constant.

The structures of benzidine, *o*-toluidine, and azobenzene are shown in Chart I. The data obtained from the sorption experiments with aromatic amines are presented in Table II. These include amine solubilities, salinities, organic carbon contents of the enriched colloidal fraction, K_{oc}, and $\Delta G_T°$ values for each compound–colloid system. The water-solubility number for *o*-toluidine is the first experimental value to be reported. The amount of amine sorbed on colloids is reported on an organic carbon basis because the amount of carbon is the most precise measure available for monitoring the amount of colloid present in both bulk water and enriched colloidal fraction. In addition, organic matter is known to be a major source of sorptive capacity of a variety of environmental substrates for hydrophobic compounds.

The equilibrium sorption isotherms of benzidine, *o*-toluidine, and azobenzene at 20 °C are shown in Figures 1, 2, and 3, respectively. The sorption isotherm of *o*-toluidine was linear, although benzidine and azobenzene yielded curvilinear isotherms over the concentration ranges tested. The sorption data were well represented by the Freundlich equation ($r^2 > 0.96$),

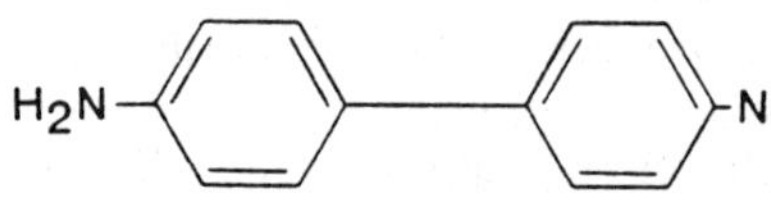

Benzidine

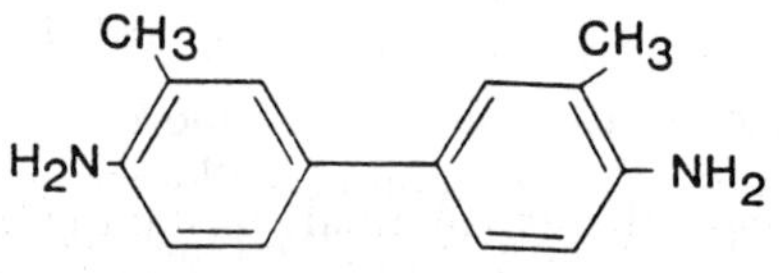

o-Toluidine

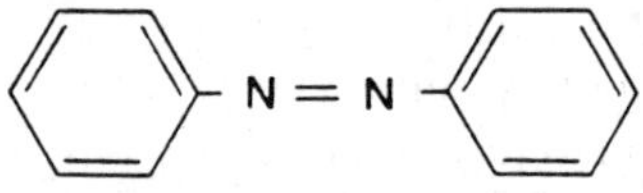

Azobenzene

Chart I. Structures of aromatic amines.

Table II. Sorption Parameters for Aromatic Amines on Estuarine Colloids

Chemical	*Solubility (ppm)*	*Salinity (‰)*	*TOC, Colloidal Fraction (mg/L)*	K_{oc}[a]	r^2	*1/n*	*Standard Free Energy (kcal/mol)*
Benzidine[b]	400	19.2	31.6	3825	0.981	0.74	−4.86
o-Toluidine[c]	75	19.3	34.5	2014	0.994	0.99	−4.83
Azobenzene[d]	300	19.4	38.3	1394	0.962	0.72	−4.72

[a]Ambient pH = 7.9.
[b]$pK_{a_1} = 4.5$, $pK_{a_2} = 3.3$.
[c]$pK_{a_1} = 4.8$, $pK_{a_2} = 3.7$.
[d]$pK_a = -2.48$.

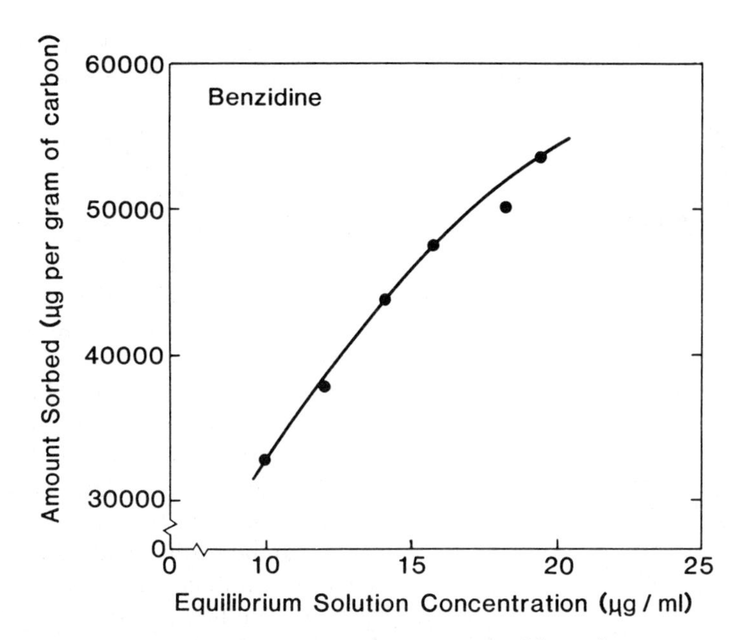

Figure 1. Equilibrium sorption isotherm at 20 °C of benzidine on estuarine colloids at pH 4.9.

where the exponential constant ($1/n$) had values of 0.74, 0.99, and 0.72 for benzidine, *o*-toluidine, and azobenzene, respectively. The curvilinear isotherm patterns of benzidine and azobenzene are in agreement with the class "L" isotherms defined by Giles et al. (*14*). They suggested that isotherm shape provides an indication of the sorption mechanism in operation for a given solute–solvent sorbent system. The L-type isotherm, the most common type, represents a relatively high affinity between the solute and the sorbent in the initial stages of the isotherm. As sorption sites are filled, the solute molecules have a decreasing probability of colliding with vacant sites. The curvilinear response may also be attributed to multiple mechanisms of sorption, as has been demonstrated with benzidine and a soil–sediment system (*15*).

The sorption experiments of aromatic amines with colloids under estuarine conditions (pH = 7.9, salinity ~19‰) yielded K_{oc} values of 3430, 2010, and 1390 for benzidine, *o*-toluidine, and azobenzene, respectively. These values, somewhat higher than corresponding K_{oc} data for soil–sediment systems, suggest that aromatic amines are more strongly bound to natural estuarine colloids. If the K_{oc} values reported here are converted to a mass basis by correcting for the fraction of carbon in the colloids (50% on average) (*15*), then they become K_p values of 1500, 1000, and 700, respectively. Zierath reported K_p values for benzidine on soils in the range of

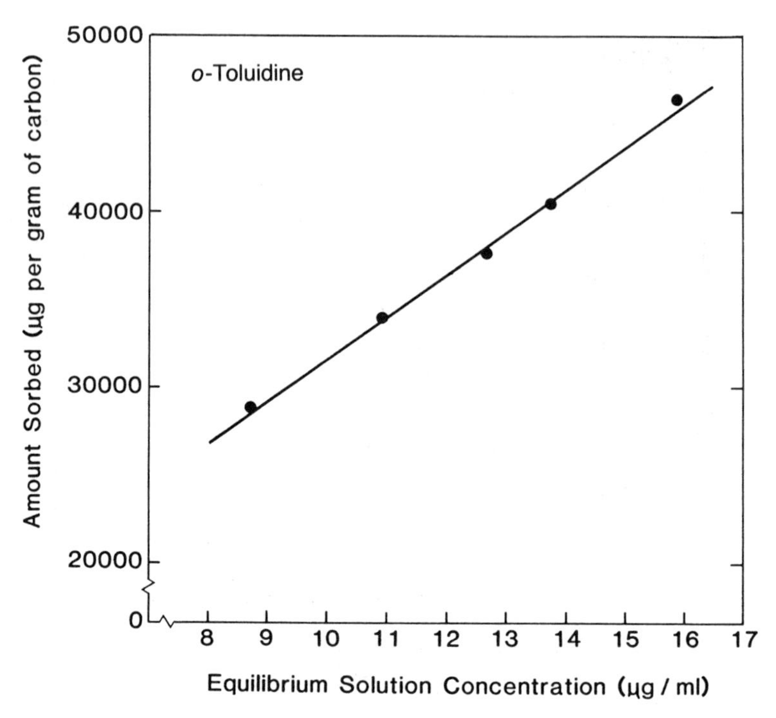

Figure 2. Equilibrium sorption isotherm at 20 °C of o-*toluidine on estuarine colloids at pH 7.9.*

50–3940 for soil with organic carbon contents of 0.15–2.4%. More recently, Johnson and Means (*16*) have reported K_p values in the range of 168–1230 for estuarine sediments having carbon contents of 0.9–3.0%. Estuarine colloids have been characterized as being composed of a carbohydrate–proteinaceous matrix in association with crystalline clay minerals and trace metals (*17, 18*).

In previous studies we demonstrated that estuarine colloids have a high affinity for PAHs (*9*) and for the herbicides atrazine and linuron (*10*). We also found that colloidal organic matter is on the order of 10 times better as sorptive substrate than soil or sediment organic matter (*11*). These apparent differences in sorption affinities exhibited by natural colloids and soils–sediments for hydrophobic organic pollutants may be related to differences in positional availability of sites for hydrophobic bonding and charge density separations on the two sorbents. Soil–sediment organic matter is itself sorbed to a highly porous and irregular inorganic matrix, often occupying or filling micropores in the structure. If it can be assumed that the soil organic matter is not a monomolecular film, the K_{oc} values calculated on a total-carbon basis may be low by 1 or 2 orders of magnitude, because all the organic carbon is not available as a sorptive surface and many charged

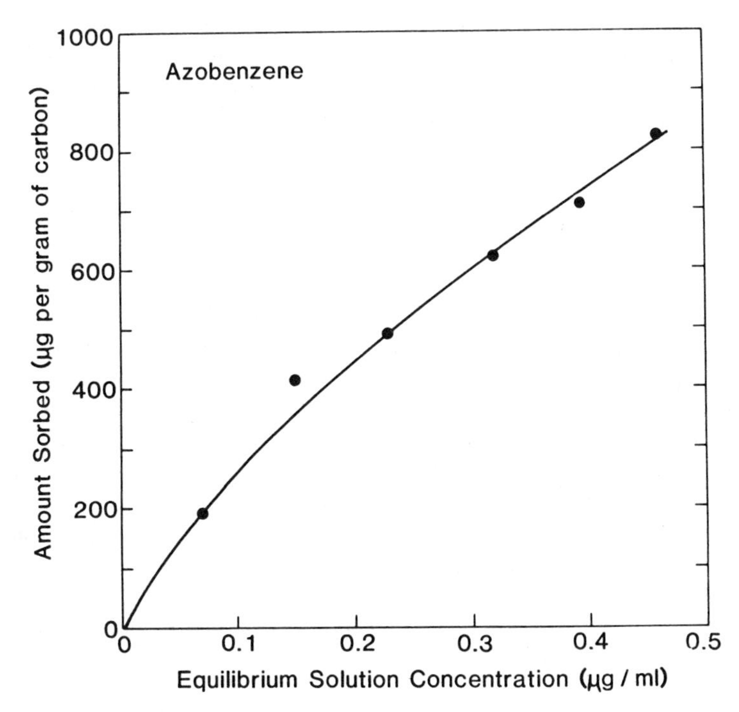

Figure 3. Equilibrium sorption isotherm at 20 °C of azobenzene on estuarine colloids at pH 7.9.

sites on the organic matter may already be involved in ion pairs with the inorganic matrix. The estuarine colloids used in the present and previously reported studies (*9–11*) have a very low inorganic content (<5% ash) and are presumably accessible for sorption from all sides in colloidal suspension. Thus, we hypothesize that the differences observed in sorptive capacities of sediment and colloidal organic matter may be explained in part by these factors.

Wijayaratne and Means (*9*) demonstrated, on the basis of a study of sorption of PAHs by natural estuarine colloids, that the solubility of a hydrophobic compound was significantly correlated with the colloid sorption constant (equation 6).

$$\log K_{oc} = -0.693 \log S(\mu g/mL) + 4.851 \qquad (r^2 = 0.985) \qquad (6)$$

The calculated K_{oc} values obtained with equation 6 for benzidine, *o*-toluidine, and azobenzene are 1116, 3561, and 1362, respectively. The calculated K_{oc} value of benzidine obtained with equation 6 is significantly lower than the observed value of 3430. This discrepancy suggests that benzidine sorption is enhanced above that expected on the basis of hydrophobic bonding alone, but the enhancement is nevertheless associated with the organic carbon content of the substrate. The mechanisms for enhanced sorption must involve one or more specific interactions of the amine functional group (e.g., ion pair) with components of either the substrate organic matter or associated clay minerals.

Benzidine can exist in solution as both a neutral species and an ionic (cationic) species by protonation of the amino groups. Zierath et al. (*15*) investigated the sorption behavior of benzidine by soils and sediments and found that sorption was highly correlated with pH because the ratio of neutral to ionized benzidine molecules is controlled by the pH of the aqueous phase. Although both species are subject to sorption, the cationic form may be sorbed to a greater extent.

Presumably, the curvilinear isotherms observed in this study are caused by multiple sorption mechanisms that are dependent on the protonation of nitrogen atoms of the diamino–biphenyl compounds in solution. The extreme pH-dependence of the colloid K_{oc} values for benzidine and *o*-toluidine, particularly as pH values approach the pK_a values of the compounds (Figures 4 and 5), further substantiates this hypothesis. The K_{oc} values of benzidine

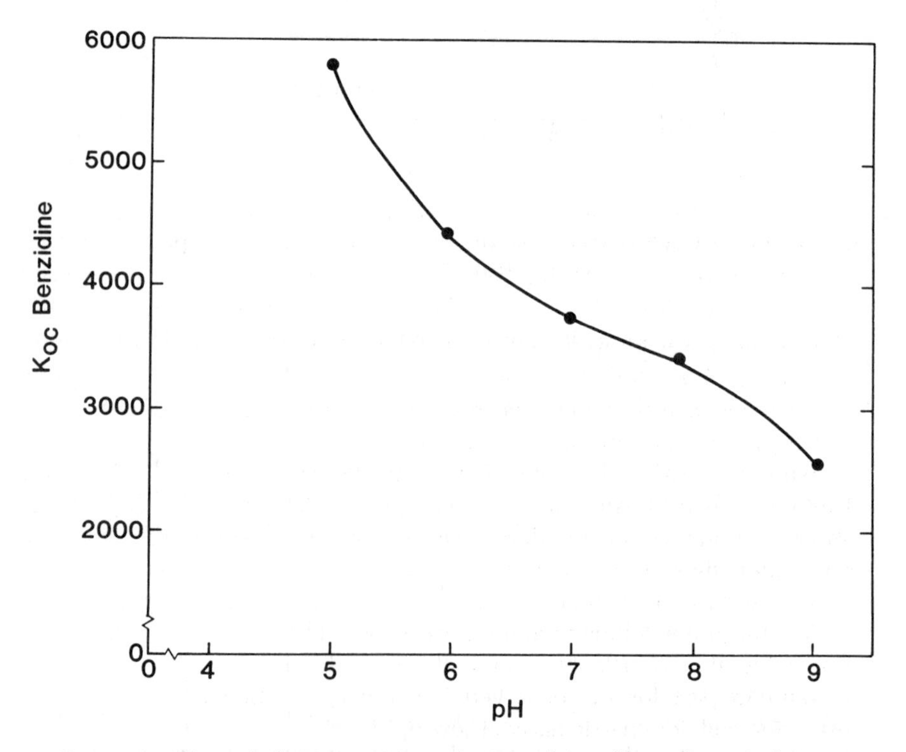

Figure 4. Effects of changes in pH on the K_{oc} *values of benzidine. Each point represents a full isotherm determined at the pH specified. Regression coefficients were all greater than 0.95.*

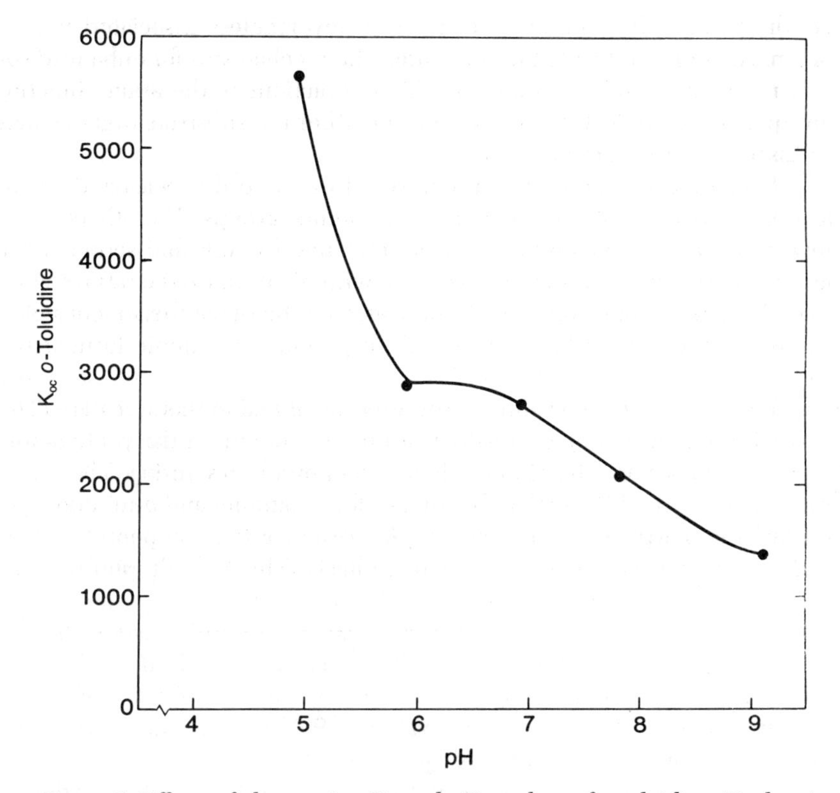

Figure 5. Effects of changes in pH on the K_{oc} *values of* o-*toluidine. Each point represents a full isotherm determined at the pH specified. Regression coefficients were all greater than 0.95.*

showed increases of up to 2000 units when the ambient pH was decreased from 7.9 to 5.0.

The mechanism for enhanced sorption of aromatic amines must involve specific interactions of the amine functional group with components such as carbonyl and carboxyl groups of the colloidal organic matter or associated clay minerals in possible modes leading to adduct formation. Parris (*8*) has shown that amines may undergo both reversible and irreversible reactions with humates and compounds that serve as models of humate functional groups (e.g., carbonyls and quinones) to yield a variety of products. Benzidine and other aromatic amines are known to react with components of clay minerals (*15, 19, 20*), especially iron. Reactions of these types may account in part for the observed increase in sorption of benzidine with estuarine colloids, particularly at low pH levels.

In contrast to the results from benzidine sorption, the K_{oc} value of 3561 for *o*-toluidine predicted by equation 6 is higher than the observed value of 2010. From thermodynamic considerations, because the $\Delta G_T°$ values of benzidine and *o*-toluidine are similar, the two compounds would be expected to show similar sorptive behavior. Furthermore, the solubility of *o*-toluidine (75 ppm) is considerably lower than that of benzidine (400 ppm). This solubility difference indicates that the escaping tendency, which is measured quantitatively by the chemical potential in solution, would be greater for *o*-toluidine. Hence, *o*-toluidine would be expected to show more sorption. This apparent discrepancy in sorptive behavior may be explained if molecular configurations are taken into consideration. The methyl substituents at the ortho positions of the phenyl ring in *o*-toluidine cause inductive and resonance effects to operate at the reaction center. Although the sorption reaction may be facilitated by a higher electron density at the amino groups, the close proximity of the methyl groups to the nitrogen atoms may sterically hinder the molecule in forming adducts with colloidal organic matter. The influence of steric factors on bonding mechanisms is particularly noticeable with aromatic amines and humates (*8*). The linearity in the sorption isotherm (Figure 2) strongly suggests that the sorption of *o*-toluidine by organic colloids is essentially a partitioning process at higher pH values. We observed similar sorption behavior with PAHs, a class of neutral aromatic hydrophobic compounds (*9*). This observation further lends support to our hypothesis that, in the case of *o*-toluidine, the aromatic rings are primarily responsible for the observed sorption characteristics with estuarine colloids at pH 7–8.

The K_{oc} value predicted for azobenzene by equation 6 was 1362. The close agreement of the observed K_{oc} value (1390) with the calculated value suggests that sorption is controlled by the neutral aromatic portion of the molecule. This observation is consistent with the fact that, unlike benzidine, the two doubly-bonded nitrogen atoms contained in azobenzene are sterically hindered with sorption reactions involving attack on the nitrogen atoms. Although the K_{oc} values increased with decreasing pH (Figure 6), the magnitude of this increase was at least 1000 units less than that observed with benzidine for corresponding pH changes. The increase in sorption observed at lower pH values cannot be attributed to protonation of the azobenzene molecule, because the pK_a value of the azo nitrogens is –2.48. Furthermore, changes occurring within the estuarine polymeric material itself may account in part for the high K_{oc} values observed at low pH (*10*).

Acknowledgments

This work was performed at the Center for Environmental and Estuarine Studies of the University of Maryland and represents portions of the doctoral dissertation of R. D. Wijayaratne, who recognizes the support by a graduate research assistantship from the Center.

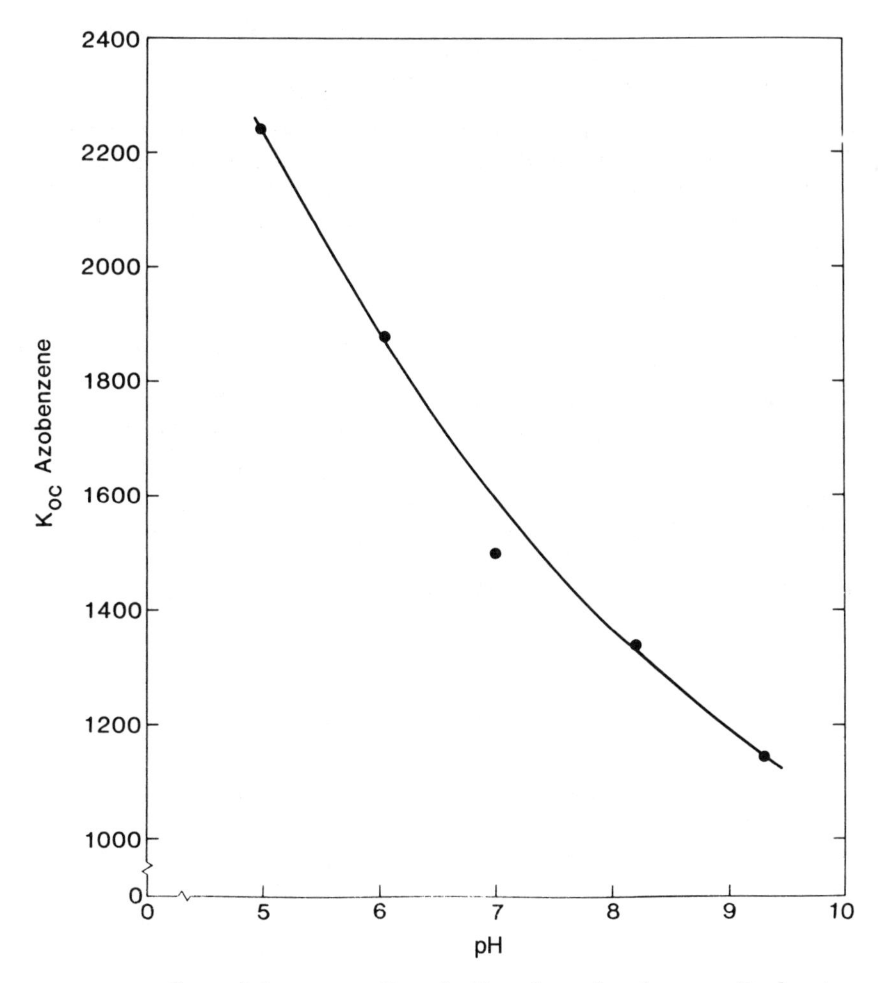

Figure 6. Effects of changes in pH on the K_{oc} values of azobenzene. Each point represents a full isotherm determined at the pH specified. Regression coefficients were all greater than 0.95.

References

1. Peters, R. H. *Textile Chemistry*; Elsevier: Amsterdam, 1975; pp 582–648.
2. *Fed. Regist.* **1973,** *48*(85), 10929.
3. Weisberger, E. K.; Russfield, A. B.; Homburger, F.; Weisburger, J. H. Boger, E.; Van Dongen, C. G.; Chu, K. C. *J. Environ. Pathol. Toxicol.* **1978,** *2*, 325.
4. Malaney, G. W.; Latin, P. A.; Cibulka, J. J.; Hickerson, L. H. *J. Water Pollut. Control Fed.* **1967,** *37*, 2020.
5. Baird, R. B.; Carmona, L. G.; Jenkins, R. L. *J. Water Pollut. Control Fed.* **1977,** *49*, 1609.

6. Nelson, C. R.; Hites, R. A. *Environ. Sci. Technol.* **1980,** *14,* 1147.
7. Hsu, T.-S.; Bartha, R. *Soil Sci.* **1974,** *116,* 444.
8. Parris, G. E. *Environ. Sci. Technol.* **1980,** *14,* 1099.
9. Wijayaratne, R. D.; Means, J. C. *Mar. Environ. Res.* **1984,** *11,* 77.
10. Means, J. C.; Wijayaratne, R. D. *Science (Washington, DC)* **1982,** *215,* 968.
11. Wijayaratne, R. D.; Means, J. C. *Environ. Sci. Technol.* **1984,** *18,* 121.
12. Menzel, D. W.; Vaccaro, R. F. *Limnol. Oceanogr.* **1964,** *9,* 138.
13. Means, J. C.; Hassett, J. J.; Wood, S. G.; Banwart, W. L. In *Polynuclear Aromatic Hydrocarbons*; Jones, P. W.; Leber, P., Eds.; Ann Arbor Science: Ann Arbor, 1979; p 327.
14. Giles, C. H.; MacEwan, T. H.; Nakhwa, S. N.; Smith, D. *J. Chem. Soc.* **1960,** *3,* 973.
15. Zierath, D. L.; Hassett, J. J.; Banwart, W. L.; Wood, S. G.; Means, J. C. *Soil Sci.* **1980,** *129,* 277.
16. Johnson, W. E.. M. S. Thesis, University of Maryland, 1986.
17. Sigleo, A. C.; Hare, P. E.; Helz, G. R. *Estuarine Coastal Shelf Sci.* **1983,** *17,* 87.
18. Means, J. C.; Wijayaratne, R. D. *Bull. Mar. Sci.* **1984,** *35,* 449.
19. Theng, B. K. G. *Clay Miner.* **1971,** *19,* 383.
20. Soloman, C. H.; Loft, B. C.; Swift, J. C. *Clay Miner.* **1968,** *7,* 389.

RECEIVED for review July 24, 1987. ACCEPTED for publication April 7, 1988.

15

Effect of Dissolved Organic Matter on Extraction Efficiencies

Organochlorine Compounds from Niagara River Water

Caryl L. Fish, Mark S. Driscoll, and John P. Hassett

Department of Chemistry, State University of New York, College of Environmental Science and Forestry, Syracuse, NY 13210

Simon Litten

New York State Department of Environmental Conservation, 50 Wolf Road, Albany, NY 12233

Two techniques for the extraction of a series of organochlorine compounds (chlorinated benzenes, polychlorinated biphenyls, DDT, DDE, and mirex) from centrifuged Niagara River water were compared. The more hydrophobic compounds were extracted more efficiently by a digestion technique than by conventional hexane extraction. Plots of the relative recovery (R = *undigested/digested*) *versus log of the octanol–water coefficient* (K_{ow}) *show* R *decreasing exponentially with log* K_{ow}. *This decrease suggests that the digestion–extraction recovers both the dissolved fraction and the fraction bound to organic matter, although conventional solvent extraction does not recover the bound fraction efficiently.*

HUMIC ACIDS, FULVIC ACIDS, AND DISSOLVED ORGANIC MATTER, which are found in all natural waters, are very complex compounds of both aquatic and terrestrial origin. Humic substances are polyelectrolytes containing both aromatic and aliphatic carbon with phenolic, alcoholic, carbonyl, acidic, and amino functional groups.

0065–2393/89/0219–0223$06.00/0

The binding of hydrophobic organic compounds to humic substances has been studied by several investigators. Mirex (1,1a,2,2,3,3a,4,-5,5,5a,5b,6-dodecachlorooctahydro-1,3,4-metheno-1*H*-cyclobuta[*cd*]pentalene) (*1*), 2,2′,5,5′-tetrachlorobiphenyl (*2, 3*), DDT (1,1′-(2,2,2-trichloroethylidene) bis[4-chlorobenzene]) (*4*), and cholesterol (*5*) bind to humic substances. This binding has been shown to affect gas exchange (*2, 3*), bioavailability (*6*), particle adsorption (*7*), photolysis (*8*), and hydrolysis (*9*) reactions. To date, very few investigators have looked at how this binding affects the extraction efficiencies of hydrophobic compounds from water. Hassett and Anderson (*5*) found that the solvent extraction efficiency of cholesterol from water was reduced in the presence of dissolved organic matter (DOM). Similarly, Carlberg and Martinsen (*10*) showed that both solvent and XAD–2 extraction efficiencies of alkanes, polycyclic aromatic hydrocarbons (PAH), chlorinated hydrocarbons, and phthalates were less than 100% in the presence of humic substances. The extraction efficiencies of these compounds decreased as the equilibrium time of the binding was increased from 4 to 60 days. Therefore, the equilibrium time between spiking and extraction is very important.

In this study, the extraction efficiencies of 23 organochlorine compounds from Niagara River water were determined. The compounds were chlorobenzenes, polychlorinated biphenyl (PCB) congeners, DDT, DDE, mirex, and photomirex. The extraction efficiencies were determined by two methods: conventional liquid–liquid extraction with hexane and digestion to break down dissolved organic matter combined with hexane extraction.

Experimental Details

Reagents. All extractions were done with pesticide-grade hexane (Baker Chemical), and digestion–extractions used chromic acid cleaning solution (90–96% sulfuric acid, 1% CrO_3). The acetone and petroleum ether (30–60 °C boiling range) used for cleaning were glass-distilled. All stock solutions were made with pesticide-grade benzene (Fisher Scientific).

Standard Compounds. Table I lists the 23 compounds used in this study in order of increasing elution time from the gas chromatograph. The chlorinated benzenes were obtained from Aldrich Chemical Co.; mirex was donated by the United States Environmental Protection Agency; photomirex was donated by Environment Canada. All other compounds were obtained from Ultra Scientific Inc.

Accurate amounts of each standard compound were prepared by weighing on a microbalance, with the exception of 1,2,4-trichlorobenzene, which was measured volumetrically. Each standard was dissolved in benzene and diluted to 10 mL in an individual volumetric flask. Mixed standards were prepared by measuring appropriate volumes of the stock solutions into a 10-mL volumetric flask and diluting to the mark with hexane. One mixed standard contained mirex and photomirex; the other contained the remaining 21 compounds.

Table I. Organochlorine Compounds Used in This Study

No.	*Compound Name*	*log* K_{ow}[a]	*Relative Recovery*[b]	*Mass Spiked*[c] *(ng)*
1	1,2,4-Trichlorobenzene	3.98	1.008	145.4
2	1,2,3-Trichlorobenzene	4.04	1.001	76.6
3	1,2,3,4-Tetrachlorobenzene	4.55	0.926	26.6
4	2-Chlorobiphenyl	4.5	0.577	984.0
5	Pentachlorobenzene	5.03	0.960	39.3
6	4-Chlorobiphenyl	4.61	0.671	2720.0
7	2,4-Dichlorobiphenyl	5.15	0.947	249.0
8	Hexachlorobenzene	5.47	0.924	41.2
9	4,4′-Dichlorobiphenyl	5.36	0.890	675.0
10	2,4,4′-Trichlorobiphenyl	5.74	0.935	103.9
11	2,2′,5,5′-Tetrachlorobiphenyl	6.26	0.938	107.0
12	2,2′,3,4-Tetrachlorobiphenyl	6.11	0.916	49.1
13	*o,p*′-DDE	—	0.879	30.5
14	2,2′,4,5,5′-Pentachlorobiphenyl	6.85	0.896	99.1
15	*p,p*′-DDE	—	0.792	34.7
16	3,3′,4,4′-Tetrachlorobiphenyl	5.62	0.707	243.0
17	2,3′,4,4′,5-Pentachlorobiphenyl	7.12	0.793	95.6
18	*o,p*′-DDT	—	0.902	12.9
19	2,2′,4,4′,5,5′-Hexachlorobiphenyl	7.75	0.830	117.0
20	*p,p*′-DDT	6.36	0.830	139.0
21	Photomirex	—	0.760	57.8
22	2,2′,3,4,4′,5,5′-Heptachlorobiphenyl	7.20	0.686	96.6
23	Mirex	6.89	0.640	63.4

[a]Data are from ref. 13.
[b]The relative recovery is the undigested fraction recovered per digested fraction recovered.
[c]Into approximately 1 L of water.

Sampling. Samples were collected biweekly from January 22 to April 30, 1986, from the outlets of continuous-flow centrifuges at sampling stations established by Environment Canada on the Niagara River. These were located at Fort Erie and Niagara-on-the-Lake, Ontario. Seven sets of samples were taken. Duplicate samples for direct solvent extraction were collected in 1-L glass Wheaton bottles with aluminum-foil-lined screw caps. Bottles were weighed on a triple-beam balance before and after sampling to determine the amount of water collected. Duplicate samples for digestion–extraction were collected in 2-L round-bottom flasks with standard taper (24/40) glass stoppers. The flasks were filled to an approximate 1200-mL mark, and the exact amount was determined by weighing as just described.

Solvent Extraction. Samples in the Wheaton bottles were spiked with 5 μL of the mirex–photomirex stock solution and 10 μL of the other combined standard solution. These samples were shaken for 24 h at 25 °C on a shaker table. To extract each spiked sample, the contents of the sample bottle was poured into a 2-L separatory funnel, the bottle was rinsed with 70 mL of hexane, and the hexane was poured into the separatory funnel. The funnel was shaken vigorously for 2 min to allow the phases to separate, and the hexane layer was removed. This process was repeated twice more, including the rinsing of the sample bottle, so that the water was extracted a total of three times. The combined hexane layers were passed through a column of 8 g of anhydrous sodium sulfate to remove traces of water and evaporated to 5–10 mL in a concentration apparatus (Kuderna-Danish).

Digestion–Extraction. The round-bottom flasks were spiked with 5 μL of the mirex–photomirex standard solution and 10 μL of the other combined standard solution. The round-bottom flasks were shaken for 24 h on a shaker table at 25 °C. Each sample was treated by addition of 5 mL of chromic acid and 200 mL of hexane. After each flask was fitted with a reflux condenser and placed in a heating mantle, the hexane was refluxed for 2 h. The samples refluxed smoothly without boiling chips or stirring. After it cooled, each sample was transferred to a 2-L separatory funnel and shaken vigorously. Once the phases had separated, the hexane layer was removed, passed through a column containing 8 g of anhydrous sodium sulfate, and evaporated to 5–10 mL in a concentration apparatus (Kuderna-Danish).

Gas Chromatography. Extracts were analyzed with a gas chromatograph (Varian 3400) with a splitless capillary column injector, auto injector, a 60-m × 0.25-mm (i.d.) SPB–1 (0.25-μm film thickness) fused silica capillary column (Supelco Inc.) and a Ni-63 electron-capture detector. Data were collected by a data aquisition system (Keithley DAS, Series 500) connected to a microcomputer (Leading Edge, Model D) and stored on floppy disks. Data aquisition was controlled with Labtech Notebook software. The output from the gas chromatograph was also sent to a strip chart recorder. Prior to injection, 100 μL of 2,2′,3,3′,5,5′,6,6′-octachlorobiphenyl was added to the extracts as an internal standard. The column temperature was held at 60 °C for 5 min, raised to 270 °C at a rate of 10 °C/min, and held at 270 °C for 30 min. Compounds were quantified by comparing peak heights to a standard curve and making volume corrections based on the internal standard response.

Results and Discussion

The 23 compounds used in this study are listed in Table I, with amounts spiked into the Niagara River water samples, log K_{ow}, and average relative recoveries [(R) = fraction recovered by solvent extraction per fraction recovered by digestion–extraction]. Unspiked samples were also analyzed, and the native concentrations were not significant relative to the spiked concentrations. Preliminary experiments indicated that all 23 compounds were recovered efficiently from distilled water by both the solvent extraction and the digestion–extraction methods. Relative recoveries in Table I demonstrate that some of the test compounds were recovered more efficiently from Niagara River water by the digestion–extraction method than by the solvent extraction method (relative recoveries less than 1). In addition, hexane–water emulsions did not form in the digested samples, but frequently caused phase separation problems in the undigested samples. Therefore, the digestion–extraction method is a superior technique for extraction of the test compounds from the Niagara River water. However, use of chromic acid digestion may not always be the method of choice for the extraction of natural waters. The oxidation of some compounds of interest is a possible problem with this method. Problems may also arise from interfering compounds formed as a result of the partial digestion of the dissolved organic matter.

Relative recovery also decreased with increasing log K_{ow} (Figure 1). This result may appear counterintuitive but can be explained if some fraction of a compound bound by DOM in Lake Erie water (dissolved organic carbon

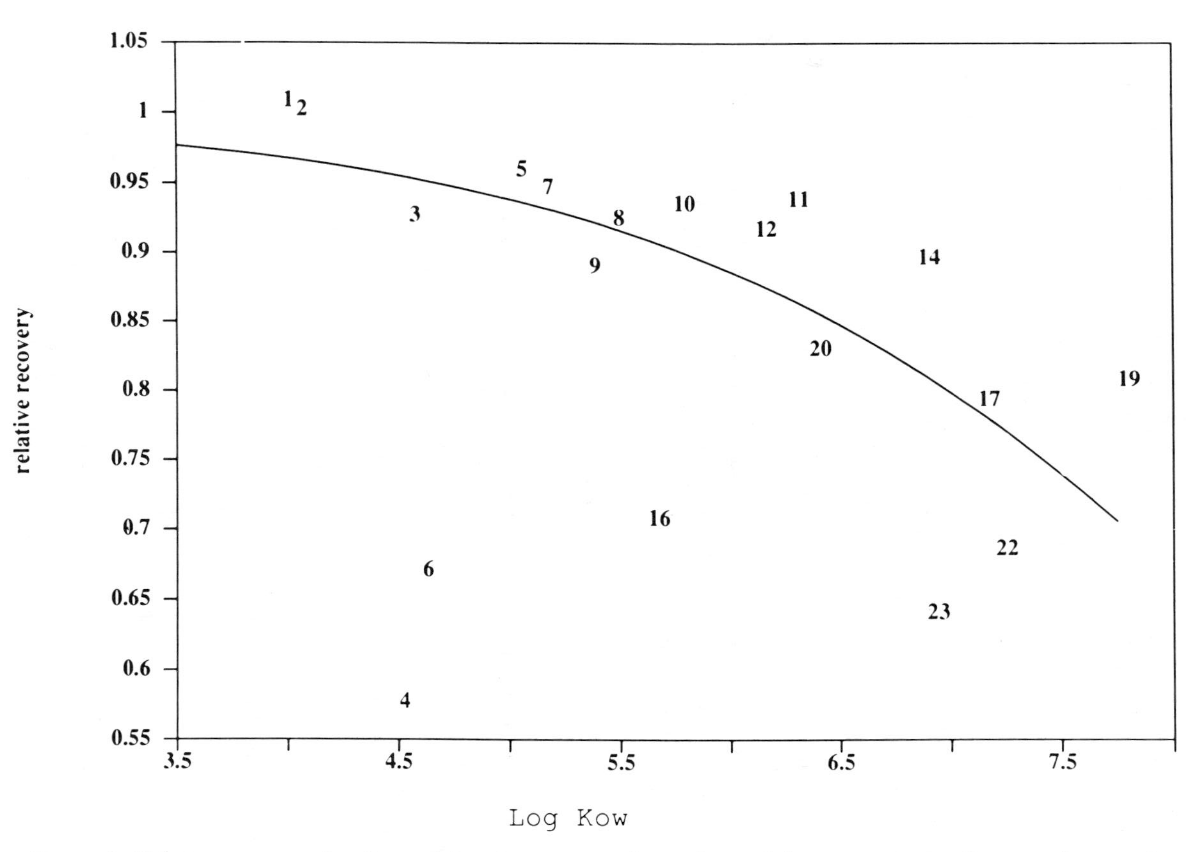

Figure 1. Relative recovery (undigested fraction recovered per digested fraction recovered) versus log K_{ow} *for each compound in Table I.*

(DOC) is 2.1–7.0 mg/L) is recovered more efficiently by digestion–extraction than by solvent extraction. This effect could occur as a result of partial or complete destruction of the DOM by reaction with chromic acid, which would release DOM-bound compounds and make them more available for solvent extraction. Compounds with higher K_{ow} values tend to be bound to a greater extent by DOM (*11*). These results can be explained quantitatively from

$$C_T = C_b + C_f \tag{1}$$

where C_T is the total concentration of the organochlorine compound in water, C_b is the bound concentration, and C_f is the free concentration. The binding constant, K_{doc}, is defined as

$$K_{doc} = \frac{C_{doc}}{C_f} = \frac{C_b}{C_f[DOC]} \tag{2}$$

where C_{doc} is the concentration of the bound compound in the dissolved organic carbon and [DOC] is the concentration of the dissolved organic carbon. Combining equations 1 and 2 yields

$$\frac{C_T}{C_f} = [DOC]K_{doc} + 1 \tag{3}$$

K_{doc} is related to K_{ow} (*12*) by an equation that takes the form of

$$\log K_{doc} = a \log K_{ow} + b \tag{4}$$

Equation 4 can be written as

$$K_{doc} = 10^{a \log K_{ow} + b} \tag{5}$$

So, from equations 3 and 5,

$$\frac{C_T}{C_f} = [DOC]10^{a \log K_{ow} + b} + 1 \tag{6}$$

Assuming that the digestion–extraction method extracts both bound and free compounds (C_T), and that solvent extraction recovers only the free compounds (C_f), then $R = C_f/C_T$, and therefore

$$R = [[DOC]10^{a \log K_{ow} + b} + 1]^{-1} \tag{7}$$

The line in Figure 1 is the nonlinear least squares fit of equation 7 to the experimental points. Compounds **4** (2-chlorobiphenyl), **6** (4-chlorobiphenyl), and **16** (3,3′,4,4′-tetrachlorobiphenyl) appear to be outliers and are not included in the regression. The use of an average DOC concentration of 4.0 mg/L for Lake Erie water yields a and b values of 0.29 and 2.76, respectively.

As can be seen from Table I and Figure 1, the ratios for compounds **4**, **6**, **16**, and **23** (mirex) are substantially lower than the line fitted to equation 7. The low ratios indicate that they are less efficiently extracted by solvent extraction than predicted by the correlation with log K_{ow}. Such deviations might occur if the driving force binding these compounds to organic matter included both hydrophobic interactions (as reflected by dependence on K_{ow}) and specific electronic interactions of some compounds with sites in the DOM.

Conclusion

A chromic acid digestion technique yields enhanced performance compared to conventional hexane solvent extraction for recovery of organochlorine compounds from Niagara River water. The effect is especially pronounced for highly hydrophobic compounds. These results may be due to digestion of natural organic matter, which otherwise binds hydrophobic compounds and inhibits their solvent extraction.

References

1. Yin, C.; Hassett, J. P. *Environ. Sci. Technol.* **1986,** *20,* 1213.
2. Hassett, J. P.; Milicic, E. *Environ. Sci. Technol.* **1985,** *19,* 638.
3. Jota, M. M.S. Thesis, State University of New York, College of Environmental Science and Forestry, 1983.
4. Carter, C. W.; Suffet, I. H. *Environ. Sci. Technol.* **1982,** *16,* 735.
5. Hassett, J. P.; Anderson, M. A. *Environ. Sci. Technol.* **1979,** *13,* 1526.
6. Boehm, P. D.; Quinn, J. G. *Estaurine Coastal Mar. Sci.* **1976,** *4,* 93.
7. Hassett, J. P.; Anderson, M. A. *Water Res.* **1982,** *16,* 681.
8. Zepp, R. G. *Chemosphere* **1981,** *10,* 109.
9. Perdue, E. M.; Wolfe, N. L. *Environ. Sci. Technol.* **1982,** *16,* 847.
10. Carlberg, G. E.; Martinsen, K. *Sci. Total Environ.* **1982,** *25,* 245.
11. Carter, C. W. Ph.D. Thesis, Drexel University, 1984.
12. Karickhoff, S. W.; Brown, D. S.; Scott, T. A. *Water Res.* **1979,** *13,* 241.
13. Rappaport, R. A.; Eisenreich, S. J. *Environ. Sci. Technol.* **1984,** *18,* 163–170.

RECEIVED for review July 24, 1987. ACCEPTED for publication February 12, 1988.

16

Effects of Humic Acid on the Adsorption of Tetrachlorobiphenyl by Kaolinite

Gregory A. Keoleian and Rane L. Curl

Department of Chemical Engineering, University of Michigan, Ann Arbor, MI 48109

Humic acids (HA) can affect the partitioning of organic contaminants between aqueous and sediment phases by complexation in solution and adsorption of the contaminant–HA complex to a mineral surface. The objective of this work was to elucidate these and other mechanisms, using ^{14}C-radiolabeled 2,2′,4,4′-tetrachlorobiphenyl (TeCB), a filtered humic acid preparation, and two natural kaolinites (KGA). The binary interactions, TeCB with KGA, HA with KGA, and TeCB with HA, were studied experimentally at 25 °C and pH 6.9. Isotherms were measured for the TeCB–HA–KGA multicomponent system to evaluate adsorption partition coefficients. The data were fitted satisfactorily, within estimated limits of uncertainty, by a model that assumes noncompetitive TeCB and HA adsorption and the same binding constant between free TeCB and dissolved or adsorbed HA.

HUMIC SUBSTANCES PLAY AN IMPORTANT ROLE as macromolecular vectors for the transport of hydrophobic organic contaminants (HOC) in groundwater and surface water. In natural waters, HOC can exist free or in a bound state associated with dissolved humic substances, aquatic organisms, suspended particulates, and other colloidal matter. The binding or association of HOC with dissolved humic substances has been measured for many systems (*1–12*).

The equilibrium binding constant, K_{AB}, relates the concentration of the pollutant (A) in the bound state, C_{AB}; the concentration of A in the free state,

0065–2393/89/0219–0231$06.00/0

C_A; and the concentration of humic substance (let B represent fulvic acid and/or humic acid) as dissolved organic carbon, C_B.

$$C_{AB} = K_{AB}C_BC_A \quad (1)$$

The humic substances can enhance the total concentration of HOC in solution by as much as the factor $K_{AB}C_B$ and, consequently, can strongly influence HOC migration from hazardous waste sites, agricultural runoff, and other sources.

The fate of this HOC–humic acid (AB) complex has not been fully investigated. McCarthy and Jimenez (*8*) studied the dissociation of the AB complex and observed that the binding of benzo[*a*]pyrene to dissolved humic material was completely reversible. Other studies were concerned with the degradation of the pollutant that is solubilized or complexed by humic substances (*1, 3*).

An important mechanism that can immobilize HOC in an aqueous environment is adsorption of a HOC–humic (AB) complex on a solid surface. The effect of humic substances and other dissolved organic matter on the sorption of HOC by natural sorbents has been studied by Hassett and Anderson (*13*), Caron et al. (*14*), Brownawell and Farrington (*15*), and Baker et al. (*16*). They observed a reduction in the solid-phase partition coefficient as dissolved organic carbon was added to HOC–natural sorbent systems. This effect was attributed to an enhancement in the aqueous phase HOC chemical activity by the factor $K_{AB}C_B$, which leads to the following relation for the apparent partition coefficient:

$$K_A^* = \frac{K_A}{1 + K_{AB}C_B} \quad (2)$$

where K_A is the linear adsorption constant in the absence of dissolved organic carbon. The mechanism of AB complex immobilization, however, was not studied experimentally by the investigators cited. It is a major objective of the present work.

Generalized Mechanistic Description of Interactions

The possible mechanisms of interaction between HOC, humic acid, and natural sorbent systems are illustrated in Figure 1. This interaction represents a complex network of equilibria further complicated by the heterogeneous nature of the sorbent and humic acids. Naturally occurring organic matter indigenous to soils and sediments can desorb or dissolve into the aqueous phase and complex with HOC (*17*). These "implicit adsorbates" (*18*) may also compete for sites on the sorbent surface; in both scenarios the true sorption behavior may be misrepresented if these effects are not identified.

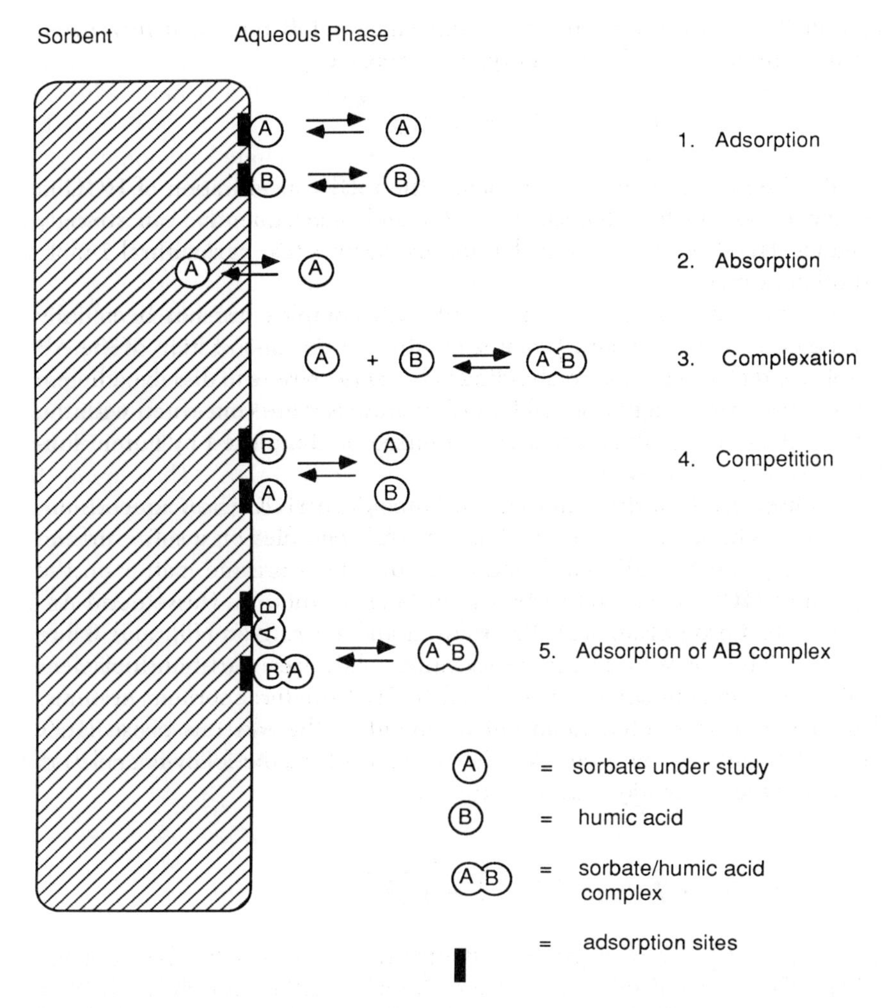

Figure 1. Possible mechanisms of interaction between a model sorbate, humic acid, and a natural sorbent.

The mechanism for the interaction of HOC with soils is still unresolved; Chiou et al. (*19*) and others support an absorption or partitioning process, whereas Mingelgrin and Gerstl (*20*) and MacIntyre and Smith (*21*) conclude that the process is adsorption. The unclear distinction between absorption and adsorption arises in part from the difficulty in distinguishing between a sorbent surface and sorbent phase for the organic matrix of soil.

Scope of This Investigation

The complete set of mechanisms that are represented in Figure 1 would be difficult to resolve through experimental investigation. Selecting a non-swelling and nonporous mineral such as kaolinite precludes absorption of the model sorbate and reduces the complexity of the investigation. In addition, implicit adsorbate effects are not possible with a mineral adsorbent free of organic carbon.

Criteria for selection of a model adsorbate included nonionic organic compound, relatively high aqueous-phase activity coefficient, chemical stability (resistance to chemical or microbial degradation), and environmental significance. On the basis of these criteria, ^{14}C-radiolabeled 2,2′,4,4′-tetrachlorobiphenyl (TeCB) was chosen for this study.

Aldrich humic acid was used after purification by membrane filtration. Malcolm and MacCarthy (*22*) have reported, on the basis of ^{13}C NMR spectra, that commercial humic acids are not representative of soil and aquatic humic acids. Gauthier et al. (*12*), however, measured partition coefficients (normalized to the fraction of organic carbon) for pyrene and 14 different humic substances, including untreated Aldrich humic acid (HA). They found that K_{AB} varied by as much as a factor of 10, depending on the source of humic material. The K_{AB} measured for the Aldrich humic acid was within the range of values reported for the other soil humic acids studied.

Sorption data at 25 °C were measured and isotherms were deduced for the interactions between the TeCB–humic acid, TeCB–kaolinite, and humic acid–kaolinite systems. Equilibrium parameters for these isotherms were used in multicomponent sorption models to describe TeCB partitioning in the TeCB–humic acid–kaolinite system. Model predictions are compared with experimental multicomponent sorption data to choose between models and provide a basis for further refinement.

Adsorption of Tetrachlorobiphenyl by Kaolinite

Anomalous results for adsorption partition coefficients have been observed for many HOC–clay mineral systems (*23, 24*). The adsorbent–concentration effect (dependence of the partition coefficient on the clay concentration of a suspension) and adsorption–desorption hysteresis are the two most significant apparent anomalies examined in the soil science and environmental science literature. In some cases that exhibit an adsorbent–concentration effect and adsorption–desorption hysteresis, experimental artifacts and other complex physicochemical phenomena may have been misinterpreted as anomalous sorption behavior (*17, 25, 26*). With this perspective in mind, the adsorption experiments were designed with a special emphasis on control and TeCB mass accountability.

Experimental Details. The ^{14}C-radiolabeled 2,2′4,4′-tetrachlorobiphenyl (TeCB) was purchased from Pathfinder Laboratories with a specific activity of 10.5 mCi/mmol and a chemical and radiochemical purity of greater than 98%. Two natural kaolinites (KGA) from the Source Clays Repository at the University of

Missouri, one well crystallized (KGA–1) and the other poorly crystallized (KGA–2), were studied. Surface properties and other characteristics of these kaolinites are given in Table I; additional physical and chemical characteristics are available from van Olphen and Fripiat (27). Type I deionized water, which was produced from a microfilter system (Millipore Corporation, Milli-R/Q), was glass-distilled from permanganate. This reagent water was buffered with 0.0025 M KH_2PO_4 and 0.0025 M $NaHPO_4$ to maintain a pH of 6.9. All glassware was baked in an oven at 600 °C to remove residual organics, except for volumetric glassware and 50-mL and 150-mL Corex centrifuge tubes used as sorption vessels. These items were cleaned with chromic–sulfuric acid cleaning solution and thoroughly rinsed with deionized water. The centrifuge tube caps were lined with aluminum foil discs that were cleaned with acetone and baked.

TeCB–KGA adsorption isotherms were measured by both a modified aqueous difference method and a direct extraction method. For both methods, TeCB material balance calculations were corrected for adsorption of TeCB by the vessel. Stock aqueous TeCB solutions were prepared by injecting TeCB dissolved in hexane into a glass bottle, allowing the hexane to evaporate, adding phosphate buffer solution, and agitating overnight. The 50-mL Corex tubes were charged with 50 mL of stock TeCB solution, with a Class A volumetric pipet. The tubes were agitated in a thermostatted shaker bath (150 cycles/min) at 25.0 ± 0.5 °C for 24 h to achieve vessel adsorption equilibrium, and then two approximately 2-mL initial TeCB concentration (C_{A0}) samples were collected. An Oxford pipetter was adapted to fit glass disposable Pasteur pipet tips, and sample volume was calibrated gravimetrically. The kaolinite was weighed and added to make a suspension of nominally m/V = 2.0 g/L for all isotherm measurements, except for the TeCB–KGA–1 isotherm measurement, where m/V = 1.0 g/L. Vessels were agitated for 24 h, although equilibrium was achieved in 2 h or less, as shown in Figure 2. The aqueous and solid phases were separated by centrifugation at 3500 × g for 2 h in a thermostatted centrifuge, and two 2-mL equilibrium (C_A) samples were taken from the supernatant liquid. For the extraction method, 35 mL of supernate was carefully removed and 20 mL of hexane added to extract both the TeCB adsorbed by the kaolinite and the vessel and the TeCB dissolved in the residual supernate (approximately 7 mL). The vessels were agitated 24 h for extraction and centrifuged for 15 min; then two 5-mL samples of the hexane phase were collected. Both aqueous and hexane samples were added to 15 mL of scintillation cocktail (Safety-Solve) and counted with a scintillation counter (either Packard 4430 or LKB 1219). Calibrations were performed for quench correction.

The adsorbed TeCB concentration, C'_A, was determined for the modified difference method by

$$C'_A = \frac{V(C_{A0} - C_A) + M_{AV0} - M_{AV}}{m} \tag{3}$$

where M_{AV0} and M_{AV} were the initial and final equilibrium TeCB mass adsorbed by

Table I. Surface and Other Characteristics of the Kaolinites

Characteristic	*KGA–1 (well crystallized)*	*KGA–2 (poorly crystallized)*
BET surface area (m^2/g)	10.05 ± 0.02	23.50 ± 0.06
Cation-exchange capacity (meq/100 g)	2.0	3.3
Median (mass basis) particle diameter (μm)	1.59	1.59

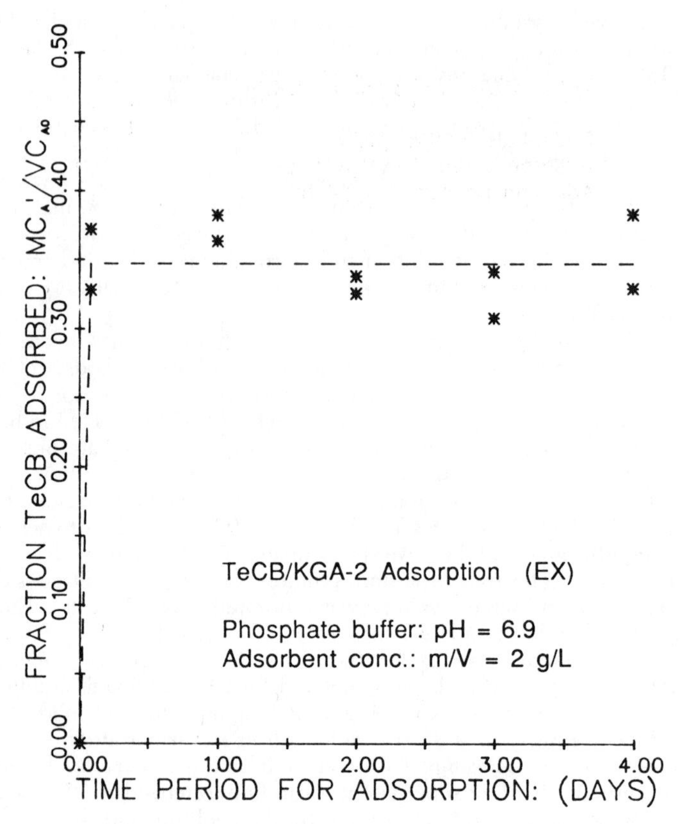

Figure 2. Demonstration of equilibrium for TeCB adsorption by kaolinite (KGA–2).

the vessel. $M_{AV} = f(C_A)$ was determined through a series of control studies. For the extraction method,

$$C'_A = \frac{M_{ex} - V_{rspnt}C_A - M_{AV}}{m} \tag{4}$$

where M_{ex} was the total TeCB mass extracted, and V_{rspnt} was the volume of residual supernatant liquid. Uncertainty estimates for the adsorbed concentration, C'_A, were propagated from uncertainties in each of the measured quantities (19 variables and parameters) used to determine C'_A. Uncertainty bars representing the 99% confidence intervals from these propagations are shown on isotherm plots.

Results and Discussion. The TeCB adsorption data for KGA–1 and KGA–2 measured by the extraction method are shown in Figures 3 and 4, respectively. The data were fitted with a one-parameter linear isotherm model (fixed at the origin) and gave equilibrium adsorption constants, K_A,

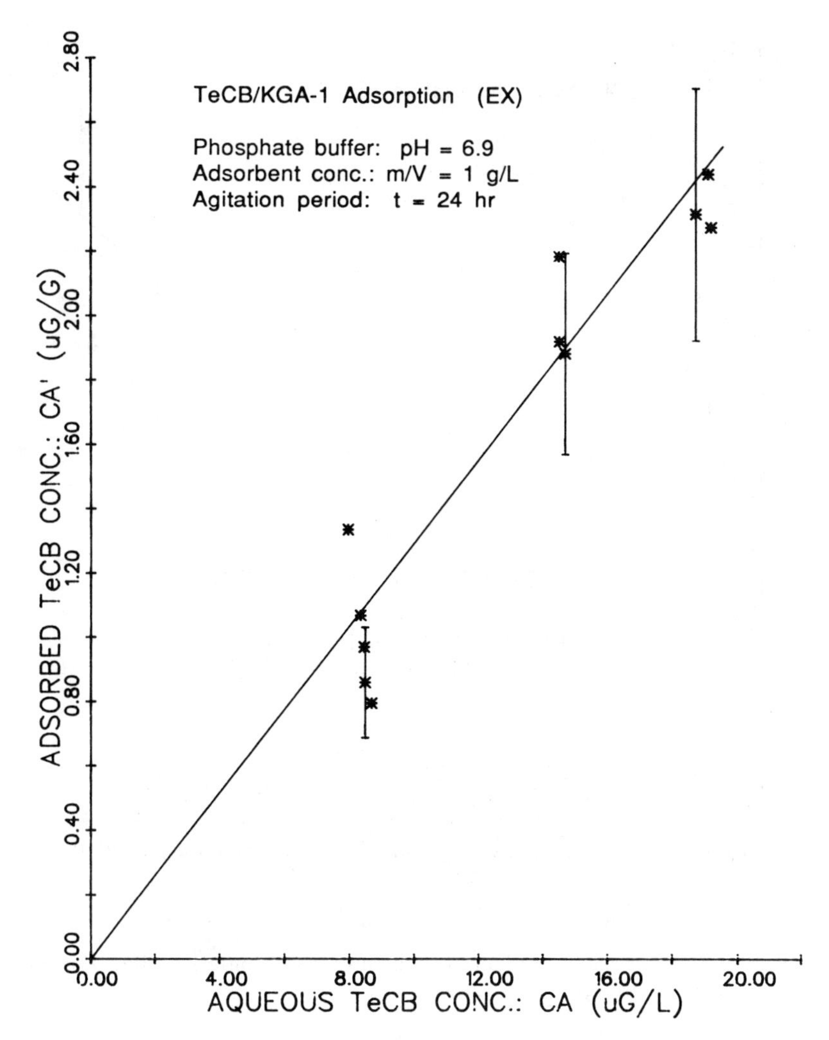

Figure 3. Adsorption isotherm for TeCB and well-crystallized kaolinite (KGA–1) measured by the extraction method.

of 129.4 ± 17.2 L/kg with KGA–1 and 224.2 ± 20.6 L/kg with KGA–2. Adsorption constants measured by the modified difference method agree within 4%, which provides verification of these two independent methods. A one-parameter model was found to be acceptable; the two-parameter least squares regression yielded intercepts not significantly different from zero.

The differences in magnitude of the adsorption constants of the two kaolinite samples may be resolved if the adsorption constants are expressed on the basis of adsorbent surface area instead of mass,

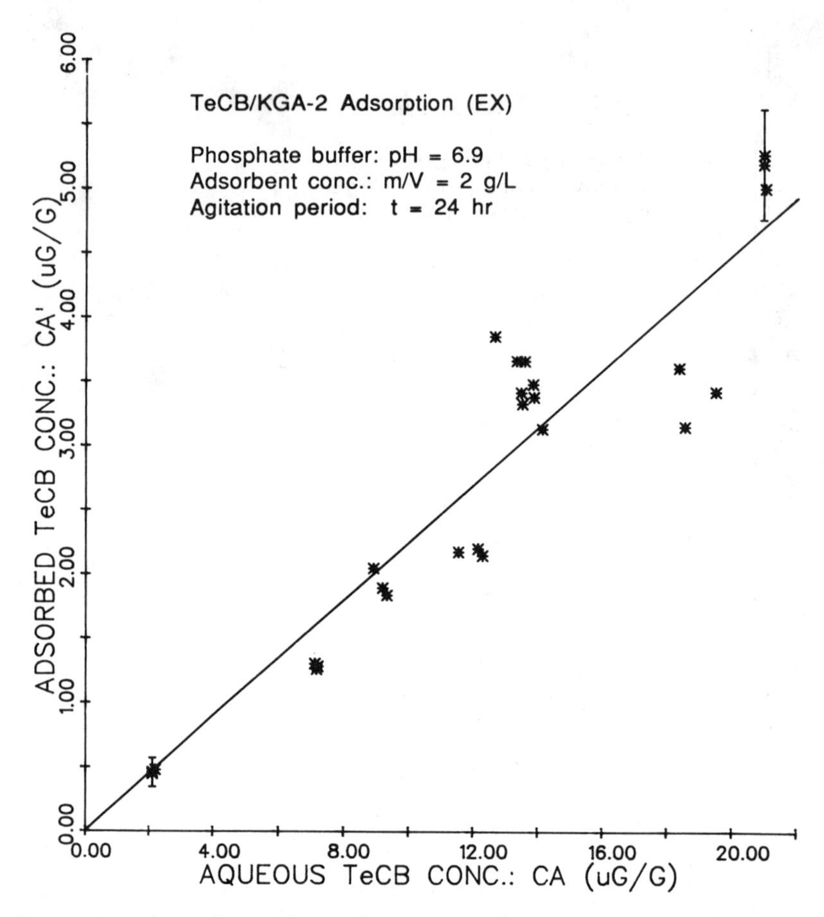

Figure 4. Adsorption isotherm for TeCB and poorly crystallized kaolinite (KGA–2) measured by the extraction method.

$$K_{A,S} = \frac{K_A}{a_S} \tag{5}$$

where a_S is the specific particle surface area (m^2/g). The BET surface area measurement with N_2 gives the total external surface area and can be used to normalize adsorption constants for nonporous, nonexpanding minerals. For expanding clay minerals and microporous soils, normalization by BET surface area is not useful. This conversion gives equilibrium adsorption constants, $K_{A,S}$ of 0.01288 ± 0.00171 L/m^2 with KGA–1 and 0.00954 ± 0.000877 L/m^2 with KGA–2. The factor of 2 difference in total specific surface area may be accounted for in the particle size distributions of these kaolinite samples. The distribution for KGA–2 includes a much higher fraction than

KGA–1 of particles less than 0.5 μm; 17.0% for KGA–2 vs 7.6% for KGA–1. Kaolinite platelets in this size range would be expected to have a higher specific surface area than particles in larger size fractions.

The low adsorption constants for kaolinite and the linearity of the isotherms indicate a nonspecific physical adsorption mechanism. The surface coverage of TeCB can be estimated from the total surface area of the TeCB molecule, reported as 259.56 Å^2 (*28*). At the maximum C'_A = 2.4378 μg/g (C_A = 19.06 μg/g) measured for KGA–2, the total projected surface coverage of TeCB is 0.0130 m^2/g. This value is orders of magnitude less than the total surface area available for monolayer coverage, 23.50 m^2/g. Apparently, the hydration of the aluminum oxide surface and hydrogen bonding of water to the silanol groups of kaolinite is energetically a more favorable interaction than the van der Waals energy of attraction between TeCB and these surface functional groups. Water can strongly compete with HOC for mineral surfaces (*29*).

Adsorption of Humic Acid by Kaolinite

Adsorption of the AB complex to the surface of kaolinite should be governed by the factors affecting humic acid adsorption by kaolinite. Humic acid adsorption by kaolinite was studied previously by Evans and Russell (*30*) and Davis (*31*). Mechanisms for the adsorption of humic substances to clay mineral surfaces have been reviewed (*32–34*).

Experimental Details. A 200-mg/L suspension of Aldrich sodium humate (lot no. 01816HH) was prepared in a 250-mL volumetric flask with phosphate buffer. The suspension was stirred with a poly(tetrafluoroethylene)-coated (Teflon) magnetic stir bar for 20 min, then filtered through a glass fiber prefilter (Gelman Type A/E) and a 0.2-μm pore size membrane filter (Gelman HT–200). The filtration removed insoluble components of the Aldrich humic acid. A "standard" humic acid solution at a concentration of 60 mg of carbon/L could be reproduced for each experiment, following this protocol. The humic acid concentration was measured by UV spectrophotometry (Varian DMS 90) at 285 nm. This method is very sensitive and can be used to analyze humic acid solutions containing radioactive TeCB, which cannot be safely treated in a total carbon analyzer. A Beer's law relation, calibrating absorbance against total carbon, was measured using a carbon analyzer (Ionics 1270) with combustion at 900 °C.

The HA–KGA adsorption experiments were conducted according to the method given for TeCB–KGA adsorption experiments. The kinetics of adsorption were studied, and a 4-day agitation period was satisfactory to achieve equilibrium. A complete isotherm was measured for HA–KGA–2. For the HA–KGA–1 system, triplicate adsorption measurements were made using the same stock humic acid solution.

Results and Discussion. The average for the adsorbed concentrations obtained from the triplicate HA–KGA–1 adsorption measurements are C'_B = 0.21 ± 0.059 mg of carbon/g at an average aqueous concentration of $C_B = 12.38 \pm 0.12$ mg of carbon/L. A Langmuir model fit of HA–KGA–2 isotherm data is presented in Figure 5. The Langmuir equation is

$$C'_B = \frac{K_B C_S C_B}{1 + K_B C_B} \tag{6}$$

where K_B is an adsorption constant and C_S is the adsorbed concentration at saturation. The adsorption data were regressed by using a nonlinear fitting program giving $K_B = 0.6844$ L/mg of carbon and $C_S = 1.3483$ mg of carbon/g. Originally a linear regression of the data was performed with the reciprocal form of the Langmuir equation, but this fitting method favors the adsorption data in the lower concentration range.

The Langmuir model assumes that individual adsorbate molecules interact at localized sites on the adsorbent surface, whereas humic acids are polyfunctional, with several segments of a macromolecule attaching at multiple surface sites.

An alternative approach is to treat humic acids as polymer molecules having an average number, ν, of adsorbable segments per molecule. An isotherm model was derived for this case by Simha, Frisch, and Eirich (*35*):

$$\left(\frac{\Theta \exp(2K_1\Theta)}{1 - \Theta}\right)^{\nu} = K_2 C_B \tag{7}$$

where Θ is the surface coverage (C'_B/C_S), K_1 is an interaction parameter, and K_2 is an effective adsorption isotherm constant. For $K_1 = 0$ and $\nu = 1$, equation 7 reduces to the Langmuir equation. The Simha, Frisch, and Eirich isotherm model requires two parameters more than the Langmuir equation, which could improve data fitting. Humic acids, however, are not well-defined polymeric molecules of repeating monomeric units. For this reason the Langmuir model was used as an empirical model, because it could fit the asymptotic approach of the adsorbed concentration to a maximum, as is observed for the data.

Another feature of humic acid systems is their wide distribution of molecular weights. Preferential adsorption of a particular molecular size fraction will depend on the conformation of the molecule at the solution–solid interface (*36*).

Several mechanisms have been proposed to describe humic acid adsorption by clay mineral and oxide surfaces (*31, 32, 34, 37, 38*). These mechanisms include cation bridging, water bridging, ligand exchange, hydrogen bonding, and van der Waals interactions. Identification of the surface functional groups and origin of the surface charge is necessary to understand the mechanism of humic acid adsorption. Kaolinite is a 1:1 layer clay mineral consisting of alternating sheets of gibbsite and silicate, with a low cation ion-exchange capacity (<10 meq/100 g).

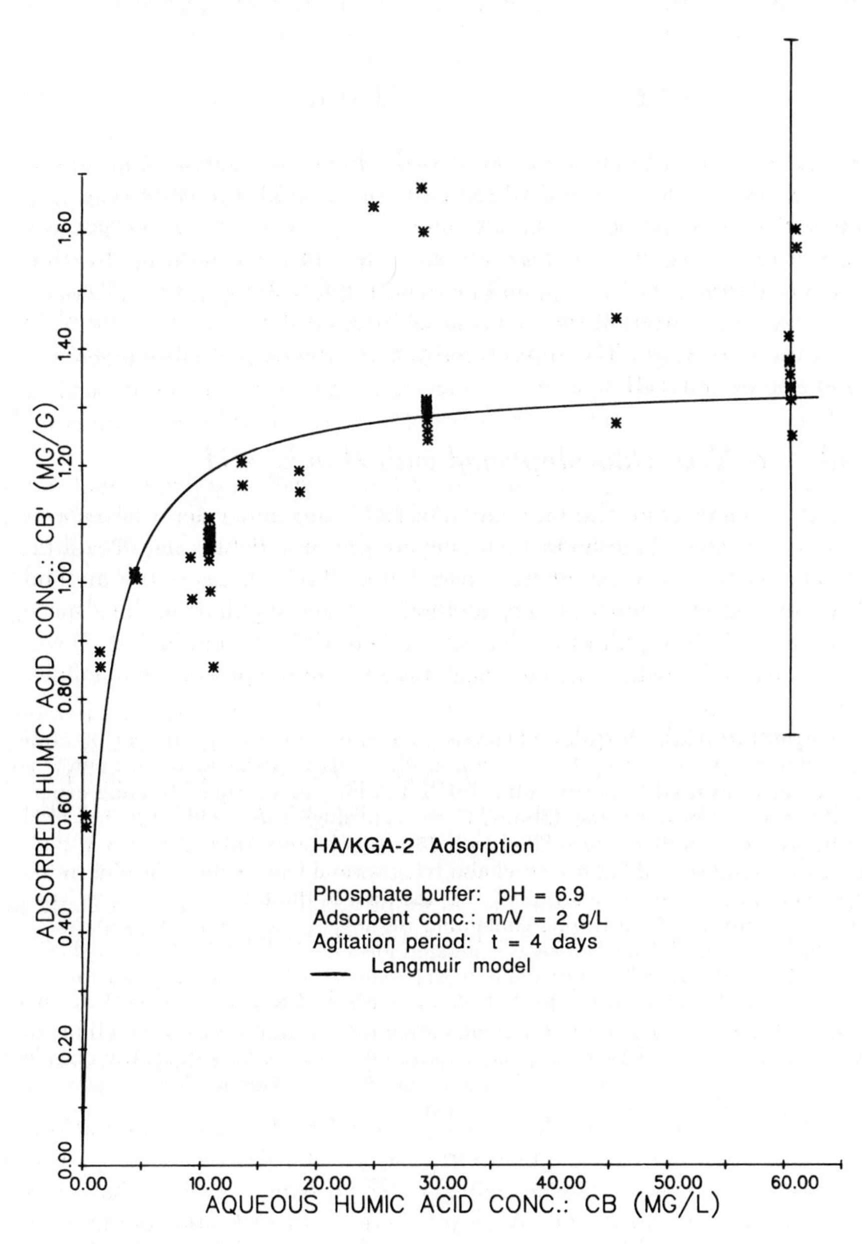

Figure 5. Humic acid and kaolinite (KGA–2) adsorption data fitted by a Langmuir isotherm model.

Davis (*31*) described the adsorption of natural organic matter by alumina with a ligand-exchange mechanism. The reaction between organic anions (A) and surface metal hydroxyls (MeOH) was written as

$$x\text{MeOH} + y\text{H}^+ + \text{A}^{z-} = [\text{MeOH}_x{}^+ - \text{HA}]^{y-z} \qquad (8)$$

where z is the charge on the anion. For kaolinite, the aluminol groups at the edges would be expected to form complexes with the carboxylic acid groups of the humic acid. The aluminol groups of the basal surface are believed to be coated with silica (*39*) and may not participate in the complexation reaction. Parfitt et al. (*40*) demonstrated that adsorption of oxalate and benzoate occurred at the edges of gibbsite crystals, although the plate faces were unreactive. The siloxane surface of kaolinite is hydrophobic and would also be unreactive.

Binding of Tetrachlorobiphenyl with Humic Acid

Binding coefficients for the interaction of HOC and humic acids have been measured by several methods, including equilibrium dialysis (*1*), ultrafiltration method (*4*), a reverse-phase separation method (*3*), gas purge method (*7*), water solubility enhancement method (*10*), and equilibrium headspace technique (*5*). The equilibrium dialysis method of Carter and Suffett (*1*) was used in this study, with some modifications given in the procedure to follow.

Experimental Details. Dialysis membrane tubing (Spectra/Por 6, 24.2-mm diameter, preserved in 0.1% sodium azide) with a molecular weight cutoff of 1000 daltons was used to measure the TeCB–HA binding constant. Binding experiments were conducted using 150-mL Corex centrifuge tubes with aluminum foil lined caps. Dialysis tubing was filled with 25 mL of humic acid and closed at both ends by knotting the tubing itself and also tying several knots with a piece of cotton string (cleaned by soaking in acetone) near the knot in the tube. The dialysis "bag" was placed into the 150-mL tube containing 100 mL of TeCB solution prepared as described previously. The tubes were agitated for 4 days in a thermostatted shaker bath at 25 °C and covered to prevent photoexcitation of the humic acid. Quadruplicate 2-mL dialysate samples were collected from the solution outside the bag, $\overline{C}'_{A,o}$, and an aliquot was taken to measure the concentration of humic acid that leaked through the membrane, $C_{B,o}$. Likewise, triplicate 2-mL samples were withdrawn from inside the bag, $\overline{C}_{A,i}$, and an aliquot was taken to determine $C_{B,i}$. The binding constant K_{AB} was calculated from the following expression:

$$K_{AB,aq} = \frac{\overline{C}_{A,i} - \overline{C}_{A,o}}{\overline{C}_{A,o}C_{B,i} - \overline{C}_{A,i}C_{B,o}} \qquad (9)$$

The term $\overline{C}_{B,o}C_{A,i}$ in the denominator corrects for the enhanced TeCB activity in the dialysate due to humic acid that has escaped from the dialysis bag. This correction assumes that the binding constant and extinction coefficient used to determine the humic acid concentration in milligrams of carbon per liter is applicable to the humic acid that has escaped from the dialysis bag.

Results and Discussion. A one-parameter regression of [^{14}C]-TeCB–HA isotherm points gave a binding constant, $K_{AB} = 48{,}300 \pm 15{,}000$ L/kg. Although the binding constant increased slightly with increasing humic acid concentration, there was no significant dependency of K_{AB} on C_B in the range of humic acid concentration studied, 11–53 mg of carbon/L. The isotherm data for $C_{B,i}$ = 27 mg of carbon/L are shown in Figure 6.

The colloidal structure of humic acid must be identified before the mechanism for the association between TeCB and HA in solution can be understood. Aggregate or micelle structures have been proposed for humic acid by Senesi et al. (*41*) and Wershaw et al. (*42*). However, a critical micelle

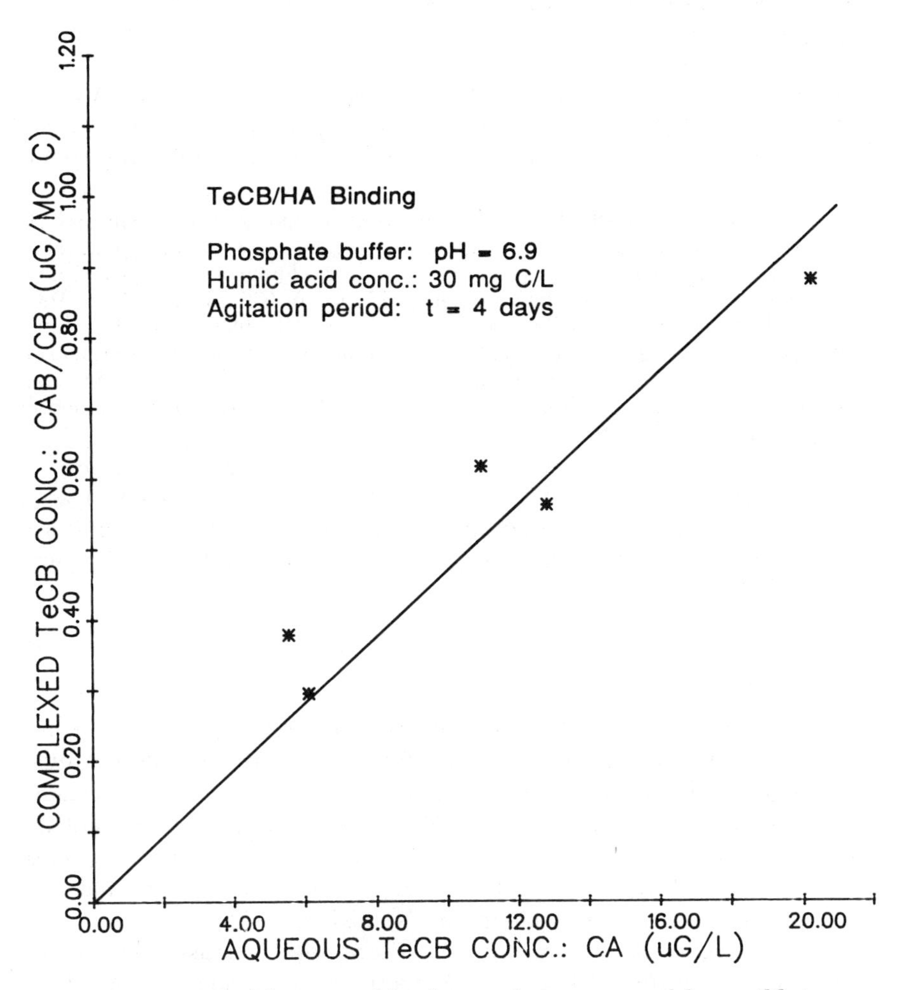

Figure 6. TeCB and humic acid binding isotherm measured by equilibrium dialysis partitioning.

concentration (cmc) has not yet been determined within the concentration range of humic acid normally encountered in natural waters (<100 mg of carbon/L). Piret (*43*) reported a cmc of 16–21 g/L, which is well above this concentration range. Chiou et al. (*10*) found that high-molecular-weight poly(acrylic acids), unlike humic acid, were unable to enhance DDT solubility. This finding was attributed to the inability of poly(acrylic acids) to form a hydrophobic environment that could solubilize DDT. On the basis of evidence of Chiou et al. and Wershaw et al., the likely mechanism for TeCB–HA association is solubilization in the hydrophobic interior of a humic micellelike structure.

Multicomponent Sorption for TeCB, Humic Acid, and Kaolinite

The measurement of the simultaneous adsorption of TeCB and HA on kaolinite is necessary to test the hypothesis of competition between TeCB and HA, and also to examine the mechanism of TeCB–HA complex adsorption.

Experimental Details. Both the adsorbed and aqueous TeCB and HA concentrations were measured for each multicomponent sorption experiment. Stock aqueous TeCB–HA solution was prepared by adding a humic acid solution (15, 30, or 60 mg of carbon/L) to a stock bottle, either 250 mL or 650 mL, containing TeCB residue left after evaporation of hexane transfer solvent. The stock TeCB–HA solution bottles were wrapped in aluminum foil to prevent photoexcitation of HA by exposure to UV light. The remaining steps are identical to those outlined for TeCB–KGA and HA–KGA single-sorbate adsorption experiments, with a 4-day agitation period for adsorption.

Results and Discussion. The effect of humic acid on the apparent TeCB–KGA–2 adsorption isotherm is shown in Figure 7 at three humic acid concentrations. The apparent adsorption partition coefficient is given by

$$K_A^* = \frac{\overline{C'_A}}{\overline{C}_A} \qquad (10)$$

where C'_A is the total TeCB adsorbed concentration and $\overline{C}_A$ is the total dissolved concentration, each including both free and HA-complexed TeCB. K_A^* is evaluated from the slope of each isotherm, with their uncertainty estimates, as given in Table II. The linearity of the multicomponent isotherms indicates that K_A^* is independent of the total $\overline{C}_A$, and consequently is independent of free C_A. The average K_A^* of the three experimental cases for TeCB–KGA–1 are also included in Table II.

The decrease in the partition coefficient with increasing humic acid concentration can be attributed to competitive adsorption between TeCB and HA or to a decrease in the free TeCB concentration due to complexation

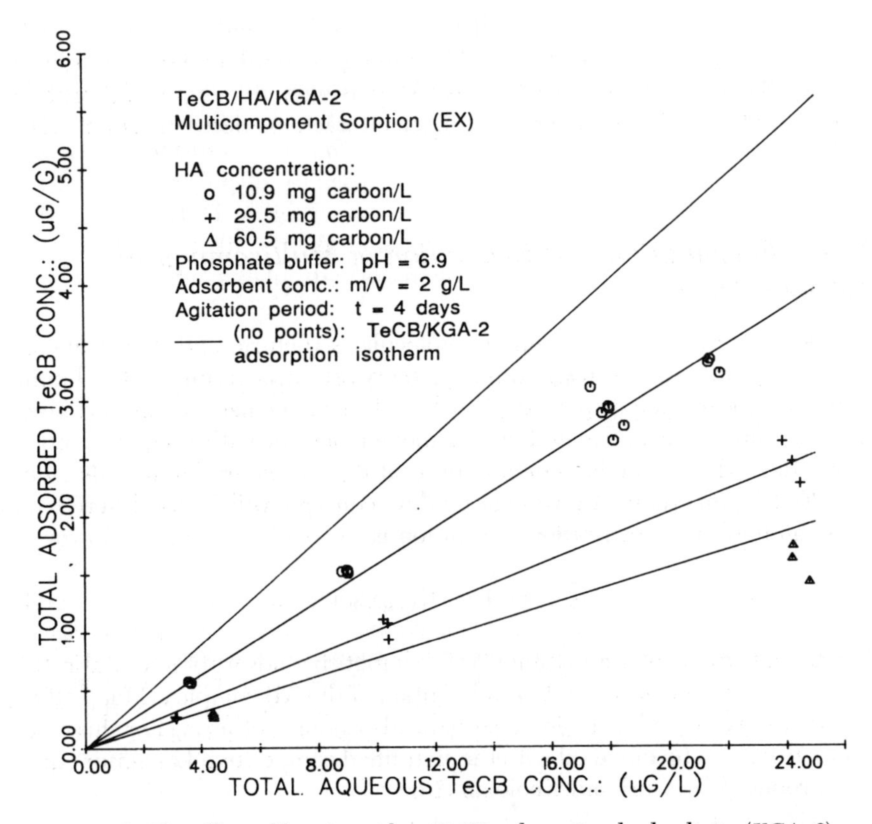

Figure 7. The effect of humic acid on TeCB adsorption by kaolinite (KGA–2). Three additional isotherm data points for C_B = 60.5 mg of carbon/L are not shown and lie outside the domain of this plot.

Table II. Comparison of Multicomponent Model Predictions of Adsorption Partition Coefficients

Kaolinite	C_B (*mg C/L*)	C_B' (*mg C/g*)	K_A^* (*observed*) (*L/kg*)	K_A^* (*eq. 17*) (*L/kg*)	K_A^* (*eq. 18*) (*L/kg*)
KGA–1	0.0	0.0	129.4 (±17.2)[a]		
KGA–1	12.4	0.21	88.2 (±12.1)[b]		87.3 (±43.5)[c]
KGA–2	0.0	0.0	224.2 (±20.6)		
KGA–2	10.9	1.06	157.8 (±7.5)	50.7 (±13.6)[c]	180.3 (±17.4)
KGA–2	29.5	1.29	100.9 (±9.4)	30.0 (±4.1)	118.1 (±16.1)
KGA–2	60.5	1.35	77.1 (±6.8)	17.9 (±2.2)	73.7 (±13.1)

[a]The 99% confidence intervals are shown for K_A^* observed, except for in footnote *b*.
[b]Three standard deviations for n = three data points.
[c]The 99% confidence intervals for predicted K_A^* from an uncertainty analysis propagation.

with HA or to both. The extent of this latter interaction is known a priori from measurements of the TeCB–HA binding in solution. The importance of A and B competitive adsorption was tested using multicomponent sorption models that include the measured binary interactions between TeCB, HA, and KGA.

Model Formulation and Evaluation of Multicomponent Sorption Data

Humic acids do not have a defined molecular structure, and their colloidal nature in water is unknown. Initially, however, it is assumed that humic acids, which are polydispersed (distributed in molecular weight and functional groups), can be treated as a homogeneous set with unique sorption properties. There may be a dependence of K_{AB} on molecular size (*44*), but the effect is uncertain. An average binding constant will be used here.

The total A concentration in solution is

$$\overline{C}_{A} = (1 + K_{AB,aq}C_{B})C_{A} \tag{11}$$

For the adsorbed phase, assume A/B competitive adsorption and also the binding of A to the adsorbent via adsorption of the AB complex. Many HOC adsorbates exhibit a linear single-sorbate adsorption isotherm. For this case, competition can be modeled with the simplified competitive Langmuir equation proposed by Curl and Keoleian (*18*):

$$C'_{A} = \frac{K_{A}C_{A}}{1 + K_{B}C_{B}} \tag{12}$$

As written here, K_{A} represents the product of the Langmuir adsorption constant and the maximum adsorbed concentration. In the absence of B, the equation will predict linear adsorption. The adsorption of the AB complex to the adsorbent, which can be thermodynamically equivalent to the binding of A to adsorbed B, can be expressed by

$$C'_{AB} = K_{AB,ads}C'_{B}C_{A} \tag{13}$$

which is analogous to equation 1 for the aqueous phase. This relation assumes that AB complex adsorption is dominated by the adsorption properties and behavior of the B macromolecules or that binding of A to adsorbed B does not change the chemical activity of adsorbed B or that both are true. For the multicomponent sorption experiments conducted in this investigation, TeCB and HA were agitated for 24 h before kaolinite was added; this favors complexation equilibrium prior to adsorption of the complex. Combining equations 12 and 13 gives the following expression for the total adsorbed concentration:

$$\overline{C'}_A = \left(\frac{K_A}{1 + K_B C_B} + K_{AB,ads} C'_B \right) C_A \tag{14}$$

The partition coefficient, found by substituting equations 11 and 14 into equation 10, is given by

$$K_A^* = \frac{\dfrac{K_A}{1 + K_B C_B} + K_{AB,ads} C'_B}{1 + K_{AB,aq} C_B} \tag{15}$$

In this model K_A^* is independent of $\overline{C}_A$, which is observed experimentally. No direct measurement of $K_{AB,ads}$ was done, so the following assumption will be made:

$$K_{AB,ads} = K_{AB,aq} \tag{16}$$

Equation 16 assumes that the conformational changes occurring in B upon adsorption do not affect A/B association. K_A^* now becomes

$$K_A^* = \frac{\dfrac{K_A}{1 + K_B C_B} + K_{AB,aq} C'_B}{1 + K_{AB,aq} C_B} \tag{17}$$

The alternative hypothesis is that A and B can adsorb on independent sites, in which case the competitive interaction coefficient K_B is zero and K_A^* becomes

$$K_A^* = \frac{K_A + K_{AB,aq} C'_B}{1 + K_{AB,aq} C_B} \tag{18}$$

Model Predictions

All of the parameters given in equations 17 and 18 can be evaluated from TeCB–HA, TeCB–KGA, and HA–KGA sorption experiments. With these parameters, the model predictions for K_A^*, considering both competitive (equation 17) and noncompetitive (equation 18) A adsorption, are given in Table II. The uncertainties in the predicted partition coefficients were propagated from the uncertainties in the parameters K_A, K_B, K_{AB}, and the concentration data C'_B and C_B. The competitive model predictions for K_A^* clearly do not agree with data for K_A^*, indicating that competition between A and B may not be significant. In all cases, the predicted K_A^* underestimates the measured K_A^* data. The competition model was not tested for KGA–1 because the complete isotherm was not measured and K_B was unknown.

A comparison of the noncompetitive model predictions with data for K_A^* shows agreement within the 99% confidence intervals. This result supports the assumption that independent surface sites on kaolinite are available for TeCB and HA adsorption. The TeCB would be expected to adsorb on the most hydrophobic surfaces of the kaolinite. The siloxane surface of kaolinite has a low affinity for water and may accommodate TeCB adsorption. Conversely, HA, which is hydrophilic in nature, probably interacts with AlOH at the edges of the kaolinite crystal lattices. A detailed description of the surface characteristics of kaolinite is given in ref. 39.

The agreement between noncompetitive model predictions and data for K_A^* supports the assumption made for K_{AB} with equation 16, that K_{AB} is the same for both the adsorbed and aqueous phases. Chiou et al. (*10*) found approximate agreement between the binding constant for a dissolved soil humic acid extract (K_{doc}) and bulk soil adsorption constants (K_{oc}) for DDT and 2,4,4′-trichlorobiphenyl. Means and Wijayaratne (*2*), however, reported that K_{doc} for colloidal organic matter was 10 to 35 times greater than K_{oc} literature values. Differences in the origin of the soil and the dissolved organic matter alone can account for the variability in K_{oc} (*45*) and K_{doc} values (*12*). Furthermore, the entire organic matrix of a soil may not be accessible to a sorbate, in which case $K_{oc} < K_{doc}$ would be expected.

If the micelle model for humic acid is accepted, it follows that TeCB solubilization in the hydrophobic interior of the micelle would not be altered upon adsorption to the kaolinite surface. The micelle structure may not be significantly disrupted during the adsorption reaction because the complexing functional groups are oriented toward the exterior of the micelle.

Summary

The binary interactions between 2,2′,4,4′ tetrachlorobiphenyl, humic acid, and two kaolinites were characterized by conducting TeCB–KGA, HA–KGA, and TeCB–HA isotherm experiments. Equilibrium sorption constants were measured for these interactions and used in alternative multicomponent sorption models to predict the partitioning of TeCB to kaolinite in the presence of humic acid. Humic acids were found to affect the partitioning by enhancing the TeCB aqueous-phase concentration through a complexation mechanism, and by immobilizing TeCB to the kaolinite via adsorption of the TeCB–HA complex to the mineral surface. This latter mechanism was previously unstudied. Adsorption of TeCB and humic acid were found to be noncompetitive. This work is applicable for the prediction of HOC mobility through clay liners of "secure" landfills and clay mineral subsurface zones.

Other environmental implications of this investigation were discussed by Keoleian (*46*).

Acknowledgments

This research was funded in part by the University of Michigan Memorial–Phoenix Project Grant. Gregory A. Keoleian was supported by the Chemodynamics of Toxic Substances Control Internship established at the University of Michigan, Department of Chemical Engineering, by the Jessie Smith Noyes Foundation.

References

1. Carter, C. W.; Suffet, I. H. *Environ. Sci. Technol.* **1982,** *16,* 735–740.
2. Means, J. C.; Wijayaratne, R. *Science (Washington, D.C.)* **1982,** *215,* 968–970.
3. Landrum, P. F.; Nelhart, S. R.; Eadie, B. J.; Gardner, W. S. *Environ. Sci. Technol.* **1984,** *18,* 187–192.
4. Wijayaratne, R. D.; Means, J. C. *Mar. Environ. Res.* **1984,** *11,* 77–89.
5. Garbarini, D. R.; Lion, L. W. *Environ. Sci. Technol.* **1985,** *19,* 1122–1128.
6. Haas, C. N.; Kaplan, B. M. *Environ. Sci. Technol.* **1985,** *19,* 643–645.
7. Hassett, J. P.; Milicic, E. *Environ. Sci. Technol.* **1985,** *19,* 638–643.
8. McCarthy, J. F.; Jimenez, B. D. *Environ. Sci. Technol.* **1985,** *19,* 1072–1076.
9. Whitehouse, B. *Estuarine Coastal Shelf Sci.* **1985,** *20,* 393–402.
10. Chiou, C. T.; Malcolm, R. L.; Brinton, T. I.; Kile, D. E. *Environ. Sci. Technol.* **1986,** *20,* 502–508.
11. Morehead, N. R.; Eadie, B. J.; Lake, B.; Landrum, P. F.; Berner, D. *Chemosphere* **1986,** *15,* 403–412.
12. Gauthier, T. D.; Seitz, W. R.; Grant, C. L. *Environ. Sci. Technol.* **1987,** *21,* 243–248.
13. Hassett, J. P.; Anderson, M. A. *Water Res.* **1982,** *16,* 681–686.
14. Caron, G.; Suffet, I. H.; Belton, T. *Chemosphere* **1985,** *14,* 993–1000.
15. Brownawell, B. J.; Farrington, J. W. *Geochim. Cosmochim. Acta* **1986,** *50,* 157–169.
16. Baker, J. E.; Capel, P. D.; Eisenreich, S. J. *Environ. Sci. Technol.* **1986,** *20,* 1136–1143.
17. Gschwend, P. M.; Wu, S. *Environ. Sci. Technol.* **1985,** *19,* 90–96.
18. Curl, R. L.; Keoleian, G. A. *Environ. Sci. Technol.* **1984,** *18,* 916–922.
19. Chiou, C. T.; Porter, P. E.; Schmedding, D. W. *Environ. Sci. Technol.* **1983,** *17,* 227–231.
20. Mingelgrin, U.; Gerstl, Z. *J. Environ. Qual.* **1983,** *12,* 1–11.
21. MacIntyre, W. G.; Smith, C. L. *Environ. Sci. Technol.* **1984,** *18,* 295–297.
22. Malcolm, R. L.; MacCarthy, P. *Environ. Sci. Technol.* **1986,** *20,* 904–911.
23. O'Connor, D. J.; Connolly, J. P. *Water Res.* **1980,** *14,* 1517–1523.
24. Di Toro, D. M.; Horzempa, L. M.; Casey, M. M.; Richardson, W. *J. Great Lakes Res.* **1982,** *8,* 336–349.
25. Bowman, B. T.; Sans, W. W. *J. Environ. Qual.* **1985,** *14,* 265–269.
26. Voice, T. C.; Weber, W. J., Jr. *Environ. Sci. Technol.* **1985,** *19,* 789–796.
27. *Data Handbook for Clay Materials and Other Non-Metallic Minerals;* van Olphen, H.; Fripiat, J. J., Eds.; Pergamon: Oxford, England, 1979; 346 pp.

28. Mackay, D.; Mascarenhas, R.; Shiu, W. Y.; Valvani, S. C.; Yalkowsky, S. H. *Chemosphere* **1980,** *9*, 257–264.
29. Chiou, C. T.; Shoup, T. D.; Porter, P. E. *Org. Geochem.* **1985,** *8*, 9–14.
30. Evans, L. T.; Russell, E. W. *J. Soil Sci.* **1959,** *10*, 119–132.
31. Davis, J. A. *Geochim. Cosmochim. Acta* **1982,** *46*, 2381–2393.
32. Greenland, D. *Soil Sci.* **1971,** *111*, 34–41.
33. Orlov, D. S.; Pivovarova, I. A.; Gorbunov, N. I. *Agrokhimiya* **1973,** *9*, 140–153.
34. Theng, B. K. G. *Formation and Properties of Clay–Polymer Complexes*; Elsevier: Amsterdam, 1979; 362 pp.
35. Simha, R.; Frisch, H. L.; Eirich, F. R. *J. Phys. Chem.* **1953,** *57*, 584–589.
36. Sato, T.; Ruch, R. *Stabilization of Colloidal Dispersions by Polymer Adsorption*; Marcel Dekker: New York, 1980; pp 1–36.
37. Parfitt, R. L.; Fraser, A. R.; Farmer, V. C. *J. Soil Sci.* **1977,** *28*, 289–296.
38. Tipping, E. *Geochim. Cosmochim. Acta* **1981,** *45*, 191–199.
39. Greenland, D. J.; Mott, C. J. B. In *The Chemistry of Soil Constituents*; Greenland, D. J.; Hayes, M. H. B., Eds.; Wiley: New York, 1978; pp 321–353.
40. Parfitt, R. L.; Fraser, A. R.; Russell, J. D.; Farmer, V. C. *J. Soil Sci.* **1977,** *28*, 40–47.
41. Senesi, N.; Chen, Y.; Schnitzer, M. In *Soil Organic Matter, Vol. 2*; International Atomic Energy Agency: Vienna, Austria, 1977; pp 143–155.
42. Wershaw, R. L.; Thorn, K. A.; Pinckney, D. J.; MacCarthy, P.; Rice, J. A.; Hemond, H. F. In *Peat and Water*; Fuchsman, C. H., Ed.; Elsevier: New York, 1986; pp 133–157.
43. Piret, E. L.; White, R. G.; Walther, H. C., Jr.; Madden, A. J., Jr. *Sci. Proc. R. Dublin Soc., Ser. A* **1960,** 69–79.
44. Hassett, J. P.; Milicic, E.; Jota, M. A. T. *Abstracts of Papers*, 188th National Meeting of the American Chemical Society, Philadelphia, PA; American Chemical Society: Washington, DC, 1984; ENVR 86.
45. Rao, P. S. C.; Davidson, J. M. In *Environmental Impact of Nonpoint Source Pollution*; Overcash, M. R.; Davidson, J. M., Eds.; Ann Arbor Science: Ann Arbor, 1980; pp 23–67.
46. Keoleian, G. A. Ph.D. Dissertation, University of Michigan, 1987; 205 pp.

RECEIVED for review December 1, 1986. ACCEPTED for publication July 24, 1987.

17

Methods for Dissolving Hydrophobic Compounds in Water

Interactions with Dissolved Organic Matter

G. R. Barrie Webster, Mark R. Servos, G. Ghaus Choudhry, and Leonard P. Sarna

Pesticide Research Laboratory, Department of Soil Science, University of Manitoba, Winnipeg, Manitoba R3T 2N2, Canada

Derek C. G. Muir

Department of Fisheries and Oceans, Freshwater Institute, 501 University Crescent, Winnipeg, Manitoba R3T 2N6, Canada

Variability inherent in the methods used to dissolve hydrophobic compounds in water is particularly problematic for compounds with solubilities <1 *mg/L* ($<1 \times 10^{-3}$ *g/L). A number of methods have been described to deal with these problems in both chemical and biological systems. Several methods make use of a generator column or fluidized bed of solid support coated with the hydrophobic compound. Water (the quality of which is often not well defined) or aqueous solutions of dissolved organic matter (DOM) can be pumped through the column or can be used to dilute the saturated solutions for experiments involving biota. Other methods use a variety of direct addition techniques to add either the compound in solution, or the solid or liquid compound itself, to water or aqueous solutions of DOM. These methods are described and critically compared.*

THE STUDY OF SOLUTION-PHASE INTERACTIONS of dissolved organic matter (DOM) with hydrophobic environmental contaminants requires that candidate compounds be dissolved in water. Dissolving hydrophobic compounds

0065–2393/89/0219–0251$06.00/0

in water is a challenging task; indeed, comparatively few numerically useful solubilities of very hydrophobic (log $K_{ow} > 6$; K_{ow} is the octanol–water partition coefficient) pesticides and related environmental contaminants have been reported. Variability inherent in the method used to prepare saturated solutions and to measure aqueous solubilities is particularly problematic at solubilities below 1 mg/L (1×10^{-3} g/L). Solubility measurements can be influenced by temperature, pH, suspended solids, salt concentration, and DOM concentration in the water. Furthermore, purity of the compound and its sorption characteristics from solution, especially with respect to container surfaces, filter materials, and membrane materials, are of great importance. Particle size of the solid or suspended compound can also affect the rate of dissolution in water. None of the methods described is ideal for all studies; however, particular approaches to individual investigations have distinct advantages.

Saturated Solutions of Hydrophobic Compounds in Water

To study the effect of the presence of DOM on the solubility and other physical properties of hydrophobic compounds, saturated solutions of the compounds in pure water are needed. A number of methods have been used to generate such solutions (Table I).

Continuous Methods: The Generator Column Approach. Several research groups have used the generator column approach originally described by Veith and Comstock (*9*). The column can be packed with glass beads or with gas chromatographic column packing. Glass beads are convenient because they are easily obtained, are easily cleaned, and have a highly consistent shape. Infusorial earth (Chromosorb W) has a higher specific surface area, but a less uniform surface of varying activity. Beads and Chromosorb W can be coated with the hydrophobic compound prior to packing the generator column (*1, 2, 8, 10, 11*) or in situ by passage of the hydrophobic liquid onto and through the prepacked column (*7*). Retention of the packing in the column with glass wool may allow particulate material to migrate from the column; stainless steel frits as used in normal HPLC columns will lessen this effect.

Preconditioning the generator column with flowing water will remove mobile particles. Reproducibility of results is good evidence of lack of particulate migration from the generator column once it has been thoroughly conditioned. The life of the generator column is determined largely by the solubility of the hydrophobic compound, the amount coated on the column packing, and the length of time that the column is used. Flow rates of 0.5–2.0 mL/min are commonly used for water flowing through the generator column for physical constant determinations (*2*). In fish exposure experiments, flows of 2–5 mL/min have been used (*3, 9*).

Table I. Methods for the Preparation of Saturated Solutions of Hydrophobic Compounds in Water

Method	*Compound*[a]	*Comments*	*References*
	Continuous techniques		
Generator column of glass beads precoated with solute	PCDDs	• for superlipophilic compounds • 2-μm frits	1, 2
		• glass wool plugs	3–6
Generator column of Chromosorb W HP 100–120 mesh coated in situ with solute	*n*-propylbenzene	• large surface area; cf. glass beads • coating may not be uniform; glass wool plugs	7
Generator column of glass beads 60–80 mesh, 1% w/w solute	PAHs	• quantities of hydrophobics very high; glass wool plugs	8
Generator column of sand, glass beads, or XAD–7 resin	OCl insecticides	• upward water flow presents plugging with particulates	9
Fluidized bed of sand or Chromosorb W 45–60 mesh precoated with solute	PCBs	• large surface area; cf. glass beads	10, 11
	Batch techniques		
Direct addition of immiscible liquid to water plus equilibration	PCBs	• low specific surface area	12
Deposit solution on inner surface	PCBs	• adsorption a problem?	13
of vessel and evaporate solvent; stir water in flask	cholesterol, T_4CB	• low specific surface area	14
for extended period	Bzs, $ArCO_2H$, OP insecticides, PCBs, XBzs, 2,4-D	• long time period required to generate saturated solution	15
Direct addition of solid (shaking)	DDT, PCBs, T_3CBz, lindane	• long time period required to generate saturated solution	16

[a]PCDDs, polychlorinated dibenzo-*p*-dioxins; PAHs, polycyclic aromatic hydrocarbons; OCl, organochlorine; PCBs, polychlorinated biphenyls; T_4CB, tetrachlorobiphenyl; Bzs, benzenes; $ArCO_2H$, benzoic acids; OP, organophosphorus; XBzs, halogenated benzenes; 2,4-D, 2,4-dichlorophenoxyacetic acid; and T_3CBz, trichlorobenzenes.

Batch Techniques. Saturated solutions of hydrophobic compounds can also be produced by direct addition followed by an extended mixing–equilibration period (*12, 16*). These techniques appear to have been simple and effective in the work described; however, they are somewhat tedious. Mixing periods may be as long as several weeks. Further, the opportunity for error may be high because of the likelihood of small quantities of undissolved material remaining in the system in the form of fine particulates or molecular aggregates. As in the case of the continuous methods, however, good reproducibility is a good indication that such problems are unimportant. Reproducibility in itself is not a guarantee of accuracy; however, it does provide evidence that the method is reliable.

Another widely used batch technique is the introduction of the hydrophobic compound in solution to the empty flask, followed by evaporation of the solvent (*13–15*). Water is then added and stirred for an extended period before the experiment. The method is simple, but solutions so generated may readily be depleted of their hydrophobic solute during transfer to another vessel by adsorption to the walls of the apparatus used for the transfer. It is difficult to avoid such problems; thus, this technique should be used only if the solution need not be transferred from its original container.

Water Quality for Studies with Hydrophobic Compounds and DOM

The solubilities in water of highly hydrophobic compounds must be determined with purified water, because even small amounts of DOM in the water will enhance the quantity of the compound carried by the water. In studies of the interaction of hydrophobic compounds with DOM, the water used has often been less than completely pure, sometimes intentionally so.

Pure water can be generated in a number of ways: distillation, deionization, charcoal filtration, filtration through a reverse-phase HPLC-type column, photooxidation, or a combination of two or more of these techniques (Table II). Friesen et al. (*1*) and Webster et al. (*2*) used strong oxidizing

Table II. Quality of Water Used in Solution Studies with Hydrophobic Compounds

Refs.	*Water Preparation*	*Needs*	*Comments*
		Abiotic studies	
7	Distilled water passed through column	Water solubilities and K_{ow}	DOM not considered
8	a. Distilled water (resin cleanup) b. Salt water	Water solubility	DOM removed, salting out effect tested
1	Ultrapure triple-distilled oxidized filtered (0.22 μm) water	Solubility studies	Without DOM
2	Ultrapure triple-distilled oxidized filtered (0.22 μm) water	Solubility studies	With or without DOM
16	Deionized water (Sybron/Barnstead Nanopure II)	Solubility enhancement study	Isolated DOM added
17	Distilled deionized water and Barnstead organic removal cartridge and filter	Effects of DOM on Henry's law constant	Natural DOM removed, commercial DOM added
14	River water	Extraction of pollutants on DOM	Natural OM sewage samples

Table II.—Continued

Refs.	*Water Preparation*	*Needs*	*Comments*
18	pH 6.4 phosphate buffer	Sorption rate constants	Water quality not given
19	Distilled water buffered to pH 8.3	Sediment sorption with DOM	Water quality not given
20	a. Dechlorinated tap water passed through activated charcoal b. Deionized distilled water (both a and b filtered through Gelman A–E glass fiber filter)	Sorption studies with PAHs	Natural DOM removed with charcoal, particulates removed with filter of 0.3 μm, commercial DOM added
21	a. River water b. Distilled water c. Filtered natural water	Partitioning of pollutants to DOM	Water quality not given
22	a. Lake water (filtered) b. Photooxidized lake water	Pollutant sorption to DOM	DOM at different depths
		Studies with biota	
10, 11	Tap water plus demineralized water	Fish habitat	DOM not considered
23	Milli-Q water and salts	Sorption and BCF of PAHs with DOM	DOM removed
24	Lake water	BCF determination	Bog water diluted with lake water and filtered (0.45 μm)
25	Charcoal filtered and UV sterilized water	BCF in fish of PAHs	Natural DOM removed, commercial DOM used (continuous flow)
3	a. Distilled water reconstituted with inorganic ions b. Natural filtered lake water	Fish habitat	Fish uptake study Fish uptake study (continuous flow)
26	a. Simulated lake water—distilled deionized water and salts b. Sediment interstitial water	Amphipod habitat	Bioavailability study
27	Deionized water	Yeast habitat	Uptake study
28	Water charcoal treated and filtered through 0.3-μm Gelman glass fiber filter	Cladoceran habitat	BCF study
29	Millipore-filtered well water	Cladoceran habitat	BCF study
5	Dechlorinated tap water	Fish habitat	BCF/depuration study
30	a. Filtered lake water b. Osmotically balanced distilled water	Fish habitat	BCF/depuration study

agents in conjunction with multiple distillation. Deionization or distillation coupled with filtration through a reverse-phase resin column (*7, 8, 16, 17*) or an activated charcoal column (*20*) also have been used. Alternatively, photooxidation of natural water can be used to destroy DOM (*22*). Commercially available purification systems using ion-exchange and organic extraction columns can provide a useful alternative (e.g., Milli-Q water). Other techniques, such as reverse osmosis, may also be suitable, but each must be tested to ensure that DOM levels in the water are acceptably low before the work proceeds.

Studies focusing on uptake of hydrophobic compounds by fish in the presence or absence of DOM must be carried out with water that is compatible with the organisms living in the system. A number of investigators have used natural waters such as lake water or sediment interstitial water (*3, 24, 26, 30*). McCarthy and Jiminez (*25*) used natural waters, charcoal filtered and UV sterilized to remove DOM, and Muir and co-workers (*3–6, 30*) and Landrum and co-workers (*23, 26*) used distilled water or Milli-Q water as the starting point, followed by addition of salts to achieve osmotically balanced water. Other studies made use of filtration through activated carbon and a 0.3-μm glass fiber filter (*28*), filtered well water (*29*) presumed to be very low in DOM, dechlorinated tap water (*5*), or deionized water plus nutrients (*27*). Earlier work was often carried out without DOM being considered (*7, 10, 11*).

Microscale Use of Nonaqueous Cosolvents

A number of other methods of dissolving organic hydrophobic compounds in water attempt to create a solution of the hydrophobic substance below its saturation point, either in pure water or in water amended with soluble organic matter and varying levels of particulate matter. The most common approach is to dissolve the hydrophobic compound in the nonaqueous solvent and then to inject a very small quantity of the solution (often in methanol) with stirring into the aqueous system being studied. The major difficulties here may be that the hydrophobic compound may come out of solution immediately upon addition of the solution and thus create a particulate hydrophobic compound, or that the presence of the organic solvent may interfere with the system and alter the effect being studied. A list of the investigations in which hydrophobic compounds have been added in nonaqueous cosolvent solution is given in Table III.

The most commonly used cosolvents have been acetone (*17, 19, 21, 23–25, 30*) and methanol (*21–23, 26, 28*). Studies have also involved the use of dioxane (*20, 25, 27, 28*), benzene (*18, 23*), ethyl ether (*23*), and toluene (*23*). In most cases, the choice of solvent has been made by the commercial supplier; that is, the hydrophobic compound has been delivered to the researcher in solution, and that solvent has become the cosolvent.

Table III. Use of Cosolvents To Dissolve Hydrophobic Compounds in Water for Studies with Humic Substances

Cosolvent	*Concentration (%)*	*Compound*	*Water*	*References*
Acetone	0.005–0.016	Mirex	Pure	17
	Undefined	DDT Lindane	pH 8.3 buffer	19
	<0.1	PAHs	Pure	25
	0.001	PAHs	a. River b. Distilled (both filtered)	21
	1.0	2,4,6-T_3CP	Lake	24
	0.002	Lindane		
	0.01	PCBs		
	0.001–0.03	Anthracene	Synthetic	23
		Pyrethroids DDT	a. Lake b. Synthetic	30
Benzene	0.02	T_3CBs	pH 6.4 buffer	18
	0.001–0.03	PAHs	Synthetic	23
Dioxane	<0.1	PAHs	Pure	20, 25
	<0.1	PAHs	Pure	28
	<0.1	PAHs	Deionized	27
Ether	0.001–0.03	Naphthalene	Synthetic	23
Methanol	<0.1	PAHs	Pure	25
	<0.1	PAHs	Pure	28
	0.0004–0.0008	PAHs	a. River b. Distilled (both filtered)	21
	0.0009	PAHs	a. Lake (filtered) b. Lake ($h\nu$ + [0])	22
	<0.04	PAHs, PCBs	a. Sediment interstitial (filtered) b. Simulated lake	26
Toluene	0.001–0.03	Benzo[*a*]pyrene	synthetic	23

The use of a cosolvent provides a very convenient route to the generation of an aqueous solution of a hydrophobic compound. It is relatively quick and easy, and the cosolvent is likely to be lost through volatilization within the first day of the experiment. The cosolvent can also be removed by sparging; however, this must be of limited duration to ensure that the hydrophobic solute is not also removed. Comparison of Henry's constants will show the degree to which this precaution must be observed.

Evaluation of Techniques

The use of the generator column technique is attractive because of the control it gives the experimenter over the quality of the solution being generated.

No aging of the solution occurs. It is easy to use the same apparatus to ensure that the solubility of the compound is constant and that the experimental conditions, including the quality of the water, have not changed. The apparatus required for this technique (*2, 9*) is logistically somewhat complex (Figure 1); however, it is quite simple to operate and allows the use of several different solvent systems with one pump.

The generator column approach is also useful in providing constant exposure conditions for biological systems (Figure 2) (*3*). It is much more convenient, however, to use the microscale addition of solutions of the hydrophobic compound in a nonaqueous cosolvent because of the simplicity of the technique and the apparatus required.

The use of a cosolvent could introduce a source of bias into the experimental system, although this possibility has not been directly tested with respect to the interaction of hydrophobic compounds of low volatility. Gossett (*31*) and Munz and Roberts (*32*) show, in the case of low-molecular-

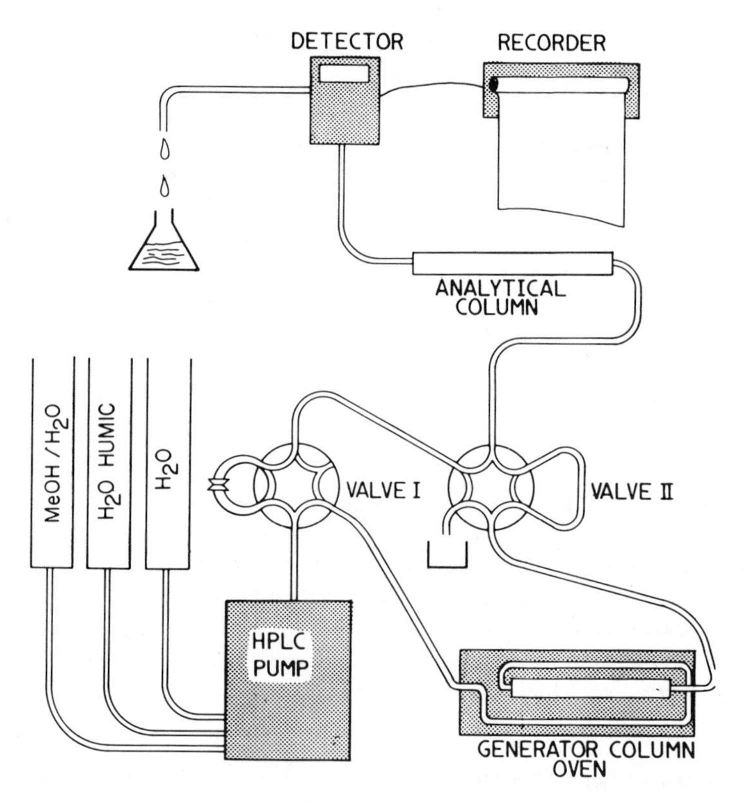

Figure 1. Generator column apparatus for the study of the interaction of hydrophobic compounds (chlorinated dioxins) with soluble humic substances. (Reproduced with permission from ref. 2. Copyright 1986 Pergamon Press.)

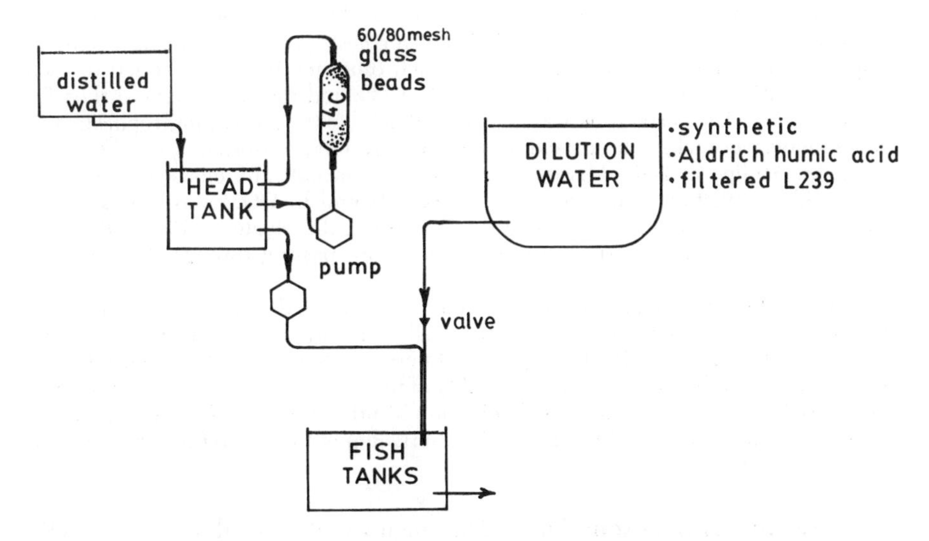

Figure 2. Continuous-flow system used to expose fish to dioxins in solutions of differing water quality: synthetic (reconstituted with inorganic ions), synthetic plus Aldrich humic acid, or filtered natural lake water.

weight halogenated hydrocarbons, that <1–2% v/v methanol used as a cosolvent does not affect the determination of Henry's constant and, therefore, the aqueous activity coefficients for these compounds. Levels typically used in hydrophobic compounds or DOM experiments are much lower than 1–2% (Table III), although concentrations as high as 1% have been used in at least one case (*24*). Higher-molecular-weight solutes, with much greater hydrophobicity, have not been examined. A short series of experiments was conducted to evaluate the effect of the use of the cosolvent methanol.

To test the effect of a common cosolvent, methanol, with respect to the partitioning and solubility enhancement of higher molecular weight hydrophobic compounds in the presence of DOM (Aldrich humic acid), the extremely hydrophobic compound 1,3,6,8-tetrachlorodibenzo-*p*-dioxin (1,3,6,8-T_4CDD) was examined in the presence of varying concentrations of methanol.

Experimental Details. Milli-Q water was used for all experiments; Aldrich humic acid was made up to yield a 5.0-mg/L solution. The ^{14}C-1,3,6,8-T_4CDD (Pathfinder Laboratories, Inc., Kansas City, MO) was purified to >99.8% by preparative thin-layer chromatography on silica plates and dissolved in methanol (HPLC grade, Caledon Laboratories, Inc., Georgetown, ON, Canada). A liquid scintillation counter (Beckman, model 7500) was used to quantify concentrations of ^{14}C-1,3,6,8-T_4CDD.

The partitioning experiment was carried out by first depositing the labeled 1,3,6,8-T_4CDD on the wall of a 2-L Erlenmeyer flask. Aldrich humic acid solution (20 mg/L in Milli-Q water) was added to each tube; methanol was added to each tube (0, 0.05, 0.1, 0.25, 2.5, 5.0, and 15.0% v/v); and the tubes were shaken for 24 h at 23 °C. After each tube was centrifuged at 20,000 g for 30 min, replicate 4.0-mL aliquots were taken and passed through reverse-phase adsorbent cartridges (C_{18}-SepPaks, Millipore, Ltd., Mississauga, Ontario, Canada). The eluate was counted by liquid scintillation counting (LSC). Percent bound to DOM was calculated by dividing the concentration of ^{14}C-1,3,6,8-T_4CDD in the eluate × 100 by the total concentration of ^{14}C-T_4CDD in the solution before passing through the cartridge (*21*).

The solubility enhancement experiment was carried out by first depositing 65 ng of the ^{14}C-1,3,6,8-T_4CDD on the wall of 25-mL centrifuge tubes (Corex). Sufficient ^{14}C-1,3,6,8-T_4CDD was present to create a solution with a concentration of 3250 ng/L, well in excess of the solubility of the dioxin in the absence or presence of DOM (*1, 2*). Aldrich humic acid solution (20 mL) and methanol were added as described and shaken for 24 h at 23 °C. A 4.0-mL aliquot from each tube was analyzed directly by LSC.

Results and Discussion. The quantity of labeled 1,3,6,8-T_4CDD that passed through the SepPak was 10–15 times greater in the presence of 20 mg/L of DOM, and this result was not affected by the presence of up to 5% methanol (Figure 3). Further, the apparent solubility of the dioxin was enhanced four- to five-fold in the presence of the DOM, even in the presence of 15% methanol. The levels of methanol typically used in recent

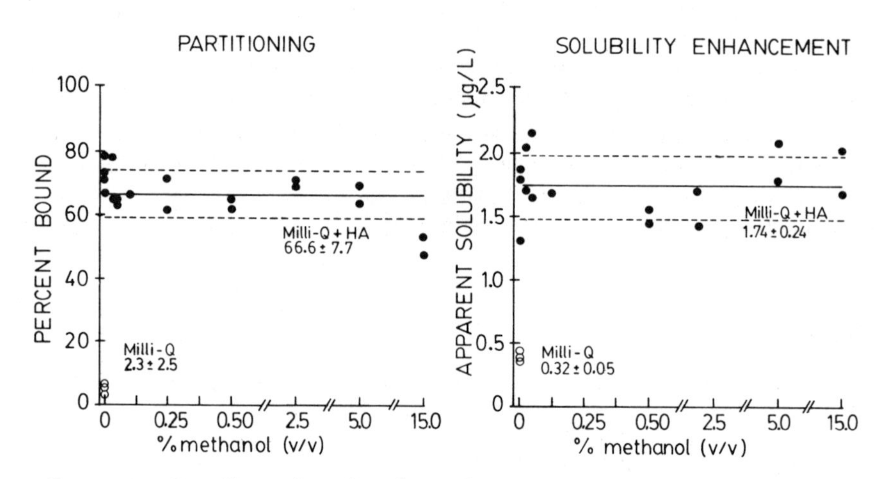

Figure 3. The effect of methanol on the interaction between ^{14}C-1,3,6,8-tetrachlorodibenzo-p-dioxin and Aldrich humic acid (20 mg/L, 0.45-µm filter). Partitioning was determined with centrifugation and C_{18}-SepPak methodology (21). *Solubility enhancement was determined as described by Chiou et al.* (16), *except that ^{14}C-dioxin was spiked onto the walls of glass tubes (Corex), the solvent was allowed to evaporate, and the tubes were shaken for 24 h after the water solution was added.*

studies with hydrophobic compounds are well below those that would affect the phenomena being studied.

Recommendations

None of the methods described is ideal for all studies concerning the interaction of DOM with hydrophobic compounds; however, each technique class has distinct advantages in individual projects. The generator column technique produces saturated solutions, provided the flow rates used are appropriate. The delivery of the hydrophobic compound solution is continuous, mixing is uniform when DOM is introduced, emulsions and aggregates are eliminated, and there is no need for a cosolvent. Further, losses via adsorption to the walls of the apparatus are much reduced, compared to methods that make use of glassware. Reduced loss is one of the main advantages of the generator column technique.

Direct addition of the liquid or solid hydrophobic compound is simple, but it involves the presence of vast excesses of the compound, and extended times are required to reach equilibrium concentration. The microscale addition of hydrophobic compound in a cosolvent is convenient, and the cosolvent effect for methanol appears to be of little practical significance. By implication, the use of other cosolvents also poses very little concern, although they have not been critically evaluated. However, in biological experiments where the cosolvent may have undesired physiological effects on the organisms, the cosolvent technique as currently used might present a problem.

Finally, the quality of the water used in any experiment examining the interaction of DOM with hydrophobic compounds must be well defined. Reference water must be organic-free; that is, DOM concentration must be lower than the detection limit of the best available technique. In addition, biota themselves have a marked effect on the quality of the water in which they are living, with respect to both DOM and particulate organic matter.

References

1. Friesen, K. J.; Sarna, L. P.; Webster, G. R. B. *Chemosphere* **1985,** *14,* 1267–1274.
2. Webster, G. R. B.; Muldrew, D. H.; Graham, N. J.; Sarna, L. P.; Muir, D. C. G. *Chemosphere* **1986,** *15,* 1379–1386.
3. Servos, M. R.; Muir, D. C. G.; Webster, G. R. B. *Aquat. Toxicol.* **1988,** in press.
4. Muir, D. C. G.; Yarachewski, A. L.; Webster, G. R. B. In *ASTM Aquatic Toxicology and Hazard Assessment: 8th Symposium*; Philadelphia, PA, 1985; ASM STP 891, pp 440–454.
5. Muir, D. C. G.; Marshall, W. K.; Webster, G. R. B. *Chemosphere* **1985,** *14,* 829–833.

6. Muir, D. C. G.; Yarachewski, A. L.; Knoll, A.; Webster, G. R. B. *Environ. Toxicol. Chem.* **1986,** *5,* 261–272.
7. DeVoe, H.; Miller, M. M.; Wasik, S. P. *J. Res. Natl. Bur. Stand. (U.S.)* **1981,** *86,* 361–366.
8. May, W. E.; Wasik, S. P.; Freeman, D. H. *Anal. Chem.* **1978,** *50,* 997–1000.
9. Veith, G. D.; Comstock, V. M. *J. Fish. Res. Board Can.* **1975,** *32,* 1849–1851.
10. Bruggeman, W. A.; Martron, L. B. J. M.; Kooiman, D.; Hutzinger, O. *Chemosphere* **1981,** *10,* 811–832.
11. Bruggeman, W. A.; Opperhuizen, A.; Wijbenga, A.; Hutzinger, O. *Toxicol. Environ. Chem.* **1984,** *7,* 173–189.
12. Murphy, T. J.; Mullin, M. D.; Meyer, J. A. *Environ. Sci. Technol.* **1987,** *21,* 155–162.
13. Haque, R.; Schmedding, D. *Bull. Environ. Contam. Toxicol.* **1975,** *14,* 13–18.
14. Hassett, J. P.; Anderson, M. A. *Environ. Sci. Technol.* **1979,** *13,* 1526–1529.
15. Chiou, C. T.; Freed, V. H.; Schmedding, D. W.; Kohnert, R. L. *Environ. Sci. Technol.* **1977,** *11,* 475–478.
16. Chiou, C. T.; Malcolm, R. C.; Brinton, T. I.; Kile, D. E. *Environ. Sci. Technol.* **1986,** *20,* 502–507.
17. Yin, C.; Hassett, J. P. *Environ. Sci. Technol.* **1986,** *20,* 1213–1217.
18. Hassett, J. P.; Milicic, E. *Environ. Sci. Technol.* **1986,** *19,* 638–643.
19. Caron, G.; Suffet, I. H.; Belton, T. *Chemosphere* **1985,** *14,* 993–1000.
20. McCarthy, J. F.; Jimenez, B. D. *Environ. Toxicol. Chem.* **1985,** *4,* 511–521.
21. Landrum, P. F.; Nihart, S. R.; Eadie, B. J.; Gardner, W. S. *Environ. Sci. Technol.* **1984,** *18,* 187–192.
22. Morehead, N. R.; Eadie, B. J.; Lake, B.; Landrum, P. F.; Berner, D. *Chemosphere* **1986,** *15,* 403–412.
23. Leversee, G. L.; Landrum, P. F.; Giesy, J. P.; Fauvin, T. *Can. J. Fish. Aquat. Sci.* **1983,** *40*(suppl. 2), 63–69.
24. Carlberg, G. E.; Martinsen, K.; Kringstad, A.; Gjessing, E.; Grande, M.; Kallgvist, T.; Sharer, J. V. *Arch. Environ. Contam. Toxicol.* **1986,** *15,* 543–548.
25. McCarthy, J. F.; Jimenez, B. D. *Environ. Sci. Technol.* **1985,** *19,* 1072–1076.
26. Landrum, P. F.; Nihart, S. R.; Eadie, B. J.; Herche, L. R. *Environ. Toxicol. Chem.* **1987,** *6,* 11–20.
27. McCarthy, J. F.; Black, M. C. In *ASTM Aquatic Toxicology and Hazard Assessment: 10th Symposium*; Adams, W. J.; Chapman, G. A.; Landis, W. G., Eds.; Special Technical Publication 971; American Society for Testing and Materials: Philadelphia, 1988; pp 233–245.
28. McCarthy, J. F.; Jimenez, B. D.; Barber, T. *Aquat. Toxicol.* **1985,** *7,* 15–24.
29. McCarthy, J. F. *Arch. Environ. Contam. Toxicol.* **1983,** *12,* 559–568.
30. Muir, D. C. G.; Hobden, B. R.; Servos, M. R. Presented at the 7th Annual Meeting of the Society of Environmental Toxicology and Chemistry, Alexandria, VA, 1986; Abstract 68.
31. Gossett, J. M. *Environ. Sci. Technol.* **1987,** *21,* 202–208.
32. Munz, C.; Roberts, P. V. *Environ. Sci. Technol.* **1986,** *20,* 830–836.

RECEIVED for review July 24, 1987. ACCEPTED for publication February 16, 1988.

18

Bioavailability and Toxicity of Metals and Hydrophobic Organic Contaminants

John F. McCarthy

Environmental Sciences Division, Oak Ridge National Laboratory, Oak Ridge, TN 37831–6036

The effect of humic substances on the availability and toxicity of organic and inorganic contaminants in the aquatic environment is reviewed. Organic contaminants associated with humic substances appear to be essentially unavailable for uptake by amphipods, daphnids, and fish. Acute toxicity of these compounds is also diminished proportionally. Because the affinity of organic solutes for binding to humic substances is related to their hydrophobicity, the effect of humic substances is significant only for compounds with octanol–water partition coefficients greater than 10^4. In most cases, association of toxic metals with humic substances reduces the uptake and toxic effects of the contaminants. However, complex interactions among the toxicants, humic ligands, other transition metals and major cations in solution, and the carrier proteins on biological membranes make it difficult to generalize and predict any reduction in accumulation and toxicity of metals. Humic substances may have secondary effects on biota uptake and accumulation of toxicants through their role in altering the transport and fate of pollutants.

BINDING OF ORGANIC OR INORGANIC CONTAMINANTS to humic substances, or to other dissolved or colloidal organic matter in aquatic systems, can alter the availability of the contaminants for uptake by biota. It is generally thought that humic substances alter bioavailability by altering the concentration of a contaminant that is in a physicochemical form capable of traversing mem-

0065–2393/89/0219–0263$06.00/0

branes. Association of metal or organic solute with the humic macromolecule masks the chemical properties of the contaminant and thus alters the normal biochemical interaction of the contaminant with the membrane. This physicochemical interaction also changes the toxicity of the contaminant, because the toxic effect of a pollutant is directly related to the dose incorporated by the organism. The relationship between dose and adverse response is not always straightforward. Contaminants can be sequestered in pharmacokinetically inactive compartments within the animal (e.g., stored in lipid deposits that are isolated from sites of toxic action) or complexes can be formed with specific proteins, such as metallothioneins, which prevent metals from entering tissues and promote excretion of toxic metals.

In this chapter, I review current knowledge of how reversible interaction of humic substances with organic and inorganic contaminants alters the biological availability and toxic effect of the pollutants. The nature of the interaction of the humic substances with the contaminants and the factors that influence the quantitative distribution of the contaminants between a bound and free form are not discussed here; these are the topics of other chapters. In much of this discussion, the general term "dissolved or colloidal organic matter" (DOM) will be used to identify organic components of natural systems that alter the physicochemical properties and the bioavailability of organic or inorganic contaminants. This term includes both aquatic humic and fulvic substances, although the affinity of different components of natural DOM to bind organic or inorganic solutes may be quite different. Metal contaminants are discussed first; then follows a discussion of the effects of humic substances on the availability and toxicity of organic contaminants.

Effect of DOM on Bioavailability and Toxicity of Metals

Because of their polyelectrolytic nature, humic and fulvic substances are capable of complex associations with metals (*1*, *2*). The presence of DOM has often been reported to alter the bioavailability and toxic effects of metals in aquatic systems; however, interactions are complex and not easily described or generalized. Some examples of the many and often conflicting reports can illustrate the difficulties in attempting to interpret how the interactions of metals with DOM affect the accumulation and toxcity of metals. Humic acids have been reported to increase the uptake of cadmium by mussels (*3*) and rainbow trout (*4*), but to decrease accumulation in phytoplankton (*5*, *6*), algae (*7*), *Daphnia magna* (*8*), and corn roots in water culture (*9*). DOM slightly increased uptake of americium and plutonium by phytoplankton (*10*), but decreased uptake of several other metals, including mercury and zinc by fish (*4*), and zinc, chromium, and cobalt by algae (*7*). DOM had no measurable effect on accumulation of copper by a polychaete

(*11*), nor cadmium by *D. magna* (*12*). The presence of DOM also altered the toxicity of metals, but not necessarily in direct relation to the accumulation of the metals in organisms (*13*). Toxicity decreased for cadmium in Atlantic salmon and algae (*14*); for copper in Atlantic salmon (*15*), algae (*5*), and *D. magna* (*12, 16*); and for zinc in daphnids (*17*). Toxicity of cadmium to *D. magna* and copper to *D. pulex* (*16*) increased in the presence of humic acid, but only in hard water (*12, 18*).

Several mechanisms have been proposed to account for the diverse and seemingly conflicting results. It is generally assumed that the free metal ion is the chemical species responsible for the biological effects of metals. The biological uptake and toxicity of the metals are expected, therefore, to be altered by interactions of the metal with inorganic and organic ligands, which alter the concentration of the ionic species (e.g., *5, 19–22*). The major processes that control chemical speciation of metals in natural waters (precipitation, formation of complexes with inorganic and organic ligands, and adsorption by particulate material) can be modeled to permit calculation of the concentration of free ionic metal (*23, 24*). However, direct measurement of the free metal ion concentration during accumulation or toxicity experiments failed to confirm a direct relationship between free ion concentration and biological effect (e.g., *4, 8, 17*).

The models and measurements of solution chemistry may fail because they neglect possible changes in solution equilibria at the interface between the water and the biological membrane (gill or gut). Competitive effects between the DOM–metal complex and the carrier proteins in biological membranes responsible for active transport of metals into the organism, or changes in binding affinity of these membrane ligands due to competitive binding of other transition metals or major cations in solution, could alter the amount of metal translocated across biological membranes into the organism. Enhanced uptake of metals could be due to passive diffusion of organic–metal complexes. Association of metals with relatively low-molecular-weight organic ligands could make the metal more lipid-soluble and thus increase permeability through membranes by a mechanism other than carrier-mediated transport of the free ionic metal species (*3, 25*).

Interpretation of uptake and toxicity results can be further complicated by biochemical processes within the test organism. *D. magna* excrete organic compounds with metal-binding activity similar to that of humic and fulvic acids (*26*). The release of agents that form complexes may affect the results of toxicity tests by reducing metal activities during incubation. Furthermore, relationships between the accumulated body burden of metal and adverse toxic effects may be obscured by the protective action of internal metal-binding proteins, such as metallothioneins, which are produced at higher levels within organisms chronically exposed to nonlethal levels of metals (e.g., *27*).

Effect of DOM on Bioavailability and Toxicity of Organic Contaminants

The effect of humic substances on the biological uptake and toxicity of organic contaminants appears much more consistent and predictable than their effect on metal toxicants. Available data suggest that association of organic contaminants with DOM reduces the uptake and toxicity of the contaminant. The interaction of the organic contaminant discussed in this chapter is limited to the reversible association of the organic solute with the humic macromolecule. This is to be distinguished from the covalent incorporation of an organic solute into the chemical structure of the humic molecule by oxidative coupling. The detoxification of a pollutant by copolymerization with humic macromolecules is discussed by Bollag in ref. 28.

The reversible association of organic solutes with DOM appears to result from the solvophobic partitioning of hydrophobic solutes from the polar aqueous environment into the more nonpolar domain of the humic macromolecule (*29*). The chemical nature of the solute does not change because of the association; the partitioning is simply an entropy-driven equilibration of the solute between a polar and nonpolar phase similar to that described for the partitioning of organic chemicals between water and octanol, or between water and the organic coatings of sediment particles (*30, 31*). The equilibrium or steady-state concentrations of solute in the two phases (freely dissolved in water or bound to DOM) can be described quantitatively by a distribution coefficient, K_{dom}, analogous to the octanol–water partition coefficient (K_{ow}) or the carbon-referenced sediment partition coefficient (K_{oc}) (*32*):

$$K_{dom} = C_{dom}/C_d[\mathrm{DOM}] \qquad (1)$$

where C_{dom} and C_d are the concentrations of solute associated with DOM (mol/kg of carbon) or freely dissolved in water (mol/L of water), respectively, and [DOM] is the concentration of DOM (kg of carbon/L). The affinity of an organic solute for associating with DOM is inversely related to the water solubility and directly related to the K_{ow} of the solute (*29, 33, 34*). Because aqueous solubility is related to the molecular surface area of the solute (*35*), contaminants with higher molecular weights, such as polycyclic aromatic hydrocarbons (PAHs) with three or more rings, or polychlorinated biphenyls (PCBs), have a high affinity for associating with the DOM. A linear relationship has been described, for example, between the log K_{ow} of a series of PAHs and their log K_{dom}s for Aldrich DOM [log K_{dom} = (1.03 log K_{ow}) – 0.5] (*36*).

Binding of Organic Contaminants to DOM. The association of a hydrophobic organic contaminant (HOC) with DOM alters the bioavailability and toxicity of the HOC. Boehm and Quinn (*37*) observed that uptake of hexadecane by the clam, *Mercinaria mercinaria*, increased significantly when the natural organic matter in seawater was removed by filtration through activated charcoal, but the uptake of phenanthrene was unaffected. The accumulation of the five-ring PAH, benzo[*a*]pyrene (BaP), by *D. magna* decreased by over 95% in the presence of 20 mg C/L of a commercial (Aldrich) humic acid (*38*). While Leversee et al. (*39*) also observed a decrease in BaP uptake by *D. magna* (25% decline with 2 mg C/L of Aldrich humic acid), they reported that the same concentration of humics had little effect on uptake of anthracene, dibenzanthracene, and dimethylbenzanthracene. Although they reported that the humic acid increased uptake of another five-ring PAH, 3-methylcholanthrene (3-MC) (*39*), this was not confirmed by a subsequent study that demonstrated that 3-MC uptake by *D. magna* decreased with increasing concentrations of Aldrich humic acid (*40*). Leversee et al. (*39*) also found that removal of natural DOM from streamwater by photooxidation (DOM decreased from 5.5 to 0.2 mg C/L) increased uptake of BaP by 40%. The uptake of 2,4′,5-trichlorophenol by Atlantic salmon fry (*Salmo salar*) was 30% lower when exposures were conducted in humic lake water (7 mg C/L), compared with exposures in water from a nonhumic lake (*41*). Bioaccumulation of the less hydrophobic compound tetrachloroguaiacol was reduced by 14% in the presence of humic water, but humics had no significant effect on accumulation of the even less hydrophobic contaminants lindane and trichlorphenol (*41*). Similarly, the addition of 10 mg/L of Aldrich DOM had only a slight effect on the accumulation of bis(tributyltin) oxide (TBT) by the mussel, *Mytilus edulus* (*42*). In general, DOM has been reported to have had an inhibitory effect on the uptake of very hydrophobic contaminants, but little effect on compounds with K_{ow}s below 10^4. This relationship may be related to the fraction of the total contaminant that binds to the DOM, which can be calculated:

$$\text{fraction of HOC bound to DOM} = \frac{K_{dom}[\text{DOM}]}{1 + K_{dom}[\text{DOM}]} \qquad (2)$$

On the basis of the direct relationship between K_{ow} and K_{dom} (*33, 36*) and equation 2 (*40*), it can be estimated that only a few percent of the phenanthrene, lindane, trichlorophenol, or TBT would have been bound to the DOM under conditions of the aforementioned studies.

Bioavailability of HOC Bound to DOM. Direct physical measurement of the amount of HOC bound to the DOM during toxicokinetic studies has confirmed that contaminant bound to DOM is essentially unavailable for uptake by aquatic organisms. Uptake is therefore reduced in proportion to the fraction of the HOC bound to the DOM. The uptake and elimination of BaP and naphthalene were measured in bluegill sunfish, *Lepomis macrochirus* (*34*), in the presence and absence of Aldrich DOM.

In the presence of 20 mg C/L of DOM, 97% of the BaP was bound to DOM (measured by equilibrium dialysis) (*43*, *36*). Uptake was reduced by about 95%, compared with uptake in the absence of DOM. Only 2% of the less hydrophobic compound naphthalene was bound to the same concentration of DOM, and uptake of naphthalene was not significantly affected by the presence of DOM (*34*). The kinetics of accumulation of a series of two- to five-ring PAHs by *D. magna* was measured in the presence of several concentrations of Aldrich DOM (*40*) (Figure 1). Uptake and accumulation decreased in proportion to the fraction of PAH bound to the DOM. Both kinetic and steady-state analyses of the data confirmed that PAH bound to DOM was not taken up by the *D. magna*.

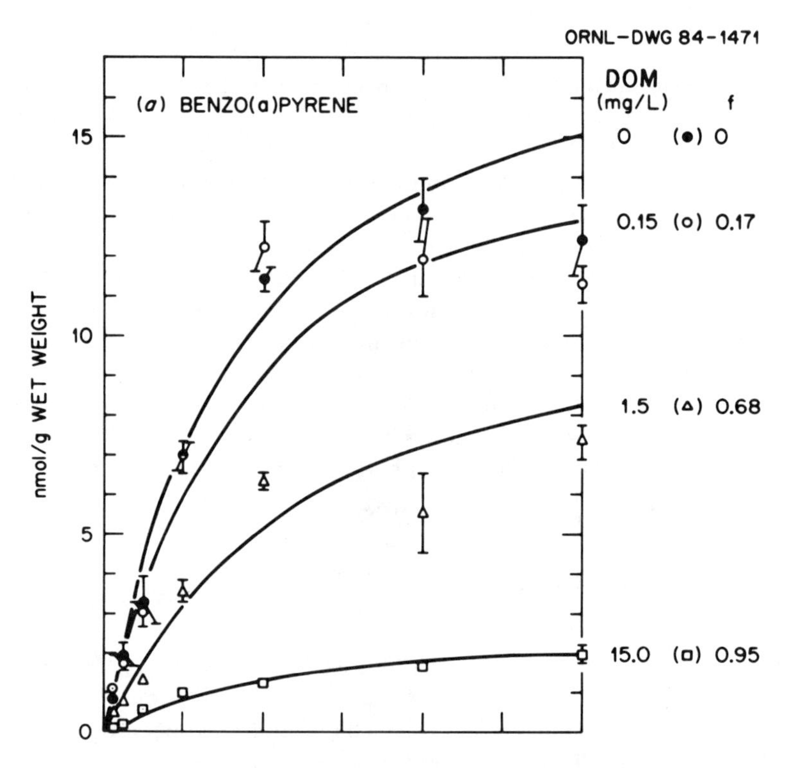

Figure 1. The time course of uptake of PAH shown for D. magna *exposed to aqueous solutions of (a) BaP, (b) benzanthracene, and (c) anthracene in the presence of different concentrations of Aldrich DOM. The presence of 15–60 mg C/L of DOM did not decrease the accumulation of naphthalene (data not shown). The mean concentrations in the animals (+SE of four replicates from each time point) are plotted. The concentration of DOM and the fraction of PAH bound to DOM (f) are indicated. The body burden of 3-methylcholanthrene accumulated after a 30-h exposure in the presence of 0.15–15 mg C/L of DOM decreased in proportion to the fraction of contaminant bound to the DOM (data not shown). (Reproduced with permission from ref. 40. Copyright 1985 Elsevier Science Publishers.)*

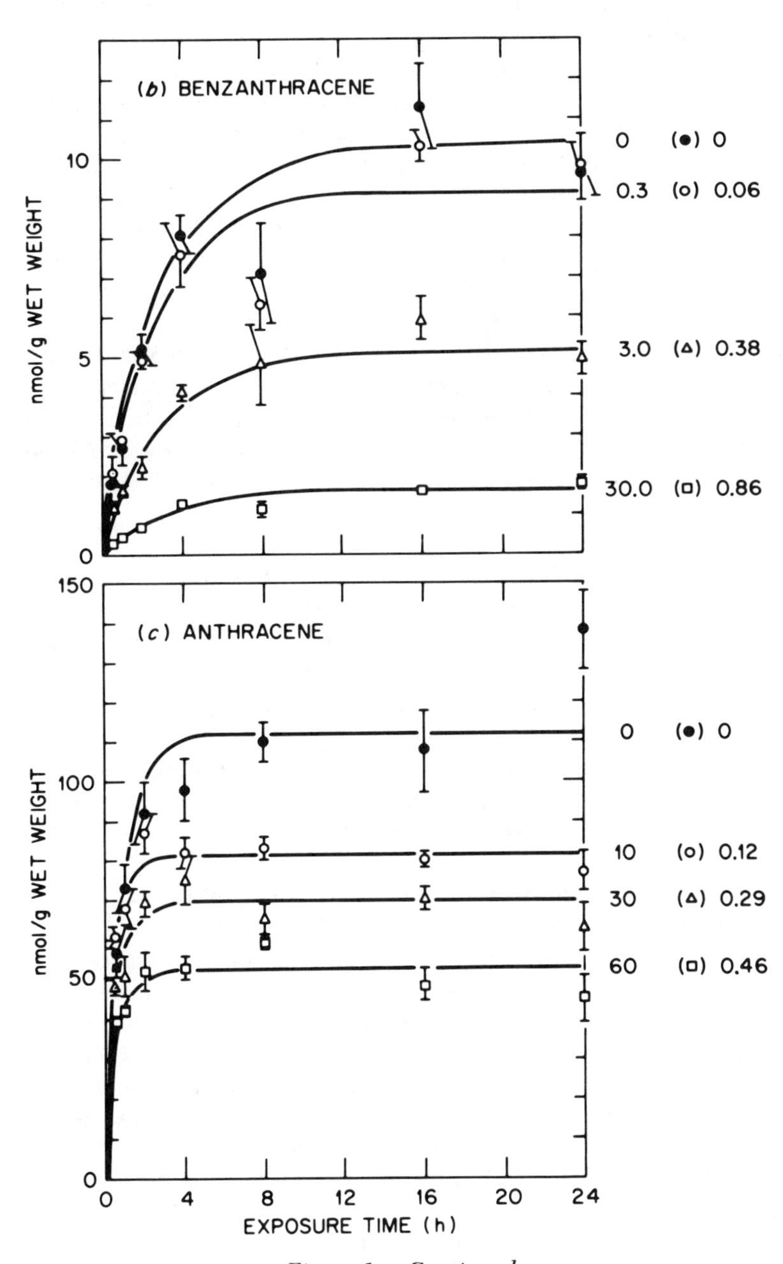

Figure 1.—Continued.

Landrum et al. (*44*) measured the change in the rate of uptake by the amphipod, *Pontoporeia hoyi*, of a variety of HOCs (several PAHs, DDT, diethylhexyl phthalate, and two PCBs) in the presence of different concentrations of Aldrich DOM. The change in the uptake rate was compared with the amount of HOC bound to the DOM (measured by a reverse-phase separation method) (*33*). Biological partition coefficients were derived from the toxicokinetics, based on an assumption that only freely dissolved HOC was being taken up by the amphipods. The partition coefficients derived from the reverse-phase method and from the toxicokinetics correlated well, supporting the conclusion that HOCs bound to DOM were not available to the organisms (*44*). Bioconcentration of tetrachloro- and octachlorodibenzo-*p*-dioxins (TCDD and OCDD, respectively) by rainbow trout and fathead minnows was orders of magnitude lower than would have been predicted from well-established regression correlations between bioconcentration and the hydrophobicity of the compounds (*45*). The low accumulation of at least the OCDD was attributed to binding of the dioxins to natural DOM in the exposure water. Analyses using the reverse-phase separation method indicated that 15% of the TCDD, but apparently all of the OCDD, was associated with the DOM.

Effect of HOC–DOM Interaction on Toxicity. The reduction in accumulation of HOCs due to their association with DOM results in a corresponding reduction in their toxicity. The presence of humic material inhibited the effect of several organic mutagens, as measured in vitro by using an Ames mutagenesis bioassay system (*46*). The in vivo toxicity of DDT to *D. magna* was reduced in the presence of natural DOM and was directly related to the lower body burden accumulated due to association of the DDT with the DOM. The toxicity and body burden of lindane, which has a low affinity for binding to DOM, were not affected (*47*).

Possible Mechanisms for Reduced Accumulation. The available evidence indicates that the association of HOCs with natural DOM reduces the uptake and accumulation of those contaminants by aquatic organisms. It appears that DOM prevents transport of the HOC through biological membranes. DOM does not reduce bioaccumulation by changing the rate of metabolism or eliminating the contaminant. In bluegill sunfish exposed to PAH, rate coefficients for the elimination of the contaminant were identical in the presence and absence of DOM, and the rate of biotransformation of the BaP within the fish was unaffected by the presence of DOM (*34*). Association of HOC with DOM has a direct effect on the efficiency with which the solute is taken up from water passing over the gills of rainbow trout. Measurements using a metabolic chamber demonstrated that the reduction in the uptake efficiency was equal to the fraction of HOC bound to the DOM and that only dissolved compound was translocated across the gill membranes

(*48*). Most likely, the impaired uptake involves a charge-exclusion or size-exclusion mechanism. HOCs are translocated across the gills by a passive diffusion through the lipophilic membranes. Polar solutes are not readily transported by this mechanism, but generally require active transport by specific ligands in the membrane. The polar character of the polyelectrolyte humic macromolecule undoubtedly overwhelms the lipophilic properties of the HOCs and controls the transport of those contaminants. Uptake of the humic macromolecule (and the associated HOCs) would be minimal because of both the DOM's polar nature and the large molecular size of the macromolecule, which would inhibit or prevent its passage through the membrane.

Environmental Significance of Association of HOCs with DOM. Figure 2 presents a conceptual model for calculating the effect of humic substances on the bioaccumulation of HOCs in aquatic systems. The total amount of contaminant in a particle-free system can be viewed as being distributed among dissolved, bound, and biotic compartments. The accumulation of the HOC by the organism can be described:

$$dC_a/dt = k_1 C_d - k_2 C_a + k_3 C_{dom} \quad (3)$$

where C_a is the concentration of HOC in the animal (mol/g), k_1 and k_3 are the rate coefficients (h^{-1}) for uptake of dissolved and bound HOC, respec-

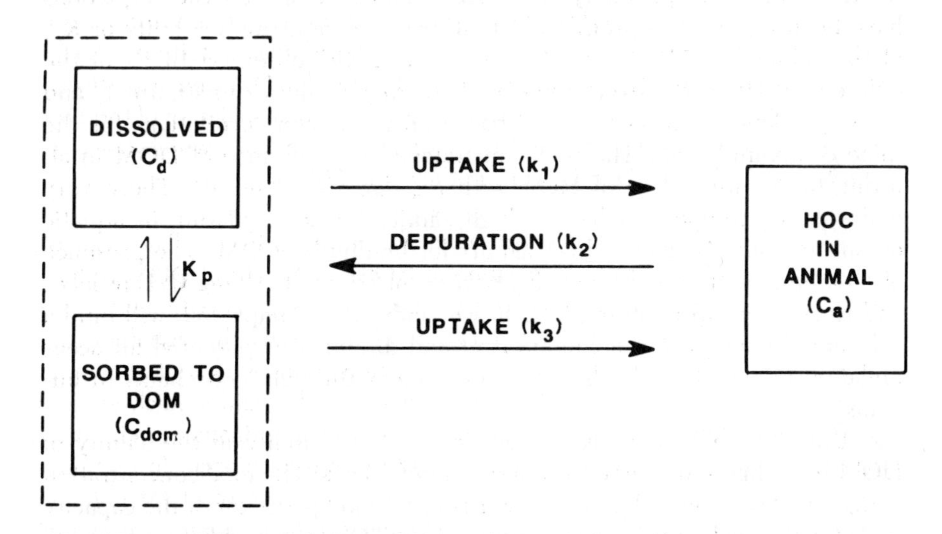

Figure 2. Conceptual model used for multicompartment analysis. The HOC in the system is viewed as being distributed among three compartments: dissolved in water (C_d)*; bound to DOM* (C_{dom})*; and incorporated by the animal* (C_a)*. Rate coefficients describe the steady-state flux of HOC among the compartments. (Reproduced with permission from ref. 40. Copyright 1985 Elsevier Science Publishers.)*

tively, and k_2 is the rate coefficient (h^{-1}) for elimination of HOC from the animal. The potential of an HOC accumulating in an aquatic organism is often quantified as a bioconcentration factor (BCF), which is the ratio of the steady-state distribution of the HOC between the animal and the water. In almost all toxicological and regulatory contexts, it is generally assumed that all of the nonparticulate contaminant in the water is dissolved and available for uptake. However, if DOM is present in the system, only a fraction of the HOC is freely dissolved, while the remainder is bound to DOM and much less bioavailable. Thus, the BCF achieved in the presence of DOM (BCF_{dom}) will be less than that reached in the absence of the sorbent. If, as the evidence suggests, HOC bound to DOM is totally unavailable for uptake ($k_3 = 0$), the BCF_{dom} will depend directly on the fraction of HOC that is freely dissolved in the water (f_{free}):

$$BCF_{dom} = f_{free} BCF_{no\ dom} \quad (4)$$

$$f_{free} = 1/(1 + K_{dom}[DOM]) \quad (5)$$

This conceptual model has been used to develop a structure–activity relationship (SAR) between a physicochemical property of the HOC (its K_{ow}), the amount of DOM present in the system, and the bioconcentration of the HOC from the water (Figure 3) (*40*). The BCF of an HOC is logarithmically related to the hydrophobicity of the HOC, and several regression equations have been reported to predict the BCF from the aqueous solubility or K_{ow} of the solute (*49–51*). Figure 3 was based on the observed BCFs of the different PAHs in the experiment with *D. magna* shown in Figure 1, and on K_{dom} calculated from the K_{ow}, based on regression correlations for the same compounds (*36*). The SAR illustrates the significance of DOM in altering the accumulation of HOC with K_{ow}s greater than 10^4. These very hydrophobic compounds have a high affinity for accumulating in aquatic organisms, but also have a high affinity for binding to DOM. The presence of environmentally realistic concentrations of DOM (1–10 mg C/L in lakes and rivers, and up to 50 mg C/L in wetlands and swamps) (*52*) will bind a substantial fraction of the contaminant and diminish its potential for accumulation by aquatic organisms and movement through food chains to humans.

Unfortunately, there is a large degree of variability in the affinity of DOM from different sources of water to bind HOCs. The total concentration of dissolved organic carbon in a water is not a good predictor of the capacity of that water for binding organic contaminants. Qualitative differences in the nature of the organic carbon have large effects on its affinity for binding HOC. The K_{dom} for binding of individual HOC (measured by the reverse-phase method) differed by 2 orders of magnitude for DOM from waters taken from different locations within the Great Lakes (*53, 54*). The affinity of different water sources for binding DDT (measured by equilibrium dialysis) also varied by orders of magnitude (*55*). There is an obvious need to un-

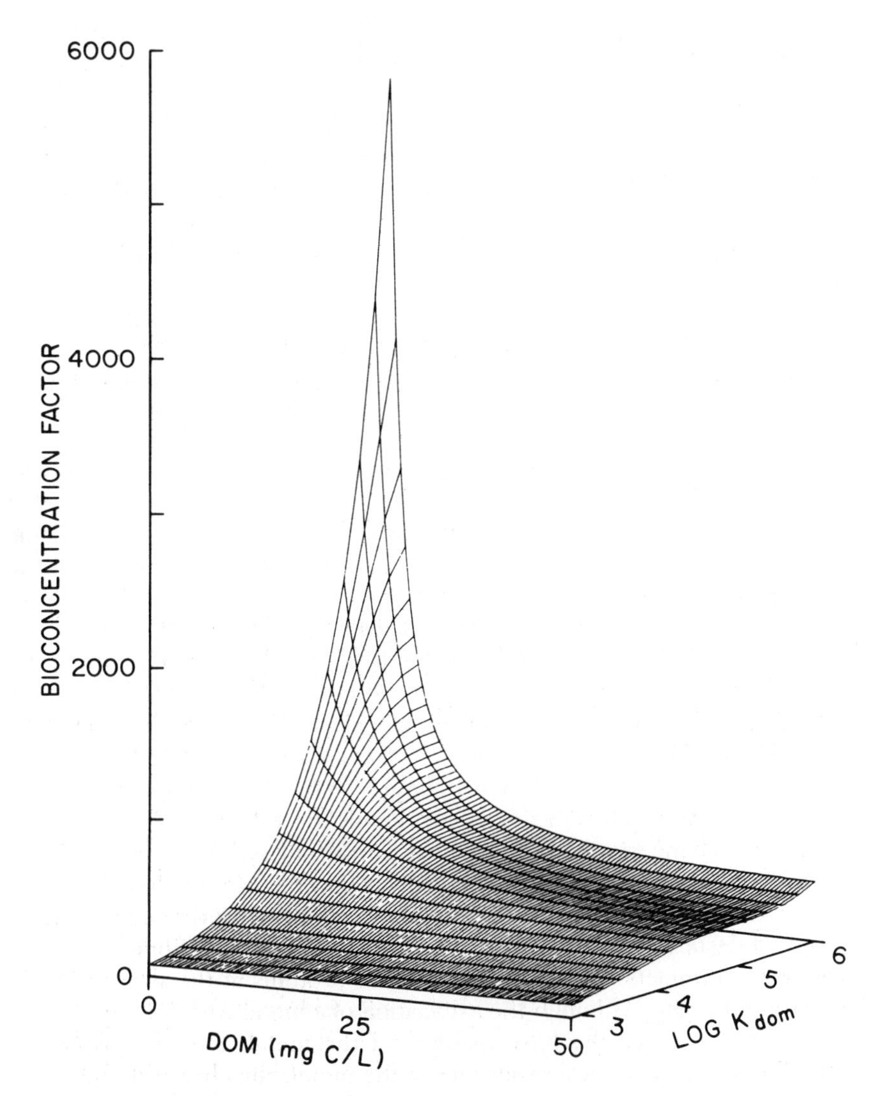

Figure 3. Structure–activity relationship (SAR) between K_{dom}, *[DOM], and predicted BCF, based on equations 4 and 5. The left edge of the figure describes the direct relationship between the hydrophobicity of the contaminant and its observed BCF in* D. magna *when no DOM is present (Figure 1). For a less hydrophobic contaminant, such as naphthalene (front edge of figure, log* K_{dom} = 3), *the BCF is low in the absence of DOM, and the presence of increasing amounts of DOM has little effect. However, for a very hydrophobic compound, such as BaP (back edge of figure, log* K_{dom} = 6), *even small amounts of DOM bind a substantial fraction of the contaminant, and this results in a large decrease in bioaccumulation. The observed BCF in the presence of DOM (Figure 1) agreed well with the BCF predicted in this figure. (Reproduced with permission from ref. 40. Copyright 1985 Elsevier Science Publishers.)*

derstand the chemical and structural properties of DOM that underlie its attraction to HOC and to develop methods for easily predicting the organic complexation capacity of natural waters. Delineation of these properties will greatly improve capabilities to predict the bioaccumulation and potential for food chain transfer of the higher molecular weight, more environmentally persistent, organic contaminants—such as PAHs, PCBs, and dioxins—that pose the greatest risks to human health.

Conclusions

Aquatic humic and fulvic substances can have a profound effect on the bioavailability and toxicity of both organic and inorganic contaminants. When contaminants interact with the DOM by chelation (for metals) or solvophobic partitioning (for HOC), the original chemical properties of the contaminant are altered and their transport and bioavailability are often dominated by the properties of the humic molecule. Available evidence indicates that HOCs that bind to humic macromolecules become essentially unavailable for uptake through biological membranes, possibly due to exclusion of the humic molecule from the membrane on the basis of the molecular size or net electrical charge of the humic substance. This interaction appears consistent and readily predictable because it is based on a thermodynamic gradient favoring the partitioning of a water-insoluble solute from an aqueous environment into the nonpolar domain of a second physicochemical phase, the humic substance.

Interactions of DOM with metals are far more complex. While the "average" properties of an "average" humic or fulvic molecule can be modeled as having two binding sites with different stability constants (*56*), a humic molecule is a heterogeneous array of specific sites capable of a range of interactions with metals or with competing cations. Experimental evidence for the effect of DOM on the uptake of metals is further complicated by competitive interactions with specific carrier proteins at the surface of biological membranes. Although the association of a metal with DOM appears, in general, to reduce the bioavailability of the metal, the effect is clearly dependent on the chemical properties of the metal, the chemical properties and amounts of competing metals or cations, the pH, and possibly the source or properties of the DOM.

The alteration of the physicochemical properties of the contaminant due to its association with the humic macromolecule can also have an indirect effect on the accumulation and toxicity of the contaminant in an ecosystem. DOM can alter the dose of toxicant to which biota are exposed by altering the transport and fate of the toxicants within the environment. Water-insoluble solutes (such as HOCs) can be stabilized in solution by association with DOM and advectively transported downstream rather than concentrating on sediment particles. For example, the presence of DOM may reduce the uptake and accumulation of a PCB near a source of contamination, but the contaminant now stabilized in the water column may be advectively transported downstream. Dilution of the contaminant concentrations downstream can favor the reversible dissociation from the DOM and make the contaminant available for accumulation far from the site of contamination. Because humics can compete with sediment organic matter for binding of HOCs, humics may also promote removal of persistent contaminants bound to sediment. Dissociation of the HOC in the water column could make the contaminants more readily available to pelagic biota that might otherwise be physically removed from the sediment-bound contaminants. To adequately ascertain the role of humic substances on the bioavailability and toxicity of environmental contaminants requires consideration of the physicochemical and hydrological, as well as the toxicological, aspects of the problem.

Acknowledgments

I thank M. C. Black, J. E. Breck, and B. D. Jimenez for reviewing this manuscript. Oak Ridge National Laboratory is operated by Martin Marietta Energy Systems, Inc., under contract DE–AC05–84OR21400 with the U.S. Department of Energy. Publication No. 3043, Environmental Sciences Division, Oak Ridge National Laboratory.

References

1. Perdue, E. M.; Lytle, C. R. In *Aquatic and Terrestrial Humic Materials*; Christman, R. R.; Gjessing, E. T., Eds.; Ann Arbor Science: Ann Arbor, 1983; pp 295–313.
2. Perdue, E. M. Chapter 19 in this volume.
3. George, S. G.; Coombs, T. L. *Mar. Biol.* **1977,** *39*, 261–268.
4. Ramamoorthy, S.; Blumhagen, K. *Can. J. Fish. Aquat. Sci.* **1984,** *41*, 750–756.
5. Sunda, W. G.; Lewis, J. A. M. *Limnol. Oceanogr.* **1978,** *23*(5), 870–876.
6. Fisher, N. S.; Frood, D. *Mar. Biol.* **1980,** *59*, 85–93.
7. Vymazal, J. *Hydrobiologia* **1984,** *119*, 171–179.
8. Poldoski, J. E. *Environ. Sci. Technol.* **1979,** *13*(6), 701–706.
9. Tyler, L. D.; McBride, M. B. *Plant Soil* **1982,** *64*, 259–262.
10. Fisher, N. S.; Bjerregaard, P.; Huynh-Ngoc, L.; Harvey, G. R. *Mar. Chem.* **1983,** *13*, 45–56.
11. Milanovich, F. P.; Spies, R.; Guram, M. S.; Sykes, E. E. *Estuarine Coastal Mar. Sci.* **1976,** *4*, 585–588.
12. Winner, R. W. *Aquat. Toxicol.* **1984,** *5*, 267–274.
13. Winner, R. W.; Gauss, J. D. *Aquat. Toxicol.* **1986,** *8*, 149–161.
14. Gjessing, E. T. *Arch. Hydrobiol.* **1981,** *91*, 144–149.
15. Zitko, P.; Carson, W. V.; Carson, W. G. *Bull. Environ. Contam. Toxicol.* **1973,** *10*(5), 265–271.
16. Winner, R. W. *Water Res.* **1985,** *19*(4), 449–455.
17. Pommery, J.; Imbenotte, M.; Erb, F. *Environ. Pollut., Ser. B* **1985,** *9*, 127–136.
18. Winner, R. W. *Aquat. Toxicol.* **1986,** *8*, 281–293.
19. Sunda, W. G.; Guillard, R. R. L. *J. Mar. Res.* **1976,** *34*, 511–529.

20. Sunda, W. G.; Engel, D. W.; Thuotte, R. M. *Environ. Sci. Technol.* **1978,** *12,* 409–413.
21. Zamuda, C. D.; Sunda, W. G. *Mar. Biol.* **1982,** *66,* 77–82.
22. Pagenkopf, G. K. *Environ. Sci. Technol.* **1983,** *17,* 342–347.
23. Parkhurst, D. L.; Thorstenson, D. C.; Plummer, L. N. *PHREEQE—A Computer Program for Geochemical Calculations*; U.S. Geological Survey: Reston, VA, 1982; Water Resources Investigations WRI 80–96.
24. Westall, J. C.; Zachary, J.; Morel, F. Report 8601, Department of Chemistry, Oregon State University: Portland, OR, 1986.
25. Blust, R.; Verheyen, E.; Doumen, C.; DeCleir, W. *Aquat. Toxicol.* **1986,** *8,* 211–221.
26. Fish, W.; Morel, F. M. M. *Can. J. Fish. Aquat. Sci.* **1983,** *40,* 1270–1277.
27. Benson, W. H.; Birge, W. J. *Environ. Toxicol. Chem.* **1985,** *4,* 209–217.
28. Bollag, J.-M. In *Chemical and Biochemical Detoxification of Hazardous Wastes*; Glaser, J. A., Ed.; Lewis Publishers: London, in press.
29. Chiou, C. T.; Porter, P. E.; Schmedding, D. W. *Environ. Sci. Technol.* **1983,** *17,* 227–231.
30. Karickhoff, S. W.; Brown, D. S.; Scott, T. A. *Water Res.* **1979,** *13,* 241–248.
31. Means, J. C.; Wood, S. G.; Hassett, J. J.; Banwart, W. L. *Environ. Sci. Technol.* **1980,** *14,* 1524–1528.
32. Karickhoff, S. W. *J. Hydraul. Eng.* **1983,** *110,* 707–735.
33. Landrum, P. F.; Nihart, S. R.; Eadle, B. J.; Gardner, W. S. *Environ. Sci. Technol.* **1984,** *18,* 187–192.
34. McCarthy, J. F.; Jimenez, B. D. *Environ. Toxicol. Chem.* **1985,** *4,* 511–521.
35. Yalkowsky, S. H.; Valvani, S. C.; Amidon, G. L. *J. Pharm. Sci.* **1975,** *64,* 1488–1494.
36. McCarthy, J. F.; Jimenez, B. D. *Environ. Sci. Technol.* **1985,** *19*(*11*), 1072–1076.
37. Boehm, P. D.; Quinn, J. G. *Estuarine Coastal Mar. Sci.* **1976,** *4,* 93–105.
38. McCarthy, J. F. *Arch. Environ. Contam. Toxicol.* **1983,** *12,* 559–568.
39. Leversee, G. J.; Landrum, P. F.; Giesy, J. P.; Fannin, T. *Can. J. Fish. Aquat. Sci., Suppl. 2* **1983,** *40,* 63–69.
40. McCarthy, J. F.; Jimenez, B. D.; Barbee, T. *Aquat. Toxicol.* **1985,** *7,* 15–24.
41. Carlberg, G. E.; Martinsen, K.; Kringstad, A.; Gjessing, E.; Grande, M.; Kallqvist, T.; Skare, J. U. *Arch. Environ. Contam. Toxicol.* **1986,** *15,* 543–548.
42. Laughlin, R. B., Jr.; French, W.; Guard, H. E. *Environ. Sci. Technol.* **1986,** *20*(9), 884–890.
43. Carter, C. W.; Suffet, I. H. *Environ. Sci. Technol.* **1982,** *16*(*11*), 735–740.
44. Landrum, P. F.; Reinhold, M. D.; Nihart, S. R.; Eadie, B. J. *Environ. Toxicol. Chem.* **1985,** *4,* 459–467.
45. Muir, D. C. G.; Yarechewski, A. L.; Knoll, A.; Webster, G. R. B. *Environ. Toxicol. Chem.* **1986,** *5,* 261–272.
46. Sato, T.; Ose, Y.; Nagase, H. *Mutat. Res.* **1986,** *162,* 173–178.
47. Henry, L. L.; Friant, S. L.; Suffet, I. H. *Environ. Sci. Technol.*, in press.
48. Black, M. C.; McCarthy, J. F. *Environ. Toxicol. Chem.* **1988,** *7,* 593–600.
49. Chiou, C. T.; Freed, V. H.; Schmedding, D. W.; Kohnert, R. L. *Environ. Sci. Technol.* **1977,** *11*(5), 475–478.
50. Chiou, C. T. *Environ. Sci. Technol.* **1985,** *19,* 57–62.
51. Neely, W. B.; Branson, D. R.; Blau, G. E. *Environ. Sci. Technol.* **1974,** *8,* 1113–1115.
52. Thurman, E. M. *Organic Geochemistry of Natural Waters*; Kluer Academic: Hingham, MA, 1985.
53. Landrum, P. F.; Nihart, S. R.; Eadie, B. J.; Herche, L. R. *Environ. Toxicol. Chem.* **1987,** *6,* 1120.

54. Morehead, N. R.; Eadie, B. J.; Lake, B.; Landrum, P. F.; Berner, D. *Chemosphere* **1986,** *15*(*4*), 403–412.
55. Carter, C. W.; Suffet, I. H. *Am. Chem. Soc.* **1983,** *11,* 215–229.
56. McKnight, D. M.; Feder, G. L.; Thurman, E. M.; Wershaw, R. L. *Sci. Total Environ.* **1983,** *28,* 65–76.

Received for review July 24, 1987. Accepted for publication December 21, 1987.

INTERACTIONS IN NATURAL WATERS WITH INORGANIC CONTAMINANTS

19

Effects of Humic Substances on Metal Speciation

E. Michael Perdue

School of Geophysical Sciences, Georgia Institute of Technology, Atlanta, GA 30332

This chapter addresses some of the problems that must be understood and solved before the effects of metal–humic substance complexation on water-treatment processes can be quantitatively addressed. The heterogeneity of ligands in a humic substance not only complicates the mathematical description of equilibrium data, but also makes the complexation capacity of a humic substance almost impossible to determine accurately. Complexation capacities (meq/g) of humic substances are widely reported to vary with pH, ionic strength, concentration of the humic substance used in the measurement, and nature of the metal being studied. By analogy with the behavior of a simple ligand (citrate), this chapter demonstrates that the reported effect of humic-substance concentration on complexation capacity is probably an artifact and that other experimental parameters affect conditional concentration quotients for metal complexation reactions. These effects create the illusion that complexation capacity is a function of pH, ionic strength, and nature of the added metal ion.

HUMIC SUBSTANCES ARE UBIQUITOUS in the aquatic environment, and their ability to form complexes with metal ions is well documented by many experimental and modeling studies. The interaction of humic substances with metal ions has been the subject of several recent review papers (*1–6*). In the context of water-treatment chemistry, the interaction of humic substances with metal ions can potentially affect removal of humic substances by coagulation–flocculation processes, removal of toxic heavy metals from polluted waters, and rates and products of reaction of humic substances with disinfectants.

0065–2393/89/0219–0281$06.00/0

Rather than process-oriented aspects of metal–humic substance complexation, the focus here is on describing some of the experimental and conceptual pitfalls that undermine our efforts to quantitatively describe metal–humic substance complexation in a predictive manner. Until such problems are clearly understood, the effects of metal–humic substance complexation on water-treatment processes will remain obscure, and trial-and-error will continue to be the most common approach toward the development of water-treatment methodologies.

Experimental studies can be conducted to examine metal complexation by humic substances at any pH, ionic strength, or combination of competing metal ions. Quantitative modeling of metal–humic substance complexation has not advanced, however, beyond single-metal complexation at constant pH and ionic strength. Even in such relatively simple systems, proper recognition of the effects of pH, ionic strength, nature of the metal, and the concentration of humic substances used in an experiment on metal–humic substance complexation equilibria is needed for interpretation of complexation-capacity measurements and interpretation of thermodynamic data on metal–humic substance complexation.

Overview of Metal Complexation Equilibria

This section presents an overview of the pertinent equations that describe metal–ligand complexation equilibria, both for a simple ligand and for a complex mixture of ligands. The distinction between concentrations and activities must be clearly developed and maintained. ***Equilibrium constants*** depend only on temperature and pressure; ***concentration quotients*** depend on temperature, pressure, and ionic strength; and ***conditional concentration quotients*** depend on temperature, pressure, ionic strength, pH, concentrations of competing metals, and ligands.

Complexation by a Single Ligand. The reactions between a metal ion (M) and a single binding site (L_i) can be described by either overall or stepwise formation constants. For example, for the 1:1 metal-to-ligand complexation reaction, $M + L_i = ML_i$, the overall and stepwise formation constants are the same (the stepwise K will be used):

$$K_i = \frac{\{ML_i\}}{\{M\}\{L_i\}} = \frac{[ML_i]}{[M][L_i]} \cdot \frac{\gamma_{ML_i}}{\gamma_M \gamma_{L_i}} = K_i^c \cdot \Gamma_i \qquad (1)$$

where M is a metal aqueous ion; L_i is a fully deprotonated binding site; ML_i is the complex formed from 1 mol each of M and L_i; braces { } and square brackets [ ] denote activities and concentrations, respectively; and γ-values are activity coefficients. K_i is a true thermodynamic constant, but the concentration quotient K_i^c and the activity coefficient ratio Γ_i are complementary functions of ionic strength. For most simple metal–ligand complexes, basic electrostatic considerations (Debye–Hückel theory) indicate that the activity coefficient ratio (Γ_i) equals 1 at zero ionic strength and increases with increasing ionic strength. K_i^c values therefore equal K_i values at zero ionic strength and tend to decrease with increasing ionic strength. In a given solution of metal and ligand, the concentration of the complex ML_i is thus expected to decrease upon addition of a background electrolyte. Experimental studies generally yield concentrations, rather than activities, of reactants and products. Consequently, K_i values cannot be directly measured, but must be obtained either by estimation of Γ_i values or by extrapolation of K_i^c values (obtained at several ionic strengths) to zero ionic strength.

Another factor that affects the extent of complexation of M by L_i is competition from side reactions, especially the hydrolysis of the metal ion to produce hydroxy complexes and the protonation of the ligand to produce its conjugate acid(s). These reactions do not actually change K_i^c as we have defined it, but they do affect the degree of complexation of M by L_i. As a general rule, ligands tend to form protonated ligands at low pH and metal ions tend to form hydroxy complexes at high pH. Consequently, the reaction between the metal ion and the ligand is often most favorable at intermediate pH values.

For mathematical convenience, a conditional concentration quotient K_i^* is often defined, in which the precise terms in equation 1 are replaced by more convenient terms:

$$K_i^* = \frac{[ML_i(\text{bound})]}{[M(\text{free})][L_i(\text{free})]} \tag{2}$$

In this equation, M(free) represents all forms of the metal ion that are not bound to the ligand of interest, L_i(free) represents all forms of the ligand that are not bound to the metal ion, and ML_i(bound) represents all complexes of 1:1 metal-to-ligand stoichiometry. Unlike K_i^c, which is a function only of ionic strength, K_i^* is a function of ionic strength, pH, concentrations of competing metal ions and ligands, and so on. If all side reactions are well understood, K_i^* is a useful parameter that can be directly related to K_i^c.

Complexation by a Multiligand Mixture. In the previous section, the use of K_i^* instead of K_i^c was a matter of mathematical and experimental convenience. In the study of metal binding by a multiligand mixture such as humic substances, however, there is no choice. It is simply not possible to fully describe the side reactions of a ligand mixture whose individual components are unknown. Conditional concentration quotients or related hybrid expressions are used exclusively, even though the users of such expressions may not always recognize their limitations. In extending the concept of a conditional concentration quotient for metal complexation by a single ligand to multiligand mixtures, an expression that formally resembles equation 2 is usually written

$$\overline{K}^* = \frac{\sum [\mathrm{ML}_i(\mathrm{bound})]}{[\mathrm{M(free)}] \sum [\mathrm{L}_i(\mathrm{free})]} \tag{3}$$

where $\Sigma[\mathrm{ML}_i(\mathrm{bound})]$ is the sum of the concentrations of all complexes formed between M and the multiligand mixture, $\Sigma[\mathrm{L}_i(\mathrm{free})]$ is the sum of the concentrations of all binding sites that are not associated with M, and [M(free)] is the sum of metal species that are not associated with the multiligand mixture.

The experimental methods that are used to study metal complexation by humic substances directly provide either none or, at best, one of the three terms in equation 3. The missing terms are always calculated from the experimental data and some stoichiometric assumptions about the system being studied. The most common assumptions involve the neglect or invocation of the existence of simple inorganic complexes (hydroxy and carbonato complexes) in the system under investigation. For example, $\overline{K}^*$ is often calculated directly from experimental data as

$$\overline{K}^* = \frac{C_\mathrm{M} - [\mathrm{M}]}{[\mathrm{M}](C_\mathrm{L} - C_\mathrm{M} + [\mathrm{M}])} \tag{4}$$

where C_M and C_L are the total stoichiometric concentrations of metal and ligand in the system under study and [M] is the concentration of free metal ion. In calculating $\Sigma[\mathrm{ML}_i(\mathrm{bound})]$ as $(C_\mathrm{M} - [\mathrm{M}])$, the presence of inorganic complexes of the metal ion has been neglected. In calculating $\Sigma[\mathrm{L}_i(\mathrm{free})]$ as $(C_\mathrm{L} - C_\mathrm{M} + [\mathrm{M}])$, an average 1:1 metal-to-ligand stoichiometry has been assumed for the mixture of binding sites. It is also assumed that C_L is known. Most of the remainder of this chapter will address the experimental determination of C_L from metal-binding data.

Although an expression can be written for $\overline{K}^*$ in equations 3 and 4 that formally resembles the conditional concentration quotient in equation 2 ($\overline{K}_i^*$), $\overline{K}^*$ is not a constant at a given pH and ionic strength. Rather, $\overline{K}^*$ will decrease steadily as the total metal-to-ligand ratio $(C_\mathrm{M}/C_\mathrm{L})$ increases. The functional nature of $\overline{K}^*$ arises from preferential reactions of stronger ligands at low metal-to-ligand ratios and has been discussed by several investigators (*1–11*). Nevertheless, the variation of $\overline{K}^*$ with the total metal-to-ligand ratio has often erroneously been cited as evidence for the existence of two binding sites in a humic substance, one reacting more favorably than the other with the metal ion. Average $\overline{K}^*$ values are ultimately functions of ionic strength, pH, and the degree of saturation of the multiligand mixture with metal ion. This latter term is loosely reflected in the $C_\mathrm{M}/C_\mathrm{L}$ ratio. Reported stability "constants" for metal–humic-substance complexation are not actually constant and should be viewed with skepticism.

In simple systems containing one metal ion and one ligand, K_i^*, K_i^c, and K_i values can be interconverted, as described in the preceding section. In metal–multiligand mixtures, however, similar interconversions of $\overline{K}^*$, $\overline{K}^c$, and $\overline{K}$ values are not practical at all. For example, to remove the pH dependence of $\overline{K}^*$ values, it would be necessary to treat all pH-dependent side reactions of the metal ion and all components of the multiligand mixture quantitatively. Although such corrections are practical for metal-ion hydrolysis, there is not yet a rigorous treatment of the acid–base chemistry of humic substances (*12*). Even if the corrections could be made, the resulting $\overline{K}^c$ values would still be functions of ionic strength and the degree of saturation of the multiligand mixture with metal ion. The conversion of $\overline{K}^c$ values into $\overline{K}$ values could theoretically be accomplished by extrapolation to zero ionic strength of $\overline{K}^c$ values obtained at constant degree of saturation of the multiligand mixture with metal ion and variable ionic strength. The resulting $\overline{K}$ values would still be functions of the degree of saturation of the multiligand mixture with metal ion. The remaining functional dependence is a fundamental characteristic of mixtures, and it cannot be eliminated by any experimental method short of total fractionation of the mixture into pure compounds that could be studied separately.

Complexation Capacity: Definitions and Measurements

Definition of Complexation Capacity. For a pure ligand reacting with divalent or trivalent metal ions, even though complexes of higher stoichiometry (1:2, 1:3, etc.) may form at low levels of bound metal, 1:1 complexes predominate at higher levels of added metal. Thus, the complexation capacity of the ligand is usually about 1 mol of metal per mole of ligand. The important point is that complexation capacity is a compositional, rather than thermodynamic, parameter. The complexation capacity of citrate ion (Cit^{3-}), for example, is about 1 mol of metal per mole of citrate, regardless of pH, ionic strength, nature of the metal, or the concentration of citrate ion used in the measurements.

Theoretically, the complexation capacity (CC) of a humic substance or other complex mixture is, to a good approximation, a weighted average of the complexation capacities of the individual ligands in the mixture:

$$CC = \frac{\sum (CC)_i[\text{weight}]_i}{\sum [\text{weight}]_i} \qquad (5)$$

where $(CC)_i$ is the complexation capacity of the ith ligand in the mixture and $[\text{weight}]_i$ is a weighting factor that reflects the relative abundance of that ligand in the multiligand mixture. The nature of the weighting factor depends on the dimensional units of CC, commonly given in milliequivalents per gram. If $(CC)_i$ values are also in milliequivalents per gram, then $[weight]_i$ is the mass of the *i*th ligand in the mixture. If the mixture is not fractionated, CC will be an average constant from which total ligand concentrations (C_L) can be computed for use in equilibrium expressions (equation 4). The metal–humic substance literature suggests that the complexation capacity of a humic substance varies considerably with almost every conceivable experimental variable: increasing at higher pH (*13–18*), decreasing at higher ionic strength (*13, 16*), increasing at higher humic-substance concentrations (*13–15, 19, 20*), and generally varying with the nature of the added metal ion (*13, 14, 17, 21*). The differences between these reported results and theoretical expectations must be resolved.

Major Misconceptions Concerning Complexation Capacities. Two fundamental misconceptions are responsible for much of the confusion over complexation capacities and their reported dependence on experimental conditions (*13–21*). First, the effects of experimental conditions on conditional concentration quotients ($\overline{K}^*$) are erroneously interpreted as modifications of the complexation capacity (CC) of the humic substance. Complexation capacity data are usually interpreted with no regard for the fact that $\overline{K}^*$ values are affected by pH, ionic strength, and the nature of the reacting metal. It is simply easier to saturate ligands with metal at near-neutral pH, low ionic strength, and with a strongly binding metal such as Cu^{2+} than otherwise. This greater ease of formation of metal–humic substance complexes leads directly to apparently higher CC values under these optimum conditions. Second, the effect of simple dilution on the position of the equilibrium in reactions of the type $M + L_i \leftrightarrows ML_i$ is erroneously interpreted as a variation of CC with humic-substance concentration. Because metal complexation equilibria shift toward free metal and ligand with dilution, it is more difficult (at a constant C_M/C_L ratio) to saturate a dilute ligand mixture than a concentrated ligand mixture, which gives rise to the illusion that the CC of a humic substance decreases with dilution. These apparent effects on CC values will be demonstrated in a later section of the chapter.

A potential source of confusion in reported CC values for metal–humic substance complexation is dimensional units. A typical humic substance has a carboxylic acid content of about 5 meq/g and contains about 50% carbon. Suppose that the CC of that humic substance is also 5 meq/g for divalent and trivalent metal ions. Then, depending on the choice of dimensional units and the charge of the metal ion, CC might be reported as 10 meq/g of C, 5.0 mmol/g of C, 3.3 mmol/g of C, 5 meq/g, 2.5 mmol/g, or 1.67 mmol/g. All of these dimensional units can be found in the literature on metal–humic substance complexation, as well as much poorer choices such as milligrams per gram or milligrams per gram of C. This potentially confusing problem is not particularly serious and, if recognized, it is easily corrected.

If C_L values in $\overline{K}^*$ expressions such as equation 4 are obtained from CC values, C_L is actually a binding-site concentration, where a binding site is defined as a group of one or more donor atoms in the humic substance that can bind a single metal ion. It is implicit in this definition that all metal–ligand complexes are of 1:1 stoichiometry (one metal ion per binding site). If a total ligand concentration is defined in this manner, metal complexation data should not be interpreted in terms of a mathematical model that assumes the occurrence of complexes of greater than 1:1 stoichiometry.

Simulation of Complexation Capacity Titrations. The integrated effects of ionic strength, pH, nature of the added metal ion, and ligand concentration on apparent CC values can be demonstrated by computer simulations of CC titrations for a system of one metal and one ligand. The additional inherent complications that are due to the multiligand nature of humic substances could be demonstrated only by analogous computer simulations of multiligand mixtures. In this chapter, complexation capacity titrations have been simulated for citrate ion, with Cu^{2+} and Ca^{2+} as titrants at several pH values (4, 5, 6, 7), ionic strengths (0, 0.1), and total ligand concentrations [$\log(C_L) = -4, -5, -6$]. Thermodynamic data were obtained from a recent tabulation (22) and are given in Table I. Experimental parameters for 10 simulated titrations are given in Table II. In all cases, titrations were simulated over a C_M/C_L range of 0.0–3.0. The choice of an upper C_M/C_L ratio of 3.0 was based on this value as typical of most complexation

Table I. Typical Thermodynamic Data for Metal–Citrate Complexation Reactions at Zero Ionic Strength and 298 K

Reaction	*log* K_{eq}
$H^+ + Cit^{3-} \rightleftarrows HCit^{2-}$	6.40
$2H^+ + Cit^{3-} \rightleftarrows H_2Cit^-$	11.16
$3H^+ + Cit^{3-} \rightleftarrows H_3Cit^0$	14.29
$Cu^{2+} + H_2O \rightleftarrows CuOH^+ + H^+$	−7.50
$Cu^{2+} + 2H_2O \rightleftarrows Cu(OH)_2^0 + 2H^+$	−16.20
$Cu^{2+} + 4H_2O \rightleftarrows Cu(OH)_4^{2-} + 4H^+$	−39.60
$2Cu^{2+} + 2H_2O \rightleftarrows Cu_2(OH)_2^{2+} + 2H^+$	−17.60
$Cu^{2+} + Cit^{3-} \rightleftarrows CuCit^-$	7.20
$Cu^{2+} + H^+ + Cit^{3-} \rightleftarrows CuHCit^0$	10.70
$Cu^{2+} + 2H^+ + Cit^{3-} \rightleftarrows CuH_2Cit^+$	13.90
$Cu^{2+} + H_2O + Cit^{3-} \rightleftarrows CuOHCit^{2-} + H^+$	2.40
$2Cu^{2+} + 2Cit^{3-} \rightleftarrows Cu_2(Cit)_2^{2-}$	16.30
$Ca^{2+} + H_2O \rightleftarrows CaOH^+ + H^+$	−12.85
$Ca^{2+} + Cit^{3-} \rightleftarrows CaCit^-$	4.70
$Ca^{2+} + 1H^+ + Cit^{3-} \rightleftarrows CaHCit^0$	9.50
$Ca^{2+} + 2H^+ + Cit^{3-} \rightleftarrows CaH_2Cit^+$	12.30

Table II. Experimental Parameters for Simulated Complexation Capacity Titrations of Citrate Ion

Curve[a]	*Symbol*[b]	*Metal*	*pH*	I	*log* C_L
1	⊠	Cu	7	0	−4
2	△	Cu	5	0	−5
3	●	Ca	7	0	−4
4	+	Cu	5	0	−6
5	×	Cu	5	0.1	−6
6	▽	Ca	7	0	−5
7	■	Ca	7	0.1	−5
8	◇	Ca	7	0	−6
9	⊕	Ca	4	0	−4
10	□	Cu	4	0	−6

[a]Curve numbers are used in Figures 1–3.
[b]These symbols are used in Figure 3.

capacity experiments. In most such studies, the bound-metal concentration is calculated as C_M − [M], which becomes experimentally indistinguishable from C_M at higher excess C_M values.

The simulated titration data sets were analyzed by three common graphical methods that are used to obtain CC values. In the first method, [M] is plotted versus C_M, with the expectation that the resulting curve will have a slope of 1:1 when the ligand is saturated. The CC of the ligand can be obtained by extrapolation of the high-C_M data to an *X*-axis intercept. In the second method, the bound-metal concentration [ML] (assuming that [ML] = C_M − [M]) is plotted versus C_M, with the expectation that [ML] will approach a constant limiting value (the CC) at high C_M values. The CC is estimated from such plots by fitting data to an empirical equation of the form: $Y = \mathrm{CC}(1 - \exp(bX))$. The second approach generally yields higher estimates of CC than the first approach. In the third method, [ML] is plotted as a function of pM, where pM = −log[M]. This type of plot is characteristically S-shaped, with an inflection point at pM = log K_i* and [ML]/C_L = CC/2.

For graphical convenience in presenting the results of simulated titrations over a wide range of ligand concentrations, the [M], [ML], and C_M data were normalized to C_L. In these simulations, where the C_L value can be expressed in molar units, the *X* and *Y* variables have dimensional units of moles of metal per mole of ligand. Similar normalizations can be done for humic substances if C_L is used in mass units (e.g., grams per liter). In this case, the *X* and *Y* variables would have dimensional units of moles of metal per gram of humic substance. The results of 10 typical simulations are shown in Figures 1–3.

Figure 1 illustrates what is probably the most common graphical approach toward estimation of complexation capacities. Curve 1 (Cu^{2+}, pH 7, ionic strength (I) = 0, log(C_L) = −4) typifies extremely favorable metal–ligand complexation (high K_i*). Below a C_M/C_L ratio of 1.0, very little

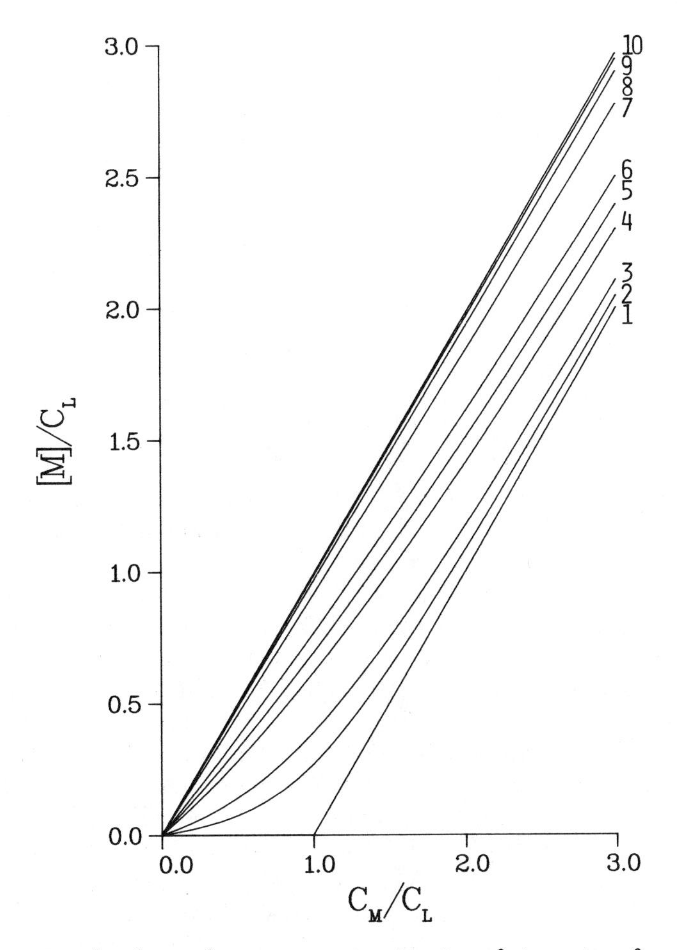

Figure 1. Simulated complexation capacity titration of citrate ion: free metal versus total metal. See *Table II for explanation of curves 1–10.*

[M] is present at equilibrium; however, at higher C_M/C_L ratios, $[M]/C_L$ increases linearly (1:1 slope) with increasing C_M/C_L. Extrapolation of the high C_M/C_L data to an *X*-axis intercept will accurately estimate CC from this titration curve. At the opposite extreme is curve 10 (Cu^{2+}, pH 4, $I = 0$, $\log(C_L) = -6$). In this case, there is no obvious indication from the graph that Cu^{2+} is being incorporated into complexes at all. The other curves (2–9) are intermediate in behavior. Thus, the common extrapolation to an *X*-axis intercept will significantly underestimate CC in almost every case.

A simple transform of Figure 1 is presented in Figure 2. Instead of plotting $[M]/C_L$ as the dependent variable, its stoichiometric complement $[ML]/C_L$ is used. The resulting figure more clearly illustrates the effects of pH, ionic strength, nature of the added metal, and ligand concentration on

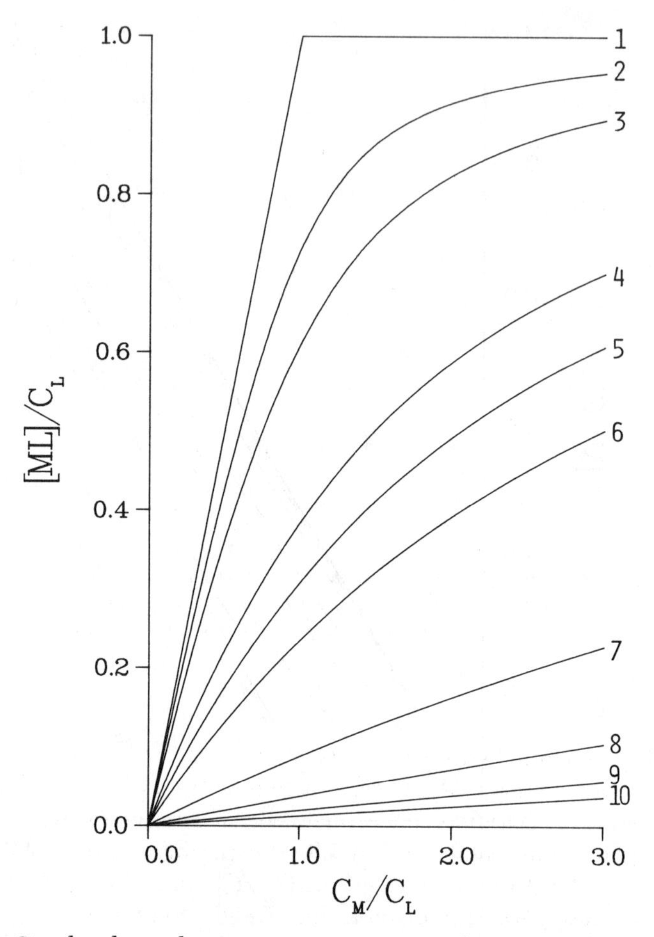

Figure 2. Simulated complexation capacity titration of citrate ion: bound metal versus total metal. See *Table II for explanation of curves 1–10.*

CC estimates. In Figure 2, curve 1 initially rises steeply (1:1 slope) until the ligand becomes saturated with added metal (at $C_M/C_L = 1.0$); then it levels off at $[ML]/C_L = 1.0$. The CC of the sample can thus be easily determined from curve 1. Curve 10 indicates the weakest metal binding, but at least Figure 2 more readily demonstrates that metal binding is taking place in that titration (compare with curve 10 in Figure 1). Curves 2–9 are again intermediate in behavior. It is doubtful that the correct CC value could be obtained by fitting data to the empirical equation: $[ML]/C_L = CC(1 - \exp(bC_M/C_L))$ for curves 4–10. Such an approach would work for curve 1 and possibly for curves 2–3.

Figures 1 and 2 present basically the same information, but Figure 2 more effectively illustrates the effects that are being evaluated in this sim-

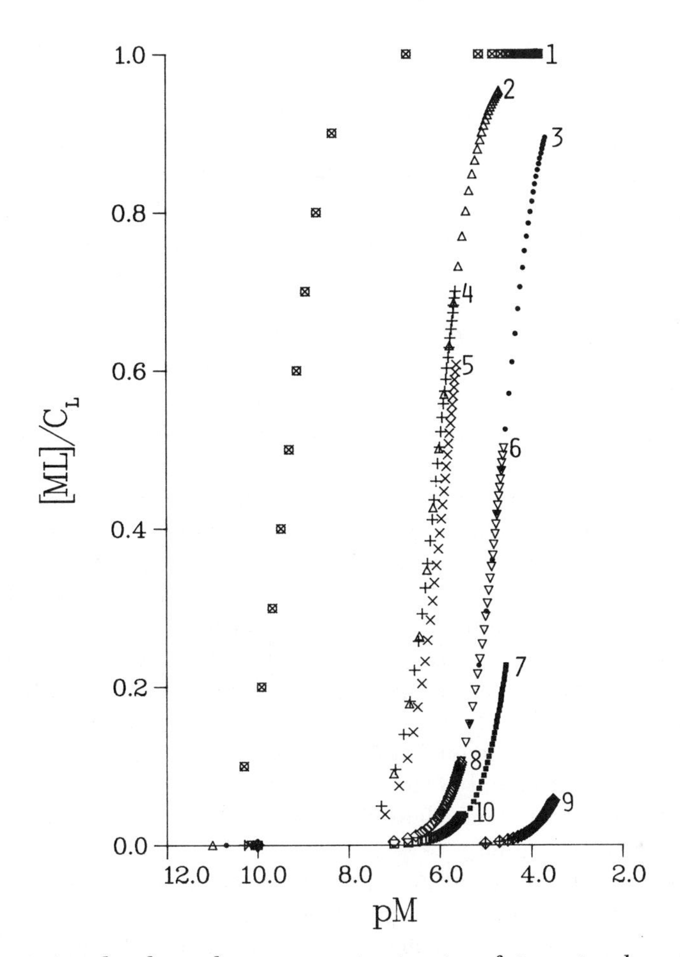

Figure 3. Simulated complexation capacity titration of citrate ion: bound metal versus log (free metal). See *Table II for explanation of curves 1–10.*

ulation. Typical effects of individual experimental parameters at otherwise-constant conditions on experimental estimates of CC can be observed in these figures by comparing the following curves. The effect of pH is exemplified by comparison of curves 4 and 10 and of curves 3 and 9. The effect of C_L is particularly evident in curves 3, 6, and 8 and in curves 2 and 4. Ionic-strength effects, which are smaller but significant, are seen in curves 4 and 5 and in curves 6 and 7. The effect of the nature of the metal ion is seen in curves 1 and 3.

One of these effects, ligand concentration (C_L), is an artifact that results from the use of C_M rather than [M] as the independent variable in Figures 1 and 2. The extent to which a ligand is saturated with a metal ion at equilibrium is a direct function of [M], not of C_M. For example, in the generalized reaction M + L $\leftrightarrows$ ML, $[ML]/C_L$ is given by the expression $[ML]/C_L = K[M]/(1 + K[M])$, where K is a conditional concentration quotient. If the simulated titration data are plotted as $[ML]/C_L$ versus pM, the curves in Figure 3 are obtained. Some of the 10 curves exactly overlap one another in this figure, so symbols were used rather than smooth lines to distinguish between curves. Curves 3, 6, and 8 and curves 2 and 4 are perfectly overlapped, although the maximum achieved level of metal binding differs from one curve to another. For these overlapping curves, all parameters other than C_L were constant. The observed overlap is not evident in Figures 1 and 2, in which C_M was used as the independent variable. Recalling that the inflection point in each curve in Figure 3 corresponds to pM = log K_i^* and $[ML]/C_L$ = CC/2, it is evident that the simulated titration curves, which appeared to indicate different CC values in Figures 1 and 2, actually indicate different log K_i^* values.

These simulated titrations demonstrate the ambiguities that can arise in interpreting complexation capacity titration experiments, even in a one-ligand system. Clearly, the effect of C_L on CC estimation is an artifact that is easily eliminated by using plots such as Figure 3 for analysis of experimental data. The remaining effects alter the energetics of the complexation (K_i^*) and thus make it difficult to experimentally realize the complexation capacity of a ligand under less-than-optimal experimental conditions. As a general conclusion, these simulated titrations suggest that higher levels of saturation of ligand at a given C_M/C_L ratio can be attained in experimental studies that are conducted at near-neutral pH, low ionic strength, with a strongly binding metal ion such as Cu^{2+}, and at high ligand concentrations.

Experimental Limitations in Complexation-Capacity Measurements. From an experimental perspective, it is extremely difficult to actually saturate a humic substance with an added metal ion, even under optimal conditions. The complex mixture of compounds in humic substances includes molecules with strong, average, and weak K_i^* values. Accordingly, not all binding sites are equally saturated with M at a given pM. A binding site with a log K_i of 5.0 will be about 90% saturated with M when log[M] = –4.0 and about 99% saturated when log[M] = –3.0, but a weaker binding site with a log K_i of 4.0 will only be half-saturated with M when log[M] = –4.0 and about 90% saturated when log[M] = –3.0.

Given the relative abundance of carboxylic acid functional groups and the very low concentrations of N- and S-containing ligands in humic substances, most metal-binding sites probably are relatively weak (as are proton-binding sites). Such binding sites are not easily saturated with metal ions in dilute aqueous solutions such as natural waters. Attempts to measure complexation capacities of unconcentrated humic substances directly in a water sample will always underestimate the true CC of the sample. For example, suppose that a dissolved humic substance in a lake water contains 10^{-5} M of a ligand with a log K_i^* of 4.0. A C_M/C_L of at least 10:1 is required just to half-saturate the binding site, and at least a 100-fold excess of metal ion is required to realize 90% of the complexation capacity of that ligand. In the presence of such large excesses of added metal ion, $[M] \sim C_M$, and changes in [ML] with increasing C_M are almost impossible to determine accurately.

If the first two graphical methods of data analysis are used, the illusion that CC values change with experimental conditions is strengthened by the heterogeneity of binding sites in humic substances. K_i^* does not change as C_M/C_L is increased in the single-ligand system used in the computer simulations. If a humic substance is titrated with a metal ion, however, the titration of very strong ligands at low C_M/C_L values gradually shifts to a titration of much weaker ligands at high C_M/C_L values. This change in the nature of reacting ligands with C_M/C_L ratio causes metal-binding curves such as those in Figure 2 to resemble curve 1 at low C_M/C_L and curve 10 at high C_M/C_L. The resulting initial steepness and later flatness of the binding curves visually enhance the illusion that the CC of a relatively strong ligand has been reached. To avoid this pitfall, metal–humic substance CC titration data should always be examined graphically, according to the approach taken in Figure 3. The resulting plots probably will not indicate more than half-saturation of the humic substance.

The optimum conditions for saturating a humic substance with metal ion (a relatively large excess of metal ion in a rather concentrated solution of the humic substance) tend to cause metal–humic substance flocculation to occur. Flocculation results in considerable ambiguity in CC estimates. Some consideration should be given to using total acidity as a surrogate parameter for the complexation capacity of a humic substance. Even though total acidity and CC are probably not equal, total acidity provides an upper limit for CC that can be measured by well-established methods (*12*).

Summary and Conclusions

The use of thermodynamic models to calculate the effect of humic substances on metal speciation requires that the complexation capacity (CC) of the humic substance be determined. If the CC of a humic substance, like that of a single ligand, is viewed as a compositional rather than thermodynamic property, then the conventional approaches toward experimental measurements and calculations of complexation capacities need to be reexamined. I propose that the CC of a humic substance is approximately equivalent to its total exchangeable acidity, and that the extent to which this CC can be realized in experimental measurements is strongly a function of pH, ionic strength, nature of the metal, and the concentration of humic substances used in the measurement. The first three parameters affect conditional concentration quotients for metal complexation, and the last parameter is simply an example of Le Chatelier's principle in metal–ligand complexation reactions.

The results of modeling studies of the effects of pH, ionic strength, and the nature of the metal ion in the complex could be more properly interpreted if the concept of CC as a compositional, rather than thermodynamic, parameter is accepted.

The current state of the art in mathematical modeling of metal–humic substance complexation is still somewhat unsophisticated. No models can quantitatively describe even the effects of pH and ionic strength on the extent of complexation of a single metal ion. Multimetal binding cannot be modeled at all. These modeling deficiencies, together with the limitations on experimental determinations of complexation capacities, present a substantial challenge to those working in this area of environmental chemistry.

Acknowledgments

Although the research described in this chapter has been funded wholly or in part by the United States Environmental Protection Agency through Cooperative Agreement No. CR813471–01 to Georgia Institute of Technology, it has not been subjected to Agency review and therefore does not necessarily reflect the views of the Agency and no official endorsement should be inferred.

References

1. Perdue, E. M.; MacCarthy, P.; Gamble, D. S.; Smith, G. C. In *Humic Substances: Volume 3—Interactions with Metals, Minerals, and Organic Chemicals*; Swift, R.; Hayes, M.; MacCarthy, P.; Malcolm, R., Eds.; Wiley: Chichester, England, 1988; in press.
2. Fish, W.; Dzombak, D. A.; Morel, F. M. M. *Environ. Sci. Technol.* **1986,** *20,* 676–683.
3. Sposito, G. *CRC Crit. Rev. Environ. Control* **1986,** *16,* 193–229.
4. Turner, D. R.; Varney, M. S.; Whitfield, M.; Mantoura, R. F. C.; Riley, J. P. *Geochim. Cosmochim. Acta* **1986,** *50,* 289–297.
5. Fish, W.; Morel, F. M. M. *Can. J. Chem.* **1985,** *63,* 1185–1193.
6. Cabaniss, S. E.; Shuman, M. S.; Collins, B. J. In *Complexation of Trace Metals in Natural Waters*; Kramer, C. J. M.; Duinker, J. C., Eds.; Junk: The Hague, 1984; pp 165–179.
7. Dempsey, B. A.; O'Melia, C. R. In *Aquatic and Terrestrial Humic Materials*; Christman, R. F.; Gjessing, E. T., Eds.; Ann Arbor Science: Ann Arbor, 1983; pp 239–273.
8. Perdue, E. M.; Lytle, C. R. In *Aquatic and Terrestrial Humic Materials*; Christman, R. F.; Gjessing, E. T., Eds.; Ann Arbor Science: Ann Arbor, 1983; pp 295–313.
9. Perdue, E. M.; Lytle, C. R. *Environ. Sci. Technol.* **1983,** *17,* 654–660.
10. Gamble, D. S.; Underdown, A. W.; Langford, C. H. *Anal. Chem.* **1980,** *52,* 1901–1908.
11. MacCarthy, P.; Smith, G. C. In *Chemical Modeling in Aqueous Systems*; Jenne, E. A., Ed.; ACS Symposium Series 93; American Chemical Society: Washington, DC, 1979; pp 201–222.

12. Perdue, E. M. In *Humic Substances in Soil, Sediment, and Water—Geochemistry, Isolation, and Characterization*; Aiken, G. R.; McKnight, D. M.; Wershaw, R. L.; MacCarthy, P.; Eds.; Wiley–Interscience: New York, 1985; pp 493–526.
13. Langford, C. H.; Gamble, D. S.; Underdown, A. W.; Lee, S. In *Aquatic and Terrestrial Humic Materials*; Christman, R. F.; Gjessing, E. T., Eds.; Ann Arbor Science: Ann Arbor, 1983; pp 219–237.
14. Weber, J. H. In *Aquatic and Terrestrial Humic Materials*; Christman, R. F.; Gjessing, E. T., Eds.; Ann Arbor Science: Ann Arbor, 1983; pp 315–331.
15. Lytle, C. R. Ph.D. Thesis, Portland State University, 1982.
16. Gamble, D. S.; Langford, C. H.; Underdown, A. W. *Org. Geochem.* **1985,** *8*, 35–39.
17. Campbell, P. G. C.; Bisson, M.; Gagné, R.; Tessler, A. *Anal. Chem.* **1977,** *49*, 2358–2362.
18. Chau, Y. K.; Gachter, R.; Lum-Shue-Chan, K. J. *Fish. Res. Board Can.* **1974,** *31*, 1515–1519.
19. Giesy, J. P.; Alberts, J. J.; Evans, D. W. *Environ. Toxicol. Chem.* **1986,** *5*, 139–154.
20. Perdue, E. M.; Lytle, C. R. In *Mineral Exploration: Biological Systems and Organic Matter*, Rubey Volume V; Carlisle, D.; Berry, W. L.; Kaplan, I. A.; Watterson, J., Eds.; Prentice–Hall: Englewood Cliffs, 1986; pp 428–444.
21. Tuschall, J. R.; Brezonik, P. L. In *Aquatic and Terrestrial Humic Materials*; Christman, R. F.; Gjessing, E. T., Eds.; Ann Arbor Science: Ann Arbor, 1983; pp 275–294.
22. Morel, F. M. M. *Principles of Aquatic Chemistry*; Wiley–Interscience: New York, 1983; pp 242–249.

RECEIVED for review July 24, 1987. ACCEPTED for publication February 29, 1988.

20

Geochemistry of Dissolved Chromium–Organic-Matter Complexes in Narragansett Bay Interstitial Waters

Gregory S. Douglas[1] and James G. Quinn

Graduate School of Oceanography, University of Rhode Island, Narragansett, RI 02882

C_{18} reverse-phase liquid chromatography and atomic absorption spectroscopy were employed to isolate and quantify dissolved interstitial chromium–organic-matter complexes in Narragansett Bay sediments. Concentration of total and organic chromium increased with depth in the core. The degree of chromium remobilization was dependent on the redox environment of the sediment and occurred mainly within the sulfate-reduction zone. Proton titration studies of interstitial water samples indicated that the chromium–organic-matter bond was exceptionally strong and that recovery could be increased significantly by protonating the normally dissociated anionic sites on the fulvic acid molecule. This protonation could decrease its polarity and increase its recovery through reverse-phase liquid chromatography. The chromium–fulvic acid association may be an important factor in the transport of reduced chromium into oxic environments in the estuarine environment.

INTERSTITIAL WATERS OF ESTUARINE SEDIMENTS provide a unique environment for the study of metal–organic-matter complexes because of the

[1]**Current address: Enseco, 205 Alewife Brook Parkway, Cambridge, MA 02138**

0065–2393/89/0219–0297$07.00/0

generally elevated concentrations of both components relative to the overlying waters. The rapid geochemical changes that occur in the upper 20 cm of these anoxic sediments provide additional information concerning the chemistry of trace metals and their organic complexes.

In this study, chromium–organic-matter complexes were isolated from estuarine interstitial waters by C_{18} reverse-phase liquid chromatography (RPLC) and atomic absorption spectroscopy (AAS) (*1–3*). In this method, dissolved organic matter (DOM) and metal–organic-matter complexes are isolated from the interstitial waters by using a C_{18} preconcentration column and subsequent elution with an organic solvent. The eluate is then measured for metal content by AAS. This method has low blank levels, requires minimal sample handling, and is rapid and reproducible (*1–3*). Chromium was selected because of its potential affinity for interstitial DOM (*4, 5*), significant input into Narragansett Bay (*6*), and the potential effects of complexation on the solubility and transport of chromium across the sediment–water interface (*3*).

The major form of chromium in reducing interstitial waters, Cr^{3+} (pH 7.4, redox electrode potential (E_h) = −200 mV), would be expected to be predominately adsorbed onto particles. Unfortunately, few interstitial chromium measurements have been reported, and no measurements of dissolved interstitial organic chromium are in the current literature. However, Brumsack and Gieskes (*4*) examined the interstitial water chemistry of laminated sediments from the Gulf of California. They showed that chromium concentrations increased with depth in the core and are strongly correlated with a Gelbstoff absorbance at 375 nm (r = 0.95). This study suggests that chromium is being remobilized during early diagenesis and may be associated with DOM in interstitial water. The strong association of interstitial chromium with DOM may decrease the particle reactivity of Cr^{3+}. The soluble Cr^{3+}–organic-matter complexes may then be transported out of the sediments and into the overlying waters. The trivalent chromium–organic-matter complex is so stable that it has been reported in oxidized marine waters (*5, 7, 8*).

In this investigation, field and laboratory studies were performed to further elucidate the chemical nature and distribution of these chromium–organic-matter complexes in the marine environment.

Experimental Details

Study Area. Sediment cores were collected from a station in mid-Narragansett Bay on May 22, 1984. McCaffrey et al. (*9*) described the Jamestown North (JN) site (Figure 1). Core samples were also collected from a series of mesocosms from the Marine Ecosystems Research Laboratory (MERL) at the University of Rhode Island in August 1983 and again 1 month later in September 1983. Pilson et al. (*10*) gave a description of the MERL facility (Figure 2). The sediments in the mesocosms were originally from the same location as the bay station and had been held in these mesocosms for 28 months.

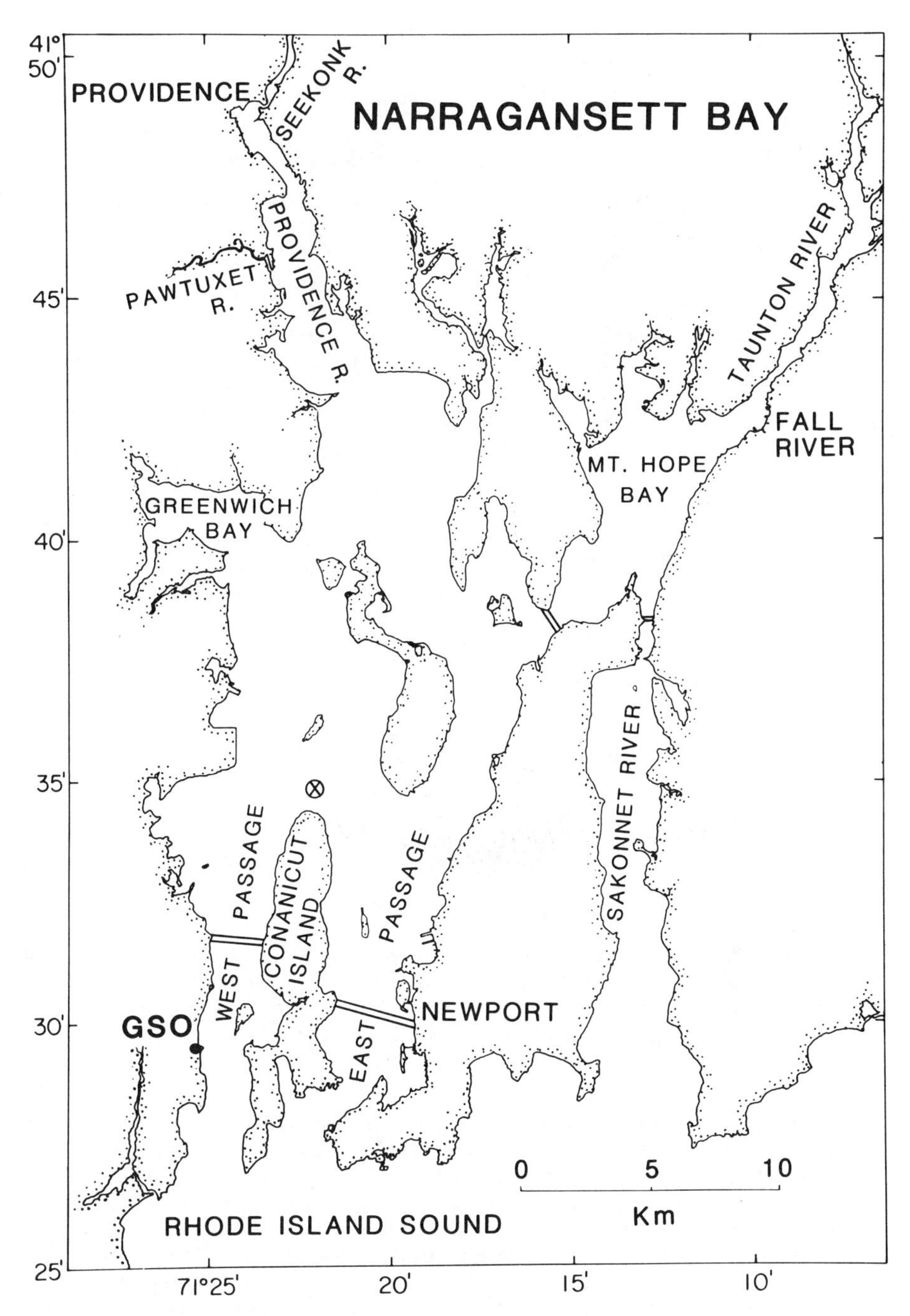

Figure 1. Map of Narragansett Bay. The circled × represents the location of the sample area.

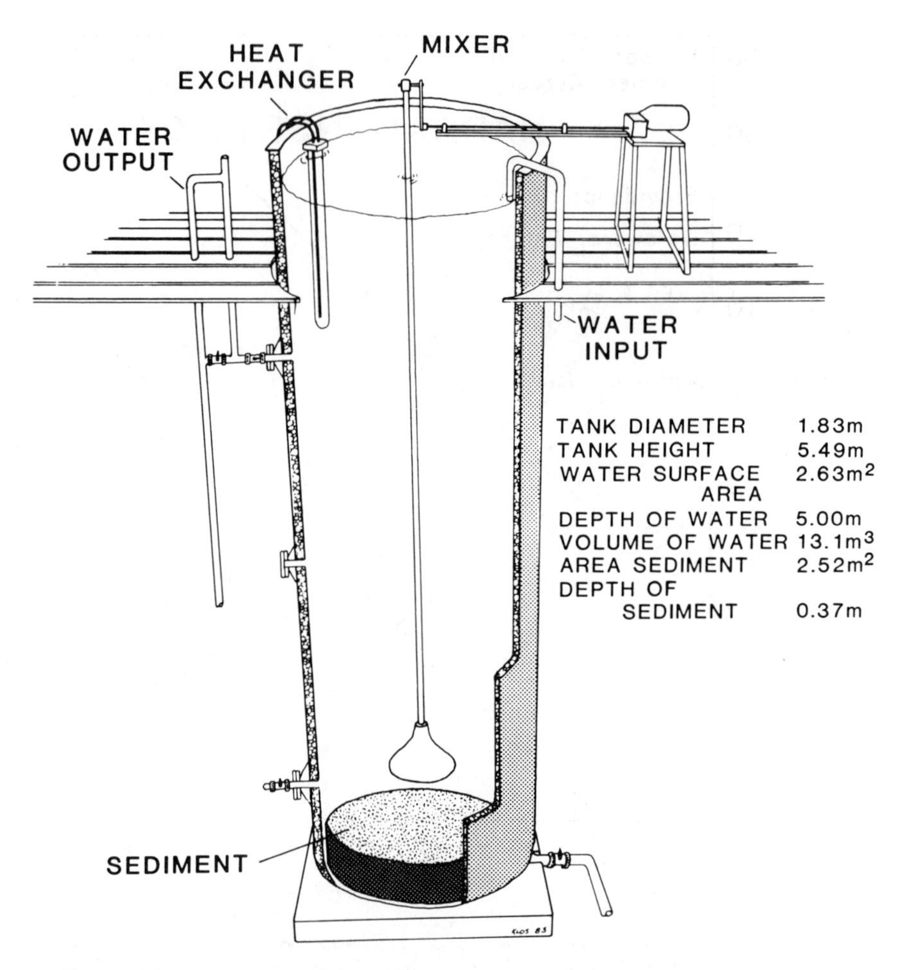

Figure 2. A cross section of MERL mesocosm tank. Seawater is pumped from the bay at a rate sufficient to replace the water in the mesocosms once every 27 days, which is the approximate flushing time of Narragansett Bay. Annual temperature range is 0 °C to about 20 °C, with a salinity range of 28–32 ‰. Reproduced from ref. 10.

Sediments for the MERL mesocosms were collected from the bay station with a large box corer designed to retain the vertical integrity of the sediments. During this period, these mesocosms were part of a nutrient-gradient experiment designed to investigate the effects of increased nutrient loading on the chemistry and biology of these systems. The nutrients NO_3, PO_4, and SiO_2 were added in fixed ratios, but at increasing amounts (Figure 3), to a series of mesocosms (*10*). These inputs were multiple factors of the average yearly inputs from sewage into Narragansett Bay. In this experiment, we examined two of these microcosms, a control (or 0×) and a 2× system (*see* Nixon et al. (*11*) for details). Ongoing studies at MERL have shown that this nutrient loading increased productivity. The new productivity level, in turn, resulted in additional carbon reaching the sediments (*12*). This increase in sedimen-

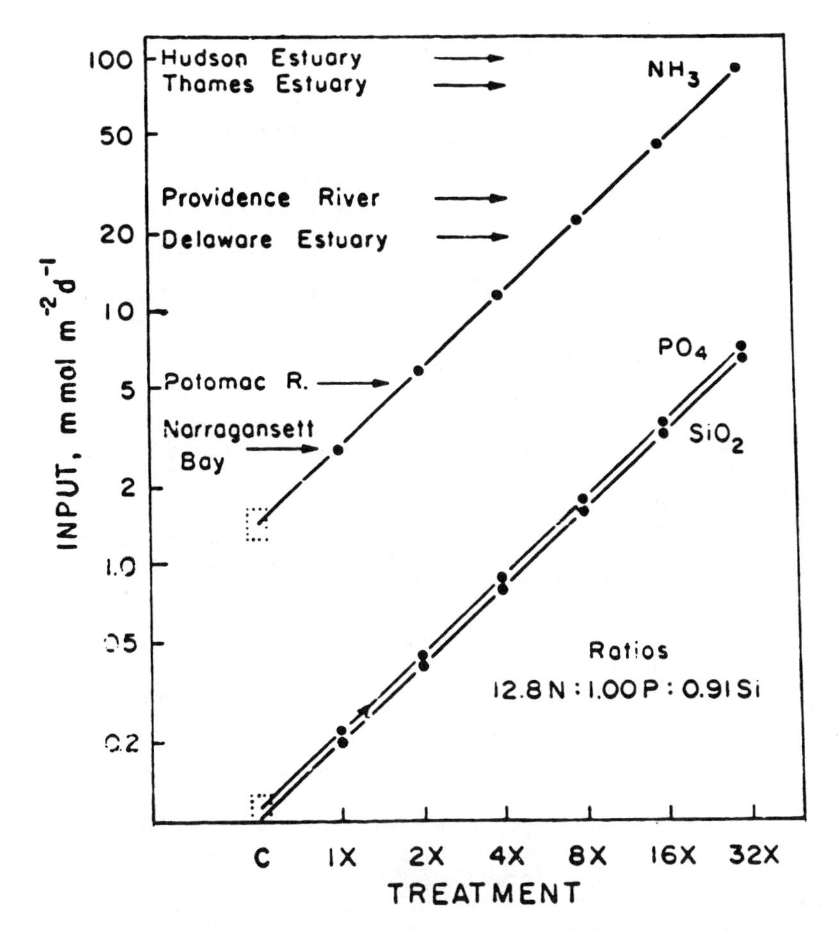

Figure 3. Additions of NH_3, PO_4, and SiO_2 versus tank treatment for seven MERL mesocosms. Reproduced from ref. 35.

tary carbon fueled sulfate reduction, which increased the reducing potential of the sediment. Thus, these mesocosms were ideal systems in which to study the distribution of interstitial total and organic chromium under a variety of redox environments.

Sediment Collection and Processing of Interstitial Water. Sample collection and processing procedures are described in Douglas et al. (3). Briefly, divers collected 16-cm-diameter cores from the study areas. The samples were returned to a nitrogen-filled glove box located within a temperature-controlled room that was programmed to operate at the in situ temperature of the sample. The overlying water was removed first, and E_h and pH measurements were made at 1-cm intervals. Each slice was placed into a 250-mL acid-washed (3.2 N HNO_3) polycarbonate bottle and centrifuged for 30 min at 10,000 rpm (16,000 × g). The samples were then returned to the glove box and the expressed interstitial fluid was filtered in a specially designed pressure-filtration unit through a 0.4-μm filter (Nuclepore).

The first 5 mL of porewater was discarded; the next 30 mL was collected for dissolved organic carbon (DOC), hydrogen sulfide, total chromium (T-Cr), iron, manganese, ammonium, phosphate, and silicate analysis. The remaining 25–50 mL of interstitial water was passed through a precleaned C_{18} sample enrichment and purification cartridge (SEP-PAK, Waters Associates) (*1–3*) to isolate the organic chromium complexes (SP-Cr). The post-SEP-PAK porewater (P-SP-Cr, interstitial water that had passed through the SEP-PAK column) was then saved in preweighed polyethylene bottles for dissolved chromium measurements.

Total Dissolved Cr, Fe, and Mn Analysis. Filtered porewater (5-mL samples) was acidified with 50 μL of nitric acid (Baker ultrex) and then analyzed for T-Cr, Fe, and Mn by direct-injection flameless AAS (Perkin-Elmer 603 with HGA 400 graphite furnace). Blanks for T-Cr, total manganese (T-Mn), and total iron (T-Fe) were 2–3 nmol/kg (residual standard deviation [RSD] = ±9%, average concentration of 20 nmol/kg), 0.1 μmol/kg (RSD = ±4%, average concentration of 20 μmol/kg), and 0.1 μmol/kg (RSD = ±5%, average concentration of 20 μmol/kg), respectively.

Dissolved Organic Chromium. RPLC was used to isolate the chromium–organic-matter complexes from the interstitial water (*3*). Briefly, the SP-Cr was collected on precleaned C_{18} SEP-PAK cartridges (10 mL of methanol, 10 mL of pH 3 HCl [Ultrex], and 20 mL of distilled–deionized water [Milli-Q, MQ]), and rinsed with 3 mL of MQ water. The organic complexes were removed with 3 mL of 50:50 methanol–water (F_1). The extract was then measured for chromium by flameless AAS. All samples were corrected for a blank that ranged between 0.1 and 1.0 nmol/kg, with RSD of ±4% at an average organic chromium concentration of 8 nmol/kg.

Dissolved Organic Carbon. DOC was measured in filtered interstitial water samples by using the persulfate oxidation method of Menzel and Vaccaro (*13*). Organic carbon in the 50:50 methanol–water SEP-PAK eluate was measured by evaporating 0.5 mL of sample in a precombusted glass ampule at 40 °C for 8 h. These samples were then analyzed in the same fashion as interstitial samples. The evolved CO_2 was measured with a carbon analyzer (Oceanography International total carbon analyzer model 0524A with a Horiba PIR–2000 IR detector). Blanks averaged 2 μg and RSD = ±5% at an average 19 μg of total carbon.

HPLC Analysis. Organic matter and associated chromium complexes, which were isolated by the SEP-PAK procedure, were further characterized by high-performance liquid chromatography (HPLC) on a liquid chromatograph (Waters Associates, ALC 224). This system was equipped with two pumps (model 6000 A), a solvent programmer (model 660), and a UV detector (model 440, 254 nm). The column was a 300 × 3.9-mm-i.d. C_{18} RPLC column (μ-Bondapack). Gradient-elution programming was employed with a 30-min linear gradient from 95% water (pH 3.2 with H_3PO_4)–5% acetonitrile to 100% acetonitrile. Flow rate was 1 mL/min, chart speed was 1 cm/min, and detector sensitivity was set for 1.0–0.01 absorbance units at full scale (AUFS).

The concentration of fulvic acid in several SEP-PAK eluates was determined by HPLC coupled with UV detection (254 nm). The area under the fulvic acid peak was determined by planimetry and compared to a standard curve prepared by using an International Humic Substance Society (IHSS) stream fulvic acid standard. The organic carbon content of the eluate was calculated by assuming that they had the same carbon to fulvic acid weight ratio as the standard (53.7% carbon). The precision of this method at 55 μmol/kg was ±9% (RSD).

Nutrients. Interstitial waters were analyzed for NH_4, PO_4, and SiO_2 with an autoanalyzer (Technicon) (*14*). Samples were diluted as necessary with low-nutrient Sargasso surface water or artificial seawater. Standards and blanks were made with this diluent water in order to eliminate errors caused by differing refractive indices (*15*). The ammonia analysis was based on Technicon Industrial Method No. 154–71W (*16*), and the phosphate technique was based on Technicon Industrial Method No. 155–71W (*17*). Silicates were analyzed by using a modification of the Brewer and Riley (*18*) method, in which oxalic acid is used to minimize phosphate interference.

Dissolved Hydrogen Sulfide Analysis. The concentration of H_2S was measured immediately after the filtration step. A mixed diamine reagent produced a blue complex that has maximum absorbance at 670 nm (*19*). The RSD of this method at a H_2S concentration of 100 μmol/kg was ±9%.

pH and E_h Measurements. The pH of the overlying water and interstitial water was measured with a combination pH electrode and a pH meter (Corning 112) equipped with an electrode switch to increase the number of electrodes that could be used with one meter. The pH electrode was calibrated in standard buffers at pH 4.0 and 7.5 at the bottom-water temperature.

After standardization, the electrode was rinsed with deionized water and equilibrated for 30 min in overlying water from the core tubes. The core was raised within the core tube until the top layer of sediment was about 0.5 cm from the lip of the core tube. A small amount of overlying water was left on the sediment so that the reference port in the pH electrode was immersed in seawater during the first measurement. The core was probed with a thin stainless steel wire to be sure that the electrode could be placed in a region where there were no shells or other obstructions. The electrode, now clamped over the core, was lowered at a rate of 1 cm per 15 min to a depth of 10 cm. The redox electrode potential between a Metrohm combination "E_h" platinum electrode and the silver–silver chloride reference electrode was also measured. The electrodes were positioned in the same way as the pH measurements. Values reported were corrected for the potential of the reference electrode at the temperature of measurement (*20*).

Results and Discussion

Artifact Studies. Within anoxic estuarine sediments, the major species of interstitial chromium is Cr^{3+} (*21*). This form of the metal is highly particle-active (*22, 23*) and may adsorb onto the C_{18} or silica material within the SEP-PAK column. Several experiments were performed to examine the possibility that chromium–organic-matter complexes were formed during the isolation procedure using SEP-PAK cartridges.

STUDY 1. $CrCl_3$ was added to a degassed interstitial water sample that had previously been passed through a SEP-PAK cartridge to remove the indigenous chromium–organic-matter complexes. The initial concentration of the porewater solution was 50 nmol/kg. It was immediately passed through a precleaned SEP-PAK and eluted with 3 mL of MQ water, 3 mL of 50:50 methanol–water (F_1), 3 mL of methanol (F_2), and 3 mL of 0.001 N HCl. The experiment was repeated with a Cr^{6+} species ($K_2Cr_2O_7$). No chromium was detected (<0.5 nmol/kg) in any of the extracts; therefore, Cr^{3+} and

Cr^{6+} were not adsorbing onto the cartridge material during the <1-s contact time in the processing of samples. These results support the interstitial budget measurements that show that the budget of SP-Cr and P-SP-Cr equals the T-Cr in these samples. This finding is strong evidence that Cr^{3+} adsorption does not occur on the SEP-PAK column.

STUDY 2. A second study was performed to determine if Cr^{3+} complexes could form on the DOM isolated by the SEP-PAK column. A 400-mL interstitial water (10–18-cm depth) sample was split into four 100-mL subsamples, and each was passed through a precleaned SEP-PAK. Two of the subsamples were processed in the normal fashion. The P-SP-Cr porewater (100 mL/sample) in the remaining two subsamples was spiked with $CrCl_3$ to a concentration of 50 nmol/kg and passed through the two DOM-loaded SEP-PAKs. These samples were then eluted with 3 mL of 50:50 methanol–water and 3 mL of methanol, and analyzed for chromium. Results (Table I) show little difference within the analytical uncertainty between spiked and unspiked samples. Therefore, no complexes were formed with the DOM-loaded SEP-PAKs.

Artifact study 2 examined the potential formation by DOM of chromium–organic-matter complexes, which were isolated with the indigenous complexes. In these samples, the chromium-complexing capacity of this SEP-PAK DOM may have been exceeded. In the present investigation, we repeated artifact study 2 with IHSS stream fulvic acid (SFA) instead of indigenous DOM because we were interested in the potential artifact formation on uncomplexed sites. The operational procedure used to isolate the SFA displaces any exchangeable chromium, and thus makes it appropriate for use in this study.

Each SEP-PAK cartridge used in this study was first processed with 5 mL of a solution containing 0.1 mg of IHSS stream fulvic acid per milliliter of filtered Sargasso seawater (salinity adjusted to 30 ‰ with MQ water). The presence of a brown ring in the SEP-PAK cartridge and the colorless nature of the post-SEP-PAK SFA solution indicated that a significant fraction of the SFA was loaded onto the SEP-PAK cartridge. Filtered Sargasso seawater (100-mL volumes, salinity adjusted to 30 ‰ with MQ water) was adjusted to 38.5-, 192-, and 962-nmol/kg concentrations of Cr^{3+} using $CrCl_3$. These solutions were then passed through the fulvic acid-loaded SEP-PAKs, eluted in the same fashion as samples in the previous two artifact studies, and analyzed for SP-Cr. In all cases, the SP-Cr was less than 2% of the initial T-Cr concentration. Again, this concentration is strong evidence that Cr^{3+} complexes do not form during the SP-Cr isolation procedure.

Colloidal Chromium Analysis. Possible colloidal artifacts were examined in two bay interstitial water samples (Table II). The samples were collected from 1–9- and 10–18-cm sections of the core. The expressed

Table I. Results from a Study of Cr^{3+} Adsorption on Indigenous DOM

Sample	*Processing*	*Mass of Porewater (g)*	*DOC Retained*[a] *(μg organic C)*	*Chromium Retained*[a] *(ng)*	*Chromium Retained*[a] *(nmol Cr/kg)*
SP–1	normal	102	96	28.9	5.42
SP–2	normal	102	105	29.9	5.61
SP–3	spiked, 50 nmol/kg	102	101	27.3	5.11
SP–4	spiked, 50 nmol/kg	102	93	30.2	5.67

NOTE: 10–18-cm bay interstitial water, collected by the SEP-PAK procedure (SP–Cr F_2 concentrations were 0 for all samples).
[a]Retained by SEP-PAK F_1.

Table II. Results from the Colloidal Chromium Study

Sample (cm)	*Filter (μm)*	*Number of Sample Replicates*	*Avg. SP-Cr of Sample Replicates (nmol Cr/kg)*	*Standard Deviation*	*Relative Standard Deviation*
1–9	0.4	3	3.37	0.10	3%
	0.4–0.05	2	3.40	NA[a]	NA
10–18	0.4	3	7.97	0.12	2%
	0.4–0.05	2	8.23	NA	NA

[a]NA means not applicable.

porewater was immediately filtered through 0.4-μm filters (Nuclepore), and three 75-mL aliquots from each sample were analyzed for SP-Cr. In addition, two 75-mL aliquots from each 0.4-μm-filtered sample were filtered through 0.05-μm filters (Nuclepore) and then analyzed for SP-Cr. The results show little difference between the 0.4- and 0.05-μm-filtered samples in both upper (1–9-cm) and lower (10–18-cm) zones of the core. This result is good evidence that the interstitial chromium–organic-matter complexes isolated by the SEP-PAK procedure are dissolved. These results agree with a similar experiment performed by Mills and Quinn (*1*) and Mills et al. (*2*) on water column samples for dissolved copper–organic-matter complexes. In their work, filtration through 0.05-μm filters (Nuclepore) removed the colloidal iron fraction but had no significant influence on the dissolved copper–organic-matter concentrations.

SEP-PAK Retention Efficiency. A study was performed to investigate the possibility that interstitial DOM could overload the SEP-PAK column capacity and therefore reduce column efficiency. Increasing volumes of filtered bay interstitial water (10–18-cm, bay combined sample) were passed through a series of SEP-PAK cartridges, eluted with 3 mL of 50:50 methanol–water, and measured for SP-Cr by AAS. The results (Figure 4) clearly show a linear relationship between total chromium complexes measured and volume of interstitial water passed through the SEP-PAK (SP-Cr [ng] = 0.24 porewater volume [mL] + 1.69, $r = 0.93$). Therefore, interstitial water samples of up to 150 mL may be processed through the SEP-PAK cartridge without any significant loss of column efficiency.

SEP-PAK Sequential Elution Experiment. To further characterize the chromium–organic-matter association, a 150-mL bay interstitial water sample was passed through a precleaned SEP-PAK cartridge and eluted with methanol–water mixtures (3 mL) of decreasing polarity. The procedure fractionates the organic carbon with the highest metal activity and can be used to determine the eluant polarity range to use for field samples (*2*). Results from the study are presented in Table III. The data show that the largest amount (45.1%) of SEP-PAK dissolved organic carbon (SP-DOC) is isolated

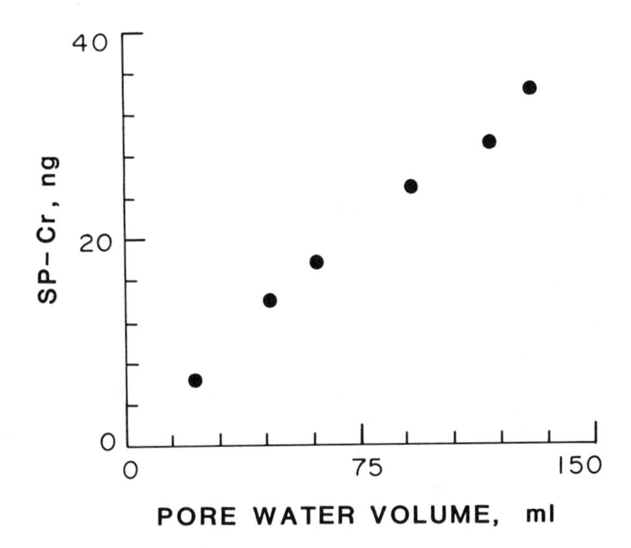

Figure 4. SP-Cr versus porewater volume passed through the SEP-PAK cartridge for a bay 10–18-cm combined interstitial water sample.

in the 10% methanol–water fraction, with smaller amounts in the less polar fractions.

Chromium recoveries followed the same general trend, although the chromium activity (nanograms of Cr per microgram of SP-DOC) was highest in the 20–40% range. The 100% methanol elution recovered 16.5% of the SP-DOC, but less than 2% of the SP-Cr. These results indicate that the 50:50 methanol–water fraction elutes about 83% of the organic carbon and

Table III. Recovery of SP-Cr and SP-DOC as a Function of Eluant Polarity

Fraction No.	*Eluant (% Methanol)*	*SP-DOC Dissolved (μg)*	*SP-DOC (%)*	*Cr (ng)*	*Total Isolated Cr (%)*	*Specific Cr Activity*[a]
		Sample No. 1				
1	0	<2	0	0	0	0.00
2	10	277	45.1	39.7	46.6	0.14
3	20	99.3	16.1	19.8	23.2	0.20
4	30	59.2	9.6	11.3	13.3	0.19
5	40	43.2	7.0	7.8	9.2	0.18
6	50	32.8	5.3	5.3	6.2	0.16
7	100	101	16.5	1.3	1.5	0.01
8	pH 3 $HCl–H_2O$	<2	0.0	0.0	0.0	0.00
Total		615.5	NA[b]	85.2	NA	NA
		Sample No. 2				
1	50	521	79.9	75.3	93.5	0.14
2	100	131	20.1	5.2	6.5	0.04
Total		652	NA	80.5	NA	NA

[a](ng of Cr)/(μg of SP-DOC).
[b]NA means not applicable.

98% of the available chromium. The negligible DOC and chromium content of the water (fraction 1) and pH 3 HCl fractions indicate that the loaded column can be rinsed with water initially to remove salts and that there is no inorganic chromium adsorption (pH 3 HCl fraction).

A second replicate sample (150 mL) was passed through a SEP-PAK cartridge and eluted with only 50:50 methanol–water and 100% methanol (Table III). The amounts of DOC and chromium recovered from the 50:50 methanol–water elution agree with the sum of the 10–50% elution experiment, ±10%. Slightly more chromium was recovered in this 100% methanol fraction than in the sequential elution experiment, a result suggesting that the sequential elution procedure was more efficient at eluting the complexes off the SEP-PAK cartridge than the 3 mL of 50:50 methanol–water. Smaller volumes of interstitial water (40–50 mL) were processed for the field samples, and no chromium was detected in any of the 100% methanol (F_2) extracts.

Interstitial Chromium Distribution. Brumsack and Gieskes (*4*) reported interstitial chromium concentrations increasing with depth in a series of cores collected from laminated sediments from the Gulf of California, Mexico. Chromium remobilization occurred below the Fe and Mn maxima and predominated within the sulfate-reduction zone. Brumsack and Gieskes (*4*) also showed a high correlation between chromium and yellow-substance absorbance (375 nm, $r = 0.95$) and suggested a strong association with DOM. The major species of chromium in these sediments is most likely Cr^{3+} (*21*), which is very particle-active. Its presence in the dissolved form is surprising, considering the large surface area of the sediments. The association with DOM may decrease particle reactivity or increase the solubility of Cr^{3+} in these systems, and thus play an important role in the geochemical release of chromium to the overlying water.

The profiles of T-Cr and SP-Cr versus depth for the bay core and a series of MERL mesocosms are shown in Figures 5a, 6a, and 7a. The T-Cr and SP-Cr concentrations generally increase with depth in the core, with SP-Cr being 23–55% of the T-Cr. The bay and control sediments (Figures 5a and 6a, respectively) reflect similar systems with relatively low dissolved chromium concentrations in the upper sediment column (0–6 cm) and increasing chromium concentrations deeper in the core (6–14 cm). This configuration suggests a deep source of chromium leading to the interstitial waters. A budget was calculated on the bay samples for T-Cr, SP-Cr, and P-SP-Cr, the dissolved chromium that was not recovered by the SEP-PAK procedure. Average budget recovery (average of all depths measured) was 92%, with a coefficient of variation of 22%.

Within the experimental mesocosms, the mean standing crop of phytoplankton increased with nutrient inputs, though not in strict proportion to loading (*11*). As a result of increased productivity, additional amounts of organic carbon entered the sediments. The organic carbon fueled elevated

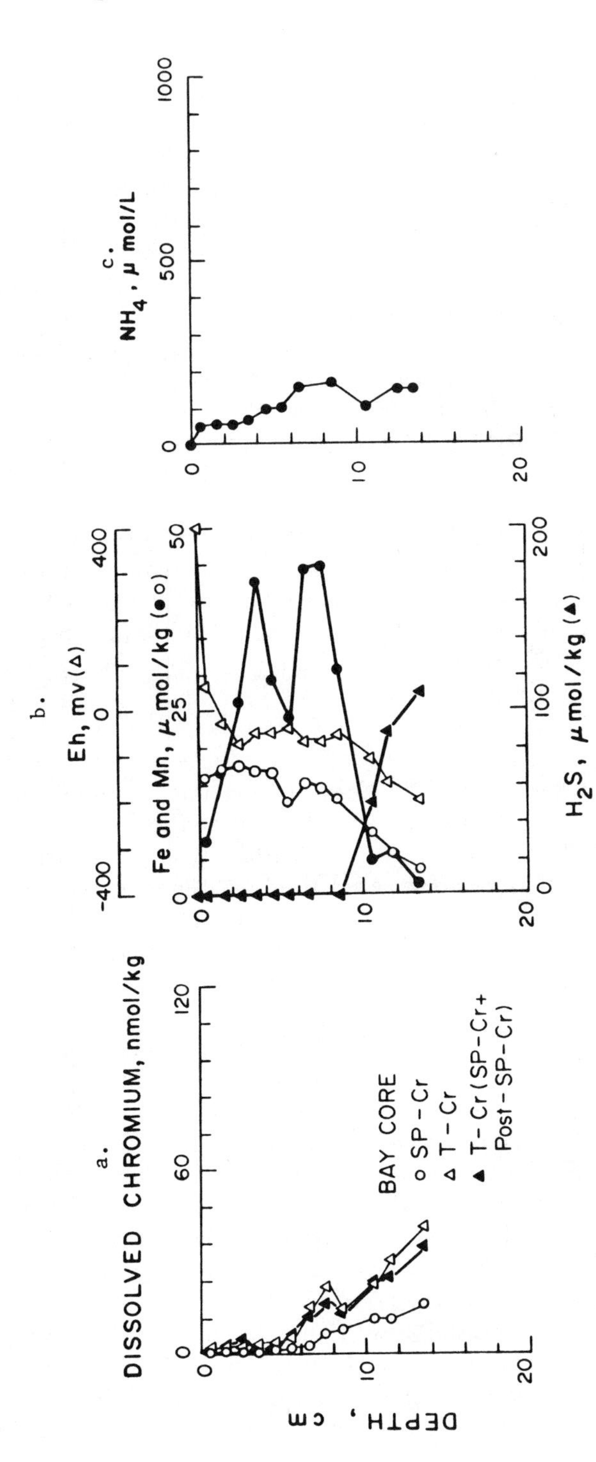

Figure 5. (*a*) *Interstitial SP-Cr, T-Cr, and budget total chromium versus depth for the bay core;* (*b*) *interstitial Fe, Mn,* H_2S, *and* E_h *versus depth for the bay core;* (*c*) *interstitial* NH_4 *versus depth for the bay core.*

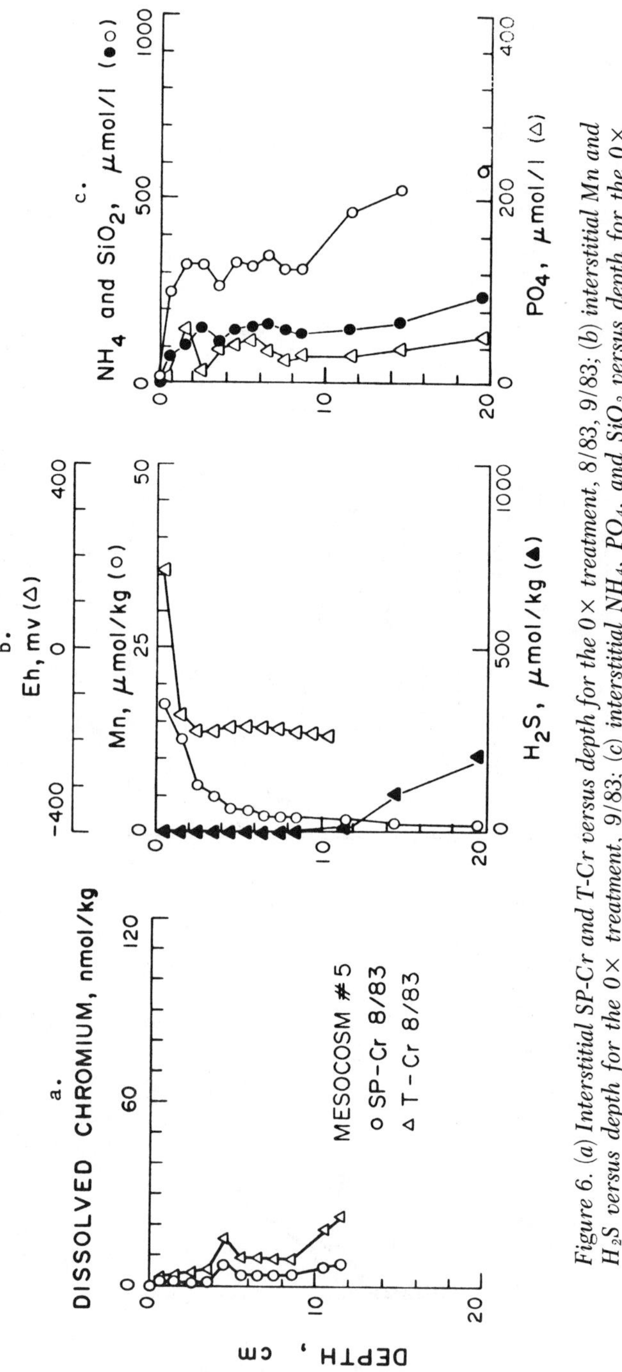

Figure 6. (a) Interstitial SP-Cr and T-Cr versus depth for the 0× treatment, 8/83, 9/83; (b) interstitial Mn and H_2S versus depth for the 0× treatment, 9/83; (c) interstitial NH_4, PO_4, and SiO_2 versus depth for the 0× treatment.

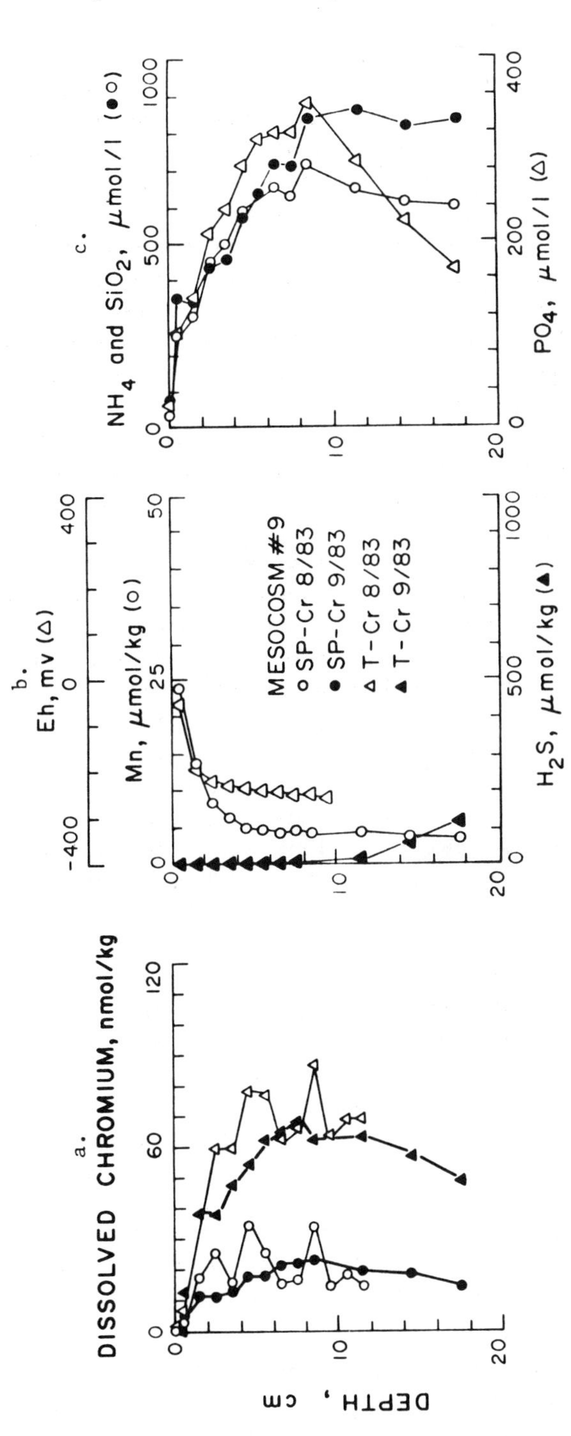

Figure 7. (a) Interstitial SP-Cr and T-Cr versus depth for the 2× treatment, 8/83, 9/83; (b) interstitial Mn, H_2S, and E_h versus depth for the 2× treatment, 9/83; (c) interstitial NH_4, PO_4, and SiO_2 versus depth for the 2× treatment.

levels of sulfate reduction, as indicated by the presence of more negative E_h values in the treatment tanks relative to the bay and control mesocosms (Figures 5b–7b). The intensity of sulfate reduction (as indicated by E_h and nutrient production, Figures 5c–7c) generally increased with increasing nutrient loading (*12*). As a result, these experimental systems were more reducing than the controls and provided a range of redox environments for the study of the interstitial trace metal remobilization processes.

Figure 7a represents dissolved porewater chromium measurements from the experimental mesocosm (2×). Concentrations of both T-Cr and SP-Cr are greatly elevated over the control and bay systems, with strong concentration gradients present up to the sediment–water interface. These gradients suggest a diffusive flux of dissolved chromium out of the sediment. Variability between sediment cores collected over a 35-day interval was small within individual mesocosms. Organic chromium ranged between 18 and 55% of T-Cr, about the same as the percentages reported for the bay and control systems (22–55%).

Within the experimental mesocosm, both T-Cr and SP-Cr reached a maximum concentration within a 4–10-cm depth range. This range suggests that chromium is remobilized from some unknown solid phase such as $Cr(OH)_3$, $FeCrO_4$ (*24*), or even possibly a chromium–silica ($Cr–SiO_2$) association (*25*). The increase in dissolved interstitial chromium in the experimental mesocosms suggests that the solid-phase release is extremely sensitive to changes in redox environments. For example, in the 2× treatment system (Figure 7), nutrients were increased only by a factor of 2 relative to the control, yet the interstitial chromium concentrations were elevated by a factor of almost 9. Once released into the dissolved phase, interstitial chromium can be transported within the sediments by diffusion and biopumping.

Organic Chromium Characterization. The organic matter isolated by C_{18} RPLC, although not operationally defined as fulvic acid (*26*), has chemical characteristics very similar to terrigenous and stream fulvic acid (SFA). The inability to completely resolve the structure of fulvic acid has not prevented significant advances in understanding the nature of this complex molecule. Three lines of evidence suggest that DOM isolated by C_{18} RPLC from marine waters is chemically similar to fulvic acid.

1. Mills et al. (*2*) showed that the protonation characteristics of SEP-PAK organic matter are similar to those of fulvic acid. This similarity indicates a diversity of active sites on the SEP-PAK DOM.
2. Electron paramagnetic resonance spectra for copper–organic-matter complexes isolated by C_{18} RPLC (*2*) have spin Hamiltonian parameter values similar to those reported by Bresnahan et al. (*27*) for a fulvic acid–copper complex.

3. Organic copper stability constants determined by Zuelke and Kester (28) on data collected by Mills et al. (2) were similar to copper–organic-matter stability constants reported by Mantoura et al. (29) for river water and seawater fulvic acid samples.

A series of experiments was designed to investigate the nature of the SEP-PAK organic matter and strength of the chromium–organic-matter bonds through a series of protonation experiments and HPLC studies of interstitial water samples and SFA.

Stream Fulvic Acid Recovery Experiment. SFA was prepared to a concentration of 0.86 mg of SFA per milliliter of filtered Sargasso seawater (salinity adjusted to 30 ‰ with MQ water, initial pH 7.2). This solution (5-mL aliquots) was then passed through a precleaned SEP-PAK cartridge and eluted with 5 mL of 50:50 methanol–water (F_1) and 5 mL of methanol (F_2) (Figure 8).

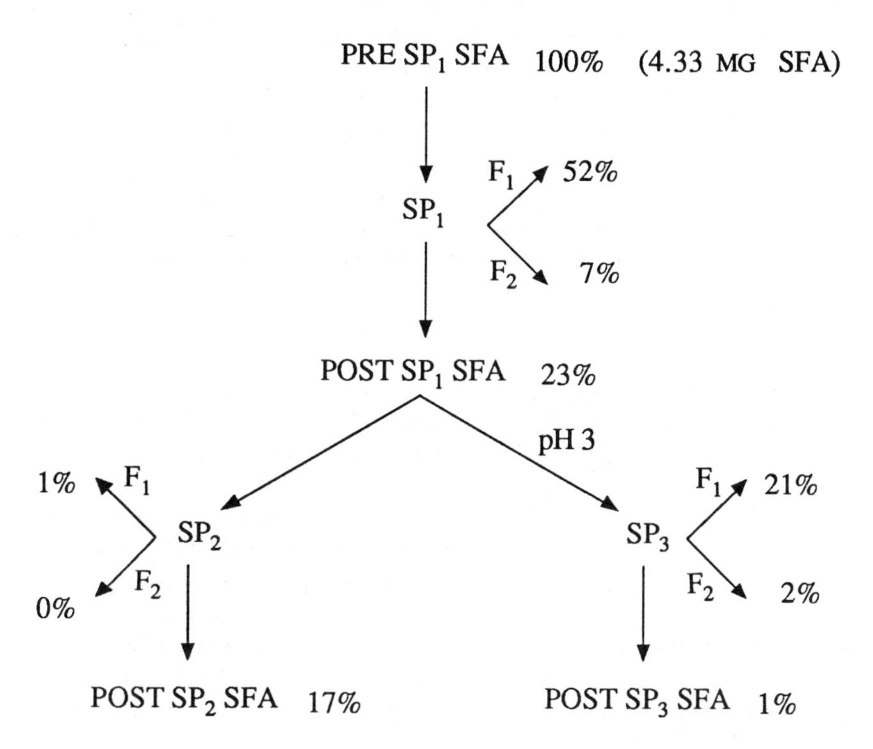

Figure 8. Flow diagram describing the SEP-PAK recovery study. All recovery percentages are based on an initial SFA–Sargasso seawater concentration of 0.86 mg of SFA per milliliter of Sargasso seawater. Total amount of SFA passed through SP_1 = 4.33 mg (5.0 mL). Total SFA recovery = SP_1 + SP_3 = 82.0%. SP_2 and SP_3 fractions were normalized to 5-mL volumes.

The post-SEP-PAK sample was then split into two 2.5-mL fractions. One fraction was passed through a second precleaned SEP-PAK column (SP_2) and eluted with 2.5 mL of 50:50 methanol–water and 2.5 mL of methanol. The second fraction was acidified to pH 3 with 6 N HCL, allowed to equilibriate for 1 h, passed through a precleaned SEP-PAK column (SP_3), and eluted with 2.5 mL of 50:50 methanol–water and 2.5 mL of methanol. The pH was lowered to 3.0 in order to suppress the ionization of weak organic acids, which decreases their solubility (and polarity) and increases retention on the C_{18} column (*30, 31*).

All fractions were analyzed for SFA by HPLC, and the results are presented in Figure 8. The SFA budget indicates that 52% of the SFA was recovered in the 50:50 methanol–water fraction, with lesser amounts in the methanol (F_2) fraction (7%). Of the SFA that passed through the SEP-PAK, less than 2% was recovered on a fresh column (SP_2), which indicated that the SEP-PAK cartridge (SP_1) was not overloaded. When the post-SP_1-SFA sample was acidified to pH 3.0, an additional quantity of SFA was recovered (23%) in the SEP-PAK eluants (SP_3, F_1 + F_2). The increased recovery most likely resulted from protonation of normally dissociated anionic sites (pH 7.2), which reduced the polarity of the SFA (*29, 32*). Total recovery of SFA by SP_1 + SP_3 was 82%.

pH 3 Experiment. The SFA study showed that increased amounts of SFA are recovered by the SEP-PAK as the pH of the fulvic acid–Sargasso seawater solution was lowered. A second experiment was performed to demonstrate this relationship on interstitial water samples and to observe the response of chromium–organic-matter complexes under acidic conditions (Table IV). A 10–18-cm-filtered (0.4-μm Nuclepore) bulk interstitial water from the bay was first analyzed for T-Cr and then passed through a SEP-PAK that was eluted with 50:50 methanol–water (F_1) and methanol (F_2). The post-SEP-PAK interstitial samples were analyzed for T-Cr, acidified to pH 3, and then passed through a second SEP-PAK cartridge from which the F_1 and F_2 were also analyzed for fulvic acid carbon (FA-C, Table IV) by HPLC.

Results from the field samples agree with the SFA experiment; they show an increase in fulvic acid and DOC in the SEP-PAK extractable fractions after a lowering of pH. A surprising result from this experiment was the increase in organic chromium (SP_2, pH 3.0), which was not recovered at the indigeneous pH (SP_1, pH 7.4). The results from this study (Table IV) indicate that SP-Cr recovery is increased as pH is lowered, with 34% of the T-Cr and 18% of the DOC recovered in the SP_1 fraction (F_1 + F_2, pH 7.4). An additional 29% of the T-Cr and 12% of the DOC was recovered by SP_2 (F_1 + F_2) after the post-SP_1 sample was adjusted to pH 3. The fulvic acid content of the SP_1 and SP_2 DOC (F_1, F_2) ranged from 30 to 48% of the SP-DOC.

Table IV. Recovery of SP-Cr, DOC, and SP-FA-C as a Function of pH and Solvent Polarity

No.	*Sample*	*Cr (nmol/kg)*	*Cr (% Total)*	*SP-DOC (μmol/kg)*	*SP-DOC (% Total)*	*SP-FA-C (μmol/kg)*	*(%SP-FA-C)/(SP-DOC)*
1	pre-SP_1[a]	29.7	100	912	100	NA[b]	NA
2	SP_1-F_1	9.42	32	143	16	57	40
3	SP_1-F_2	0.57	2	17	2	7	41
4	post-SP_1	19.8	66	807	88	NA	NA
5	SP_2-F_1 pH 3	7.51	25	96	10	46	48
6	SP_2-F_2 pH 3	1.21	4	20	2	6	30
7	post-SP_2 pH 3	10.8	36	682	75	NA	NA

[a]Initial pH of the interstitial water was 7.4.
[b]NA means not applicable.

Previous studies (2, 33) have shown that for other transition metals that form organic complexes, the complexed metal is replaced by protons as the pH is lowered. The pH at which the metal is released is indicative of the strength of metal–ligand bond. The results from this experiment suggest that chromium forms exceptionally strong complexes with organic matter.

pH 1.5 Experiment. To further characterize the strength of the chromium–organic-matter bonding, an experiment was designed to measure the chromium complexes at pH 1.5 (Table V). A pH of 1.5 was selected as the lowest possible pH because the silica backbone in the SEP-PAK begins to degrade seriously below this value. A filtered bay porewater sample (1–20 cm) was split into three 80-g samples. The pH of Sample 1 was 7.4. Sample 2 was adjusted to pH 3, and Sample 3 was adjusted to pH 1.5; both samples were pH-adjusted with 6 N HCl. These samples were allowed to equilibrate for 1 h, filtered (0.4-μm Nuclepore), and passed through precleaned SEP-PAK cartridges. Filtration studies showed that the chromium complex did not precipitate after acidification. Lack of a precipitate is additional evidence that the organic material responsible for complexation is the "acid-soluble" fulvic acid, as opposed to acid-insoluble humic acid.

The results (Table V) are consistent with the SFA and pH 3 experiments, with increased recoveries of SP-Cr, DOC, and fulvic acid (FA) with decreasing pH (7.4–1.5). The pH 3 experiment (Table IV) showed substantial increases in the recovery by SEP-PAK in SP-Cr, SP-DOC, and SP-FA-C when the sample pH was lowered to 3 (SP-Cr/T-Cr at pH 3 = $[SP_1 (F_1 + F_2) + SP_2 (F_1 + F_2)]$/T-Cr = 63%, Table IV). The pH 1.5 study reaffirmed these results (SP-Cr/T-Cr at pH 3 = $[F_1 + F_2]$/T-Cr = 83.6%, Table V) and provided additional information concerning the response of SP-Cr to higher proton concentrations.

The maximum amount of SP-Cr was recovered in the pH 3 sample (Table V), with 54.3% of the T-Cr in F_1 and 29.3% in F_2. When the pH in Sample 3 was lowered to 1.5, there was an expected increase in F_1 (61.7%

Table V. Results of SP-Cr, SP-DOC, and SP-FA-C from pH-Adjusted 1–20-cm Combined Interstitial Bay Porewater Samples

Sample	*SP-Cr (nmol/kg)*	*% T-SP-Cr[a]/T-Cr[b]*	*SP-DOC (μmol/kg)*	*SP-FA-C (μmol/kg)*
pH 7.4–F_1	4.89	31.0	94.0	39
pH 7.4–F_2	0.00	00.0	NA[c]	0
pH 3.0–F_1	8.56	54.3	132	62
pH 3.0–F_2	4.62	29.3	NA	22
pH 1.5–F_1	9.72	61.7	152	73
pH 1.5–F_2	2.18	13.8	NA	13

[a]Total SEP-PAK chromium.
[b]15.8 nmol/kg.
[c]NA means not applicable.

T-Cr); however, there was also an unexpected decrease in F_2 (13.8% T-Cr). A decrease in F_2, also observed in the FA-C, suggests that one possible mechanism for the observed reduction of SP-Cr in the pH 1.5 F_2 is that protonation at the pH level has rendered a significant fraction of this DOM so nonpolar that it is no longer eluted with methanol. The presence of organically complexed chromium at pH 1.5 suggests that the chemical bonding between chromium and fulvic acid is very strong and relatively nonlabile because of the formation of inner-sphere complexes.

The increase in SP-Cr with lowered pH was also accompanied by corresponding increases in the amount of SP-FA-C. These results support the previous pH studies and provide additional detail concerning the SP-Cr:SP-FA-C relationship. The strength of this relationship is demonstrated in Figure 9, which is a plot of SP-Cr versus SP-FA-C for the pH 3 and 1.5 studies. The strong correlation between SP-Cr and SP-FA-C (SP-Cr [nmol/kg] = 0.14 SP-FA-C [μmol/kg] + 0.43, $r = 0.98$) provides additional evidence that the main complexing agent for chromium in interstitial waters is fulvic acid. The results from these experiments also suggest that additional amounts of organic chromium are present in interstitial waters that would not be recovered by the SEP-PAK procedure at ambient pH. Up to 84% of the T-Cr in interstitial waters (SP-Cr at pH 3.0, $F_1 + F_2$, [Table V]) may be complexed with "fulvic acidlike" organic matter. The increased recoveries obtained through pH adjustments should be viewed cautiously because possible equilibrium shifts have not been examined.

The effects of pH on SEP-PAK extraction efficiency must be considered when interpreting SP-Cr interstitial water profiles and their geochemical significance. In this study, the ambient pH (bay and mesocosm sediments)

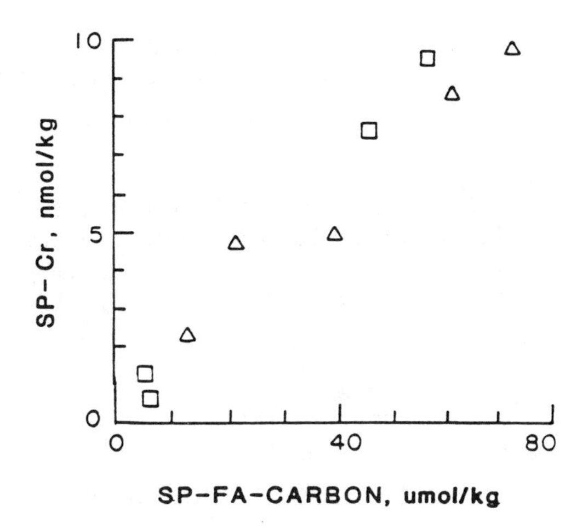

Figure 9. SP-Cr versus SP-FA-C for the pH 3–DOC study (□), pH 1.5 experiment (Δ). SP-Cr (nmol/kg) = 0.14 SP-FA-C (μmol/kg) + 0.43, r = 0.98.

varied with depth over a narrow range (pH 7.4 ± 0.2). Acid titration studies of Narragansett Bay interstitial waters (*34*) showed that the increase in SEP-PAK extraction efficiency from pH 7.4 to 7.0 was significantly less (<10%) than the observed increase of interstitial SP-Cr with depth in the core. This finding suggests that the observed interstitial SP-Cr distribution was not an artifact of the extraction method.

The previous laboratory studies have all shown a strong relationship between SP-Cr and SP-FA-C. In another study, the F_1 fraction of the bay core (Figure 5a) was further analyzed by HPLC for SP-FA-C. These values (SP-FA-C) were plotted versus SP-Cr and are presented in Figure 10. The results support the previous protonation studies and reveal a strong linear relationship between SP-FA-C and SP-Cr (SP-Cr [nmol/kg] = 0.33 SP-FA-C [μmol/kg] − 3.55, r = 0.97). The correlation of SP-Cr with DOC and SP-DOC were always poor (r = <0.5) because of the bulk property of these measurements. The SP-Cr:SP-FA-C relationship observed in the bay sample at ambient pH and the various supporting protonation studies are strong evidence that fulvic acid plays a major role in the geochemistry of interstitial chromium.

Conclusion

The chromium–organic-matter complexes are tightly bound and have chemical characteristics similar to fulvic acid. Increased recovery of both fulvic acid and chromium was achieved by lowering the pH of the sample prior to the SEP-PAK extraction. By using the pH adjustment procedure, up to 84% of the total dissolved chromium may be associated with DOM.

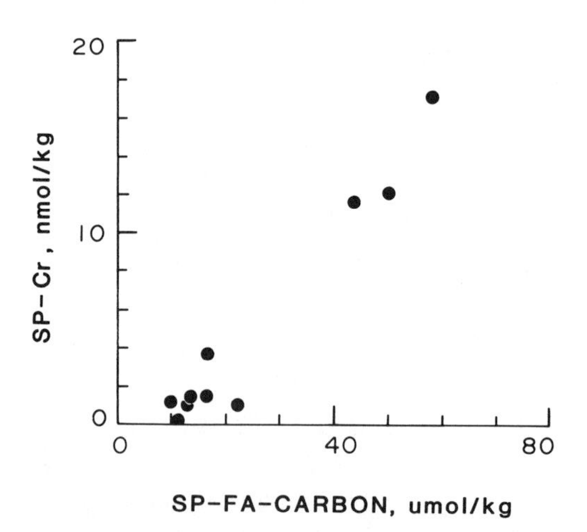

Figure 10. SP-Cr versus SP-FA-C for the bay core, 5/83 (pH 7.4). SP-Cr (nmol/kg) = 0.33 SP-FA-C (μmol/kg) − 3.55, r = 0.97.

Dissolved organic complexes of chromium in the interstitial water of Narragansett Bay sediments represent a significant amount of the total dissolved metal (22–53%). The concentrations of total and organic chromium increase with depth in the core; the degree of remobilization is dependent on the redox environment of the sediment and predominates in the sulfate-reduction zone. The formation of the reduced-organic-matter complex (Cr^{3+}) increases the solubility of the particle-active species and may be an important factor in the release of dissolved chromium from anoxic sediments.

Acknowledgments

We would like to thank Gary Mills for his help and advice in this study. This investigation was supported by a grant from the Marine Chemistry Program of the National Science Foundation (OCE–8200150).

References

1. Mills, G. L.; Quinn, J. G. *Mar. Chem.* **1981,** *10,* 93–102.
2. Mills, G. L.; Hanson, A. K.; Quinn, J. G.; Lammela, W. R.; Chasteen, D. N. *Mar. Chem.* **1982,** *11,* 355–377.
3. Douglas, G. S.; Mills, G. L.; Quinn, J. G. *Mar. Chem.* **1986,** *19,* 161–174.
4. Brumsack, H. J.; Gieskes, J. M. *Mar. Chem.* **1983,** *14,* 89–106.
5. Nakayama, E.; Kuwamoto, T.; Tsurubo, S.; Tokoro, H.; Taitiro, F. *Anal. Chim. Acta* **1981,** *130,* 289–294.
6. Olsen, S.; Lee, V. *A Summary and Preliminary Evaluation of Data Pertaining to the Water Quality of Upper Narragansett Bay*; technical report; Coastal Resources Center, University of Rhode Island: Narragansett, RI, 1979.
7. Nakayama, E.; Kuwamoto, T.; Tsurubo, S.; Fujinaga, T. *Anal. Chim. Acta* **1981,** *130,* 401–404.
8. Nakayama, E.; Kuwamoto, T.; Tokoro, H.; Fujinaga, T. *Anal. Chim. Acta* **1981,** *131,* 247–254.
9. McCaffrey, R. J.; Meyers, A. C.; Davey, E.; Morrison, G.; Bender, M.; Leudtke, N.; Cullen, D.; Froelich, P.; Klinkhammer, G. *Limnol. Oceanogr.* **1980,** *25,* 31–44.
10. Pilson, M. E. Q.; Oviatt, C. A.; Nixon, S. W. In *Microcosms in Ecological Research*; Giesy, J. P., Ed.; U.S. Department of Energy Symposium Series; U.S. Department of Energy: Washington, DC, 1980; pp 724–741; Symposium: Microcosms in Ecological Research, Augusta, GA, 1978; CONF 7811–1, NTIS.
11. Nixon, S. W.; Pilson, M. E. Q.; Oviatt, C. A.; Donaghay, P.; Sullivan, B.; Seitzinger, S.; Rudnick, S.; Frithsen, J. In *Flows of Energy and Materials in Marine Ecosystems: Theory and Practice*; Fasham, M., Ed.; Plenum: London, 1984; pp 105–135.
12. Sampou, P. personal communication, 1986.
13. Menzel, D. W.; Vaccaro, R. F. *Limnol. Oceanogr.* **1964,** *9,* 138–142.
14. Hager, S. W.; Gordon, L. I.; Park, P. K. *A Practical Manual for Use of the Technicon Auto Analyzer in Seawater Nutrient Analysis*; final report to the Bureau of Commercial Fisheries for Contract No. 14–17–0001–1759, Department of Oceanography, Oregon State University: Corvallis, 1968; Ref. No. 68–33.

15. Froelich, P. N.; Pilson, M. E. Q. *Water Res.* **1978,** *12,* 599–603.
16. *Ammonia in Water and Seawater*; Technicon Industrial Systems; Industrial Method No. 154–71w, 1978.
17. *Phosphate in Water and Seawater*; Technicon Industrial Systems; Industrial Method No. 155–71w, 1973.
18. Brewer, P. G.; Riley, J. P. *Anal. Chim. Acta* **1966,** *35,* 514.
19. Cline, J. D. *Limnol. Oceanogr.* **1969,** *14,* 454–458.
20. Sawyer, D. T.; Roberts, J. L. *Experimental Electrochemistry for Chemists*; Wiley: New York, 1974.
21. Hem, J. D. *Geochim. Cosmochim. Acta* **1977,** *41,* 527–538.
22. Amdurer, M.; Adler, D.; Santchi, P. H. In *Trace Metals in Seawater*; Wong, C. S.; Boyle, E.; Bruland, K.; Burton, J. D.; Goldberg, E. D., Eds.; Plenum: New York, 1983; pp 537–562.
23. Mayer, L. M.; Schinck, L. L.; Chan, C. A. *Geochim. Cosmochim. Acta* **1984,** *48,* 1717–1722.
24. Elderfield, H. *Earth Planet. Sci. Lett.* **1970,** *9,* 10–16.
25. Cranston, R. E. *Mar. Chem.* **1983,** *13,* 109–125.
26. Stuermer, D. H.; Harvey, G. R. *Nature (London)* **1974,** *250,* 480–481.
27. Bresnahan, W. T.; Grant, C. L.; Weber, J. H. *Anal. Chem.* **1978,** *50,* 1675–1679.
28. Zuehlke, R. W.; Kester, D. R. In *Trace Metals in Seawater*; Wong, C. S.; Boyle, E.; Bruland, K.; Burton, J. D.; Goldberg, E. D., Eds.; Plenum: New York, 1983; pp 537–562.
29. Mantoura, R. F. C.; Riley, J. P. *Anal. Chim. Acta* **1975,** *76,* 97–106.
30. Karger, B. L.; Giese, R. W. *Anal. Chem.* **1978,** *50,* 807–822.
31. Mills, G. L.; McFadden, E.; Quinn, J. G. *Mar. Chem.* **1986,** *20,* 313–325.
32. Thurman, E. M.; Malcolm, R. L. *Environ. Sci. Technol.* **1981,** *14,* 463–466.
33. Raspor, B.; Nurnberg, H. W.; Valenta, P.; Branica, M. *Mar. Chem.* **1984,** *15,* 231–249.
34. Douglas, G. S. Ph.D. Thesis, University of Rhode Island, 1986.
35. Pilson, M. E. Q. *Marine Ecosystem Research Laboratory: Fates and Effects of Nutrients Along a Simulated Estarine Gradient*; proposal submitted to U.S. Environmental Protection Agency, Office of Exploratory Research; Cr. No. 810265–01–0.

RECEIVED for review October 15, 1987. ACCEPTED for publication March 21, 1988.

ENVIRONMENTAL REACTIONS IN NATURAL WATERS

21

Influences of Natural Organic Matter on the Abiotic Hydrolysis of Organic Contaminants in Aqueous Systems

Donald L. Macalady

Department of Chemistry and Geochemistry, Colorado School of Mines, Golden, CO 80401

Paul G. Tratnyek and N. Lee Wolfe

Environmental Research Laboratory, U.S. Environmental Protection Agency, College Station Road, Athens, GA 30613

This chapter reviews investigations that attempt to provide a systematic understanding of the role of natural organic matter in aqueous hydrolytic reactions of anthropogenic organic chemicals. In particular, the suggestion that humic substances exert a catalytic effect on a wide variety of hydrolytic reactions is evaluated in the light of experimental evidence. With the possible exception of the hydrolytic dechlorination of the chloro-1,3,5-triazine herbicides, available data do not support the idea of a general catalytic effect. On the contrary, more evidence exists for an inhibitory role than for one of enhancement. The (limited) experimental evidence that may lead to an understanding of the possible role of natural organic matter in mediating abiotic hydrolytic reactions is discussed. Several models that imply a general effect of natural organic matter on the kinetics of hydrolysis are outlined.

HYDROLYTIC DEGRADATION OF CONTAMINANT ORGANIC CHEMICALS is one of the most important and widely investigated pollutant transformation proc-

0065–2393/89/0219–0323$06.00/0

esses in aqueous systems. Microorganisms play an important role in the hydrolysis of organic chemicals, either directly (*1*, *2*) or indirectly through their molecular components, such as soil enzymes (*3*). For many contaminants, however, abiotic hydrolytic reactions are more important under conditions found in most natural waters. Therefore, abiotic hydrolyses of anthropogenic organic chemicals, especially pesticides, have been extensively investigated (*4–8*). In particular, there has been considerable focus on the factors that enhance or retard the rates of these abiotic hydrolytic reactions.

This review will deal only with hydrolysis, as distinguished from other reactions between organic chemicals and water, such as acid–base, addition, elimination, and hydration. In the hydrolytic process, an organic molecule, RX, reacts with water to form a new carbon–oxygen bond with the water molecule and cleave a C–X bond in the original molecule. The net reaction is commonly a direct displacement of X by OH (*9*). Important classes of organic molecules that undergo hydrolytic reactions include aliphatic halides; esters of carboxylic, phosphoric, phosphonic, sulfonic, and sulfuric acids; carbamates; epoxides; amides; and nitriles. Hydrolytic half-lives for abiotic reactions at pH 7 and 25 °C vary from millions of years to about 1 s (*8*).

Under pH-buffered reaction conditions, many hydrolytic reactions can be represented by a pseudo-first-order rate law $-d[P]/dt = k[P]$, where $[P]$ represents the concentration (activity) of the pollutant molecule and k is the pseudo-first-order rate constant, valid for a fixed temperature and pH only. The rate constant, k, is system-specific. It may be composed not only of contributions from specific acidic or basic and neutral hydrolyses, but also from catalytic processes such as general acid–base catalysis. A profile of the pH dependence of the hydrolytic rate constant for the methyl ester of (2,4-dichlorophenoxy)acetic acid (2,4-DME), typical of many carboxylic acid esters, is represented in Figure 1. This figure does not include generalized acid or base catalysis, and either specific acid or base catalysis may be the rate-controlling process at pH values commonly encountered in natural waters (*8*, *9*). A few metal ions also catalyze specialized hydrolytic reactions (*10*, *11*). Under most reaction conditions that exist in natural waters, however, neither general acid–base nor metal-ion catalysis is expected to contribute significantly to the overall rate of hydrolysis (*12*).

This review addresses the effects of natural organic matter in aquatic systems (humic substances) on the rates of hydrolytic reactions. In a recent review of the interactions of humic substances with environmental chemicals, Choudhry (*13*) stated that humic substances can play an important role in hydrolytic transformations of a wide variety of organic chemicals in the aqueous environment. We will reevaluate the evidence for this generalization, take into account studies published since 1983, and offer alternate suggestions that may lead to a more general understanding of the influences of natural organic matter on hydrolytic processes. Except in a few cases

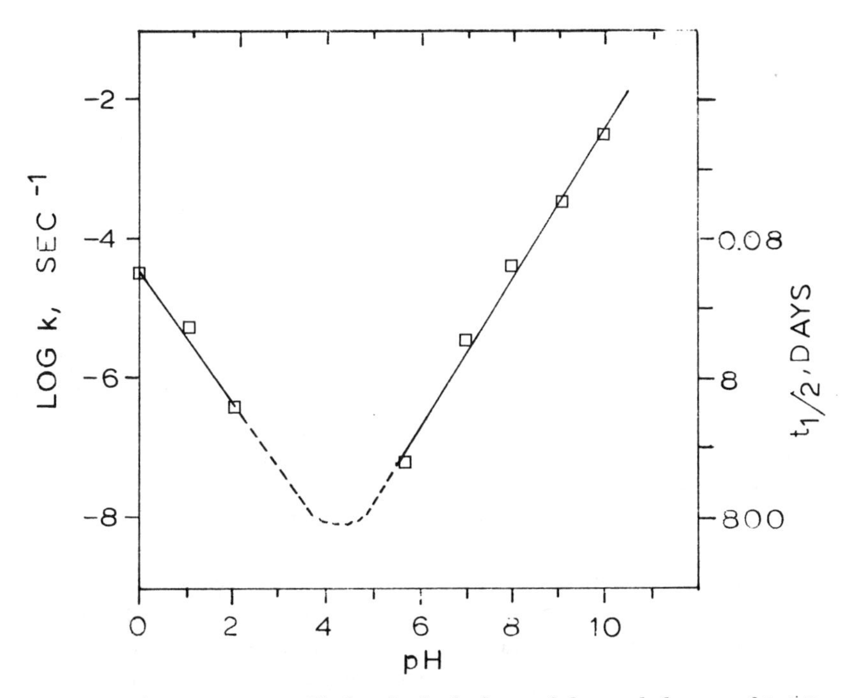

Figure 1. The pH–rate profile for the hydrolysis of the methyl ester of 2,4-D in buffered distilled water at 30 °C. Data are from ref. 28. The dashed lines represent extrapolations of the experimental data.

included for comparison, this discussion refers only to abiotic reactions. The means by which sterility is ensured in the various experimental systems will not be discussed. Though considerations of these methods may be important, the present state of the literature does not warrant further discussion at this time (*14*).

Triazines

The catalytic effects of humic acids (HAs) on the dechlorohydroxylation of the chloro-1,3,5-triazine herbicides in soils and aqueous solutions are well documented. Harris (*15*) first reported that the decomposition of simazine, atrazine, and propazine in soils proceeds predominantly by way of the acid-catalyzed hydrolysis at the 2-position of the triazine ring to produce the hydroxy derivatives, as shown in Reaction I.

Evidence for the catalytic effect on this reaction of solution-phase HAs and soil organic matter has been presented by (among others) Armstrong and Chesters (*16*), Li and Felbeck (*17*) and Khan (*18*). Nearpass (*19*) reported similar evidence for propazine. Although these catalytic effects are well

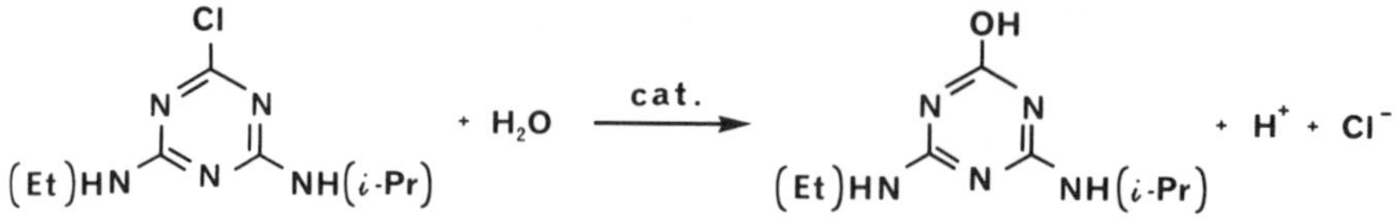

Reaction I. Hydrolysis of atrazine.

documented, the mechanism is poorly understood. Attempts to establish a pathway for the observed enhancement have produced some contradictory evidence. The effect of HAs on the activation energy for this hydrolysis, for example, has been reported to be substantial by Li and Felbeck (*17*) and negligible by Kahn (*18*). Indeed, Metwally and Wolfe (*20*) failed to observe rate enhancement of acid-catalyzed hydrolysis of atrazine in sediment–water versus buffered distilled water systems.

Attempts to explain these observations in terms of general acid catalysis are not convincing, and specific interaction between HAs and the triazines is more likely. Hydrogen bonding between HA surface acid groups and the ring nitrogens of the triazines has been suggested as a mechanism for reduction of the activation energy barrier for hydrolytic cleavage of the C–Cl bond (*13*).

Interpretation of the effects of HAs on triazine hydrolyses is complicated by the complexity of the reaction pathways for these reactions at lower pH values. Plust et al. (*21*) demonstrated that the complex pH dependence of the atrazine rate of hydrolysis (Figure 2) is due to the protonation of atrazine at lower pH values. They also verified the absence of a pH-independent hydrolytic pathway for atrazine.

Under certain reaction conditions, the surface acidity of clays will facilitate the hydrolysis of adsorbed atrazine (*22, 23*). This activity indicates that the catalytic effects observed in soils may not be entirely due to natural organic matter. Where clean surfaces are available, they may contribute a catalytic effect confounded with that from other sources (*24*).

Solution-Phase Reactions

For convenience, one can divide the role of natural organic matter in hydrolytic reactions into solution-phase and sediment-phase interactions. Fulvic and humic acids in water are known to interact with many (generally hydrophobic) contaminants (*25*). Investigations of the effect, if any, of such associations on rates of hydrolysis are few. Struif et al. (*26*) considered the role of fulvic acids in the hydrolyses of a series of *n*-alkyl esters of 2,4-D and concluded that the presence of this form of organic matter accelerated

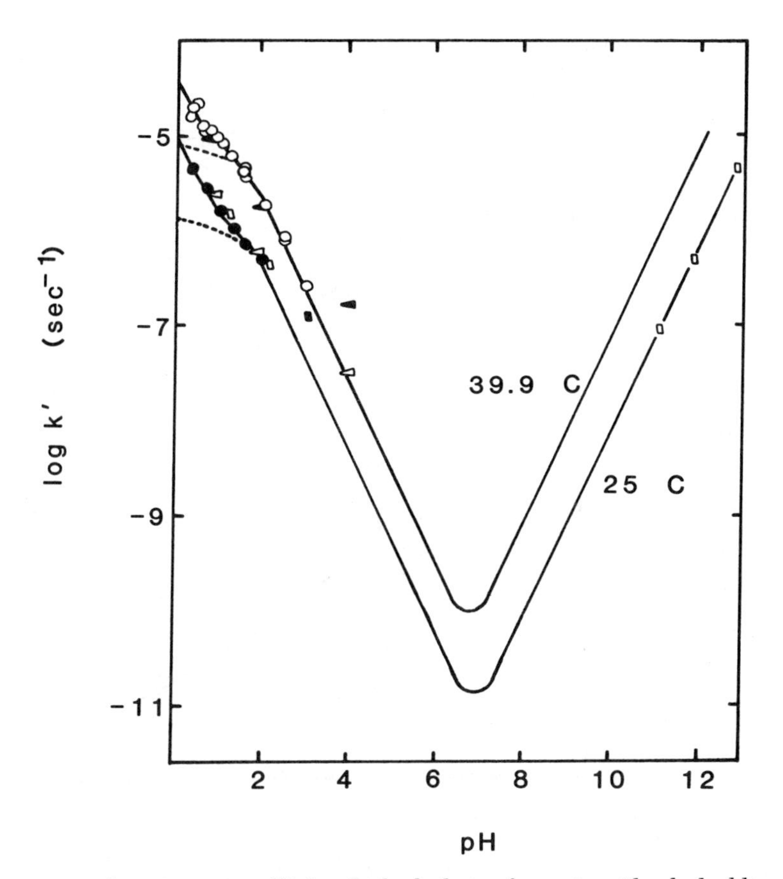

Figure 2. The pH–rate profile for the hydrolysis of atrazine. The dashed lines represent the rates that would result if there were no contribution from diprotonated atrazine. (Reproduced from ref. 21. Copyright 1981 American Chemical Society.)

the reactions. However, Perdue and Wolfe (27) later demonstrated that, at least for base-catalyzed hydrolyses, the opposite is true. Solution-phase humic substances (those that pass through a 0.2-μm filter) were shown to retard the rate of the base-catalyzed hydrolysis of the *n*-octyl ester of 2,4-D (Figure 3). Perdue and Wolfe attributed the rate enhancements observed by Struif et al. to microbially catalyzed processes. No reports of similar investigations of the effect of solution-phase organic matter on acid-catalyzed or pH-independent hydrolytic reactions have been published.

Heterogeneous Systems

In sediment–water or soil–water systems, associations of hydrophobic contaminant molecules with the organic phase of the solids ("sorption") can also

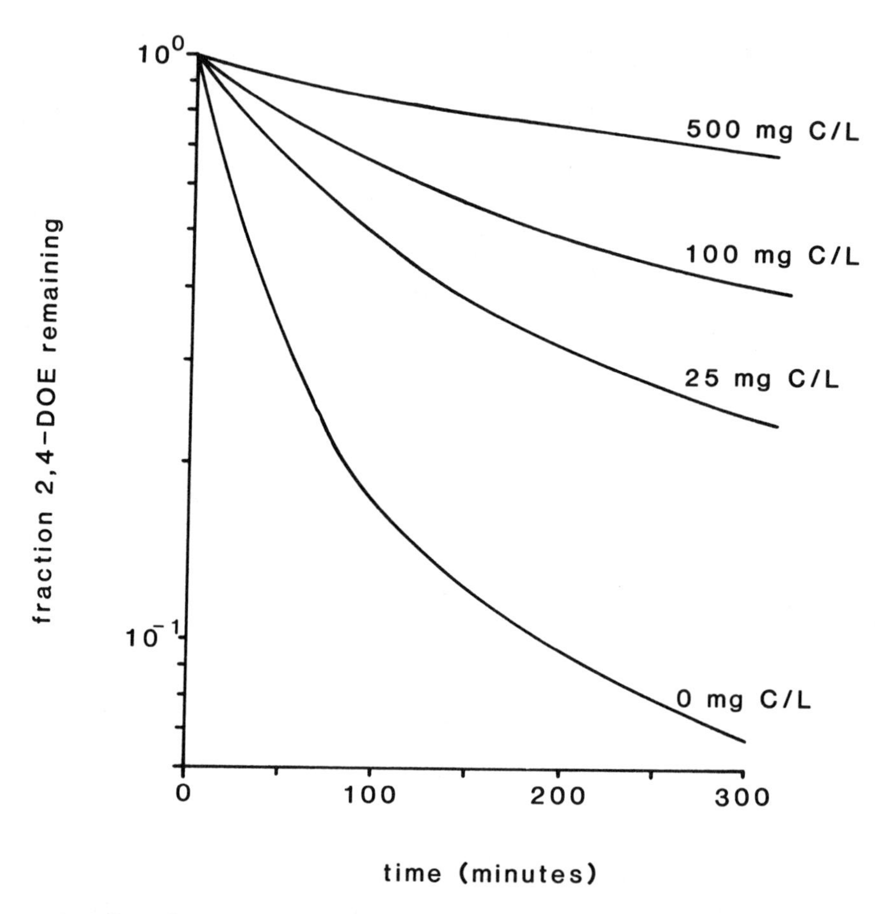

Figure 3. Effect of partitioning on the hydrolysis of 2,4-DOE at pH 10 with varying amounts of "dissolved humic substances". (Reproduced from ref. 27. Copyright 1982 American Chemical Society.)

be considered in terms of effects on hydrolytic reactions. Studies of rates of hydrolysis in the presence of soils and sediments are more numerous than solution-phase studies, and considerable progress has been made toward a general understanding of the effects of sorption on abiotic hydrolyses. In particular, a large and growing body of evidence supports the idea that sorption has little or no effect on the rates of pH-independent (i.e., non-acid/base catalyzed) hydrolyses. For a wide variety of organic molecular structures and a concomitant variety of hydrolytic mechanisms, sediment organic matter does not have a measurable effect on pH-independent rates of hydrolysis.

Macalady and Wolfe (*28*, *29*) demonstrated that the hydrolyses of chlorpyrifos, diazinon, and ronnel (Chart I), all organophosphorothioate esters, proceed at rates that are independent of sediment sorption. These reactions presumably proceed via an S_N2 pathway involving attack by water at the ester carbon atoms (*30*). The S_N1 hydrolysis of benzyl chloride (Chart I) is

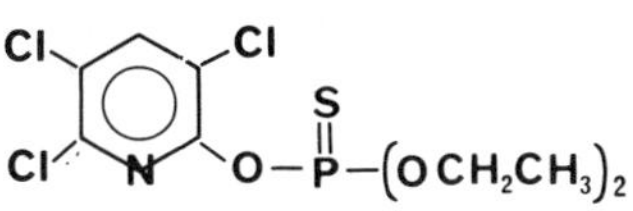

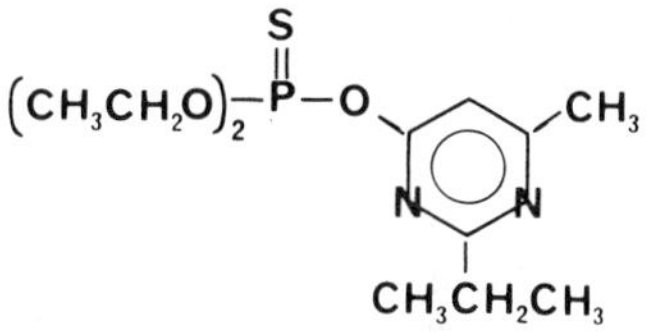

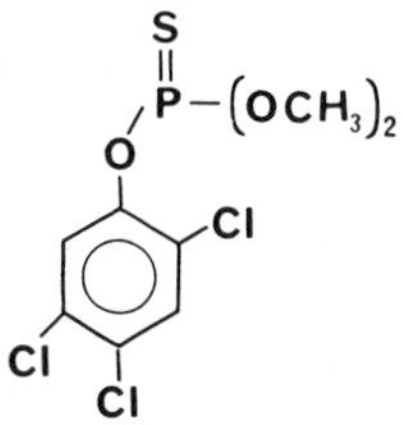

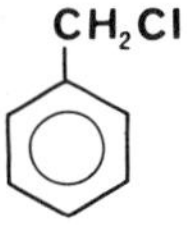

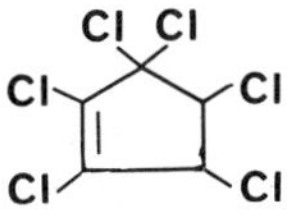

Chart I. Molecular structures of compounds mentioned in the text. A, chlorpyrifos; B, diazinon; C, ronnel; D, benzyl chloride; E, hexachlorocyclopentadiene; F, 4-(p-chlorophenoxy)butyl bromide; and G, chlorostilbene oxide.

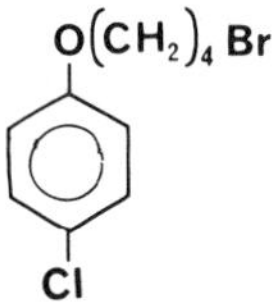

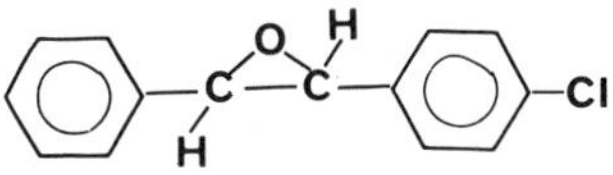

also unaffected by the presence of sediments (*28*). The S_E2 hydrolysis of hexachlorocyclopentadiene (*31*, Chart I) and the nucleophilic substitution reaction of 4-(*p*-chlorophenoxy)butyl bromide (*32*, Chart I) are also not measurably affected by sorption of the substrate to sediment organic matter (*28*). Recently, El-Ammy and Mill (*33*) have demonstrated a similar lack of influence of aquifer materials on the disappearance kinetics of halogenated alkenes, which probably involves hydrolytic reactions. Metwally and Wolfe (*20*) have recently demonstrated that the hydrolysis of chlorostilbine oxide (Chart I) at near-neutral pH values proceeds at experimentally equivalent rates in the presence and absence of sediments.

A similarly convincing array of evidence exists that base-catalyzed hydrolytic reactions are retarded by natural organic matter. The homogeneous retardation of 2,4-DOE hydrolysis by dissolved HAs is matched by influences of solid-phase organic matter (*34*). The alkaline hydrolysis of chlorpyrifos is characterized by pseudo-first-order kinetics with rate constants that, in the presence of sediments, are lower by factors of 0.01 to 0.001 than those in buffered distilled water at the same pH (*29*).

For acid-catalyzed hydrolyses, the picture is considerably less resolved. An attractive working hypothesis is that acid-catalyzed hydrolyses should be enhanced in the presence of natural organic matter. This hypothesis is consistent with the rationalization of rate retardation observed for the alkaline case. The retardation of base-catalyzed hydrolysis can be envisioned in terms of the changes in solution chemistry in the vicinity of the negatively charged (at neutral and alkaline pH values) surfaces of HA molecules and sediment particles. To the extent that such a simplistic picture is reasonable, one should be able to predict, at least in a semiquantitative way, the effects of organic matter on acid-catalyzed hydrolytic reactions. Such predictions might be based on the pH of the reaction medium, the known or assumed surface charge of the solution or sediment organic matter at that pH, and the pH–rate profile for the abiotic hydrolysis in buffered distilled water.

Although bits of evidence seem to support such a scenario, the entire body of experimental evidence does not. At pH values considerably above the zero point of charge (zpc) or isoelectric point of the sediments, for example, one would predict enhancements of acid-catalyzed rates of hydrolysis compared to rates at similar pH values in buffered distilled water. Such effects are seen in some cases, but not in others. Using the methyl ester of 2,4-D (2,4-DME) as a substrate and two standardized EPA sediments (*35*) as the source of solid-phase organic matter, Coleman (*36*) has recently observed enhanced hydrolysis rates at pH values around 4.0 and unaffected reaction rates at pH values near 2.0–2.5.

However, evidence from investigations of the hydrolysis of chlorostilbene oxide (*20*) and several aziridine derivatives (*37*) in a broader series of sediments and at pH values well above the expected zpc's of sediment organic matter (about 2.0) is not supportive of the proposed model for acid-catalyzed hydrolyses. For chlorostilbine oxide (Chart I), rate enhancements of acid-catalyzed hydrolyses were not observed in the presence of sediments at pH values around 4.0. Aziridine derivatives have the general form shown in Structure I and react via acid-catalyzed and pH-independent hydrolyses to produce the corresponding ring-opened amino alcohols. Hydrolyses of aziridine derivatives in four different sediments at pH values between 4 and 7 produced hydrolysis rate constants that differed from constants at equivalent pH values in buffered distilled water by factors of 0.4 to 60. Enhancement factors were not related to pH or sediment organic matter content, nor were they in any way consistent with the proposed model of the effects of sediment organic matter. Metal catalysis at the sediment–water interfaces is a hypothesis presently under consideration to account for the results.

R_1–N (ring) with R_3(H)C–C(H)R_2

Structure I. Generalized aziridine derivatives.

The search for a general model by which one can predict the effects of HAs and sediment or soil organic matter on the rates of hydrolyses of contaminant organic chemicals is thus incomplete. The proposed model is clearly over-simplified and inadequate. Nor is it the only model. Additional mechanisms for the involvement of HAs in the mediation of hydrolysis have been postulated. Perdue (*38*) proposed a general mechanism for the effects of HAs on hydrolysis kinetics, based on an analogy to the effects on detergent micelles. This model is interesting, but largely untested. Finally, inhibition by HAs of hydrolytic enzymes in soils has been indicated by another mode of interaction that may operate in certain reactions (*39, 40*). A comprehensive understanding of the role of natural organic matter in hydrolytic reactions is critical for efforts to predict the behavior of organic pollutants in soils and aqueous systems. An incomplete knowledge of the effects of natural organic matter on acid-catalyzed hydrolytic reactions remains a principle gap in this knowledge. The validity of a micelle-based model should also be evaluated.

References

1. Paris, D. F.; Steen, W. C.; Baughman, G. L.; Barnett J. T., Jr. *Appl. Environ. Microbiol.* **1981,** *41*, 603–609.
2. Paris, D. F.; Wolfe, N. L.; Steen, W. C. *Appl. Environ. Microbiol.* **1984,** *47*, 7–11.
3. Burns, R. G.; Edwards, J. A. *Pestic. Sci.* **1980,** *11*, 506–512.
4. Wolfe, N. L.; Zepp, R. G.; Doster, J. C.; Hollis, R. C. *J. Agric. Food Chem.* **1976,** *24*, 1041–1045.
5. Wolfe, N. L.; Zepp, R. G.; Paris, D. F. *Water Res.* **1978,** *12*, 561–563.
6. Wolfe, N. L.; Burns, L. A.; Steen, W. C. *Chemosphere* **1980,** *9*, 393–402.

7. Camilleeri, P. *J. Agric. Food Chem.* **1984,** *32*, 1122–1124.
8. Mabey, W.; Mill, T. *J. Phys. Chem. Ref. Data* **1978,** *7*, 383–415.
9. Harris, J. C. In *Handbook of Chemical Property Estimation Methods*; Lyman, W.; Reehl, W.; Rosenblatt, D., Eds.; McGraw–Hill: New York, 1982; Chapter 7.
10. Mortland, M. M.; Raman, K. V. *J. Agric. Food Chem.* **1967,** *15*, 163–167.
11. Banks, S.; Tyrell, R. J. *J. Org. Chem.* **1985,** *50*, 4938–4943.
12. Perdue, E. M.; Wolfe, N. L. *Environ. Sci. Technol.* **1983,** *18*, 635–642.
13. Choudhry, G. G. In *Humic Substances*; Gordon and Breach: New York, 1984; Vol. 7, Current Topics in Environmental and Toxicological Chemistry Series, pp 143–169.
14. Block, S. S. *Disinfection, Sterilization and Presevation*; Lea and Febiger: Philadelphia, 1983; 1053 pp.
15. Harris, C. I. *J. Agric. Food Chem.* **1967,** *15*, 157–162.
16. Armstrong, D.; Chesters, G. *Environ. Sci. Technol.* **1968,** *2*, 683–689.
17. Li, G.-C.; Felbeck, G. T., Jr. *Soil Sci.* **1972,** *114*, 201–209.
18. Khan, S. U. *Pestic. Sci.* **1978,** *9*, 39–43.
19. Nearpass, D. C. *Soil Sci. Soc. Am. J.* **1972,** *36*, 606–611.
20. Metwally, M.; Wolfe, N. L., submitted to *Environ. Toxicol. Chem.*.
21. Plust, S. J.; Loehe, J. R.; Feher, F. J.; Benedict, J. H.; Hebrandson, H. F. *J. Org. Chem.* **1981,** *46*, 3661–3665.
22. Brown, C. B.; White, J. L. *Soil Sci. Soc. Am. J.* **1969,** *33*, 863–869.
23. Terce, M.; LefebureDrouet, E.; Calvert, R. *Chemosphere*, **1977,** *6*, 753–758.
24. Davis, J. A. *Geochim. Cosmochim. Acta* **1982,** *46*, 2381–2393.
25. Carter, C. W.; Suffet, I. H. *Environ. Sci. Technol.* **1982,** *16*, 735–740.
26. Struif, B.; Weil, L.; Quentin, K. E. *Vom Wasser* **1975,** *45*, 53–73.
27. Perdue, E. M.; Wolfe, N. L. *Environ. Sci. Technol.* **1982,** *16*, 847–852.
28. Macalady, D. L.; Wolfe, N. L. In *Treatment and Disposal of Pesticide Wastes*; Krueger, R. F.; Seiber, J. N., Eds.; American Chemical Society: Washington, DC, 1984; ACS Symposium Series No. 259, pp 221–244.
29. Macalady, D. L.; Wolfe, N. L. *J. Agric. Food Chem.* **1985,** *33*, 167–175.
30. March, J. *Advanced Organic Chemistry*, 3rd ed.; Wiley–Interscience: New York, 1985; Chapter 10.
31. Wolfe, N. L.; Zepp, R. G.; Schlotzhauer, P.; Sink, M. *Chemosphere* **1982,** *11*, 91–102.
32. Sanders, P.; Wolfe, N. L. U.S. Environmental Protection Agency, Athens, GA, unpublished data.
33. El-Ammy, M. M.; Mill, T. *Clays Clay Miner.* **1984,** *32*, 67–73.
34. Wolfe, N. L. U.S. Environmental Protection Agency, Athens, GA, unpublished data.
35. Hassett, J. J.; Means, J. C.; Banwort, W. L.; Wood, S. G. *Sorption Properties of Energy Related Pollutants*, U.S. Environmental Protection Agency: Athens, GA, 1980; EPA–600/3–80–041.
36. Coleman, K. D. M.S. Thesis, Colorado School of Mines, 1987.
37. Mani, J.; Wolfe, N. L. U.S. Environmental Protection Agency, Athens, GA, unpublished results.
38. Perdue, E. M. In *Aquatic and Terrestrial Humic Matter*; Christman, R. F.; Gjessing, E. T., Eds.; Butterworths: Stoneham, MA, 1983; pp 441–460.
39. Malini de Almeida, R.; Popisil, F.; Vockova, K.; Kutacek, M. *Biol. Plant* **1980,** *22*, 167–175.
40. Mulvaney, R. L.; Bremner, J. M. *Soil Biol. Biochem.* **1978,** *10*, 297–302.

RECEIVED for review October 15, 1987. ACCEPTED for publication December 29, 1987.

22

Sunlight-Induced Photochemistry of Humic Substances in Natural Waters: Major Reactive Species

William J. Cooper

Drinking Water Research Center, Florida International University, Miami, FL 33199

Rod G. Zika, Robert G. Petasne, and Anne M. Fischer

Marine and Atmospheric Chemistry, Rosenstiel School of Marine and Atmospheric Sciences, University of Miami, Miami, FL 33149

This chapter presents a review of the photochemical formation of reactive species, the aqueous electron, singlet oxygen, organoperoxy radicals, hydroxyl radicals, superoxide, hydrogen peroxide, and other phototransients, and the effect of these reactive species on the fate of pollutants is discussed. Most likely, the primary source of most of these reactive species is photochemical reactions involving humic substances.

SUNLIGHT-INDUCED PHOTOCHEMICAL PROCESSES that occur in surface waters have a profound effect on the redox chemistry of these waters (*1–11*). Photochemically mediated processes may also affect the chemistry of pollutants in natural waters. Humic substances are initiators of photoreactions, some of which involve reactive secondary products. The primary and secondary photoprocesses have important ramifications with respect to modifications of pollutants, regulation of the redox properties of natural waters, and the decomposition of humic substances.

Natural organic substances, in part referred to as humic substances, are found in most natural waters (*12–18*). For the purpose of this review, the

0065–2393/89/0219–0333$08.50/0

term "humic substances" (HS) is used only when fulvic and humic acids were isolated from soils or water. The term "dissolved organic matter" (DOM) is used when natural waters were studied without further characterization. Humic substances play a major role in photoprocesses of surface waters, as they are ubiquitous and are among the most highly absorbing compounds found in natural waters, accounting for about half of the organic matter in fresh waters (*16*).

Figure 1 shows the absorption spectra of waters that contain different concentrations of organic carbon, 3.0–15.4 mg L^{-1} (dissolved organic carbon, DOC). In general, the absorption gradually decreases with increasing wavelength, with essentially no absorption above 550 nm. As the concentration of DOC increases, the integrated absorption between 200 and 550 nm also increases. In some cases this generalization does not hold, as shown for samples C and D. Each has a DOC of 13.4 mg L^{-1} (obtained from two different aquifers), but it appears that the chromophores differ in absorbance characteristics (*19*).

The solar spectrum has been described in detail elsewhere (*20, 21*), and a simplified example is shown in Figure 1. Figure 1 shows that most of the solar energy absorbed by DOC occurs between 300 and 500 nm. The primary excitation step therefore has to be between 95 and 58 Kcal mol^{-1}. This energy is sufficient to initiate a number of different photochemical processes.

To extrapolate from laboratory experiments to natural environments, it is necessary to take into account solar intensity, which in natural waters is affected by numerous variables (*8, 9, 20, 22, 23*). The photochemistry of molecules in natural water will depend upon their distribution in the water column. This distribution may exhibit both spatial and temporal variability, which in turn may be a function of biological properties and physical mixing. Therefore, it is also necessary to understand mixing in water bodies (*24*). In some cases, it is also necessary to assess the biological sources and sinks of the individual reactive species and the dry deposition to surface waters.

The field of humic-substance photochemistry has been the subject of several reviews (*25–27*). A number of primary processes have been tentatively identified as resulting from humic-substance chromophores. Our understanding of these primary photoprocesses of humic substances is limited (*28, 29*); these processes, such as charge transfer, energy transfer, and photoincorporation, will be discussed later in this chapter. It may be possible to obtain a better understanding of humic-substance photochemistry by studying polymeric materials similar to those investigated by Rabek and Ranby (*30*).

Three transients have been observed with laser flash photolysis methods. One is believed to be the aqueous (hydrated) electron, e_{aq}^{-}. The other transients have radical and triplet properties (*31–34*). Once formed, these reactive species may undergo secondary thermal reactions. In most cases the reactive species that have been identified and studied in most detail are

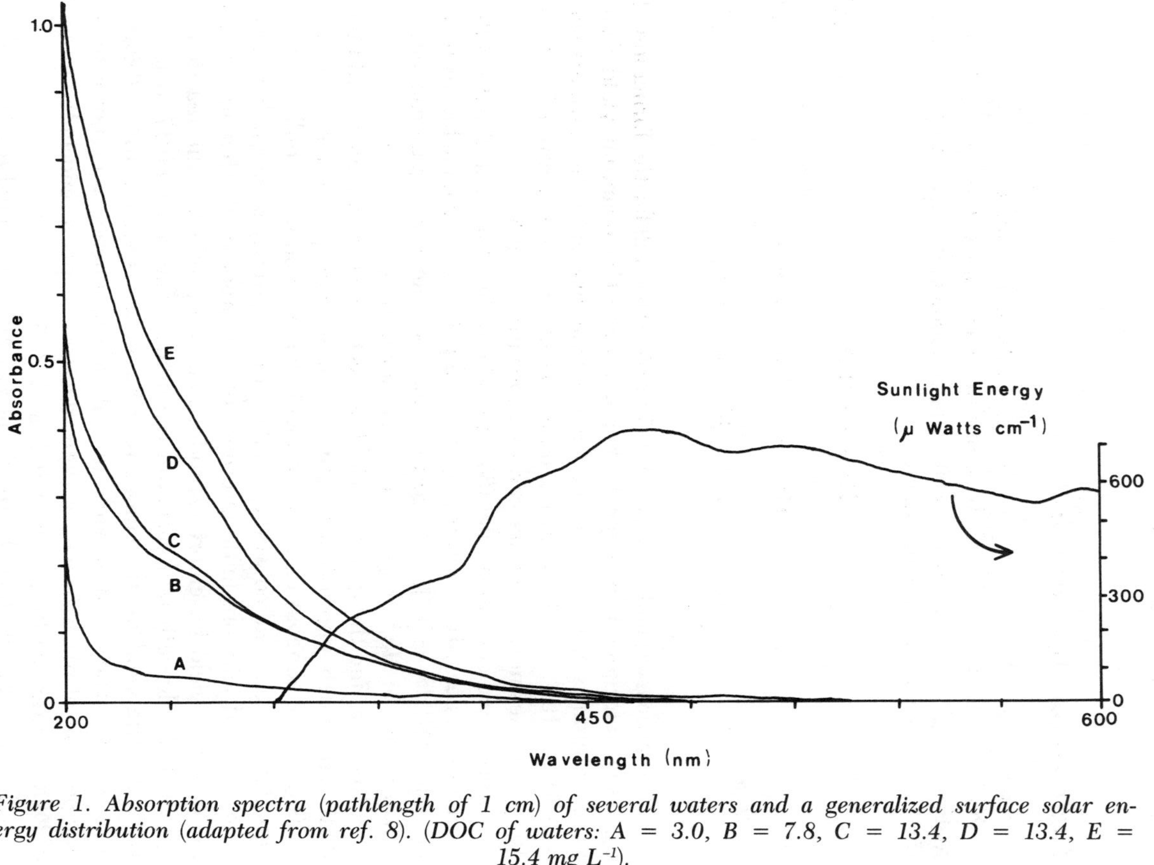

Figure 1. Absorption spectra (pathlength of 1 cm) of several waters and a generalized surface solar energy distribution (adapted from ref. 8). (DOC of waters: A = 3.0, B = 7.8, C = 13.4, D = 13.4, E = 15.4 mg L^{-1}*).*

secondary products. These include singlet oxygen (*10, 35–44*), peroxy radicals, $RO_2\cdot$ (*44, 45*), hydroxy radicals, $HO\cdot$ (*44–51*), superoxide, $O_2^{-}\cdot$ (*52–55*), and unidentified redox-active species, including excited-state and humic-substance organic radicals (*57*).

Hydrated or Aqueous Electron

The formation of e_{aq}^- from irradiated solutions of humic substances and DOC has been reported in laboratory investigations (*31–34, 58*). The formation of e_{aq}^- is thought to result from photoejection of an electron from the excited-state humic substances.

$$\mathrm{HS} \xrightarrow{\text{light}} \mathrm{HS}^* \tag{1}$$

$$\mathrm{HS}^* \longleftrightarrow [\mathrm{HS}^{+}\cdot + e^-] \tag{2}$$

$$[\mathrm{HS}^{+}\cdot + e^-] \underset{-\mathrm{H_2O}}{\overset{+\mathrm{H_2O}}{\rightleftharpoons}} \mathrm{HS}^{+}\cdot + e_{aq}^- \tag{3}$$

The primary quantum yield at 355 nm (equation 2) for the formation of e_{aq}^- was determined by laser flash photolysis. The quantum yield from purified humic substances isolated from several different natural waters was $4.6\text{–}7.6 \times 10^{-3}$, and from two commercial humic acids it was $1.7\text{–}4.0 \times 10^{-3}$ (all values normalized for carbon concentration) (*58*).

Recent and current studies (*58, 59*) indicate that the excited state of the humic substances, HS*, forms a tight caged pair, somewhat analogous to an ion pair, $[\mathrm{HS}^{+}\cdot + e^-]$, with formal charges. The caged pair can either collapse back to HS* or eject the e^- and form the e_{aq}^-.

To determine the quantum yield of ejected electrons under conditions that more closely approximate natural surface water photochemical conditions (i.e., the aqueous electrons available for reaction in the bulk solution, equation 3), Zepp and co-workers (*58*) conducted steady-state irradiations of deoxygenated solutions in the presence of 2-chloroethanol. They found that the quantum yield for chloride production, one product of the reaction of e_{aq}^- with 2-chloroethanol, varied in the natural waters from 0.17 to 1.2×10^{-4}, and the yield for the commercial humic acids varied from <0.08 to 0.23×10^{-4}. The quantum yield for chloride production was shown to be independent of pH 6–8, with a slight decrease at pH 4, and independent of DOC (Suwannee River) up to 100 mg L^{-1} (*58*).

Thus, it appears that the quantum yield of the ejected e_{aq}^- is approximately 2 orders of magnitude lower than the primary quantum yield in reaction 2 (*58*), suggesting that the role of the e_{aq}^- may be minimal compared to other phototransients in reactions with $^3O_2$ (*58, 59*).

The reaction of e_{aq}^- with O_2 (k_4 = 1.9–2.2 × 10^{10} L mol^{-1} s^{-1} (as reviewed in refs. 60 and 61) is the predominant reaction in aerobic surface waters:

$$O_2 + e_{aq}^- \longrightarrow O_2^- \quad (4)$$

With a reaction rate for quenching reaction 4 of 1.9 × 10^{10} L mol^{-1} s^{-1} and a production rate typical of Zepp and co-workers (58), a steady-state concentration of e_{aq}^-, $[e_{aq}^-]_{ss}$, of 2 × 10^{-17} mol L^{-1} is reasonable. By using the $[e_{aq}^-]_{ss}$ and rate constants (*60, 61*), it is possible to estimate the half-lives of several organic pollutants (Table I). Because the half-lives calculated assume continuous noontime sunlight irradiation, they severely underestimate the real half-lives of the compounds in surface waters, at least with respect to the reaction with e_{aq}^-. However, the half-lives can be used to determine the relative contribution of the photochemical processes to the decomposition of the compounds.

Table I. Compilation of Estimated Sunlight Half-Lives of Organic Pollutants for Their Reaction with e_{aq}^-

Compound	*pH*[a]	k^b *(L mol⁻¹ s⁻¹)*	$t_{1/2}^c$ *(days)*
Benzene	7	<7 × 10^6	5.7 × 10^4
Benzophenone	7 ± 1	(3.0 ± 0.5) × 10^{10}	13
Bromoacetate	10	(6.2 ± 0.7) × 10^9	65
Carbon tetrachloride	7	3.0 × 10^{10}	13
Chloroacetate ion	7	1.1 × 10^9	3.7 × 10^2
Chlorobenzene	11	5.0 × 10^8	8.0 × 10^2
1-Chlorobutane	7.3	4.5 × 10^8	8.9 × 10^2
Chloroform	7.0	3.0 × 10^{10}	13
Chloromethane	—	5 × 10^8	8.0 × 10^2
1-Chloropropane	6.3	6.9 × 10^8	5.8 × 10^2
Dichloroacetate ion	7.5	1.0 × 10^{10}	40
1,1-Dichloroethene	—	2.3 × 10^{10}	17
1,2-Dichloroethene[d]	—	7.5 × 10^9	54
Methanol	—	<10^4	4.0 × 10^7
Methylene chloride	10	(6.3 ± 0.5) × 10^9	64
Nitrobenzene	7	3.0 × 10^{10}	13
Phenol	6.3–6.8	(1.8 ± 0.2) × 10^7	2.2 × 10^4
Trichloroacetate ion	6.6	2.1 × 10^{10}	19
1,1,2-Trichloroethene	—	1.9 × 10^{10}	21
Vinyl chloride	—	2.5 × 10^8	1.6 × 10^3

[a]pH at which the reaction rate was determined.
[b]Reaction rates were obtained from refs. 60 and 61.
[c]Assuming $[e_{aq}^-]_{ss}$ = 2 × 10^{-17} mol L^{-1}, which is the midday value at water body surface.
[d]cis or trans isomer was not specified.

Singlet Oxygen

The presence of $^1O_2$ has been established in natural waters, and rate constants for reactions between several organic chemicals and $^1O_2$ have been reported (*35–44, 62–67*). The data have recently been summarized and extended to include other pollutants (*68*).

The formation of $^1O_2$ by energy transfer from excited-state humic substances (HS*) to O_2 is shown in the following equations:

$$^1HS \xrightarrow{\text{light}} {}^1HS^* \xrightarrow{\text{ISC}} {}^3HS^* \quad (5)$$

$$^3HS^* + {}^3O_2 \longrightarrow {}^1HS + {}^1O_2 \quad (6)$$

where ISC is intersystem crossing. Only 22.7 kcal mol^{-1} is required for the $^3\Sigma > {}^1\Delta_g$ transition in oxygen. Hence, it is not surprising that singlet oxygen is a major product resulting from triplet sensitizers in natural waters (*35, 42, 62*). This reaction has been shown to be pH-independent in the range 4.5–9.0 (*38, 42*).

Frimmel and co-workers (*69*), using laser flash photolysis, have recently shown that $^1O_2$ is formed with a quantum yield of 1–3% from humic substances of different origin. The irradiations were conducted at 366 nm, under conditions that simulated natural conditions. The data provide support for direct energy transfer from triplet-state humic substances to $^3O_2$, as shown in equation 6. The quantum yields of Frimmel and co-workers (*69*) are very similar to results obtained using broad-band irradiation, where it was reported that the efficiency was 0.5–2.6% (*41*).

The major pathway for the loss of $^1O_2$ in natural waters is physical quenching by water:

$$^1O_2 + H_2O \longrightarrow {}^3O_2 + H_2O \quad (7)$$

with a first-order rate constant of $k_7 = 2.3 \times 10^5\ s^{-1}$ (*70*).

Steady-state concentrations of $^1O_2$, $[^1O_2]_{ss}$, have been determined in numerous waters (*35, 36, 41, 42*). The $[^1O_2]_{ss}$ varies over 2 orders of magnitude, $0.4–94 \times 10^{-14}$ mol L^{-1}, apparently related primarily to the amount of light-absorbing species in the water (under similar irradiation conditions). Figure 2 shows the relationship of $[^1O_2]_{ss}$ and DOC for waters examined in these studies. These results are uncorrected, drawn as reported. In all three studies (*35, 36, 41*), the same trend of increasing $[^1O_2]_{ss}$ with increasing DOC was observed.

More recently the apparent discrepancies between the work of Zepp and co-workers (*35*) and Haag and co-workers (*41*) have been corrected. The initial work of Zepp and co-workers (*35*) did not take into account the geometric effects of tube measurements relative to flat surfaces. When these

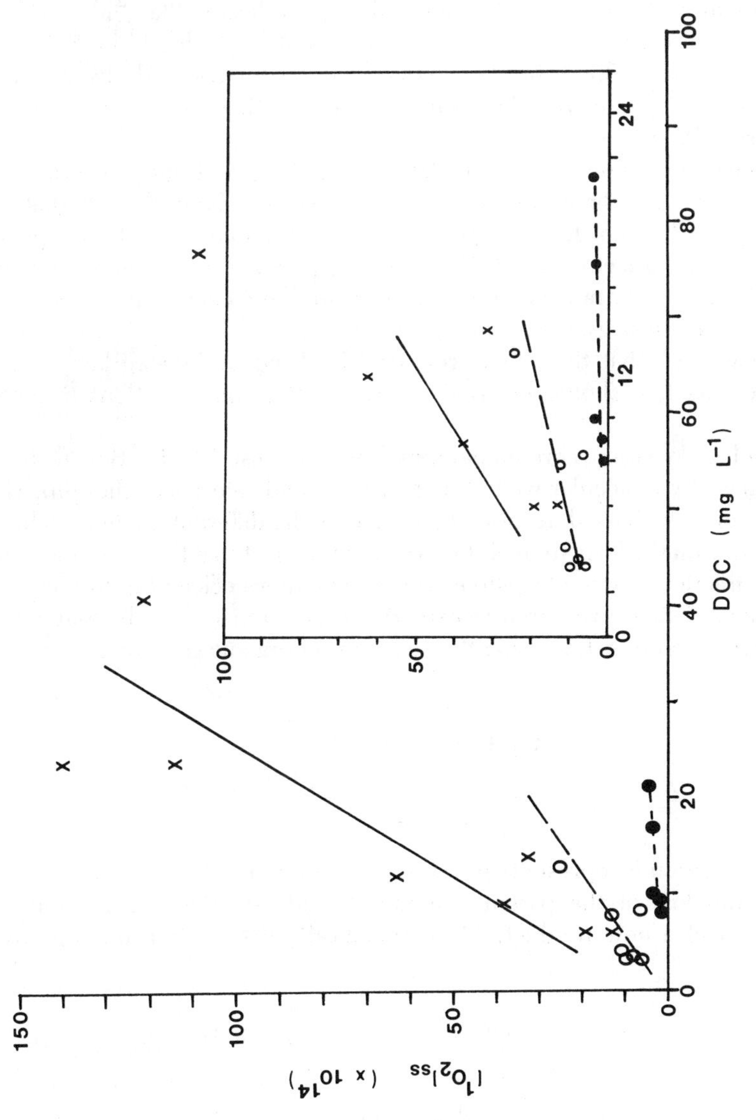

Figure 2. Uncorrected steady-state concentration of $^1O_2$ *in surface waters with different concentrations of DOC (see text for details). Insert is the lower portion of the graph enlarged. Key:* ×, *ref.* 35; ○, *ref.* 41; ●, *ref.* 36.

effects are considered, it lowers the factor by about 1.5–2 for a flat water body. Further, a recent reevaluation of the reaction rate of $^1O_2$ with 2,5-dimethylfuran (used as a $^1O_2$ chemical trap) indicates that the value that Zepp and co-workers used (*35*), 4×10^8 L mol^{-1} s^{-1}, should be 8.2×10^8 L mol^{-1} s^{-1} (*68*). This value also reduces the estimates made by Zepp and co-workers (*35*), and the data are consistent with those of Haag and co-workers (*42*).

The differences between the data reported by Wolff and co-workers (*36*) and the results of the other studies (*35, 41, 42*) may lie in the fact that they used up to 10^{-3} mol L^{-1} dimethylfuran. This high concentration represses the $[^1O_2]_{ss}$ by a factor of 4.3 (W. R. Haag, personal communication). It is also difficult to extrapolate their data to sunlight conditions because the pyranometer they used was not specified.

It appears that the $[^1O_2]_{ss}$ reported by Haag and co-workers (*41, 42*) are the data that should be used to predict $^1O_2$ concentrations in natural waters.

Gel permeation chromatography has been used to fractionate humic substances by molecular weight (e.g., ref. 71 and references therein). Haag and Hoigné (*42*) have determined the effect of the different fractions of humic substances on the formation of $^1O_2$. The data suggest that the lower molecular weight fractions, 100–500 daltons, seem to be more efficient in forming $^1O_2$. Additional data are required to extend and generalize this phenomenon.

Two equations describing the reactions of organic compounds with $^1O_2$ are

$$^1O_2 + \text{org} \longrightarrow {}^3O_2 + \text{org} \tag{8}$$

$$O + \text{org} \longrightarrow \text{product} \tag{9}$$

where equation 8 represents physical quenching of $^1O_2$ by an organic compound (no loss of the parent compound) and equation 9 represents the formation of a new product. More specifically (*64*, section on superoxide ion):

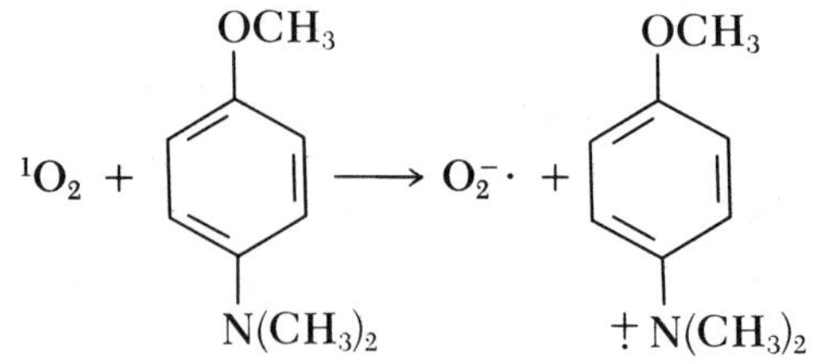

$$^1O_2 + \text{[4-methoxy-}N,N\text{-dimethylaniline: ring with OCH}_3\text{ and N(CH}_3)_2] \longrightarrow O_2^{-}\cdot + \text{[ring with OCH}_3\text{ and }\overset{+}{\cdot}\text{N(CH}_3)_2] \tag{10}$$

$$^1O_2 + \text{[cyclohexadiene]} \longrightarrow \text{[bicyclic endoperoxide with O–O bridge]} \tag{11}$$

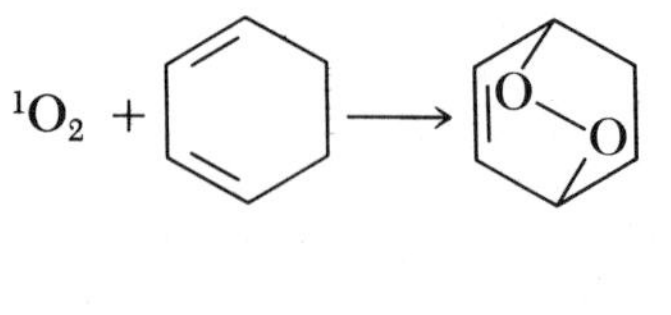

Data for reaction rates of $^1O_2$ with numerous organic chemicals have been compiled (72). For the most part, the data are not applicable to environmental systems because they were obtained in nonaqueous solvents. However, for several organic compounds the half-life can be estimated in natural waters (Table II). Thus, with a $[^1O_2]_{ss}$ of 4×10^{-14} mol L^{-1}, half-lives of organic compounds vary from 0.02 to ≥ =5000 days (*see* Table II).

Thus, the reaction of $^1O_2$ with organic pollutants may be significant when considering the fate of compounds in surface waters. Because of the wide variability in reaction rates, the relative importance of these processes is dependent on the specific pollutant. Considerable data gaps are apparent, and additional chemicals should be studied to better understand the potential importance of reactions with $^1O_2$ in natural waters. Experiments should be conducted in natural waters under the variable conditions found in natural environments and water-treatment conditions.

Organoperoxy Radicals

Three studies have reported the possible presence of organoperoxides in surface waters exposed to sunlight (*44, 45, 57*). The formation of organic

Table II. Compilation of Estimated Sunlight Half-Lives of Organic Pollutants for Their Reaction with $^1O_2$

Compound	*pH*[a]	k^b *(L mol^{-1} s^{-1})*	$t_{1/2}^c$ *(days)*
9,10-Dimethyl-1,2-benzanthracene	—	1.1×10^{10}	0.02
1-Anthracenesulfonate ion	—	5.0×10^{8}	0.4
2-Anthracenesulfonate ion	—	3.0×10^{8}	0.7
		4.5×10^{7}	4.5
9,10-Dimethylanthracene	—	9.1×10^{8}	0.2
2-Methylfuran	7	1.0×10^{8}	2
2,5-Dimethylfuran	—	1.6×10^{9}	0.13
1,3-Diphenylisobenzofuran	—	$(4.5–28) \times 10^{9}$	0.007–0.04
Histamine	7.1	$(2.8–20) \times 10^{7}$	1–7.2
Imidazole	7.1	$(2–4.0) \times 10^{7}$	5–10
1,5-Anthracenedisulfonate ion	—	7.0×10^{6}	29
Furfuryl alcohol[d]	—		40
Histidine[d]	8		50
Methionine[d]	6–11		~200
Lysozyme[d]	6		≥500
Arginine, adenine, thymine[d]	—		≥5000
Thiobencarb[e]	—		no reaction

[a]pH at which the reaction rate was determined.
[b]Reaction rates were obtained from ref. 72.
[c]Assuming $[^1O_2]_{ss} = 4 \times 10^{-14}$ mol L^{-1}, which is the midday value at water body surface.
[d]Data from ref. 42.
[e]Data from ref. 73.

peroxy compounds could be accounted for through the following reaction sequence (e.g., ref. 8):

$$HS^{+}\cdot + {}^{3}O_2 \longrightarrow HSO_2{}^{+}\cdot \qquad (12)$$

$$HSO_2{}^{+}\cdot + RH \longrightarrow HSO_2H + R^{+}\cdot \qquad (13)$$

where $HS^{+}\cdot$ is a photochemically generated cation radical, such as shown in reaction 3. The $HS^{+}\cdot$ reacts with ground-state ${}^{3}O_2$ to give a peroxy-cation radical, followed by hydrogen abstraction from another organic substrate (RH).

Mill and co-workers, using product analysis of known reactions, have estimated the concentration of $RO_2\cdot$ in three waters exposed to sunlight. Their estimates varied from 0.45 to 9.5×10^{-9} mol L^{-1} (*45*). For a more detailed discussion, *see* Chapter 23.

Reactions leading to the formation of these radicals in natural waters are poorly understood, in part because the structures of the humic substances are not well defined. Recent compilations of aliphatic, alkoxy (RO·), acyloxyl (RCOO·), $HO_2\cdot$, and $RO_2\cdot$ radical rate constants reveals considerable variation in expected reaction half-lives (*74, 75*). Therefore, although considerable data are available, there is no convenient means of summarizing the data set. *See* references 74 and 75 for specific rate constants.

Mill and co-workers (*76*) have estimated half-lives for various classes of compounds when reacting with $RO_2\cdot$. They used a steady-state concentration of $RO_2\cdot$ of 1×10^{-8} mol L^{-1} and estimated half-lives (in days) of the following classes of organic chemicals:

alkane	2×10^5
olefin	9×10^2
benzyl	8×10^3
aldehyde	0.4
alcohol	9×10^4
chlorocarbon	15

Although the half-lives are generally long, reactions involving $RO_2\cdot$ may be important for some pollutants.

Hydroxyl Radicals

Hydroxyl radicals, HO·, have been implicated in several studies of natural water photochemistry (*10, 44–51*). Because of its highly reactive nature, direct observation of this intermediate is difficult in complex solutions. Thus, most of the evidence for the existence of HO· comes from product analysis studies of photolytic reactions.

The formation mechanism of HO· in solutions is not clear. One possibility is the direct sunlight photolysis of H_2O_2 (*8*):

$$H_2O_2 \xrightarrow{\text{light}} 2HO\cdot \quad (14)$$

Because of its small absorption cross section in sunlight (*46*, *77*), the direct photolysis of H_2O_2 is not considered to be an important HO· source in surface waters.

Other possible sources of HO· have been considered. For example, it is possible that the reaction of H_2O_2 and Fe(II) salts, Fenton's reagent (*78* and references therein), could lead to the formation of HO· in surface waters (*73*). The presence of Fe(II) in surface waters has been established (*79–81*) and appears to be related to photochemical processes in those waters. From the recent work of Zepp and co-workers (*82*), it appears that a major source of HO· in some waters may be nitrate ion photolysis (*46*). In marine waters, the oxidation of Cu(I) and Fe(II) by H_2O_2 may be a significant source of HO· (*83*).

Several groups have estimated the steady-state concentration of HO·, $[HO\cdot]_{ss}$, in natural waters with remarkably similar results. Mill and co-workers calculated a concentration of $0.15–1.8 \times 10^{-17}$ mol L^{-1} (*45*). Zepp and co-workers have calculated a $[HO\cdot]_{ss}$ of 2.5×10^{-17} mol L^{-1} (*82*). Haag and Hoigné have calculated $[HO\cdot]_{ss}$ for a Swiss lake, Greifensee, during summer noon sunlight to be 20×10^{-17} mol L^{-1} (*46*). Russi and co-workers estimated a $[HO\cdot]_{ss}$ of 50×10^{-17} mol L^{-1}, and noted that the primary factor controlling formation was nitrate ion concentration (*49*).

Several compilations of reaction rate constants of chemicals with HO· exist in the radiation-related literature (*84*, *85*). Further discussion of this topic is also found in Chapter 23.

Table III summarizes data on the half-life of several pollutants with the reaction of HO·. Reaction rates are rapid, but probably negligible in surface waters because of the low steady-state concentrations of HO·.

Superoxide

The presence of $O_2^{-}\cdot$ in natural waters has also been reported (*55*, *57–59*). It has been established with the aid of superoxide dismutase (SOD). This enzyme is known to catalyze the disproportionation of $O_2^{-}\cdot$ to H_2O_2 (*86–88*):

$$2O_2^{-}\cdot + 2H^+ \xrightarrow{\text{SOD}} H_2O_2 + O_2 \quad (15)$$

with an overall reaction rate $k_{15} = 2 \times 10^9$ L mol^{-1} s^{-1}. The measurement of H_2O_2 can be used as a probe for $O_2^{-}\cdot$. SOD inhibits reactions mediated by $O_2^{-}\cdot$ and increases the H_2O_2 yield. This method (the addition of SOD) is commonly used. For example, it has been used to show that the oxidant

Table III. Compilation of Estimated Sunlight Half-Lives of Organic Pollutants for Their Reaction with HO·

Compound	*pH*[a]	k^b (*L mol*$^{-1}$ *s*$^{-1}$)	$t_{1/2}{}^c$ (*days*)
Acetonitrile	9	3.5×10^6	9.2×10^4
Acetophenone	7	$(6.5 \pm 0.7) \times 10^9$	49
Benzene	6–7	7.5×10^9	43
Benzophenone	—	9×10^9	36
Bromoacetate ion	9	4.4×10^7	7.3×10^3
Chloroacetate ion	9	5.5×10^7	5.8×10^3
Chlorobenzene	9	4.5×10^9	71
Chloroform	9	1.4×10^7	2.3×10^4
m-Chlorophenol	9	7.2×10^9	45
o-Chlorophenol	9	8.2×10^9	39
5-Chlorouracil	7	5.2×10^9	62
o-Cresol	9	1.1×10^{10}	29
p-Cresol	9	1.3×10^{10}	25
1,1-Dichloroethene	—	$(4.1 \pm 0.4) \times 10^9$	78
1,2-Dichloroethene[d]	—	$(4.4–5.0) \times 10^9$	64–73
m- and *o*-Hydroxyphenol	9	$(1.1–1.2) \times 10^{10}$	27–29

[a]pH at which reaction rate was determined.
[b]Reaction rates were obtained from refs. 84 and 85.
[c]Assuming $[HO\cdot]_{ss} = 2.5 \times 10^{-17}$ mol L^{-1}, which is the midday value at water body surface.
[d]cis or trans isomer was not specified.

arising from the photodecomposition of tryptophan in oxygenated waters was $O_2^{-}\cdot$ (*89, 90*).

The formation of $O_2^{-}\cdot$ can result from direct electron transfer from triplet states of organic compounds. The formation of e_{aq}^{-} occurs with the subsequent reduction of O_2 to $O_2^{-}\cdot$. The relative contribution of these processes is unknown.

A generalized photochemical reaction mechanism leading to the formation of $O_2^{-}\cdot$, from sunlight-induced reactions involving humic substances, is the following sequence:

$$^1HS \xrightarrow{\text{light}} {}^1HS^* \xrightarrow{\text{ISC}} {}^3HS^* \quad (5)$$

$$HS^* \text{ or } {}^3HS^* \longrightarrow [HS^{+}\cdot + e^{-}] \quad (2)$$

$$[HS^{+}\cdot + e^{-}] \longrightarrow HS^{+}\cdot + e_{aq}^{-} \quad (3)$$

$$O_2 + e_{aq}^{-} \longrightarrow O_2^{-}\cdot \quad (4)$$

$$^3HS^* + {}^3O_2 \longrightarrow HS^{+}\cdot + O_2^{-}\cdot \quad (16)$$

$$^3HS^* + {}^3O_2 \longrightarrow {}^1HS^* + {}^1O_2 \quad (6)$$

$$^1HS + {}^1O_2 \longrightarrow HS^{+}\cdot + O_2^{-}\cdot \quad (17)$$

$$^3HS^* + RNH_2 \longrightarrow HS^{-}\cdot + RNH_2{}^{+}\cdot \quad (18)$$

$$HS^{-}\cdot + O_2 \xrightarrow{?} HS + O_2{}^{-}\cdot \quad (19)$$

Reactions 4, 16, 17, and 19 all lead to the formation of $O_2{}^{-}\cdot$.

Reaction 4 involves the reduction of O_2 with $e_{aq}{}^{-}$. This reaction has a rate constant of $k_4 = 1.9\text{–}2.2 \times 10^{10}$ L mol^{-1} s^{-1} (*60, 61, 91–93*). It would be the dominant reaction in aerated surface waters, provided that reaction 3 occurs under sunlight-induced reaction conditions.

This mechanism, reaction 3 and 4, is supported by data on the direct spectroscopic observation of $e_{aq}{}^{-}$ in solutions of humic substances (*31–34*). The quantum yield for the formation of the $e_{aq}{}^{-}$ is considerably reduced from that of the caged complex reaction 2 (*59*). This observation has been investigated by Zepp and co-workers (*58*), and they suggest that the second mechanism, reaction 16, may be significant.

The third mechanism involves $^1O_2$, reaction 17. Singlet oxygen may be important in some cases in the formation of $O_2{}^{-}\cdot$. Although charge-transfer reactions involving $^1O_2$ are thought to result in quenching (*94*), the reaction of $^1O_2$ with *N,N*-dimethyl-*p*-anisidine, reaction 10, leads to the formation of $O_2{}^{-}\cdot$ (e.g., *95*).

Electron transfer between the primary light absorber and the organic substances, reaction 18, has been shown to be involved when flavin and its analogues (*96*) or methylene blue and alkylamines (*97, 98*) are present. Subsequent electron transfer between the secondary organic radicals and O_2, reaction 19, leads to $O_2{}^{-}\cdot$. However, this mechanism has not been established in humic-substance photochemistry.

All of these reactions probably occur in sunlight-induced photochemical reactions involving humic substances. Additional studies are required to determine the relative importance of these various processes occurring in surface waters, leading to the formation of $O_2{}^{-}\cdot$.

Photochemically induced redox cycling of transition metals may also be important in natural-water photochemistry and the formation of $O_2{}^{-}\cdot$. Moffett and Zika (*99*) have discussed the formation of H_2O_2 in sea water, resulting from the redox cycling of copper. They suggest that the following reaction may be involved:

$$Cu_{aq}{}^{+} + O_{2aq} \longrightarrow Cu_{aq}{}^{2+} + O_{2aq}{}^{-}\cdot \quad (20)$$

with disproportionation of the $O_2{}^{-}\cdot$ to H_2O_2. H_2O_2 accumulated upon oxidation of Cu(I), consistent with reaction 20. Other studies of the oxidation of various metals—iron (*79–82, 100*), manganese (*101*), and chromium (*102*)—suggest that these may also play a role in either the formation or the decomposition of $O_2{}^{-}\cdot$ and H_2O_2. The relative contribution of these processes in fresh waters is not known, and additional studies are necessary to establish this information.

The rate of formation of $O_2^{-}\cdot$ under midday sunlight has been estimated in one coastal sea water at 5×10^{-7} L mol^{-1} h^{-1}, and the $[O_2^{-}\cdot]_{ss}$ has been estimated at 1×10^{-8} mol L^{-1} (*56*). Although this study was in salt water, it provides a framework from which to extrapolate to freshwater environments. Because the rate depends upon DOC concentration, formation rates probably range between 10^{-7} and 10^{-5} L mol^{-1} h^{-1} in surface waters.

Superoxide is known to react with organic chemicals (*103, 104*). However, little is known regarding reactions of $O_2^{-}\cdot$ and pollutants. Table IV summarizes the applicable reaction rate data and half-lives of several chemicals. Because of the relatively high $[O_2^{-}\cdot]_{ss}$ and the estimated lifetimes in surface water, superoxide might be important in the degradation of some compounds. The reported reaction half-lives with benzidine, benzo[*a*]pyrene, and catechol are fast, and they indicate that additional data should be obtained for other pollutants.

Hydrogen Peroxide

This section thus far has dealt with rather unstable reactive species, generated via sunlight-induced reactions. Hydrogen peroxide, H_2O_2, is also thought to form as a secondary product of sunlight-induced reactions. However, it is relatively stable in most surface waters and may accumulate to concentrations that far exceed these reactive species.

The first reports of hydrogen peroxide in fresh water were those of Sinel'nikov and co-workers (*105–108*) in Russian lakes and streams, where the concentration was 7.1–32.3 $\times$ 10^{-7} mol L^{-1}. Other reports of the presence of H_2O_2 in surface waters have appeared more recently (*54, 55, 109, 110*). Several pathways result in H_2O_2 in natural waters. These different pathways include redox cycling of metals, dry and wet deposition of atmospherically formed oxidants, biological activity, and in situ photochemical formation.

Metal redox cycling has been mentioned in the section on $O_2^{-}\cdot$ and will not be discussed further. The contribution of wet and dry deposition are covered in the appendix to this chapter. The focus of this section is the in

Table IV. Compilation of Estimated Sunlight Half-Lives of Organic Pollutants for Their Reaction with $O_2^{-}\cdot$

Compound	*pH*[a]	k^b (*L mol*$^{-1}$ *s*$^{-1}$)	$t_{1/2}^c$ (*days*)
Benzidine	—	$>2.5 \times 10^7$	$<3.2 \times 10^{-5}$
Benzo[*a*]pyrene	—	$<1 \times 10^7$	$>8.0 \times 10^{-5}$
Catechol	0.5–9	$(2.3 \pm 0.3) \times 10^5$	3.5×10^{-3}
Resorcinol	5.0–8.5	2.0 ± 1	4.0×10^2

[a]pH at which the reaction rate was determined.
[b]Reaction rates were obtained from ref. 103.
[c]Assuming $[O_2^{-}\cdot]_{ss} = 1 \times 10^{-8}$ mol L^{-1} (*56*), which is the midday value at water body surface.

situ photochemical formation of H_2O_2, with some discussion of biological formation.

The photochemical formation of H_2O_2 is thought to result from the reduction of oxygen to the superoxide radical, as shown in reactions 4, 16, 17, and 19. In aqueous solution, the pK_a of superoxide:

$$HO_2\cdot + H_2O \longrightarrow O_2^-\cdot + H_3O^+ \quad (21)$$

has been reported to be 4.5 ± 0.15 (*111–114*), 4.8 (*115*), 4.88 (*116*), and 4.69 (*117*). A value of 4.8, given in a review (*103*), will be used in this chapter. In most natural waters with pH 6–8, $O_2^-\cdot$ is the major species.

The disproportionation of superoxide in aqueous solution occurs via three reactions (*111*):

$$HO_2\cdot + HO_2\cdot \longrightarrow O_2 + H_2O_2 \quad (22)$$

$$HO_2\cdot + O_2^-\cdot \longrightarrow O_2 + HO_2^- \xrightarrow{H_2O} O_2 + H_2O_2 + OH^- \quad (23)$$

$$O_2^-\cdot + O_2^-\cdot \longrightarrow O_2 + O_2^{2-} \xrightarrow{2H_2O} O_2 + H_2O_2 + 2OH^- \quad (24)$$

Reaction 24 is negligible, with a rate constant of k_{24} = 0.3–6 L mol^{-1} s^{-1} (*117* and references therein) and therefore can be ignored. The disproportionation rate of $O_2^-\cdot$ at pH 7 is $k = 6 \times 10^5$ L mol^{-1} s^{-1} (*115, 116*). Over a pH range of 7–9, encountered in most natural waters, the disproportionation rate has been evaluated as log k = 12.681–0.998 (pH) (*118*).

The addition of superoxide dismutase is known to catalyze the disproportionation of $O_2^-\cdot$, reaction 15 (*86–88*). Thus, the addition of SOD to natural waters should increase the formation of H_2O_2 (55, 56). Figure 3 presents the results of the addition of SOD to a nonfiltered water with 7.35 mg L^{-1} DOC. The rate of H_2O_2 production in the irradiated samples with SOD present was substantially higher than in the same natural water sample without SOD. (The addition of SOD to organic free water with subsequent exposure to sunlight resulted in no H_2O_2 formation.) These results, similar to those reported by Petasne and Zika (*56*), confirm the involvement of $O_2^-\cdot$ in the formation of H_2O_2 in this nonmarine water.

Petasne and Zika showed that 24–41% of the $O_2^-\cdot$ formed did not form H_2O_2 (*56*). Other pathways for the destruction (consumption) of $O_2^-\cdot$ were not investigated. However, it does indicate that other mechanisms for consumption (or destruction) may exist, and these may indeed be important in considering the fate of pollutants in surface waters.

In natural waters it is appropriate to define an experimentally determined rate of H_2O_2 formation, that is, the accumulation rate of H_2O_2. We use the term "accumulation rate" because in natural waters the reactions

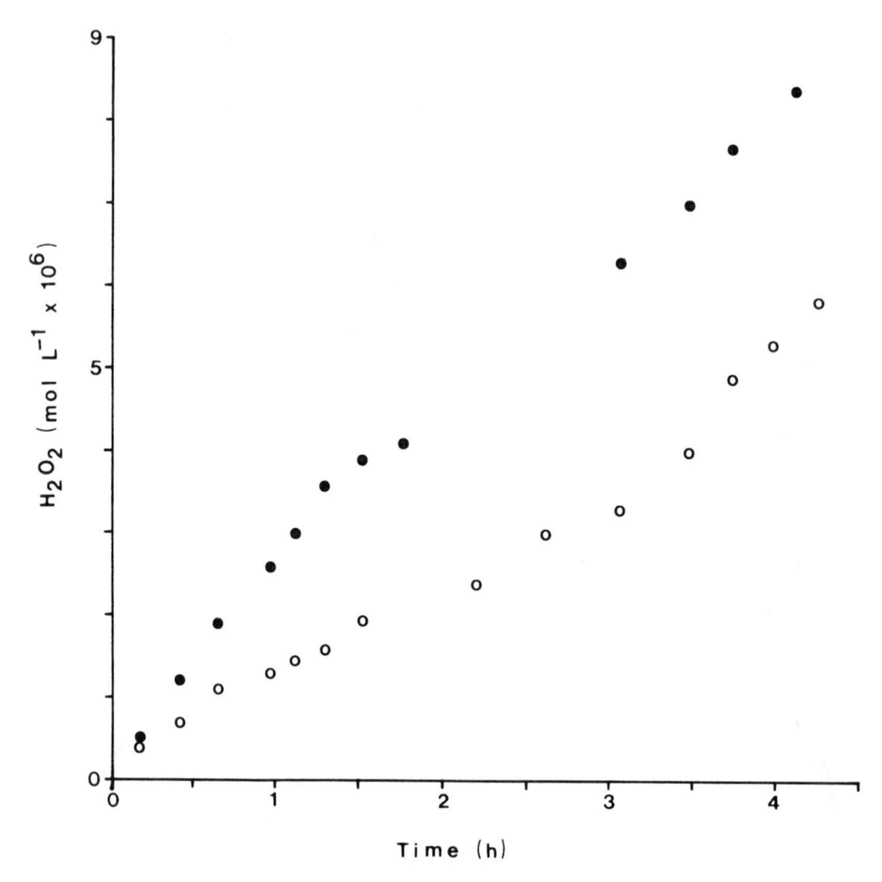

Figure 3. Photochemical formation of H_2O_2 in ground water, of 7.35 mg L^{-1} of DOC, exposed to sunlight in the presence (●) and absence (○) of superoxide dismutase.

leading to the formation and decomposition of H_2O_2 are, or may be, occurring simultaneously. Thus, the experimentally determined concentration of H_2O_2 is the sum of formation and decomposition.

Relatively little is now known about the details of the processes that result in H_2O_2. However, it is generally accepted that the rate at which H_2O_2 is formed is related to the concentration of humic substances in the water (*55*, *110*). From Figure 4, the calculated accumulation rate of H_2O_2 is 1.5×10^{-7} mol L^{-1} h^{-1} DOC–C^{-1} through the range 0–10 mg L^{-1} DOC, and 5.5×10^{-7} mol L^{-1} h^{-1} DOC–C^{-1} through the range 10–18 mg L^{-1} DOC. The reason(s) for the apparent change above 10 mg C L^{-1} is unknown, but may reflect a reduction of back reactions in the solution.

More appropriately, the accumulation rate of H_2O_2 should be related to the integrated area of the absorption spectra of natural waters, 300–800 nm. This step has been qualitatively demonstrated by determining the concentration of H_2O_2 with time (solar irradiation) in waters of different DOC

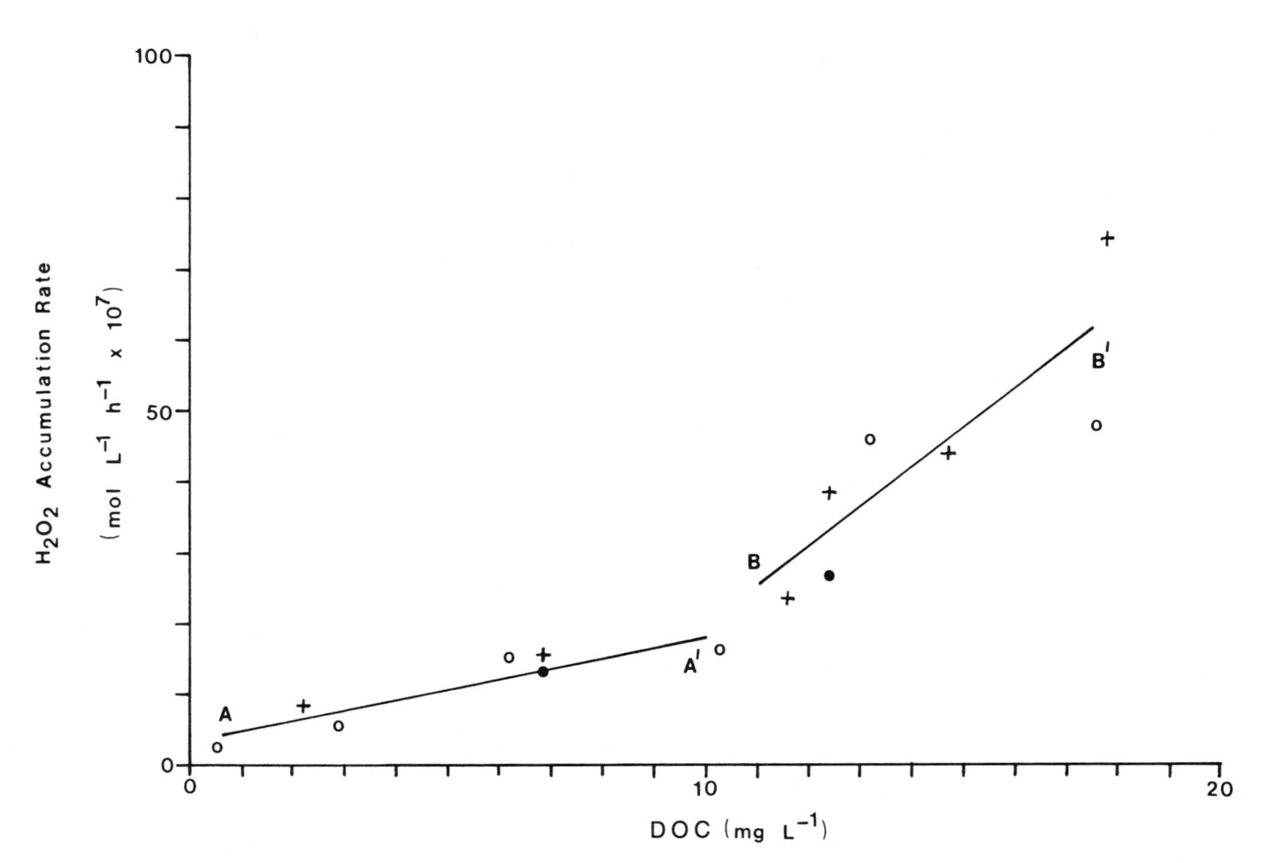

Figure 4. Accumulation rate of H_2O_2 in natural waters with different concentrations of dissolved organic material (110). The slope and correlation coefficient for A–A′ were 1.53 and 0.93, and for B–B′ were 5.53 and 0.84, respectively. Key: ○, ground water; +, filtered surface water; ●, unfiltered surface water.

concentrations. These light-absorbing organic compounds are thought to be primarily humic substances. However, only dissolved organic carbon determinations were performed, and no detailed characterization was undertaken to confirm this identification.

Other chemicals of biological origin may contribute to the formation of H_2O_2 in natural waters, e.g., flavins (*119*). The amino acids tryptophan (*89, 90*) and tyrosine (*54*) may give rise to H_2O_2 in oxygenated aqueous solution. Further, phenylalanine has been shown to form e_{aq}^- upon irradiation at 313 nm (*120*), which would result in the formation of $O_2^{-}\cdot$ in surface waters and hence produce H_2O_2.

Peroxides other than H_2O_2 may be formed in sunlight-initiated photochemical reactions. Evidence that H_2O_2 is the major peroxide formed comes from studies using the peroxide-decomposing enzyme catalase (*121*). Figure 5 shows the results of one such experiment. Ground water was exposed to sunlight in two 300-mL quartz flasks for approximately 1.5 h. The samples were then analyzed for peroxide concentration (*see* ref. 110 for experimental details and H_2O_2 methodology). To a third flask, 0.5×10^{-5} mol L^{-1} reagent-grade H_2O_2 was added. Following the addition of catalase to two of the flasks, the one exposed to sunlight and the one to which reagent-grade H_2O_2 had been added, the H_2O_2 concentration was determined with time (in the dark).

Figure 5 indicates that decomposition rates were identical within experimental error. If a significant portion of the photochemically formed peroxides were low-molecular-weight peroxides other than H_2O_2, an apparent decomposition rate intermediate between the control and the H_2O_2 solution would have been observed. The conclusion is that H_2O_2 is the major peroxide formed under the conditions used in our study.

Additional data to support the photochemical formation of H_2O_2 in natural waters can be found in a limited investigation in which Klockow and Jacob (*109*) reported a limited diel (24-h day–night cycle) study of hydrogen peroxide concentration in the surface layer (1 cm) of a pond on a clear day. At 800 h they observed a concentration of 2.9×10^{-7} mol L^{-1}, which increased to a maximum of 8.8×10^{-7} mol L^{-1} at 1300 h, followed by a decrease to 1.5×10^{-7} mol L^{-1} at 2400 h.

To develop a more quantitative understanding of the in situ photochemical formation of hydrogen peroxide in surface waters, the quantum yield must be determined at several wavelengths throughout the solar spectrum. However, the complications imposed by natural water samples require that the classical concept of a quantum yield for a well-characterized system be qualified. The major complications are that both formation and decomposition may be occurring simultaneously, possibly through complex secondary reactions, and that an exact molar absorptivity cannot be obtained because the structure and molecular weight of the humic substances are not known. In fact, the chromophores that are responsible for the absorption of the light have not yet been determined. For these reasons, it is necessary

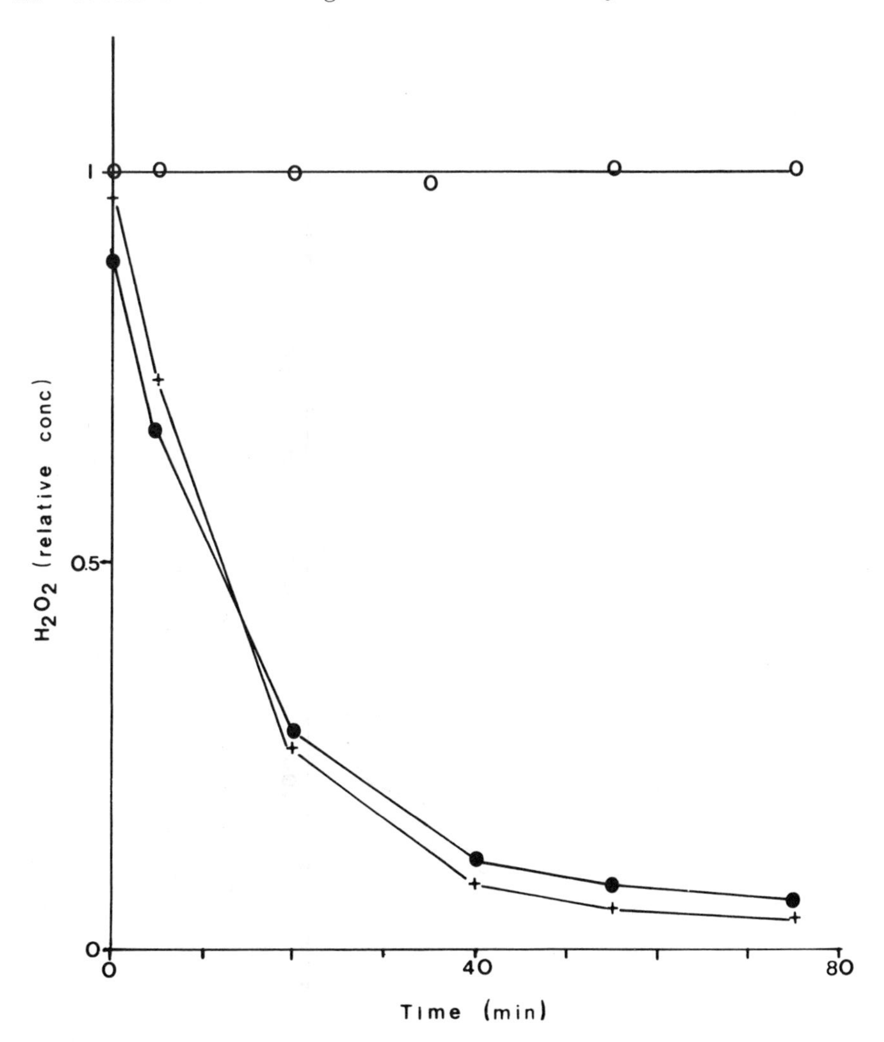

Figure 5. Hydrogen peroxide decay in ground water in the presence and absence of catalase. Key: ○, exposed to sunlight and no catalase added; +, exposed to sunlight and catalase added; ●, not exposed to sunlight and 0.5 × 10^{-5} mol L^{-1} H_2O_2 added.

to define an apparent quantum yield for H_2O_2 (*110*), from which accumulation rates can be determined.

Apparent quantum yields for six waters have been determined from 284 to 400 nm, and for three waters from 284 to >500 nm (*110*). An average quantum yield for the formation of H_2O_2 is shown in Figure 6. Although some slight differences occur from one water to the next, the data at any one wavelength vary by less than a factor of 2. Therefore, the data in Figure 6 could be used as a first approximation in studies where an accumulation

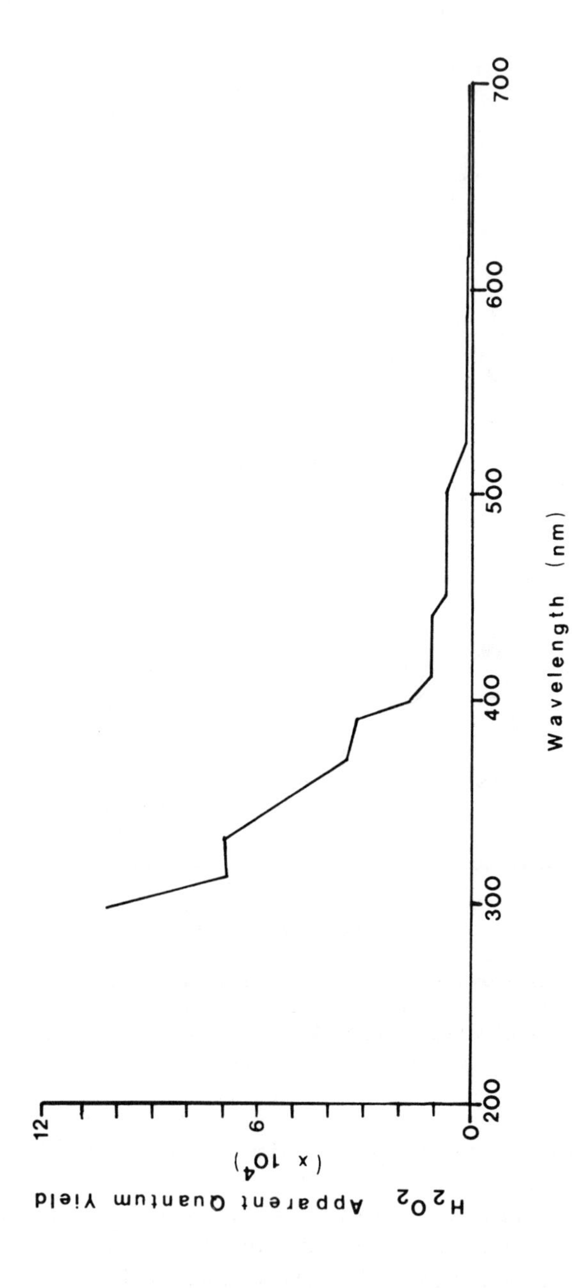

Figure 6. Average apparent quantum yield of H_2O_2 in several waters. Adapted from ref. 110.

rate of H_2O_2 is required. Photochemical processes involving humic substances are thought to be the primary route of formation.

In natural waters, the observed H_2O_2 could result from processes related to biota. Studies have also been conducted on the formation of H_2O_2 in the presence of several algal cultures (*122*). The average rate of formation of H_2O_2 was 0.4–8 × 10^{-7} L mol^{-1} h^{-1}. The production of H_2O_2 was found to be related to light exposure and algal concentration. However, the different algal cultures formed H_2O_2 with different efficiencies, a result suggesting that considerably more work is necessary to understand the role of algae (and microorganisms) in the formation of H_2O_2 in fresh waters.

Very little work has been reported on the decomposition of hydrogen peroxide in natural waters. Several pathways are possible, such as dark reactions involved in redox cycling, photodecomposition through either primary or secondary processes, and biologically mediated processes.

Draper and Crosby (*54*) determined a half-life of 1 h for eutrophic (fresh) water. This value was obtained by irradiating the water for 5 h, removing it from light, and recording the change in concentration with time.

Zepp and co-workers (*122*) conducted H_2O_2 decay experiments using five freshwater green algae and four cyanobacteria. They found no decay in the supernatant portion of the algal suspensions, and concluded that the decay in their system was therefore heterogeneous. The overall decay rate equation was second order and is as follows:

$$\text{rate} = k_1[H_2O_2] = k_{\text{bio}}C_a[H_2O_2] \qquad (25)$$

where k_1 is the pseudo-first-order rate constant for H_2O_2 decay, k_{bio} is the second-order rate constant for H_2O_2 decay, and C_a is the concentration of chlorophyll *a* in milligrams of Chl *a* per cubic meter.

The second-order decay rates ranged from 0.18 ± 0.04 to 8.8 ± 2.0 × 10^{-3} m^3 (mg of Chl *a*)$^{-1}$ h^{-1}. Estimating the decay rate in a eutrophic lake with a chlorophyll *a* concentration of 100 mg of Chl *a* m^{-3}, Zepp and coworkers report a half-life of 1.5 h. From this data, they concluded that (1) most of the microorganisms studied were very efficient in catalyzing the decay of H_2O_2 and (2) in fresh waters, this pathway is probably significant for H_2O_2 decay.

Petasne and Zika (*123*) examined the decomposition or dark decay of H_2O_2 in several environments. The decay rates they reported for oligotrophic waters were on the order of 130 h (half-life). In coastal and near-shore marine environments, the dark decay rate is substantially faster, sometimes as fast as 10 h. Apparently, the decay rate in marine systems is largely an effect of microorganisms.

The two studies that have reported decay rates of H_2O_2 in fresh water (*54*, *122*) agree quite well, even though they appear to be studying different processes. There is a need for more extensive data on the decay rates and

Table V. Aqueous Reaction of Photochemically Formed Reactive Species with Hydrogen Peroxide

Reactive Species Reaction	*pH*[a]	k *(L mol^{-1} s^{-1})*	$t_{1/2}$[b] *(days)*
e_{aq}^- (*60, 61*)			
$[e_{aq}^-]_{ss} = 2 \times 10^{-17}$ mol L^{-1}			
$e_{aq}^- + H_2O_2 \rightarrow OH\cdot + OH^-$	7	$(1.2 \pm 0.1) \times 10^{10}$	33
	—	1.4×10^{10}	29
$^1O_2$			
none reported			
$OH\cdot$ (*45, 46, 82, 84, 85*)			
$[OH\cdot]_{ss} = 2.5 \times 10^{-17}$ mol L^{-1}			
$OH\cdot + H_2O_2 \rightarrow HO_2\cdot + H_2O$	—	2×10^7	1.6×10^4
$O_2^-\cdot$ (*56, 85, 103*)			
$[O_2^-\cdot]_{ss} = 1 \times 10^{-8}$ mol L^{-1}			
$O_2^-\cdot + H_2O_2 \rightarrow OH^- + OH\cdot + O_2$	7.0–9.0	0.13 ± 0.07	6.2×10^3
	—	3.7 ± 1.6	2.2×10^2
	—	16.0 ± 3.3	50

[a]pH at which the reaction rate was determined.
[b]$t_{1/2}$ of H_2O_2 at given steady-state concentration of reactive intermediate.

factors affecting decay rates in fresh waters. The one study that reports decay rates in oceanic environments should be extended in an attempt to develop a more detailed understanding of the processes.

Table V lists the reactions of H_2O_2 with these reactive species. From the $t_{1/2}$ calculations, it is obvious that these reactive intermediates do not limit the H_2O_2 concentration to any extent under natural conditions.

Because of the relative inertness of H_2O_2 in surface waters, $[H_2O_2]_{ss}$ of between 10^{-7} and 10^{-5} mol L^{-1} have been reported (*110*, and references therein). Of the reactive species studied, H_2O_2 appears to have the highest steady-state concentrations. This result is presumably due to its relative inertness when compared to the other reactive species under discussion. Although thermodynamically it is a very powerful oxidant, it is kinetically limited in most cases and does not affect the oxidation of many organic compounds directly.

Very little work has been reported on the direct involvement of H_2O_2 with pollutants. Draper and Crosby (*73*) have suggested the involvement of H_2O_2 in the decomposition of pesticides. Additional studies are needed to better define the role of H_2O_2 in the fate of pollutants in surface waters.

Humic-Substance Parent Molecule Phototransients

The previous sections have discussed the fairly well-defined phototransient byproducts produced upon irradiation of humic substances. The humic-substance parent macromolecule is also photochemically reactive. Three major reaction types of the photoexcited humic-substance macromolecule are energy transfer (or photosensitization), charge transfer, and photoincorporation. Photoincorporation involves reactions such as radical combination or cycloaddition, and it results in the incorporation of smaller reactive constituents into the humic-substance moiety. The reactions discussed here are a summary of much data in a diverse literature. The references cited herein contain typical examples of each reaction type and additional references to similar studies (e.g., *124*).

In a discussion of the photochemical reactivity of humic substances and other molecules over 500 daltons, many workers hesitate to use classical photochemical nomenclature, such as singlet and triplet. In humic-substance photochemical work, these and similar terms are used as qualitative descriptions of molecular properties and do not imply a detailed understanding of humic-substance electronic energy levels.

The parent humic-substance molecule is known to contain conjugated structures that include ketonic and quinoid functional groups (*125–127*). These groups can photoreact by several processes. In the first process, indirect photolysis, the photoexcited humic substance can donate its excitational energy to an acceptor molecule, which does not absorb the exciting radiation:

$$\text{donor}^* + \text{acceptor} \rightarrow (\text{donor} \ldots \text{acceptor})^* \rightarrow (\text{donor}) + (\text{acceptor}^*) \qquad (26)$$

This process would enhance photochemical degradation of some environmental chemicals that have lower excited-state energies, but do not absorb sunlight directly. A tripletlike excited state of humic substance would be likely to participate in this energy-transfer collisional process, because organic triplet lifetimes are sufficiently long for good yields of energy-transfer reactions. If the humic substance has a lower energy than the xenobiotic under study, the humic substance becomes the acceptor and photochemical degradation is quenched, i.e., retarded by the humic substance. This process works both ways with environmental chemicals (*20, 91, 92, 128, 129*). It depends on relative excited-state energies and structural properties.

The second way the parent humic substance can react with substances of environmental interest is by photoinduced electron transfer (*130*). In the irradiated humic-substance system a radical cation is produced in a cage with the aqueous electron. The aqueous electron can back-react to form a neutral excited-state humic-substance molecule or can react with another humic-substance group to form a radical anion. Many studies (e.g., *131*) by different methods have shown that humic substances form donor–acceptor complexes with substances of environmental interest.

At least three classes of compounds have been studied as reactive with humic substances in these charge-transfer reactions. Polycyclic aromatic hydrocarbons have been studied by UV-vis, fluorescence, phosphorescence, and laser studies (*124, 130*). The photosynthesis inhibitor triazine (a herbicide) has been studied by electron spin resonance (*132–134*). The reactivity of paraquat has been studied by excited-state absorption spectroscopy (*33, 124*).

A third type of humic-substance photochemical reactivity is photoincorporation of the environmental reactant directly into the covalent structure of the humic substance. Very recent laboratory experiments have shown that polychlorobenzenes photoincorporate with humic-substance monomers (e.g., *125*).

Summary

Reactive species formed photochemically in surface waters may play a major role in the degradation of many pollutants. However, in general, very little work has been conducted with these reactive intermediates that is directly relevant in determining the fate of pollutants in surface waters. Additional studies should be conducted, and the reactive species likely to be of most importance in the fate and transport of pollutants are $^1O_2$ and $O_2^{-}\cdot$. Hydrogen peroxide may not react directly with many organic pollutants, but may be important in metal speciation and can be used as a probe for $O_2^{-}\cdot$.

Another area that has not been addressed to any extent is the direct effect of triplet excited states of humic substances and their potential reactions with pollutants. Studies conducted with dyes (*91, 92*) showed that these longer-lived excited-state molecules participate directly in reactions with various chemicals. Therefore, humic-substance triplet states may be important in better understanding the fate of pollutants in surface waters, and further research is indicated.

Acknowledgments

We thank Werner R. Haag and Richard G. Zepp for helpful discussions during the preparation of this manuscript. We also thank the Drinking Water Research Center, Florida International University, Miami, FL, for support of this project.

Appendix

Wet Deposition. Rain has been shown to have relatively high concentrations of H_2O_2 (*135*, and references therein). As such, it can be a significant source of H_2O_2 in surface waters. This subject has recently been reviewed for marine environments (*135*). Except for somewhat lower rain H_2O_2 concentrations over areas affected by urban pollution, the same discussion is valid for freshwater environments. The data available are sparse, and more information is needed to determine the relative importance of this source of H_2O_2.

Dry Deposition. Dry deposition may be a source of H_2O_2 in surface waters. To determine the extent of this source, it is necessary to consider three species: $HO_2\cdot$, H_2O_2, and O_3. Thompson and Zafiriou (*136*) calculated the relative magnitude of dry deposition of these three species to the ocean surface. These data are relevant to freshwater surfaces as well.

The air–sea flux of $HO_2\cdot$ was calculated to be 1×10^8 cm^{-2} s^{-1}. In aqueous solution the predominant form of $HO_2\cdot$ would be $O_2^{-}\cdot$, pK_a = 4.5. To approximate the in situ concentration of $O_2^{-}\cdot$, it is assumed that it serves as the source of all of the observed H_2O_2. Thompson and Zafiriou (*136*) assumed an in situ production of 10^{12}–10^{13} cm^{-2} s^{-1}, or approximately 10^4 times the estimated atmospheric input. The conclusion is that the dry deposition of $HO_2\cdot$ is not significant to the ocean surface H_2O_2 concentration.

The air–sea flux of H_2O_2 was estimated to be 1.4×10^{10} mol cm^{-2} s^{-1} (*136*). With a midday H_2O_2 production rate of 2×10^{-8} mol L^{-1} h^{-1} in the top 2 m of a water body, then the dry deposition rate would lead to an increase of 2% in the H_2O_2 in the surface layer. These calculations and the data presented on surface water H_2O_2 concentrations show that air–sea flux is an insignificant source of H_2O_2 in most surface waters.

The dry deposition of O_3 was estimated to be 2.0×10^{10} cm^{-2} s^{-1} (*136*). Ozone is known to disproportionate in aqueous solutions, and the rate of this reaction is base catalyzed. If all of the O_3 were to produce H_2O_2, it would be approximately the same as H_2O_2 dry deposition. However, ozone may react with numerous species in surface water without leading to the formation of H_2O_2. Thus, dry deposition of O_3 is not a significant source of H_2O_2 in the surface waters.

In summary, none of the three atmospheric reactants, $HO_2\cdot$, H_2O_2, and O_3, contribute significantly through dry deposition to surface water H_2O_2 concentrations in remote areas. No data are available to assess the relative contribution of dry deposition in areas adjacent to urbanization.

References

1. *Photochemistry of Environmental Aquatic Systems*; Zika, R. G.; Cooper, W. J., Eds.; ACS Symposium Series 327; American Chemical Society: Washington, DC, 1987; 288 p.
2. Zafiriou, O. C. *Mar. Chem.* **1977**, *5*, 497–522.
3. Sundstrom, G.; Ruzo, L. O. In *Aquatic Pollutants: Transformation and Biological Effects*; Hutzinger, O.; Van Lelyveld, I. H.; Zoeteman, B. C. J., Eds.; Pergamon Press: New York, 1978; pp 205–222.
4. Zepp, R. G.; Baughman, G. L. In *Aquatic Pollutants: Transformation and Biological Effects*; Hutzinger, O.; Van Lelyveld, I. H.; Zoeteman, B. C. J., Eds.; Pergamon Press: New York, 1978; pp 237–263.
5. Mill, T. In *The Handbook of Environmental Chemistry: Reactions and Proc-*

esses, Vol. 2, Part A; Hutzinger, O., Ed.; Springer–Verlag: New York, 1980; pp 77–105.
6. Zepp, R. G. In *Dynamics, Exposure and Hazard Assessment of Toxic Chemicals*; Haque, R., Ed.; Ann Arbor Science: Ann Arbor, 1980; pp 69–110.
7. Zepp, R. G. In *Handbook of Environmental Chemistry: Reactions and Processes*, Vol. 2, Part B; Hutzinger, O., Ed.; Springer–Verlag: New York, 1980; pp 19–41.
8. Zika, R. G. In *Marine Organic Chemistry: Evolution, Composition, Interactions and Chemistry of Organic Matter in Seawater*; Duursma, E. K.; Dawson, R., Eds.; Elsevier Science: Amsterdam, 1981; pp 77–105.
9. Zepp, R. G. In *The Role of Solar Ultraviolet Radiation in Marine Ecosystems*; Calkins, J., Ed.; Plenum Press: New York, 1982; pp 293–307.
10. Zafiriou, O. C. *Chem. Oceanogr.* **1983,** *8*, 339–379.
11. Zafiriou, O. C.; Joussot-Dubien, J.; Zepp, R. G.; Zika, R. G. *Environ. Sci. Technol.* **1984,** *18*, 358A–371A.
12. Schnitzer, M.; Khan, S. U. *Humic Substances in the Environment*; Marcel Dekker: New York, 1972; 327 p.
13. Gjessing, E. T. *Physical and Chemical Characteristics of Aquatic Humus*; Ann Arbor Science: Ann Arbor, 1976; 120 p.
14. Stevenson, F. J. *Humus Chemistry, Genesis, Composition Reactions*; John Wiley and Sons: New York, 1982; 443 p.
15. Christman, R. F.; Gjessing, E. J. *Aquatic and Terrestrial Humic Materials*; Ann Arbor Science: Ann Arbor, 1983; 538 p.
16. Thurman, E. M. *Organic Geochemistry of Natural Waters*; Martinus Nijhoff/Junk: Boston, 1985; 497 p.
17. Aiken, G. R.; McKnight, D. M.; Wershaw, R. L.; MacCarthy, P. *Humic Substances in Soil, Sediment and Water: Geochemistry, Isolation and Characterization*; John Wiley: New York, 1985; 692 p.
18. Choudhry, G. G. *Toxicol. Environ. Chem.* **1981,** *4*, 209–260.
19. Cooper, W. J.; Young, J. C. In *Water Analysis: Organic Species*, Vol. III; Minear R. A.; Keith, L. H., Eds.; Academic: New York, 1984; pp 41–82.
20. Zepp, R. G.; Cline, D. M. *Environ. Sci. Technol.* **1977,** *11*, 359–366.
21. Finlayson-Pitts, B. J.; Pitts, J. N., Jr. *Atmospheric Chemistry*; John Wiley and Sons: New York, 1986, 1098 p.
22. Bener, P. *Approximate Values of Intensity of Natural Ultraviolet Radiation for Different Amounts of Atmospheric Ozone*; U.S. Army Report DAJA 37–68–C–1017; Physikalisch-Meteorologisches Observatorium Davos: Davos Platz, Switzerland, 1972; 59 p.
23. Oliver, B. G.; Cosgrove, E. G.; Carey, J. H. *Environ. Sci. Technol.* **1979,** 9, 1075–1077.
24. Plane, J. M. C.; Zika, R. G.; Zepp, R. G.; Burns, L. A. In *Photochemistry of Environmental Aquatic Systems*; Zika, R. G.; Cooper, W. J., Eds.; ACS Symposium Series 327; American Chemical Society: Washington, DC, 1987; pp 250–267.
25. Choudhry, G. G. *Toxicol. Environ. Chem.* **1981,** *4*, 261–295.
26. Zepp, R. G. Presented at the Dahlem Workshop on Humic Substances and Their Role in the Environment, Dahlem Konferenzer, Wallatstrasse 19, 1000 West Berlin 33, 1987.
27. Zika, R. G. *Rev. Geophys. Space Phys.* **1987,** 25, 1390–1394.
28. Slawinska, D.; Slawinski, J.; Sarna, T. *J. Soil Sci.* **1975,** *26*, 93–99.
29. Ziechman, W. Z. *Pflanzeneraehr. Bodenkd.* **1977,** *140*, 133–150.
30. Rabek, J. F.; Ranby, B. *Polym. Eng. Sci.* **1975,** *15*, 40–43.

31. Fischer, A. M.; Kliger, D. S.; Winterle, J. S.; Mill, T. *Chemosphere* **1985,** *14,* 1299–1306.
32. Power, J. F.; Sharma, D. K.; Langford, C. H.; Bonneau, R.; Joussot-Dubien, J. *Photochem. Photobiol.* **1986,** *44,* 11–13.
33. Fischer, A. M.; Winterle, J. M.; Mill, T. In *Photochemistry of Environmental Aquatic Systems*; Zika, R. G.; Cooper, W. J. Eds.; ACS Symposium Series 327; American Chemical Society: Washington, DC, 1987; pp 141–156.
34. Power, J. F.; Sharma, D. K.; Langford, C. H.; Bonneau, R.; Joussot-Dubien, J. In *Photochemistry of Environmental Aquatic Systems*; Zika, R. G.; Cooper, W. J., Eds.; ACS Symposium Series 327; American Chemical Society: Washington, DC, 1987; pp 157–173.
35. Zepp, R. G.; Wolfe, N. L.; Baugman, G. L.; Hollis, R. C. *Nature (London)* **1977,** *267,* 421–423.
36. Wolff, C. J. M.; Halmans, M. T. H.; van der Heijde, H. B. *Chemosphere* **1981,** *10,* 59–62.
37. Zepp, R. G.; Baughman, G. L.; Schlotzhauer, P. F. *Chemosphere* **1981,** *10,* 109–117.
38. Zepp, R. G.; Baughman, G. L.; Schlotzhauer, P. F. *Chemosphere* **1981,** *10,* 119–126.
39. Baxter, R. M.; Carey, J. H. *Freshwater Biol.* **1982,** *12,* 285–292.
40. Haag. W. R.; Hoigné, J.; Gassmann, E.; Braun, A. *Chemosphere* **1984,** *13,* 631–640.
41. Haag, W. R.; Hoigné, J.; Gassmann, E.; Braun, A. *Chemosphere* **1984,** *13,* 641–650.
42. Haag, W. R.; Hoigné, J. *Environ. Sci. Technol.* **1986,** *20,* 341–348.
43. Haag, W. R.; Hoigné, J. *Water Chlorination: Chem. Environ. Impact Health Eff. Proc. Conf., 5th, 1984* **1985,** 1011–1020.
44. Hoigné, J.; Faust, B. C.; Haag, W. R.; Scully, R. E.; Zepp, R. G. Chapter 23 in this volume.
45. Mill, T.; Hendry, D. G.; Richardson, H. *Science (Washington, DC)* **1980,** *207,* 886–887.
46. Haag, W. R.; Hoigné, J. *Chemosphere* **1985,** *14,* 1659–1671.
47. Zafiriou, O. C. *J. Geophys. Res.* **1974,** *79,* 4491–4497.
48. Zafiriou, O. C.; True, M. B. *Mar. Chem.* **1979,** *8,* 9–32.
49. Russi, H.; Kotzias, D.; Korte, F. *Chemosphere* **1982,** *11,* 1041–1048.
50. Kotzias, D.; Parlar, H.; Korte, F. *Naturwissenschaften* **1982,** *69,* 444–445.
51. Hoigné, J.; Bader, H.; Nowell, L. H. Presented at the 193rd National Meeting of the American Chemical Society, Denver, CO; American Chemical Society: Washington, DC, 1987; paper ENVR 78.
52. Baxter, R. M.; Carey, J. H. *Nature (London)* **1983,** *306,* 575–576.
53. Draper, W. M.; Crosby, D. G. *J. Agric. Food Chem.* **1983,** *31,* 734–737.
54. Draper, W. M.; Crosby, D. G. *Arch. Environ. Contam. Toxicol.* **1983,** *12,* 121–126.
55. Cooper, W. J.; Zika, R. G. *Science (Washington, DC)* **1983,** *220,* 711–712.
56. Petasne, R. G.; Zika, R. G. *Nature (London)* **1987,** *325,* 516–518.
57. Larson, R. A.; Smykowski, K.; Hunt, L. A. *Chemosphere* **1981,** *10,* 1335–1338.
58. Zepp, R. G.; Braun, A. M.; Hoigné, J.; Leenheer, J. A. *Environ. Sci. Technol.* **1986,** *21,* 485–490.
59. Sturzenegger, V. Ph.D. Thesis, Swiss Federal Institute of Technology, EAWAG Dubendorf, Switzerland, in preparation.
60. Anbar, M.; Bambenek, M.; Ross, A. B. *Natl. Stand. Ref. Data Series (U.S. Natl. Bur. Stand.* **1973,** *43 (NSRDS–NBS43),* 54 pp.

61. Ross, A. B. *Natl. Stand. Ref. Data Series* (*U.S. Natl. Bur. Stand.* **1975,** *43 Suppl.* (*NSRDS–NBS43*), 43 pp.
62. Zepp, R. G.; Schlotzhauer, P. F.; Simmons, M. S.; Miller, G. C.; Baughman, G. L.; Wolfe, N. L. *Fresenius Z. Anal. Chem.* **1984,** *319*, 119–125.
63. Sluyterman, L. A. *Biochim. Biophys. Acta* **1962,** *60*, 557–561.
64. Kearns, D. R. *Chem. Rev.* **1971,** *71*, 395–427.
65. Seely, G. R. *Photochem. Photobiol.* **1977,** *26*, 115–123.
66. Matsura, T. *Tetrahedron* **1977,** *33*, 2869–2905.
67. Thomas, M. J.; Foote, C. S. *Photochem. Photobiol.* **1978,** *27*, 683–693.
68. Scully, F. E., Jr.; Hoigné, J. *Chemosphere* **1987,** *16*, 681–694.
69. Frimmel, F.; Bauer, H.; Putzien, J.; Murasecco, P.; Braun, A. M. *Environ. Sci. Technol.* **1987,** *21*, 541–545.
70. Rodgers, M. A. J.; Snowden, P. T. *J. Am. Chem. Soc.* **1982,** *104*, 5541–5543.
71. Fuchs, F.; Raue, B. *Vom Wasser* **1981,** *57*, 95–106.
72. Wilkinson, F.; Brummer, J. G. *J. Phys. Chem. Ref. Data* **1981,** *10*, 809–999.
73. Draper, W. M.; Crosby, D. G. *J. Agric. Food Chem.* **1981,** *29*, 699–702.
74. Ross, A. B.; Neta, P. *Natl. Stand. Ref. Data Series* (*U.S. Natl. Bur. Stand.* **1982,** *70* (*NSRDS–NBS70*), 96 pp.
75. Howard, J. A.; Scaiano, J. C. *Radical Reaction Rates in Liquids*. Subvolume d. Oxyl-, Peroxyl-, and Related Radicals. Landolt-Borstein: Numerical Data and Function Relationships in Science and Technology; Springer–Verlag: Berlin, 1984; Vol. 13, 431 pp.
76. Mill, T.; Richardson, H.; Hendry, D. G. In *Aquatic Pollutants: Transformation and Biological Effects*; Hutzinger, O.; Van Lelyveld, I. H.; Zoeteman, B. C. J., Eds.; Pergamon: New York, 1978; pp 223–236.
77. Volman, D. H. *Adv. Photochem.* **1963,** *1*, 43–82.
78. Walling, C. *Acc. Chem. Res.* **1975,** *8*, 125–131.
79. McMahon, J. W. *Limnol. Oceanogr.* **1967,** *12*, 437–442.
80. McMahon, J. W. *Limnol. Oceanogr.* **1969,** *14*, 357–367.
81. Collienne, R. H. *Limnol. Oceanogr.* **1983,** *28*, 83–100.
82. Zepp, R. G.; Hoigné, J.; Bader, H. *Environ. Sci. Technol.* **1986,** *21*, 443–450.
83. Moffett, J. W.; Zika, R. G. *Environ. Sci. Technol.* **1987,** *21*, 804–810.
84. Dorfman, L. M.; Adams, G.E. *Natl. Stand. Ref. Data Series* (*U.S. Natl. Bur. Stand.* **1973,** *46* (*NSRDS–NBS46*), 59 pp.
85. Farhatazis; Ross, A. B. *Natl. Stand. Ref. Data Series* (*U.S. Natl. Bur. Stand.* **1977,** *59* (*NSRDS–NBS59*), 113 pp.
86. Klug, D.; Rabani, J.; Fridovitch, I. J. *Biol. Chem.* **1972,** *247*, 4839–4842.
87. Fridovitch, I. *Acc. Chem. Res.* **1972,** *5*, 321–326.
88. Fridovitch, I. A. *Rev. Biochem.* **1975,** *44*, 147–159.
89. McCormick, J. P.; Fischer, J. R.; Pachlatko, J. P.; Eisenstark, A. *Science* (*Washington, DC*) **1976,** *191*, 468–469.
90. McCormick, J. P.; Thomason, T. *J. Am. Chem. Soc.* **1978,** *100*, 312–313.
91. Gordon, S.; Hart, E. J.; Matheson, M. S.; Rabani, J.; Thomas, J. K. *Discuss. Faraday Soc.* **1963,** *36*, 193–205.
92. Keene, J. P. *Radiat. Res.* **1964,** *22*, 1–13.
93. Hentz, R. R.; Farhataziz; Hansen, E. M. *J. Chem. Phys.* **1972,** *56*, 4485–4488.
94. Kacher, M. L.; Foote, C. S. *Photochem. Photobiol.* **1979,** *29*, 765–769.
95. Saito, I.; Matsura, T.; Inoue, K. *J. Am. Chem. Soc.* **1981,** *103*, 188–190.
96. Massey, V.; Strickland, S.; Mayhew, S. G.; Howell, L. G.; Engel, P. C.; Mathews, R. G.; Schuman, M.; Sullivan, P. A. *Biochem. Biophys. Res. Commun.* **1969,** *36*, 891–897.
97. Kayser, R. H.; Young, R. H. *Photochem. Photobiol.* **1976,** *24*, 395–401.
98. Kayser, R. H.; Young, R. H. *Photochem. Photobiol.* **1976,** *24*, 403–411.

99. Moffett, J. W.; Zika, R. G. *Mar. Chem.* **1983,** *13,* 239–251.
100. Miles, C. J.; Brezonik, P. L. *Environ. Sci. Technol.* **1981,** *15,* 1089–1095.
101. Sunda, W. G.; Huntsman, S. A.; Harvey, G. R. *Nature (London)* **1983,** *301,* 234–236.
102. van der Weijden, C. H.; Reith, M. *Mar. Chem.* **1982,** *11,* 565–572.
103. Bielski, B. H. J.; Cabelli, D. E.; Arudi, R. L.; Ross, A. B. *J. Phys. Chem. Ref. Data* **1985,** *14,* 1041–1100.
104. *Superoxide and Superoxide Dismutase in Chemistry, Biology and Medicine*; Rotilio, G., Ed.; Elsevier Science: New York, 1986; 688 p.
105. Sinel'nikov, V. E. *Gidrobiol. Zh.* **1971,** *7,* 115–119 (*Chem. Abstr.* **1971,** *75,* 25016a).
106. Sinel'nikov, V. E. *Tr. Inst. Biol. Vnutr. Vod Akad. Nauk SSR* **1971,** *20,* 159–171 (*Chem. Abstr.* **1971,** *75,* 121176u).
107. Sinel'nikov, V. E.; Liberman, A. Sh. *Tr. Inst. Biol. Vnutr. Vod Akad. Nauk SSR* **1974,** *29,* 27–40 (*Chem. Abstr.* **1976,** *85,* 25170y).
108. Sinel'nikov, V. E.; Demina, V. I. *Gidrokhim. Mater.* **1974,** *60,* 30–40 (*Chem. Abstr.* **1976,** *83,* 151980j).
109. Klockow, D.; Jacob, P. In *Chemistry of Multiphase Atmospheric Systems*; Jaeschke, W., Ed.; Springer–Verlag: New York, 1986; pp 117–130.
110. Cooper, W. J.; Zika, R. G.; Petasne, R. G.; Plane, J. M. C. *Environ. Sci Technol.* **1988,** *22*(10), 1156–1160.
111. Czapski, G.; Bielski, B. H. J. *J. Phys. Chem.* **1963,** *67,* 2180–2184.
112. Czapski, G.; Dorfman, L. M. *J. Phys. Chem.* **1964,** *68,* 1169–1177.
113. Rabani, J.; Mulac, W. A.; Matheson, M. S. *J. Phys. Chem.* **1965,** *69,* 53–70.
114. Sehester, K.; Rasmussen, O. L.; Fricke, H. *J. Phys. Chem.* **1968,** *72,* 626–631.
115. Rabani, J.; Nielsen, S. O. *J. Phys. Chem.* **1969,** *73,* 3736–3744.
116. Behar, D.; Czapski, G.; Rabani, J.; Dorfman, L. M.; Schwarz, H. A. *J. Phys. Chem.* **1970,** *74,* 3209–3213.
117. Bielski, B. H. J. *Photochem. Photobiol.* **1978,** *28,* 645–649.
118. Millero, F. M. *Geochim. Cosmochim. Acta* **1987,** *51,* 351–353.
119. Mopper, K.; Zika, R. G. In *Photochemistry of Environmental Aquatic Systems*; Zika, R. G.; Cooper, W. J., Eds.; ACS Symposium Series 327; American Chemical Society: Washington, DC, 1987; p 174–190.
120. Mossoba, M. M.; Makino, K.; Riesz, P. *J. Phys. Chem.* **1982,** *86,* 3478–3483.
121. Chance, B. J. *J. Biol. Chem.* **1949,** *179,* 1311.
122. Zepp, R. G.; Skurlatov, Y,I.; Pierce, J. T. In *Photochemistry of Environmental Aquatic Systems*; Zika, R. G.; Cooper, W. J., Eds.; ACS Symposium Series 327; American Chemical Society: Washington, DC, 1987; pp 213–224.
123. Petasne, R. G.; Zika, R. G. *Limnol. Oceangr.* submitted.
124. Fischer, A. M. Ph.D. Thesis, University of California—Santa Cruz, 1985.
125. Choudhry, G. G.; van den Broecke, J. A.; Webster, G. R. B.; Hutzinger, O. *Chemosphere* **1987,** *16,* 495–504.
126. Choudhry, G. G. *Humic Substances: Structural, Photophysical, Photochemical, and Free Radical Aspects and Interactions with Environmental Chemicals*; Gordon and Breach Science: New York, 1984; 185 pp.
127. Wilson, S. A.; Weber, J. H. *Anal. Lett.* **1977,** *10,* 75–84.
128. Zepp, R. G.; Schlotzhauer, P. F.; Sink, R. M. *Environ. Sci. Technol.* **1985,** *19,* 74–81.
129. Herman, M.; Kotzias, D.; Korte, F. *Chemosphere* **1987,** *16,* 523–585.
130. Kress, B. M. *Chem. Erde* **1978,** *37,* 80–100.
131. Senesi, N.; Testini, C.; Miano, T. Presented at the 2nd International Conference of the International Humic Substances Society; Hayes, M. H. B.; Swift, R. S., Eds.; University of Birmingham, England, 1984; pp 234–238.

132. Senesi, N.; Chen, Y.; Schnitzer, M. *Soil Biol. Biochem.* **1977**, *9*, 397–403.
133. Senesi, N. *Z. Pflanzenernaehr. Bodenkd.* **1981**, *144*, 580–586.
134. Senesi, N.; Steelink, C. In *Proceedings of the International Humic Substances Society Conference*; Birmingham, July 23–28, 1984.
135. Cooper, W. J.; Saltzman, E. S.; Zika, R. G. *J. Geophys. Res.* **1987**, *92*, 2970–2980.
136. Thompson, A. M.; Zafiriou, O. C. *J. Geophys. Res.* **1983**, *88*, 6696–6708.

RECEIVED for review July 24, 1987. ACCEPTED for publication January 26, 1988.

23

Aquatic Humic Substances as Sources and Sinks of Photochemically Produced Transient Reactants

Jürg Hoigné, Bruce C. Faust[1], Werner R. Haag[2], Frank E. Scully, Jr.[3], and Richard G. Zepp[4]

Swiss Federal Institute of Water Resources and Water Pollution Control (EAWAG), 8600 Dübendorf, Switzerland

In sunlit surface waters aquatic humic substances and nitrate act as sensitizers or precursors for the production of photoreactants such as singlet oxygen, humic-derived peroxy radicals, hydrogen peroxide, solvated electrons, and ˙OH radicals. Lifetimes of the various reactants are controlled by their reactions with aquatic humic substances (˙OH radicals), by solvent quenching (singlet oxygen), by reactions with molecular oxygen (solvated electron), or by other processes (peroxy radicals). The steady-state concentration of each transient formed during solar irradiation was determined from the apparent first-order disappearance rate of added organic probe compounds. The probe compounds used had selective reactivities with the individual transient species of interest. Effects of the photoreactants on the elimination of micropollutants and on chemical transformations of DOM are discussed.

[1]Current address: School of Forestry and Environmental Sciences, Duke University, Durham, NC 27706
[2]Current address: SRI International, Menlo Park, CA 94025
[3]Current address: Department of Chemical Sciences, Old Dominion University, Norfolk, VA 23508–8503
[4]Current address: Environmental Research Laboratory, U.S. Environmental Protection Agency, Athens, GA 30613

0065–2393/89/0219–0363$07.00/0

DISSOLVED ORGANIC MATERIAL IN SURFACE WATERS HAS A ROLE in producing or consuming different types of photoreactants. The conclusions stated in this chapter are based on data from a series of our recent publications. Extensive literature reviews are given in these publications and are not repeated here. This chapter will focus on the reactions for which humic materials act as sources or sinks. We make no attempt to include all other possible photochemical processes. For example, no discussion of heterogeneous processes is included, although there is evidence that they are important (e.g., in the redox cycling of metals). Moreover, photochemical processes mediated by superoxide and hydrogen peroxide are discussed in Chapter 22 by Cooper et al.

During a cloudless summer noon hour, surface waters receive approximately 1 kW/m^2 of sunlight, or about 20 $einsteins/m^2$ (20 mol of $photons/m^2$) (Figure 1). Within 1 year about 1300 times this dose is accumulated (*1*). A large portion of these photons is absorbed by dissolved organic material (DOM) present in natural water. In addition, a rather small fraction of short-wavelength light is absorbed by nitrate (Figure 2).

From a chemist's viewpoint, the resulting rate of interactions between photons and absorbers is very high. Assuming that most of the photons are absorbed in a well-mixed 1-m water column, we estimate that about 20 mmol/(L·h) of interactions occur between photons and absorbing sub-

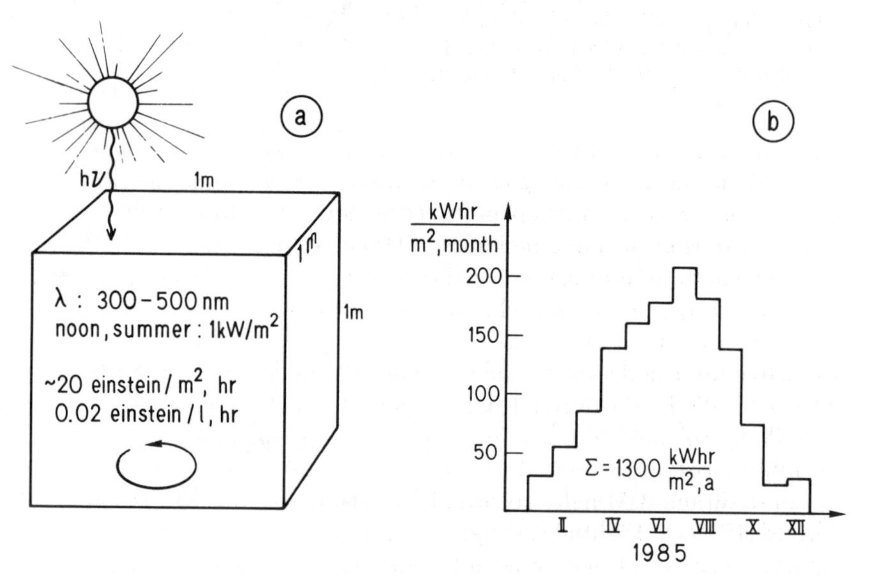

Figure 1. Solar radiation. (a) Mean dose intensity in a mixed 1-m water column in which all light is absorbed. (b) Monthly solar flux ($280 < \lambda < 2800$ nm) in Dübendorf (47.5° N), 1985.

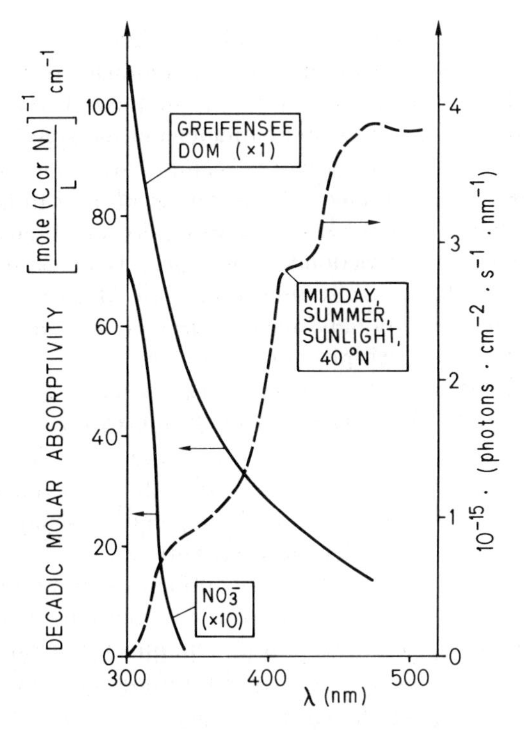

Figure 2. Decadic molar absorptivities of Greifensee DOM and NO_3^- anion. The values for NO_3^- have been multiplied by a factor of 10. Typical DOC and $[NO_3^-]$ values for Greifensee are 4 mg/L (300 μM of carbon units) and 100 μM, respectively. Solar irradiation data are for sea level, after ref. 1.

strates (Figure 1a). Assuming an average chromophore unit weight of 120 for DOM in water containing 4 mg of dissolved organic carbon (DOC) per liter, we arrive at a chromophore concentration of 0.033 mM. Thus, each chromophore is excited at a high rate of 600 times per hour.

Some of these interactions lead to direct photochemical transformations of DOM and aqueous micropollutants to secondary products. But in addition, aquatic humic materials act as sensitizers or precursors for the production of reactive intermediates (so-called "photoreactants") such as singlet oxygen ($^1O_2$) (*1*, *3*, *4*), DOM-derived peroxy radicals ($ROO^{\cdot}$) (*5–7*), hydrogen peroxide (*8*, *9*), solvated electron (e_{aq}^-) (*10–12*), superoxide anion (O_2^-) (*13*, *14*) and humic structures excited to triplet states (*15*).

In addition, UV light absorbed by nitrate and nitrite produces $OH^{\cdot}$ radicals (*16*, *17*). Light-absorbing redox-active metal species may also be important sources of photoreactants, such as metals in lower valence states (*18*, *19*).

Of these photoreactants only H_2O_2, because of its relative inertness, accumulates and decomposes during extended illumination periods (hours).

All other species are highly reactive and short-lived; they are present only at very low concentrations and only during illumination. The role of DOM as a source and a sink of photoreactants is of interest because these photoreactants can chemically transform pollutants and, in many cases, the DOM itself.

In principle the role of DOM as source and sink for photoreactants can be discussed without detailed knowledge of particular kinetic models (*see* **Conclusions**). However, a reaction kinetic approach is required for designing experiments that yield generalizable results. The main ideas of the model are summarized here (for details, *see* ref. 20).

The steady-state concentration of relatively short-lived photoreactants ($[X]_{ss}$) is given by the rate with which these reactants are produced (r_X), relative to the pseudo-first-order rate constant with which they become consumed k_X'):

$$[X]_{ss} = r_X \times \frac{1}{k_{X'}} \qquad (1)$$

The formation rate (r_X) is proportional to the rate of light absorption by the photochemical source substance (i.e., proportional to $k_A[A]$) and to the quantum efficiency (Φ). As shown in equation 2, the rate of X consumption or quenching can be controlled by solvent quenching (k_q), reaction with DOM acting as a scavenger (S) for the photoreactant (X) ($k_{X,S}[DOM]$), reaction with oxygen ($k_{X,O_2}[O_2]$), reaction with other scavengers, and possibly by bimolecular reactions with itself ($k_{X,X}[X]$).

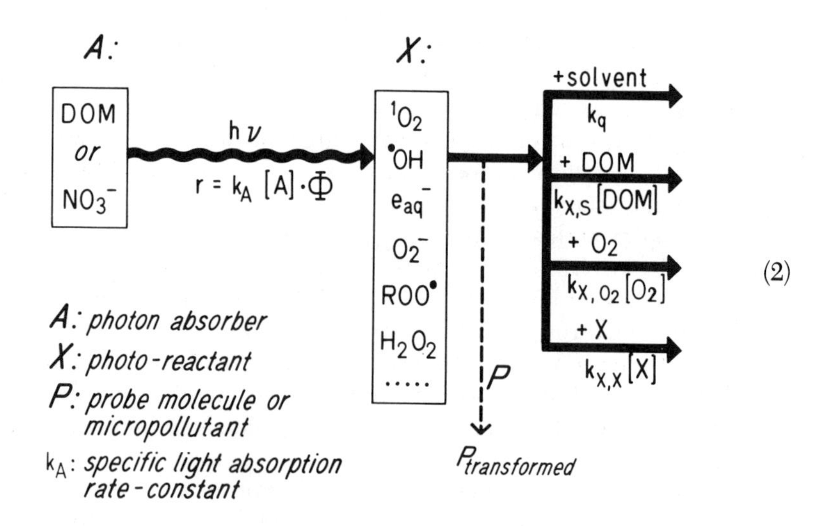

(2)

Rate constants of reactions controlling the fate of transient reactants cover a wide range. Therefore, in most cases only one of these reactions dominantly controls the lifetime of a specific photoreactant in a given system.

To quantify production rates and steady-state concentrations of the main photoreactants ($^1O_2$, $^{\cdot}OH$, $ROO^{\cdot}$, and e_{aq}^{-}), rates of their selective reactions with added probe molecules (P) were determined (equation 2). Highly selective probe molecules were chosen to discriminate between different types of photoreactants. Whenever possible, probe compounds with structures similar to those of the micropollutants of interest were applied.

To probe for $^{\cdot}OH$, $^1O_2$, and $ROO^{\cdot}$, experiments were performed in a way that produced a simple second-order rate law. The rate of transformation of P was first order in concentration of both P and X.

$$-\frac{d[P]}{dt} = k_{P,X}[X]_{ss}\,[P] \tag{3}$$

where $k_{P,X}$ is the second-order rate constant for the reaction of X with P. For a closed parcel of water, and if the concentration of P is low enough not to change $[X]_{ss}$ significantly, equation 3 integrates to

$$-\ln\frac{[P]}{[P]_0} = k_{P,X}\,[X]_{ss} \times t \tag{4}$$

Therefore, the logarithm of the relative residual concentration of P declines linearly with time (t) with a slope of $k_{P,X}[X]_{ss}$ (apparent first-order kinetics). Then $[X]_{ss}$ can be calculated directly from the experimental elimination rate constants in cases where $k_{P,X}$ of a selected probe molecule is known. All kinetic experiments used in this overview for deducing rate constants yielded very good first-order plots.

Transient production rate decreases with depth (z) in a surface water because of light screening by DOM (Figure 2). In waters that contain particles, light-scattering terms and absorption by particles must also be accounted for. Assuming complete mixing, measured surface rates ($r^{(0)}$) for production of photoreactants are normally converted to depth-averaged rates ($r^{(z)}$) by multiplying by the light-screening factor $S_{\lambda}^{(z)}$ (*15*).

$$r^{(z)} = r^{(0)} \times S_{\lambda}^{(z)} \tag{5}$$

where

$$S_{\lambda}^{(z)} = \frac{1 - 10^{-\alpha_{\lambda}z}}{2.3 \times \alpha_{\lambda} \times z} \tag{6}$$

Here α_λ is the decadic absorption coefficient of the water at wavelength λ and z is the average water depth. In principle, the choice of wavelength for the corrections must account for the overlap (product) between the action spectrum (quantum efficiency times molar absorptivity) of the reaction considered and the spectrum of the light. Application of the screening factor to waters of medium depth may be highly complicated. However, for large depths (such that $\alpha_\lambda \times z > 1$), the light-screening factor approximates

$$S_\lambda^{(z)} = \frac{1}{2.3 \times \alpha_\lambda \times z} \tag{7}$$

and the depth-averaged production rate becomes

$$r^{(z)} = \frac{r^{(0)}}{2.3 \times \alpha_\lambda \times z} \tag{8}$$

Given the assumption of vertical mixing, equation 8 corresponds essentially to a dilution of the photochemical effect with increasing depth. For example, most UV light is absorbed within the top meter of even slightly eutrophic lakes (*1*). When most light is absorbed within the considered depth, z, $r^{(z)}$ is independent of the concentration of DOM (DOC) if both α and $r^{(0)}$ increase proportionally to DOC. However, it decreases proportionally with the depth of the water body, because a higher absorbance at the surface is directly compensated by lower light penetration. If $r^{(0)}$ is independent of DOC (e.g., is proportional to NO_3^-), then $r^{(z)}$ decreases with DOC and z.

Finally, diurnal and seasonal variations in light intensity (Figure 1b) must be taken into account in any generalization of results.

Characteristics of Various Photooxidants

Singlet Oxygen. It is possible to measure the steady-state concentration of singlet oxygen by following the oxygenation of selective probe molecules such as dimethylfuran and furfuryl alcohol (*1*, *3*). Furfuryl alcohol is less volatile and leads to products that are highly specific for $^1O_2$ reactions. Dimethylfuran, however, needs lower exposure times.

From the literature and our own studies (*1*) we conclude that the formation of $^1O_2$ from ground-state oxygen ($^3O_2$) is sensitized by DOM, and that in natural waters its destruction rate is generally controlled by solvent (water) quenching. The sequence of reactions is given in equation 9.

SINGLET OXYGEN FORMATION :

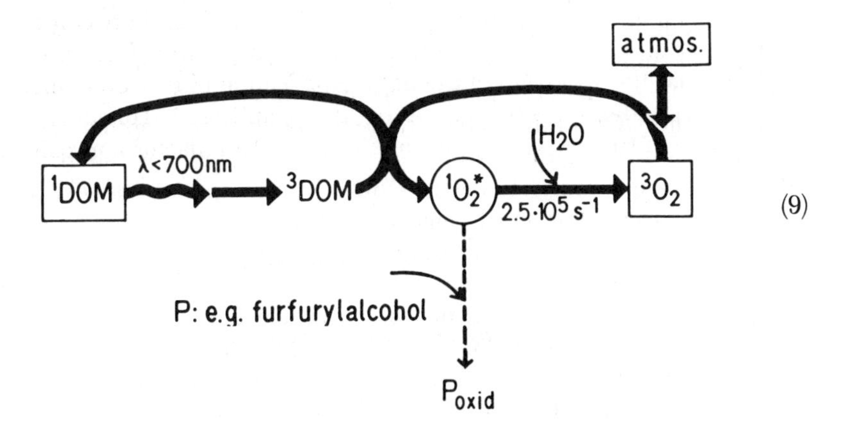

(9)

At DOM concentrations typical for surface waters (DOC < 20 mg/L), quenching of singlet oxygen by DOM can be neglected.

Different types of aquatic DOM exhibit different quantum efficiencies for $^1O_2$ production. Selected examples are presented in Table I. In summarizing phenomenological studies using a variety of different humic substances, we conclude that DOM with higher specific light absorption exhibits somewhat lower quantum efficiencies (*1*). No significant relationship between quantum efficiency and molecular weight fraction was found (*1*).

As an example, the steady-state concentration of singlet oxygen at the surface of the somewhat eutrophic pre-alpine Lake Greifensee in Switzerland (DOC ~ 4 mg/L) during June, with summer midday sunshine (1 kW/m^2) is 8×10^{-14} M. This number results from the observed furfuryl alcohol elimination rate of 3%/h.

A light-screening factor can be estimated on the basis of action and absorption spectra typical for this lake water. Accounting for this, equations

Table I. Half-Lives of Added Furfuryl Alcohol and Corresponding Steady-State Concentrations of $^1O_2$

Water Source	*DOC (mg/L)*	$t_{1/2}$ *FFA*[a] *(h)*	$[^1O_2]_{ss}^{0}$ [a,b] $(M \times 10^{14})$	$[^1O_2]_{ss}^{1m}$ [a,c] $(M \times 10^{14})$
Greifensee	3.5	20	8	4.6
Etang de la Gruyère	13	6	28	1.5
Rhine, Basel	3.2	27	5.6	3.6
Secondary effluent[d]	15	14	11	2.2

[a]Summer midday sunlight, 1 kW/m^2.
[b]Concentration at the surface.
[c]Concentration averaged over 1-m depth.
[d]Communal wastewater.

1 and 8 indicate that a depth-averaged steady-state concentration for an ideally mixed top meter of the lake ($[^1O_2]_{ss}^{1m}$), within which most of the photoactive light is absorbed, is 5×10^{-14} M. At greater depth, the average $[^1O_2]_{ss}^{z}$ will simply be 5×10^{-14} M divided by the depth in meters. Surface values of $[^1O_2]_{ss}^{0}$ increase proportionally with the rate of light absorption (i.e., with the DOC of the water) (Table II). For a water depth within which all light is absorbed, the area-based production of $^1O_2$ is independent of DOC (*see* last column in Table II). However, a comparison of very different types of surface waters must consider that quantum efficiences and action spectra for producing $^1O_2$ vary somewhat with the type of DOM (*1*).

The occurrence of singlet oxygen in sunlit waters can be important for the elimination of cyclic 1,3 dienes, polynuclear aromatic hydrocarbons, organic sulfides, and phenolic compounds when the latter are significantly dissociated into the reactive phenolate anions (e.g., chlorophenols) (Figure 3). For phenols with high phenolic pK_a values, correspondingly low reactivities are found. (For comprehensive literature, *see* ref. 21).

OH Radical. Because OH radicals react with most organic substrates at nearly diffusion-controlled rates (H atom abstraction or ·OH addition reactions), most organic substances can be applied as ·OH probe molecules. Butyl chloride, in addition to other compounds, was often used as our probe substance of choice because of its easy analysis and its inertness against direct photolyis and other photoreactants (except e_{aq}^-)(*20*).

In fresh surface waters a large part of ·OH is produced from slow photolysis of nitrate (*17, 20*). The reaction is shown in equation 10.

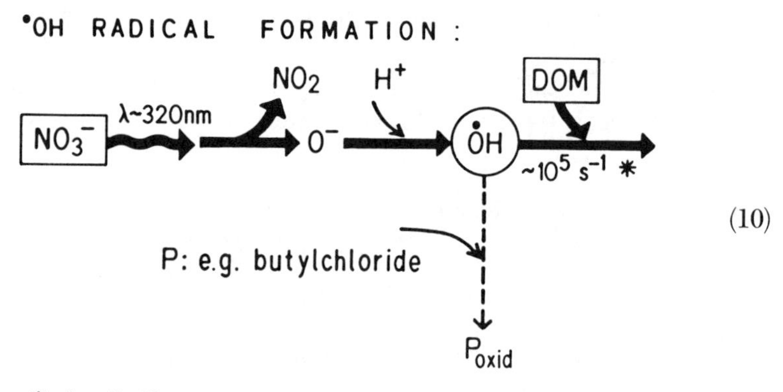

(10)

The absorption of the relevant solar UV light by nitrate (at about 310 nm) is much smaller than that by DOM (Figure 2). As has been shown in preceding studies in which decomposed ozone has been used as an OH

Table II. Photochemical Transient Reactants Produced in Lake Greifensee and Functional Dependencies of Their Steady-State Concentrations

Reactant	*Probe Molecule,* $k_{P,i}$ $(M^{-1}s^{-1})$	$k_{P,i}[X_i]_{ss}^0, [X]_{ss}^0$ (%/h)	$[X]_{ss}^0$ (*M*)	$[X]_{ss}^{1m}$ [a] (*M*)	*Functionalities of* $[X]_{ss}$ *Surface*	*Functionalities of* $[X]_{ss}$ *1-m Layer*[a]
$^1O_2$	furfuryl alcohol (1.2×10^8)	3	8×10^{-14}	5×10^{-14}	$\propto[DOM]$	independent[b]
$OH^{\cdot}$	butyl chloride (3×10^9)	0.2	2×10^{-16}	4×10^{-17}	$\propto[NO_3^-]/[DOM]$	$\propto[NO_3^-]/[DOM]^2$
$ROO^{\cdot}$	trimethylphenol (?)	15	(?)[c]	(?)[c]	$\propto[DOM]$[d]	independent[b, d]
e_{aq}^-	CCl_4 (3×10^{10})	0.13	1.2×10^{-17}	5.2×10^{-18} [e]	$\propto[DOM]/[O_2]$	$\propto 1/[O_2]$

[a] Light screening by suspended sediments is neglected.
[b] Independent of [DOM].
[c] Not determined for reasons given in the text.
[d] Only for DOC < 5 mg/L.
[e] Estimated for an assumed screening factor, $S_{355}^{1m} = 0.43$.

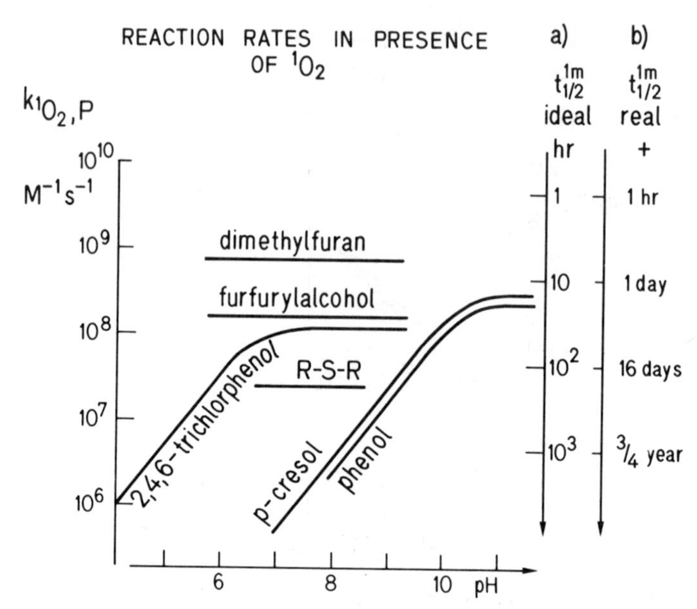

Figure 3. Compilation of rate constants for reactions of $^1O_2$ (left scale), and sunlight irradiation times required for solute eliminations (right scales, 1-m average depth) vs. pH. Data are from ref. 21. (a) Half-life of selected pollutants in Greifensee water during exposure to June midday sunlight (1 kW/m^2). For the case of $^1O_2$ this yields a $[^1O_2]_{ss}^{1m}$ value of 4×10^{-14} M. (b) Scale of times for achieving the irradiation dose required for the reduction of the concentration to 50% of its initial value. These times are an estimate based on the real sum-curve of measured solar irradiations in Dübendorf, Switzerland, when starting on a clear summer day (i.e., June 2, 1985, 11 a.m.).

radical source (*22, 23*), $^{\cdot}OH$ is primarily consumed by fast scavenging by DOM or, in waters with low DOM, even by carbonate anions. Different types of aquatic DOM exhibit comparable rate constants for trapping $^{\cdot}OH$. Thus, we can assume that the value of $[^{\cdot}OH]_{ss}^{0}$ increases with the ratio of the concentration of NO_3^- to DOC, as shown in Figure 4.

For deep water columns, within which most photoactive light is absorbed, the light-screening factor increases with DOC (equation 8). Therefore depth-averaged $[^{\cdot}OH]_{ss}$ values for water columns in which all light is absorbed decrease with the square of the DOM concentration (*see* entry in Table II).

On the basis of detailed measurements of OH information quantum yield, absorption spectrum of NO_3^-, and experimentally calibrated rate constants for consumption of $^{\cdot}OH$ by DOM, we estimate the rate constant for elimination of compounds with reactivity similar to that of butyl chloride for the surface of Lake Greifensee ($[NO_3^-{-}N]$/DOC = 0.4 mg/mg) to be

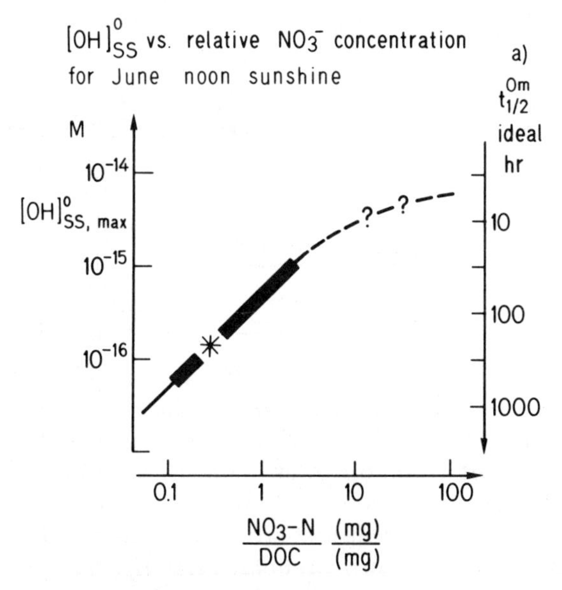

Figure 4. Surface values of $[{}^{\cdot}OH]_{ss}$ during summer midday sunshine (1 kW/m²) vs. the ratio of $[NO_3^-]$/DOC. Values are given (right scale) for a compound P of mean reactivity ($k_{OH,P} = 6 \times 10^9\ M^{-1}\ s^{-1}$) exposed to ideal midday summer sunshine (1 kW/m²). Star indicates that the experiment was performed in Greifensee water. Bold line defines range within which most medium-sized Swiss rivers lie. Nonlinear relationships could prevail at high nitrate concentrations.

0.2%/h in June midday sunshine (1 kW/m²). This corresponds to a $[{}^{\cdot}OH]_{ss}^{0}$ value of 2×10^{-16} M.

Rates of reaction of $OH^{\cdot}$ with many compounds in natural waters may be estimated from the calibrations performed with probe molecules and with information from tables of relative rate constants (Figure 5).

Organic Peroxy Radicals. Compounds that are classified as antioxidants (such as alkylphenols, aromatic amines, thiophenols, and imines) generally exhibit moderately high reactivity toward organic peroxy radicals ($ROO^{\cdot}$) (*26*). 2,4,6-Trimethylphenol, the most water-soluble representative of the antioxidant class of 2,4,6-trialkylphenols, was our preferred probe molecule for characterizing the organic peroxy radical reactivity of natural waters (*7*). Experiments have demonstrated that its direct sunlight photolysis is slow, and that possible interfering reactions with singlet oxygen are negligible.

A plausible reaction sequence for the formation of $ROO^{\cdot}$ is presented in equation 11.

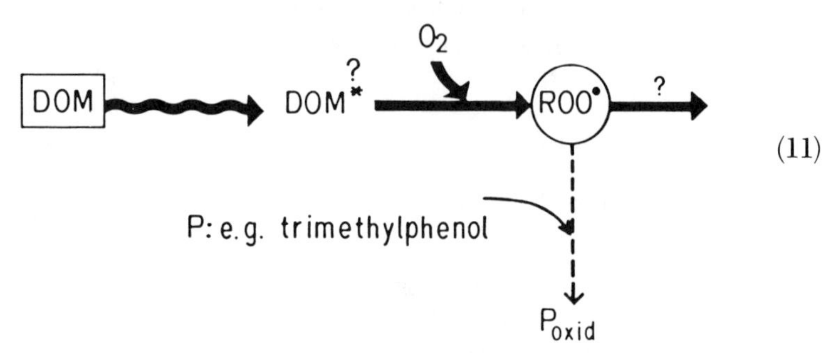

(11)

Experiments show that some electron-rich phenols are rapidly oxdized by a transient oxidant postulated to be an organic peroxy radical. All phenols exhibited apparent first-order oxidation kinetics, consistent with the assumption of the kinetic model described previously. 2,4,6-Trimethylphenol is photooxidized with an apparent first-order rate constant of 15% per hour in sunlit (midday, June, 1 kW/m^2) Greifensee water. Other phenols exhibit lower reactivity toward this transient oxidant, as shown in Figure 6.

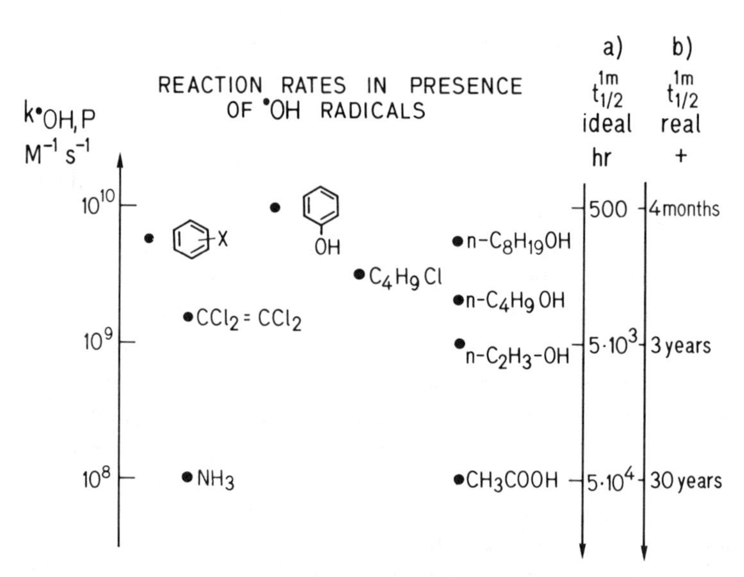

Figure 5. Compilation of rate constants for reactions of OH radicals (left scale), and sunlight irradiation times required for solute elimination (right scales) as in Figure 3. $k_{OH,P}$ *values for tetrachloroethylene and butyl chloride are from our own measurements of competition kinetics (22); other data are from ref. 23. The mean half-life at the surface would be about 20 times shorter because of the absence of light screening by DOM.* $[{}^{\cdot}OH]_{ss}{}^{1m} = 1 \times 10^{-17}$ *M.*

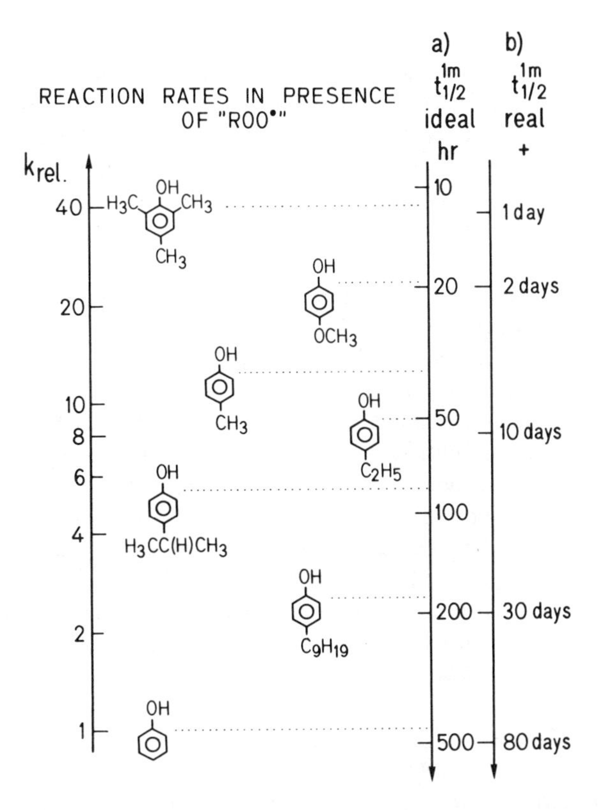

Figure 6. Compilation of experimentally determined relative rate constants for ROO˙ radicals (left scale), and sunlight irradiation times required for solute eliminations in Greifensee (right scales; see Figure 3 for sunlight intensity). Data are from ref. 7. Half-life scales are depth-averaged values for a well-mixed 1-m water column, calculated by using equation 6 with α_{366} *= 0.01 cm*$^{-1}$*. Calculations using* α_{313} *= 0.022 cm*$^{-1}$ *predict values to be ~1.8 times the half-lives given.*

4-Alkylphenols exhibit the following relative reactivity sequence: methyl > ethyl > isopropyl > nonyl > H. The relatively low reactivity of nonylphenol might be caused by steric hindrance or hydrophobic interactions of the alkyl group with parts of the DOM, which could inhibit attack of the phenol by the transient oxidant. Benzoate and hydroxybenzoates (ortho, meta, and para forms) exhibited no measurable reactivity toward ROO˙.

From diagnostic tests and kinetic analyses, we conclude that DOM is the source of this transient oxidant, and we definitely exclude singlet oxygen, hydroxyl radical, and DOM triplets as responsible for the oxidation of these phenols (7).

Steady-state concentrations of the transient oxidant (postulated ROO˙) are linearly related to DOC (for DOC < 5 mg/L) and to sunlight intensity.

This relationship indicates that the fate of this transient oxidant is neither controlled by DOM scavenging (for DOC < 5 mg/L) nor by reaction with itself. The absence of $ROO^\cdot$ scavenging by DOM is not surprising because $ROO^\cdot$ exhibits negligible reactivity toward benzoate and hydroxybenzoates. Therefore, for low DOC we assume that DOM is a source, but not a sink, of the transient oxidant. In such a case the depth-averaged $[ROO^\cdot]_{ss}$ for a well-mixed water column ($\alpha_\lambda \cdot z > 1$) is independent of the DOC concentration (Table II).

Additionally, the area-based production rate of postulated $ROO^\cdot$ is independent of DOC concentration (for DOC < 5 mg/L), provided that most photoactive light is absorbed within the water column. However, at higher DOC (DOC > 5 mg/L) scavenging reactions of DOM may begin to control the fate of the transient oxidant (proposed $ROO^\cdot$). Under these conditions DOM may function as both source and sink of the postulated $ROO^\cdot$.

In contrast to the other photoreactants discussed here, $ROO^\cdot$ are DOM-derived radicals and contain parts of the DOM. Consequently the $ROO^\cdot$ speciation, like that of DOM, is complicated, unknown, and dependent on water source. Therefore, DOM-derived $ROO^\cdot$ may well exhibit a wide range of reactivities. This variation introduces considerable uncertainty into the determination of their steady-state concentration and production rate by any competing kinetic method (7).

Solvated Electron. Flash photolysis combined with kinetic spectroscopy of natural-water DOM has shown high concentrations of a transient whose kinetic and spectroscopic characteristics correspond with those of the solvated electron (*10–12*), e_{aq}^-. However, experiments with added probe compounds at lower light intensities, which are more representative of continuous (sunlight) irradiations, show that the chemical effect of such species is rather low.

We assume that the discrepancy between the high apparent production rates of transient absorbers representative of solvated electrons and the low efficiency observed for typical chemical reactions of solvated electrons might be due to the fact that most of the transients remain trapped in the macromolecular structure. Therefore, transient recombination may occur before reaction with compounds of interest.

The postulated formation sequence of e_{aq}^- and its reactions with probe molecules are described in equation 12.

Probe molecules of choice are chlorinated organic compounds. These rapidly scavenge free solvated electrons to produce chloride anion. However, the formation rate of solvated electrons appears so low, and the rate with which solvated electrons in aerobic water are scavenged by dissolved oxygen is so high ($\sim 5 \times 10^6$ s^{-1}), that their steady-state concentration is very low. Consequently, probe molecules (present in low concentrations) are not significantly reduced by solvated electrons within feasible irradiation times.

e_{aq}^- FORMATION :

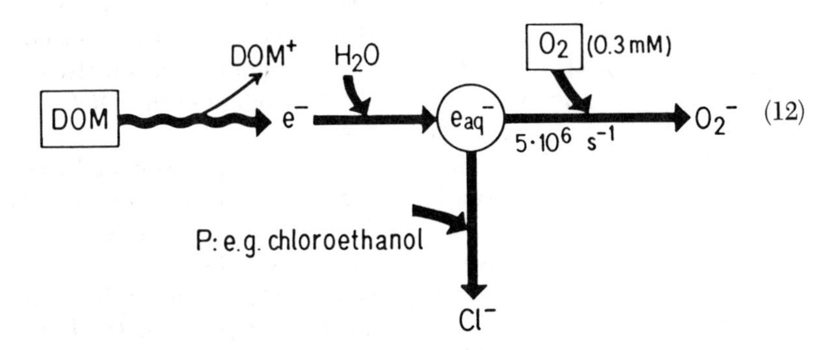

Therefore, for the determination of these photoreactants, high concentrations of probe molecules were used to trap all solvated electrons before they reacted with oxygen. The total production rate of solvated electrons was determined from the formation rate of Cl^- in such experiments. Then the steady-state concentration of solvated electrons was determined by combining the results with the known rate constant of solvated electrons with oxygen.

From these results we could deduce that the value of $[e_{aq}^-]_{ss}^0$ for Lake Greifensee water during June midday sunshine (1 kW/m²) is about 1.2 × 10^{-17} M. The effect of such a concentration on reactive micropollutants is exemplified in Figure 7 and in Table II. Compounds with the kinetic characteristics of trichloroacetic acid and chloroform, which exhibit maximal rate constants, would be degraded at a rate of only 0.13%/h at the water surface.

Although compounds present in DOM are presumably the main precursor of solvated electrons, reactions with oxygen constitute the major sink. The value of $[e_{aq}^-]_{ss}^0$ will therefore increase with the water concentration ratio of DOC to O_2. However, the mean concentration, when averaged over a water depth in which all photoactive light is absorbed, again becomes independent of the DOC (*see* last column in Table II).

Most solvated electrons will add to oxygen and form O_2^-. Part of the O_2^- seems to be converted to H_2O_2 (*14*). However, because added electron scavengers did not affect the formation rate of hydrogen peroxide in illuminated Greifensee water, the solvated electron itself is not considered a main precursor for this species (*27*).

Humic-derived cation radicals must also be considered photoionization products occurring when electrons are released. Such cations will eventually lead to oxygenated products and presumably could be an additional source of O_2^- or even hydrogen peroxide. But because the yield of solvated electrons

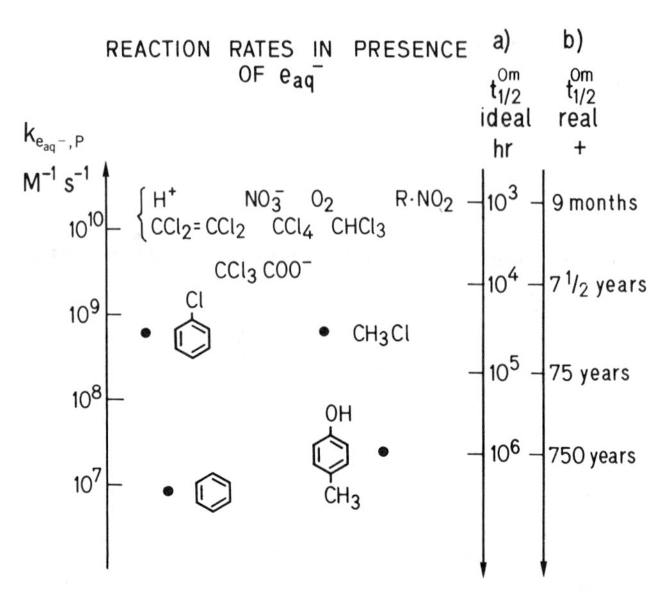

Figure 7. Compilation of rate constants for surface-level reactions of solvated electrons (left scale) and sunlight irradiation times required for solute elimination (right scales; sunlight intensity as in Figure 3). $k_{e_{aq}^-,P}$ values from ref. 25. The mean half-life within a steadily mixed 1-m layer would be about 5 times longer because of light screening by DOM. $[e_{aq}^-]_{ss}^{0m} = 1.2 \times 10^{-17}$ M.

is so low, such reactions can be of only minor importance in producing further photooxidants.

Conclusions

Table II summarizes the apparent rate constants with which appropriate probe or reference substances are transformed by the various photoreactants produced in eutrophic lake water under summer midday sunshine (1 kW/m^2). Steady-state concentrations of the photoreactants are calculated by using absolute or estimated reaction-rate constants for reactions of photoreactants with the listed probe substances. The functions in the last two columns describe the dependence of the photoreactant steady-state concentration on DOC or other relevant water parameters.

The dimensions of the functions describing steady-state concentrations are different for different photoreactants; they depend on the photolytic source and on the lifetime-controlling sinks to be considered. But in spite of this circumstance, which makes comparisons difficult, the results show that $^1O_2$ and ROO$^{\cdot}$ radicals are efficient photoreactants for transformations of some specific types of micropollutants. However, both of them react highly selectively and will degrade only a few types of chemical structures within a reasonable time.

On the basis of steady-state concentrations given in Table II and known lifetimes of the various photoreactants in these waters, we can also estimate the total photoreactant production per kilowatt-hour of absorbed light or per integral amount of sunshine throughout 1 year (about 1300 kW-h/m^2). The resulting values reported in Table III for the yearly production of photoreactants per square meter are remarkably high. Comparisons between these yearly production rates and the amount of DOM present per unit of area in the water column indicate that these reactants may also be important for the aging of DOM. For example, a lake containing 4 mg/L of DOC as "DOM molecular units" (molecular weight = 120 g of C/mol of such units) contains 0.3 mol of DOM molecular units per square meter over a 10-m depth. Comparison of this concentration with entries in Table III shows that 10% of the DOM molecular units would react, for example, with $^{\cdot}OH$ radicals in 1 year.

Table III. Production Rates of Photoreactants (r_X) in Greifensee Water

Reactant	*mol/kW*	*mol/(m²·yr)*
$^1O_2$	50×10^{-3}	60
$^{\cdot}OH$	20×10^{-6}	0.025
$ROO^{\cdot}$	—	—
e_{aq}^-	70×10^{-6}	0.1

NOTE: For water containing 4 mg of DOC/L, we calculate 0.03 mol of DOM units per square meter and per meter depth, assuming a molecular weight of 120 g of C/mol. This means that DOM/m^2 = 0.03 mol $\times$ z(m) (molecular units: 120 g of C/mol).

Aging effects of DOM induced by the other photoreactants cannot be quantified as easily because these do not significantly react with the DOM under the conditions considered in this overview. However, for each photoproduced solvated electron, a corresponding DOM cation radical is also produced. We expect the DOM cation radical to undergo molecular rearrangements that may either split out a cationic entity (possibly a proton) or add an anion (e.g., OH^-) and O_2 (a biradical). Therefore, DOM might also undergo significant transformations by such reactions.

On the basis of the data given in Table III for the production rate of solvated electrons, we deduce that about 40% of the DOM present in a 10-m-deep water layer can be converted in 1 year, if we accept the same assumptions. Although these latter ideas are somewhat speculative, they show that photoreactants could have large effects on the chemical transformations of humic substances. Data are not available to make additional comparisons of these effects with other possible direct photolytic reactions of DOM.

References

1. Haag, W. R.; Hoigné, J. *Environ. Sci. Technol.* **1986,** *20,* 341–348.
2. Zepp, R. G.; Cline, D. M. *Environ. Sci. Technol.* **1977,** *11,* 359.
3. Zepp, R. G.; Wolfe, N. L.; Baughman, G. L.; Hollis, R. C. *Nature (London)* **1977,** *267,* 421–423.
4. Wolff, C. J. M.; Halmans, M. T. H.; van der Heijde, H. B. *Chemosphere* **1981,** *10,* 59–62.
5. Mill, T.; Hendry, D. G.; Richardson, H. *Science (Washington, DC)* **1980,** *207,* 886–887.
6. Smith, J. H.; Mabey, W. R.; Bohonos, N.; Holt, B. R.; Lee, S. S.; Chou, T.-W.; Bomberger, D. C.; Mill, T. *Environmental Pathways of Selected Chemicals in Freshwater Systems: Part II. Laboratory Studies*; U.S. Environmental Protection Agency: Athens, GA, 1978; EPA Report EPA–600/7–78–074.
7. Faust, B. C.; Hoigné, J. *Environ. Sci. Technol.* **1987,** *21,* 957–964.
8. Cooper, W. J.; Zika, R. G. *Science (Washington, DC)* **1983,** *220,* 711–712.
9. Draper, W. M.; Crosby, D. G. *J. Agric. Food Chem.* **1981,** *29,* 699–702.
10. Fischer, A. M.; Kliger, D. S.; Winterle, J. S.; Mill T. *Chemosphere* **1985,** *14,* 1299–1306.
11. Power, J. F.; Sharma, D. K.; Langford, C. H.; Bonneau, R.; Joussot-Dubien, J. In *Photochemistry of Environmental Aquatic Systems;* Zika, R. G.; Cooper, W. J., Eds.; ACS Symposium Series 327, American Chemical Society: Washington, DC, 1987; pp 157–173.
12. Zepp, R. G.; Braun, A. M.; Hoigné, J.; Leenheer, J. A. *Environ. Sci. Technol.* **1987,** *21,* 485–490.
13. Baxter, R. M.; Carey, J. H. *Nature (London)* **1983,** *306,* 575–576.
14. Petasne, R. G.; Zika, R. G. *Nature (London)* **1987,** *325,* 516–518.
15. Zepp, R. G.; Schlotzhauer, P. F.; Sink, R. M. *Environ. Sci. Technol.* **1985,** *19,* 74–81.
16. Zafiriou, O. C.; True, M. B. *Mar. Chem.* **1979,** *8,* 9–42.
17. Zepp, R. G.; Hoigné, J.; Bader, H. *Environ. Sci. Technol.* **1987,** *21,* 443–450.
18. Faust, B. C.; Hoffmann, M. R. *Environ. Sci. Technol.* **1986,** *20,* 943–948.
19. Waite, T. D. In *Geochemical Processes at Mineral Surfaces;* Davis, J. A.; Hayes, K. F., Eds.; ACS Symposium Series 323, American Chemical Society, Washington DC, 1986; pp 426–445.
20. Haag, W. R.; Hoigné, J. *Chemosphere* **1985,** *14,* 1659–1671.
21. Scully, F. E.; Hoigné, J. *Chemosphere* **1987,** *16,* 681–694.
22. Hoigné, J.; Bader, H. *Ozone Sci. Eng.* **1979,** *1,* 357–372.
23. Hoigné, J.; Bader, H. In *Organometals and Organometalloids, Occurrence and Fate in the Environment;* Brinckman, F. E.; Bellama, J. M., Eds.; ACS Symposium Series 82; American Chemical Society, Washington, DC, 1978; pp 292–313.
24. Farhatazis; Ross, A. B. In *Selected Specific Rates of Reactions of Transients from Water in Aqueous Solution. III. Hydroxyl Radical and Perhydroxyl Radical and their Radical Ions;* National Bureau of Standards, Report No. NSRDS–NBS 59; Washington DC, 1977.
25. Howard, J. A.; Scaiano, J. C. In *Landolt Börnstein New Series, Kinetic Rate Constants of Radical Reactions in Solution: Part d, Oxyl-, Peroxyl- and Related Radicals*; Springer Verlag: Berlin, 1984; Series II/13d.
26. Anbar, M.; Bambenek, M.; Ross, A. B. *Selected Specific Rates of Reactions of Transients from Water in Aqueous Solution. 1. Hydrated Electron*; National

Bureau of Standards, U.S. Department of Commerce: Washington DC, 1973; National Standard Reference Data Series Report NSRDS–NBS 43.
27. Sturzenegger, V. Ph.D. Thesis, Federal Institute of Technology, Zurich; in preparation.
28. Zepp, R. In *Humic Substances and Their Role in the Environment*; Frimmel, F. H.; Christman, R. H., Eds.; Wiley: New York, 1988; pp 193–214.

RECEIVED for review July 24, 1987. ACCEPTED for publication July 26, 1988.

INFLUENCES OF COAGULATION PROCESSES ON WATER TREATMENT

24

Coagulation Process for Removal of Humic Substances from Drinking Water

Eilen Arctander Vik

Aquateam, Norwegian Water Technology Centre A/S, P.O. Box 6326, Etterstad, 0604 Oslo 6, Norway

Bjørnar Eikebrokk

Norwegian Hydrotechnical Laboratory, SINTEF, 7034 Trondheim–NTH, Norway

Humic substances adversely affect the quality of drinking water in many ways. For instance, they impart color, serve as precursors to the formation of chlorinated compounds, possess ion-exchange and complexing properties that include association with toxic elements and micropollutants, and precipitate in distribution systems. This chapter reviews the coagulation process, in which aluminum sulfate is the most commonly used coagulant aid, along with ferric salts and some organic polyelectrolytes. Aluminum chemistry and coagulation mechanisms are reviewed in detail. Factors that are critical to the design of processes, such as the importance of the rapid-mixing process, are discussed. The results of several recent studies on conventional and direct filtration treatment are given. These results illustrate, from an operational perspective, the chemical theory that is presented. In spite of the development of new separation techniques for water treatment, the coagulation process will continue to be the most important water-treatment process worldwide for removal of humic substances. Further research is needed to improve the coagulation process, optimize coagulant dose, minimize residual aluminum, and integrate separation, disinfection, taste-, odor-, and corrosion-control treatment into the water-treatment program.

0065–2393/89/0219–0385$07.00/0

HUMIC SUBSTANCES IMPART A YELLOW OR BROWN COLOR to waters with high organic content. These substances are undesirable in a potential water supply for a number of reasons, ranging from aesthetics to the fact that they are precursors of potentially carcinogenic compounds (trihalomethanes). The health effects of high concentrations of humic substances are unknown. Several factors make aquatic humic substances an important constituent of natural water systems:

- Humic substances, 40–60% of the dissolved organic carbon (DOC) in natural waters, are the largest fraction of natural organic matter in water. They are normally present in concentrations >1 mg of C/L, although identifiable organic compounds can be present at concentrations that are orders of magnitude lower (*1*, *2*).
- Humic substances possess ion-exchange and complexing properties that are associated with most constituents of water, including toxic elements and organic micropollutants (*1–4*).
- Humic substances act as a vehicle for transport of toxic, water-insoluble elements and organic micropollutants (*1–4*).
- Chlorine combines with aquatic humic substances to form chlorinated organic compounds, such as chloroform (*5–8*) and complex chlorinated compounds (*9–13*), that may have negative effects on health (*9–14*).
- Humic substances precipitate in the distribution system, where they lead to deterioration of tap water quality and increase the need for interior cleaning of the pipes.

Concentrations of DOC in natural waters vary over a wide range. The world-stream-volume weighted mean, according to Meybeck (*15*), is 5.7 mg of C/L as DOC. Particulate organic carbon, including planktonic organisms, generally accounts for about 10% of the total organic matter (*3*, *16*). DOC consists predominantly of humic substances. The lowest concentrations vary from 0.05 to 0.60 mg of C/L. Streams, rivers, and lakes contain from 0.50 to 4.0 mg of C/L. Colored rivers and lakes have much larger concentrations of humic substances, 10–30 mg of C/L (*2*).

Characteristics of Humic Substances

The yellowish-brown color of natural surface water comes from humic substances leached from plant and soil organic matter. This color is seen in bogs and wetlands along streams. Organic acids dissolve into the stream. This gives the yellow color and contributes protons to the soil weathering process (*2*). DOC, according to an operational definition, is organic carbon smaller than 0.45 μm in diameter (Figure 1). At the molecular level, most of the dissolved organic carbon comes from polymeric organic acids called humic substances. These yellow organic acids (1000–2000 MW) are polyelectrolytes of carboxylic, hydroxyl, and phenolic functional groups. They compose 50–75% of the dissolved organic carbon and are the major class of organic compounds in natural waters. Figure 1 shows the continuum of organic carbon in natural waters.

Dissolved molecules of fulvic acid are approximately 2 nm in diameter and at least 60 nm apart. There might be five inorganic ions (calcium, sodium, bicarbonate, chloride, and sulfate) between each pair of these organic molecules.

There is also some colloidal organic matter in the water. These colloids are large aggregates of humic acids, 2–50 nm in diameter. They are commonly associated with clay minerals or oxides of iron or aluminum. In most natural waters the colloidal organic matter is approximately 10% of the DOC. Colloidal organic carbon is the humic acid fraction of humic substances. This fraction is larger in molecular weight (2000–100,000) (Figure 1) and contains fewer carboxylic and hydroxyl functional groups than the fulvic acid fraction. Humic acid adsorbs and chemically binds to the inorganic colloids, modifying their surface.

Dissolved humic substances compose 5–10% of all anions in streams and rivers. This anionic character gives humic substances their aqueous solubility, binding sites for metals, buffer capacity, and other characteristics. The anionic character comes from the dissociation of carboxylic acid functional groups:

$$\text{R—COOH} \rightleftharpoons \text{R—COO}^- + \text{H}^+ \tag{1}$$

Carboxylic functional groups occur on aquatic humic substances with a frequency of 5–10 per molecule. At the pH of most natural waters, 6–8, all carboxylic groups are anionic or dissociated. These charged groups repulse one another and spread out the molecule. The counterions balancing the charge of the negative groups are mostly calcium and sodium. However, trace metals may be bound to some carboxylic groups that have favorable steric location. That is, a carboxylic group in association with a phenolic group may form a chelate or a ring structure and bind metal ions.

Because of the complex nature of humic substances, they have traditionally been classified according to the operational procedures needed to separate them. Oden's classification (*17*) of these substances has been used since its introduction in 1919. Several methods have been used for determining the molecular size of humus; results have varied according to method. Newer separation techniques have improved the results.

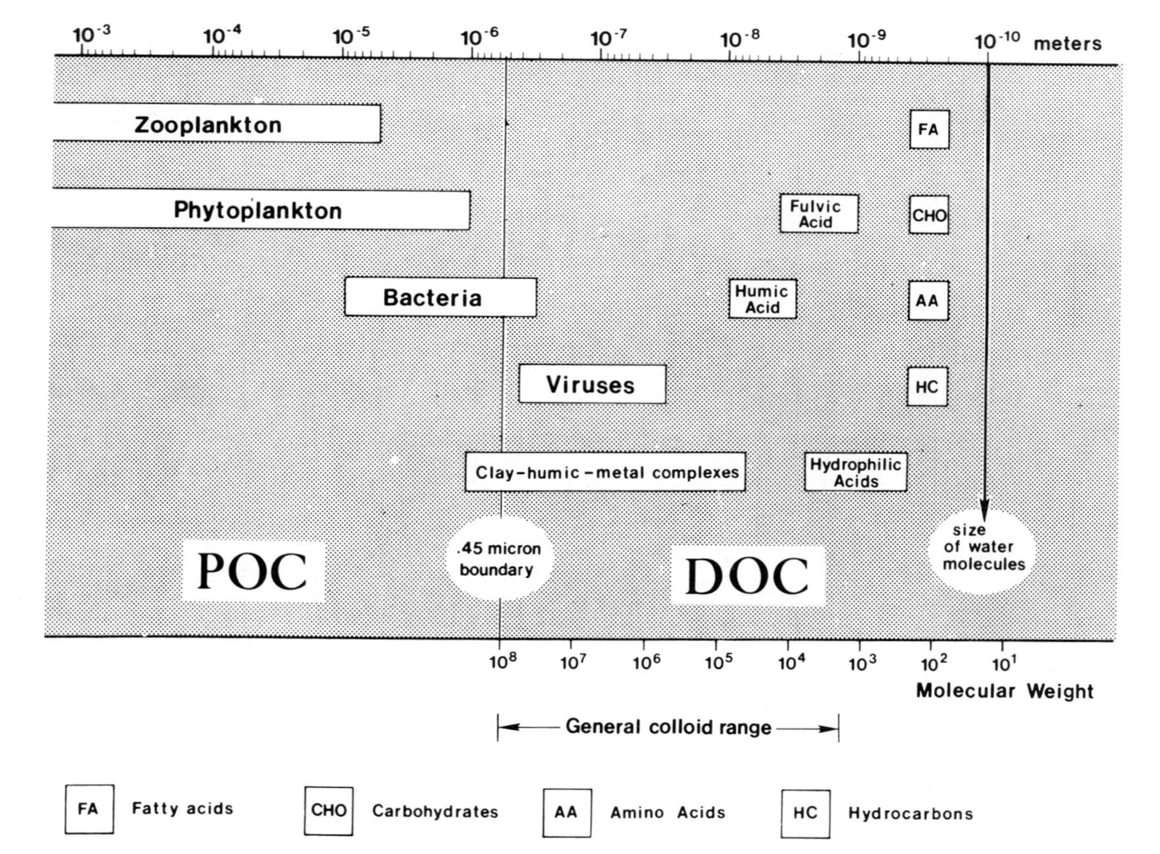

Figure 1. Continuum of particulate and dissolved organic carbon in natural waters. (Reproduced with permission from ref. 2. Copyright 1985 Kluwer Academic Publishers.)

Thurman (2), in a broad discussion of literature on the molecular weight of humic substances, concluded that aquatic fulvic acid is generally less than 2000 MW. This low molecular weight means that most aquatic humic substances are dissolved, rather than colloidal. Because humic matter in water is associated with various metal ions, clays, and amorphous oxides of iron and aluminum, they may have larger molecular weight than determined with the purified free acid. Humic acid is 2000–5000 MW or greater, and is therefore considered to be colloidal.

Thurman (2) also discussed the possible structures and structural units that may be present in aquatic humic substances. The separation of humic acid from fulvic acid by precipitation at pH 1 is important. Humic acid is less soluble than fulvic acid. The lower amount of carboxylic acid lowers the aqueous solubility of humic acid and is the main reason that most natural waters contain 5–25 times more fulvic acid than humic acid. The second factor is the size of the humic acid, 2–10 times larger than fulvic acid. This large size lowers the aqueous solubility of humic acid. Also, the phenolic content of the humic acid is somewhat greater than that of the fulvic acid, and there are more color centers on the humic acid molecule (*18*). Humic acid does contain longer-chain fatty-acid products than fulvic acid (*19*). This finding suggests that humic acid is more hydrophobic because of the longer-chain fatty acids (C_{12}–C_{18}).

According to Stevenson (*20*), the biochemistry of the formation of humic substances is one of the least understood and most intriguing aspects of humus chemistry. Pathways suggested for the formation of humus in soil are the two lignin-degradation models, the polyphenol theory, and the sugar–amine condensation theory. Because aquatic humic substances are different from soil humic substances, those originating in water may have other mechanisms of formation (*21*). New analytical techniques have been developed within the last few years. Nuclear magnetic resonance (NMR) spectroscopy, and especially ^{13}C NMR spectroscopy, has become important in elucidation of the chemical structures of humic materials (*22*). New analytical methods are also giving a new perspective to a detailed understanding of the coagulation process for humic-substances removal.

The Coagulation Process

Coagulation in Water Treatment. Coagulation with aluminum sulfate, a standard procedure for removing suspended colloids, is also an effective method for removing aquatic humus. Coagulation is one of the most important processes used in treatment of surface waters (*23*).

The conventional water-treatment process (Figure 2) includes coagulation, flocculation, sedimentation, and final filtration. Direct filtration (Figure 3) is an important option in conventional water treatment. A status report (*24*) discussed results of a worldwide survey of direct filtration plants. Color

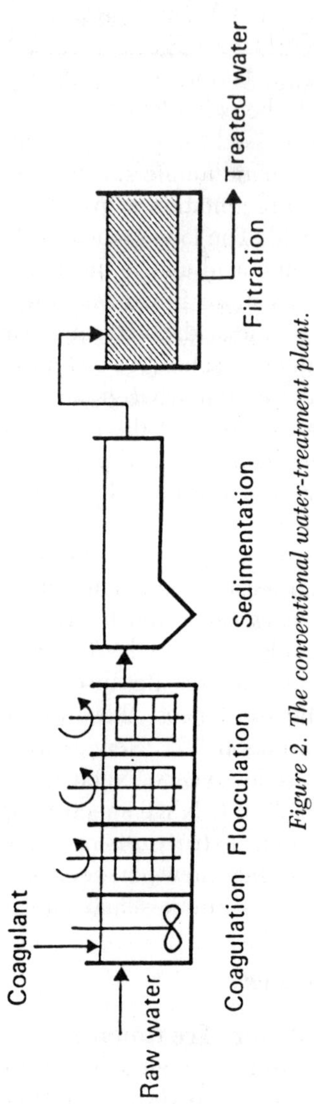

Figure 2. The conventional water-treatment plant.

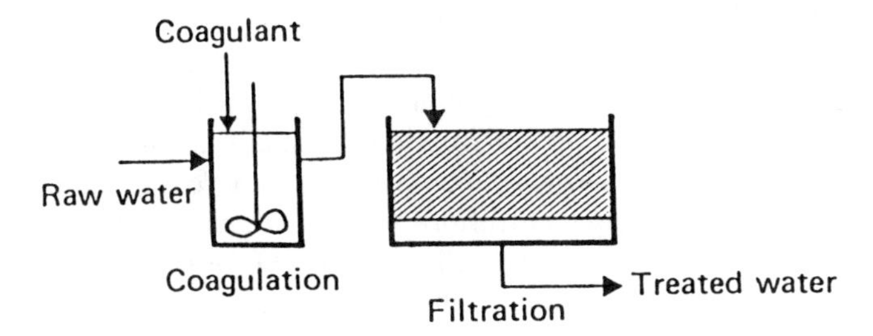

Figure 3. The direct filtration plant.

exceeding 30–40 Hazen units and continuing turbidity greater than 15 FTU were pointed out as problem situations. Polyelectrolytes were proposed as a short-term substitute for all or part of the required primary coagulant; thus the amount of sludge was reduced. Full-scale experiments (*25*) indicate that this process is suitable for high concentrations of humic substances when turbidity is low. Further improvement may be obtained if the coagulation process is carried out at lower pH values (≃4.7) and low alum concentrations (≃0.3 mg of Al/L).

Coagulants in Use. Aluminum is the most important coagulant in use. Ferric compounds are applied to some extent. Alum and ferric salts are, however, not the only reactants used in coagulation. Magnesium carbonate, hydrolyzed with lime, has been shown to be as effective as alum (*26, 27*).

Various organic polyelectrolytes have been used for organic substance removal (*28–31*). All of them show good humus removal, especially at low-to-moderate DOC concentrations. Most of the studies using polymers are related to the use of direct filtration (*30–35*). Synthetic organic cationic polyelectolytes are widely used as coagulants in direct filtration. Their use as primary coagulants in direct filtration of humic matter has also been studied (*30–32, 36*). Although cationic polymers showed good removal efficiency, they are not as effective as alum (aluminum sulfate). For waters containing a relatively high concentration of humic substances, direct filtration with cationic polyelectrolytes produced less head loss development and longer filter runs when a flocculation period was included (*31*).

Many studies have been conducted to evaluate the performance of coagulation with aluminum and ferric salts (*38*). Up to 90% removal of the humic acid fraction has been achieved with both Al(III) and Fe(III) (*39–41*).

Alum reacts differently in water than the chloride, nitrate, or perchlorate salts of aluminum (*42*). Polyaluminum chloride (PACl), a coagulant that consists of partially hydrolyzed aluminum chloride, is produced by the addition of base to concentrated $AlCl_3$ solution. Dempsey and co-workers (*42*) found PACl to be a promising alternative to alum as a coagulant for removing humic substances from water supplies.

The removal of soluble organic contaminants by lime softening has been studied (*37*). The precipitate in this case is mainly $CaCO_3$, but if magnesium is present and the pH is sufficiently high, $Mg(OH)_2$ will also be formed. The process effectively removes several humic substances, including a fulvic acid. Removal efficiency increased with increasing pH, increasing amount of precipitate, and decreasing concentrations of total organic carbon (TOC). Removal efficiency was significantly enhanced by the presence of magnesium or phosphate, especially when the amount of precipitate formed was small.

ALUMINUM CHEMISTRY. The aqueous chemistry of aluminum is complex and diverse because of the numerous hydrolysis intermediates formed before precipitation of aluminum hydroxide, $Al(OH)_3(s)$. The aluminum ion, Al^{3+}, behaves very much like Fe^{3+} in solution, except that it has a greater tendency to form polynuclear species. Several authors have reported (*16, 43, 44*) a stepwise conversion of the positive aluminum hydrated ion to the negative aluminum ion. When aluminum salts are dissolved in water, the metal ion Al^{3+} hydrates, coordinates six water molecules, and forms an aquometal ion, $Al(H_2O)_6{}^{3+}$. For simplicity, the H_2O ligands attached to the Al ions are omitted (e.g., $Al(H_2O)_5OH^{2+}$ is written as $AlOH^{2+}$). Several hydrolysis species are possible:

$$Al^{3+} \rightarrow AlOH^{2+} \rightarrow Al(OH)_2{}^{+} \rightarrow Al(OH)_3\,(aq) \rightarrow Al(OH)_4{}^{-} \rightarrow Al(OH)_5{}^{2-} \rightarrow Al(OH)_6{}^{3-} \quad (2)$$

These species can form polymers with several of the hydrolysis products. The following polymers have been suggested: $Al_{13}(OH)_{34}{}^{5+}$; $Al_7(OH)_{17}{}^{4+}$; $Al_8(OH)_{20}{}^{4+}$; $Al_6(OH)_{15}{}^{3+}$; $Al_2(OH)_2{}^{4+}$.

For more than 20 years, scientists have argued about the predominance and existence of aluminum polymers. Baes and Mesmer (*45*) indicated that aqueous aluminum equilibrium chemistry can be explained accurately (but not uniquely) by considering three polymeric species [$Al_2(OH)_2{}^{4+}$, $Al_3(OH)_4{}^{5+}$, and $Al_{13}O_4(OH)_{24}{}^{7+}$], five monomers [$Al^{3+}$, $AlOH^{2+}$, $Al(OH)_2{}^{+}$, $Al(OH)_3$, and $Al(OH)_4{}^{-}$], and a solid precipitate [$Al(OH)_3(s)$]. Amirtharajah and Mills (*46*) emphasized that two important deductions follow from the existence of hydrolysis Al(III) species:

1. Hydroxy metal complexes readily adsorb on surfaces, and the charges they carry may cause charge reversals of the surfaces on which they adsorb (*44, 47, 48*). The hydrolysis products of aluminum in aqueous solution are adsorbed more readily than the free Al^{3+} ion. Matijevic (*49*) showed that the hydrolyzed aluminum species reversed the charge of the originally negative hydrolyzed halide ions, whereas the simple hydrated Al^{3+} did not. The greater the degree of hydrolysis, the more extensive is the adsorption (*49*).

2. The sequential hydrolysis reactions release H^+ ions, which lower the pH of the solution in which they are formed. Additionally, the concentration of the various hydrolysis species will be controlled by the final concentration of H^+ ions, (i.e., by pH value) (*46*).

In addition to the hydroxy compounds, aluminum forms other complexes such as the fluorides and sulfates. Although fluoride is a minor constituent of most natural waters, the complexing action is strong enough to have considerable influence on the form of dissolved aluminum, even when very little fluoride is present (*50, 51*).

REACTION RATES FOR ALUMINUM COMPOUNDS. A few attempts have been made to establish the rate of the reactions between Al(III) and colloidal suspensions. Bratby (*52*) summarized data on reaction rates and calculated the time needed for adjusting the structure of the double layer at around 10^{-8} s and for a Brownian collision (diffusion) around 10^{-7}–10^{-3} s. According to Amirtharajah and Mills (*46*), sweep coagulation, which involves formation of $Al(OH)_3(s)$ and entrapment of the colloid amid the precipitate, is a slower process that occurs in the range of 1–7 s. In contrast, the adsorption destabilization process occurs in approximately 10^{-4} s. Formation of polymerized aluminum hydrolysis products is a slower process, on the order of 1 s.

Reaction between aluminum and fulvic acid is faster than the reaction between aluminum and inorganic particles (*53*). Hahn and Stumm (*54*) found that the rate of coagulation of silica particles by hydrolyzed aluminum was limited by collision frequency and efficiency, rather than by the hydrolysis of aluminum.

Coagulation and Flocculation. Humic-substance colloids retain a dispersed or dissolved (stable) state in surface water. In the pH of most natural waters, humic materials are negatively charged macromolecules in the colloidal size range. The coagulation process is used to overcome the factors promoting the stability of humic-substance molecules in water. Alum is frequently used to destabilize humic molecules, and is included in this chapter's focus on the coagulation process.

Destabilization is the conversion of the stable state of a given dispersion or solution to an unstable state. Such a process could alter the surface properties of particulate material and thereby increase the adsorptivity of the particles to a given filter medium or generate a tendency for small particles to aggregate into larger units. Alternatively, destabilization could precipitate dissolved material and thereby create particulate material for which separation by sedimentation or filtration is feasible.

Coagulation achieves destabilization of a given suspension or solution. That is, the function of coagulation is to overcome the factors promoting the stability of a given system.

Flocculation is the process whereby destabilized particles, or particles formed as a result of destabilization, are induced to come together, make contact, and thereby form larger agglomerates.

Coagulation and flocculation involve aggregation of particles into larger, more readily removable, aggregates. Aggregation of colloidal particles can be considered to involve two separate and distinct steps: particle transport to achieve interparticle contact (fluid and particle mechanisms) and particle destabilization to permit attachment when contact occurs (colloid and surface chemistry).

COAGULATION MECHANISMS. There are four distinct mechanisms for destabilization of colloids: compression of the diffuse layer; adsorption to produce charge neutralization; enmeshment in the precipitate; and adsorption to permit interparticle bridging. O'Melia (*47*) discussed the four mechanisms as follows:

1. Compression of the diffuse layer involves purely electrostatic interactions. The phenomenon is described by the theoretical Verwey–Overbeek model and the empirical Schulze–Hardy rule. The coagulants, called indifferent electrolytes, are of limited interest in water and wastewater treatment processes. The concentrations of Na^+, Ca^{2+}, and Al^{3+} required to destabilize a negatively charged colloid vary approximately in the ratio of 1 to 10^{-2}–10^{-3}. Ionic strength is of major importance, but the salt concentration required for effective destabilization is too high for practical water-treatment applications. Restabilization is not possible because of a chemical overdose.
2. Adsorption and charge neutralization produce destabilization at low dosages; overdosing may lead to restabilization and charge reversal. If coulombic interaction were the only driving force for destabilization, it would not be possible to adsorb excess counterions to produce charge reversal and restabilization. Aluminium salts used in coagulation have been shown to follow this mechanism at low dosages.
3. Enmeshment in a precipitate (also called sweep coagulation) can occur at high dosages of metal salts. The concentration of $Al_2(SO_4)_3$ must be sufficiently high to cause rapid precipitation of $Al(OH)_3(s)$ before the colloid particle can be enmeshed in these precipitates. To the first approximation, the rate of precipitation of a metal hydroxide is dependent upon the extent to which the solution is oversaturated. The extent of oversaturation is described by $(Al^{3+})\cdot(OH^-)^3/K_s$, where K_s is the solubility constant for the $Al(OH)_3(s)$ product. For very rapid

precipitation, the ratio must be 100 or even larger. In neutral and acid water the rate of precipitation is also increased by the presence of anions in solution, particularly sulfate ions. The colloidal particles serve as nuclei for the formation of the precipitate, so the rate of precipitation increases with increasing concentration of the colloidal particles to be removed. The greater the concentration of colloidal material, the lower is the amount of metal coagulant required.

4. Adsorption and interparticle bridging usually occur when polymers are used as destabilizing agents in the treatment of water and wastewater. This mechanism will not be further discussed here.

Bratby (*52*) illustrated the three first mechanisms in Figures 4 and 5. Amirtharajah and Mills (*46*) did an extensive literature review, summarizing important coagulation studies. Figure 6 summarizes in schematic form the proposed predominant mechanism of coagulation with alum. Figure 7 presents a design and operational diagram for alum coagulation, including specific areas where coagulation would occur and the major mechanisms causing coagulation. The skeletal forms of the diagram were previously presented by other workers. Minor changes in the restabilization zone are expected, depending on the type of colloid. The area for optimum sweep coagulation

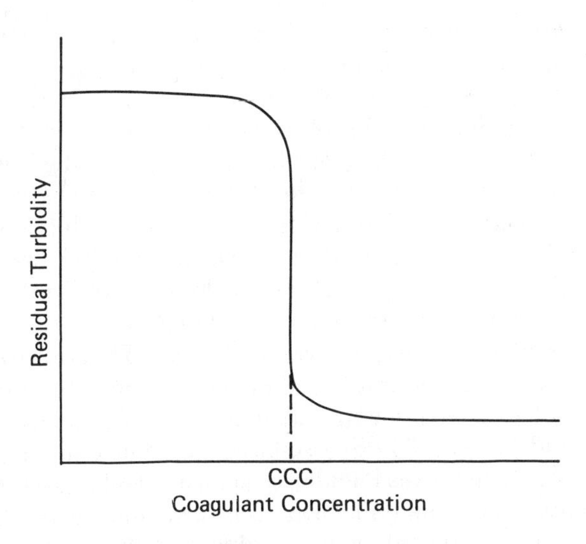

Figure 4. Destabilization characteristics where an electrical double-layer repression mechanism is predominant. Increasing the coagulant concentration beyond the critical coagulation concentration (CCC) has no substantial effect.

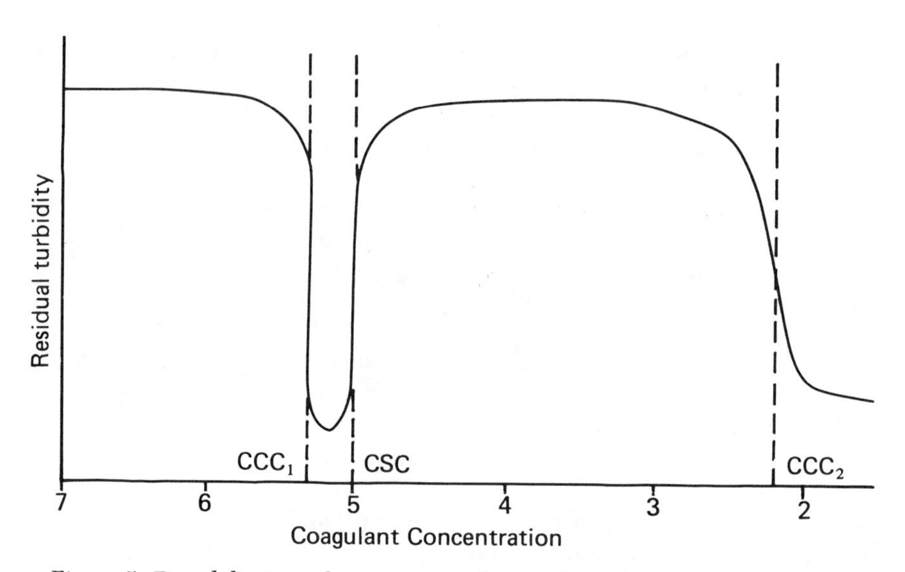

Figure 5. Destabilization characteristics that involve adsorption of coagulant species to colloidal particles. CCC_1 and CSC signify the concentration, C_t, necessary to destabilize and restabilize the dispersion, respectively. A further critical coagulation concentration, CCC_2, indicates double-layer repression or enmeshment mechanisms of higher coagulant dosages.

(i.e., the area for best settling flocs) is specifically defined by an alum dose of 20–50 mg/L, with a final pH of 6.8–8.2.

RAPID-MIXING PROCESS. Several researchers have discussed the importance of rapid-mix parameters. The major consideration in rapid mixing has been uniform distribution of the coagulant in the raw water in order to avoid over- and undertreatment of water. Amirtharajah and Mills (*46*) discussed the two modes of destabilization in relation to the influence of the rapid mix, based on the reaction rates previously discussed (Figure 6).

For adsorption–destabilization, Amirtharajah and Mills (*46*) emphasized that the coagulants have to be uniformly distributed in the raw water stream as rapidly as possible (<0.1 s). This distribution allows the hydrolysis products, which develop in 0.01–1 s, to cause destabilization of the colloid. Sweep coagulation reactions occur in the range of 1–7 s. The extremely short dispersion times and high intensities of mixing are thus not so crucial in this case, compared to adsorption–destabilization as the predominant mechanism. Vråle and Jorden (*57*) discussed the importance of rapid mixing and recommended instantaneous blender-type mixing (<1 s), based on chemical theories of adsorption–destabilization. They also discussed the different destabilization theories regarding the mixing conditions. Amirtharajah and Mills (*46*) established boundaries under which conditions for each coagulation mechanism predominate (Figure 7). They did experiments to determine

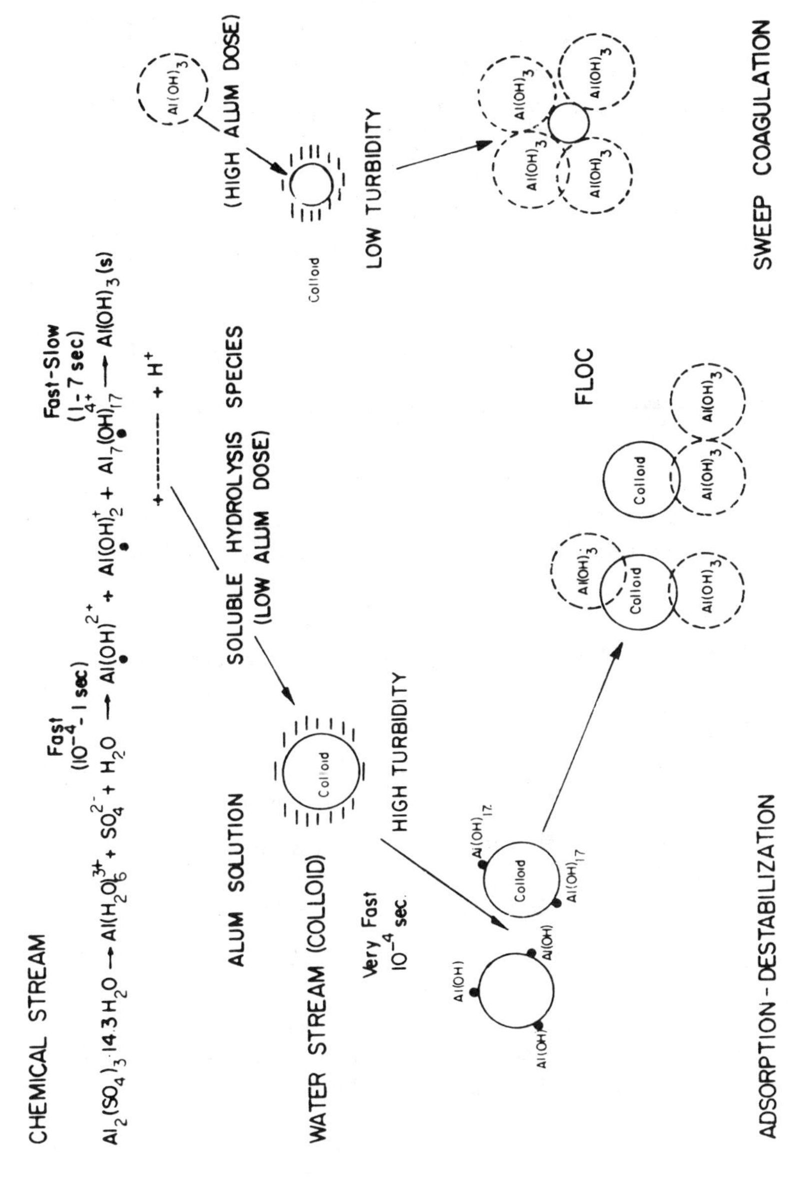

Figure 6. Reaction schemes of coagulation. (Reproduced with permission from ref. 46. Copyright 1982 American Water Works Association.)

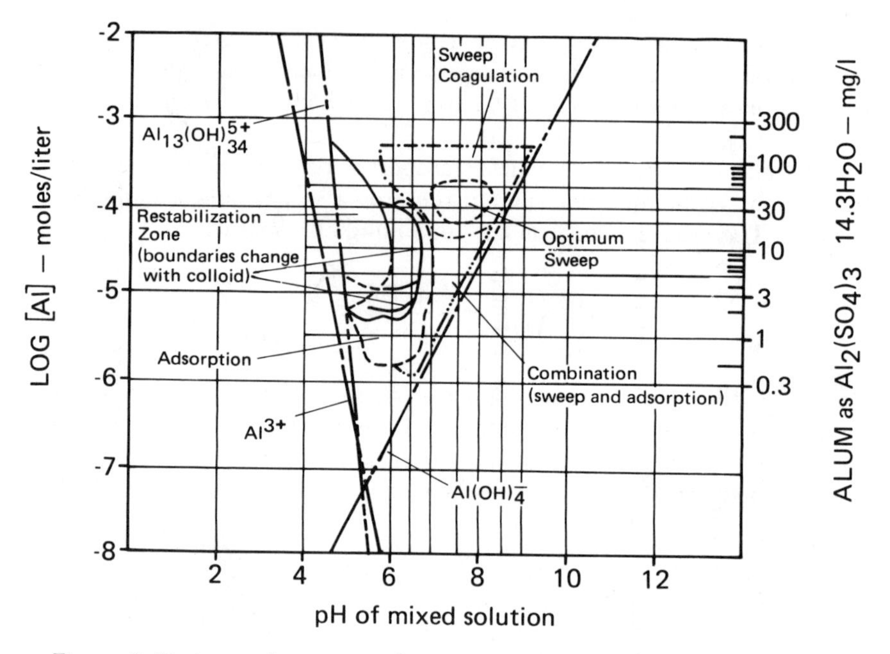

Figure 7. Design and operation diagram for alum coagulation. (Reproduced with permission from ref. 46. Copyright 1982 American Water Works Association.)

whether the kinetics of coagulation were influenced by the coagulant transport step needed for hydrolysis and precipitation of alum to occur. Their conclusions are as follows:

- Under optimum sweep conditions there is no difference in results for various mixing intensities (G values).
- Under adsorption–destabilization conditions, the high-intensity blender-type rapid mixing (high mixing intensity) is superior to the backmix reactor.
- In the restabilization region, there is no difference in the results.
- In the region defined as combination coagulation, the rapid-mixing mode produces no significant difference in the rate of coagulation–flocculation–sedimentation.

The velocity gradient or G-value concept is a simplistic and totally inadequate parameter for rapid-mixer design. However, until a more detailed understanding of rapid mixing is possible, this parameter will continue to be used as the only available means for designing the hardware of a rapid-mixer unit. Particle transport (flocculation) occurs either by Brownian motion (perikinetic) or by agitation-induced orthokinetic velocity gradients. Particle-transport efficiency depends on temperature, velocity gradients, volumetric particle concentration, flocculation time, and hydraulic properties of the reactor.

O'Melia (47) discussed particle transport as perikinetic and orthokinetic flocculation. Perikinetic flocculation requires a suspension of uniform-size particles. The role of change in the total number of particles is

$$\frac{dN}{dt} = \frac{-4kTN^2}{3\upsilon} \tag{3}$$

where N is the total number of particles at time t, T is the absolute temperature, k is Boltzmann's constant, and υ is the fluid viscosity. In orthokinetic flocculation, a homogeneous suspension is subjected to agitation that causes velocity gradients. The rate of change in the total number of particles is

$$\frac{dN}{dt} = -\frac{2}{3} G d^3 N^2 \tag{4}$$

where G is a mean velocity gradient and d is the diameter of the particles.

Equations 3 and 4 assume 100% efficiency of particle destabilization (coagulation). The relative importance of these mechanisms is

$$\frac{\text{orthokinetic rate}}{(\text{perikinetic rate})\ kT} = \frac{\upsilon G d^3}{2kT} \tag{5}$$

This ratio is unity at 25 °C, 1-μm particle diameter, and a velocity gradient of 10 s^{-1}.

Camp and Stein (58) presented an equation for estimating the mean velocity gradient, G (s^{-1}), in terms of the power input, P (watts), to the system as

$$G = \left(\frac{P}{\upsilon V}\right)^{1/2} \tag{6}$$

where υ is the dynamic viscosity of the liquid ($N\ s/m^2$) and V is the volume of the system (m^3).

In conventional jar tests, the rapid-mixing procedure is carried out at around 100 rpm, which gives a velocity gradient around 100 s^{-1}. This quantity is not sufficient to obtain good treatment results in the region where the adsorption–destabilization mechanisms dominate. Alternative laboratory testing techniques have been studied (59). The water to be treated was added through a syringe (250 mL) to the beaker containing the coagulant in the dose wanted. Water samples were taken at different time frequences, centrifuged, and analyzed. This method was found to be very good for simulating the direct filtration process (*59*). Chemical engineering approaches to studies of micromixing in the rapid-mixing process have been used (*60*). The authors described a new test reaction technique for studying the micromixing environment in a variety of mixing vessels. These results seem very promising and represent the first step in making a bridge between rapid-mixing problems and chemical-engineering-based models of flow and reaction.

Practical Experiences with Coagulation

Humic Substances Removal. Coagulation is one of the most important processes used in treatment of surface waters (*22*). Relatively few investigations reported in the literature deal specifically with the coagulation of naturally occurring organic substances (*38*), but much has been done with regard to color (*61*) and turbidity removal. Relatively few investigations deal with coagulation of humic and fulvic acids. The humic acid fraction of humic substances contributes more color per unit weight than the fulvic acid fraction, undoubtedly because of its larger size (*62*). Babcock and Singer (*40, 63*) showed that coagulation of the humic acid fraction resulted in 86% TOC removal, compared to only 22% TOC removal from the fulvic acid fraction. They also showed that chlorination of the fulvic acid fraction yields less than half as much chloroform as the humic acid fraction. This difference indicates that the coagulation process is an effective way of reducing trihalomethanes (THM).

Kavanaugh (*64*) discussed the wide range of TOC removal by coagulation that is reported in the literature. Symons et al. (*65, 66*) reported an average of 30% TOC removal in 63 plants using conventional water-treatment processes. About 60% TOC removal has been reported from Ohio River water (*65*). Kavanaugh (*64*) described studies from the Rhine and Alpine lakes in Germany that measured 25–40% removal of dissolved and colloidal organic carbon by alum at pH 7.0. Other authors (*39, 40*) have shown that up to 90% of the humic acid fraction can be removed by using Al(III) and Fe(III), although the removal of the fulvic acid fraction is more difficult. Color removal from highly colored surface waters with ferric sulfate was shown to range between 87 and 98% (*41*). Kavanaugh (*64*) showed that the optimum pH values for turbidity and TOC removal are not necessarily the same. Semmens and Field (*76*) pointed out the differences between turbidity and organic-substance removal, even though good turbidity always coincided with good organic-substance removal. Several scientists have studied the alum coagulation of aquatic humic substances (*25, 42, 59, 63, 67–70*). Vik showed that maximum TOC removal varied from 45 to 81% for three different lake waters. The water with the highest amount of lower-molecular-weight organic substances showed the highest coagulant demand and the lowest TOC removal. In most cases, however, TOC removal in the 60–80% range was achieved.

Edzwald (*31*) studied the use of cationic polymers as the sole coagulant in direct filtration and achieved approximately 40% TOC removal.

Coagulation Mechanisms. The main parameters affecting the coagulation of humic substances with alum include the initial concentration of humic substances in the raw water, the coagulant dosage, and the pH at which the coagulation process is carried out. A number of studies (*30, 38, 39, 53, 66–71*) have shown that optimum color removal with alum occurs in a pH range of 5–6.5.

Mangravite and co-workers (*42*) studied the mechanisms involved in the coagulation of humic substances when alum was used. Destabilization of the humic substances was observed in two regions, one at low coagulant dosage and low pH values and one at high alum dosages and higher pH values.

According to an American Water Works Association Committee Report (*38*), coagulation of humic substances with aluminum can be accomplished through two destabilization mechanisms: charge neutralization (adsorption) and precipitation. The destabilization accomplished by charge neutralization (adsorption) results from a specific chemical interaction between positively charged aluminum species and negatively charged groups of the humus colloids. This interaction could be accomplished over a narrow pH range (4–6). A stoichiometric relationship between the raw-water humic-substances concentration and the optimum coagulant dosage would be observed. Humic substances can form water-soluble and water-insoluble complexes with metal ions. As the alum dosage is increased, precipitation may occur. However, destabilization by this mechanism may incorporate the humic matter within the aluminum hydroxide floc (sweep coagulation) or coprecipitate it as aluminum humate. The optimum operation region for water-treatment plants worldwide is the optimum sweep area shown in Figure 7.

Vik (*67, 70*) studied the coagulation mechanisms involved in TOC removal from natural humic substances obtained from several Norwegian lakes, with raw-water TOC concentration varying from 6 to 15 mg of C/L. The coagulation process was studied using jar tests. For the sources studied, no restabilization was observed. The efficiency of the coagulation process depended on the efficiency of $Al(OH)_3(s)$ precipitation. Vik concluded that the poor mixing intensities achieved in the jar test might have reduced the possibility of destabilizing the humus by the adsorption mechanism (charge neutralization).

Other researchers (*42, 46*) have shown that optimum-color-removal areas tend toward lower pH ranges and higher alum dosages as the concentration of humic acid increases, but that color removal occurs in two areas, regardless of the concentration of color in the raw water. Results from Eikebrokk's work (*25*, *68*, *69*) are shown in Figure 8. Natural Norwegian humic water, blended with tap water to give a color of 30–45 mg of Pt/L and TOC of 3.3–4.2 mg of C/L, was used in the study. Alum in the dose range of 2–3 mg of Al/L gave good results in the pH ranges of 4.5–4.9 and 6.5–7.0. For direct-filtration purposes the study focused on the possibility of reducing the alum dosage in the low pH range. Figure 9 illustrates that at pH 4.7 and an alum dosage of 0.2 mg of Al/L, the color was reduced from 35 mg of Pt/L to <10 mg of Pt/L. These studies are being continued. The pilot-scale studies of direct filtration also showed that sweep coagulation at pH 6.5–6.8 resulted in early breakthrough in the filter if the flocs were formed at low velocity gradients (*G*-values) ($<100\ s^{-1}$), which means that they actually were suitable for removal by sedimentation rather than filtration.

Removal of humic substances by alum coagulation may turn out to be a particle separation problem. Turbidity created by the destabilization process may be difficult to remove, even in filters. Figure 9 also shows that the use of sulfate (H_2SO_4) instead of chloride (HCl) widens the optimum pH range. This finding has been reported by others (*73*).

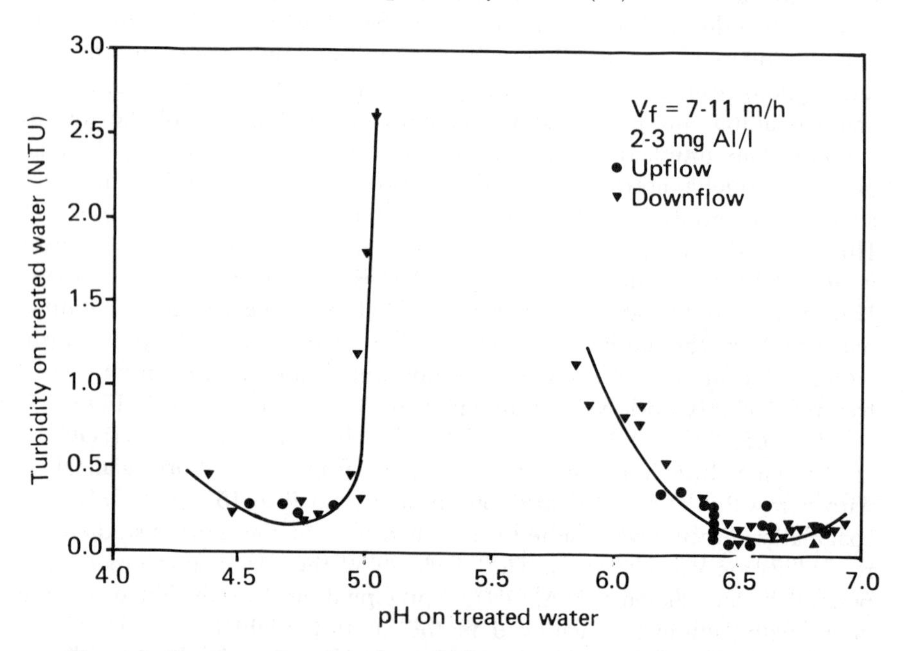

Figure 8. Optimum pH ranges for coagulation. Results are from direct filtration studies using upflow and downflow filters. (Reproduced with permission from ref. 68. Copyright 1984 Royal Norwegian Council of Scientific Sanitary Engineering Research.)

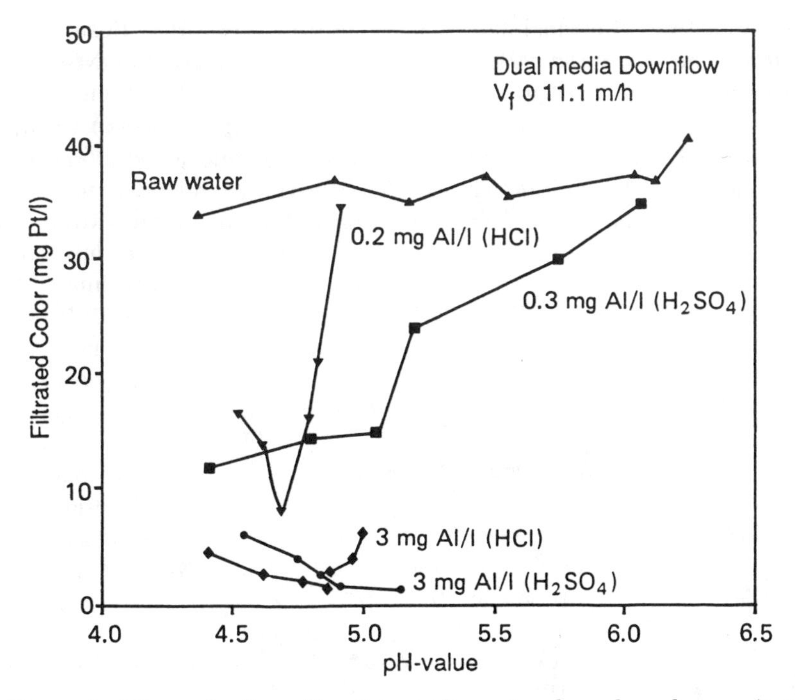

Figure 9. Pilot-scale direct-filtration studies using low alum dosages (0.2–0.3 mg of Al/L) and HCl and H_2SO_4 for pH adjustment to 4.7. The residual color was reduced from 35 to 10 mg of Pt/L. (Reproduced with permission from ref. 25. Copyright 1986 Royal Norwegian Council for Scientific and Industrial Research.)

Operational Aspects. The coagulation process is used all over the world for removal of turbidity and color. Many operators have recognized elevated levels of aluminum in finished waters, especially in low-temperature periods (winter). High aluminum content in finished water was found in a United States Environmental Protection Agency (U.S. EPA) 1983 survey, in Norway in a 1985 survey, and in several other countries. The content varies considerably, even up to a few milligrams per liter. U.S. EPA found residual aluminum concentrations varying from 0.01 to 2.7 mg/L; in Norway the residual aluminum concentration varied from 0.01 to 5 mg/L.

Aluminum does not provide any known benefit in water, and levels higher than 0.1 mg/L in coagulated water have been reported to cause precipitation of aluminum, with pH changes in the distribution system. The resulting sediment buildup creates sites for bacterial growth and concentration of micropollutants. Higher levels of aluminum concentration tend to bring increased galvanic corrosion caused by the contact of dissimilar metal oxides. The hydraulic capacity of the pipes is influenced by the buildup of interior coatings. Residual aluminum concentrations higher than expected could also be attributed to the metal complexation capacity of the humus molecule.

Vik (69) did jar-test experiments on the coagulation process for removal of humic substances from naturally occuring organic material from lake Tjernsmotjern. The residual color, UV absorbance, and TOC were measured, as well as the residual aluminum concentration. Figure 10 shows the

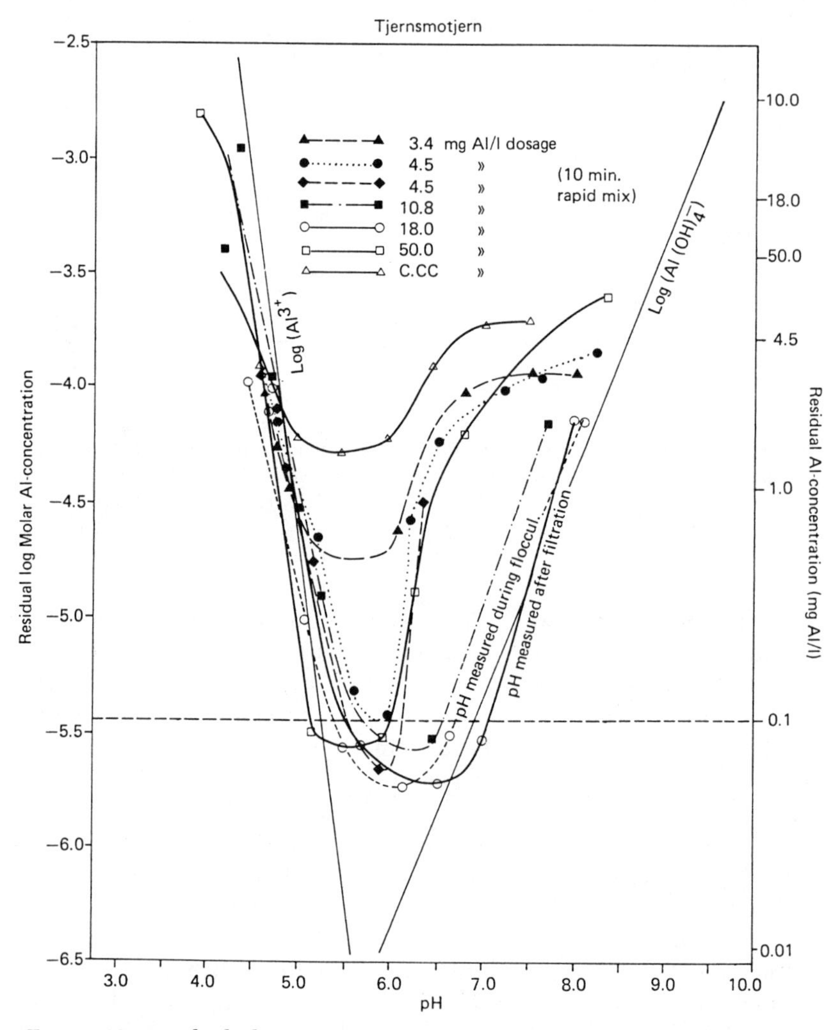

Figure 10. Residual aluminum concentration (log scale) as a function of pH value at various aluminum dosages. The jar-test experiments were carried out in the laboratory with rapid mixing at 100 rpm and temperature around 20 °C. (Reproduced with permission from ref. 70. Copyright 1982 University of Washington in Seattle.)

residual aluminum concentration related to the pH value for various dosages. A dosage greater than 10 mg of Al/L was necessary to achieve acceptable removal of humic substances under these conditions. The residual aluminum concentrations were higher than those theoretically calculated, possibly because of complexation of the Al ion and the humus molecule. The poor results achieved for humus removal at low alum dosages was suspected to be the result of the poor rapid mixing in the jar test. For full-scale conventional water-treatment plants, poor rapid-mixing conditions are probably common. In low-temperature winter conditions, the rapid-mixing effect is decreased with the same energy input. Decreasing the temperature from 20 to 0 °C increases the viscosity 50%. When the viscosity increases, the possibility of obtaining an even distribution of the coagulants in the water decreases (*74*).

Research Needs

Except for filtration, coagulation will continue to be the most important water-treatment process used for humus removal around the world. The relatively low cost of this process, compared to any other equally effective treatment for humic-substance removal (*75*), will make it even more important in the future, especially when the need for removal of specific organic compounds in the nanogram per liter range becomes more and more important. The coagulation and filtration processes will be used in the future as pretreatment methods, and membrane or adsorption processes (e.g., activated carbon) will be included for removal of specific organic compounds.

There are, however, still many possible ways to make the process more efficient and cheaper, especially in the area of operational aspects. Ability to use the direct filtration plant even for high concentrations of humic substances requires a more efficient coagulation process, one that will create destabilization at low coagulant dosages. More reliable processes that give low residual coagulant concentrations in the effluent water will be important. Other ways to improve the mixing process, especially rapid mixing and floc separation, are being studied. Better laboratory testing techniques are needed, including modified jar tests and analytical tools for determining the colloidal charge in the solution, simple tools for determining the actual mixing intensities being used, and better floc separation test equipment. There is an increasing public demand for alternative coagulants, like iron products or lime products. We need to develop integrated processes, such as coagulation and corrosion control.

In spite of all the experiments undertaken, we still do not fully understand all aspects of this process, such as the importance of the characteristics of the humic substances being used (e.g., molecular weight distribution, colloidal charge, pH for zero charge) and the influence of the turbidity and ion content of the water. The influence of the metal-complexing capacity of the humus molecule on the process of coagulation is not fully understood. Futhermore, the high residual aluminum concentrations found in treated water from alum coagulation plants could be related to this phenomenon. Because of the low rapid-mixing intensities used in full-scale treatment plants, it is impossible to achieve acceptable treatment results with low-range alum dosages at low pH values. This combination is necessary in order to make the direct-filtration process suitable even for high raw-water concentrations of humic substances. Future research into the coagulation process should include basic chemical engineering approaches.

References

1. Gjessing, E. T. *Physical and Chemical Characteristics of Aquatic Humus*; Ann Arbor Science: Ann Arbor, 1976; Chapters 2–4.
2. Thurman, E. M. *Organic Geochemistry of Natural Waters*; Martinus Nijhoff/ Dr. W. Junk Publishers (Kluwer): Boston, 1985; Chapters 10 and 11.
3. Wetzel, R. G. *Limnology*; W. B. Saunders Company: Philadelphia, 1976; pp 583–621.
4. Schnitzer, M.; Khan, S. V. *Humic Substances in the Environment*; Marcel Dekker: New York, 1972.
5. Stevens, A. A.; Slocum, C. J.; Seeger, D. R.; Robeck, G. G. *J. Am. Water Works Assoc.* **1976,** *68*(11), 615.
6. Rook, J. J. *Water Treat. Exam.* **1974,** *23*(2), 234.
7. Bellar, T. A.; Lichtenberg, J. J.; Kroner, R. C. *J. Am. Water Works Assoc.* **1979,** *71*(11), 703.
8. Stevens, A. A. Presented at Karlsruhe Conference on Oxidation Techniques, CCMS 111; U.S. Environmental Protection Agency, U.S. Government Printing Office: Washington, DC, 1979; EPA–570/9–79–020.
9. Organization for Economic Cooperation and Development (OECD). *Control of Organochlorinated Compounds in Drinking Water*; Paris, 1980.
10. Oliver, B. G. *Can. Res.* **1978,** *11*(6), 21.
11. Stevens, A. A.; Slocum, C. J.; Seeger, D. R.; Robeck, G. G. U.S. Environmental Protection Agency: Washington, DC, 1980; Quarterly Report, Task 19.
12. Christman, R. F.; Johnson, J. D.; Pfaender, F. K.; Norwood, D. L.; Webb, M. R.; Haas, J. R.; Babenrieth, M. J. In *Water Chlorination Environmental Impact and Health Effects*; Ann Arbor Science: Ann Arbor, 1980; Volume 3.
13. Peters, C. J.; Perry, R. Presented at the International Association on Water Pollution Research conference, Edmonton, Alberta, 1980.
14. Eidness, F. *Willing Wat.* **1974,** *18,* 12.
15. Meybeck, M. In *Flux of Organic Carbon by Rivers to the Ocean*; Likens, G. E., Ed.; U.S. Department of Energy, 1981, National Technical Information Service Report, Conf.–8009140, UC–11, Springfield, VA.
16. Stumm, W.; Morgan, J. J. *Aquatic Chemistry*; Wiley and Sons: New York, 1981.
17. Oden, S. *Kolloidchem. Beih.* **1919,** *11,* 75–260.
18. Thurman, E. M.; Wershaw, R. L.; Malcolm, R. L.; Pinckney, D. J. *Org. Geochem.* **1982,** *4,* 27–35.
19. Liao, W.; Christman, R. F.; Johnson, J. D.; Millington, D. S.; Hass, J. R. *Environ. Sci. Technol.* **1982,** *16,* 403–410.
20. Stevenson, F. J. *Humus Chemistry*; Wiley and Sons: New York, 1982.
21. Christman, R. F.; Gjessing, E. T. *Aquatic and Terrestrial Humic Materials*; Ann Arbor Science: Ann Arbor, 1983.

22. Wershaw, R. L; Mikita, M. A. *NMR of Humic Substances and Coal: Techniques, Problems and Solutions*; Wiley: New York, 1987.
23. Sontheimer, H. In *The Scientific Basis of Flocculation*; Ives, K. J., Ed.; Sijthoff and Noordhoff: Alphenan den Rijn, Netherlands, 1978.
24. AWWA Committee Report. *J. Am. Water Works Assoc.* **1980,** *72*(7), 405.
25. Eikebrokk, B. Technical Report No. 55/86; Royal Norwegian Council for Scientific and Industrial Research (NTNF): Trondheim, Norway, 1986; Program on Sanitary Engineering Research (VAR).
26. Thompson, C. G.; Singley, J. E.; Black, A. P. *J. Am. Water Works Assoc.* **1972,** *64*(1), 11.
27. Thompson, C. G.; Singley, J. E.; Black, A. P. *J. Am. Water Works Assoc.* **1972,** *64*(2), 93.
28. Kisla, Thomas C.; McKelvey, Ronald D. *Environ. Sci. Technol.* **1978,** *12,* 207.
29. Edzwald, J. K.; Haff, J. D.; Boak, J. W. *J. Environ. Eng. Div.* **1977,** *103*(12), 989.
30. Glaser, Harold T.; Edzwald, James K. *Environ. Sci. Technol.* **1979,** *13*(1), 631.
31. Edzwald, J. K.; Becker, W. C.; Tambini, S. J. *J. Environ. Eng. Div.* (*Am. Soc. Civ. Eng.*) **1987,** *113*(1), 167.
32. Edzwald, J. K. In *Organic Carcinogens in Drinking Water;* Ram, N. M.; Calabrese, E.; Christman, R. F., Eds.; Wiley and Sons: New York, 1986.
33. Scheuch, L. E.; Edzwald, J. K. *J. Am. Water Works Assoc.* **1981,** *73*(9), 497–502.
34. Tate, C. H.; Trussell, R. R. *J. Am. Water Works Assoc.* **1980,** *72*(3), 165–169.
35. Rebhun, M.; Fuhrer, Z.; Adin, A. *Water Res.* **1984,** *18*(8), 963–970.
36. Amy, G. L.; Chadik, P. A. *J. Am. Water Works Assoc.* **1983,** *75*(10), 527–531.
37. Johnson, D. E.; Randtke, S. J. *J. Am. Water Works Assoc.* **1983,** *75*(5), 249.
38. AWWA Committee Report. *J. Am. Water Works Assoc.* **1979,** *71*(10), 558.
39. Hall, E. S.; Packham, R. F. *J. Am. Water Works Assoc.* **1965,** *57*(9), 1149.
40. Babcock, D. S.; Singer, P. C. *J. Am. Water Works Assoc.* **1979,** *71*(3), 148.
41. Black, A. P.; Christman, R. F. *J. Am. Water Works Assoc.* **1963,** *55*(6), 753.
42. Dempsey, B. A.; Ganho, R. M.; O'Melia, C. R. *J. Am. Water Works Assoc.* **1984,** *76*(4), 141–150.
43. Matijevic, E.; Janauer, G. E.; Kerker, M. *J. Colloid. Sci.* **1964,** *19,* 333.
44. Matijevic, E.; Stryker, L. J. *J. Colloid Sci.* **1966,** *22,* 68.
45. Baes, C. F., Jr.; Mesmer, R. E. *The Hydrolysis of Cations*; Wiley and Sons: New York, 1976.
46. Amirtharajah, A.; Mills, K. M. *J. Am. Water Works Assoc.* **1982,** *74*(4), 210–216.
47. O'Melia, C. R. In *Physicochemical Processes for Water Quality Control*; Weber, W. J., Ed.; Wiley–Interscience: New York, 1972.
48. Hayden, P. L.; Rubin, A. J. In *Aqueous Environmental Chemistry of Metals*; Rubin, A. J., Ed.; Ann Arbor Science: Ann Arbor, 1974.
49. Matijevic, E. *J. Colloid Interface Sci.* **1973,** *43*(2), 217.
50. Seip, H. M. Central Institute for Industrial Research: Oslo, Norway, unpublished results.
51. Hem, J. D. *Graphical Methods for Studies of Aqueous Aluminum Hydroxides, Fluorides and Sulfate Complexes*; U.S. Geological Survey. U.S. Government Printing Office: Washington, DC, 1968; Water Supply Paper, 1068, No 1827 B.
52. Bratby, J. *Coagulation and Flocculation*; Upland Press: Croydon, England, 1980.
53. Narkis, N.; Rebhun, M. *J. Am. Water Works Assoc.* **1975,** *62*(2), 101.
54. Hahn, H. H.; Stumm, W. *J. Colloid Interface Sci.* **1968,** *28*(1), 134.
55. Rubin, A. J.; Kovac, T. W. In *Chemistry of Water Supply Treatment and Distribution*; Rubin, A. J., Ed.; Ann Arbor Science: Ann Arbor, 1974; Chapter 8.
56. McCooke, N. J.; West, J. R. *Water Res.* **1978,** *42,* 793.
57. Vråle, L.; Jorden, M. *J. Am. Water Works Assoc.* **1971,** *63, 1,* 52.
58. Camp, T. R.; Stein, P. C. *J. Boston Soc. Civ. Eng.* **1943,** *30,* 219.

59. Dolejs, P. In *Proc. Int. Conf. Stud. Environ. Sci., Leuwen, Belgium;* Pawlowski, L.; Gond Lacy, W. J., Eds.; Elsevier: Amsterdam, 1985.
60. Clark, M. M.; David, R.; Wiesner, M. R. In *Proc. AWWA Annu. Conf.* **1987,** 1957–1975.
61. AWWA Committee Report. *J. Am. Water Works Assoc.* **1967,** *69*(8), 1023.
62. Ghassemi, M.; Christman, R. F. *Limnol. Oceanogr.* **1968,** *13*, 583.
63. Reckhow, D. A.; Singer, P. C. *J. Am. Water Works Assoc.* **1984,** *76*(4), 151.
64. Kavanaugh, M. C. *J. Am. Water Works Assoc.* **1978,** *70*(11), 613.
65. Symons, J. M.; Bellar, T. A.; Carswell, J. K.; DeMarco, J.; Kropp, K. L.; Robeck, G. G.; Seeger, D. R.; Slocum, C. J.; Smith, B. L.; Stevens, A. A. *J. Am. Water Works Assoc.* **1975,** *67*(11), 634.
66. Symons, J. M.; Stevens, A. A.; Clark, R. M.; Geldreich, E. D.; Love, O. T., Jr.; Demarco, J. *Treatment Techniques for Controlling Trihalomethanes in Drinking Water;* U.S. Environmental Protection Agency: Cincinnati, OH, 1981; EPA–600/2–81–156.
67. Vik, E. A.; Carlson, D. A.; Eikum, A. S.; Gjessing, E. T. *J. Am. Water Works Assoc.* **1985,** *77*(3), 58.
68. Eikebrokk, B. Project Report 6/84; Royal Norwegian Council of Scientific Sanitary Engineering Research: Trondheim, Norway, 1984.
69. Eikebrokk, B. Ph.D. Thesis; Technical University of Trondheim, Norway, 1982.
70. Vik, E. A. Ph.D. Thesis; University of Washington in Seattle, 1982.
71. Kawamura, S. *J. Am. Water Works Assoc.* **1976,** *68*(6), 328.
72. Mangravite, F. C., Jr.; Buzzell, T. D.; Cassell, E. A.; Matijevic, C. E.; Saxton, G. B. *J. Am. Water Works Assoc.* **1975,** *67*(2), 88.
73. de Hek, H.; Stol, R. J.; Bruun, P. L. *J. Colloid Interface Sci.* **1978,** *64*(3), 72.
74. Hedberg, T. Technical Report No. 75:2, 1975; Chalmers Technical University of Gothenburg, Gothenburg, Sweden.
75. Hem, L. J. Project Report 22/87, Royal Norwegian Council for Scientific Drinking Water Research: Oslo, Norway, 1987.
76. Semmens, M.; Field, T. K. *J. Am. Water Works Assoc.* **1989,** *72*(8), 476.

RECEIVED for review January 4, 1988. ACCEPTED for publication April 19, 1988.

25

Reactions Between Fulvic Acid and Aluminum

Effects on the Coagulation Process

Brian A. Dempsey

Department of Civil Engineering, Pennsylvania State University, University Park, PA 16802

The effects of fulvic acids on the speciation of aluminum are measured by timed colorimetric analyses. Precipitates of $Al(OH)_3(s)$ (pK_{sp} = *32.8 for pH 4.5–6.5) form in every region where alum has been shown to be successful for the removal of fulvic acids. Stability functions (average log K = 3.39) are reported for the formation of soluble aluminum–fulvic acid complexes. Adsorption functions (fulvic acid on freshly precipitated* $Al(OH)_3(s)$*) are more than 10 times larger than the stability functions for complexation of fulvic acids with dissolved aluminum.*

FULVIC ACID (FA) IS USUALLY REMOVED from raw water by the coagulation process followed by sedimentation and rapid sand filtration. Filter alum ($Al_2(SO_4)_3 \cdot 14.3H_2O$) is the coagulant most commonly used in the United States. However, ferric chloride, organic polyelectrolytes, and other salts of aluminum(III) or iron are also used. I have previously reported on the removal of FA with salts of aluminum (*1–3*).

The objectives of the work presented here are to determine the effects of FA on the speciation of aluminum (especially when filter alum is used as the coagulant) and to use this information to predict the mechanism of FA removal during water treatment. The experimental time frame in this study (minutes to days) corresponds to the hydraulic residence time of conventional water-treatment and distribution systems. The relatively short time for re-

0065–2393/89/0219–0409$06.00/0

action and the presence of fulvic acid means that analytically identifiable crystalline aluminum oxides or hydroxides are not expected to be important in these systems.

Removal of Contaminants Using Aluminum Salts

Amirtharajah and Mills (*4*) studied the removal of suspended solids from water with alum as the coagulant. They identified pH values and alum doses that resulted in the formation of a voluminous precipitate of $Al(OH)_3(s)$. Combinations of pH and coagulant dose that result in a heavy floc that settles are usually identified as the "sweep-floc" zone. The use of these conditions for the coagulation process results in good removal of clays, FA, and other contaminants.

Amirtharajah and Mills (*4*) also identified pH and coagulant dose values at which removal of the contaminants occurs by the charge-neutralization mechanism. In this case, the negative charge of the contaminants is just neutralized by the cationic coagulant species. An equivalent coagulant-to-contaminant dosage must be used to produce charge neutralization. Overdosing results in restabilization. Lower doses of chemical are required for charge neutralization than for the sweep-floc zone. Although using lower doses has potential advantages, contaminant removal rates are often lower and operational control may be more difficult than for the sweep-floc zone.

Amirtharajah (*5, 6*) explained his experimental results for coagulation with filter alum by assuming that $Al(OH)_3(s)$ forms whenever coagulation is successful. However, he noted that the results are also consistent with destabilization by soluble polymeric species of aluminum. Edwards and Amirtharajah (*7*) reported that the removal zones for humic acid involve pH and alum doses similar to those required for the removal of clays, except that the stability zone shifts slightly toward lower pH values (*7*).

Sricharoenchaikit (*8*) also assumed that the precipitation of $Al(OH)_3(s)$ precedes the destabilization of contaminant species. Packham and associates (*9–11*), on the other hand, suggested that FA and the dissolved aluminum from alum can react directly to form a precipitate without the preliminary formation of $Al(OH)_3(s)$. They stated that FA is precipitated by soluble hydrolyzed species of aluminum when salts of aluminum are used as coagulants (*9*), whereas they suggested that removal of clays is dependent on the preliminary precipitation of $Al(OH)_3(s)$ (*10, 11*). Rebhun and Narkis (*12*) suggested that removal of humic materials by alum near pH 6.7 is due to direct precipitation by the polymeric species $Al_8(OH)_{20}{}^{4+}$. Matijevic and co-workers (*13, 14*), Hayden and Rubin (*15*), and Stumm and associates (*16, 17*) also suggested that polymers of hydrolyzed aluminum are active reagents when filter alum is used as a coagulant.

Dempsey and co-workers (*1–3*) used various coagulants that contain aluminum to investigate the conditions required for the coagulative removal of fulvic acids. Some of their results are shown in Figure 1A. The black area designated I in Figure 1A corresponds to conditions in which a sweep-floc can be generated when aluminum is added in the form of filter alum. Removal of fulvic acid apparently occurs by charge neutralization with conditions designated by black area II. Under conditions within the gray dotted region, fulvic acids can be removed by membrane filtration, but not by sedimentation. Dempsey and co-workers (2) suggested that removal using alum at pH above 6 is due to adsorption of FA on $Al(OH)_3(s)$, but that some removal by alum at lower pH values is due to the direct precipitation of FA by polymers or even monomers of aluminum hydroxide.

Polyaluminum chloride (PAC) is a commercially available coagulant with the formula $Al(OH)_x(Cl)_y(SO_4)_z$. Typically, x is 1.2–2.0, and z is usually 0.16 or less. Both direct and indirect evidence indicates that PAC contains thermodynamically stable polymers of aluminum, especially $AlO_4(Al(OH)_2)_{12}$

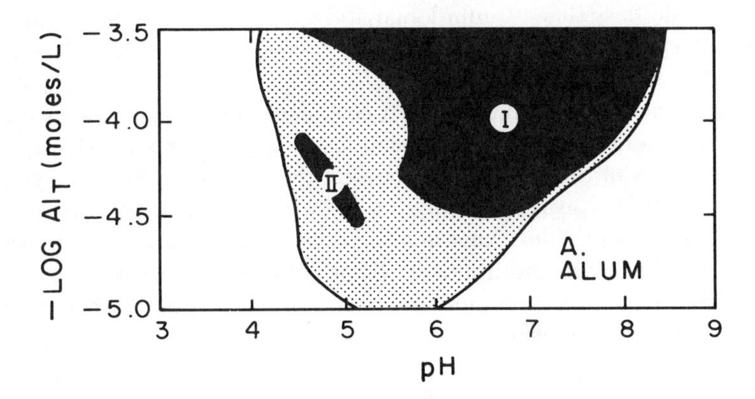

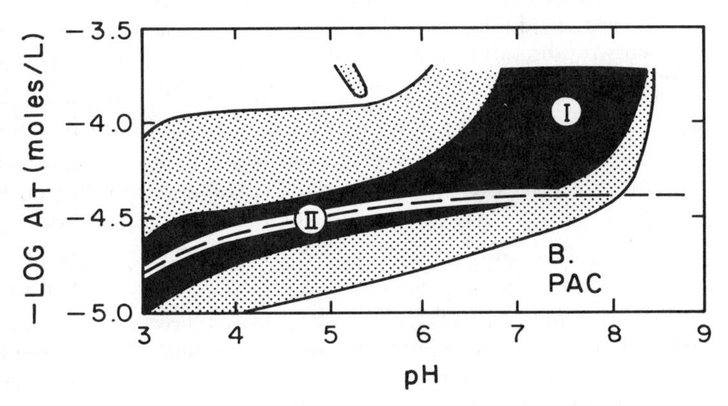

Figure 1. Aluminum doses and pH values that permit removal of fulvic acid from water when (A) alum or (B) polyaluminum chloride is the coagulant.

(subsequently abbreviated Al_{13}); this radical has a +7 charge. PAC achieves good removal of contaminants even at pH less than 4 and Al_t less than $10^{-4.5}$ M), conditions under which neither $Al(OH)_3(s)$ nor Al_{13} are thermodynamically stable (part of the black areas in Figure 1B). This coagulative behavior, analogous to that of synthetic organic polyelectrolytes (*12, 18*), indicates that the polymeric species in PAC are relatively inert.

Some of the investigators cited have measured both the removal of contaminants and the speciation of hydrolyzed aluminum. Evidence for the presence or activity of a certain species of hydrolyzed aluminum is usually based on indirect experimental evidence and logical argument (*2, 9, 13–15, 19, 20*). Hundt (*21*) characterized the species that form after the addition of alum, aluminum chloride, or PAC to water in the absence of FA or clays. He showed that when alum was added and the pH was above 4.5, most of the aluminum was retained by a membrane filter. In contrast, filterable species persisted until higher pH values for PAC. Tambo (*22*), using alum, isolated labile (less than 10-min longevity) species of aluminum with highly positive electrophoretic mobilities. These species have not been determined to be either solid or polymeric.

The speciation of the aluminum coagulants, after addition to the raw waters, is uncertain. Critical compilations of thermodynamic data (*23, 24*) and evaluations of the literature regarding the hydrolysis of aluminum (*24, 25*) are available. However, reports disagree regarding the best free energy values, and enthalpy values have not been obtained for many of the hydrolysis reactions of aluminum. Additionally, the short time frame of the coagulation process and the presence of contaminants during coagulation make the prediction of speciation difficult.

Experimental Methods

The collection, extraction, cleanup, preservation, and characterization of FAs have been previously described (*26*). Two FAs, designated FA1 and FA4, were used in these experiments. Both of these materials are derived from Lake Drummond, VA. Solid-sample ^{13}C NMR studies indicate that approximately 80% of the organic carbon is aliphatic. FA1 has 11.4 meq and FA4 has 11.6 meq of carboxyl functional groups per g of organic carbon; more than a third of these acidic groups are ionized at pH 3. Aldrich humic acid (HA) was also used in a few experiments. Water was distilled and then treated by a microfilter (Milli-Q) system. All other chemicals were reagent grade or better.

Experiments were run in 40-mL glass sample bottles with poly(tetrafluoroethylene) (Teflon)-lined caps at room temperature (23–25 °C). FA solution and stock acetate were added to water to give between 0 and 84 mg/L of FA or HA organic carbon and total acetate of 8×10^{-4} M. For comparison, 2.1 to 84 mg/L of FA4 contains 2.4×10^{-5} to 9.7×10^{-4} M of carboxylic functional groups. The pH was adjusted to the desired value by using HCl or NaOH. Then the appropriate amount of alum was injected and the solution was mixed. The stock alum solution contained 0.0018 M total Al(III). The final concentration of aluminum varied, but it was 1.05 mg/L (3.9×10^{-5} M) in most cases. Samples were taken at 5 min, 1 h,

4 h, 1 day, and sometimes 7 days for the analysis of aluminum. The pH was measured for each sampling period. Except for the 5-min samples and when filtration was used as a separation process, the sample bottles were placed in an ultrasonic bath for 10 min prior to sampling in order to break up any particulates that may have formed.

The concentration and speciation of aluminum was determined by the ferron (8-hydroxy-7-iodoquinolinesulfonic acid) method. A 2-mL aliquot of the analyte was placed in a cuvette (path length = 1 cm) and 0.8 mL of the ferron reagent as described by Bersillon (*20*) was added. The cap was inserted, and macromixing was completed within 5 s of the ferron addition. The pH after addition of ferron must be very consistent (*27*); addition of acid for digestion of aluminum species should be avoided unless the sample is back-titrated or the molar absorptivity is shown to be unchanged. Absorptivity was 7.78 mM^{-1} cm^{-1}, with a standard deviation of 0.03 at 370 nm. Monomeric aluminum reacts very rapidly, with a reported pseudo-first-order k = 2.3 min^{-1} (*28*).

This value was confirmed in our experiments for times greater than 15 s, but the reaction was even faster before 15 s, so that 85% of the total aluminum in dilute alum solutions had reacted within 30 s. Reaction rates for most other species of aluminum are considerably slower (*19, 27, 28*) and the blank-corrected absorbance at 30 s is claimed to represent inorganic monomeric aluminum. Although the rate of nonmonomeric color development increased with increasing concentrations of either FA or total aluminum, the extrapolated contribution of these nonmonomeric species to the 30-s reading was typically less than 10% of the 30-s reading.

This contribution was determined on the basis of the slopes (absorbance versus time) at 2 min, when the monomeric aluminum was 99% reacted. Reagent blanks were analyzed by measuring the absorbance of samples (minus aluminum) against distilled–deionized (DDI) water. Ferron reagent was added to both cuvettes. Sample blanks were analyzed by comparing the absorbance of the whole samples (including aluminum and FA) against DDI water; ferron reagent was not added.

Solutions of polyaluminum chloride that contain the polymer $AlO_4(Al(OH)_2)_{12}{}^{7+}$ (abbreviated Al_{13}) have an initial rate of color development of 0.071 min^{-1} (our data) to 0.075 min^{-1} (*28*); thus, only 3.5% of such aluminum is reacted with ferron at 30 s. The rate of color development from the PAC that was used in these experiments slows substantially, however, so that only 57% of the Al_{13} is reacted after 22 h. On the other hand, the aluminum in suspensions that are predominantly $Al(OH)_3(s)$ is sometimes totally reacted in less than 1 h, a result indicating greater lability with respect to the ferron reagent than for Al_{13} or other polymeric materials. As a result, definitions of the aluminum that reacts with ferron in 2 h as monomeric plus polymeric (Al^a plus Al^b) and the remaining aluminum (Al^c) as $Al(OH)_3(s)$ cannot be justified for the situations that we have studied.

Some plots of absorbance versus time are shown in Figure 2 for diluted alum (very rapid reaction), Al_{13} (very slow and monotonic reaction), amorphous $Al(OH)_3(s)$ (S-shaped curve), and aluminum that is complexed by FA (very rapid reaction for inorganic monomeric aluminum and very slow, monotonic reaction for organically bound aluminum). The S-shaped curve occurred at some sample time (typically at 5 min, 1 h, and 4 h after coagulation) in every case in which $Al(OH)_3(s)$ was visually observed. The inflection point in the S-shaped curve always occurred within 2 h after the addition of ferron reagent, and the incremental absorbance that occurred after the inflection point could often be removed by membrane filtration. In this work the ferron test is used to determine monomeric (30-s) aluminum and as evidence for the presence of $Al(OH)_3(s)$. These two uses have been corroborative in every case.

Some data for filterable aluminum are presented in this chapter. Filterable aluminum is defined as the fraction that passes membrane filters with 0.2-μm pores.

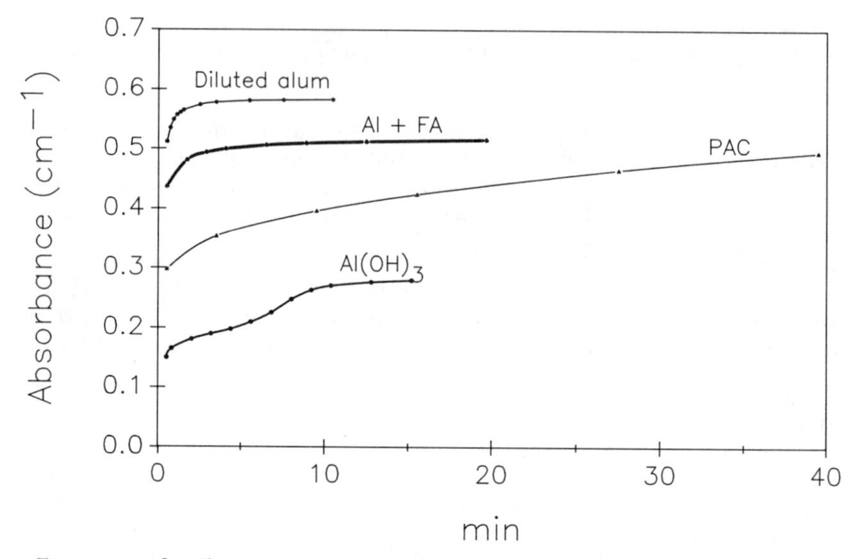

Figure 2. Absorbance versus time after addition of the ferron reagent for diluted alum (no base added), polyaluminum chloride (C_0 = 1.3 M), amorphous $Al(OH)_3(s)$, and aluminum complexed with FA.

Aliquots of 10 mL of filtrate per cm^2 of exposed filter were discarded before keeping filtrate for analysis. The filtrate was analyzed by electrothermal atomic absorption spectroscopy (*29*) or by ferron colorimetric technique.

The monomeric speciation of aluminum was calculated by using data from the 30-s ferron reactions and equilibrium constants taken from Smith and Martell (*23*) or Baes and Mesmer (*24*). Thus the overall stability constants (log β^* at ionic strength I = 0) were –4.99, –9.3, –15.0, and –23.0. Activity coefficients were calculated with the extended Debye–Hückel equation (*30*) and I = 0.001. The logarithmic values of the apparent overall constants for the sequential hydrolysis of Al^{3+} were therefore –5.07, –9.42, –15.13, and –23.12.

Precipitation of $Al(OH)_3(s)$

One objective of this work is to determine the effects of FA on the speciation of aluminum during coagulation processes. Experiments have been run to determine whether $Al(OH)_3(s)$ forms during the coagulation process. A test for the presence of this phase can be performed by observing whether consistent values exist for the ion activity product ($\{Al^{3+}\}\cdot\{OH^-\}^3$); a consistent value occurs when the slope is –3.0 for the plot log $[Al^{3+}]$ versus pH. Additional tests for the presence of $Al(OH)_3(s)$ are an S-shaped ferron trace and the removal of nonmonomeric aluminum by membrane filtration.

Data are presented first for experiments run for pH values between 3.5 and 7. The calculated concentrations of Al^{3+} (based on the 3-s ferron test and the apparent hydrolysis constants) are shown in Figure 3. These data are for analyses run at 4 h and 1 day. The total aluminum concentration was either 3.9×10^{-5} or 7.8×10^{-5} M, and the concentration of humic substances (FA1, FA4, or Aldrich humic acid) was either 0 or 4.2 mg/L of organic carbon (i.e., 4.8×10^{-5} eq/L of total carboxylic acid groups).

The concentration of the Al^{3+} species and the shape of the ferron trace are constant until a pH of about 4.75. At all higher pH values and for 0 or 4.2 mg/L of DOC, the concentration of monomeric aluminum is diminished and the ferron trace assumes the S-shape that has been identified with amorphous $Al(OH)_3(s)$. Nonmonomeric aluminum can be removed by membrane filtration. A linear least-squares analysis of the data for pH 4.75–6.5 reveals that

$$\log [Al^{3+}] = 9.06 - 2.96 \text{ pH} \qquad (r^2 = 0.987) \tag{1}$$

where $[Al^{3+}]$ refers to concentration. The theoretical slope is –3.00 if aluminum speciation is controlled by $Al(OH)_3(s)$. The pK_{sp} (corrected to zero ionic strength) was calculated for each datum within the given range of pH values; the average and standard deviation for pK_{sp} was 32.82 ± 0.17. These data are consistent with the formation of microcrystalline gibbsite. Nordstrom and Ball (*31*) and Driscoll and co-workers (*32*) discovered similar relationships between $[Al^{3+}]$ and pH for stream waters affected by acid-mine drainage and by acid precipitation, respectively.

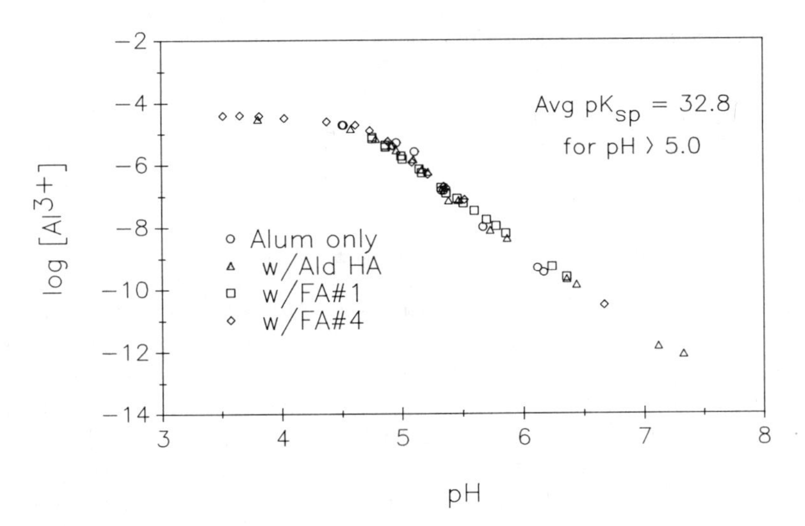

Figure 3. log $[Al^{3+}]$ versus pH under several experimental conditions. The $[Al^{3+}]$ is based on the 30-s ferron result (this research) and stability constants (23, 24).

Moderate concentrations of humic materials (up to at least 5 mg/L of DOC) enhance the precipitation of $Al(OH)_3(s)$ at the slightly acidic pH values (4.75–5). This effect of humic materials on the precipitation of $Al(OH)_3(s)$ is illustrated in Figure 4, where the inorganic monomeric aluminum (30-s ferron results) or Al^a is plotted as a function of increasing concentration of humic material.

Changes in pH account for a minor part of the decreases in inorganic monomeric aluminum; the range of pH values was 0.01 for the Aldrich HA (pH 5.0) tests and 0.09 for the FA1 (pH 5.0) tests. The sample blanks for these tests were identical to the reagent blanks, a result indicating that the changes in absorbance that are shown in Figure 4 are not due to analytical anomalies, such as unaccounted light scattering. The minimum value for inorganic monomeric aluminum occurs when the FA normality (in total carboxylic groups) is equivalent to the molar concentration of aluminum (1.05 mg/L). The charge density of FA1 is greater than that for Aldrich HA. These conditions have been designated zone II in Figure 1A.

At much higher concentrations of humic materials (e.g., the curve in Figure 4 for FA1 and pH 5.0 for DOC greater than 20 mg/L), the ferron trace assumes completely monotonic response at all reaction times. The concentration of inorganic monomeric aluminum decreases with increasing DOC. No filterable precipitate is formed. These observations are consistent with the formation of a soluble complex between aluminum and the humic materials.

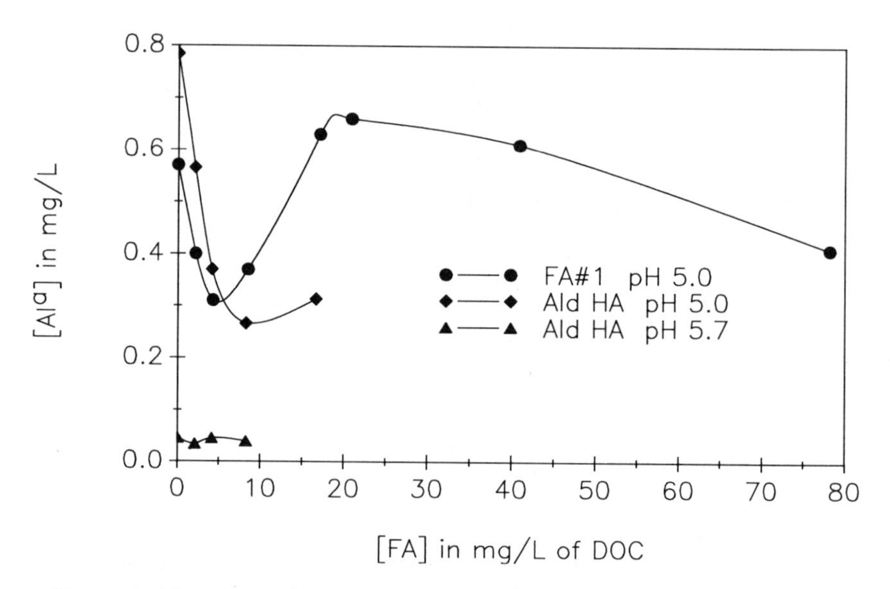

Figure 4. Monomeric aluminum as determined by the 30-s ferron test versus the concentration (DOC) of fulvic acid. See text for explanation of the inflection points.

The concentration of inorganic monomeric aluminum increases between DOC of 4.2 and 20 mg/L of FA1 and pH 5.0 (Figure 4). These conditions are analogous to the gray-hatched area below zone II in Figure 1A (i.e., a region in which the ratio of coagulant to humic material is inadequate for charge neutralization). The increase in inorganic monomeric aluminum may be due to an increasing disorder (and specific surface area) for $Al(OH)_3(s)$ with excessive and increasing concentration of the humic materials. The rate of the increase in absorbance for the nonmonomeric fraction of aluminum is directly proportional to the concentration of humic material. In addition, the ferron trace is S-shaped, and some aluminum can be removed by filtration (but not by sedimentation) in this zone.

The part of this study that deals with the speciation of aluminum at slightly alkaline pH values is an extension of prior work (*2*), which showed a linear relation at pH 7.2 between the concentration of humic materials and the required dose of alum for the removal of the FA by coagulation and filtration. Thus, removal of 3.5 mg/L of DOC required at least 1.3 mg/L of aluminum, 10 mg/L of DOC required at least 3.6 mg/L of aluminum, and 35 mg/L of DOC required at least 13 mg/L of aluminum before any removal of FA occurred. Some of these results are displayed by the curves in Figure 5. These doses of aluminum are orders of magnitude in excess of the solubility of $Al(OH)_3(s)$ at pH 7.2 and in the absence of humic materials.

These experiments were repeated to determine the speciation of the aluminum. Experiments were performed for 0, 3.5, and 10.0 mg/L of DOC (FA1) and for five concentrations of total aluminum. The DOC that passes through a membrane filter is shown as a function of the coagulant dose in Figure 5. In addition, the ion activity product for $Al(OH)_3(s)$ has been determined (on the basis of the 30-s ferron results and the hydrolysis constants for monomeric aluminum), and these values are shown in Figure 5. These values were affected neither by the total concentration of aluminum nor by the extent of removal of FA.

The consistency of the calculated ion activity products at a pH of 7.2 and the S-shaped ferron traces are strong indications of the presence of $Al(OH)_3(s)$ and of the control of aluminum speciation by this precipitate. This evidence indicates that $Al(OH)_3(s)$ has formed, although neither aluminum nor FA can be removed by membrane filtration. Prior research has demonstrated that FA adsorbs strongly on $Al(OH)_3(s)$ at this pH, with 3 mol of FA carbon (e.g., 0.4 eq of carboxylic functional groups) adsorbed on every mole of $Al(OH)_3(s)$ when the residual FA DOC is only 1 mg/L. The resultant negative charge density of these aggregates causes electrostatic repulsion between particles (*33, 34*) and, for these experiments, has resulted in stabilization of the submicrometer-sized particles of FA and $Al(OH)_3(s)$.

The zones in which FA can be removed by coagulation with alum and filtration by membrane filters are compared with the conditions required for the precipitation of $Al(OH)_3(s)$ in Figure 6. The zone of precipitation of $Al(OH)_3(s)$ is indicated by the solid line. At slightly acidic values of pH the

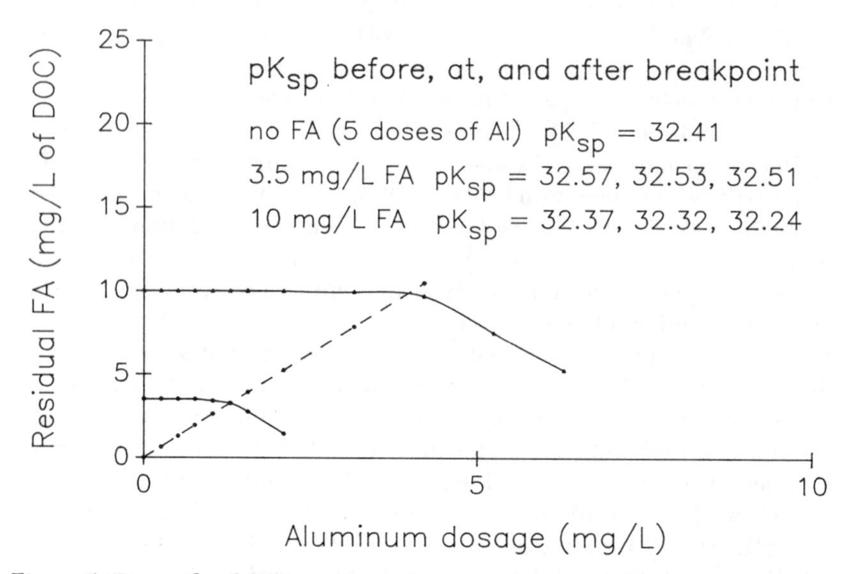

Figure 5. Removal of fulvic acid at pH 7.2 as a function of the initial fulvic acid DOC and the dose of alum. The ion activity products for $Al(OH)_3(s)$ are indicated for various experimental conditions.

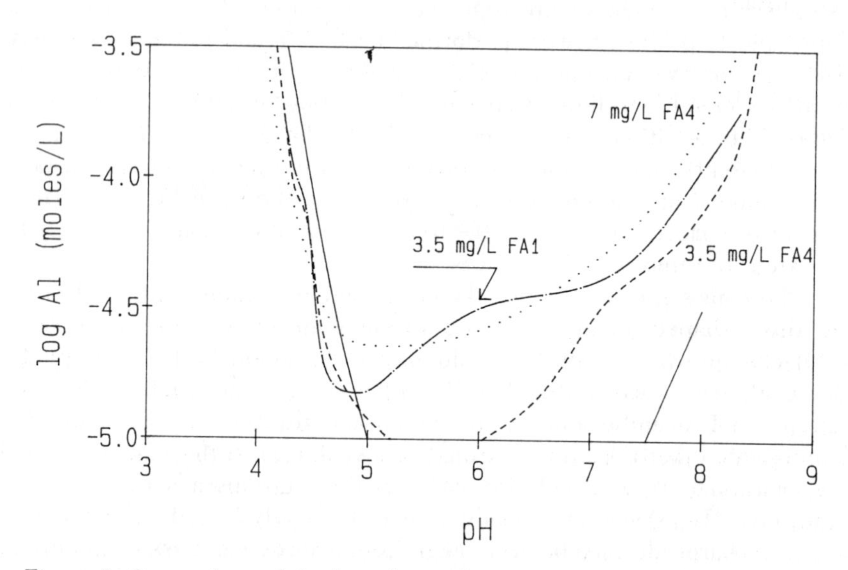

Figure 6. Comparison of the boundaries for removal (after coagulation and filtration) of various fulvic acids and the boundaries for the formation of $Al(OH)_3(s)$. Alum is the coagulant in all of these experiments.

removal of FA occurs at pH values and alum doses that are appropriate for the precipitation of $Al(OH)_3(s)$. At higher pH values the formation of $Al(OH)_3(s)$ is not sufficient for the removal of FA. According to the hypothesis that has been proposed here, the particles are dispersed until higher ratios of positively charged $Al(OH)_3(s)$ to negatively charged FA are obtained. The coherence of removal zones and solubility zones for alum is an indication that $Al(OH)_3(s)$ is present when FA is removed. It is not necessarily a negation of the simultaneous presence of reactive polymers of aluminum.

Complexation Between Al(III) and FA

Stability functions for the complexation of aluminum by FA have been calculated for zones where $Al(OH)_3(s)$ is not present (according to criteria that have been discussed). The data are shown in Table I. All stability functions are in the form

$$K = \frac{[\text{complex}]}{[\text{free Al}] \times [\text{free FA}]} \tag{2}$$

where [complex] is the molarity of complexed aluminum (i.e., the total aluminum minus [free Al]); [free Al] is the molarity of aluminum that reacts with ferron in 30 s; and [free FA] is the total normality of carboxylic acid groups minus [complex]. Thus an unsupported assumption is made that each complexed aluminum reacts with only one carboxylate group. The stability functions that correspond to data where the ion activity product for $Al(OH)_3(s)$ exceeds $10^{-32.8}$ are marked in Table I with an asterisk. The presence of very high concentrations of FA may inhibit the precipitation of $Al(OH)_3(s)$; nevertheless, the marked data should be viewed with skepticism. Some stability functions for the complexation of calcium with FA are also shown in Table I. These functions are based on charge-balance calculations (*26*). The numerators for the Ca data have been changed from normality (as reported in ref. 26) to molarity to obtain greater compatibility with other data in Table I.

The values for complexation of aluminum with FA that are shown in Table I are close to those that have been reported by Backes and Tipping (*35*) (log K = 2.90), who used a dialysis technique and concentrations for total aluminum and humic material similar to those reported here. Their pH values ranged from 3 to 4.8. Backes used a high-molecular-weight fraction of humic acid to minimize losses through the dialysis membrane.

Schnitzer and Hansen (*36*) used two techniques (method of continuous variations and the ion-exchange equilibrium method) to determine stability functions between aluminum and soil FA at pH 2. Both techniques gave log K values of 3.7. The K value was expressed in terms of molarity of the FA, and values would be somewhat lower if expressed in normality of carboxylic

Table I. Complexation Functions for FA and Stability Constants for Analogues of FA with Aluminum or Calcium

Ligand	*Metal*	*Log of Function or Constant*	*Conditions*
FA1	Al	3.59	pH 4.30, DOC 17.1 mg/L
	Al	3.62[a]	pH 5.07, DOC 17.1 mg/L
	Al	3.32[a]	pH 9.07, DOC 17.1 mg/L
	Al	3.49	pH 4.15, DOC 20.9 mg/L
	Al	3.52[a]	pH 5.07, DOC 20.9 mg/L
	Al	3.27[a]	pH 8.90, DOC 20.9 mg/L
	Al	3.23	pH 4.12, DOC 40.9 mg/L
	Al	3.32[a]	pH 4.98, DOC 40.9 mg/L
	Al	3.15[a]	pH 8.72, DOC 40.9 mg/L
	Al	3.36	pH 4.03, DOC 78.2 mg/L
	Al	3.27	pH 4.77, DOC 78.2 mg/L
	Al	3.58[a]	pH 8.08, DOC 78.2 mg/L
FA1[b]	Ca	1.72	pH 5, pCa 3, TOC 666 mg/L
	Ca	2.09	pH 7, pCa 3, DOC 666 mg/L
	Ca	2.25	pH 9, pCa 3, DOC 666 mg/L
	Ca	1.64	pH 5, pCa 2.4, DOC 666 mg/L
	Ca	1.91	pH 7, pCa 2.4, DOC 666 mg/L
	Ca	2.10	pH 9, pCa 2.4, DOC 666 mg/L
Acetate	Al	1.51	$I \rightarrow 0$
	Ca	1.18	$I \rightarrow 0$
Oxalate	Al	6.1	$I = 1.0$
	Ca	1.66	$I = 1.0$
Tartrate	Al	5.32	$I = 1.0$
	Ca	1.80	$I = 0.1$
Salicylate	Al	12.9	$I = 0.1$
	Ca	0.15	$I = 0.16$
Phthalate	Al	3.18	$I = 0.5$
	Ca	1.07	$I = 0.1$

[a]Ion activity products ($\{Al^{3+}\}\{OH\}^3$) greater than or equal to $10^{-32.8}$, but ferron traces without indication of $Al(OH)_3(s)$.
[b]*See* ref. 28 for more data regarding complexation of Ca with FA.

functional groups on the FA. Schnitzer used higher concentrations of FA and aluminum than for data reported in Table I, but used FA-to-aluminum ratios similar to those used here. Pott and co-workers (37) used a cation-exchange technique to measure complexation functions between aluminum and humic materials. They discovered functions 3 orders of magnitude stronger than those reported in Table I.

Because humic materials are mixtures, the metal-to-ligand ratios are important. Pott and co-workers (37) used lower aluminum concentrations than were used for the experiments summarized in Table I, but the ratios of humic material to aluminum fall within the same range. Young and Bache (38) also modeled the complexation reaction, but only used a limited set of data for validation of the model. Both Backes and Pott found that complexation of aluminum with humic materials was insignificant compared to the hydrolysis of aluminum at pH 6 or higher. This behavior, which agrees with the results reported here, has substantial impact on the mechanism of removal of FA during coagulation processes.

Some stability constants for the complexation of aluminum with specific organic ligands are also shown in Table I. Monocarboxylic organic acids (e.g., acetic acid), strong chelating agents (e.g., salicylic acid), and weaker chelating agents (e.g., phthalic acid) are often proposed as analogues for the reactive complexing groups on molecules of FA. When we compare FA and specific organic ligands as complexing agents, we must emphasize that FA is a mixture of materials and that the stability function cannot have the same meaning as the stability constants for complexation of aluminum by well-defined ligands (*39, 40*).

The following observations are made with respect to the stability functions and constants listed in Table I. First, the formation functions for FA and the selectivity of the ligand for Al versus Ca are both greater for FA than for acetic acid. Second, the stability functions and the selectivity for Al versus Ca are less for FA than for salicylate (and other chelators that form five- or six-membered chelate structures). Finally, the selectivity for Al versus Ca and the stability constant for complexation of aluminum with phthalate are quite close to the values for FA.

The best analogue for these experimental conditions is phthalate, which forms a seven-membered chelate ring with metals. Fettes (*41*) showed that chelate rings of greater than 30 members are sometimes stable for complexation of metals with macromolecules. Dempsey and O'Melia (*26*) suggested that such chelation may be important when metals are complexed with FA. The comparisons described support this hypothesis.

Comparison of Complexation and Adsorption

Dempsey, Ganho, and O'Melia (*2*) reported isotherms for the adsorption of FA on $Al(OH)_3(s)$ that is formed during the coagulation process. The isotherms were performed at pH 7 and for total concentrations of FA and aluminum similar to those used in the experiments reported here. For a residual soluble FA concentration of 1 mg/L, over 3 mol of FA carbon was adsorbed per mole of aluminum. If there were 11.24 meq of carboxylic functional groups per g of FA DOC, then the adsorption function would be 3.9×10^4 ($\log K = 4.6$).

The terms and derivation of the adsorption function are identical to the complexation functions described here and shown in Table I. Both are expressed in terms of complexed over free FA functional groups divided by the free concentration of Al ions (in monomeric form for complexation and in solid form for adsorption). In both cases a 1:1 stoichiometry is assumed between carboxylic functional groups on the FA and aluminum atoms (a simplification that does not take possible chelation into account).

The very interesting result is that the adsorption reaction is 16 times stronger than the complexation reaction. This result seems reasonable. The weight-averaged molecular weight of FA is at least 600, and there are over 11 meq of carboxylic functional groups per g of FA carbon, yielding more than six carboxylic groups on the average molecule of FA. Many investigators (e.g., *42*) have shown that the strength of adsorption is a function of the molecular weight and the number of adsorbing groups on the adsorbate.

This quantitative comparison corroborates the evidence reported in Figure 5, which indicates that $Al(OH)_3(s)$ has formed at pH 7.2, even when the precipitate cannot be removed by membrane filters with 0.2-μm pores. There should be similar strength of adsorption at pH 5; the FA is less negatively charged, but the $Al(OH)_3(s)$ is more positively charged. This qualitative argument strengthens allegations that FA is removed by the adsorption mechanism rather than by direct precipitation at pH values close to 5.

Conclusions

An important conclusion from this work is that $Al(OH)_3(s)$ forms in the presence of FA over the entire range of pH and total aluminum conditions that lead to removal of FA during coagulation using alum. The solubility of $Al(OH)_3(s)$ for times ranging from minutes to days and for pH values less than 6.5 is defined by $pK_{sp} = 32.8$. Solubility is greater than this when the pH exceeds 6.5.

This conclusion does not necessarily indicate that polymeric species of aluminum are unimportant in coagulation with alum, even after several minutes have elapsed, because a mass balance on the aluminum is not reported here. Most likely, polymeric species form when alum is added to water, because polymers may be intermediates between the initial coagulant (monomeric aluminum) and one of the final products ($Al(OH)_3(s)$) (*22*, *43*). It is not certain whether such intermediate species are necessary for effective coagulation. Effective coagulation can be accomplished by using relatively inert polymeric species of aluminum (e.g., polyaluminum chloride).

FA affects the nature of the $Al(OH)_3(s)$. The effects seem to be caused by the relatively strong adsorption of FA and by the moderately high negative charge densities for the FA. As a result, FA enhances the formation of $Al(OH)_3(s)$ at pH ≤ 5. This behavior is similar to that observed for sulfate. At pH 7.2, the FA causes the stabilization of submicrometer-sized particles of $Al(OH)_3(s)$. The high and variable charge density and strong adsorption of FA affect the shape of the typical removal diagram for FA. This effect causes removals at lower pH than for clay for a given coagulant dose and requires a greater coagulant dose at higher pH values than would be needed for clay.

The complexation of soluble aluminum by these FAs is moderate in strength—less than for model chelators with five- or six-membered chelate rings, much greater than for monoprotic organic acids, similar in strength (and selectivity for aluminum versus calcium) to phthalic acid. This provocative similarity is a clue (but not a confirmation) to the structure of the reactive functional groups on FA. FA is a mixture, and the reported stability functions represent the dominant complexing groups at the degree of experimentally observed complexation.

Adsorption of FA on $Al(OH)_3(s)$ is a much stronger reaction than the complexation of monomeric aluminum with FA. This fact is very useful for a mechanistic understanding of the removal of FA during coagulation with salts of aluminum. The conclusion is reasonable, but not necessary, that the most important mechanism for removal of FA with alum is by adsorption of FA on $Al(OH)_3(s)$. This prediction is made for the entire pH range for which removal occurs and for the low-to-moderate concentrations of FA that have been tested.

References

1. O'Melia, C. R.; Dempsey, B. A. *Proc. Annu. Public Water Supply Eng. Conf.*; University of Illinois at Urbana–Champaign, 1982; pp 5–14.
2. Dempsey, B. A.; Ganho, R. M.; O'Melia, C. R. *J. Am. Water Works Assoc.* **1984,** *76,* 141–150.
3. Dempsey, B. A.; Sheu, H.; Ahmed, T. M. T.; Mentink, J. *J. Am. Water Works Assoc.* **1985,** *77,* 74–80.
4. Amirtharajah, A.; Mills, K. M. *J. Am. Water Works Assoc.* **1982,** *74,* 210–216.
5. Amirtharajah, A. *Coagulation and Filtration: Back to the Basics*; from *Am. Water Works Assoc. Semin.*; 1981; pp 1–22.
6. Amirtharajah, A. *J. Environ. Eng.* (*N.Y.*) **1986,** *112,* 1085–1108.
7. Edwards, G. A.; Amirtharajah, A. *J. Am. Water Works Assoc.* **1985,** *77,* 50–57.
8. Sricharoenchaikit, P. Ph.D. Thesis, Syracuse University, 1984.
9. Hall, E. S.; Packham, R. F. *J. Am. Water Works Assoc.* **1965,** *57,* 1149–1166.
10. Packham, R. F. *J. Appl. Chem.* **1962,** *12,* 564–568.
11. Packham, R. F. *Proc. Soc. Water Treat. Exam.* **1963,** *12,* 15.
12. Narkis, N.; Rebhun, M. *J. Am. Water Works Assoc.* **1977,** *69,* 325–328.
13. Matijevic, E.; Mathai, K. G.; Ottewill, R. H.; Kerker, M. *J. Phys. Chem.* **1961,** *65,* 826–830.
14. Matijevic, E.; Janauer, G. E.; Kerker, M. *J. Colloid Sci.* **1964,** *19,* 333–346.
15. Hayden, P. L.; Rubin, A. J. In *Aqueous–Environmental Chemistry of Metals*; Rubin, A. J., Ed.; Ann Arbor Science: Ann Arbor, 1974; p 317.
16. Hahn, H. H.; Stumm, W. *J. Colloid Interface Sci.* **1968,** *28,* 134–144.
17. Stumm, W.; Morgan, J. J. *J. Am. Water Works Assoc.* **1962,** *54,* 971–992.
18. Edzwald, J. K.; Becker, W. C.; Tambini, S. J. *J. Environ. Eng.* (*N.Y.*) **1987,** *113,* 167–185.
19. Smith, R. W. In *Non-Equilibrium Systems in Natural Water Chemistry*; Gould, R. F., Ed.; American Chemical Society: Washington, DC, 1971; p 250.
20. Bersillon, J. L.; Hsu, P. H.; Fiessinger, F. *Soil Sci. Soc. Am. J.* **1980,** *44,* 630–634.
21. Hundt, T. R. Ph.D. Thesis, The Johns Hopkins University, 1985.
22. Tambo, N.; Kamei, T. Chapter 28 in this volume.

23. Smith, R. M.; Martell, A. E. *Critical Stability Constants*; Plenum: New York, 1974.
24. Baes, C. F., Jr.; Mesmer, R. E. *The Hydrolysis of Cations*; Wiley: New York, 1976; p 112.
25. Davis, J. A.; Hem, J. D. In *The Environmental Chemistry of Aluminum*; Sposito, G., Ed.; CRC Press: Boca Raton, FL, 1988.
26. Dempsey, B. A.; O'Melia, C. R. In *Aquatic and Terrestrial Humic Materials*; Christman, R. F.; Gjessing, E. T., Eds.; Ann Arbor Science: Ann Arbor, 1983; p 239.
27. Schonherr, S.; Gorz, H.; Gessner, W. *Z. Chem.* **1980,** *20*, 422.
28. Gessner, W.; Winzer, M. *Z. Anorg. Allg. Chem.* **1979,** *452*, 151–156.
29. Pagoni, P. M.S. Thesis, University of Missouri–Rolla, 1986.
30. Stumm, W.; Morgan, J. J. *Aquatic Chemistry*, 2nd ed.; Wiley–Interscience: New York, 1981.
31. Nordstrom, D. K.; Ball, J. W. *Science (Washington, D.C.)* **1986,** *232*, 54–56.
32. Driscoll, C. T.; Baker, J. P.; Bisogni, J. J.; Schofield, C. L. In *Geological Aspects of Acid Precipitation*; Bricker, O. P., Ed.; Butterworths: London, 1984; p 55.
33. Gibbs, R. J. *Environ. Sci. Technol.* **1983,** *17*, 237–240.
34. Neihof, R. A.; Loeb, G. I. *Limnol. Oceanogr.* **1972,** *17*, 7–16.
35. Backes, C. A.; Tipping, E. *Water Res.* **1987,** *21*, 211–216.
36. Schnitzer, M.; Hansen, E. H. *Soil Sci.* **1970,** *109*, 333–340.
37. Pott, D. B.; Alberts, J. J.; Elzerman, A. W. *Chem. Geol.* **1985,** *48*, 293–304.
38. Young, S. D.; Bache, B. W. *J. Soil Sci.* **1985,** *36*, 261–269.
39. MacCarthy, P.; Smith, G. C. In *Chemical Modeling in Aqueous Systems*; Jenne, E. A., Ed.; American Chemical Society: Washington, DC, 1979; p 201–222.
40. Perdue, E. M.; Lytle, C. R. In *Aquatic and Terrestrial Humic Materials*; Christman, R. F.; Gjessing, E. T., Eds.; Ann Arbor Science: Ann Arbor, 1983; pp 295–313.
41. Fettes, E. M. *Chemical Reactions of Polymers*; Interscience: New York, 1964; p 14.
42. Howard, D. B.; Woods, S. J. In *Adsorption from Solution at the Solid–Liquid Interface*; Parfitt, G. D.; Rochester, C. H., Eds.; Academic: London, 1983.
43. Teagarden, D. L.; White, J. L.; Hem, S. L. *J. Pharm. Sci.* **1981,** *70*, 808–810.

RECEIVED for review July 24, 1987. ACCEPTED for publication March 2, 1988.

26

Aluminum and Iron(III) Chemistry

Some Implications for Organic Substance Removal

J. Y. Bottero

Equipe de Recherche sur la Coagulation–Floculation, UA 235 du Centre National de la Recherche Scientifique et GS Traitement Chimique des Eaux, BP40 54501 Vandoeuvre Cedex, France

J. L. Bersillon

Laboratoire de Recherche de la Lyonnaise des Eaux, 38 rue du Président Wilson, 78230 Le Pecq, France

Removal of organic substances by metallic salts is one of the most important processes in the water-treatment industry. Some of the subsequent treatment phases depend on the efficiency of the coagulation–flocculation phase. Development of new coagulant–flocculant agents is one way to improve the removal of organic compounds by coagulation. A new generation of aluminum (Al_{13}) and iron hydroxides is being developed by prehydrolyzing aluminum or iron chloride salts to form active species. These polymers have greater reactivity than alum in removing organic material when the solution pH is lower or higher than the sweep flocculation zone. In acidic solutions (pH <5.5–6) coagulation–flocculation is achieved by charge neutralization of organic acid functions, followed by precipitation. Removal is achieved at pH >6.5 by adsorption onto large particles. The polycations are stable over pH and time. Their structure is open and loose (fractal), ensuring a large available surface area for complexing or adsorbing molecules. The stability constants of organic compound–Al polycation complexes are higher than the stability constants of organic compound–Al or Fe monomer complexes. Prehydrolyzed coagulant–flocculant agents can be designed to adapt the treatment to specific raw-water characteristics.

0065–2393/89/0219–0425$06.00/0

NATURAL ORGANIC MATERIALS ARE NOW THE PRIMARY TARGET of water treatment (*1*). These compounds (responsible for color, taste, and odor in water) have a potential impact on public health. Generally, these compounds originate in the soil and appear as humic or fulvic acids (*2, 3*). ^{13}C NMR spectroscopy reveals that these complexes contain ~50% aromatic –COOH groups, ~20% carbon bonded to phenolic OH, polysaccharides, proteins, and fatty acids. Their major building blocks are formed by structures like benzenecarboxylic acid (*4*). The chemistry of humic substances in water is somewhat different; carbohydrates and proteins are more prominent (*5*).

Coagulation–flocculation, the most widely used treatment process for removal of organic substances in surface water, has received little attention as a potentially powerful process for removing soluble organic contaminants. Aluminum or iron flocculants are used in drinking-water treatment. From an engineering point of view, a better understanding of the interactions involved in organic-substance removal should lead to models that can be used in optimization of the process. Because coagulation is the first step in water treatment, its improvement will have large consequences on subsequent treatments (activated-carbon adsorption, ozonation, chlorination).

The mechanisms involved (neutralization, flocculation, and adsorption on flocs) and the efficiency of organic-compound removal by coagulation depend on the coagulant chemistry and structure (which vary with pH) and the oxidation pretreatments (ozonation). This chapter reviews the chemistry and structure of the mineral coagulants used (Al and Fe) and their effects on complexation, adsorption, and the relative kinetics of interaction with the organic functions in humic substances.

Mineral Coagulant–Flocculant

The aluminum coagulants in use are essentially alum ($Al_2(SO_4)_3 \cdot nH_2O$). However, alternative coagulants have been developed in recent years. Aluminum chloride salts partially neutralized before use (polybasic aluminum chloride or PBAC) and hydrolyzed ferric salts are extensively studied in Europe. Some of them were successfully implemented in full-scale plants 10 years ago. In Japan, a partially hydrolyzed aluminum chloride ($[OH]_{bound}/[Al]$ ~1.9) containing sulfate ions (PACS) has been developed (Taki Fertilizer Company). PBAC can be optimized with respect to the quality of the raw water (temperature, organic content) by changing the hydrolysis ratio $[OH]_{added}/[Al]$ (*6*).

The ferric products are relatively more recent. Fe ions can be partially hydrolyzed before use, but the control of hydrolysis during preparation is a major problem (*7*). A new generation of well-controlled partially hydrolyzed iron coagulant is under study.

Aluminum Coagulant. The use of metallic coagulants presents a problem because little is known about the nature of the substance that precipitates when the solution is rapidly diluted at near-neutral pH. Many transient polymeric species have been suggested (8, 9) to explain the transformations from Al^{3+} ions to solid-phase compounds. Most of these data on transient polymeric species were obtained from highly indirect methods of investigation. Spectroscopic methods and radiation scattering have revolutionized aluminum chemistry for systems that are far from equilibrium, as in the case of water treatment.

Chemistry of Coagulants and Flocs. Solution-state ^{27}Al NMR patterns of aluminum sulfate (or alum) dissolved in water exhibit two resonance peaks. The first, at 0 ppm, is relative to the monomers (Al^{3+}, Al^{2+}, and Al^{+}) in octahedral symmetry; the other corresponds to the complex $Al(SO_4)^+$ (Figure 1a). The chemistry of PACS is somewhat different. The matrix solution (~2.2 mol/L) contains more than 60% of the aluminum present as Al^{3+} and 2–3% as dimers (Figure 1b). The rest of the aluminum is layered octahedral medium-sized polymers. Dilution or a sharp increase in pH triggers the formation of large particles or precipitates with the same local range order (i.e., aluminum atoms in octahedral coordination, as alum or the medium polymers in the matrix solution, as shown by ^{27}Al NMR in solid-state pattern in Figure 2). Alum makes a direct transition between monomers and precipitates. The sulfate ions form inner complexes with aluminum and inhibit the formation of transient polymeric species.

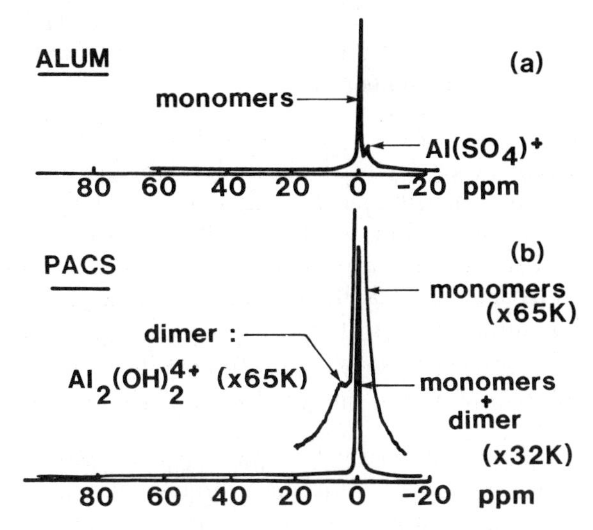

Figure 1. Solution-state ^{27}Al NMR spectra of alum (a) and PACS (b).

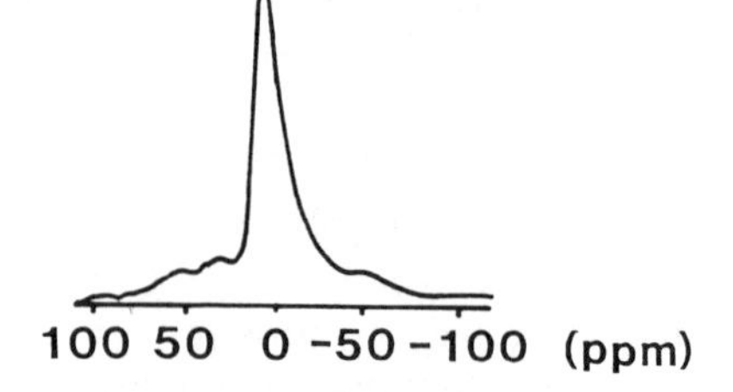

Figure 2. Solid-state ^{27}Al NMR spectra of alum or PACS after precipitating at neutral pH.

The products formed by hydrolyzing pure aluminum chloride solution are quite different. In these solutions the hydrolysis of Al^{3+} with sodium hydroxide or sodium carbonate (*10–12*) yields the formation of soluble polycations. The most important parameters that govern the synthesis are $R = [OH]_{added}/[Al]$, which varied from 0.5 to 3; [Al] from 10^{-4} to 0.2 mol/L; and T, the age of the solution.

Solution-state ^{27}Al NMR at 23.45 MHz (*10, 12*) and 93.84 MHz (*13*) shows spectra that correspond to octahedral and tetrahedral symmetry of aluminum (Figure 3). From R_{added} = 0.5 to 2.3–2.4, a signal appears at 79.9 ppm from the reference (AlO_4^-), corresponding to aluminum ions in octahedral coordination. This resonance peak, which corresponds to monomers $Al(H_2O)_6{}^{3+}$, $Al(OH)(H_2O)_5{}^{2+}$, and $Al(OH)_2(H_2O)_4{}^{+}$ (*14*), does not shift, but its width at half-maximum increases with R. The peak located at 17 ppm from the reference was attributed to the tetrahedral aluminum of the polymer $Al_{12}{}^{VI}(OH)_X Al^{IV}O_4(H_2O)_{12-X}{}^{(31-X)+}$. Its structure was described by Johansson in 1962 (*15*). The central Al atom is in tetrahedral coordination, whereas the 12 other aluminum atoms are octahedral (Figure 4). They are not observed because they do not have a symmetrical environment (*12*). The relative concentrations of each one of the species can be calculated from the areas under the peaks; the precision of the measurements is ±10%. For example, for Al = 10^{-1} mol/L, Figure 5 shows that Al_{13} concentration corresponds approximately to 90% Al for R_{added} = 2.0 or 2.2.

Beyond these values, some aluminum is not detected by solution-state NMR because colloids are formed. The turbidity of the suspensions increases from 1 nephelometric turbidity unit (NTU) (R = 2.2) to ~50 NTU (R = 2.5); they precipitate rapidly for $R > 2.6$. ^{27}Al solid-state magic angle spinning (MAS) NMR spectra at 104 MHz relative to the particles formed from $R \geq 2.5$ (Figure 6) look like the spectra obtained in solution state.

The resonance band at ~63 ppm is related to the tetrahedra of Al_{13}. The relative content of Al_{13} in the particles is obtained by comparing normalized areas of the peak at 63 ppm with the peak resonance at ~0 ppm (octahedral symmetry) (*16*). The data (Table I) show that the Al_{13} content in the small particles (R = 2.5) and large particles (R = 2.6–2.8) correspond nearly to 100%, at least for $T \leq 1$ h (*13, 16*). The Al_{13} content does not vary with time T for R = 2.6–2.8. At R = 3 the precipitated solid is bayerite.

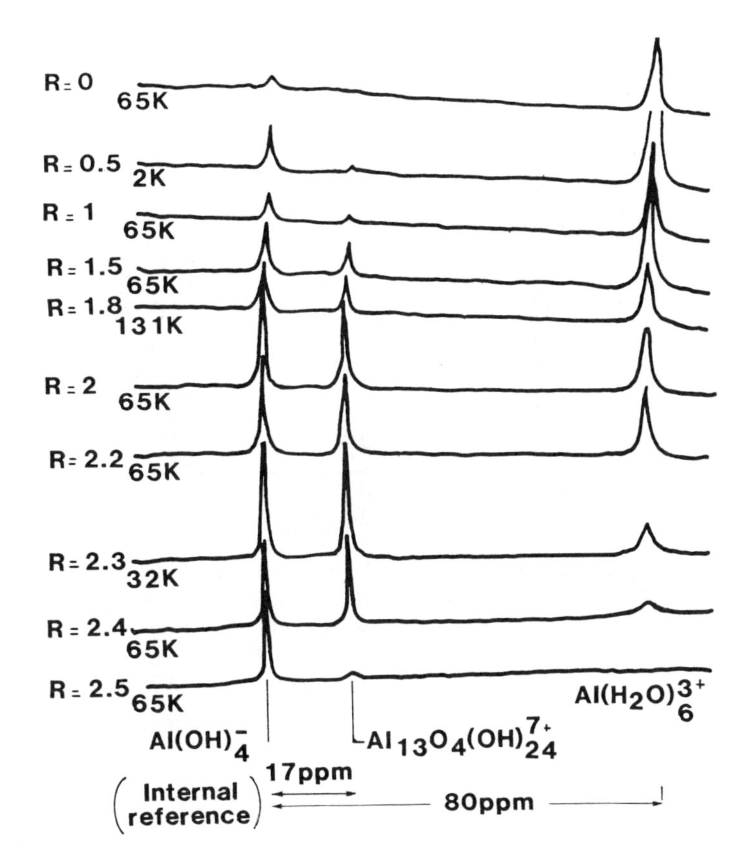

Figure 3. Solution-state ^{27}Al NMR spectra typical of PACl as a function of R = [OH]/[Al]. The peak at 80 ppm from $Al(H_2O)_6^{3+}$ is an integration reference needed to calculate the relative concentration of each one of the species. (Reproduced from ref. 12. Copyright 1980 American Chemical Society.)

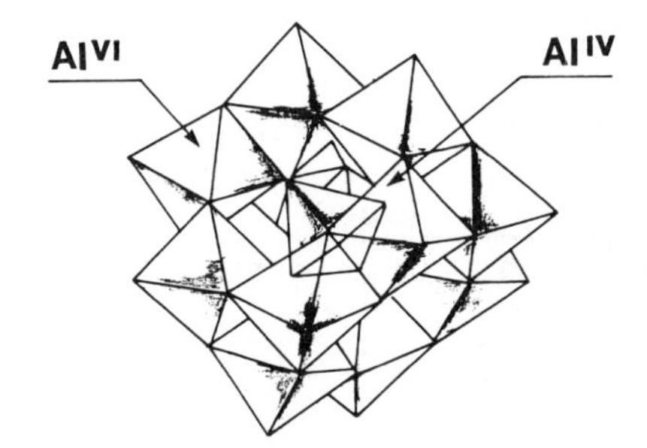

Figure 4. Structure of the $Al^{IV}O_4Al_{12}^{VI}(OH)_{24}(H_2O)_{12}^{7+}$ polycation. (Reproduced with permission from ref. 16. Copyright 1987 Academic.)

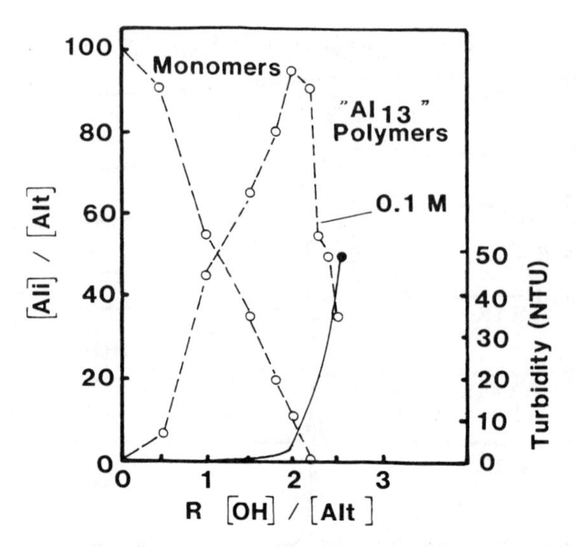

Figure 5. Distribution of the different hydroxy aluminum species (monomers and Al_{13}) in PACl solutions from NMR data, versus R = *[OH]/[Al]* (- - - - -) *and subsequent turbidity variation* (——).

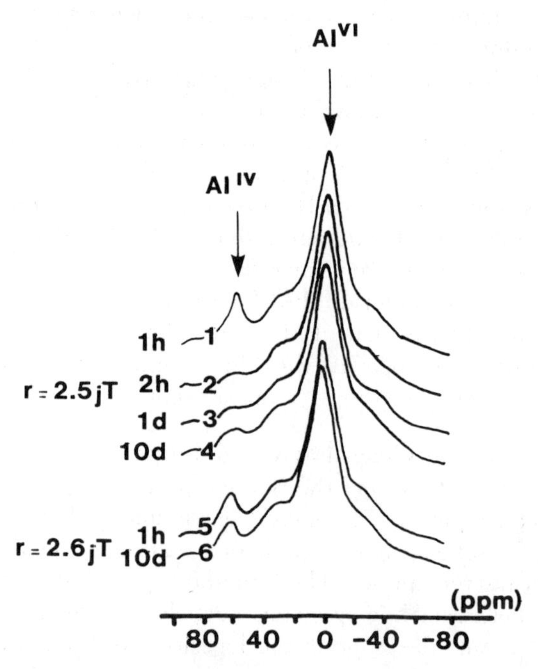

Figure 6. ^{27}*Al MAS NMR spectra of colloids in* R = *2.5 and 2.6 solutions versus time. The peak at 63 ppm is relative to the tetrahedral coordinated Al of Al_{13}. The large peak at ~0 ppm is the octahedral coordinated Al of Al_{13} and other species when the age of the solutions increases.*

Table I. Concentration of Al_{13} in Colloids

Spectrum	*Sample*	Al_{13}[a]
1	$R = 2.5$; $T = 1$ h	8.8
2	$R = 2.5$; $T = 2$ h	4.7
3	$R = 2.5$; $T = 1$ day	4.2
4	$R = 2.5$; $T = 1$ week	3.9
5	$R = 2.6$; $T = 1$ h	8.5
6	$R = 2.6$; $T = 1$ week	6.7
7	$R = 2.8$; $T = 1$ h	6.7
8	$R = 2.8$; $T = 1$ week	6.6
9	$R = 3.0$; $T = 1$ h	1–2 (bayerite)
10	$R = 3.0$; $T = 1$ week	0 (bayerite)

NOTE: Colloids obtained from ^{27}Al MAS NMR patterns and calculated from $Al^{IV}/(Al^{IV} + Al^{VI})$ ratio. The theoretical ratio is $1/13 = 7.7\%$. From $R = 2.8$, the flocculation occurs in the solution.

[a] $Al^{IV}/\Sigma(Al^{IV} + Al^{VI})$.

The Al_{13} content does not change very much (~80–100%) from $R = 2.2$ to 2.8. This behavior implies that the charge of Al_{13} polymers decreases as pH or R increases.

The modifications attributable to dilution and pH shock that characterize the use of aluminum salts in water treatment have been evaluated. The dilution, in demineralized water, of $R = 2$ (~90% of isolated Al_{13}), $R = 2.5$ (~40% of isolated Al_{13} and 60% of aggregated Al_{13}) from $Al = 10^{-1}$ to 10^{-3} mol/L at pH 6.5, 7.5, and 8.5 led to formation of flocs. Solid-state MAS ^{27}Al NMR spectra of these flocs precipitated at pH 7.5, $[Al] = 10^{-3}$ mol/L, and from $R = 2.0$ and 2.5 (Figure 7) are similar to those of flocs precipitated in acidic solutions (Figure 6). Table II summarizes the Al_{13} content of the flocs at various precipitation pH and aging times. The Al_{13} content decreases as precipitation pH and aging time increase.

The polyaluminum chloride (PACl) coagulants are different from sulfate or polyaluminum chlorosulfate (PACS) coagulants because Al_{13} exists only in the PACl solutions and because the dilution and pH increase do not modify the local range order (i.e., the Al_{13} ions are not immediately destroyed).

Structure of the Coagulants and Flocs. The structure of active species in matrix solutions and the relative precipitated flocs at neutral pH have a great consequence on removal of organic matter and solids. Such structures can be investigated by radiation scattering (small-angle X-ray or neutron scattering) techniques. These methods permit definition of the size, shape, and structure of homogeneous particles or agregates. If fractallike particles are present, the slope of the logarithm of the scattering intensity versus the logarithm of the wave vector can be expressed as

$$S = \frac{2\Theta}{\lambda} \tag{1}$$

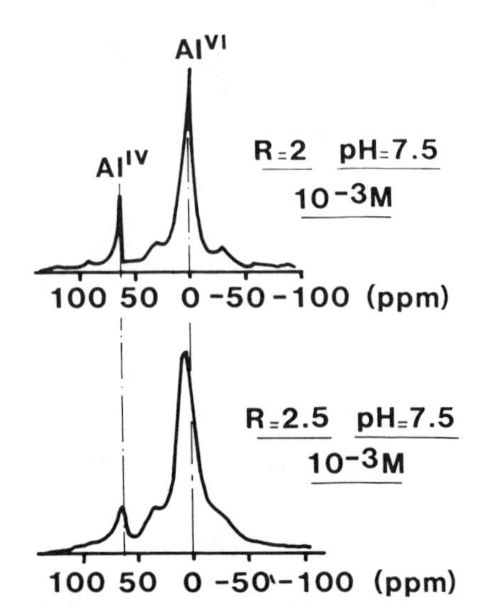

Figure 7. ^{27}Al MAS NMR spectra of PACl flocs, R = 2 and 2.5, precipitated at pH 7.5 after diluting at 10^{-3} mol/L.

Table II. Al_{13} Content in Precipitated Flocs

	pH 6.5			*pH 7.5*			*pH 8.5*		
Flocculant	*10 min*	*1 h*	*24 h*	*10 min*	*1 h*	*24 h*	*10 min*	*1 h*	*24 h*
R = 2.0	80 | 100%	80	N/A	80 | 100%	80	80	60	45	35
R = 2.5	80 | 100%	75	N/A	80	60	60	60	35	25

NOTE: Flocs precipitated from $R = 2$ and 2.5 after diluting Al solutions from 10^{-1} to 10^{-3} M at pH 6.5, 7.5, and 8.5, as calculated by ^{27}Al MAS NMR experiments.

where 2Θ and λ are the scattering angle and the wavelength, respectively. This relationship yields the fractal dimension, D_f. D_f, which varies between 1 and 3, characterizes the structure of the aggregates (i.e., their density). Low, medium, and high D_f values correspond to linear, branched, and dense aggregates, respectively. PBAC coagulants have been extensively studied. In $R = 2$ solution, the Al_{13} polymers have a radius of 12.6 Å, which corresponds to the radius of the polymers plus the thickness of the hydration shell (*17*). In $R = 2.5$ solutions, Al_{13} aggregates have a radius of gyration (r) of ~90 Å and a fractal geometry with a fractal dimension D_f ~1.43 (*16*, *18*). This coagulant solution is made of isolated and aggregated Al_{13} (Figure 8). In $R = 2.6$ solutions, the Al_{13} aggregates are bigger ($r \geq 300$ Å) and have a fractal geometry with a fractal dimension D_f ~1.7–1.8. After flocculating at ~pH 7, these aggregates form large flocs of Al_{13} with a fractal dimension D_f ~1.8 (*19*). A very loose structure that provides a large surface area for adsorption of organic molecules is certainly more useful than a dense microporous aggregate on which a large part of the surface area is not accessible.

Specific Surface Area of Flocs and Surface Charge. Alum and PACl floc particles obtained at pH >6 present large specific surface areas that have been measured by using adsorption of anionic surfactants (*20* and Rakotonarivo, CNRS 1987, unpublished data). The specific surface area obtained with PACl depends on pH of precipitation and on initial R_{added} = [OH]/[Al]. The specific surface areas decrease as pH and R increase (Table III). The specific surface area of alum flocs is much lower. The surface charge distribution can be evaluated by electrophoresis and potentiometric titration. From electrokinetic potential (ζ) and potentiometric titration data in NaCl

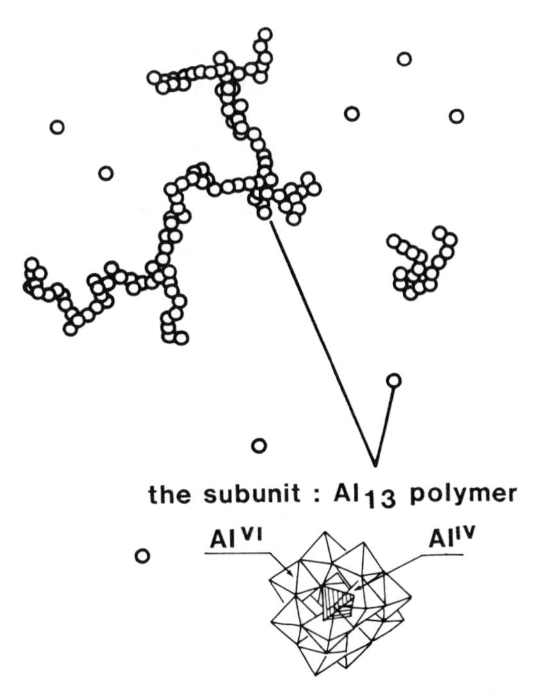

Figure 8. Schematic representation of PACl R = 2.5 *solution.*

Table III. Specific Surface Area Values

pH	R = *2.5*	R = 2	R = *0*
6.5	587	774	1100
7.5	540	601	800

NOTE: All values are in square meters per gram; calculated from the adsorption isotherms of long-alkyl-chain soaps (*21*) or long-alkyl-chain sulfonates (*20*).

electrolyte, the diffuse-layer charge σ_d and the surface charge σ_0 can be calculated. Then the active site number N_S and the surface reaction constant K_i can be evaluated. Four equilibria between surface sites and liquid could be postulated for the purpose:

$$Al(OH_2)^+ \overset{K_1}{\rightleftarrows} Al(OH) + H^+ \tag{2}$$

$$Al(OH) \overset{K_2}{\rightleftarrows} AlO^- + H^+ \tag{3}$$

$$Al(OH_2)^+Cl^- \overset{K_3}{\rightleftarrows} Al(OH_2)^+ + Cl^- \tag{4}$$

$$AlO^-Na^+ \overset{K_4}{\rightleftarrows} AlO^- + Na^+ \tag{5}$$

Reactions 4 and 5 correspond to the complexation of positive and negative sites by Cl^- and Na^+ ions. The variables N_S and K_i are optimized to fit ζ–pH and σ_0–pH curves (*21*). The active-site-number density and surface-reaction pK of flocs precipitated from PACl solutions $R = 2.5$ or 2 in NaCl electrolyte have been calculated by computer (*21*). The results (Figure 9) show that at neutral pH the sites are predominantly neutral: 60% is Al(OH); 20–40% of positive sites are $Al(OH_2)^+$ complexed by Cl^- ions.

The physical and chemical relations between PACl solutions and PACl flocs are summarized in Figure 10. Alum and unhydrolyzed aluminum chlo-

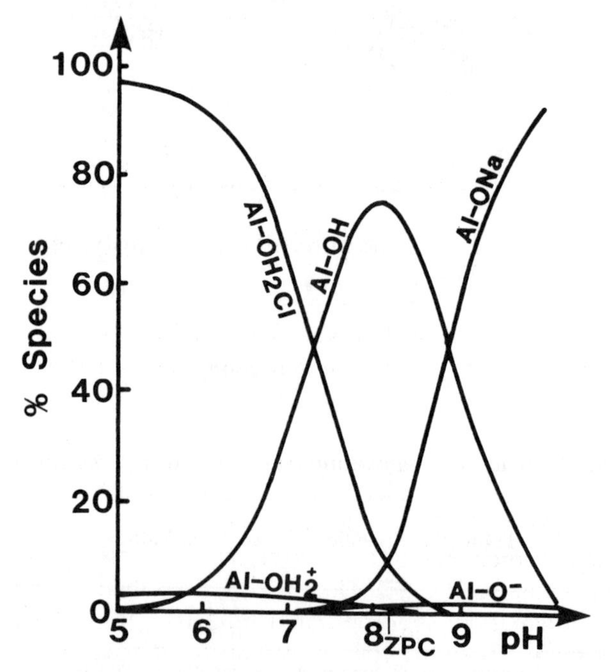

Figure 9. Distribution of the active surface sites of PACl flocs versus pH in deionized water.

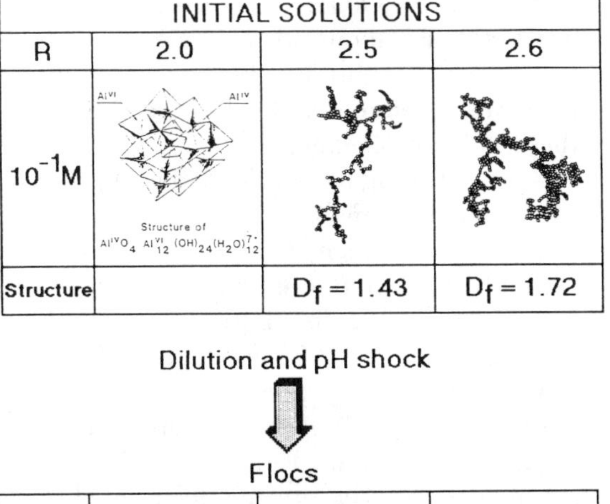

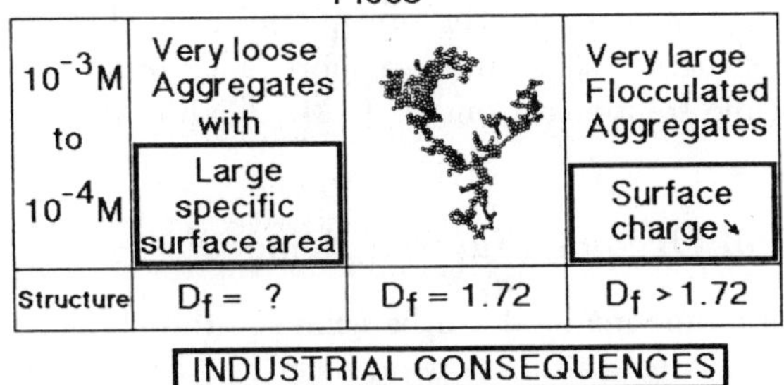

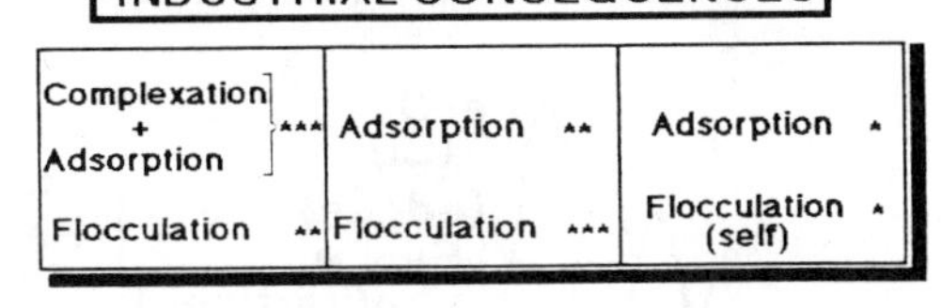

Figure 10. Representation of the physical and chemical evolution of PACl solutions versus R = *[OH]/[Al], dilution, and pH increase.*

ride coagulant solutions are extensively modified by dilution and pH increase during treatment; however, this does not happen with PACl and only partly with PACS.

Iron Coagulant. The chemistry of iron ions is not as developed as that of aluminum because no local probe exists except X-ray absorption fine structure (EXAFS), which can play the same role as NMR. Nevertheless, different-sized colloids ($Fe(H_2O)_6^{3+}$, $Fe(H_2O)_5^{2+}$, $Fe(H_2O)_4^{+}$) exist in equilibrium with monomers (*22–25*). The size of these colloids (5–40 nm) depends on the time of aging. In supersaturated iron(III) nitrate solutions, studied by methods such as base titration, ultracentrifugation, and IR spectroscopy (*26–28*), the precursor for the formation of the organized phase is amorphous. The critical nucleus is postulated to contain 16–32 atoms. Small polycations such as $[Fe_3(OH)_4]^{5+}$ and $[Fe_4O(OH)_4]^{6+}$, in which octahedra share edges and corners, control both nucleation and growth steps by associating in larger polymers (29).

Direct methods were used to follow the size and the growth of iron(III) clusters during the first steps of aging, just after the addition of base. Data from small-angle X-ray scattering were recorded every 50 s. Small polymers were detected (from 0.5- to 0.9-nm diameter as time increased). They formed linear clusters by aggregating the small polymers (30). The local range order as revealed by the EXAFS technique is similar to that of goethite, as $R = [OH]/[Fe] \geq 1.5$ (31). The radial distribution function (Figure 11) reveals the presence of a first distance at ~0.2 nm, characteristic of Fe–O or Fe–Cl distances. As hydrolysis increases, a second and a third distance appear to indicate Fe–Fe distances. The Fe octahedra are shared either by edges (0.295 nm) or by vertices (0.330 nm). From $R = 1.5$ the spectra exhibit the same shape as that of goethite. The precipitates formed at near-neutral pH have the same structure as goethite, a short range order similar to that of polymers in acidic solutions (Combe, J. M., CNRS, 1988, personal communication).

Mechanisms of Organic Material Removal

The removal of humic material can be achieved through complexation and adsorption. These mechanisms depend on pH and the physical chemistry

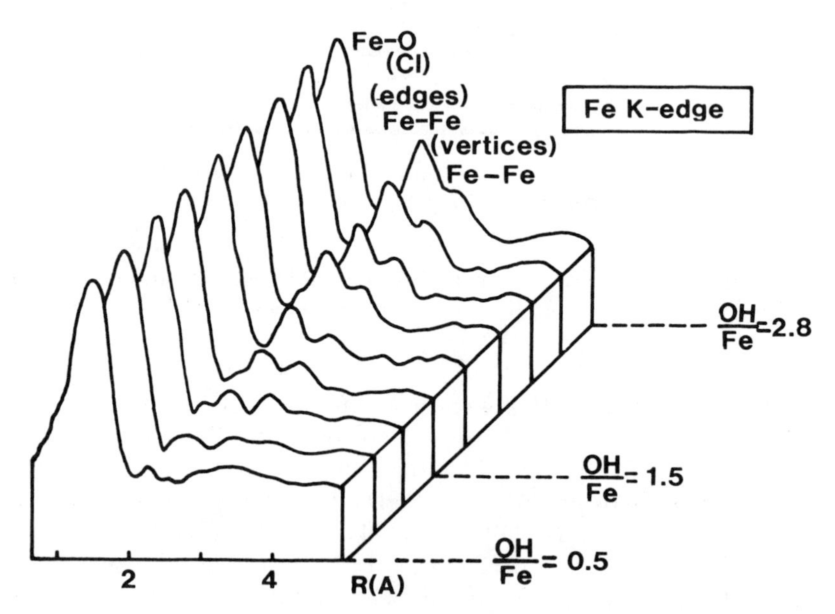

Figure 11. Radial distribution function of iron(III) hydroxide solutions versus R = *[OH]/[Fe] through EXAFS technique.*

of the mineral coagulant. They are detected by indirect evidence. Extensive projects have measured color and total organic carbon removal after treatment with alum, PACl, or PACS. Jar-test experiments with organic raw water have led to a design and operation diagram with alum as coagulant (*32, 33*). The diagram shows restabilization zones and two regions of removal that are located approximately at pH <7 and $-6 < \log Al < -4$ and at pH >5.5 and $-4.5 < \log Al < -3.2$. The first zone corresponds to charge neutralization (complexation) by small or medium polymers, followed by sedimentation. The second zone corresponds to a sweep coagulation (*34*) or adsorption on floc particles. A similar study compared alum and PACl (*32*).

As a result of this research (*32*), it appears that alum is the preferred coagulant within the sweep floc zone, whereas the PACl is a better coagulant at all pH values more basic or more acidic than the alum sweep floc zone. With PACl at pH < 5, removal can occur by charge neutralization and precipitation of fulvic acids (FA). An FA stoichiometry exists between the concentration of organic material and of coagulant. If the FA type is changed, the coagulant dosage will be changed. At pH >7 adsorption occurs on large aluminum flocs; the stoichiometry no longer exists, and the floc surface reacts the same as any oxide surface (*21*).

The choice of coagulant for the removal of fulvic acids depends on the FA type and concentration, pH, mineral complexing concentration (SO_4^{2-} or PO_4^{3-}) (*32*). The same kind of result has been obtained with PACS; this shows a clear relationship between the size and charge of the flocs formed (*35*).

Organic-substance removal by coagulation in the presence of montmorillonite and kaolinite clay particles has been investigated by Dempsey et al (*36*). The benefits of treatment with PACl instead of alum have been investigated as a function of pH, raw water composition, and mixing conditions. At low contaminant concentration, PACl is a better coagulant than alum at pH values <5 or >7. The relative benefit of PACl over alum decreases at higher contaminant concentration. Turbidity slightly influences color removal. The presence of fulvic acids increases the dose of coagulant required to remove turbidity.

Ozone treatment has been investigated as a way to modify the nature of raw organic material and increase its affinity for coagulants. Low concentrations of ozone (0.2–0.6 mg/mg of dissolved organic carbon) are sufficient to oxidize the adsorbable high-molecular-weight fraction (*37*). An excess of ozone dosage decreases the affinity of humic substances for flocs, presumably because of small units humic formation. In the presence of aluminum the metal–humate complexes are more soluble after ozonation, and removal efficiency can be lower (*38*). The 1:1 compounds formed through ozonation treatment excluded the formation of dicarboxylic acids (*39*). The removal of dibutyl phthalate and diethyl phthalate (one of the principal groups of ozonation products of humic materials) by Al or Fe coagulation (Figure 12) is

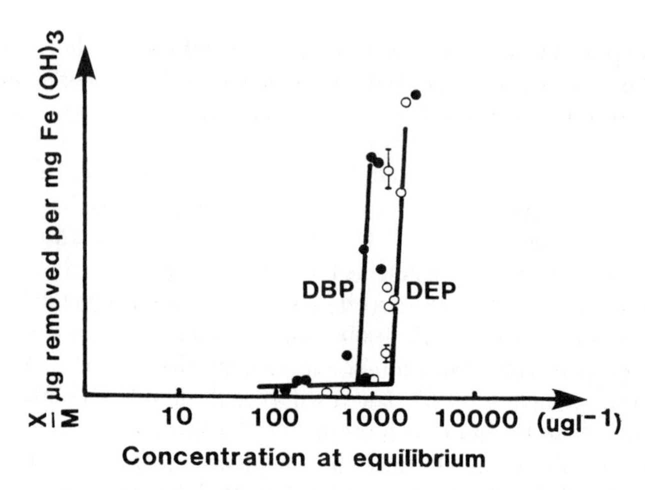

Figure 12. Adsorption isotherms of DBP (dibutyl phthalate) and DEP (diethyl phthalate) onto Fe (R = 2.5) *and PACl hydroxide* (R = 2.5) *flocs at pH 8 and* T = *20 °C. (Reproduced with permission from ref. 40. Copyright 1981 Pergamon.)*

organic-size dependent. The amount adsorbed increased with the molecular weight because of the increasing strength of the lateral bonds between molecules in the adsorbed phase (cooperative adsorption) (*40*).

In a laboratory study that used representative low-molecular-weight species such as salicylic acid, glycyltyrosine, cellobiose, and *N*-acetylglucosamine, low-dosage ozone yielded polymerization into macromolecules. Jar tests performed with alum showed improved organic removal (*41*). The large discrepancies between research reports show that more basic investigations are needed. The mechanisms of the ozonation–coagulation process are not well understood.

To approach the physical origin of such differences, it is necessary to evaluate the origin of the relationship between organic functions and aluminum cations or polycations (complexation), as well as the parameters governing the adsorption. In the literature that deals with complexation and adsorption of organic substances by aluminum, the major data concern aluminum monomeric species.

Complexation. The origin of the humic-substance affinity for aluminum flocs is certainly substituted hydroxybenzoic acids (*42*). Many studies have been done of the environmental implications of the aluminum– or iron–organic substance complexation. The aluminum–complexation reaction by salicylate (*43*), sulfosalicylate (*44*), acetate (*45*), and fulvic acids (*46*) is of metal–ligand complex origin, as proposed by the Eigen theory (*47*). Kinetically, the first step is an outer-sphere complex formation. The second step, much slower, is an inner-sphere complex formation accomplished by displacing a water molecule from the hydration shell. The complexation of Al monomers (pH <4–4.5) by small organic acids that simulate humic substances follows a monoexponential kinetic law (*48*). The kinetic curves have been interpreted with a reaction scheme that takes the three monomers into account:

$$\text{monomer} + A^- \overset{K_i}{\rightleftharpoons} \text{monomer}{\cdot}A + H^+ \qquad (6)$$

where the acid AH is dissociated to form A^- and H^+ and $K_i = k_i/k_{-i}$. The rate constants k_i that characterize the complexation with salicylate ions are of the order of 1, 1000, and 3000 $mol^{-1}\ s^{-1}$ with Al^{3+}, $Al(OH)^{2+}$, and $Al(OH)_2{}^+$, respectively. The stoichiometry is always 1:1. The stability constant of the complex is $K \sim 10^5$. The results are quite different with a mixture of monomers and Al_{13} and with a pure Al_{13} solution (*48*). The formation kinetic constant between Al_{13} and the salicylate ions is $k_p \sim 4000\ mol^{-1}s^{-1}$, $k_{-p} \sim 10\ s^{-1}$, and the stability constant $K = 10^{30}$. These data show that the complexation kinetic with Al_{13} is larger than with monomers and that the complexes Al_{13}–organic acids are more stable. The stoichiometry of such complexes is six salicylate ions per Al_{13}; the charge of the complex is ~0.

In the presence of minerals, more coagulant is needed to ensure a good removal of organic materials (*36*). Two kinds of reaction take place. The coagulant is adsorbed onto solids and organic compounds are adsorbed onto coagulants. The kinetic of coagulation–flocculation of minerals by aluminum is fast, because of the high affinity of Al for solid surfaces present in water (clays, silica). The half-time of silica aggregation by Al is less than 3×10^{-3} s (*49*). From studies of complexation (*43–45*), the half-time reaction between small organic acids and Al is higher than 0.07 s. This could mean that a large part of the coagulant, because it is first adsorbed onto solids, is not available for the adsorption or complexation of organic substances.

Adsorption. In many water-treatment plants the pH is rarely lower than 6.5, so the first step in the organic-substance removal process is adsorption onto large particles of coagulant. Three factors influence the uptake of organic material: the specific surface area, the surface charge, and the extent of the most active (energetic) sites (*50*).

As shown in Table III, the specific surface area of PACl flocs depends on the neutralization ratio *R* and the pH as measured from the adsorption of surfactants (*20, 51*). For alum flocs in water, the specific surface area is lower: 200–400 m^2/g of floc (Bottero, CNRS, unpublished data). Few data exist concerning iron coagulants. It seems that 600-m^2/g floc particles can be obtained (*52*), but generally the specific surface area of hydrous iron oxyhydroxides is 100–200 m^2/g. The specific surface area in water has to be the highest. In PACl flocs the presence of an Al_{13} subunit (~1000 m^2/g) leads to a very large available surface area for the uptake of organic compounds. If the surface area of flocs obtained from $R = 0$ or alum salts can be important, the affinity is lower for the reactions of monomers with small organic acids (*43–46*).

The surface charge of hydroxylated mineral flocs is generally positive at the pH of water (pH 6–8). In this case the adsorption is partly of electrostatic origin. The value of the isoelectric point (IEP) or zero point of charge (ZPC) is important, and the extent of the removal can be determined at constant surface area. Differences exist between the coagulants. The surface charge of PACl flocs is positive until pH ~8.5 (*21*). The point of zero charge of alum flocs or hydrous iron oxyhydroxides is ~7 and 6.5 to ~8.5, respectively (*21, 52*). In the case of alum, the affinity of SO_4^{2-} (nonindifferent electrolyte) for aluminum can partly mask the adsorption sites; because of this affinity, the sulfate ions would not be totally exchanged by organic substances. On PACl flocs the weak affinity of Cl^- ions for the surface increases the available site number compared with alum.

Some researchers report that the interactions of humic acids and aluminum salts at neutral pH are primarily a surface complexation on flocs, the anionic humate molecules are exchanged against OH, and the stoichiometry is 1:1 (one acidic group complexed by one Al) (*38*). This result is not obtained from the adsorption of sodium decanoate and sodium dodecanoate onto PACl flocs at $6.5 < pH < 7.5$. For this kind of molecules the adsorption mechanism is mainly of electrostatic origin without OH exchange (*51*). If cooperative effects exist between molecules in the adsorbed phase (van der Waals at-

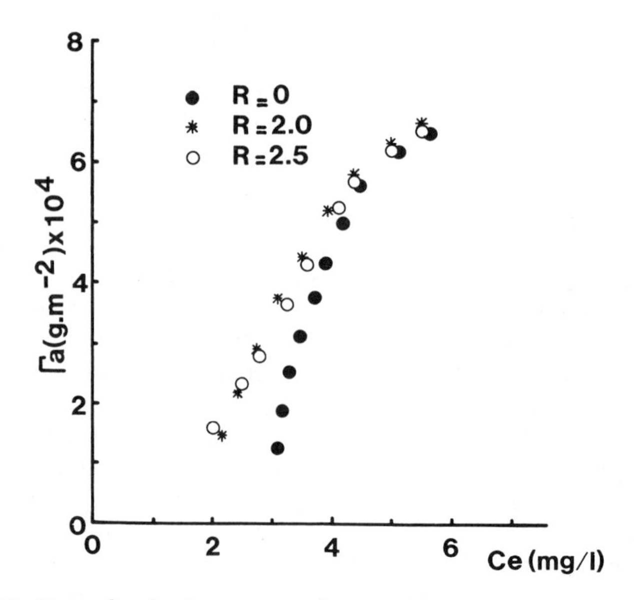

Figure 13. Normalized adsorption isotherms of fulvic acids onto PACl flocs versus R = *[OH]/[Al] at pH 7.2 ± 0.15. The specific surface area values were obtained from refs. 20 and 21.*

traction between alkyl chains), they do not generate the adsorption when the adsorbate–adsorbent interaction energy is large (*50*).

The surface of flocs is energetically heterogeneous, and the distribution or extent of this heterogeneity depends on the pH and the nature of the coagulant (*20, 51*). A study of the removal of a river humic acid by PACl flocs at pH = 7.2 ± 0.15 (*53*) showed that the surface of flocs precipitated from R = 2.0 and 2.5 is more efficient than the one from R = 0. For a same adsorbed amount, the equilibrium concentration decreases with R = $[OH]_{added}/[Al]$ (Figure 13). It means that the more energetic sites are more extended on R = 2.0 and R = 2.5 flocs than on R = 0 flocs.

Few data exist on in situ adsorption of humic substances onto iron flocs. At pH ~7, polymerized iron chloride improves humic-substance removal as compared with iron chloride, especially at low temperature (*54*).

References

1. Commitee report. *J. Am. Water Works Assoc.* **1979,** 588–603.
2. Schnitzer, M.; Khan, S. U. In *Humic Substances in the Environment*; Marcel Dekker: New York, 1972.
3. Schnitzer, M. In *Soil Organic Matter*; Schnitzer, M.; Khan, S. U., Eds.; Elsevier: New York, 1978; pp 1–64.
4. Schnitzer, M. In *Interactions of Soils Minerals with Natural Organics and Microbes*; Huang, P. M.; Schnitzer, M., Eds.; Special Publication No. 17; Madison, WI, 1986; pp 77–105.
5. Mallevialle, J.; Anselme, C.; Marsigny, O. Presented at the 193rd National Meeting of the American Chemical Society, Denver, CO, April 1987.
6. Richard, Y.; Bersillon, J. L.; Poirier, J. E. *Eau Ind.* **1979,** *34,* 29–35.
7. Thebault, P. Technical Report. Sociétè Lyonnaise des eaux, 1982.
8. Sillen, L. G. *Acta Chem. Scand., Series A* **1964,** *15,* 1981–1990.
9. Bersillon, J. L.; Hsu, P. H.; Fiessinger, F. *Soil Sci. Soc. Am. J.* **1980,** *44,* 630–637.
10. Akitt, J. W.; Farthing, A. *J. Chem. Soc. Dalton Trans.* **1981,** *12,* 1606–1610.
11. Akitt, J. W.; Greenwood, W.; Khandelwal, B. L.; Lester, G. R. *J. Chem. Soc. Dalton Trans.* **1972,** *9,* 604–612.
12. Bottero, J. Y.; Cases, J. M.; Fiessinger, F.; Poirier, J. E. *J. Phys. Chem.* **1980,** *84,* 2933–2937.
13. Parthasarathy, N.; Buffle, J. *Water Res.* **1985,** *19*(1), 25–36.
14. Bottero, J. Y.; Marchal, J. P.; Cases, J. M.; Fiessinger, F. *Bull. Soc. Chim. Fr.* **1982,** *11,* 439–443.
15. Johansson, L. *Acta Chem. Scand.* **1960,** 771–773.
16. Bottero, J. Y.; Axelos, M.; Tchoubar, D.; Cases, J. M.; Fripiat, J. J.; Fiessinger, F. *J. Colloid Interface Sci.* **1987,** *117*(1), 47–57.
17. Bottero, J. Y.; Tchoubar, D.; Cases, J. M.; Fiessinger, F. *J. Phys. Chem.* **1982,** *86,* 3667–3670.
18. Axelos, M.; Tchoubar, D.; Bottero, J. Y.; Fiessinger, F. *J. Phys. (Les Ulis, Fr.)* **1985,** *46,* 1587–1593.
19. Bottero, J. Y.; Tchoubar, D.; Cases, J. M.; Fripiat, J. J.; Fiessinger, F. In *Interfacial Phenomena in Biotechnology and Materials Processing*; Attia, Y., Ed.; Elsevier, 1988 (in press).
20. Rakotonarivo, E.; Bottero, J. Y.; Cases, J. M.; Fiessinger, F. *Colloids Surf.* **1984,** *9,* 273–293.

21. Rakotonarivo, E.; Bottero, J. Y.; Thomas, F.; Cases, J. M.; Poirier, J. E. *Colloids Surf.* **1988,** in press.
22. Stumm, W.; Morgan, J. J. *Aquatic Chemistry,* 2nd ed.; Wiley Interscience: New York, 1962.
23. Posner, J.; Quirk, R. A. *J. Colloid Interface Sci.* **1973,** *8,* 273–280.
24. Baes, R.; Mesmer, J. J. *The Hydrolysis of Polycations*; Wiley: New York, 1978.
25. Johnsson, P. N.; Amirtharajah, A. *J. Am. Water Works Assoc.* **1983,** 232–239.
26. Van der Woude, J. H. A.; de Bruyn, P. L. *Colloids Surf.* **1983,** *8,* 55–78.
27. Van der Woude, J. H. A.; de Bruyn, P. L. *Colloids Surf.* **1983,** *8,* 79–92.
28. Van der Woude, J. H. A.; de Bruyn, P. L.; Pieters, J. *Colloids Surf.* **1984,** *9,* 173–188.
29. Schneider, W. *Comments Inorg. Chem.* **1984,** *3,* 205–223.
30. Tchoubar, D.; Bottero, J. Y.; Axelos, M. A. V.; Quienne, P. *Abstracts of Papers,* 6th Meeting of the European Clay Groups, Seville, 1987; pp 521–523.
31. Combe, J. M.; Mancean, A.; Bottero, J. Y.; Calas, G. *Abstracts of Papers,* 6th Meeting of the European Clay Groups; Sociedad Española de Arcillas, Seville, Spain, 1987; pp 178–179.
32. Dempsey, B. A.; Ganho, R. M.; O'Melia, C. R. *J. Am. Water Works Assoc.* **1984,** *76*(4), 141–150.
33. Edwards, G. A.; Amirtharajah, A. *J. Am. Water Works Assoc.* **1985,** *77*(3), 49–57.
34. Hall, E. S.; Packham, R. F. *J. Am. Water Works Assoc.* **1965,** *57,* 1149–1166.
35. Wiesner, M.; Turcaud, V.; Fiessinger, F. *Abstracts of Papers,* Annual Conference of the American Water Works Association; American Water Works Association: Denver, CO; 1986.
36. Dempsey, B. A.; Sheu, H.; Tanser Ahmed, T. M.; Mentink, J. *J. Am. Water Works Assoc.* **1985,** *77*(3), 74–80.
37. Jekel, M. R. *Sci. Eng.* **1983,** *5,* 21–35.
38. Rechkow, D. A. Presented at the SVW Water Symposium, Brussels, Belgium, 1985; pp 411–431.
39. Jekel, M. R. *Water Res.* **1986,** *20*(12), 1535–1542.
40. Thebault, P.; Cases, J. M.; Fiessinger, F. *Water Res.* **1981,** *15,* 183–189.
41. Brette, B.; Duguet, J. P.; Mallevialle, J. Presented at the Association Générale des Hygienistes et Techniciens Municipaux, Nice, France, 1987.
42. Stumm, W.; Kummert, R.; Sigg, L. *Croat. Chem. Acta* **1980,** *53*(2), 291–312.
43. Secco, F.; Venturini, M. *Inorg. Chem.* **1974,** *14*(8), 1978–1981.
44. Periemuter, B.; Tapuhi, H. E. *Inorg. Chem.* **1977,** *16*(11), 2742–2745.
45. Harrada, S.; Uchida, Y.; Kuo, H. L.; Yasunaga, T. *Int. J. Chem. Kin.* **1980,** *12,* 387–392.
46. Plankey, B. J.; Patterson, H. H. *Environ. Sci. Technol.* **1987,** *21,* 595–601.
47. Eigen, M. In *Advances in the Chemistry of Coordination Compounds*; Kirshner, S., Ed.; Macmillan: New York, 1961; p 371.
48. Rakotonarivo, E.; Tondra, D.; Bottero, J. Y.; Mallevialle, J. *Water Res.,* in press.
49. Hahn, H. H.; Stumm, W. *J Colloid Interface Sci.* **1968,** *28*(1), 132–142.
50. Cases, J. M. *Bull. Mineral.* **1979,** *102,* 469–477.
51. Rakotonarivo, E.; Bottero, J. Y.; Cases, J. M.; Leprince, A. *Colloids Surf.* **1985,** *16,* 153–173.
52. Davis, J. A. *Geochim. Cosmochim. Acta* **1982,** *46,* 2381–2385.
53. Bersillon, J. L.; Benedek, A. Presented at the Annual American Water Works Association Conference, Houston, TX, 1984.
54. Leprince, A.; Fiessinger, F.; Bottero, J. Y. *J. Am. Water Works Assoc.* **1984,** 93–97.

RECEIVED for review January 26, 1988. ACCEPTED for publication May 4, 1988.

Effects of Humic Substances on Particle Formation, Growth, and Removal During Coagulation

Gary L. Amy, Michael R. Collins, C. James Kuo, and Zaid K. Chowdhury

Environmental Engineering Program, Department of Civil Engineering, University of Arizona, Tucson, AZ 85721

Roger C. Bales

Department of Hydrology and Water Resources, University of Arizona, Tucson, AZ 85721

This chapter describes research examining (1) the effect of natural organic matter (NOM) on the formation, growth, and removal of particles during coagulation using aluminum sulfate and (2) the effect of particles on NOM removal. Several sources of NOM and several types of mineral particles were studied under water treatment conditions. Particle formation, growth, and removal were found to be significantly affected by initial particle type and concentration, NOM type and concentration, pH, and aluminum sulfate dose.

SURFACE WATER CONTAINS PARTICLES ranging in size from submicrometer to supramicrometer dimensions, as well as aquatic natural organic matter (NOM), including humic substances. Important objectives of both conventional surface-water treatment and direct filtration are removal of both turbidity-causing particles and NOM. Total particle number is reduced during coagulation / flocculation by inducing particle growth, but subsequent solids–liquid separation (i.e., sedimentation and filtration) removes floc composed of aggregated particles and precipitated aluminum hydroxide.

0065–2393/89/0219–0443$06.00/0

Experimental Design

Four independent series of experiments (A through D) were conducted with solutions containing a model particle and either aquatic dissolved organic matter (DOM) or soil fulvic acid. All experiments evaluated aluminum sulfate coagulation by using a conventional jar test apparatus, but mixing conditions (Table I) and "postmixing" sample processing were different. The principal objective of series A experiments was to evaluate particle removal by passing flocculated water through a laboratory sand filter functioning as a batch-mode direct-filtration apparatus. The dimensions and operating conditions of the laboratory sand filter are described elsewhere (*1*). In contrast, the major objective of series B through D experiments was to study particle formation and growth. The filtrate was characterized in series A. In series B–D, the flocculated suspension was analyzed. For each series, a preliminary set of experiments was run to provide a basis for selecting aluminum doses.

The organic substances examined were a soil-derived fulvic acid (series A and D) and DOM isolated from two surface waters: the Grasse River in New York (series B) and the Edisto River in South Carolina (series C). DOM was defined as organic matter passing a 0.45-μm membrane filter.

Two parameters used for particle characterization were the total particle number (TPN), which represents the summation of particle concentrations observed in all channels of the particle counter, and the volume average diameter (d_v), which is a volume-weighted average. TPN was used to indicate overall particle removal, and d_v indicated particle-size changes. Although values of TPN and d_v for the initial conditions reflect only the model particle, these same parameters for the final water reflect aluminum–particle–humic aggregates in the filtered water (series A) or in the flocculated suspension (B through D).

Experimental Details

The hydrophobic (humic) fraction of the DOM, which was used in the experiments, was isolated by adsorption onto resin (XAD–8, Rohm and Haas), according to the method of Thurman and Malcolm (*2*). The average molecular weight of the DOM was determined by ultrafiltration (*3*). Carboxylic acidity was measured by potentiometric titration of the hydrophobic fraction from pH 3.0 to 8.0 (*4*). Non-purgeable organic carbon (NPOC) was determined with an organic carbon analyzer (Dorhmann DC–80), and UV absorbance was measured with a UV–visible spectrophotometer (Perkin–Elmer Model 200).

Three different particles were studied: (1) kaolinite clay that contained significant amounts of submicrometer material, series A; (2) Min–U–Sil–5 (Pennsylvania Glass Sand Corporation), series B and C; and (3) Min–U–Sil–15, series D. The Min–U–Sil materials represent crystaline silica, SiO_2. Particle enumeration and size characterization were done with an optical particle counter (Hiac 4100, Pacific Scientific Inc.), using a 2.5–150-μm sensor (series A and B) and a 1–60-μm sensor (series C and D). The more sensitive sensor was acquired midway through the research.

Electrophoretic mobilities were measured on the coagulant suspensions with a microelectrophoresis apparatus (Rank Brothers Mark II). An equilibration time of 12 h was used for DOM-coating of particles.

Results

The Grasse River contained slightly more DOM than the Edisto River (Table II). Average NPOC-based molecular weights for the two rivers were comparable, although that for the fulvic acid was much higher. Carboxylic acidities differed by a factor of 2, but are comparable to values reported in the literature. The characteristics summarized in Table II for the natural waters reflect the original water, as well as actual experimental conditions.

The Min–U–Sil–5 (SiO_2) had average d_v values of 4.2 and 3.4 μm, using the 2.5–150 and 1–60-μm sensors, respectively. Addition of the lower range sensor midway through the experiments provided more information on smaller-particle removal. The d_v for Min–U–Sil–15 (SiO_2) was 6.3 μm, using the 1–60-μm sensor.

Electrophoretic mobilities of the model particles as a function of pH are shown in Figure 1. Without the presence of humic substances and DOM, mobilities for SiO_2 and kaolinite ranged from positive to negative, with observed pH values at the isoelectric point (pH_{iep}) of 2.6 and 4.6, respectively. The DOM-coated silica exhibited a more negative mobility than the uncoated silica over the pH range of about 4.0 to 7.0, while the mobility of the DOM-coated kaolinite remained negative over the pH range of about 3.0 to 6.0. The effects of DOM on mobility were more pronounced for kaolinite than silica. Also shown in Figure 1 are mobilities for $Al(OH)_3(s)$ formed under conditions of homogeneous nucleation, indicating that the aluminum hydroxide precipitate formed can exhibit a positive or negative charge with a pH_{iep} of about 6.5.

The statistical significance of observed differences was evaluated by using the Duncan's multiple range (DMR) test, which has inherent advantages over the more commonly used *t*-test when a number of comparisons are made within a data set (5). The DMR test also permits a statistical comparison of pairs of parameter values. The data shown in Table I were subjected to a statistical analysis by the DMR test (6–8). Comparison of initial and final water characteristics shows an increase in d_v, a decrease in TPN, and a decrease in NPOC for most experiments (Table I). The following specific data comparisons are based on statistically significant differences in parameter levels at the 95% confidence level.

In Series A, fulvic acid more adversely affected particle removal at the lower pH. Higher fulvic acid concentrations generally resulted in higher TPN levels in the filtrate at pH 5.5; at a higher pH of 8.5, no statistically significant differences were found for TPN levels in experiments with higher versus lower initial fulvic acid levels. This finding may be due to the greater

Table I. Summary of Initial and Operational Conditions for Series A, B, C, and D Experiments

	Initial and Operating Conditions						Initial Water			Final Water		
Exp. ID	*Particle Type*	*Organic Source*	*Mixing Cond.*[a]	*pH*	*Turb. (NTU)*	*Al (mg/L)*	*TPN*[b] *(mL^{-1})*	d_v *(μm)*	*NPOC (mg/L)*	*TPN*[b] *(mL^{-1})*	d_v *(μm)*	*NPOC (mg/L)*
A-1	none	fulvic	I	5.5	0	2.0	1,100	—	1.9	90	4.9	1.0
A-2	kaolin	fulvic	I	5.5	10	2.0	14,000	4.4	1.9	170	4.6	1.1
A-3	none	fulvic	I	8.5	0	4.0	1,100	—	1.9	90	5.1	0.9
A-4	kaolin	fulvic	I	8.5	10	4.0	14,000	4.4	1.9	80	5.4	0.8
A-5	none	fulvic	I	5.5	0	2.0	1,100	—	7.7	200	4.8	3.4
A-6	kaolin	fulvic	I	5.5	10	2.0	14,000	4.4	7.7	230	4.7	3.5
A-7	none	fulvic	I	8.5	0	4.0	1,100	—	7.7	130	5.0	3.0
A-8	kaolin	fulvic	I	8.5	10	4.0	14,000	4.4	7.7	110	5.0	3.3
B-1	none	Grasse	II	7.0	0	2.0	210	—	5.0	930	6.5	2.8
B-2	silica	Grasse	II	7.0	5	2.0	140,000	4.2	5.0	10,000	8.8	3.0
B-3	silica	Grasse	II	7.0	10	2.0	290,000	4.2	5.0	5,100	1.2	2.7
B-4	none	Grasse	II	8.5	0	2.0	210	—	5.0	70	8.5	3.7
B-5	silica	Grasse	II	8.5	10	2.0	290,000	4.2	5.0	5,700	10.7	3.4
B-6	silica	Grasse	II	5.5	5	2.0	140,000	4.2	5.0	3,500	18.5	2.0
B-7	silica	Grasse	II	5.5	5	4.0	140,000	4.2	5.0	6,200	14.9	1.7

C-1	none	Edisto	II	7.0	0	2.0	140	—	4.4	3,500	5.5	2.7
C-2	silica	Edisto	II	7.0	5	2.0	410,000	3.4	4.4	2,200	25.6	2.6
C-3	silica	Edisto	II	7.0	10	2.0	820,000	3.4	4.4	1,800	30.6	2.5
C-4	none	Edisto	II	8.5	0	2.0	140	—	4.4	1,900	6.2	2.9
C-5	silica	Edisto	II	8.5	10	2.0	820,000	3.4	4.4	1,500	30.8	2.8
C-6	silica	Edisto	II	5.5	5	2.0	410,000	3.4	4.4	3,200	25.9	2.1
C-7	silica	Edisto	II	5.5	5	4.0	410,000	3.4	4.4	1,700	26.6	1.9
D-1	silica	fulvic	III	7.5	0.4	4.0	2,100	6.3	4.9	110,000	2.9	—
D-2	silica	fulvic	III	7.5	0.4	1.0	2,100	6.3	4.9	38,000	3.6	—
D-3	silica	none	III	7.5	0.4	4.0	2,100	6.3	0	130,000	3.4	—
D-4	silica	none	III	7.5	0.4	1.0	2,100	6.3	0	11,000	6.5	—
D-5	silica	fulvic	III	5.5	0.4	1.0	2,100	6.3	4.9	8,200	5.3	—
D-6	silica	none	III	5.5	0.4	1.0	2,100	6.3	0	8,600	6.0	—
D-7	none	fulvic	III	7.5	0	4.0	13	—	4.9	150,000	1.5	—
D-8	none	none	III	7.5	0	4.0	13	—	4.9	190,000	1.3	—

[a] Mixing conditions I, rapid mix G of 300 s^{-1} for 1 min followed by slow mix G of 60 s^{-1} for 15 min; mixing conditions II, rapid mix G of 300 s^{-1} for 1 min followed by slow mix G of 30 s^{-1} for 20 min; and mixing conditions III, single velocity gradient G of 100 s^{-1} for 20 min.

[b] Series A and B are based on a 2.5–150-μm sensor; a 1.0–60-μm sensor was used in series C and D.

Table II. Characteristics of Dissolved Organic Matter and Soil-Derived Fulvic Acid

Parameter	*Grasse*	*Edisto*	*Fulvic Acid*
NPOC (mg/L)	4.96	4.35	1.94 or 7.71
UV absorbance (cm^{-1}, 254 nm, pH 7)	0.186	0.164	0.085 or 0.334
Hydrophobic NPOC (% of total NPOC)	63	60	73
Carboxylic acidity (meq/g)	6.4	8.5	13.5
Average molecular weight	3290	3570	9700
Bicarbonate alkalinity (mg/L as $CaCO_3$)	31	24	50

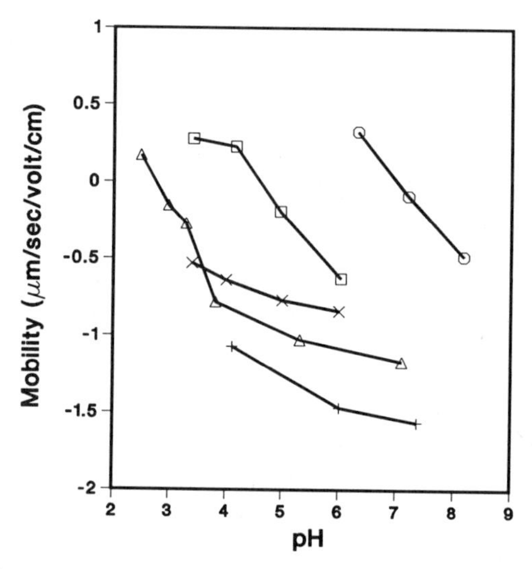

Figure 1. Electrophoretic mobilities of particles with and without organic coatings: □, kaolinite; ○, $Al(OH)_3$; △, Min–U–Sil–15; +, FA-coated Min–U–Sil–15; and ×, FA-coated kaolinite.

stabilizing effects of fulvic acid on kaolinite clay at the lower pH, as evidenced by the divergence of the mobility curves below pH 6 (Figure 1). At a pH of 5.5, the fulvic acid would have a lower charge density and the kaolinite clay would be near its pH_{iep}, thus enhancing adsorption of fulvic acid. The presence or absence of kaolinite did not influence TPN levels in the filtrate at the higher pH, although TPN trends were less clear at the lower pH. No significant differences were seen in d_v for series A; finished-water particle size was controlled to a large extent by the sand filtration.

Results of series B and C (with the objective of particle growth) indicate that d_v was greater in the presence than in the absence of silica, and was also greater when more silica was present. The effects of silica on TPN levels were less clear; with Edisto River DOM, increasing the initial silica concentration resulted in no statistically significant differences in final TPN. For Grasse River DOM, the final TPN concentration was generally highest; 5 NTU of initial silica was present. Also, with Grasse River DOM, higher pH and higher aluminum dose lead to increases in TPN and lower values of d_v. Conversely, for Edisto River DOM, higher pH and higher aluminum dose resulted in lower TPN levels, with no significant change in d_v.

In series D, the presence of silica, with all other factors equal, resulted in lower TPN levels and higher values of d_v. The presence of fulvic acid led to larger particles and a smaller TPN when silica was absent; and to smaller particles, yet no systematic change in DOM when silica was present. This result suggests that an aluminum fulvate system (without particles) provides slightly better precipitation than a homogeneous aluminum hydroxide system. Snodgrass et al. (9) also observed that the average size of particles formed in an aluminum–fulvic acid solution was larger than those observed with aluminum alone. In an aluminum fulvate–silica system, more competing (pH-dependent) reactions occur; the extent of particle aggregation is not expected to be a simple function of aluminum dose, fulvic acid concentration, and so on.

The presence of kaolinite generally inhibited NPOC removal in series A. This result is apparently due to the competing reactions between adsorption and precipitation for particle destabilization and aggregation, as opposed to aluminum fulvate precipitation. In contrast, in series B and C the presence or absence of silica had little effect on NPOC removal, a result probably reflecting less adsorption of NOM onto silica than onto kaolinite. The mobility data reflect a greater effect of NOM on kaolinite than on silica (Figure 1). This effect is consistent with observations that NOM adsorbs to aluminum (hydr)oxide surfaces (a portion of kaolinite) to a greater extent than to silica (*10, 11*). For series B and C, better NPOC removals were observed at pH 5.5 than at 7.0; at pH 5.5, one would expect a lesser charge density associated with carboxylic groups of the NOM, thus affecting charge neutralization.

A further analysis of data was conducted to delineate a possible stoichiometry between NPOC removal and aluminum added. For the seven experiments conducted with Grasse River DOM, the ratio of NPOC removed to aluminum added was 0.99 ± 0.28 mg/mg, while the corresponding ratio for the seven experiments conducted with the Edisto River DOM was 0.84 ± 0.16 mg/mg.

Discussion

Particle growth occurred in most of the series B and C experiments, probably because mixing conditions were conducive to particle growth. Although different particle-counter sensors were used in the series B and C experiments, qualitative comparisons between the two are still possible. Particles formed with Edisto River DOM in the presence of silica were generally larger than those in Grasse River DOM; TPN levels were correspondingly lower for Edisto River versus Grasse River. The Edisto River DOM was present at a lower concentration (as NPOC) than Grasse River DOM, and it had a slightly greater carboxylic acidity and molecular weight.

Significant increases in d_v (particle growth) did not occur in either series A or D. Series D experiments used the largest particle and had a high-velocity gradient, approaching rapid-mix conditions. Further, substantial numbers of small aluminum hydroxide particles were produced in these experiments. Recall that in series A, the objective was particle removal and the flocculated water was filtered prior to analysis. Although TPN levels were significantly reduced, final d_v values (3.6–4.7 μm) were comparable to those in initial waters (d_v = 4.3 μm). This result is not surprising, as clean sand filters are most effective in removing particles larger than a few micrometers (*12*). In all experiments, one would expect the initial formation of submicrometer precipitates of aluminum hydroxide, aluminum fulvate, or aluminum–particle aggregates, with the subsequent growth of these precipitates until eventually they pass through the detection limit (1.0 or 2.5 μm) of the optical particle counter. In Series D experiments, it is likely that an observation of particle growth of the original silica particles was outweighed by the appearance of small aluminum hydroxide floc.

Results show qualitative consistency with the notion of differing removals or growth, due to competing dissolved-aluminum complexation and precipitation reactions that differ in rate and extent as a function of pH. In the absence of foreign particles or humic substances, homogeneous nucleation and precipitation of aluminum hydroxide occur. In the presence of foreign particles, there is heterogeneous nucleation of aluminum hydroxide onto the initially present "seed" particles. Either homogeneous or heterogeneous nucleation can be the first step in particle formation and growth. When aluminum hydrolysis species predominate, particle destabilization occurs via adsorption of the hydrolysis species, creating conditions suitable for subsequent particle aggregation. Thus, two principal mechanisms involved in particle removal by chemical coagulation are *sweep coagulation*, in which particles are enmeshed in an aluminum hydroxide precipitate, and *charge neutralization*, whereby particles are destabilized by adsorbed, positively charged hydrolysis species.

When humic substances are present at lower pH levels (e.g., 5.5), cationic aluminum species react with negatively charged humic and fulvic acids to form aluminum–organic complexes and precipitates (*13, 14*). The observed greater NPOC removal in the current work at lower pH is consistent with adsorption versus coprecipitation; the results do not permit an unambiguous conclusion, however. Although these aluminum fulvate–humate precipitates are too small to settle effectively, they can be removed by filtration. At higher pH levels (e.g., 7.5), humic-substance adsorption onto an aluminum hydroxide precipitate is more important. Humic substances also adsorb onto materials such as clays and metal oxides, and thereby change their surface identity (*10, 11, 15*) and stabilize the particles.

Humic substances can affect aluminum–particle interactions, and particles can affect aluminum–humic interactions. Because of the higher charge density associated with the fulvic acid, as opposed to the clay found in a typical surface water, the coagulant dosage required for charge neutralization will normally be controlled by the fulvic acid (*16*).

Summary and Conclusions

On the basis of TPN levels, the presence of fulvic acid resulted in a poorer quality of sand filtrate, and the most adverse effect was observed at lower pH conditions. In the presence of DOM, greater amounts of initial seed particles resulted in greater values of d_v in flocculated suspensions. The presence of fulvic acid and the absence of initial particles led to an increase in d_v in flocculated suspensions, although the presence of both fulvic acid and initial particles led to a decrease in d_v.

Particle formation, growth, and removal were significantly affected by initial particle and organic matter or humic substance concentrations, as well as pH and aluminum dose. The effect of particles on DOM removal differs for kaolinite versus silica.

Acknowledgment

Partial financial support for this research was provided by the U.S. Environmental Protection Agency (EPA) under Grant R–812325–01–0. The contents do not necessarily reflect the views and policies of the EPA.

References

1. Collins, M.; Amy, G.; Bryant, C. *J. Environ. Eng. Div. (Am. Soc. Civ. Eng.)* **1987,** *112*(2), 330–344.
2. Thurman, E.; Malcolm, R. *Environ. Sci. Technol.* **1981,** *15*, 463–466.
3. Amy, G.; Collins, M.; Kuo, C.; King, P. *J. Am. Water Works Assoc.* **1987,** *79*, 43–49.
4. Collins, M.; Amy, G.; Steelink, C. *Environ. Sci. Technol.* **1986,** *20*, 1028–1032.
5. Dowdy, S.; Wearden, S. *Statistics for Research*; John Wiley: New York, 1983.
6. Collins, M. Ph.D. Dissertation, University of Arizona, 1985.
7. Kuo, C. Ph.D. Dissertation, University of Arizona, 1986.
8. Chowdhury, Z. Ph.D. Dissertation, University of Arizona, 1987.
9. Snodgrass, W.; Clark, M.; O'Melia, C. *Water Res.* **1984,** *18*, 479–488.
10. Davis, J. *Geochim. Cosmochim. Acta* **1982,** *46*, 2381–2391.
11. Davis, J.; Gloor, R. *Environ. Sci. Technol.* **1981,** *15*, 1223–1229.
12. O'Melia, C. *J. Environ. Eng. Div. (Am. Soc. Civ. Eng.)* **1985,** *111*, 874–891.

13. Wiesner, M.; Lahoussine-Turcaud, V.; Fiessinger, F. *Proc. AWWA Annu. Conf.* American Water Works Association Annual Conference, American Water Works Association: Denver, CO, 1986; p 1685.
14. Hundt, T. *Proc. AWWA Annu. Conf.* American Water Works Association Annual Conference, American Water Works Association: Denver, CO, 1986; p 1199.
15. Gibbs, R. *Environ. Sci. Technol.* **1983,** *17,* 237–240.
16. Dempsey, B.; Ganho, R.; O'Melia, C. *J. Am. Water Works Assoc.* **1984,** *76,* 141–150.

RECEIVED for review July 24, 1987. ACCEPTED for publication December 7, 1987.

28

Evaluation of Extent of Humic-Substance Removal by Coagulation

Norihito Tambo and Tasuku Kamei

Department of Sanitary and Environmental Engineering, Faculty of Engineering, Hokkaido University, N13 W8, Sapporo 060, Japan

Hideshi Itoh

Hokkaido Research Institute for Environmental Pollution, N19 W12, Sapporo 060, Japan

A functional relationship that quantitatively evaluates the extent of removal of general organic matter from water has been developed, on the basis of gel chromatographic analysis of raw water. This functional relationship was derived by the comparison of anion-exchange gel chromatograms before and after coagulation with the water-quality indices of dissolved organic carbon and absorbance at 260 nm. Water quality after coagulation can be predicted by the use of such functions when gel chromatograms of raw water have been evaluated. The mechanism of organic-substance removal by aluminum coagulation is discussed in connection with the extent of removal.

HUMIC SUBSTANCES ARE THE MAJOR ORGANIC CONSTITUENT of waters and wastewaters. Therefore, any evaluation of the treatability of various types of water and wastewaters must consider the behavior of humic substances as a key factor.

Many studies have been conducted since the 1960s in an effort to understand the behavior of humic substances in the coagulation and flocculation processes. Black and his co-workers (*1–6*) were among the first to discuss

0065–2393/89/0219–0453$06.00/0

these behaviors in light of the electrokinetic nature and other chemical characteristics of colored substances. They determined the optimal coagulation pH at around 5.5 and 4.0, respectively, for aluminum and iron coagulants. Letterman et al. (*7*), Edwards and Amirtharajah (*8*), and others investigated the phenomena further.

A stoichiometric relationship between coagulants and humic substances at the optimum coagulation condition was found through the research of Black et al. (*4*), Hall and Packham (*9*), Narkis and Rebhun (*10*), and others. Current theories of humic-substance coagulation are discussed from the point of view of aluminum–humate production and aluminum precipitation in the publications of Snodgrass et al. (*11*) and Dempsey et al. (*12*).

Many researchers have recognized the effect of organic-matter particle size on coagulation and other processes. Zuckerman and Molof (*13*) used gel chromatography (Sephadex G-15) to fractionate soluble organic substances in wastewater and thus identify the relationship between the molecular-size distribution of the contaminants and wastewater treatability. They showed that low-molecular-weight organic substances are more biodegradable than larger compounds. Gjessing and Lee (*14*) and Brodsky and Prochaza (*15*) carried out humic-substance research on river water with gel chromatography (Sephadex G-75). They revealed that the high-molecular-weight humic compounds are removed effectively. Tambo and Kamei (*16*) reported comprehensive but qualitative treatability evaluation of wastewater and of unpolluted and polluted natural waters on the Sephadex G-15 gel chromatogram. They revealed that small humic substances (<1500 MW) are not removed effectively by alum coagulation.

Semmens and Ayers (*17*) fractionated Mississippi River water and its coagulated water by ultrafiltration. They reported that coagulation can remove intermediate ($1–10 \times 10^3$) and high-molecular-weight substances ($10–100 \times 10^3$) effectively, but that low-molecular-weight substances (<1000 MW) cannot be removed well. Hall and Packham (*9*) stated that there is no distinction between the coagulation mechanism of humic and fulvic acids from the viewpoint of complex formation, but that fulvic acid (usually <1500 MW) cannot be removed effectively by alum and iron coagulation.

Behaviors of humic substances in the coagulation process are discussed in this chapter, with and without coexisting water-quality components, to establish a practical scheme for treatability evaluation. The mechanism of humic-substance removal in connection with the treatability evaluation scheme is also discussed.

Organic-Compound Removal by Aluminum Coagulation

Categorization of Major Dissolved Organic Compounds. Suspended organic matter is removed effectively by both water purification and wastewater-treatment processes. Therefore, practical discussion is concerned mainly with dissolved organic compounds. Gel chromatography can fractionate almost any amount of organic compounds, except trace volatile and hydrophobic organic substances, in order of apparent molecular size. Most of the gel-chromatographable organic compounds are nonvolatile substances. They are regarded as pseudobinary groups of organic materials for the purpose of treatability evaluation (*18*, *23*).

The first group consists of biodegradable organic compounds such as carbohydrates and aliphatic organic acids. It is characterized as dissolved organic carbon (DOC), with UV extinction at 260 nm (E_{260}). Dissolved organic compounds that show biological oxygen demand (BOD) roughly correspond to this group.

Humic substances make up the major part of the refractory organic material in unpolluted natural water and secondary effluent. These organic compounds, with similar infrared spectra, have a DOC/E_{260} ratio of about 30–50 (*16*, *18*, *19*, *23*). The value of DOC in this ratio is expressed in milligrams per liter, and E_{260} is expressed by absorbance through a 1-cm quartz cell.

These compounds, widely found in natural water and secondary effluent, are the end products of aerobic biological processes. E_{260}-sensitive refractory organic carbon, generated through aerobic biological wastewater-treatment processes, accounts for 3–5% of the carbon in the original biologically degradable organic substances (*19*). This group of UV-sensitive nonvolatile organic compounds is the major precursor of trihalomethane (THM) and total organic halide (TOX) formation by water chlorination (*20*). A good linear relationship exists between E_{260} and the formation potential for both THM and TOX, as shown in Figure 1. THM or TOX formation potential is defined

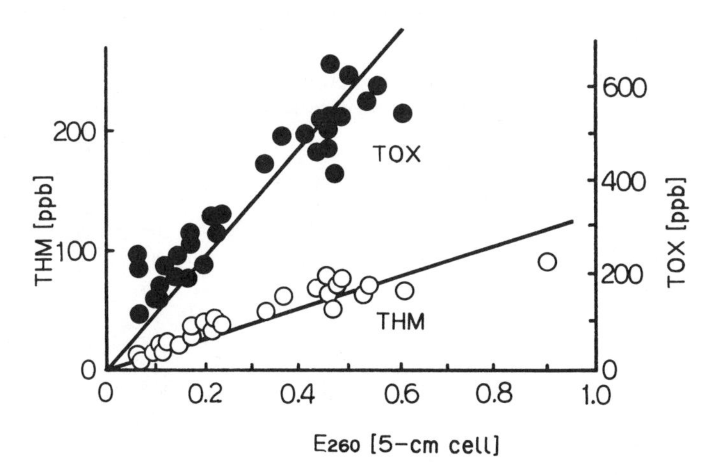

Figure 1. Correlation between E_{260} *and THM or TOX.*

as the amount of THM or TOX formed in 24 h of reaction time at 20 °C and pH 7, with 1 mg/L of free residual chlorine.

Analysis of various types of water samples has shown that the minimum value of DOC/E_{260} is approximately 30 for large E_{260}-sensitive organic compounds (>1500 MW) and approximately 50 for smaller E_{260}-sensitive organic compounds (<1500 MW) (*16, 18, 19, 23*). These values are probably characteristic of E_{260}-sensitive natural colored organic materials and might correspond to humic acid and fulvic acid. These characteristic values and observed DOC/E_{260} values can be used to calculate the amount of E_{260}-sensitive DOC in the samples.

Table I shows an example in which the characteristic DOC/E_{260} ratio for the high-molecular-weight humic substances (i.e., group 1) is taken as 33.3; in this example the extinction coefficient of 100 mg of E_{260}-sensitive DOC is 3. For the low-molecular-weight humic substances (i.e., groups 2–4), the characteristic value is usually taken as DOC/E_{260} = 50 (Figure 2, Table I).

The ratio of E_{260}-sensitive DOC to total DOC can be estimated by using the E_{260}/DOC value observed in each chromatographically separated group. This process will be discussed in the following section. A log–log plot of the relationship between the ratio of refractory E_{260}-sensitive DOC and the individual DOC/E_{260} values is presented in Figure 2 for both high- and low-molecular-weight humic substances.

Chromatographic Grouping of Major Organic Constituents. For more accurate discussion of the treatability evaluation, the two categories of organic constituents were subdivided by gel chromatography into several groups of compounds according to their apparent molecular size.

Figure 3 shows a schematic gel chromatogram of soluble organic compounds in sewage. The groups were separated in a gel column (Sephadex

Table I. Relationship Between DOC/E_{260} and Amount of E_{260}-Sensitive DOC

E_{260}	DOC/E_{260}	E_{260}-*Sensitive DOC (mg)*
3.0	33.3	100
2.7	37	90
2.4	42	80
2.1	48	70
1.8	56	60
1.5	67	50
1.2	83	40
0.9	111	30
0.6	167	20
0.3	333.3	10
0.15	667	5
0.03	3333	1

NOTE: In all cases, the total DOC was 100 mg/L.

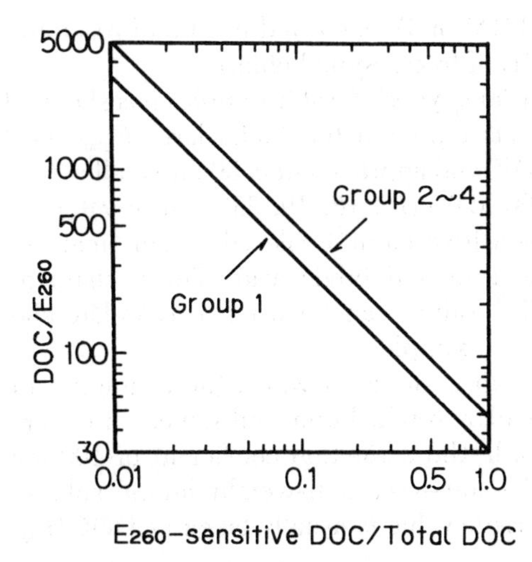

Figure 2. Relationship between E_{260}*-sensitive DOC and DOC/*E_{260}*. (Reproduced with permission from ref. 23. Copyright 1978 Journal of the Japanese Water Works Association.)*

G-15) and detected by DOC/E_{260}. We chose G-15 (Pharmacia Fine Chemicals) from among several types of Sephadex gel because of its good separation at the molecular-weight region where coagulation, carbon adsorption, and aerobic biological processes coexist in a system. In biologically treated wastewater effluent and natural colored water, a significant amount of organic material is larger than the upper fractionation limit of Sephadex G-15 (i.e., >1500 MW). However, these substances are effectively removed by coagulation and flocculation, so that no further gel fractionation is necessary (*16, 21*).

Elution of the soluble part (0.45-μm filter pass) of the sample was carried out by H_2O and 0.1 N NH_4OH. The samples were concentrated under reduced pressure in a rotary vacuum evaporator at 40 °C (20-fold concentration). Then 10-mL aliquots of these concentrated samples (>100 mg/L of DOC) were applied to the Sephadex G-15 columns (90 × 2.5 cm i.d.) and eluted with distilled water, followed by 0.1 N NH_4OH solution. Use of a single alkaline eluent, such as 0.1 N NH_4OH solution or alkaline boric acid buffer, can minimize adsorption onto the gel. However, two successive eluents were employed in an attempt to separate colored substances from synthetic organic compounds, which are much more easily adsorbed onto the gel.

Elution order can be described in terms of the distribution coefficient, K_d, defined by

$$K_d = \frac{V_e - V_0}{V_i} \tag{1}$$

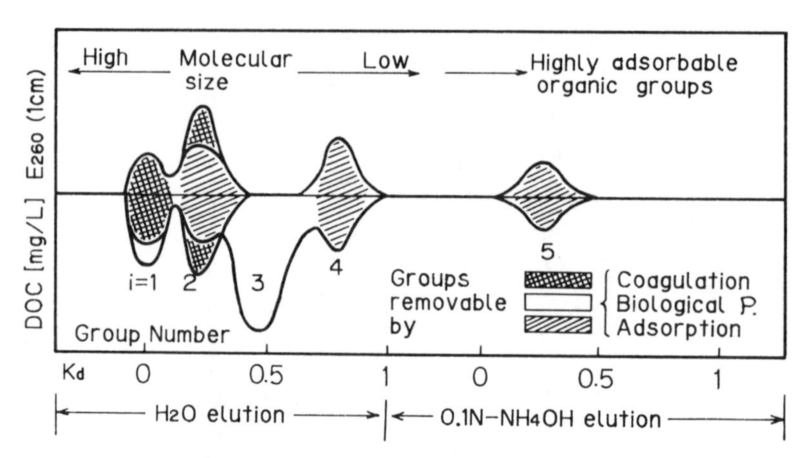

Figure 3. Group fractionation on Sephadex G-15 of typical domestic wastewater and corresponding treatment processes.

where V_e is the elution volume of the organic compounds (mL), V_0 is the external aqueous void volume of the gel (mL), and V_i is the inner void volume of the gel (mL).

These soluble organic compounds can be classified into five groups by elution order. The first four groups, eluted in water, are pseudobinary compounds characterized as E_{260}-sensitive refractory organic materials and E_{260}-insensitive biodegradable organic materials. Organic compounds in group 5, eluted by 0.1 N NH_4OH, are usually E_{260}-sensitive and have high affinity to the gel. The compounds are grouped according to the K_d value as follows: for group 1, <0.15; for group 2, 0.15–0.45; for group 3, 0.45–0.6; and for group 4, 0.6–1.2. The K_d value roughly corresponds to the apparent molecular weight of the compounds eluted by H_2O. An elution profile of model organic solutes and model reference compounds on the Sephadex G-15 gel chromatogram is shown in Figure 4.

Elution-order grouping, useful for categorizing sewage before and after biological treatment, also applies to many natural waters. Outflow from a well-conserved forest includes organic compounds (mainly fulvic substances) in groups 2, 4, and 5, with $DOC/E_{260} = 50$. Outflow from peaty swamps adds the group 1 compounds (humic substances), with $DOC/E_{260} = 20$–30. These high-molecular-weight humic substances should be removed when water passes through soil layers. Very often compounds in groups 1, 2, and 4 show a DOC/E_{260} ratio much higher than 50. This result means that some biodegradable E_{260}-insensitive organic compounds have been added to the humic and fulvic substances. These E_{260}-insensitive group 1 organic compounds are probably polysaccharides generated as an intermediate metabolic product of microbiological activity (*20, 22*). Almost all petrochemical micropollutants, such as hydrophobic and low-molecular-weight insecticides and pesticides, appear in group 5 (*25*).

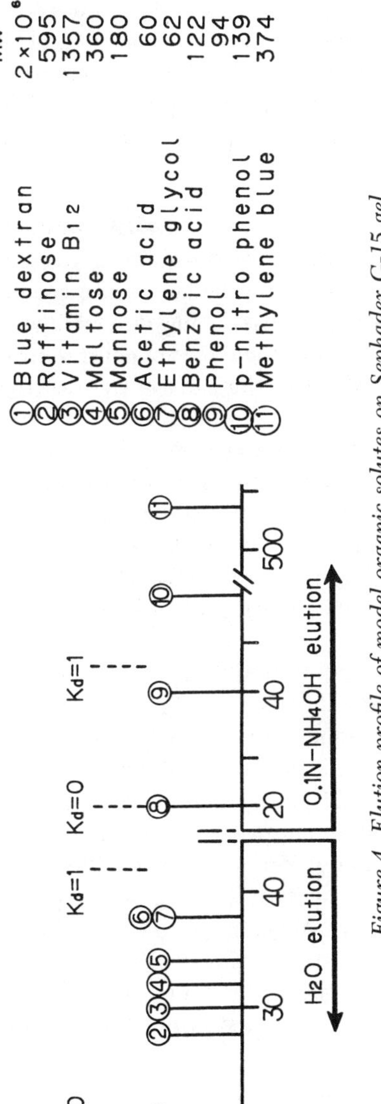

Figure 4. Elution profile of model organic solutes on Sephadex G-15 gel.

These groups of compounds are removed by successive use of conventional water-treatment processes such as biological decomposition, coagulation, and activated-carbon adsorption (*see* Figure 3) (*16*). Effectiveness of coagulation for high-molecular-weight E_{260}-sensitive compounds and of biological decomposition for E_{260}-insensitive low-molecular-weight organic substances have been reported in our previous publications (*16, 18, 23, 24*). The reports of Zuckerman and Molof (*13*), Hall and Packham (*8*), Brodsky and Prochazka (*15*), and others support our results. Adsorption removes low-molecular-weight E_{260}-sensitive organic compounds effectively (*15, 23, 24*).

Jar Test and Optimum Coagulation Conditions. The coagulation test for quantitative treatability evaluation was carried out by the standard jar test method, using aluminum sulfate as the coagulant for many waters (lake water, river water, peat water, and domestic wastewater) that contain humic substances. Constant coagulant concentration with variable pH was used for each run by adjusting the sample pH with 0.1 N NaOH or 0.1 N HCl solution.

After coagulation, flocculation, and sedimentation in a test jar, the supernate was filtered through a 0.45-μm membrane filter. Measurements of DOC concentration and UV absorbances at 260 nm (E_{260}) in a 1-cm cell were carried out for the filtrates. A TOC analyzer (Toshiba–Beckman, model 102) was used for the DOC measurements. Gel chromatography (Sephadex G-15) was carried out with samples taken before and after aluminum coagulation.

Examples of the results of the constant coagulant concentration and variable pH jar tests are shown in Figures 5 and 6. In these jar tests, pH 5.5–6.0 is an optimum zone for humic-substance coagulation with minimal aluminum dosage. Figure 7 plots residual DOC after coagulation with added aluminum concentration at the optimum pH.

The DOC removal ratio improves as aluminum concentration increases. Beyond a certain point, however, DOC is not effectively removed even with a further increase in the dosage. Thus, practically, there is a minimum to the DOC level attainable by coagulation. This minimum level suggests that a portion of the total DOC resists removal by aluminum coagulation, perhaps because of the nature of the compounds involved. In this study, removal efficiency was defined at the lowest attainable DOC level in the following equation.

$$R = \frac{[C_{\text{raw}} - C_{\text{treated}}]}{C_{\text{raw}}} \tag{2}$$

where R is removal efficiency, C_{raw} is the DOC level before treatment, and C_{treated} is the lowest DOC level attainable by coagulation.

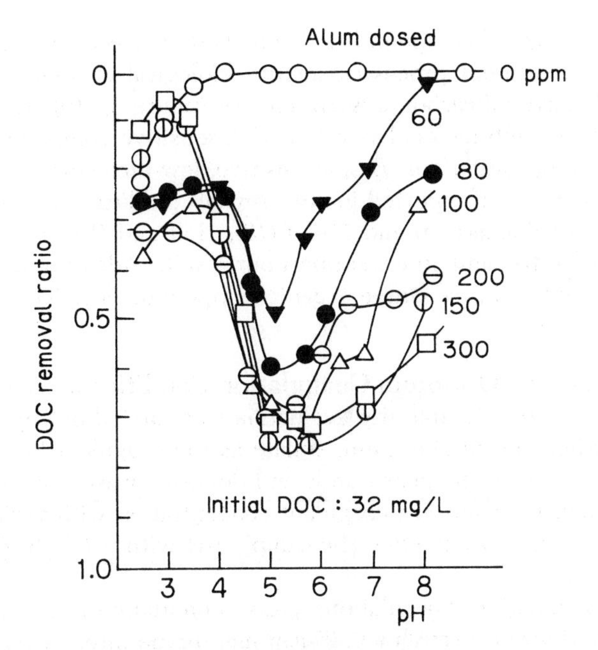

Figure 5. Coagulation pattern of peat water.

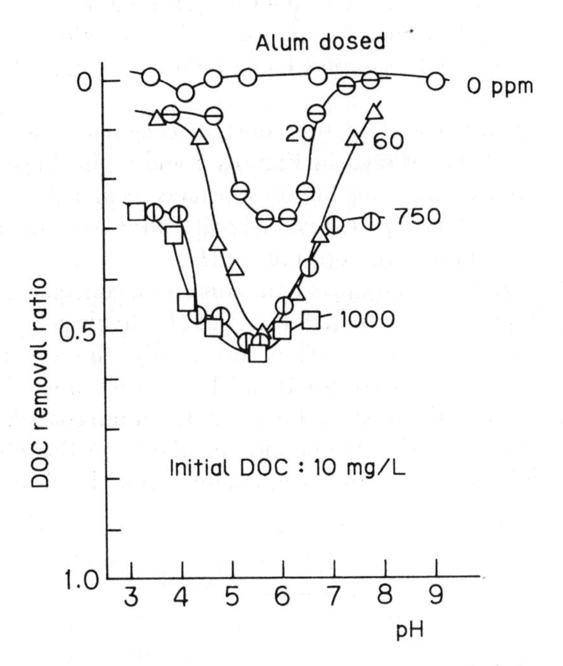

Figure 6. Coagulation pattern of effluent of an activated sludge sewage-treatment plant.

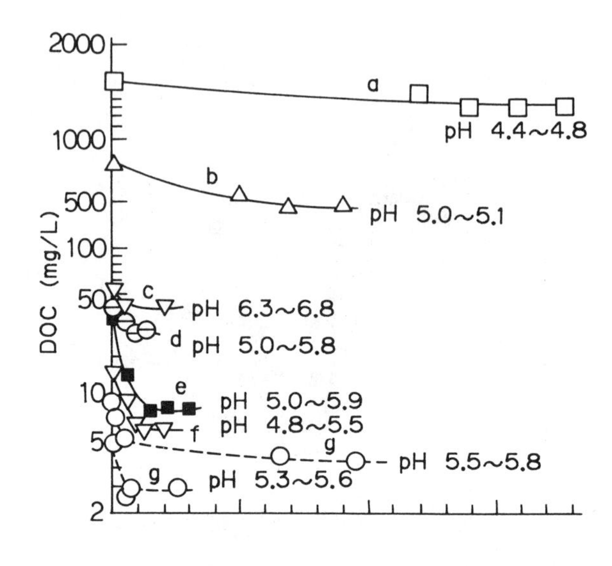

Figure 7. Improvement of DOC removal with alum dosage at the optimum pH. a, sulfite pulp wastewater; b, Kraft pulp wastewater; c, daily wastewater; d, settled raw sewage; e, peat water; f, activated sludge process effluent of excrement digester supernatant; and g, activated sludge process effluent of sewage.

Characteristic Treatability. By comparison of the G-15 gel chromatograms of raw water and coagulated water, the DOC/E_{260} ratio can be used to calculate the DOC removal efficiency in each group of organic substances. Figure 8 shows the removal efficiency of compounds in each group, characterized by the DOC/E_{260} ratio. A high DOC/E_{260} ratio indicates low removal efficiency with all groups of compounds.

If we accept the pseudobinary nature of the major organic compounds in each K_d-value group, equation 3 can be proposed to characterize their removal efficiency.

$$r_i = \frac{a_i E_{260}\text{-sensitive DOC} + b_i E_{260}\text{-insensitive DOC}}{\text{total DOC}} \qquad (3)$$

where r_i is the removal efficiency of group i, a_i is the removal efficiency of E_{260}-sensitive organic substances, and b_i is the removal efficiency of E_{260}-insensitive organic substances in group i.

Removal-efficiency values for both E_{260}-sensitive and E_{260}-insensitive compounds in each water-elution group, evaluated from many kinds of experimental data, are shown in Table II.

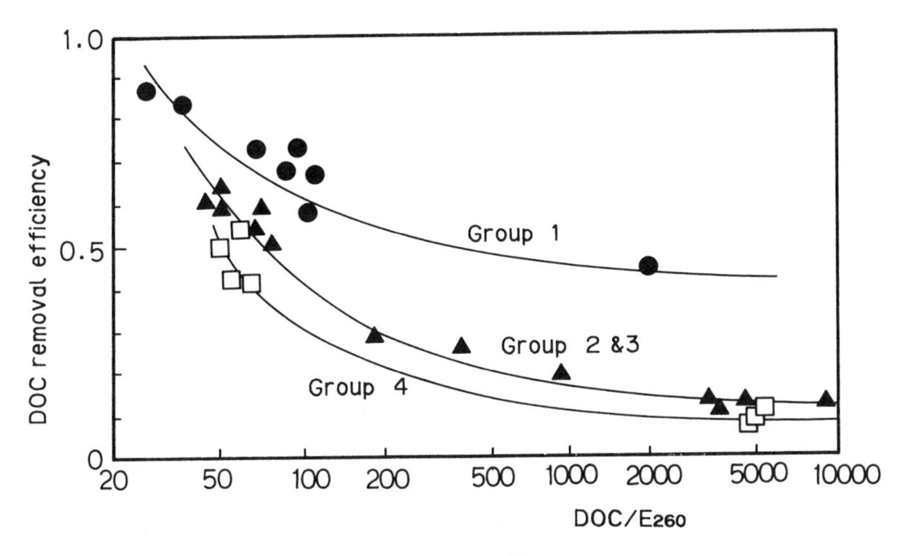

Figure 8. Removal efficiency.

Table II. Removal Efficiency of Component Organic Compounds (DOC)

	Chemical Coagulation		*Aerobic Bioprocess*	
Group	E_{260}-*Sensitive*	E_{260}-*Insensitive*	E_{260}-*Sensitive*	E_{260}-*Insensitive*
1	0.90	0.45		0.70
2	0.65	0.15	0.05	
3				0.97
4	0.50	0.10		

For prediction of the removal efficiency of group 5 organic compounds, the relationship between DOC/E_{260} and removal efficiency is plotted as shown in Figure 9 (*23*, *25*). Regardless of the differing composition of the organic substances, a high DOC/E_{260} ratio tends to produce low removal efficiency.

The experimentally determined removal efficiency of the group 5 compounds (removed by aluminum coagulation) is expressed as equation 4.

$$r_s = 47.1 \left(\frac{\text{DOC}}{E_{260}} \right)^{-1.32} \tag{4}$$

Equation 2 also applies for evaluation of the removal efficiency of aerobic biological processes, as we reported in a previous paper (*18*). Table II shows that the biological process is ineffective for the removal of humic substances.

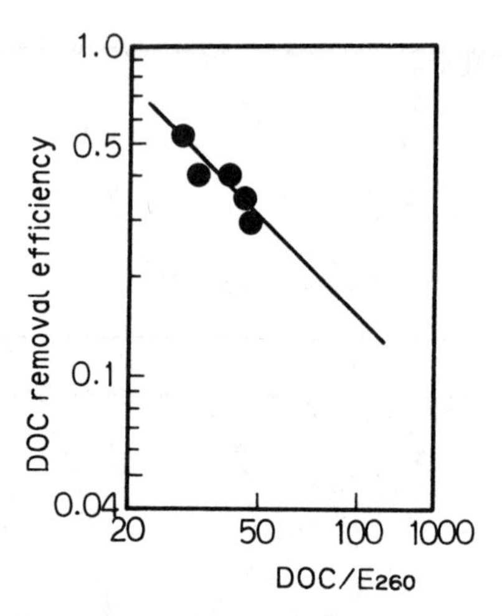

Figure 9. Removal efficiency in group 5. (Reproduced with permission from ref. 23. Copyright 1978 Journal of the Japanese Water Works Association)

Characteristic Coagulation Pattern

High-molecular-weight humic substances (i.e., mainly humic acid) are efficiently removed by coagulation. On the contrary, low-molecular-weight substances (i.e., mainly fulvic acid) are not removed effectively. The difference in removal efficiency and the characteristic coagulation pattern are discussed in this chapter in connection with the nature of the aluminum coagulant and humic substances.

Coagulation Pattern and Electrophoretic Mobility. We carried out two kinds of jar tests to illustrate the coagulation pattern of humic substances with alum. Sample colored water with negligible turbidity was taken from a shallow well in a peaty area near the city of Sapporo, Japan. General raw-water quality of this sample is as follows: pH 6.0; color, 80 color units; TOC, 20 mg/L; electrical conductivity, 73.7 μs/cm.

The tests were carried out with normal jar test procedures for constant aluminum concentration and variable pH. For method A, raw colored water was filtered through a 0.45-μm membrane filter and then diluted to 20 color units with distilled water. Coagulated and settled water was filtered through a 12-μm membrane filter. This 12-μm filter size was selected to simulate separation in a rapid sand filter. The filtrate was analyzed for color unit and aluminum concentration at the same time. The solid lines in Figure 10 show the ratio of color or aluminum concentration before and after 12-μm membrane filtration. The dotted lines show the results of the same jar test procedures with aluminum coagulant (alum) alone.

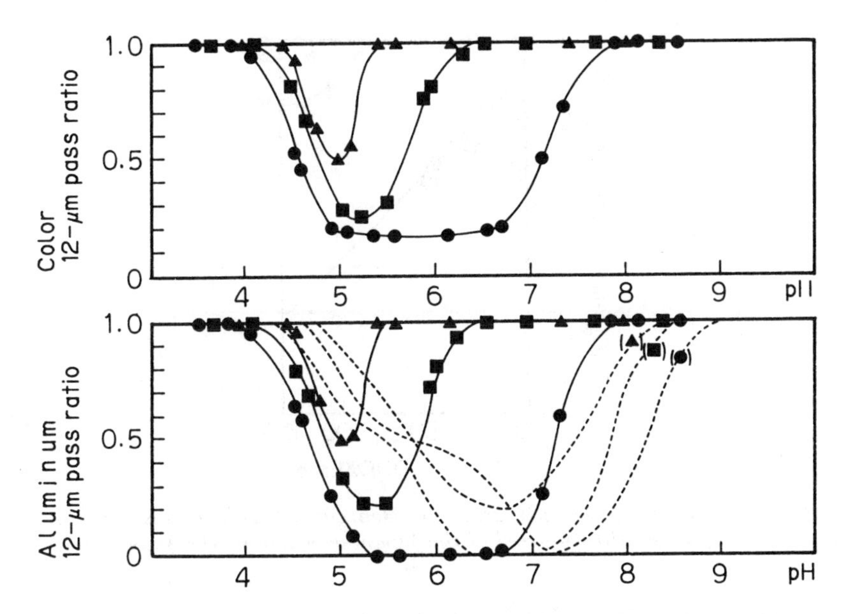

Figure 10. Filtration pattern by method A. Color: 20 units; aluminum (mg/L): (▲) 0.8; (■) 2.4; (●) 4.8. Key: —, results of jar tests; and - -, results of same jar-test procedures with aluminum coagulant alone.

In method B, equal volumes of alum solution and colored water were separately pH-controlled in two different vessels. After a 5-min reaction time, the two solutions were poured into a jar and treated according to the ordinary jar test procedure.

The results attained with method B are shown in Figure 11. Separate concentrations of aluminum and color were twice as high before mixing as those concentrations measured in the test jars.

Electrophoretic mobility was measured by the U-tube method (i.e., the electrophoretic mass-transfer method) (*26*, *27*). Measurements were carried out with two indices, color and aluminum concentration. The results are shown in Figures 12 and 13. Solid lines show gross mean mobility of aluminum species, humic substances, and aluminum–humic-substance complexes or aggregates, as measured by color unit and aluminum concentration under various combinations of pH and alum concentration. The dotted lines show the mobility of aluminum alone. The dotted broken line shows the mobility of humic substances in raw water.

If mobility curves measured by color and aluminum concentration coincided with each other, it would mean that almost all humic substances and coagulants exist as aluminum–humic-substance complexes or aggregates. This agreement is seen at pH >4.7–4.8. On the contrary, mobility as measured by color and aluminum differs greatly at pH <4.5–4.7. This variation

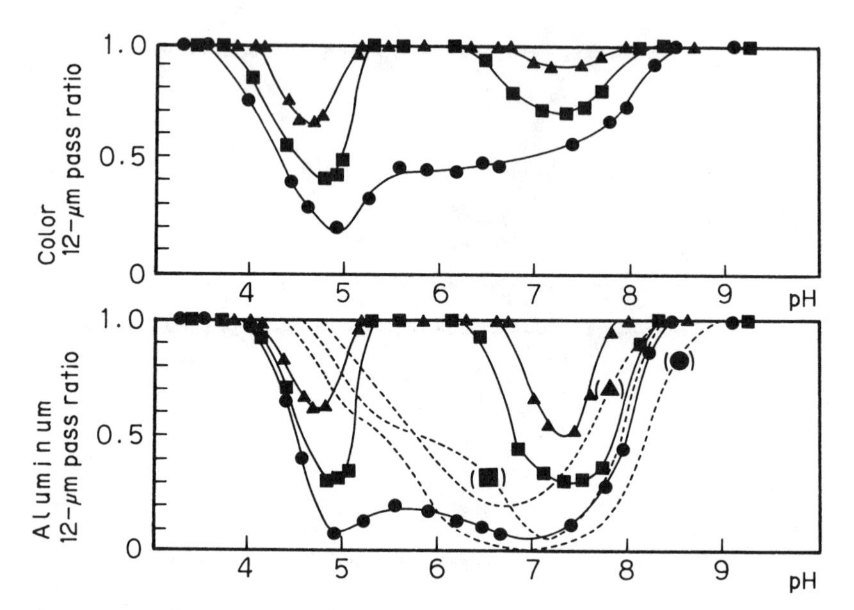

Figure 11. Filtration pattern by method B. Color: 20 units; aluminum (mg/L): (▲) 0.8; (■) 2.4; (●) 4.8.

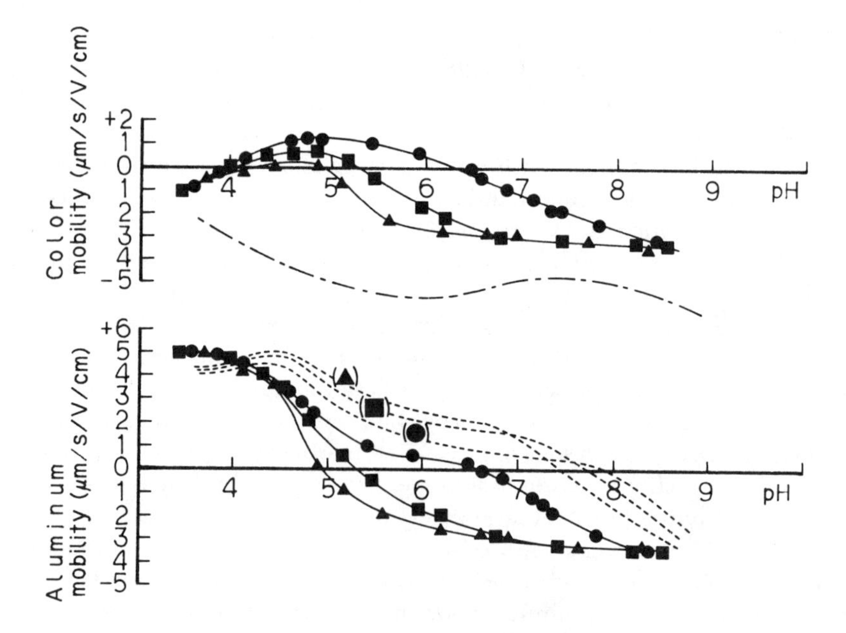

Figure 12. Mobility pattern by method A. Color: 20 units; aluminum (mg/L): (▲) 0.8; (■) 2.4; (●) 4.8.

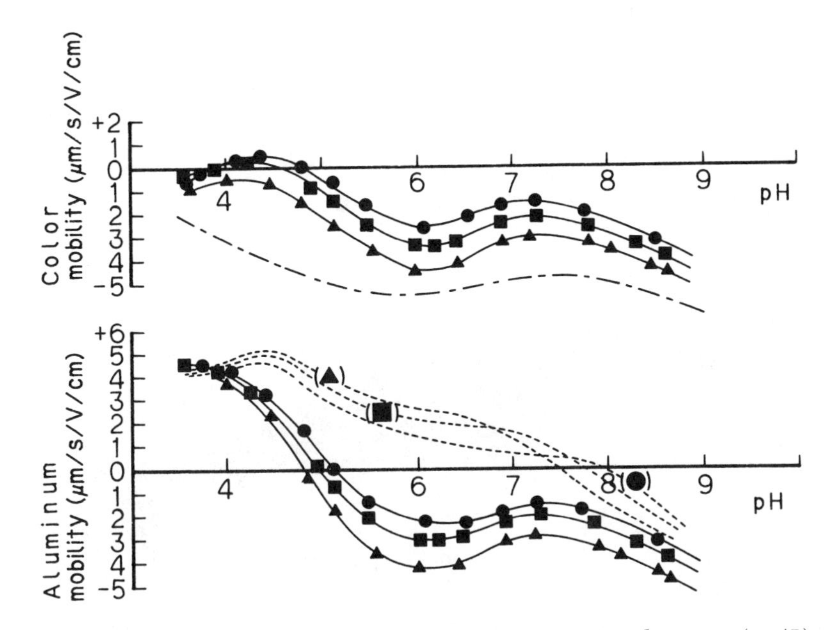

Figure 13. Mobility pattern by method B. Color: 20 units; aluminum (mg/L): (▲) *0.8;* (■) *2.4;* (●) *4.8.*

indicates that aluminum–humic-substance complexes or aggregates are not effectively generated at pH <4.5.

A comparison of Figures 10 and 11 with Figures 12 and 13 shows good coagulation in the same pH zone with good mobility, as measured by color unit and aluminum concentration. Figures 10 and 11 show optimum coagulation of humic substances at pH 5.0–5.5, as many researchers have reported (*1, 6, 9, 11, 12*).

Electrophoretic mobilities of hydrolyzed aluminum species were also measured by the U-tube method with respect to aluminum concentration and duration of electrophoresis (i.e., hydrolysis reaction time). The results are shown in Figure 14. Figure 15 shows particle-size distribution of the hydrolyzed aluminum species by ultrafiltration with respect to pH (*27*). Those measurements were carried out after 5 min reaction time with 0.8 mg of aluminum per liter of solution.

Figures 14 and 15 reveal that small, highly cationic aluminum polymers exist at pH 4.5. Weakly cationic aluminum hydroxides precipitate from pH 5.0 to 8.5. Positive mobility changes rather slowly at about pH 5.0–5.5. Therefore, competitive reactions occur between aluminum and both OH^- in water and similar kinds of functional groups on the surface of humic substances.

The hydrolyzed aluminum species generated in method B is less effective for coagulation of humic substances. However, apparent coagulation

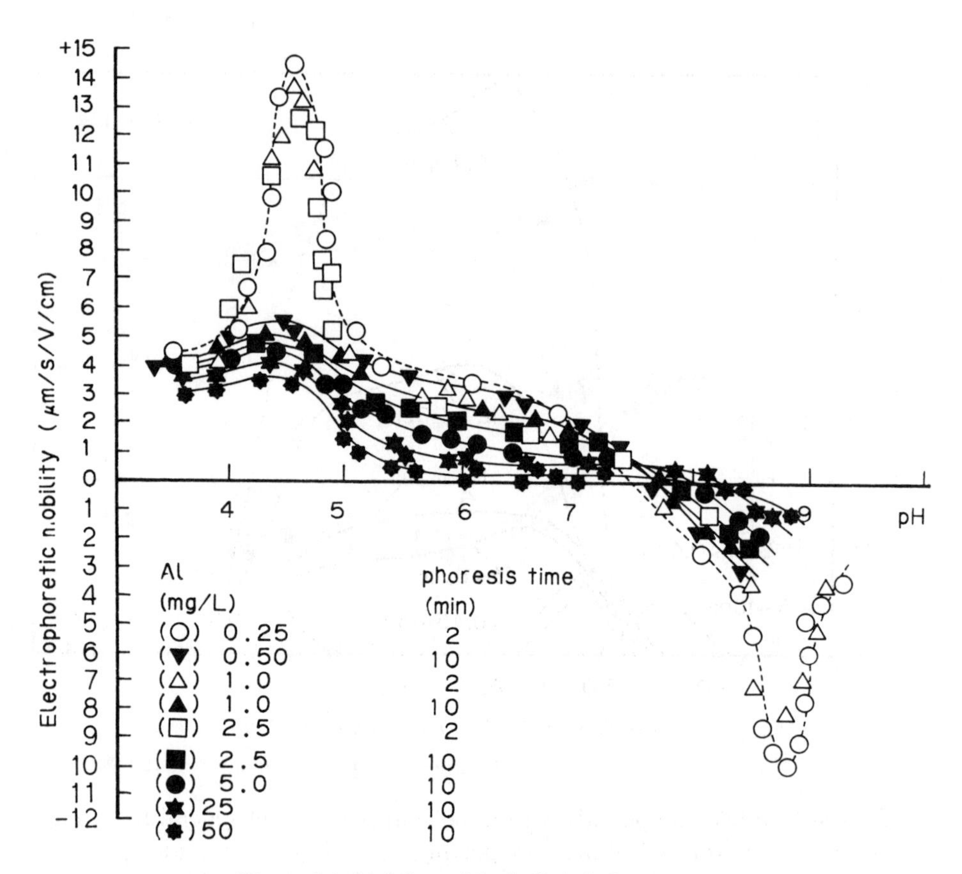

Figure 14. Mobility of hydrolyzed aluminum.

patterns at optimum pH (i.e., pH = 5.0–5.5) are similar for both methods. Although aluminum humate production and hydrolysis of aluminum compete in the first step of coagulation (*11, 12*), hydrolysis of aluminum largely controls the key coagulation pattern in the optimum pH zone. In addition, aluminum humate production in the first step explains the much higher removal efficiency of method A (*11, 12*). Coagulation patterns differ in the neutral pH zone. The aluminum hydroxide that precipitates in method B may adsorb humic substances on its surface. In the case of method A, the hydrolyzed soluble aluminum polymer complexes significantly with humic substances to produce a precipitate.

Humic substances are efficiently removed in the zone where the absolute value of electrophoretic mobility is less than 1.0 μm/s/V/cm (*28*). The maximum color removal was about 80% in this case.

Schematic Coagulation Model. Gel chromatograms show that humic substances range from an apparent molecular weight of a few thousand

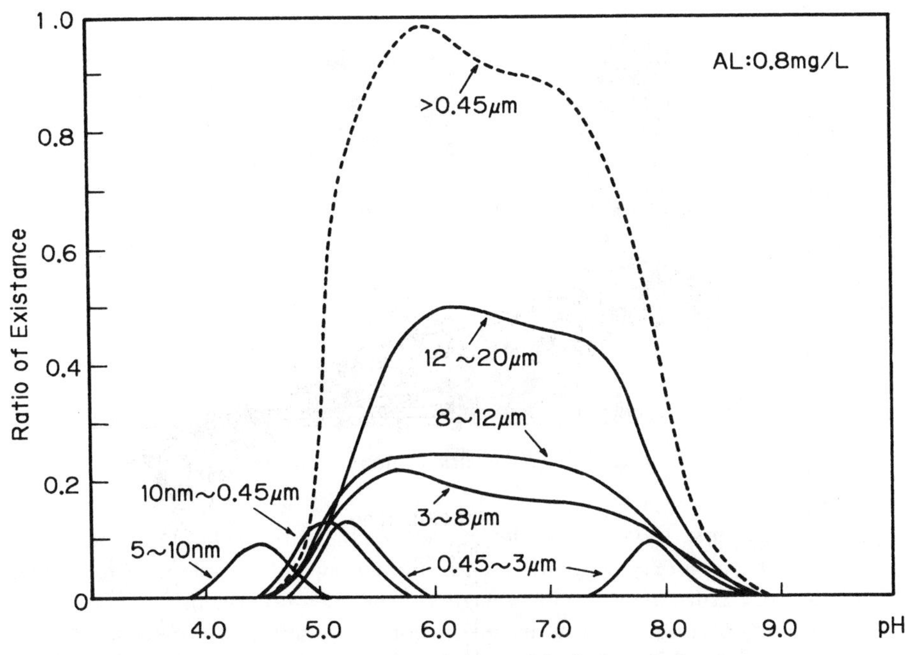

Figure 15. Particle-size distribution of hydrolyzed aluminum.

or more to fulvic compounds with a molecular weight of several hundred. The active hydrolyzed aluminum polymer at pH 5.0–5.5 could be such species as $Al_8(OH)_{20}{}^{4+}$, $Al_2(OH)_2{}^{4+}$, and $Al_{13}O_4(OH)_{24}{}^{7+}$ (*12, 29, 30*). A polymer such as $Al_8(OH)_{20}{}^{4+}$ (MW <1000) can effectively neutralize the negative charge of both colloidal substances and humic substances (*8, 29, 30*). At the same time, some aluminum species that precipitate at pH >5 can be as large as 1 μm or more.

The combination of such cationic and anionic species is the basis for the schematic floc illustrated in Figure 16. Highly charged small aluminum polymers effectively neutralize their charge and form complexes with high-molecular-weight humic substances (MW >>10^3). The neutralized aluminum–humic-substance complexes are bridged by precipitated aluminum to form a settleable floc. Both of the required conditions (formation of neutral aluminum–humic-substance complexes and bridging aluminum hydroxide precipitation) can be satisified at pH >5.

The mobility of aluminum hydroxide increases in the range from neutral to pH 4.5, the mobility of humic substances decreases toward the acidic side, and precipitated aluminum species can make settleable flocs at pH >5. These characteristics make pH 5.0–5.5 an optimum condition for coagulation and flocculation. At this point the relative sizes of humic substances and aluminum polymer that bring about complex formation might be very im-

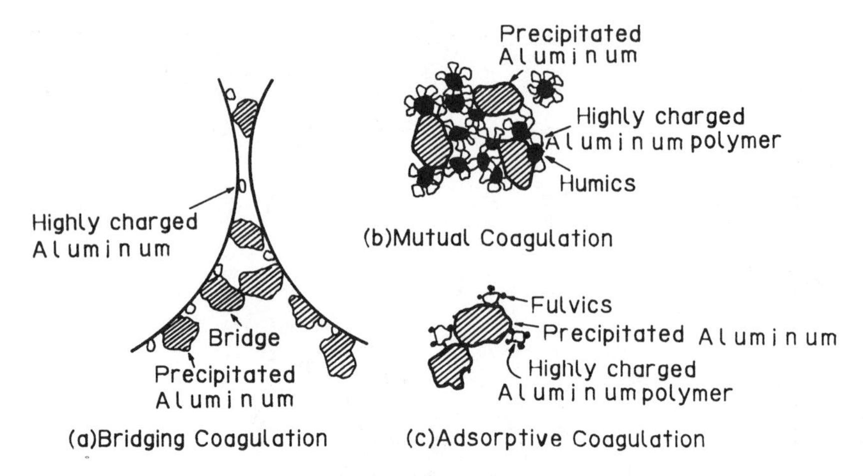

Figure 16. Schematic of floc formation.

portant. Because fulvic molecules are so much smaller relative to the active aluminum polymer, coagulation with alum may not be effective for fulvic substances. Still, small fulvic substances can become attached to the active aluminum polymer, such as in adsorption. The schematic pattern for this type of reaction can be seen in Figure 16. For these small E_{260}-sensitive compounds, carbon adsorption is much more effective (*16*).

For clay suspensions, both charge neutralization and bridging action can satisfactorily occur through reactions of weakly cationic aluminum precipitates at the neutral pH zone. When humic substances and clay colloids coexist, coagulation proceeds effectively at the weakly acidic region (*8, 10, 17*). In this case aluminum–humic-substance agglomerates may behave as bridging precipitates for the agglomeration of clay particles.

References

1. Black, A. P.; Willems, D. G. *J. Am. Water Works Assoc.* **1961,** *53*(5), 581–604.
2. Black, A. P.; Christman, R. F. *J. Am. Water Works Assoc.* **1963,** *55*(6), 753–770.
3. Black, A. P.; Christman R. F. *J. Am. Water Works Assoc.* **1963,** *55*(7), 897–912.
4. Black, A. P. et al. *J. Am. Water Works Assoc.* **1963,** *55*(10), 1347–1366.
5. Tambo, N. *J. Jpn. Water Works Assoc.* **1964,** *361*, 2–12.
6. Tambo, N. *J. Jpn. Water Works Assoc.* **1965,** *365*, 25–37.
7. Letterman, R. D. et al. *J. Am. Water Works Assoc.* **1982,** *74*(1), 45–51.
8. Edwards, J. K.; Amirtharajah, A. *J. Am. Water Works Assoc.* **1985,** *77*(3), 50–57.
9. Hall, E. S.; Packham, R. F. *J. Am. Water Works Assoc.* **1965,** *57*, 1149–1166.
10. Narkis, N.; Rebhum, M. *J. Am. Water Works Assoc.* **1977,** *69*(6), 325–328.
11. Snodgrass, W. J. et al. *Water Res.* **1984,** *18*, 479–488.
12. Dempsey, B. A. *J. Am. Water Works Assoc.* **1984,** *76*(4), 141–150.
13. Zuckerman, M. M.; Molof, A. H. *J. Water Pollut. Control Fed.* **1970,** *42*(30), 437–456.

14. Gjessing, E.; Lee, G. F. *Environ. Sci. Technol.* **1967,** *1,* 631–638.
15. Brodsky, A.; Prochazka, J. *J. Am. Water Works Assoc.* **1975,** *67*(1), 23–26.
16. Tambo, N.; Kamei, T. *Water Res.* **1978,** *42,* 931–950.
17. Semmens, M. J.; Ayers, K. *J. Am. Water Works Assoc.* **1985,** *77*(5), 79–84.
18. Tambo, N.; Kamei, T. *J. Water Pollut. Control Fed.* **1980,** *52,* 1019–1028.
19. Tambo, N.; Kamei, T. *Proc. 2nd World Cong. Chem. Eng.* Vol. VI; Montreal, Canada, 1981; pp 106–109.
20. Tambo, N.; Kamei, T. *Abstracts of Papers*; International Chemical Congress of Pacific Basin Societies, Honolulu, HI; American Chemical Society: Washington, DC, 1984; Abstract 03110.
21. Kamei, T.; Tambo, N. *J. Jpn. Water Works Assoc.* **1977,** *519,* 24–41.
22. Okuyama, H.; Kamei, T.; Tambo, N. *Gesuido Kyokaishi* **1982,** *19,* 19–32.
23. Tambo, N.; Kamei, T. *J. Jpn. Water Works Assoc.* **1978,** *530,* 8–18.
24. Tambo, N.; Kamei, T. *J. Jpn. Water Works Assoc.* **1978,** *532,* 37–44.
25. Kamei, T. Doctor of Engineering Thesis, Hokkaido University, 1979.
26. Kruyt, H. R. *Colloid Sci.* **1962,** *1,* 213–214.
27. Tambo, N.; Itoh, H. *J. Jpn. Water Works Assoc.* **1977,** *508,* 38–50.
28. Tambo, N. *Memoirs of Faculty of Engineering,* Hokkaido University, 1965, XI, pp 585–611.
29. Matijevic, E. et al. *J. Phys. Chem.* **1961,** *65,* 826–829.
30. Stumm, W.; Morgan, J. J. *J. Am. Water Works Assoc.* **1962,** *54*(8), 971–992.

RECEIVED for review July 24, 1987. ACCEPTED for publication March 21, 1988.

Characteristics of Humic Substances and Their Removal Behavior in Water Treatment

[illegible]

Characteristics of Humic Substances and Their Removal Behavior in Water Treatment

J. S. Kim[1], E. S. K. Chian, F. M. Saunders, E. Michael Perdue[2], and M. F. Giabbai[1]

School of Civil Engineering, Georgia Institute of Technology, Atlanta, GA 30332

The characteristics of naturally occurring aquatic humic substances and their removal behavior in water treatment were investigated in a plant operation and in alum coagulation with conventional laboratory jar-test experiments. Specific characteristics of humic substances affected the performance of alum coagulation in removing both humic substances and turbidity. Up to 50% of the humic substances were removed in both the treatment plant and alum coagulation. High-molecular-weight humic substances were preferentially removed. This preference resulted in substantial decreases in color intensity and trihalomethane formation potential per unit mass of humic substances. The characteristics and removal behavior by alum coagulation of commercial humic acid were significantly different from those of the source-water humic substances.

REMOVAL OF HUMIC SUBSTANCES has long been of concern in water treatment because of their abundance in natural waters and their potential adverse effects on public health and esthetics. They impart color to water and are able to form complexes with other inorganic and organic species (*1–4*) that frequently prevent the removal of these pollutants in treatment processes. More importantly, humic substances are major precursors in the formation

[1]Current address: HazWaste Industries, Inc., 2264 Northwest Parkway, Suite F, Marietta, GA 30067
[2]Current address: School of Geophysical Sciences, Georgia Institute of Technology, Atlanta, GA 30332

0065–2393/89/0219–0473$07.25/0

of trihalomethanes (THM) upon chlorination (*5, 6*). The ubiquity of THMs in drinking water (*7*), together with their potential carcinogenic activity (*8*), prompted the U.S. Environmental Protection Agency (EPA) to promulgate the maximum contaminant level of 100 μg/L of total THMs in water systems serving more than 10,000 persons (*9*).

Most traditional water-treatment plants have been designed and operated to maximize removal of turbidity and pathogens through the use of coagulation–flocculation–sedimentation, filtration, and chlorination processes. Now that humic substances can be removed in water treatment, a considerable amount of effort has been focused on the optimization of existing treatment processes, especially coagulation, for effective removal of humic substances (*10–17*). However, there are still questions relating to the performance of these processes in removing humic substances because the nature of humic substances is not fully understood. Because the nature of humic substances can be an important factor influencing the performance of the water-treatment processes and the formation of THMs upon chlorination, this study was initiated to investigate the specific characteristics of humic substances in natural water sources and their removal behavior in water treatment.

Humic substances in natural waters are unresolvably complex mixtures of organic matter; their physical and chemical properties are difficult to characterize. They are generally described as yellow, acidic, chemically complex, and polyelectrolytelike materials that range in molecular weight from a few hundred to several thousands (*18*). Recent studies (*19–21*) have indicated that humic substances in natural waters from different sources are relatively similar to each other in nature, but are significantly different from the commercial model humic substances commonly used in laboratory studies.

Most information on the removal of humic substances in water treatment and their impact on water quality has been obtained from studies with model humic substances (*11, 13, 14*). A clear distinction should be made between data acquired from model humic substances and those from natural water sources.

In order to address this concern, the characteristics of humic substances and their removal behavior in water treatment were investigated at a full-scale operating plant and in alum coagulation with conventional laboratory jar tests. The Chattahoochee Water Treatment Plant (CWTP) of Atlanta, GA, was selected for this study.

CWTP is a conventional plant that uses the Chattahoochee River as its source of raw water. The treatment sequences are coagulation–flocculation–sedimentation, filtration, and chlorination. Humic substances isolated from the CWTP source water and the model Aldrich humic acid were used in alum coagulation. The specific objectives of this study were to characterize the humic substances isolated from the source water and from the treated effluents of each subsequent treatment process of the CWTP; to evaluate the removal behavior of humic substances by alum coagulation with conventional jar-test procedures; and to compare the nature of both types of humic substances and their removal behavior by alum coagulation.

Experimental Materials and Methods

CWTP and Water Sampling. The CWTP is a conventional plant that employs coagulation–flocculation–sedimentation, filtration, and chlorination (Figure 1). Alum is used as a coagulant in coagulation–flocculation after pH adjustment with lime. Prechlorination is practiced in both plant-intake and rapid-mix units. The supernate from the sedimentation basins is filtered through a dual-media filter (anthracite and sand); this process is followed by postchlorination. The sludge produced is chemically conditioned and dewatered with a filter press. The resulting sludge cakes are disposed of in a sanitary landfill.

The CWTP, designed to treat 230 $\times$ 10^6 L/day, was treating an average of 150 $\times$ 10^6 L/day. It uses the Chattahoochee River as a water source; general water characteristics are shown in Table I. The average color and turbidity values were 28 Pt–Co units and 27 nephelometric turbidity units (NTU), respectively. These values represent relatively low color and high turbidity, compared to many surface waters (22).

The water-sampling points, selected for the isolation of humic substances and investigation of their characteristics and removal behavior in the CWTP, included the source water (SW), effluent from coagulation–flocculation–sedimentation (ACFS), filtration effluent (AF), and the clear well (CW). The water volumes and sampling dates taken for the isolation of humic substances from each sampling point are shown in Table II, along with the amount of humic substances isolated. Because a large volume of water from each sampling point was required to isolate sufficient quantities of humic substances for subsequent laboratory experiments, water samples were taken during several periods of time rather than at intervals chosen according to the hydraulic retention time of each unit process. Approximately 200–700 L of water samples per day were taken from a sampling point.

The source-water quality and the performance of the CWTP were considered to influence the removal behavior of humic substances in the treatment processes. Therefore, various water-quality parameters were measured at each sampling point according to the hydraulic retention time of each unit process during the water-sampling period, as shown in Table II.

Figure 2 illustrates the results of the water-quality parameters for the four sampling points at the CWTP. Instantaneous THMs were produced after prechlorination and increased to 29.1 μg/L in CW. Trihalomethane formation potential (THMFP) is a measure of the maximum amount of THMs that can be formed by the reaction of free residual chlorine with humic substances. A comparison of SW THMFP (228.5 μg/L) with CW THMFP (95.2 μg/L) shows a 69% reduction of THM precursors. The nonvolatile TOC (NVTOC) was reduced by 50% (i.e., from 4.3 to 2.1 mg/L), whereas UV absorbance at 254 nm and color were removed by 71 and 100%, respectively, through the treatment processes.

All water-quality parameters for humic substances decrease significantly during the treatment processes (Figure 2), especially after alum coagulation–flocculation–sedimentation. NVTOC decreased to a lesser extent than THMFP and humic substances (as measured by color). This preferential removal of THMFP and humic substances corroborates the results of Babcock and Singer (23), who observed

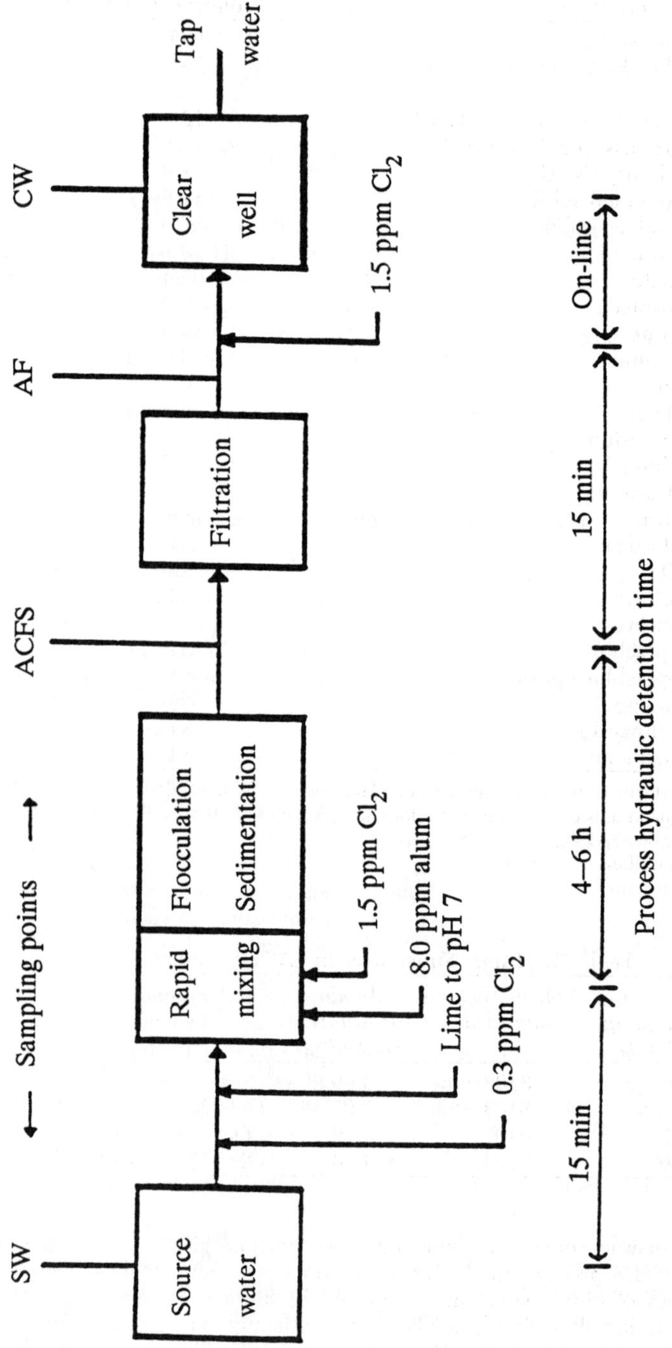

Figure 1. Flow diagram and operational conditions of the Chattahoochee Water Treatment Plant, along with the four sampling points for the isolation of humic substances. Chemical doses and process detention time are approximate annual averages for a flow of approximately 150 × 10^6 L/day.

Table I. General Characteristics of CWTP Source Water

Parameters	*Concentrations*[a]
Dissolved oxygen	8.6
Color (Pt–Co unit)	28.0
pH	6.8
Alkalinity (mg/L as $CaCO_3$)	12.0
Hardness (mg/L as $CaCO_3$)	10.8
Turbidity (NTU)	27.0
Suspended solids	17.6
Dissolved solids	37.5
Fluoride	0.1
Sulfate	5.4
Phosphate	0.2
Nitrate	0.04
Aluminum	0.6
Iron	0.4
Calcium	2.5
Magnesium	0.9
Sodium	2.2
Potassium	1.4
Aldrin	NF[b]
Chlordane	NF
DDT	NF
Dieldrin	NF
Endrin	NF
Heptachlor	NF
Heptachlor epoxide	NF
Lindane	NF
Methoxychlor	NF
Toxaphene	NF

NOTE: Data were obtained from plant records. Values presented are averages of two determinations sampled on October 5 and November 5, 1984.
[a]In milligrams per liter.
[b]Not found.

Table II. Humic Substances in CWTP Water

Water Sampling Points	*Vol. of Water Sampled (L)*	*Humic Substances Isolated (g)*	*Sampling Dates (1984)*
SW	3102	3.99	Nov. 4–Nov. 15
ACFS	3143	2.92	Oct. 25–Nov. 3
AF	6467	4.79	Oct. 1–Oct. 14
CW	7164	4.94	Oct. 15–Oct. 24

that alum coagulation selectively removed those portions of organic matter most responsible for THM production. These observations allowed us to investigate more closely the types of humic substances removed by alum coagulation. The humic substances remaining after alum coagulation are ultimately chlorinated, and chlorination results in the formation of THMs.

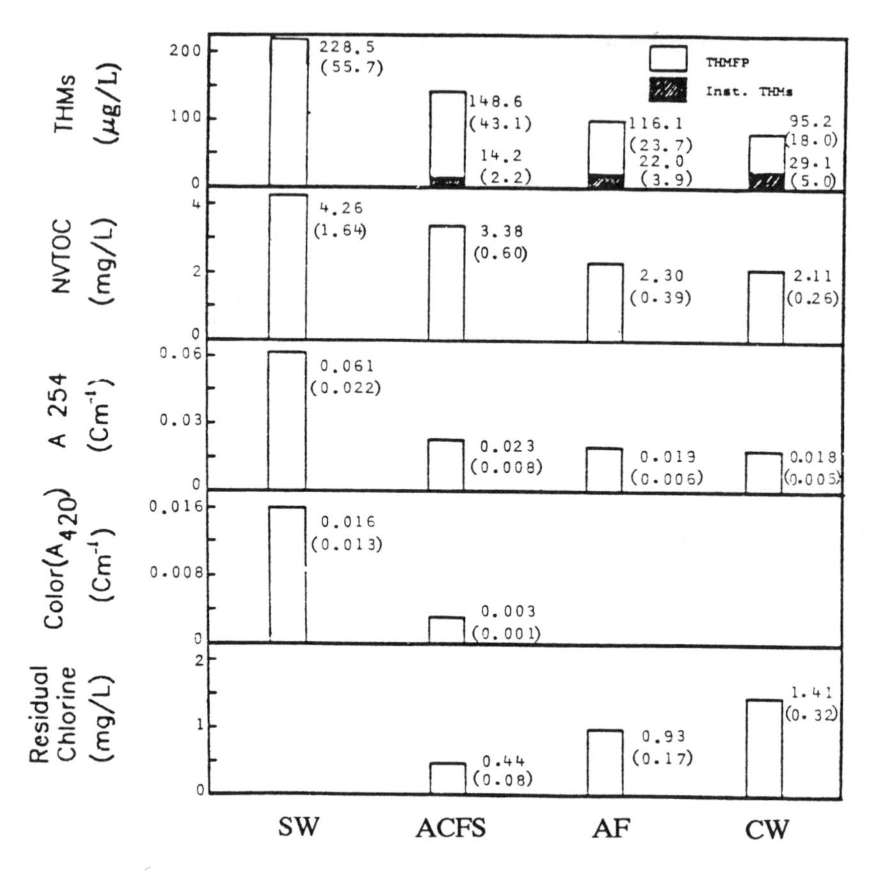

Figure 2. Monitoring the water-quality parameters of the CWTP during the water-sampling period. Values presented are daily averages and those in parentheses are standard deviations.

Isolation of Humic Substances. The humic substances in water samples from the four CWTP sampling points were isolated as a hydrophobic acid fraction of dissolved organic matter by methyl methacrylate resin (XAD–8) adsorption procedures (*24*) from the concentrate of reverse osmosis (RO). The overall scheme for the isolation of humic substances from water samples is depicted in Figure 3. The organic matter in water samples was concentrated at the plant by using the membrane process facility shown in Figure 4, which was equipped with a spiral-wound composite membrane module for seawater desalination (FT–30, FilmTec Corp., Minneapolis, MN).

Water samples were continuously pumped to the sample reservoir at the same flow rate as the RO permeate, and the result was a concentration of organic matter in the sample reservoir. The operating pressure and feed flow rate of the RO module were maintained at 4140–4830 kPa and 910 L/h, respectively. Sodium azide (0.5 mg/L) was added to the water samples to prevent algal and microbial growth, and a stoichiometrically determined amount of sodium sulfite was added to prevent possible oxidation of organic matter by residual chlorine. The initial sample, even-

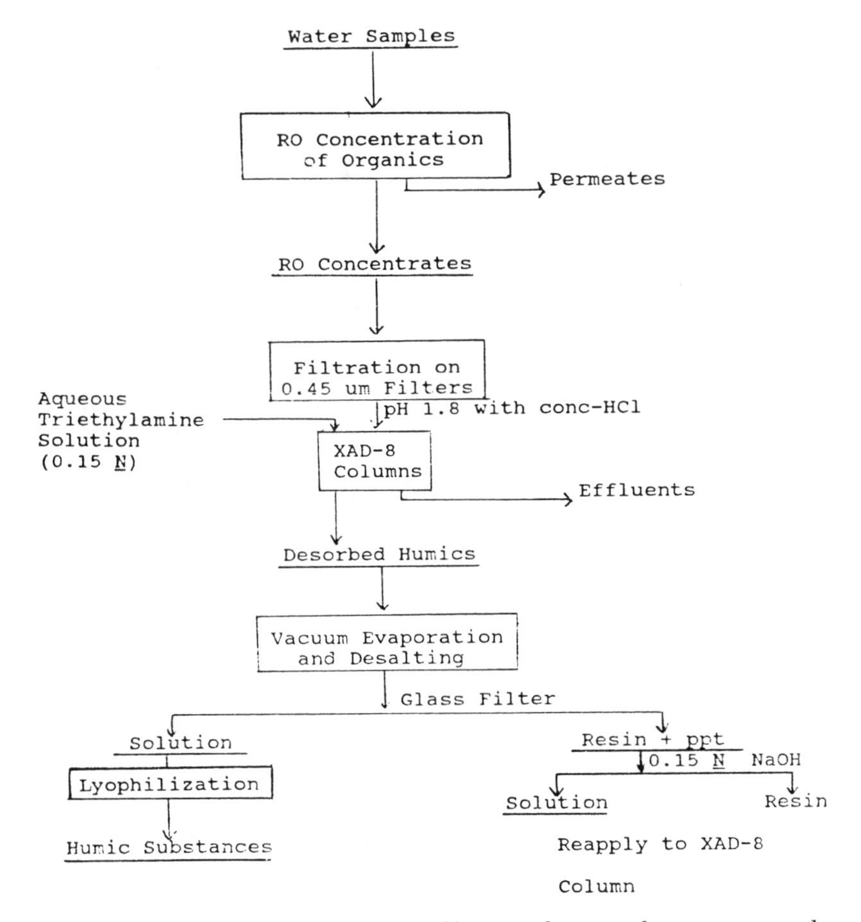

Figure 3. Overall scheme for isolation of humic substances from water samples.

tually concentrated from approximately 3000–7000 L to 20–40 L, was transferred to an 80-L Pyrex glass bottle. The final RO concentrates were filtered through 0.45-μm membrane filters (Gelman Sciences, Ann Arbor, MI), acidified to pH 1.8 with HCl, and passed over XAD–8 resin columns to adsorb the humic substances.

The filtered and acidified yellow-to-brown RO concentrates were pumped to two XAD–8 resin columns connected in series at 3 bed volumes per hour. Humic-substance adsorption was monitored by measuring the effluent absorbance at 420 nm. The humic substances adsorbed on XAD–8 resins were subsequently eluted with a 0.15 M aqueous triethylamine solution until no color was apparent in the eluates.

Three different aqueous base solutions have been commonly used to desorb humic substances from XAD–8 resins. Mantoura and Riley (*32*) used 2 M ammonium hydroxide solution, Perdue (*33*) used 0.15 M triethylammonium hydroxide solution, and Malcolm and Thurman (*24*) used 0.1 M sodium hydroxide solution. All three base solutions are effective in desorbing humic substances from XAD–8 resins. However, triethylammonium hydroxide solution has advantages over the other two so-

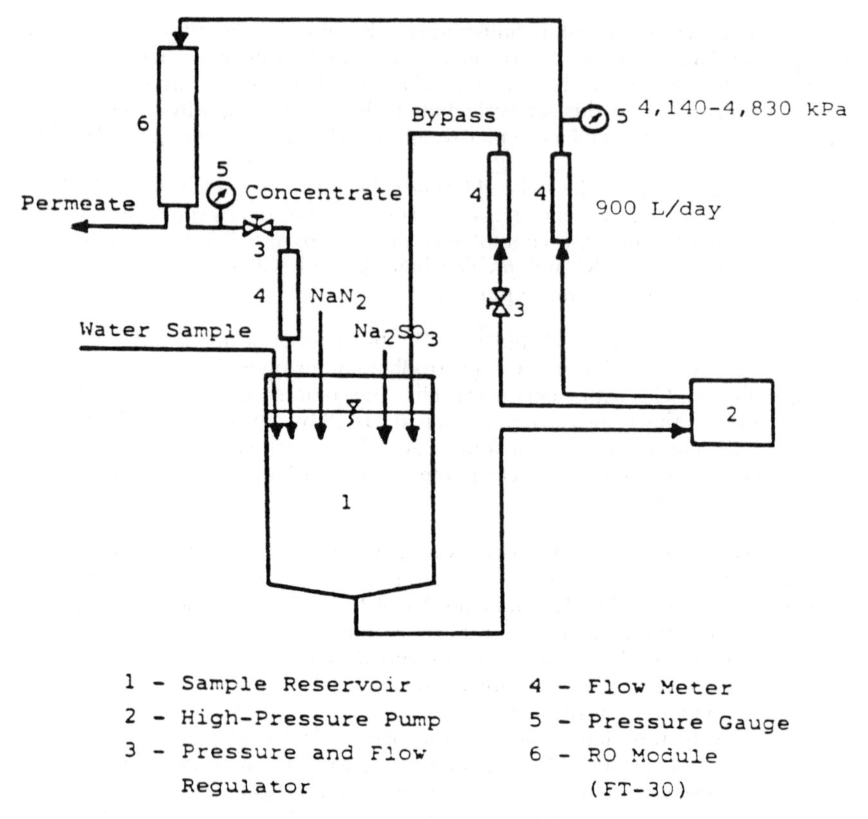

Figure 4. Schematic of reverse osmosis system.

lutions with respect to minimal possible reactions that may alter the chemical structures of humic substances and the ease of removal of excess base.

The excess triethylamine in the eluates was removed by slight concentration in a rotary vacuum evaporator. The resulting triethylamine salts of humic substances were desalted with 20–40-mesh cation-exchange resins (H^+-form AG 50W × 8, Bio-Rad Labs, Richmond, CA) by a batch operation. The desalted humic substances were filtered through glass filters to remove the resins and lyophilized to yield brown fluffy powders, which were readily soluble in water.

This isolation procedure provided an 81.5% recovery of humic substances after RO concentration of organic matter when tested with the humic substances isolated in this study. Hereafter the humic substances isolated from SW, ACFS, AF, and CW are referred as SW–HS, ACFS–HS, AF–HS, and CW–HS, respectively.

A commercially available humic acid (sodium salt) was supplied by Aldrich Chemical Co. (Milwaukee, WI). One gram of sodium humate was purified by dissolving it in 2 L of distilled water and filtering it through 0.45-μm membrane filters. The sodium humate solution was desalted with cation-exchange resins. The resulting solution was lyophilized to yield a dark brown fluffy powder. The final yield was 0.627 g of humic acid (i.e., a 62.7% recovery). Hereafter the purified Aldrich humic acid is referred as AL–HA.

Characterization of Humic Substances. Because humic substances are a heterogeneous mixture of many compounds with generally similar chemical properties, they need to be characterized on the basis of several chemical and physical properties. The characterization in this study included elemental composition, color, acidic functional groups, apparent-molecular-weight (AMW) distribution, and THMFP.

ELEMENTAL ANALYSIS. The isolated humic substances were analyzed for C, H, and N by an automatic analyzer. Halogen as chlorine was analyzed by combustion followed by subsequent titration. Ash content was analyzed by crucible combustion at 850 °C. Acetanilide (C, H, N) and *p*-chlorobenzoic acid (Cl) were used as standard compounds for quality assurance purposes.

SPECTROSCOPIC ANALYSIS. Visible and UV absorbances of humic substances were measured in aqueous solution on a spectrophotometer (model 26, Beckman Instruments, Fullerton, CA) with 1-cm quartz cells. The color intensity of humic substances was expressed as the absorbance at 420 nm at pH 10 by adjusting with 0.1 M NaOH (A_{420}), and UV absorbance was measured at 254 nm without pH adjustment (A_{254}). IR spectra were recorded on a spectrophotometer (AccuLab 6, Beckman Instruments) by using KBr pellets.

ACIDIC FUNCTIONAL GROUPS ANALYSIS. Carboxylic groups of humic substances were determined by direct titration (*25*). Accurately weighed samples of approximately 10 mg were dissolved in 100 mL of aqueous 0.1 M NaCl. The solution was titrated with 0.1 N NaOH by using a digital microburet (Gilmont). A reagent blank was titrated parallel to the sample. Net titration curves were derived by subtracting reagent blank titration data from the sample titration. The carboxylic groups were calculated from the net titration curves at pH 7.0.

Total acidity was determined by the barium hydroxide method (*18*). Thirty milliliters of 0.5 M barium hydroxide was reacted with 10–20 mg of humic substances for 24 h at room temperature. The solution was filtered through a UM05 membrane ultrafilter. The filtrates were then titrated with 0.1 N HCl. All operations were performed under nitrogen atmosphere and a blank was determined simultaneously. The total acidity was calculated as follows:

$$\text{total acidity (meq/g)} = \frac{[\text{titer}_{\text{blank}} - \text{titer}_{\text{sample}}] \times N_{\text{acid}} \times 1000}{\text{milligrams of sample}}$$

where titer is volume (mL) of 0.1 N HCl consumed to titrate the filtrate to pH 8.4 and N_{acid} is normality of the standardized HCl solution used for titration. The weak acidic groups were calculated as the difference between total acidity and carboxylic groups.

AMW DISTRIBUTION. Both gel permeation chromatography (GPC) and membrane ultrafiltration (UF) were used to determine the AMW distribution of humic substances. GPC provided an elution chromatogram that enabled comparison of the AMW distribution of humic substances; UF gave discrete AMW fractions based on the molecular weight cutoff values of specific membranes.

The AMW distribution by GPC was obtained by using a gel (Sephadex G-25, Pharmacia Fine Chemicals, Piscataway, NJ) with a MW exclusion limit of 500–5000 daltons by dextrans. A 1.25-cm i.d. glass column filled with preswollen gel to a bed height of 36 cm was used. A 0.5-mL sample (2.5 mg of humic substances) was placed on the column and eluted with a 0.01 M NaCl solution at a flow rate of 0.5 mL/min. The effluent from the column was connected to the UV detector to monitor absorbance at 254 nm. The column was calibrated with phenol (MW 94), polypropylene (MW 1000), polystyrene (MW 3600), and dextran blue (MW 2,000,000).

Fractionation of humic substances was achieved by using two UF membranes (XM50 and UM10, Diaflo Membranes, Amicon, Lexington, MA). The XM50 and UM10 membranes have reported MW cutoff values of 50,000 and 10,000, respectively. The retention of any solute by a specified membrane can be expressed by the local rejection coefficient (*R*).

$$R = \frac{\ln (C_f/C_o)}{\ln (V_o/V_f)}$$

where V_o and V_f are the volumes of original solution and retentate, and C_o and C_f are the solute concentrations in original solution and retentate, respectively. For solute entirely excluded by the membrane, *R* is equal to 1; for freely permeable solute, the value is 0.

Approximately 400 mg of humic substances was dissolved in 200 mL of distilled water and fractionated by XM50 and UM10 membranes in series. The solutions were concentrated by fourfold (v/v) with XM50 membrane under nitrogen in an Amicon model TCF10 cell at a pressure of 240 kPa. Repetitive dilution and concentration of the retentates were performed for maximal separation until *R* became 90% or more. The humic substances in the resulting retentate of the XM50 membrane are defined as an AMW fraction of >XM50 (MW >50,000). The composite XM50 filtrates were further concentrated with UM10 membrane according to the procedures used for the XM50 membrane. The UM10 retentate of the XM50 filtrates is defined as an AMW fraction of UM10–XM50 (10,000 < MW < 50,000) and the composite UM10 filtrates of the XM50 filtrates is defined as <UM10 (MW <10,000).

THMFP. THMFP is a measure of the maximum amount of THM that can be formed by the reaction of free residual chlorine with the humic substances. Samples of humic substances (5 mg/L, i.e., 2.5 mg C/L) were adjusted to pH 6.8 with a 0.1 M phosphate buffer. Chlorination of samples for the THMFP assays was conducted in 40-mL glass sample vials, which were filled with sample to exclude headspace. The chlorine dose was 42 mg/L as chlorine with NaOCl (molar ratio of Cl_2/C = 2.8). The samples were stored at 20 °C for 5 days, after which they were dechlorinated with 0.1 M sodium sulfite and analyzed for THM. All samples were run in duplicate with two reagent blanks. The standard deviation for a single analysis, including both chlorination procedure and THM determination, was 5.8% at THM concentration levels from 5 to 500 μg/L.

THMs were analyzed by headspace-free, liquid–liquid extraction techniques (*26*) with 1,1,1-trichloroethane as a surrogate and tetrachloroethylene as an internal standard. A gas chromatograph (Hewlett Packard 5830-A, Avondale, PA) with a split–splitless capillary injection system and an electron-capture detector was employed for the quantitative THM analysis. The GC conditions were as follows: column, 15-m × 0.3-mm i.d. glass capillary column coated with SE–54; injection mode, splitless; oven temperature, isothermal at 35 °C.

Coagulation of Humic Substances. The humic substances used in the coagulation experiments were SW–HS, three AMW fractions of SW–HS (obtained by UF), and AL–HA. Turbidity was added to samples by using laboratory-grade kaolin (Fisher Scientific Company, Fair Lawn, NJ), and an aluminum sulfate stock solution of 0.1 M as Al(III) was used as a coagulant.

Coagulation was performed by using a conventional six-place jar-test apparatus (Phipps & Bird, Richmond, VA) with 1-L beakers. Suspensions with humic-substance concentrations of 5 and 10 mg/L (2.5 and 5 mg of C/L, respectively) and kaolin concentration of 50 mg/L were prepared in distilled water containing 5×10^{-3} M $NaHCO_3$. The pH was adjusted by addition of HCl and NaOH so that subsequent addition of the coagulant would result in the desired final pH value.

The coagulant was added during a rapid-mix period of 1 min at 100 rpm. This was followed by a slow-mix period at 25 rpm for 30 min and gravity settling for 30 min. After settling, residual turbidity was measured by an analytical nephelometer (model 2100A, Hach Chemical Corp., Ames, IA). The residual concentrations of humic substances in the settled solutions were determined by UV absorption at 254 nm after filtration with a 0.45-μm membrane filter.

Results and Discussions

Characteristics of Humic Substances. The results of the characterization of the humic substances isolated from the four sampling points of the CWTP and AL–HA are summarized in Table II. The concentration of dissolved organic carbon (DOC) in the CWTP source water was 2.1 mg/L, of which approximately 30% was humic substances. The concentrations of DOC and humic substances were decreased through the CWTP treatment process.

The elemental compositions of the humic substances isolated from the CWTP were similar (i.e., 56.2–56.9% C, 6.0–6.1% H, 0.8–1.0% N, and 36.2–36.7% O). The values were comparable with those of surface-water humic substances (*6*, *20*, *27*), but were different from those of AL–HA and soil humic substances (*28*).

Color and THMFP of humic substances are the important parameters in evaluating water quality. In this study, color equals 50× absorbance at 420 nm at pH 10. The color and THMFP of SW–HS were 0.051 and 67.8 g of THM/mg of humic substances (i.e., 135.6 g of THM/mg of TOC), respectively, and were similar to those of surface-water humic substances (*6*). Ranges or surface-water humic substances are 0.092–0.137 for color (measured at 400 nm at pH 7 and the same unit) and 36.0–77.5 g of THM/mg of humic substances for THMFP. The values also decreased through the treatment processes of the CWTP. However, AL–HA showed approximately 7 times higher in color and 2 times higher in THMFP, as compared with SW–HS.

The existence of acidic functional groups in humic substances is evidenced by IR spectra shown in Figure 5. The carboxylic groups are responsible for the strongest IR adsorption features: an intensive broad-bend centered at 1720 cm^{-1} (C=O stretching) and a weak absorbance at 2500–2700 cm^{-1} (C–O stretching or O–H deformation). A broad and intensive absorption centered at 3400 cm^{-1} indicates weak acidic groups (O–H stretching of H-bonded hydroxyls).

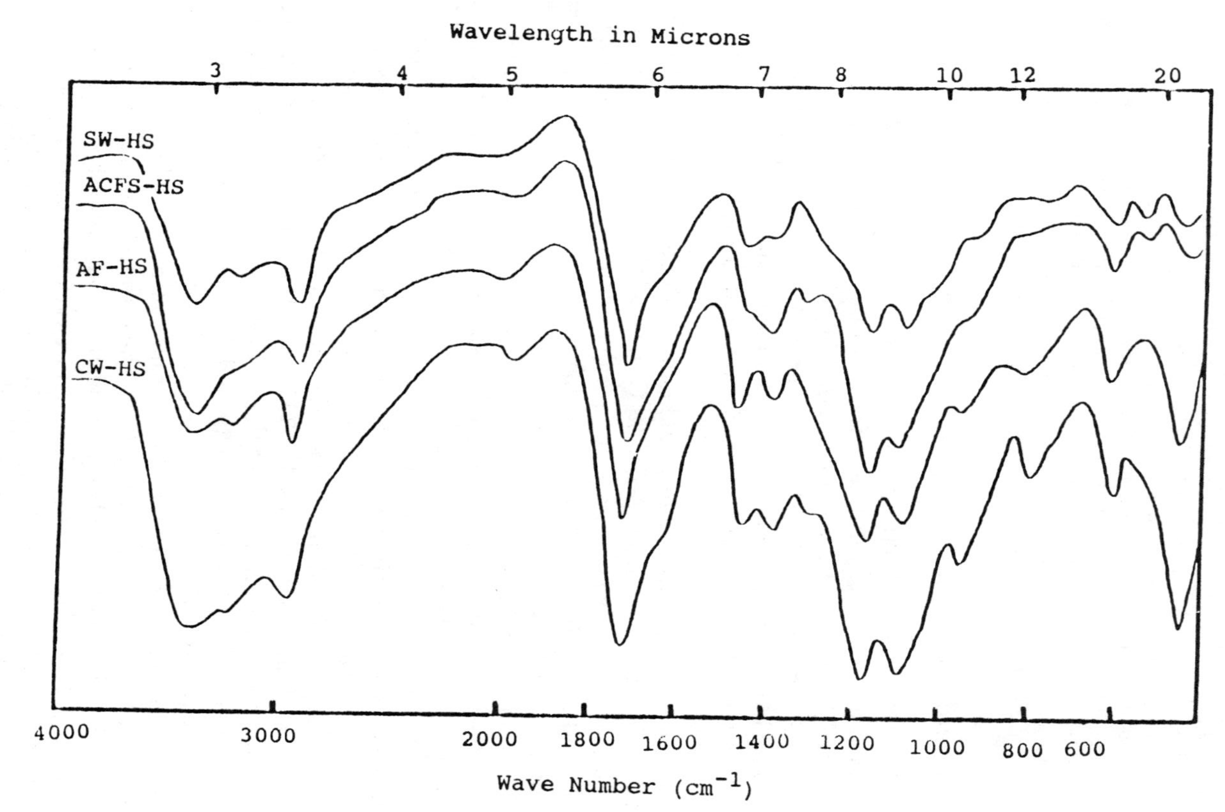

Figure 5. Infrared spectra of humic substances from the four sampling points of the CWTP.

Quantitative estimation of the carboxylic groups was made by the net titration curves shown in Figure 6. From the shape of the net titration curves, it is apparent that the acidic functional groups are titrated beyond pH 10. Although the curves did not have sharp inflection points, carboxylic groups were simply estimated at pH 7 because the curves were highly linear within the pH 7–9 intervals.

SW–HS contained 3.8 and 2.6 meq/g of carboxylic and weak acidic groups (gram basis of humic substances), respectively, which are similar to

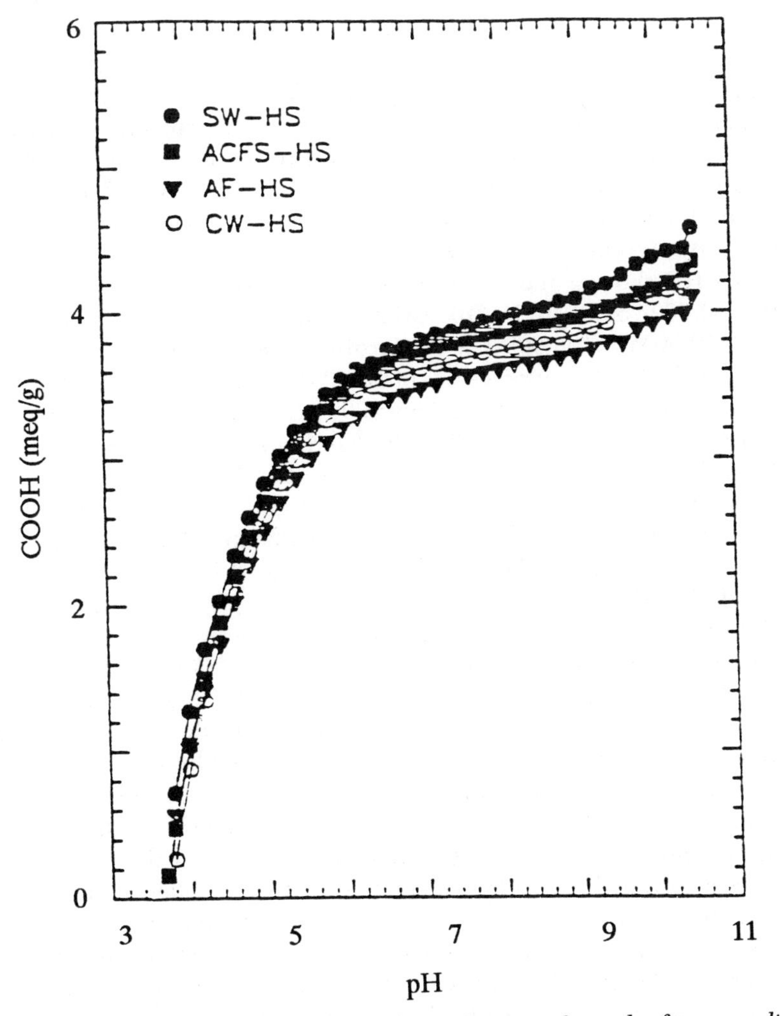

Figure 6. Net titration curves of humic substances from the four sampling points of the CWTP.

those of surface-water humic substances (*20*). There are no significant differences between the humic substances isolated from the four sampling points of the CWTP, as shown in Table III; however, AL–HA contained a lower acidic functional group content than SW–HS. The majority of the acidic functional groups of AL–HA were weak acidic functional groups.

One of the most important characteristics governing the physicochemical properties of humic substances is their molecular weight. GPC and UF were used to estimate the molecular weight of humic substances. Because both techniques separate the humic substances based on molecular size rather than molecular weight, the results are expressed as AMW distribution. The GPC chromatograms of the humic substances from the CWTP are illustrated in Figure 7. The UV absorbance responded proportionally with the molecular weight of humic substances (Table III), and thus the chromatograms represent only a relative distribution as a function of apparent molecular weight. As in most GPC studies of humic substances (*27*, *29*), there are generally two peaks in the chromatograms: MW >5000 (which is totally excluded by the column) and 1000 < MW < 5000 (which was totally included by the column).

SW–HS showed the following AMW distribution: 30.3% in >XM50 (MW >50,000), 51.2% in UM10–XM50 (10,000 < MW < 50,000) and 18.5% in <UM10 (MW <10,000). The AMW distribution of humic substances was affected by the treatment processes of the CWTP. The high-molecular-weight fractions were removed preferentially in the treatment processes. The AMW distribution of AL–HA is significantly different from that of SW–HS. Most of AL–HA were found in >XM50 fraction; this distribution indicates that AL–HA are composed of higher-molecular-weight compounds than SW–HS.

The UF was reasonably successful in fractionating humic substances on the basis of their molecular weights, as evidenced by the GPC chromatograms of SW–HS and their three AMW fractions (Figure 8). However, the MW values assigned for each fraction of SW–HS by the UF method are higher than those assigned by the GPC method. The >XM50 (MW >50,000), UM10–XM50 (10,000 < MW < 50,000), and <UM10 (MW <10,000) fractions indicated by the UF method corresponded to the MW >5000, 2000 < MW < 5000, and 1000 < MW < 5000 fractions by the GPC method, respectively. Similar observations were reported by Thurman et al. (*30*), who concluded that the MW values determined by the UF method for humic substances are 2 to 10 times higher than those shown by the GPC method.

Because the three AMW fractions of SW–HS obtained by UF were used in the jar-test experiments, they were also characterized, and the results are shown in Table III. Trends linking higher molecular weight with higher color and THMFP and lower acidic functional groups were observed. No intuitively obvious reason exists for these correlations, and more studies are needed to explain these phenomena.

Table III. Characteristics and Removal Behavior of Humic Substances

Humic Substances	*DOC*[a] *(mg C/L)*		*Elemental Composition (%)*[b]					*Color*[c] *50 · L/mg · cm*	*THMFP*[d] *(μg THM)/(mg HS)*	*Acidic Functional Groups (meq/g)*		*AMW Distribution by UF (%)*[e]		
			C	*H*	*N*	*O*	*Ash*			*COOH*	*Phenolic OH*	*>XM50*	*UM10–XM50*	*<UM10*
SW–HS	2.1	(0.7)	56.4	6.0	1.0	36.6	1.9	0.051	67.8	3.8	2.6	30.3	51.4	18.3
>XM50								0.117	79.9	3.3	2.4			
UM10–XM50								0.029	72.9	3.7	2.5			
<UM10								0.008	58.2	4.0	2.7			
ACFS–HS	1.9	(0.5)	56.2	6.1	1.0	36.7	3.7	0.029	57.0	3.7	1.7	26.0	43.8	30.2
AF–HS	1.4	(0.4)	56.4	6.1	0.8	36.7	5.1	0.029	51.4	3.4	1.3	22.8	38.5	38.7
CW–HS	1.5	(0.4)	56.9	6.1	0.8	36.2	3.1	0.015	50.2	3.7	1.6	14.4	41.9	47.0
AL–HA			51.5	4.8	0.8	43.0	9.4	0.345	133.4	3.3	2.5	94.0	6.0	–

[a] DOC of water samples for the isolation of humic substances. Numbers in parentheses indicate the portions of DOC contributed by humic substances.
[b] Ash-free basis.
[c] Absorption at 420 nm at pH 10.
[d] Chlorination conditions for THMFP assays are $Cl_2/C = 2.8$, pH 7.0, 20 °C, and 5 days reaction time.
[e] >XM50 is the retentate of XM50 membrane (MW cutoff 50,000). UM10–XM50 is the retentate of UM10 membrane (MW cutoff 10,000) from the permeate of XM50. <UM10 is the permeate of UM10 from the permeate of XM50.

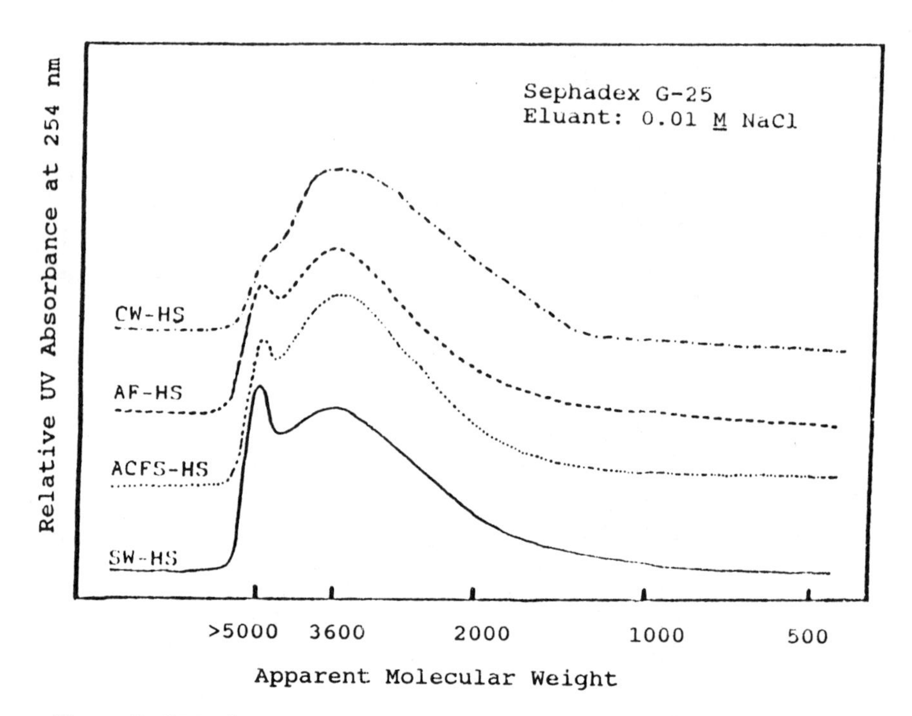

Figure 7. GPC chromatograms of humic substances from the four sampling points of the CWTP.

Removal Behavior of Humic Substances in the CWTP. Figure 9 illustrates the removal behavior of humic substances in the CWTP. The treatment processes decreased the concentration of DOC from 2.1 to 1.5 mg of C/L and the concentration of humic substances from 0.9 to 0.5 mg/L. These concentration changes indicate that the plant is able to remove some DOC, including humic substances. Overall, 44% of humic substances were removed (i.e., from 0.9 to 0.5 mg of C/L) through the treatment processes of the CWTP. Specifically, 74% of the >XM50 fraction (i.e., from 30.3% of 0.9 mg of C/L to 14.1% of 0.5 mg of C/L) and 55% of the UM10–XM50 fraction (i.e., from 51.2% of 0.9 mg of C/L to 41.9% of 0.5 mg of C/L) were removed, but the <UM10 fraction was increased by 40% (i.e., from 18.5% of 0.9 mg of C/L to 47.0% of 0.5 mg of C/L).

The selective removal of high-molecular-weight fractions of humic substances is reflected in a substantial decrease in color intensity and THMFP per unit mass of humic substances. Color intensity was decreased by 71% and THMFP was decreased by 26% through the treatment processes. The concentration of acidic functional groups of humic substances was decreased from 6.4 to 4.7 meq/g during coagulation–flocculation–sedimentation and filtration, but increased to 5.2 meq/g by chlorination. The slight increase

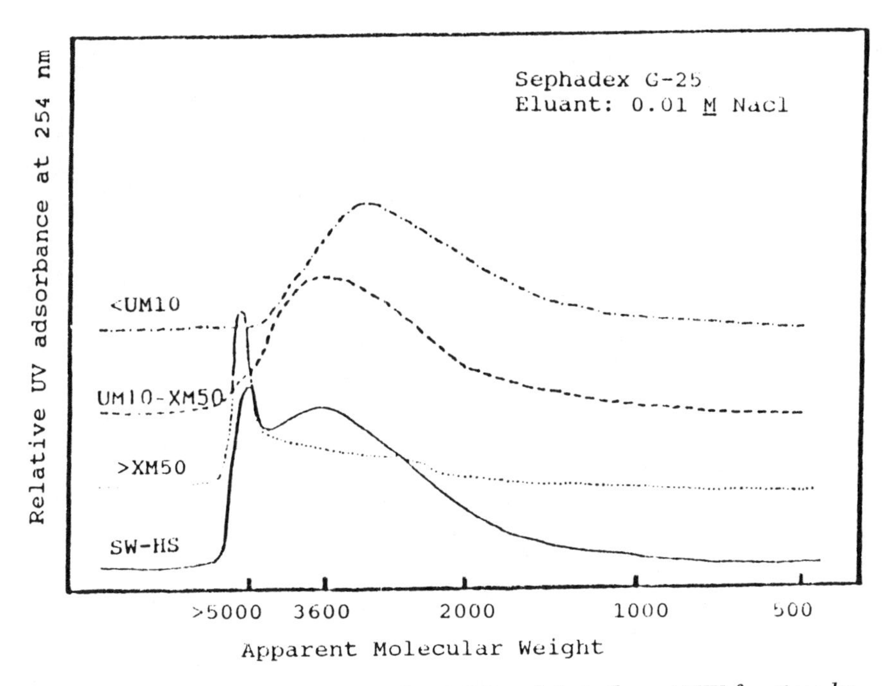

Figure 8. GPC chromatograms of SW–HS and their three AMW fractions by UF.

in the acidic functional group content may have resulted from oxidation reactions of chlorine with humic substances.

The data clearly demonstrated not only that humic substances are removed, but also that their characteristics are affected by CWTP treatment processes. The preferential removal of the high-molecular-weight fraction of humic substances through the CWTP treatment processes resulted in substantial decreases in color intensity and THMFP per unit mass of humic substances. This removal behavior is attributed to the coagulation–flocculation–sedimentation, filtration, and prechlorination practiced at plant-intake and rapid-mix units. However, alum coagulation is considered to be the major process responsible for this removal because alum coagulation of SW–HS provided similar removal behavior of humic substances observed in the CWTP.

This finding has a special significance in water-treatment practice. Because the humic substances remaining after coagulation are chlorinated, the performance of alum coagulation in removing humic substances can be an important variable in the reduction of THMs in finished drinking water. Therefore, proper operation of the water-treatment plant for maximal removal of high-molecular-weight fractions of humic substances by alum co-

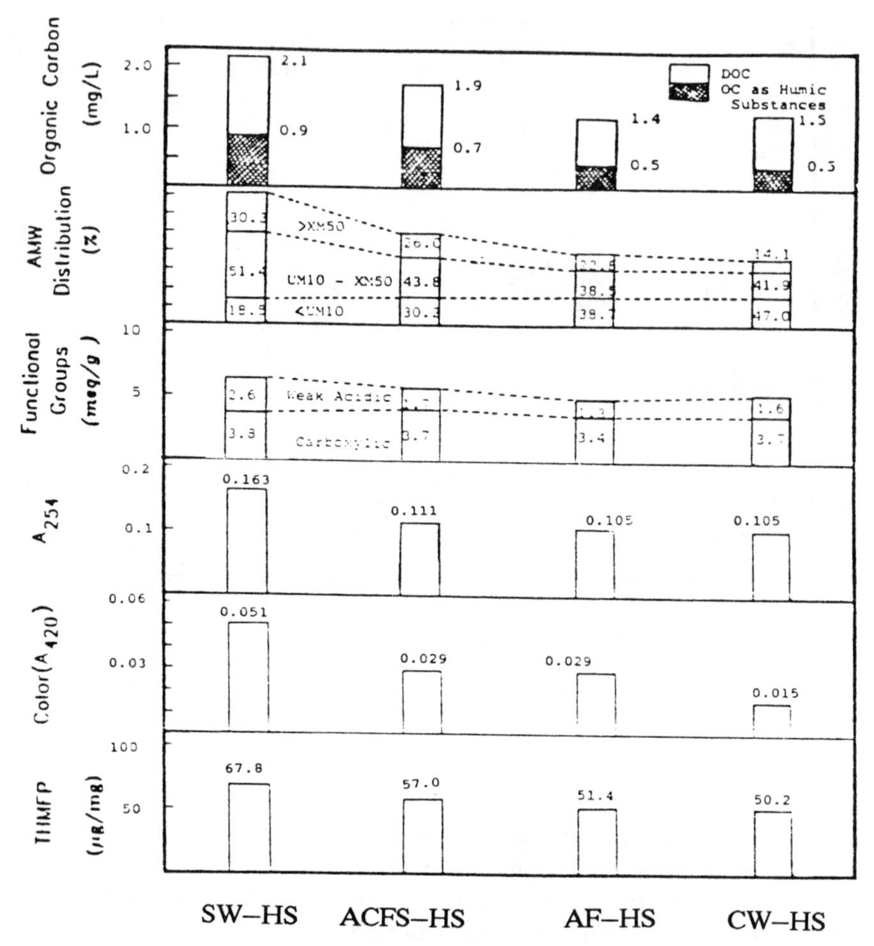

Figure 9. Summary of the removal behavior of humic substances in the CWTP water-treatment processes.

agulation can significantly reduce the color and THMs in finished drinking water.

Alum Coagulation of Humic Substances. The majority of the alum coagulation experiments were run with increasing aluminum dosages at the various initial pH values to define the removal areas of humic substances. pH values measured after the slow-mix period were presented in the removal areas of humic substances. However, the alum coagulation at initial pH 6.5 was routinely used to investigate the removal behavior of humic substances and the effects of various chemical parameters. Both increasing aluminum dose and loss of carbon dioxide during flocculation and settling periods caused the pH to decrease up to 0.2 unit at the highest aluminum dose (0.2 meq/L) at initial pH 6.5. No efforts were made to keep the pH constant during the experiments.

REMOVAL CHARACTERISTICS. A typical set of results obtained from alum coagulation of humic substances is shown in Figure 10. SW–HS were removed gradually, with increasing aluminum dose approaching a maximum of approximately 40% removal at high aluminum dosages. The removal behavior of AL–HA was quite different from that of SW–HS. At a 0.06 meq/L aluminum dose the removal of AL–HA is trivial, but removal is essentially complete at higher dosages (>0.1 meq/L).

This differential removal behavior between SW–HS and AL–HA may be caused by their different characteristics, especially AMW distribution. Most of AL–HA were found in the >XM50 fraction, although only 30.3% of SW–HS were found in that fraction. To test this hypothesis, the three AMW fractions of SW–HS were coagulated; the results are shown in Figure 11. Although the >XM50 fraction of SW–HS was not removed to the same degree as AL–HA, it was removed by 62%. In contrast, UM10–XM50 and <UM10 fractions were removed by 42 and 18%, respectively. This preferential removal of high-molecular-weight fractions of humic substances by alum coagulation is consistent with the results observed in the CWTP.

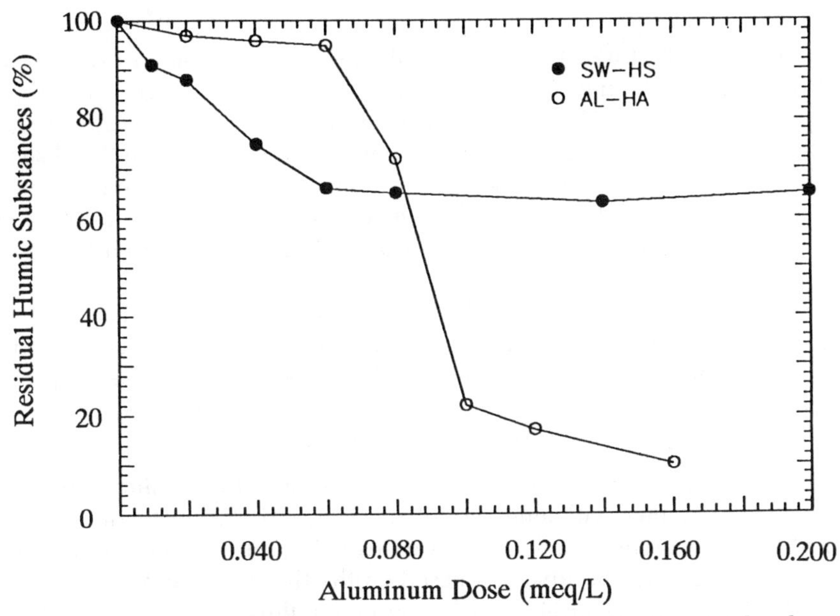

Figure 10. Removal characteristics of SW–HS and AL–HA (5 mg/L) by alum coagulation at initial pH 6.5.

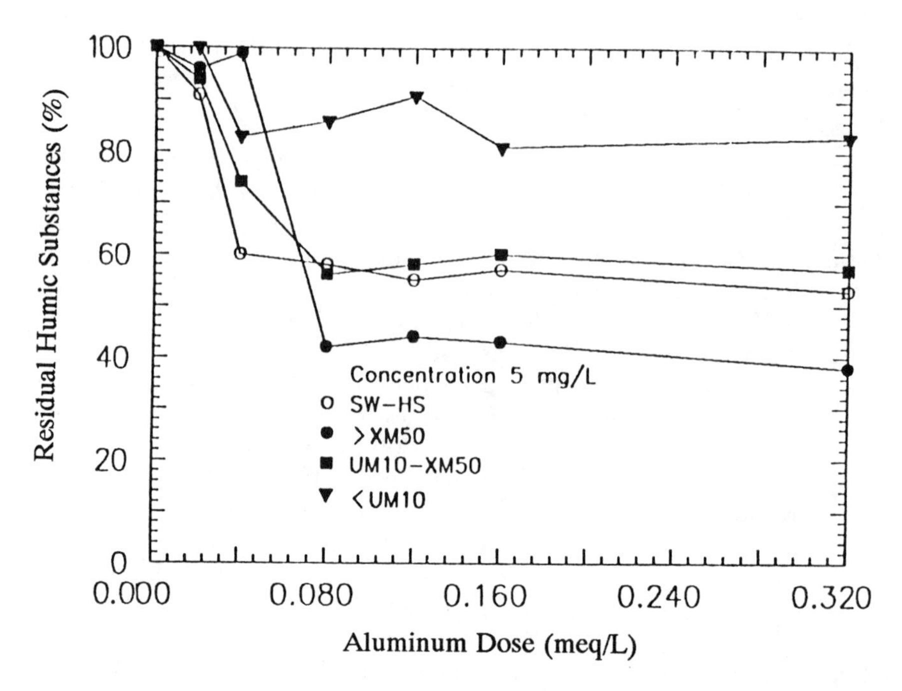

Figure 11. Removal characteristics of SW–HS and their AMW fractions by alum coagulation at initial pH 6.5.

EFFECTS OF TURBIDITY. The results of alum coagulation experiments on solutions containing 5 mg/L of SW–HS or AL–HA together with 50 mg/L of kaolin (35 NTU) are shown in Figure 12. The removal of humic substances in the presence of kaolin resembled their removal in the absence of kaolin, except for a slight increase in aluminum dose requirement for an equivalent removal of humic substances. In contrast, the presence of humic substances considerably increased the aluminum dose required for kaolin removal.

Although the behavior of kaolin is somewhat different from that of natural turbidity, the results are of practical significance. Because alum coagulation has long been operated to maximize turbidity removal, a proper improvement of the process may increase the removal of humic substances, thus reducing the color and THM in finished drinking water.

EFFECTS OF pH AND STOICHIOMETRY. Because the maximum removal of SW–HS obtained with alum coagulation was 50%, the 30% removal areas of SW–HS by alum coagulation at various pH values are shown in Figure 13 in an Al(III) stability diagram based on the thermodynamic equilibria of Al(III) hydrolysis products (*14*). The optimum aluminum dose and pH for the removal of 5 mg/L SW–HS were approximately 0.02 meq/L (2 mg/L as alum) at pH 6 and those of 10 mg/L were 0.04 meq/L (4 mg/L as alum)

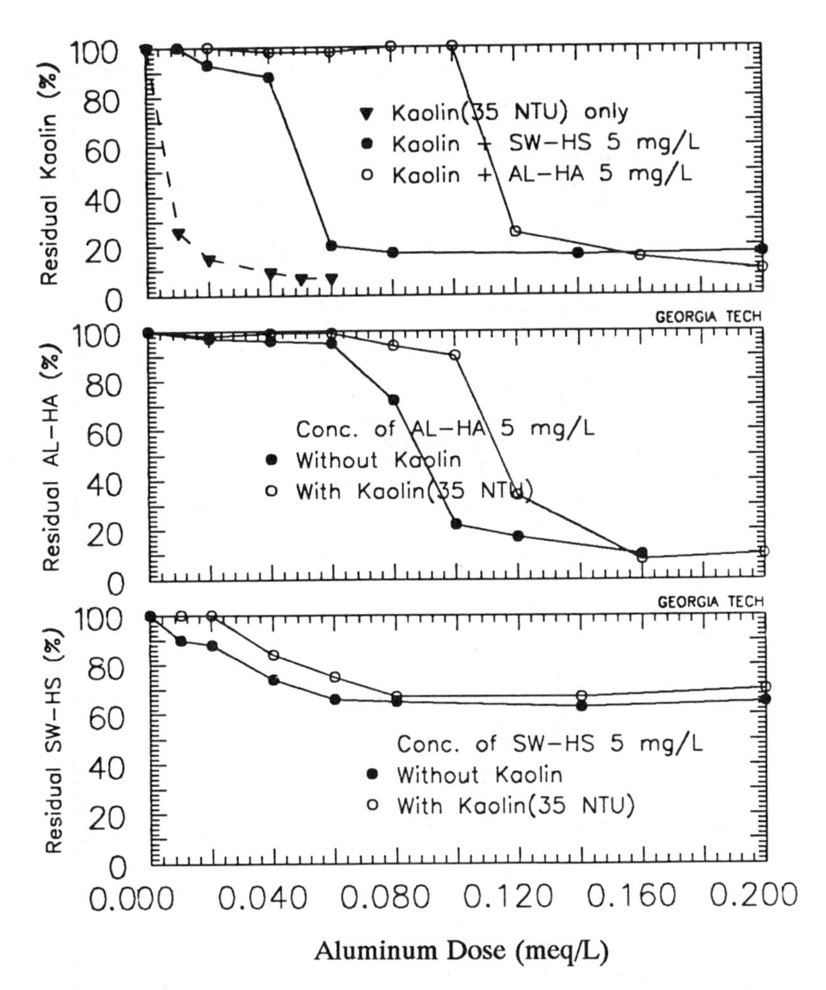

Figure 12. Effect of kaolin on the removal of humic substances by alum coagulation at initial pH 6.5.

at pH 5.5. In general, the 30% removal areas follow a trend toward higher aluminum dosages and lower pH as the concentration of SW–HS increases.

A stoichiometric relationship was observed between the concentration of SW–HS and the aluminum dose required to initiate the removal of SW–HS. This stoichiometry is consistent with the following observations. SW–HS appear to form a complex with aluminum so that $Al(OH)_3(s)$ cannot form until the aluminum dose exceeds the available complexation sites of SW–HS. SW–HS contain 3.8 meq/g of caboxylic groups and 2.6 meq/g of weak acidic groups. If only the carboxylic groups are ionized at pH 6.5 and if an initial concentration of SW–HS of 5 mg/L is used, at least a 0.02-meq/L dosage of aluminum will be required to initiate the formation

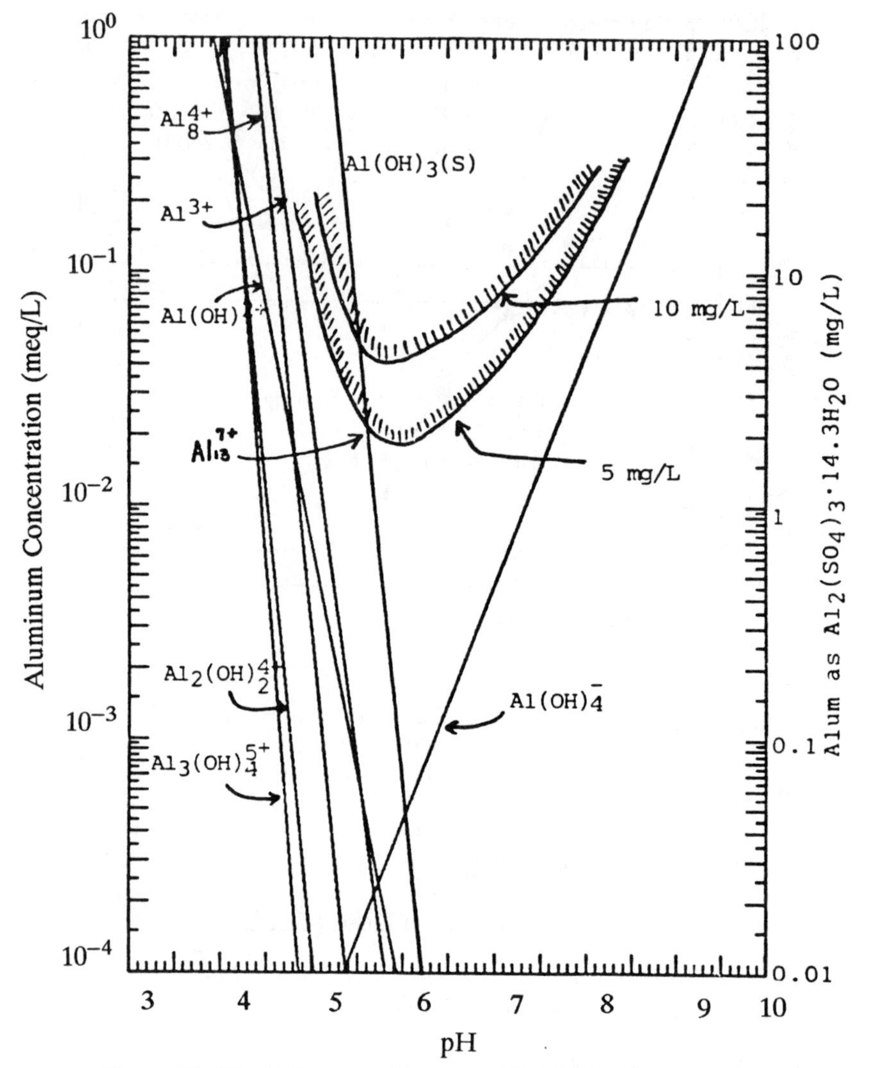

Figure 13. The 30% removal areas of SW–HS by alum coagulation.

of $Al(OH)_3(s)$. This dosage corresponds precisely to the aluminum dosage required for the removal of SW–HS shown in Figure 13.

The 50% removal areas of AL–HA by alum coagulation are shown in Figure 14, which substantiates the differences between the removal behavior of AL–HA and that of SW–HS. Two distinct 50% removal areas were observed for 5 mg/L of AL–HA; the two removal areas were connected to

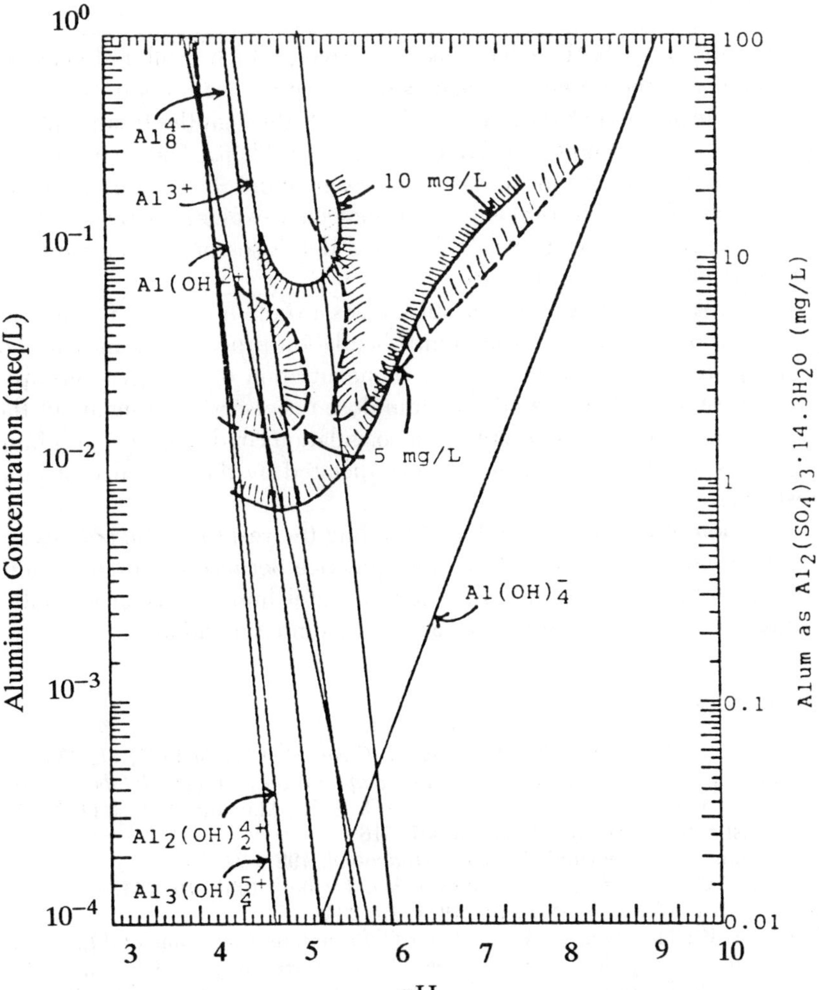

Figure 14. The 50% removal areas of AL–HA by alum coagulation.

form a single band with an initial AL–HA concentration of 10 mg/L. These removal areas of AL–HA correspond with the regions of removal observed in the literature (*14, 31*).

Considerable insight into the removal of humic substances has been gained with studies of AL–HA. However, caution must be exercised in extending the results to water-treatment practice because the characteristics and removal behavior of AL–HA are significantly different from those of SW–HS.

Conclusions

Humic substances in the Chattahoochee River, which exhibited characteristics similar to those found in many surface waters, were not only removed but also transformed physically and chemically through the treatment processes of the Chattahoochee Water Treatment Plant. The major process responsible for this removal was alum coagulation. The alum coagulation process was effective in removing high-molecular-weight fractions of humic substances, and the result was substantial decreases in the color intensity and THMFP of humic substances.

The characteristics of humic substances are the important factors influencing the performance of the alum coagulation process in removing both turbidity and humic substances. The concentration of carboxylic groups of humic substances determined the alum dose necessary to bring about their removal in the process. The optimum coagulation conditions for both humic substances and turbidity removal were dictated by the presence of humic substances.

Caution must be exercised in extending the results of the studies with Aldrich humic acid to water-treatment practice because commercial humic substances differ significantly from natural-water humic substances, both in characteristics and in removal behavior in alum coagulation.

References

1. Reuter, J. H.; Perdue, E. M. *Geochim. Cosmochim. Acta* **1977,** *41,* 325.
2. Buffle, J.; Tessier, A.; Haerdi, W. In *Complexation of Trace Metals in Natural Waters*; Kramer, C. J. M.; Duinker, J. C., Eds.; Martinus Nijhoff/Dr W. Junk Publishers: The Hague, 1984; pp 301–316.
3. Gjessing, E. T.; Berlind, L. *Arch. Hydrobiol.* **1981,** *92,* 24.
4. Carter, C. W.; Suffet, I. H. *Environ. Sci. Technol.* **1982,** *16,* 735.
5. Rook, J. J. *Water Treat. Exam.* **1974,** *23,* 234.
6. Oliver, R.; Thurman, E. M. In *Water Chlorination: Environmental Impact and Health Effects*; Jolley, R. L.; Brungs, W. A.; Cortruvo, J. A.; Cumming, R. B.; Mattice, J. S.; Jacobs, V. A., Eds.; Ann Arbor Science: Ann Arbor, 1983; Vol. 4, p 231.
7. Symons, J. M.; Bellar, T. A.; Carswell, J. K.; Demarco, J.; Kropp, K. L.; Robeck, G. G.; Seeger, D. R.; Slocum, C. J.; Smith, B. L.; Stevens, A. A. *J. Am. Water Works Assoc.* **1975,** *67,* 634.
8. *Carcinogenesis Bioassay of Chloroform*; National Cancer Institute: Washington, DC, 1976.
9. *Fed. Regist.* National Interim Primary Drinking Water Regulations; Control of Trihalomethanes in Drinking Water; Final Rule, Vol. 44, 1979; No. 231.
10. Hall, E. S.; Packham, R. F. *J. Am. Water Works Assoc.* **1965,** *57,* 1149.
11. Narkis, N.; Rebhun, M. *J. Am. Water Works Assoc.* **1975,** *67,* 101.
12. Kavanaugh, M. C. *J. Am. Water Works Assoc.* **1978,** *70,* 613.
13. Edzwald, J. K. *AIChE Symp. Ser.* **1979,** *75,* 54.
14. Amirtharajah, A.; Mills, K. *J. Am. Water Works Assoc.* **1982,** *74,* 210.

15. Dempsey, B. A.; Ganho, R. M.; O'Melia, C. R. *J. Am. Water Works Assoc.* **1984,** *76,* 141.
16. Collins, M. R.; Amy, G. L.; King, P. H. *J. Environ. Eng. (N.Y.)* **1985,** *111,* 850.
17. Sinsabaugh, R. L.; Hoehn, R. C.; Knocke, W. R.; Linkins, A. E. *J. Environ. Eng. (N.Y.)* **1986,** *112,* 139.
18. Schnitzer, M.; Kahn, S. U. *Humic Substances in Environment*; Marcel Dekker: New York, 1972.
19. Chian, E. S. K.; Giabbai, M. F.; Kim, J. S.; Reuter, J. H.; Kopfler, F. C. In *Organic Pollutants in Water*; Suffet, I. H.; Malaiyandi, M., Eds.; Advances in Chemistry Series 214; American Chemical Society: Washington, DC, 1987; p 181.
20. Malcolm, R. L. In *Humic Substances in Soil, Sediment, and Water*; Aiken, G. R.; McKnight, D. M.; Wershaw. R. L.; MacCarthy, P., Eds.; Wiley: New York, 1985; p 181.
21. Malcolm, R. L.; McCarthy, P. *Environ. Sci. Technol.* **1986,** *20,* 904.
22. Edzwald, Z. K. In *Control of Organic Substances in Water and Wastewater*; Berger, B. B., Ed.; Office of Research and Development, U.S. Environmental Protection Agency: Washington, DC, 1983; p 26.
23. Babcock, D. B.; Singer, P. C. *J. Am. Water Works Assoc.* **1979,** *71,* 149.
24. Thurman, E. M.; Malcolm, R. L. *Environ. Sci. Technol.* **1981,** *15,* 463.
25. Perdue, E. M. In *Humic Substances in Soil, Sediment, and Water*; Aiken, G. R.; McKnight, D. M.; Wershaw. R. L.; MacCarthy, P., Eds.; Wiley: New York, 1985; p 493.
26. Richard, J. J.; Junk, G. A. *J. Am. Water Works Assoc.* **1977,** *69,* 60.
27. Plechanov, N.; Josefsson, B.; Dyrssen, D.; Lundquist, K. In *Aquatic and Terrestrial Humic Materials*; Christman, R. F.; Gjessing, E. T., Eds.; Ann Arbor Science: Ann Arbor, 1983; p 387.
28. Schnitzer, M. In *Soil Organic Matter*; Schnitzer, M.; Kahn, S. U., Eds.; Elsevier: New York, 1978; p 1.
29. Gjessing, E.; Lee, G. F. *Environ. Sci. Technol.* **1967,** *1,* 631.
30. Thurman, E. M.; Wershaw, R. L.; Malcolm, R. L.; Pinckney, D. *J. Org. Geochem.* **1982,** *4,* 27.
31. Mangravite, F. J., Jr.; Buzzell, T. D.; Cassell, E. A.; Matijevic, E.; Saxton, G. B. *J. Am. Water Works Assoc.* **1975,** *67,* 88.
32. Mantoura, R. F. C.; Riley, J. P. *Anal. Chim. Acta* **1975,** *76,* 97.
33. Perdue, E. M. In *Chemical Modeling in Aqueous Systems*; Jenne, E. A., Ed.; ACS Symposium Series 93; American Chemical Society: Washington, DC, pp 94–114.

RECEIVED for review March 24, 1988. ACCEPTED for publication September 9, 1988.

SORPTION ONTO ACTIVATED CARBON: INFLUENCES ON WATER TREATMENT

30

Effects of Humic Background on Granular Activated Carbon Treatment Efficiency

Walter J. Weber, Jr., and Edward H. Smith

Environmental and Water Resources Engineering Program, The University of Michigan, Ann Arbor, MI 48109

Uncharacterized background organic matter can impair the effectiveness and complicate the design and operation of adsorption treatment processes directed at the removal of specific target organic compounds from waters and wastewaters. Mathematical models calibrated with system-specific information may facilitate process design and operation by allowing quantification of the effects of such background matter on adsorption efficiency. In this work, a two-resistance homogeneous surface diffusion adsorption model was used to simulate and predict fixed-bed adsorber breakthrough behavior for two specific solutes in background waters from various sources. Independent measurements of requisite model coefficients were made for the two target solutes directly in the presence of the background organic matter, which in turn was treated as an unspecified class of components quantified only in terms of the lumped analytical parameter of total organic carbon. This approach suitably incorporated the effects of the background matter in model forecasts of fixed-bed adsorber performance for the target compounds.

ORGANIC CONTAMINANTS OF CONCERN in the application of adsorption processes for water and waste-treatment practice can be divided into two major classes: (1) potentially hazardous specific compounds, generally of anthropogenic origin; and (2) relatively uncharacterized background dissolved organic matter (DOM), frequently of humic and fulvic character.

0065–2393/89/0219–0501$09.00/0

Granular activated carbon (GAC) adsorption in fixed-bed reactor (FBR) systems constitutes an established technology for effective removal of even trace quantities of an extensive range of the first category of compounds. However, the presence of uncharacterized background DOM in the solution matrix of many waters and wastewaters complicates the design and operation of adsorption treatment processes directed at removing specific *target* organic compounds. This hindrance stems from complexities introduced as a result of DOM interactions with both the adsorbent and target species present in solution. The results of prior efforts to address this problem have suggested that mathematical models and approaches, if properly formulated and structured, offer potential for quantifying the impact of background DOM on the adsorption of target contaminants for purposes of process design and performance forecasts (*1–6*).

Mathematical models incorporating film and intraparticle mass-transfer resistance terms have enjoyed broad application for describing and predicting the dynamics of FBR adsorbers. One such model, the homogeneous surface diffusion (HSD) version of the Michigan Adsorption Design and Applications Model (MADAM), requires determination of equilibrium and rate parameters that can be estimated or, preferably, measured in bench-scale laboratory experiments that closely simulate particular systems of interest (*7, 8*). Mass-transport parameters have traditionally been evaluated by using data derived from completely mixed batch reactor (CMBR) measurements in combination with dimensionless-group correlation techniques. Although virtually untested for complex mixtures of organic substances, the short-bed adsorber (SBA, defined as a bed of sufficiently short length that immediate contaminant breakthrough occurs) more closely approximates the hydrodynamics of full-scale columns and allows simultaneous determination of both film and intraparticle mass-transport parameters and thus offers potential as an alternative design tool (*9*).

Objectives and Approach

The research described here was designed to (1) examine the effects of different sources and concentrations of DOM background on adsorption capacities and rates for typical target organic compounds, (2) evaluate the capabilities of relatively straightforward models such as the MADAM–HSD for description and prediction of fixed-bed adsorber breakthrough characteristics for these target compounds in the presence of the various DOM backgrounds, and (3) evaluate several alternative approaches for estimating the adsorption-rate parameters required for modeling the behavior of the target compounds in the systems studied.

The systems selected for investigation were chosen to simulate a broad range of conditions typically encountered in field applications, from a relatively simple background composed of a few target species typical of a ground water supply to the more complex matrix of target compounds and humic-type DOM characteristic of leachates emanating from hazardous-waste disposal or spill sites. For each compound modeled, the MADAM–HSD approach required input relative to (1) a film diffusion parameter, k_f, characterizing mass transfer of solute to the exterior carbon particle surface; (2) an intraparticle diffusion coefficient, D_s, quantifying diffusional transport along the interior carbon particle pore surfaces; and (3) appropriate isotherm coefficients to characterize the functional dependence of the solid-phase concentration of adsorbed solute on the solution-phase concentration at equilibrium. In the modeling approach employed, these coefficients were evaluated for the target compounds directly in the presence of complex leachate material, which was considered as uncharacterized but system-specific background organic matter quantified only in terms of total organic carbon (TOC).

The rationale for this approach is that many field-scale applications of adsorption involve waters that are complex in composition. The adsorbent–solute–solution interactions associated with such complex matrices are typically site-specific, and correspondingly site-specific design criteria are needed. Further, the operation of fixed-bed adsorbers is often governed by the breakthrough pattern(s) of one or two specific compounds, either by virtue of these compounds being the only identified hazardous substances in the treatment stream or because they are the first to exceed prescribed limits in the adsorber effluent. Such an approach is easier to implement and affords reduced model input and computational requirements over methodologies that seek to identify and predict the individual adsorptive behaviors of all components present.

Experimental Details

A two-stage experimental program was implemented to obtain equilibrium and mass-transfer coefficients for various matrices containing two different target organic compounds and for background waters having different types and amounts of DOM measured as TOC. In Phase I, CMBR isotherm and rate studies were conducted for one- and two-component solutions of the target compounds. Phase II experiments used information from Phase I to formulate two different column-type adsorber experiments, namely, short-bed adsorber (SBA) measurements as an alternate to CMBR and correlation techniques for kinetic-parameter estimation and model calibration, and deep-bed adsorber (DBA) experiments for parameter–model verification.

Activated Carbon. The adsorbent used in all these experiments was activated carbon (Filtrasorb–400, Calgon Corporation). The general physical properties of F–400 are documented elsewhere (*10*). For isotherm studies, the carbon was used in powdered form to facilitate rapid attainment of equilibrium and limit interferences due to biological activity. Powdered carbon was prepared by crushing random bag samples, sieving, and retaining the 200–325 U.S. standard sieve size fraction. CMBR rate and fixed-bed adsorber experiments were conducted with a 30–40 sieve size fraction of the same carbon. The sieved carbon was washed with deionized–distilled water, oven-dried at 104 °C, and stored in airtight glass containers. Carbon for immediate use was dried to a constant weight and then cooled to ambient temperature in a desiccator.

Solutes. Two organic compounds, trichloroethylene (TCE) and *p*-dichlorobenzene (*p*-DCB), both designated as priority pollutants by the U.S. Environmental Protection Agency, were chosen as target solutes. These compounds were selected because they exhibit different adsorption characteristics, have been identified in contaminated surface and ground waters, are relatively straightforward to analyze, and embody a wide range of properties and behavioral characteristics. The majority of multicomponent adsorption studies to date have involved compounds from similar organic class groupings, whereas the intent here was to examine compounds having different structural and solution characteristics. TCE is a straight-chain, unsaturated aliphatic of relatively high volatility and solubility, compared to the aromatic *p*-DCB.

Background waters included (1) a baseline solution of deionized–distilled water (DDW) containing no background organic material (this baseline solution also served as a make-up or dilution water for all other solutions); (2) a commercial humic acid (HA, Aldrich Chemical Co.); (3) a leachate from a hazardous-waste landfill cell (HWL, Wayne Disposal, Rawsonville, MI); (4) a three-solute mixture of known organic contaminants (TRISOL); and (5) a solution composed of the commercial humic acid in conjunction with the trisolute mixture (TRISOL + HA).

Humic acid stock solutions were prepared by first dissolving an appropriate mass of dried Aldrich humic acid (Lot No. 3061–KE) in DDW at pH 11. Following readjustment of the pH to 7, the solution was filtered through a prewashed glass-fiber filter (Whatman 934 AH) to remove undissolved solids. The TOC of the filtrate was then measured. A typical stock solution had a DOM concentration of 250 mg/L as TOC. Working concentrations of humic acid were obtained by dilution of the stock with DDW to initial background concentrations of 0, 5, 15, and 25 mg/L as TOC.

Raw leachate was prefiltered through a glass-fiber filter prior to use to remove suspended particles that might adsorb pollutants in competition with activated carbon. The leachate was characterized as a high-strength waste (TOC = 10,000 mg/L; total hardness = 2100 mg/L as $CaCO_3$), and its color and adsorptive characteristics suggested the presence of significant amounts of humic material. Appropriate amounts of HWL were diluted with buffered DDW to achieve background DOM concentrations of 0, 16, 60, and 200 mg/L as TOC. (Note: HWL background with an initial concentration of *XX* mg/L as TOC is expressed as HWL(*XX*), with a similar notation for HA; for example, HWL(60), HA(15).

The three solutes constituting the TRISOL mixture (lindane, tetrachloroethylene, and carbon tetrachloride) were chosen with selection criteria similar to those for the two target compounds. The concentrations of these solutes used in all studies were approximately 1000, 575, and 525 μg/L, respectively. This amount corresponds to an equimolar amount of each compound of about 3.4 μmol/L, and a total TOC of 0.37 mg/L. Identical concentrations of the TRISOL components were used in the TRISOL + HA(15) background.

Working solutions consisted of background water spiked with TCE or *p*-DCB, prepared as methanol-based stock solutions, to the desired concentration. A slight variation of this procedure employed for application of TCE to fixed-bed adsorbers is noted in the section on column methods. Experiments were conducted at room temperature (22 ± 2 °C), and all solutions were buffered at pH 6.5 ± 0.2 with 10^{-4} M phosphate.

Mass concentration determinations for TCE and *p*-DCB were by gas chromatography, with a liquid–liquid extraction procedure for sample preparation and an external standard calibration procedure. Background organic analysis was performed by direct TOC measurement or by UV spectroscopy correlated with TOC measurements.

Data Collection and Analysis

Isotherm Studies. Equilibrium data were collected by using the completely mixed batch reactor (CMBR) bottle-point technique. Varying amounts of powdered carbon were carefully measured and added to a series of 0.16-L glass vials, followed by contact with solutions containing the adsorbates. After a 5-day contact period, a sample was taken from each reactor, filtered through a prewashed glass-fiber filter to separate the carbon, and extracted or diluted according to the appropriate analytical technique. Upon measurement of the equilibrium solute concentration, C_e, the corresponding equilibrium solid-phase concentration, q_e, was calculated from a mass balance. Control measures employed in capacity experiments included elimination of headspace in reactors to limit volatilization losses and evaluation and accomodation of losses encountered in the filtration step.

For single-solute systems of *p*-DCB and TCE, the Freundlich isotherm model was found to adequately describe the equilibrium liquid–solid-phase relationship over the concentration ranges of interest. The Freundlich equation is a semiempirical, nonlinear expression of the form

$$q_e = K_F C_e^n \tag{1}$$

Freundlich model coefficients, K_F and n, were determined with a nonlinear geometric mean functional regression algorithm that recognizes errors encountered in measurement and calculation of both liquid- and solid-phase equilibrium concentrations (*11, 12*).

Multicomponent adsorption equilibria were described by the ideal adsorbed solution theory (IAST), by using an empirical modification similar to that employed by others (*13, 14*) to provide a more precise fit of the data. The generalized equations to be solved for the modified IAST are

$$\pi_i = \pi \tag{2}$$

$$\sum_{i=1}^{N} z_i = 1 \tag{3}$$

$$\frac{1}{q_T} = \sum_{i=1}^{N} \frac{z_i}{q_i^*} \tag{4}$$

$$q_i = z_i q_T = \frac{C_{0,i} - C_i}{M/V} \tag{5}$$

$$q_i^* = f(C_i^*) \tag{6}$$

$$C_i = P_i z_i C_i^* \tag{7}$$

where P_i is incorporated as an empirical coefficient to account for competitive interactions between the target solutes in a specific system. When equation 1 is used to characterize $f(C_i^*)$ and to evaluate the spreading pressure, π_i, equations 2 through 7 reduce to the expression

$$C_{0,i} - q_i\left(\frac{M}{V}\right) = -\frac{P_i q_i}{\sum_{j=1}^{N} q_j}\left(\frac{\sum_{j=1}^{N} q_j \eta_j}{\eta_i K_{F,i}}\right)^{\eta_i} \tag{8}$$

where η is the inverse of the Freundlich exponent, n. The competition coefficient values, P_i, are searched from bisolute equilibrium data by using a minimization parameter. Equation 8 can be solved numerically by implementing a Newton–Raphson algorithm. A detailed development of the solution is given elsewhere (*12*).

Kinetic Studies. Two techniques for mass-transport parameter estimation were compared. The more traditional approach of the two involved CMBR rate experiments in conjunction with a batch-reactor MADAM algorithm to obtain D_s from time–concentration data. The 2.3-L reactor was sealed and headspace-free to prevent volatilization losses. Initial target compound and background leachate concentrations approximated the influent concentrations used in subsequent column experiments. Film diffusion coefficients for corresponding FBR model predictions were estimated from literature correlations.

The second approach used in this work involved calibrating an FBR version of MADAM with SBA experimental data to simultaneously determine k_f and D_s. In the particular experimental design employed, the SBA was placed in series with a second adsorber. The combined depths of the two columns were sufficient to fully contain the adsorption wave front and, for the purposes of these experiments, compose a DBA. The merit of this experimental design is that identical influent flow and concentration are ensured for complementary short and long column runs for purposes of model calibration and verification, respectively. The columns were constructed of borosilicate glass with an i.d. of 1.3 cm. For 30–40-mesh carbon, this size gives a column-to-particle-diameter ratio of 25 and essentially eliminates hydrodynamically related wall effects (*15*). Columns were packed with successive layers of glass microbeads (of the same size as the carbon), GAC, and more beads to establish a consistent flow pattern in and out of the bed. Stainless steel screens were placed at the entrance and exit of the columns to retain the media. The beds were carefully packed in distilled water to eliminate air, and then rinsed for several minutes at a higher-than-design flow rate to wash out remaining fines.

Columns were operated in upflow mode. Bulk influent solutions were prepared in and delivered from a 45-L glass container by a variable-speed peristaltic pump. To avoid volatilization, TCE was pumped to the system from a concentrated stock solution prepared in a well-mixed, headspace-free cylinder. A stirred 1.2-L glass chamber with baffles was inserted upstream of the influent to the SBA to provide adequate mixing of incoming volatile compound stock with the bulk solution. An air trap was employed between the bulk feed pump and mixing chamber to prevent air from passing to the carbon beds. All tubing and valves were made of glass to minimize adsorption of organic substances onto reactor surfaces. The system was fitted with sampling ports before and after column 1 and after column 2 for influent and SBA and DBA effluent sampling, respectively. Samples were collected at discrete time intervals and extracted immediately for analysis.

Results and Discussion

Equilibrium Studies. A tabulation of Freundlich isotherm coefficients for single-solute solutions of *p*-DCB and TCE in the presence of various background waters and DOM concentrations is given in Table I.

Table I. Freundlich Isotherm Coefficients for Target Compounds in Single-Solute Systems

Background Water	K_F[a]	*95% Confidence Interval* (K_F)	n[a]	*95% Confidence Interval* (n)	r[b]
With *p*-DCB as solute					
DDW	31.55	28.91–34.43	0.330	0.316–0.344	0.996
HA(5)	28.84	26.35–31.91	0.338	0.326–0.350	0.997
HA(15)	24.19	22.28–26.28	0.356	0.343–0.369	0.999
HA(25)	23.19	20.66–26.04	0.338	0.319–0.357	0.996
HWL(16)	27.07	23.00–31.86	0.323	0.296–0.349	0.995
HWL(60)	14.43	11.62–17.92	0.374	0.333–0.415	0.991
HWL(200)	8.15	6.45–10.33	0.368	0.333–0.403	0.995
TRISOL(0.37)[c]	25.25	20.56–31.02	0.337	0.304–0.370	0.993
TRISOL + HA(15)	18.97	17.17–20.96	0.358	0.342–0.374	0.999
With TCE as solute					
DDW	1.61	1.51–1.72	0.515	0.499–0.531	0.998
HA(5)	1.47	1.36–1.58	0.493	0.474–0.513	0.998
HA(15)	1.17	1.09–1.25	0.499	0.482–0.516	0.998
HA(25)	1.00	0.91–1.09	0.489	0.469–0.508	0.999
HWL(16)	1.37	1.27–1.49	0.435	0.418–0.452	0.999
HWL(60)	0.79	0.73–0.86	0.398	0.378–0.417	0.998
HWL(200)	0.43	0.38–0.49	0.391	0.368–0.414	0.998
TRISOL(0.37)[c]	1.04	0.92–1.19	0.508	0.484–0.531	0.998
TRISOL + HA(15)	0.92	0.83–1.02	0.482	0.463–0.500	0.999

[a]Based on C_e in micrograms per liter and q_e in micrograms per milligram.
[b]Correlation coefficient.
[c]Refers to milligrams per liter as TOC.

Experimental data and Freundlich model calibrations for representative cases of p-DCB and TCE are presented in log–log format in Figures 1 through 4. Initial concentrations employed for the isotherms illustrated were 6000–12,000 μg/L for p-DCB and 1000–2000 μg/L for TCE. The results indicate that the capacity of F–400 for p-DCB is much greater than for TCE

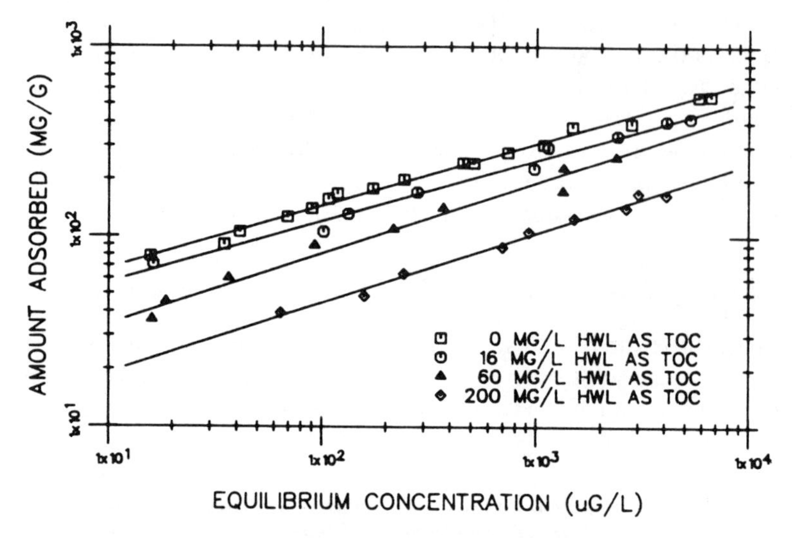

Figure 1. Single-solute isotherms for p-DCB *in the presence of varying concentrations of leachate background.*

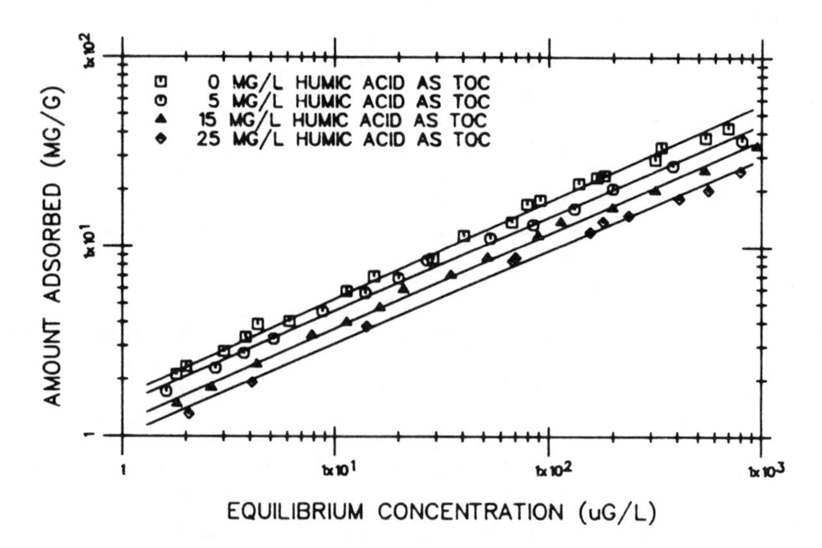

Figure 2. Single-solute isotherms for TCE in the presence of varying concentrations of humic acid background.

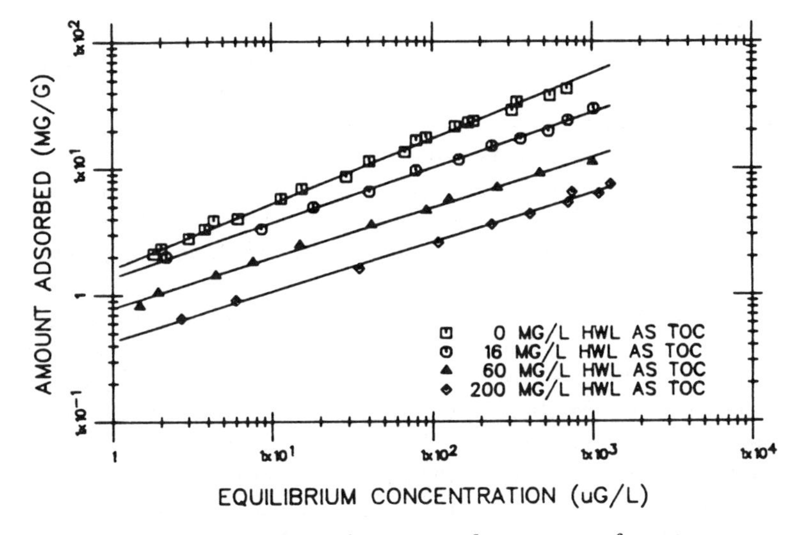

Figure 3. Single-solute isotherms for TCE in the presence of varying concentrations of leachate background.

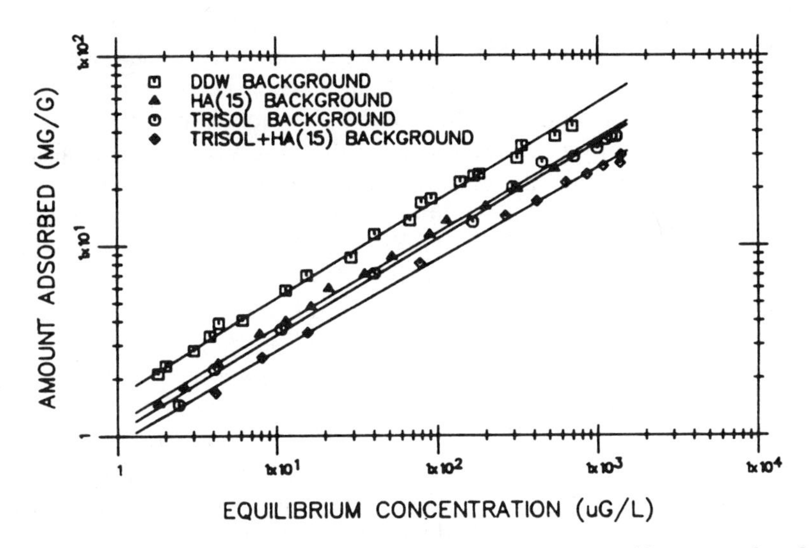

Figure 4. Single-solute isotherms for TCE in the presence of humic acid and TRISOL background.

and that the presence of background DOM reduces the adsorption capacity for both target compounds. Moreover, the decrease in adsorption capacity is proportional to the background DOM concentration expressed in terms of TOC for a given background water. These capacity effects are reflected in the magnitude of the Freundlich capacity factor, K_F.

For background DOM represented by a commercial humic acid, the impact on adsorption of *p*-DCB appears to be less than for TCE. In the case of HA(25), for example, adsorption capacities for *p*-DCB and TCE are reduced by approximately 25 and 40%, respectively. This difference may be due in part, however, to the fact that initial concentration values for *p*-DCB are higher than those for TCE; thus the *p*-DCB can compete more favorably at a given background TOC. The capacities for both target compounds are affected to a similar degree in the leachate background; for example, for HWL(60), adsorption capacities are reduced to 45–50% of the values for the case of no background DOM. The TRISOL background has a similar impact on *p*-DCB adsorption capacity as HA(15). Their combined effect on the impact of the TRISOL + HA(15) background is essentially additive.

In contrast, the combined effect of the TRISOL compounds and HA(15) background on TCE is only slightly greater than their individual impacts based upon K_F values. However, the slope of the TRISOL + HA(15) isotherm is not as steep and, as shown in Figure 4, the displacement is greater at higher liquid-phase concentration values. Overall, the impact of the TRISOL and TRISOL + HA(15) backgrounds was less for *p*-DCB than for TCE. The slope terms, n, for the *p*-DCB isotherms exhibit little variation, a result consistent with other studies conducted for target compounds in background DOM (*1*, *6*). In the case of TCE, however, n values in leachate background are approximately 20% lower than for the case of no background DOM. Because of the semiempirical nature of the Freundlich model, it is difficult to attach absolute physical significance to the observed behavior, although the slope term is roughly an indicator of adsorption energy or intensity (*16*). As illustrated in Figure 3, this condition translates for this system to relatively more favorable adsorption of TCE at the very low end of the concentration range than at higher concentrations when HWL is present.

The precise mechanisms and solid–solution characteristics responsible for reduction by DOM of target-compound adsorption are not readily apparent. Aquatic humic substances have demonstrated a tendency to form complexes with certain organic micropollutants and thus alter such solution properties as solubility and hydrophobicity (*17–19*) and, therefore, adsorption characteristics. Solution preparation procedures for the isotherm experiments allowed only a few minutes between injection of concentrated target solute into background DOM and contact with activated carbon. Thus, if the association of a compound with humic material is rate-limited, the results presented may not reflect the total magnitude of potential complexation effects. Direct competition for adsorption sites between target compounds and the strongly adsorbing components of the background DOM is certainly another potential mechanism for reduction of the apparent adsorption capacity of the target compounds. This competition is especially evident in the case of the TRISOL background, which exhibits an impact as great or greater than other backgrounds having TOC levels 50 times higher. Similarly, although HA(15), HWL(16), and TRISOL + HA(15) backgrounds all have approximately equal TOC content, they exhibit significantly different impacts on the adsorption capacities of the target species.

A related factor in the case of the leachate background is the presence of detectable levels of Ca^{2+} and other inorganic salts in the leachate, substances known to enhance the adsorption of humic substances on activated carbon (*20*). Physical and chemical properties specific to a particular activated carbon can also play determinative roles in the nature of adsorption of various types of DOM. More precise characterization of the background DOM and investigation of isolated solute–DOM–adsorbent interactions and their relative contributions to observed adsorption phenomena are required to identify operative mechanisms.

Bisolute isotherm data and modified IAST model calibrations (solid lines) are presented in Figures 5 through 8. Figure 5 shows the adsorption of *p*-DCB in HA(15) background for variable concentrations of TCE. The effect of the presence of TCE on *p*-DCB adsorption capacity is evident for small concentrations of the competing solute, but increases little for larger $C_{0,\mathrm{TCE}}$. Figure 6 is the matching plot for TCE for varying $C_{0,p\text{-DCB}}$. Data for TCE for two-component isotherms in various background humic acid and leachate concentrations are shown in Figures 7 and 8, respectively.

Comparison of these data and model traces with the dashed lines, which represent single-solute isotherms for TCE in the presence of the denoted backgrounds, illustrates the significant reduction in TCE adsorption capacity effected by the presence of *p*-DCB. The presence of background DOM in solution further suppresses TCE adsorption. The displacement of the bisolute isotherms from their corresponding single-solute curves is greatest in the upper concentration range. In fact, the tail of the isotherm bends down-

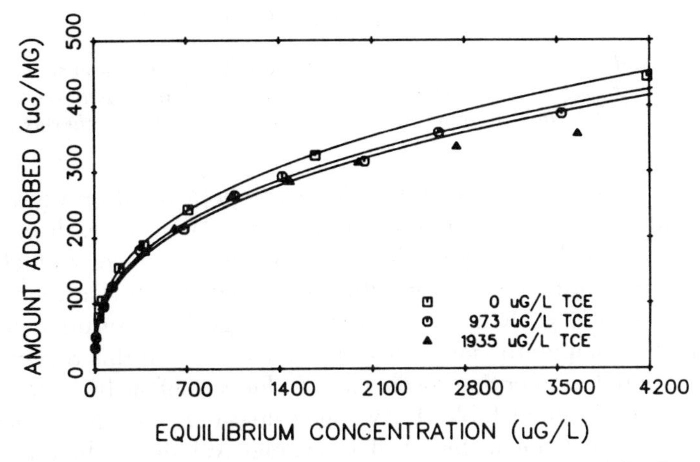

Figure 5. Isotherms for p*-DCB in the presence of TCE in HA(15) background.*

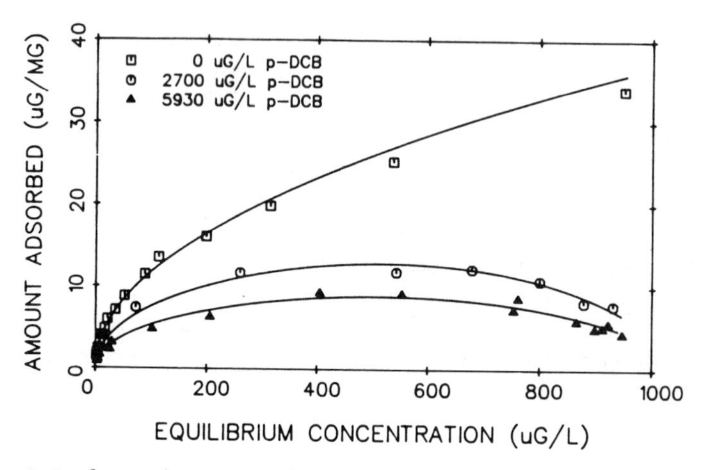

Figure 6. Isotherms for TCE in the presence of p-DCB *in HA(15) background.*

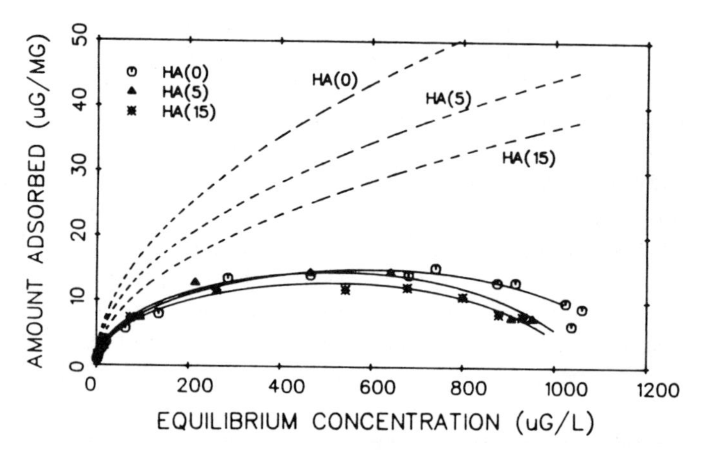

Figure 7. Isotherms for TCE ($C_0 \sim$ *1000 μg/L) in the presence of* p-DCB ($C_0 \sim$ *2700 μg/L) for various concentrations of humic acid background. (Dashed lines are corresponding TCE single-solute isotherms.)*

ward, a common feature of isotherms for competitive adsorptions between compounds which sorb strongly and weakly relative to each other. This region corresponds to very small carbon dosages and involves high surface coverage and more intense competition for available sites. The more strongly adsorbing solute, *p*-DCB, has a greater energy for adsorption than does TCE, and consequently dominates surface coverage in this region.

Computed competition coefficients, P_i, for each of the bisolute isotherms conducted are listed in Table II. Average values of P_i for a given background water have also been included. The average values are based upon the observation that estimated P_i values exhibit no significant variation for vari-

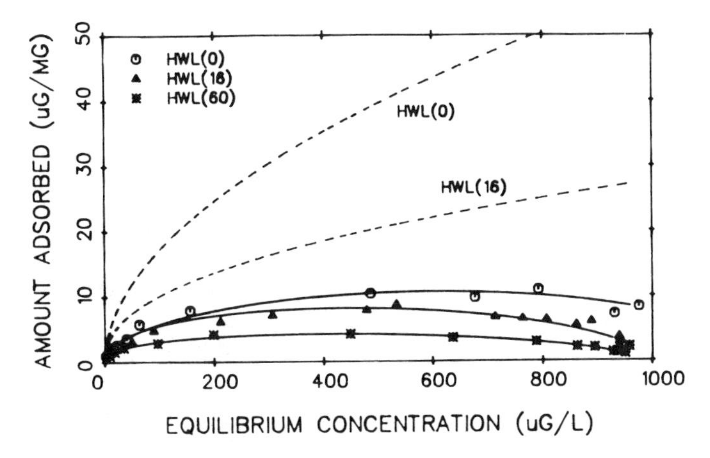

Figure 8. Isotherms for TCE (C_0 ~ *1000 μg/L) in the presence of* p-DCB (C_0 ~ *5000 μg/L) for various concentrations of leachate background. (Dashed lines are corresponding TCE single-solute isotherms.)*

able initial concentration ratios within the concentration ranges examined. Computed competition coefficients varied significantly from 1.0 for both *p*-DCB and TCE for every background type–concentration of DOM studied. The term P_i is essentially a measure of unequal or so-called nonideal competition between adsorbates. As a phenomenological coefficient, it also accounts for any hysteresis effects on adsorption equilibria.

Ideal competition (P_i = 1.0) implies that the solutes adsorb independently in the sense that there are no solute–solute, solute–DOM, or solute–sorbent interactions or energetics operative other than those manifest in their respective single-solute isotherm patterns and the stoichiometry of the bisolute system. Inherent also is the assumption of a homogeneous adsorbent with respect to active sites. This assumption is not met by most activated carbons, the majority of which possess numerous surface functional groups and variable pore-size distributions. These functional groups and pore-size distributions can promote selective adsorptions and steric exclusions of certain solutes (*13*). It is difficult to correlate the nonidealities reflected by the competition coefficient values with specific system parameters to make a priori estimates of adsorption equilibria and thereby circumvent the need for multicomponent isotherm data. The fact that P_{TCE} values are less than 1.0 and $P_{p\text{-DCB}}$ values greater than 1.0 indicates that, for the stated conditions, TCE competes more favorably with *p*-DCB for adsorption on F–400 than predicted by ideal competition.

The modeling approach has several limitations. First, the competition coefficients are valid only over a limited initial concentration range. Moreover, P_i must approach 1.0 as the concentration of one (or both) of the sorbates becomes very small (e.g., the model must predict single-solute

Table II. Competition Factors, P_i, for Bisolute Isotherm Model for Various Background Waters

Experiment Number	*Background Water*	C_0 (μg/L)		P_i		*Average* P_i			
		p-*DCB*	*TCE*	p-*DCB*	*TCE*	p-*DCB*	σ_n[a]	*TCE*	σ_n[a]
BI1	DDW	4127	843	1.23	0.73				
BI2	DDW	2868	1115	1.21	0.71	1.24	0.02	0.73	0.01
BI3	DDW	5857	1130	1.24	0.73				
BI4	DDW	5774	1897	1.27	0.74				
BI5	HA(5)	2609	998	1.19	0.43	1.19		0.43	
BI6	HA(15)	2699	978	1.22	0.39				
BI7	HA(15)	5931	973	1.18	0.38	1.21	0.02	0.37	0.02
BI8	HA(15)	5858	1935	1.24	0.35				
BI9	HWL(16)	4813	969	1.15	0.22	1.15		0.22	
BI10	HWL(60)	1258	966	1.14	0.16				
BI11	HWL(60)	5776	972	1.09	0.16	1.10	0.03	0.15	0.01
BI12	HWL(60)	5899	1860	1.06	0.13				
BI13	TRISOL	6089	2150	1.09	0.31	1.09		0.31	
BI14	TRISOL + HA(15)	5751	2020	1.22	0.20	1.22		0.20	

[a]Standard deviation based on *N* events.

equilibria when only one solute is present). In addition, the IAST formulation used in this study is valid because the single-solute isotherm data was well-fitted by the Freundlich model. If significant curvature exists in the log–log trace of single-solute equilibrium data, use of the Freundlich equation in the IAST solution will result in substantial errors in calculation of the spreading pressure. In such situations, a multiparameter isotherm model that better describes (log–log) nonlinearity must be applied.

Kinetic Studies. The questionable accuracy of model inputs that must be obtained experimentally or from empirical correlations is always a concern for modeling efforts employing dynamic models of relatively high levels of mathematical sophistication. This concern is particularly true relative to mass-transport parameter evaluations for multiple-resistance adsorption models. Such evaluations are commonly done by subjecting CMBR rate data to a statistical parameter search for estimation of the $k_{f,b}$ (i.e., k_f for CMBR conditions) and D_s values. CMBR rate data and MADAM batch model calibrations for *p*-DCB in HWL(0) and HWL(60) are illustrated in Figure 9. Because the hydrodynamic characteristics of an FBR adsorber are significantly different from those of a CMBR, correlation techniques are used to estimate k_f values for column modeling. Associated errors in both search and correlation techniques can result in the compounding of common-direction errors.

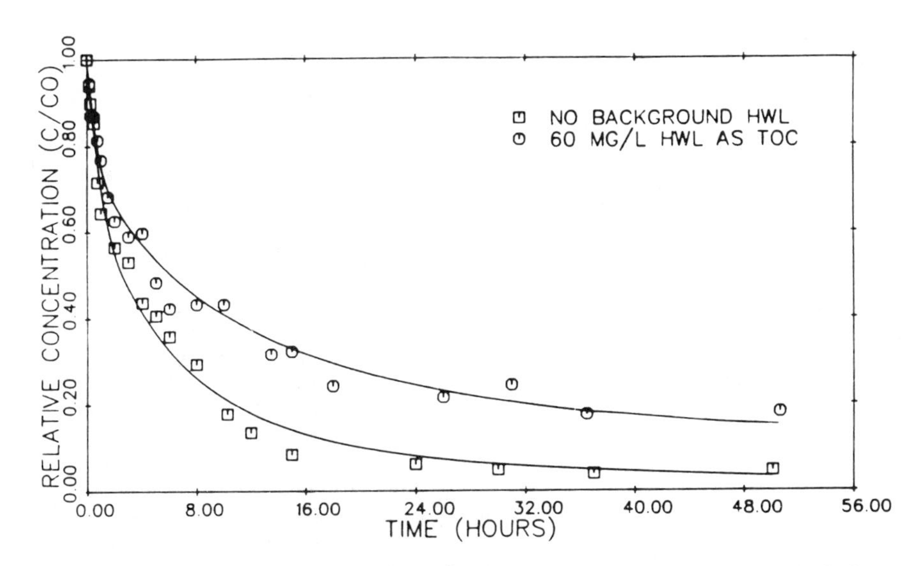

Figure 9. CMBR rate data and model simulations for p-*DCB in HWL(60) background (PB1/PB6).*

The SBA technique used here more closely approximates the hydrodynamics of full-scale columns than does a CMBR, allows simultaneous determination of k_f and D_s, and minimizes error-compounding by mutual compensation of errors in k_f and D_s during parameter search–regression analysis. Figure 10 shows typical SBA data and MADAM calibration results for TCE as a single-target solute in HA(15) background. Film transport is assumed to control at the initial breakthrough stage. The value of k_f is searched by using a single-compound column version of MADAM to fit the first 20–30 min of data, followed by calibration of D_s over the entire profile. Both search routines use a minimization function based on the sum of the squares of the errors between data and model calculations.

Desorption data collected after the influent concentration of TCE was reduced are included in Figure 10. The model calculations for this portion of the profile are based on an assumption of complete reversibility; that is, equilibrium and rate parameters are those obtained from the adsorption region. Discrepancies between the model and the data in this region may be due to some irreversibility or hysteresis. Other factors that might contribute to the differences include the difficulty in obtaining a representative value for the influent concentration during the initial stage of the desorption period, and the precision of the initial boundary condition for the modeled portion of the profile, which is preestablished by the profile for the adsorption region.

Numerous film mass-transport correlations are reported in the literature, each distinguished by a particular functional relationship between the

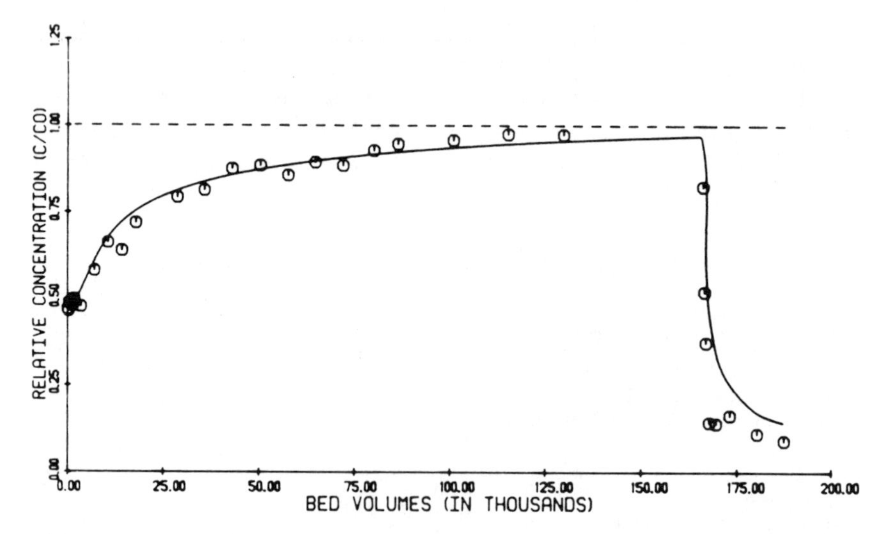

Figure 10. SBA data and model simulation for TCE in HA(15) background (TS2).

dimensionless Reynolds (Re) and Schmidt (Sc) numbers in the expression for the Sherwood number (Sh), which is related to k_f by

$$\mathrm{Sh} = \frac{k_f d}{D_l} \tag{9}$$

Re and Sc are defined as

$$\mathrm{Re} = \frac{v_z d}{\nu} \tag{10}$$

$$\mathrm{Sc} = \frac{\nu}{D_l} \tag{11}$$

Sample k_f values for single-solute cases of *p*-DCB and TCE in HWL(60) computed from six different correlations are presented in Table III. Separate calculation of k_f is required for each experiment because film diffusion is a function of flow conditions and bed void fraction. Bulk liquid diffusivities, D_l, for *p*-DCB and TCE were determined with the Wilke–Chang equation (*26*). Table III also includes k_f values estimated by the SBA–MADAM calibration technique. The different correlation values for k_f were observed to vary over a significant range, and most differed substantially from those obtained by the SBA experimental technique, depending upon the type of background DOM.

Mass-transport coefficients for single-solute solutions of *p*-DCB and TCE in the several DOM backgrounds, using both parameter determination techniques, are given in Table IV. The SBA calibrations of k_f and D_s and CMBR determinations of D_s reflect the decreased rates of film and intraparticle diffusion in the presence of background DOM observed in the data. Literature correlation calculations of k_f do not account for interactions between

Table III. Comparison of Film Mass-Transfer Coefficients

	p-*DCB in HWL(60)*		*TCE in HWL(60)*	
Method of Determination	*Sh*[a]	k_f (*cm/s* × 10^3)[a]	*Sh*[b]	k_f (*cm/s* × 10^3)[b]
Experimental (SBA)	—	5.8	—	5.0
Williamson et al. (*21*)	39.6	6.6	36.8	7.3
Wilson and Geankoplis (*22*)	46.1	7.7	43.2	8.6
Gnielinski (*23*)	45.1	7.5	42.5	8.5
Ohashi et al. (*24*)	34.0	5.6	32.1	6.4
Kataoka et al. (*15*)	48.5	8.1	45.5	9.1
Dwivedi-Upadhyay (*25*)	49.0	8.1	45.9	9.1

[a]Reynolds number, 10.4; Schmidt number, 1116.3.
[b]Reynolds number, 10.3; Schmidt number, 931.4.

Table IV. Single-Solute Rate Parameters Estimated by Two Calibration Techniques

Experiment Number	*Background Water*	k_f (*cm/s* × *10*^{3})		D_s (*cm*^{2}*/s* × *10*^{10})	
		SBA	*Correlation (Best–Worst)*[a]	*SBA*	*CMBR*
With p-DCB as solute					
PS1	DDW	7.9	8.0–5.9	1.6	2.9
PS2	HA(15)	7.2	7.8–5.8	0.72	2.2
PS3	HWL(60)	5.8	5.6–8.1	0.57	2.6
PS4	TRISOL	7.5	7.6–5.4	1.5	2.7
PS5	TRISOL + HA(15)	6.8	7.2–5.4	1.3	2.4
With TCE as solute					
TS1	DDW	6.6	6.3–9.0	5.6	9.5
TS2	HA(15)	5.8	6.5–9.5	4.8	5.2
TS3	HWL(60)	5.0	6.4–9.1	4.2	5.9
TS4	TRISOL	6.1	6.0–8.8	5.6	8.8
TS5	TRISOL + HA(15)	5.7	6.1–9.1	4.7	4.7

[a]Best and worst based on deviation from k_f (SBA).

target compounds and therefore cannot reflect the rate decreases. Moreover, no standard criteria exist for a priori selection of a particular correlation for a specific system. Table IV lists "best" and "worst" correlation values, where "best" and "worst" are based upon deviations from k_f determined from the SBA approach.

For p-DCB, correlation k_f values bracketed the experimentally determined coefficients. For TCE, however, correlation values were usually higher than SBA values, especially when leachate was present in the background. Intraparticle diffusion coefficients estimated by the CMBR method are higher than those determined from the SBA for both compounds; the largest percentage deviation exists in the case of p-DCB in leachate background and the smallest for TCE in TRISOL + HA(15) background. The discrepancy in parameter values between the two techniques is assumed to be due to differences in the hydrodynamic conditions of the experimental systems. For instance, the sequential loading of multiple adsorbing species is likely to be different in the two reactor configurations. It is anticipated that solid and solute interactions associated with the presence of additional sorbates, especially humiclike substances, may accentuate such differences, particularly in the case of intraparticle mass transport within microporous adsorbents. Both techniques predict that TCE diffuses more rapidly than p-DCB along the internal surfaces of F–400, whether background DOM is present or not.

Figures 11 and 12 illustrate model verifications for p-DCB and TCE, respectively, in HWL(60). Predicted MADAM profiles were generated with the physical parameters listed in Table V, SBA rate parameters, "best" correlation k_f with CMBR D_s, and "worst" correlation k_f with CMBR D_s. Lines 2 and 3 bound the profiles generated by using k_f values estimated from the six literature correlation procedures. Clearly, the SBA approach provides a

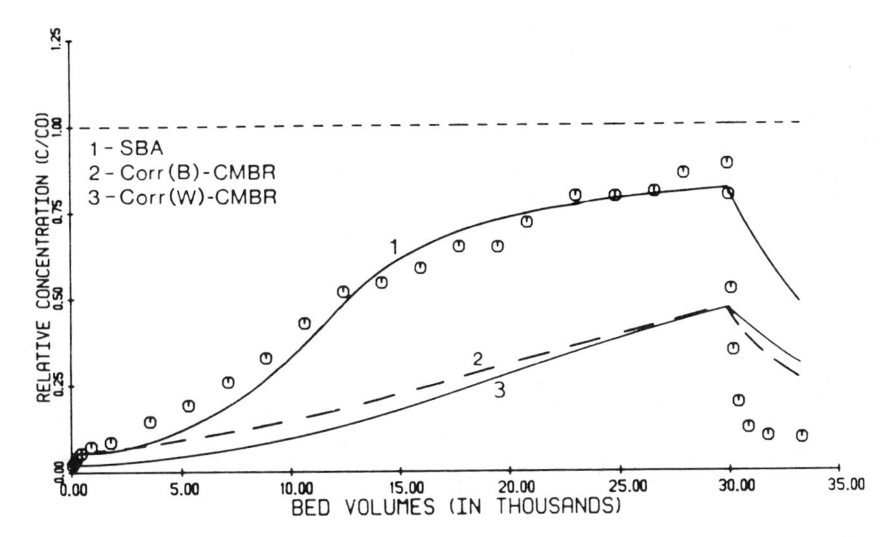

Figure 11. DBA data and model predictions for p-*DCB in HWL(60) background (PD3).*

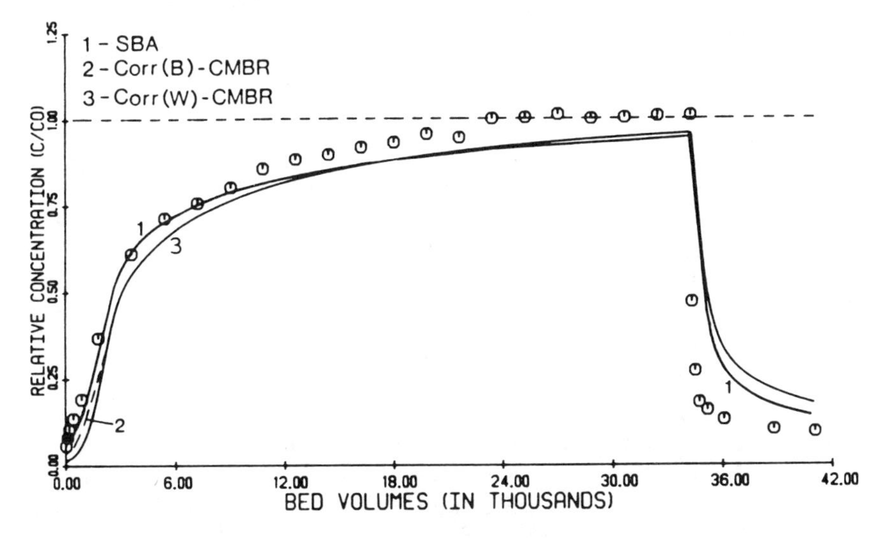

Figure 12. DBA data and model predictions for TCE in HWL(60) background (TD3).

better prediction of DBA breakthrough data than the correlation–CMBR profiles for the case of *p*-DCB. The SBA-based prediction is adequate over the entire profile, while the correlation–CMBR method significantly overestimates the performance of the column. This result is depicted numerically in Table VI, which compares experimentally observed numbers of bed vol-

Table V. Physical Parameters for DBA-Mode Verification Studies for Single-Solute Systems

Experiment Number	*Background Water*	C_0 (μg/L)	ε	v_z (cm/s)	*L* (cm)
With *p*-DCB as solute					
PD1	DDW	3484	0.356	2.11	4.5
PD2	HA(15)	3347	0.371	2.02	5.0
PD3	HWL(60)	3352	0.391	1.93	6.1
PD4	TRISOL	3425	0.343	1.82	6.5
PD5	TRISOL + HA(15)	3487	0.372	1.69	5.0
With TCE as solute					
TD1	DDW	641	0.399	1.89	5.6
TD2	HA(15)	643	0.380	1.98	8.0
TD3	HWL(60)	660	0.395	1.92	6.0
TD4	TRISOL	655	0.382	1.64	6.0
TD5	TRISOL + HA(15)	658	0.363	1.70	8.5

NOTE: *See* Table IV for coincident rate parameters.

Table VI. Comparison of Breakthrough Levels for Single-Solute Systems

		Predicted		
% Breakthrough	*Measured*	*SBA*	*Corr(B)-CMBR*[a]	*Corr(W)-CMBR*[b]
With *p*-DCB as solute				
10	1.8	3.7	4.6	9.1
25	7.1	8.5	16.6	18.7
50	12.2	12.8	32.0	32.0
With TCE as solute				
10	0.2	0.5	0.9	1.2
25	1.2	1.4	1.7	1.9
50	2.5	2.6	3.1	3.1
75	6.9	7.0	8.4	8.3

NOTE: All values are thousand bed volumes treated. They were generated by using different rate parameter determination techniques; background: HWL(60).
[a] k_f from best correlation; D_s from CMBR rate data.
[b] k_f from worst correlation; D_s from CMBR rate data.

umes treated to prescribed effluent levels to values predicted with models calibrated by the various parameter-estimation techniques.

All of the modeling approaches provide reasonably good prediction of the data for TCE in leachate background, although the SBA technique gives a better fit to the initial portion of the breakthrough profile (Figure 12 and Table VI). In this region, discrepancies between the data and model predictions are attributable to errors associated with the estimated value of k_f. The differences in D_s values obtained by the two calibration techniques are much greater for *p*-DCB than for TCE; correspondingly greater discrepancies result between MADAM profiles generated for *p*-DCB and TCE.

Model predictions in the desorption region were also more accurate for TCE than for *p*-DCB, a result suggesting that a significant portion of *p*-DCB adsorption may be irreversible. This contrast may also help to explain the difference between the intraparticle diffusion coefficients for *p*-DCB obtained by the SBA and CMBR techniques. The irreversibility observed in the *p*-DCB–HWL(60) system was greater than for *p*-DCB in distilled water background and indicated that certain solute interactions with components of the leachate background enhanced the established hysteresis pattern. The observed response to these interactions in the alternate reactor systems is anticipated to be different, because the sequential loading of multiple adsorbing species (the target compound and background water components) is also likely to be different. Further, several investigators have noted that the adsorption capacity determined by CMBR bottle-point studies may in certain systems overestimate the capacity that can be achieved in column adsorbers (*10, 27*). If this is so, then D_s values determined from SBA data and analysis will indeed be less than if equilibrium parameters representative of lower adsorption capacity were used. This condition may be especially critical if inherent irreversibility and hysteresis patterns exhibited by such a compound as *p*-DCB are augmented by background solution components (i.e., DOM).

Figures 13 and 14 briefly summarize the impact of the presence of background DOM on the removal of single-solute systems of *p*-DCB and TCE, respectively, in fixed beds. System conditions are identical for a given set of MADAM profiles, and rate parameters are those determined by the SBA technique, because these were shown to provide good prediction of DBA breakthrough data. As expected, the combined impacts of reduced equilibrium capacities and rates effected a significant shift to the left in the

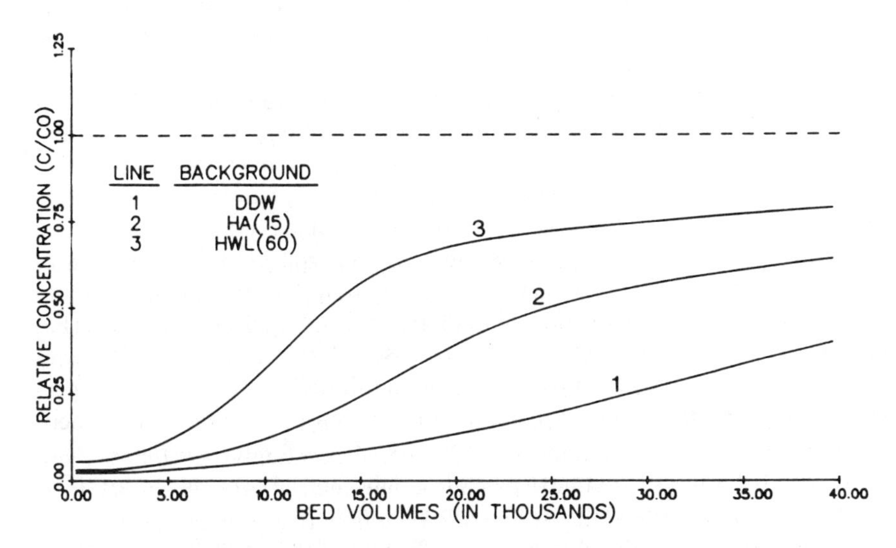

Figure 13. MADAM–DBA breakthrough profiles for p-*DCB in the presence of various background waters.*

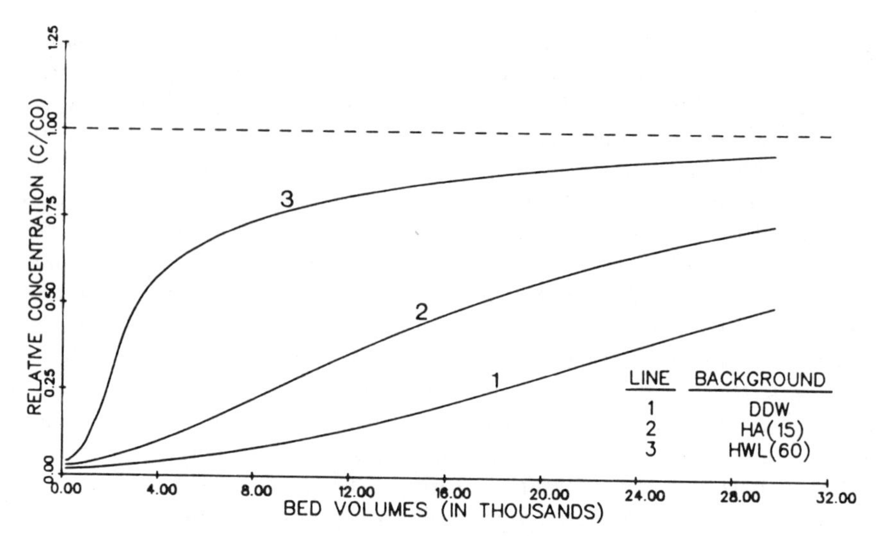

Figure 14. MADAM–DBA breakthrough profiles for TCE in the presence of various background waters.

breakthrough profiles, indicative of decreasing treatment effectiveness by the carbon column.

Typical SBA data and MADAM calibrations for a bisolute mixture in background DOM are presented in Figure 15. The small overshoot of the TCE profile is evidence of a competitive effect exerted by the more strongly adsorbed compound, *p*-DCB, on the more weakly but more rapidly adsorbed TCE. This type of displacement phenomenon is a common characteristic of multicomponent systems in which the adsorbing species have substantially different energies and rates of adsorption.

Table VII contains rate coefficients for the bisolute studies. Film mass-transport values computed from literature correlations are essentially the same as in single-solute systems, as this approach assumes that the sorbates diffuse independently of one another. As noted earlier, however, competitive adsorption equilibria in the systems studied were observed to be of nonideal character, and it is reasonable to assume that adsorption kinetics were similarly affected. Observations regarding the rate parameters for the bisolute analysis are consistent with those for single-solute modeling, namely: (1) k_f and D_s for both target compounds decrease with increasing background DOM concentrations; (2) values of k_f obtained for TCE from literature correlations are significantly higher than those determined by the SBA in leachate backgrounds; (3) D_s values obtained from CMBR data are higher than corresponding values obtained from SBA data; and (4) D_s values for TCE are higher than D_s values for *p*-DCB in each case.

Comparison of the single-solute coefficients presented in Table IV to the bisolute coefficients in Table VII reveals that D_s values for *p*-DCB are

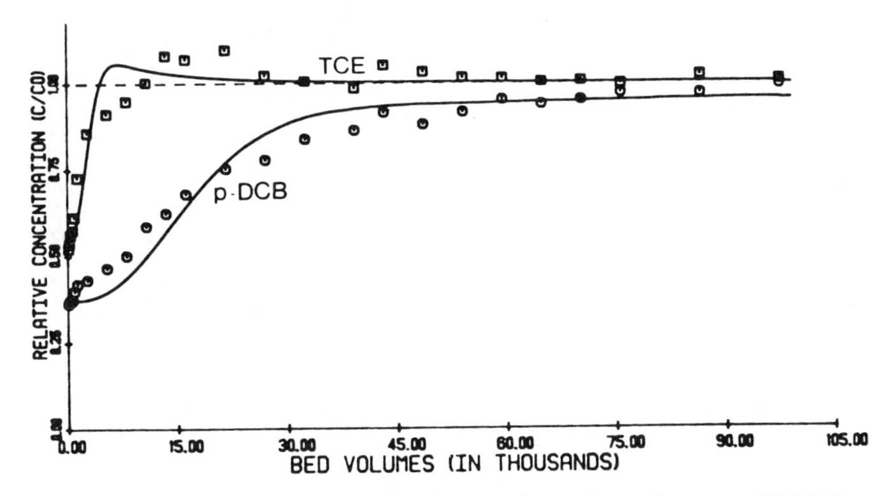

Figure 15. SBA data and model simulations for p-DCB *and TCE in HWL(60) background (BS8).*

similar, but the values determined for TCE in the bisolute systems are much higher than the corresponding single-solute value. This peculiarity is due, in part, to the fact that the nonideal equilibrium behavior of *p*-DCB and TCE in bisolute systems represented by the values of P_{TCE} is appreciably less than 1.0. This nonideality is reflected in the intraparticle rate coefficients in response to a numerical feature of MADAM relating to the value of the solute distribution parameter, D_g. The solute distribution parameter, defined in the MADAM algorithm as

$$D_g = \frac{\rho(1-\epsilon)\sum_{i=1}^{N} q_{0,i}}{\epsilon \sum_{i=1}^{N} C_{0,i}} \qquad (12)$$

is used in the expression that dedimensionalizes a term that includes D_s for solution of the solid-phase material-balance equation. Thus, abrupt changes in D_g may result in a similar discontinuity in the same direction for D_s for variable calibration runs. Initial experimental conditions and relative equilibrium partitioning of *p*-DCB and TCE are such that D_g values in the bisolute system are approximately 5 times higher than the D_g values in single-solute calibrations for TCE, a situation resulting in the corresponding shift in D_s values determined by MADAM (6).

The dependence of surface diffusivity on loading in multicomponent systems could also be interpreted in terms of the bonding energies of the

Table VII. Rate Coefficients for Bisolute Systems

Experiment Number	*Solute*	*Background*	k_f (*cm/s* × 10^3)		D_s (*cm²/s* × 10^{10})	
			SBA	*Best–Worst*[a]	*SBA*	*CMBR*
BS1	*p*-DCB	DDW	7.7	7.7–6.5	1.8	3.0
	TCE		6.7	6.4–9.3	20.0	57.0
BS2	*p*-DCB	HA(15)	6.2	6.0–7.9	0.46	2.3
	TCE		5.3	6.1–8.9	16.0	36.0
BS3	*p*-DCB	HWL(16)	7.5	7.8–5.8	0.62	2.3
	TCE		5.3	6.5–9.7	16.0	52.0
BS4	*p*-DCB	HWL(60)	5.4	5.6–8.0	0.36	2.2
	TCE		3.3	6.3–9.0	10.5	32.0
BS5	*p*-DCB	TRISOL	7.4	7.4–5.2	1.8	2.5
	TCE		5.7	5.9–8.3	18.0	54.0
BS6	*p*-DCB	TRISOL + HA(15)	5.8	5.9–8.3	1.6	2.3
	TCE		4.4	6.2–9.3	9.0	35.0

[a]Best and worst correlation-derived values based on deviation from k_f (SBA).

competing solutes (*28*). The effect would be greatest for the more weakly adsorbed solute, in this case TCE. As more strongly (and perhaps specifically) adsorbing *p*-DCB molecules displace TCE molecules at surface sites, a higher mobility of TCE in the adsorbed phase is expected.

Physical parameters for the various bisolute DBA runs are listed in Table VIII. Typical DBA breakthrough data and MADAM predictions for the target compounds in selected DOM backgrounds are illustrated in Figures 16 through 18. Numerical comparisons of model predictions to DBA data, such as that given for the case illustrated by Figure 17 and Table IX, yield conclusions that closely parallel those obtained from comparisons of single-solute data and model predictions. Breakthrough curves determined with the SBA technique predict DBA data much more closely than correlation–CMBR procedures for *p*-DCB, with model deviations most attributable to variations in D_s. Prediction of TCE breakthrough is not as sensitive to the parameter estimation methodology used, although the SBA technique is more accurate in the earlier portion of the profile and provides a better overall prediction of DBA behavior. The competitive displacement of TCE is more evident in the DBA than in the SBA profiles. Although the extent of TCE desorption is not predicted precisely by MADAM, the general pattern is obtained, particularly when SBA-estimated coefficients are used.

The regions bounded below and above the $C/C_0 = 1.0$ line show that a large fraction of the TCE adsorbed initially is displaced into the effluent by the more strongly adsorbing *p*-DCB. This phenomenon is further evident in Figure 19, which shows MADAM profiles for TCE in bisolute systems for varying concentrations of leachate background (based on SBA rate coefficients). Figure 20 presents corresponding profiles for *p*-DCB. These several examples illustrate the decrease in adsorber performance observed with increasing background DOM concentrations. The combined impact of re-

Table VIII. Physical Parameters Used in DBA Mode Verification Studies for Bisolute Systems

Experiment Number	*Background Water*	C_0 ($\mu g/L$) p-*DCB*	*TCE*	ϵ	v_z (*cm/s*)	*L* (*cm*)
BD1	DDW	3438	637	0.388	1.94	7.0
BD2	HA(15)	3539	634	0.375	1.68	10.0
BD3	HWL(16)	3537	676	0.370	2.03	7.1
BD4	HWL(60)	3425	684	0.386	1.94	11.0
BD5	TRISOL	3569	673	0.403	1.56	5.0
BD6	TRISOL + HA(15)	3662	625	0.360	1.74	5.5

NOTE: *See* Table VII for coincident rate parameters.

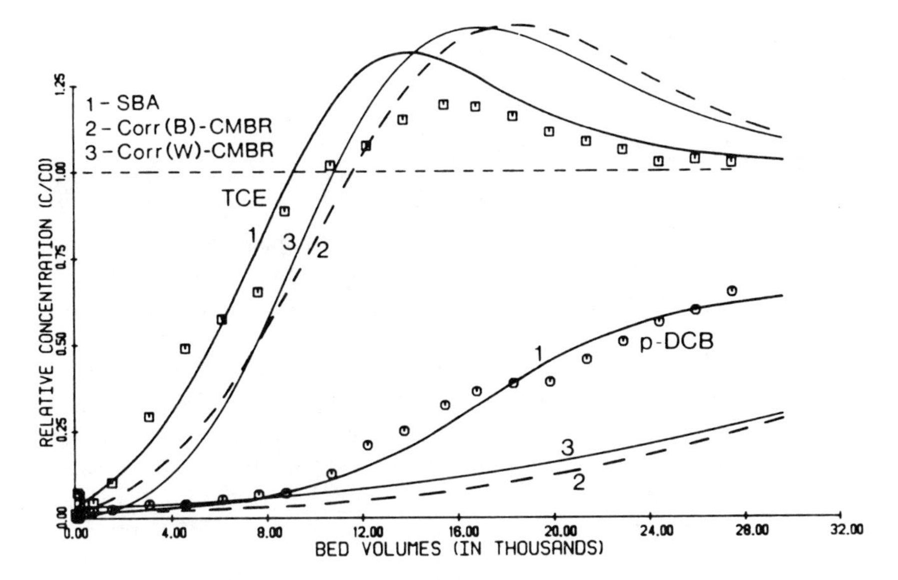

Figure 16. DBA data and model predictions for p-*DCB and TCE in HWL(60) background (BD6).*

duced equilibrium capacities and rates on adsorber performance is significant at leachate concentrations as low as 16 mg/L as TOC. Figures 21 and 22 compare the single- and bisolute breakthrough profiles in the TRISOL and TRISOL + HA(15) backgrounds for TCE and *p*-DCB, respectively. From Figure 21, *p*-DCB obviously effects a substantial decrease in TCE removal in both backgrounds. In contrast, Figure 22 illustrates that the presence of TCE at the stated concentration had no noticeable impact on the breakthrough of *p*-DCB in TRISOL and TRISOL + HA(15) backgrounds, as the respective single- and bisolute curves are essentially superimposed one upon the other.

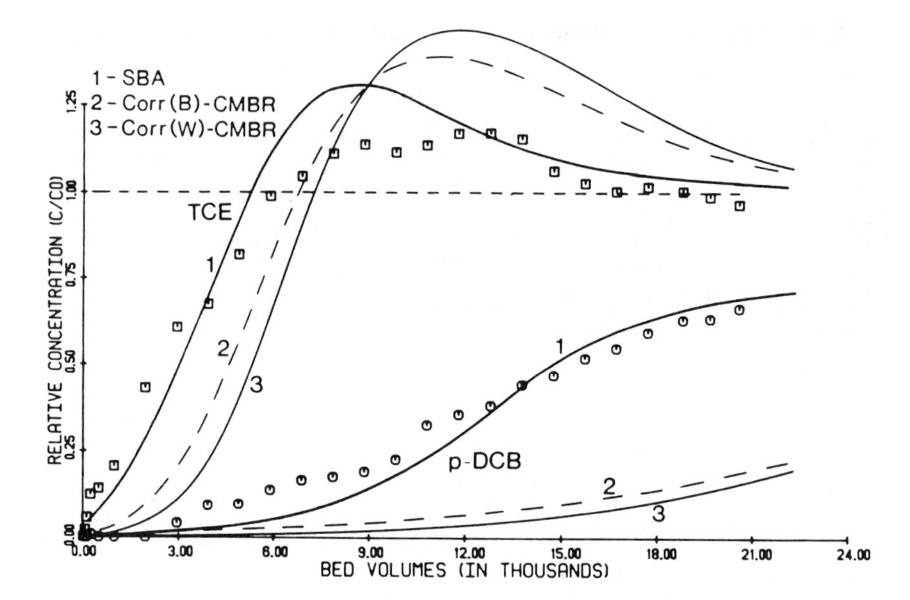

Figure 17. DBA data and model predictions for p-*DCB and TCE in HWL(60) background (BD8).*

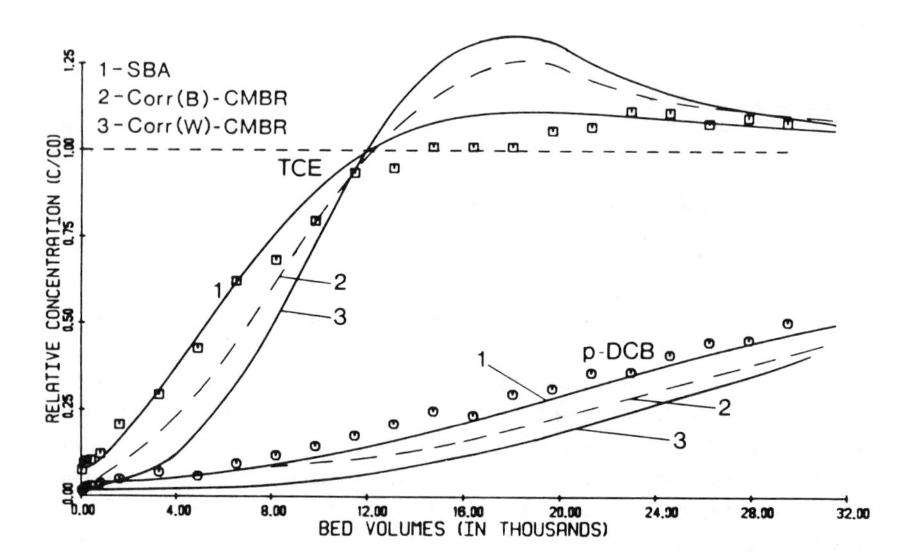

Figure 18. DBA data and model predictions for p-*DCB and TCE in TRISOL+HA(15) background (BD12).*

Table IX. Comparison of Breakthrough Levels for Bisolute Systems

		Predicted		
% Breakthrough	*Measured*	*SBA*	*Corr(B)-CMBR*[a]	*Corr(W)-CMBR*[b]
With *p*-DCB as solute				
10	9.8	10.2	17.4	14.1
25	13.7	14.9	27.6	26.5
50	22.8	21.0	>32.0	>32.0
With TCE as solute				
10	1.5	1.5	2.6	3.5
25	2.7	3.4	5.0	5.5
50	4.6	5.4	7.5	7.5
75	8.0	7.4	9.6	9.2

NOTE: All values are thousand bed volumes treated. They were generated by using different rate parameter determination techniques (Run DB6); background: HWL(16).
[a] k_f from best correlation; D_s from CMBR rate data.
[b] k_f from worst correlation; D_s from CMBR rate data.

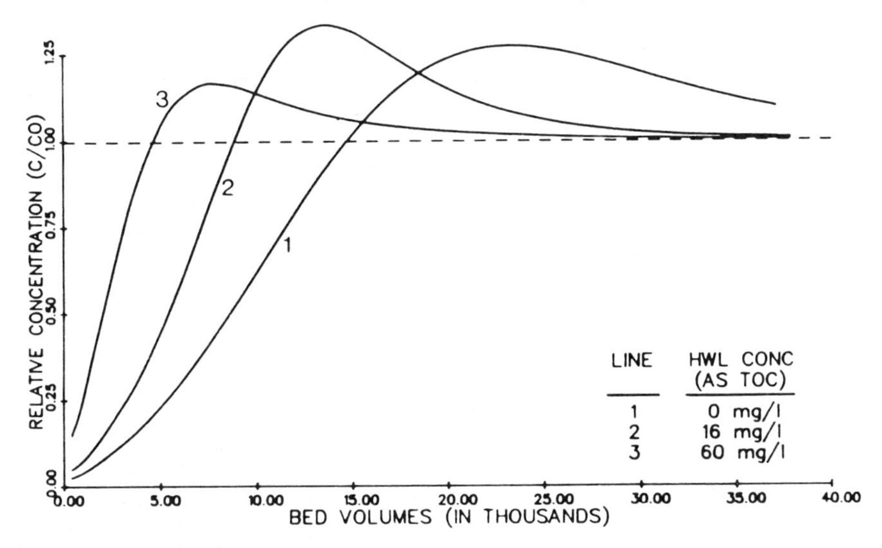

Figure 19. MADAM–DBA breakthrough profiles for TCE in the presence of p-DCB for varying concentrations of background HWL.

Summary and Significance

GAC adsorption of two target organic compounds in the presence of various sources and amounts of background DOM was analyzed with an existing mathematical adsorption model, MADAM. The modeling approach applied was one in which model parameters were experimentally determined and calibrated for the target compounds only, with the leachate considered as an unspecified background characterized only in terms of a lumped-param-

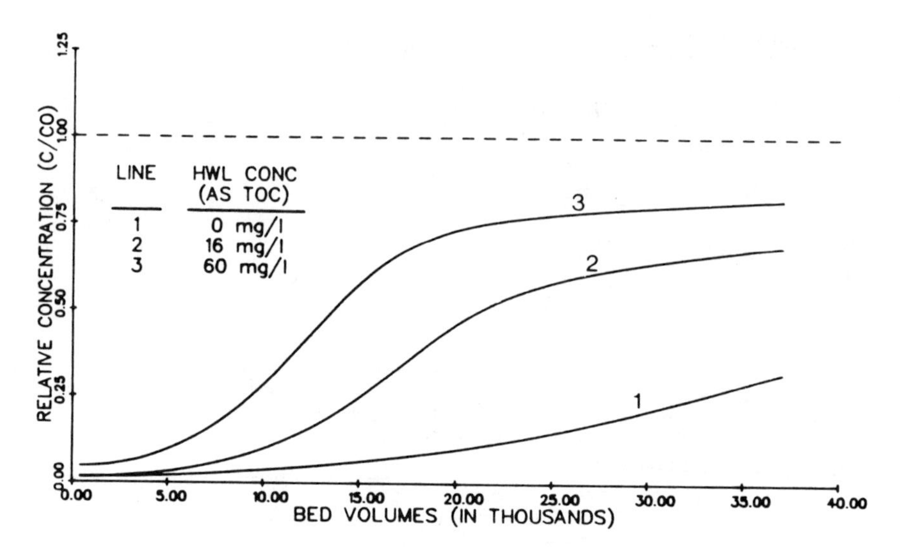

Figure 20. MADAM–DBA breakthrough profiles for p-*DCB in the presence of TCE for varying concentrations of background HWL.*

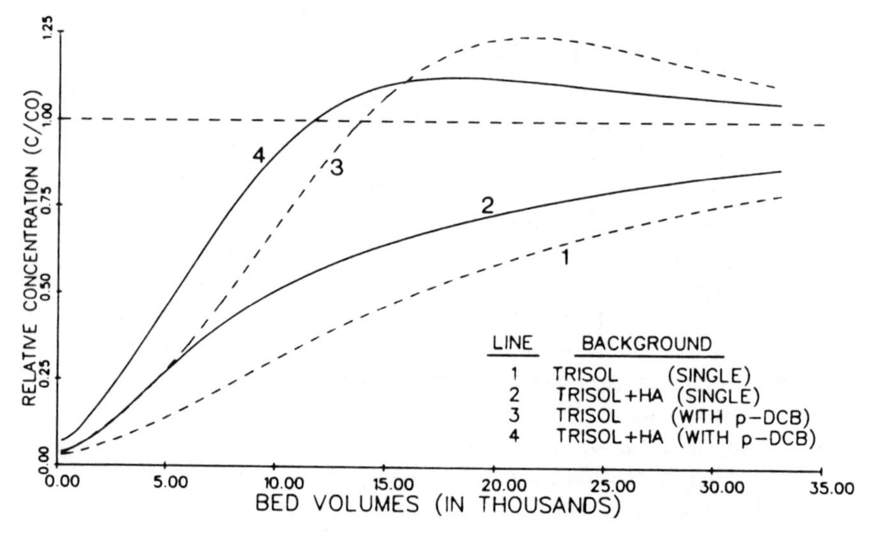

Figure 21. MADAM–DBA breakthrough profiles for TCE in TRISOL and TRISOL+HA(15) backgrounds for single- and bisolute systems.

eter measure, TOC. Findings and conclusions derived from the study are summarized as follows.

In general, the presence of leachate DOM reduced adsorption capacities and rates for both target compounds (*p*-DCB and TCE), compared to their respective values in water having no background DOM. These impacts, reflected quantitatively in equilibrium and kinetic model coefficients deter-

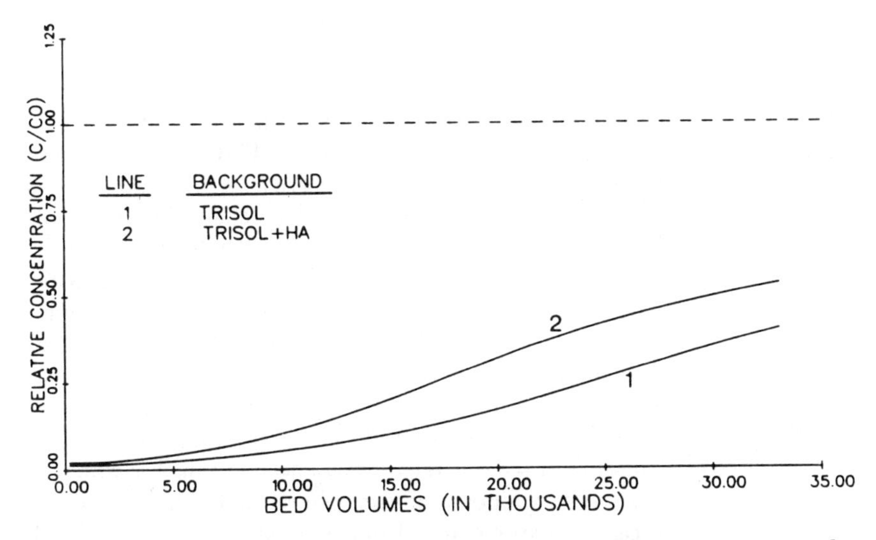

Figure 22. MADAM–DBA breakthrough profiles for p-*DCB in TRISOL and TRISOL + HA(15) backgrounds for single- and bisolute systems.*

mined by calibration of bench-scale experimental data, are proportional to the strength of the background water, as quantified by TOC.

The SBA technique was verified as a useful method for determination of system-specific kinetic parameters for complex mixtures of organic compounds. Film and surface-diffusion coefficients estimated by use of this method resulted in adequate predictions of the performance of deeper, bench-scale FBRs for single- and multicomponent cases for the range of system conditions studied. These predictions were as accurate and, in some cases, more accurate than those deriving from a more traditional methodology for estimating rate parameters through literature correlations and CMBR rate data. Discrepancies that result from application of the correlation–CMBR technique are more pronounced for *p*-DCB than for TCE, and most are attributable to differences in values for the internal diffusion coefficient. Values of D_s determined from CMBR rate data were significantly higher than those estimated from SBA analysis, presumably because of the different hydrodynamic conditions that prevail in the two reactor systems and to differences in the extent to which differences in these conditions were reflected in the corresponding k_f values. Overprediction of fixed-bed adsorber performances resulted for certain cases when the correlation–CMBR parameters were used in the MADAM model projections. Differences in film mass-transport coefficient values for fixed-bed adsorbers result from the fact that literature correlations do not account for interactions between target compounds and background DOM, although these interactions appear to be reflected in k_f values derived by the SBA technique.

Beyond the specific conclusions summarized, the findings of this study bear on an issue of more general significance in terms of demonstrating the merits of a phenomenological modeling approach for target solutes in complex systems: namely, the implicit incorporation of many of the impacts of adsorbent–solute, adsorbent–DOM, and solute–DOM interactions into nondescript model coefficients. Independent characterization and experimental verification of all of the reactions and associated mechanisms in complex systems is an enormous and, in many cases, prohibitively expensive task. A phenomenological approach focuses the study by first identifying critical model coefficients, followed by calibration of these coefficients according to perturbations in system composition. Other advantages of this approach are that: (1) it streamlines pilot-study requirements, because it enables primary information gathering and hypothesis testing to be done on the bench scale, with the pilot plant used as more of a verification tool; and (2) it allows for further investigation, at least on a macroscopic scale, into mechanistic processes and their impacts on model parameters, where the modeling approach (if not the results) can be extrapolated from one case to another and the system-specific results compared to provide meaningful scientific insights. For example, future studies might explore correlations between experimentally determined model parameters for given compounds and various system–solution characteristics, and then verify the consistency of such correlations for different types of background DOM.

The current study also illuminates the need to investigate carbon adsorption of complex mixtures in connection with other treatment processes. For instance, the removal of a weakly adsorbed compound such as TCE may be more effectively achieved by air stripping. Similarly, if the presence of humic DOM significantly reduces the performance of GAC adsorbers with respect to a particular target compound (or compounds), adjustments in the coagulation step may be warranted to effect a greater preremoval of humic substances. In addition, the potential formation of metal–DOM complexes during the course of treatment may result in substances that compete with target organic compounds for adsorptive sites on activated carbons. Therefore, the impact of prior treatment steps in the flow scheme on the removal of target organic species also requires examination.

Nomenclature

C_e equilibrium liquid-phase solute concentration (μg/L)

C_i liquid-phase concentration of species i, IAST (μM)

C_i^* single-solute concentration of species i evaluated at the bisolute mixture spreading pressure, IAST (μM)

$C_{o,i}$ initial solute concentration of species i (μM or μg/L)

d carbon particle diameter (cm)

D_g solute distribution parameter (dimensionless)

D_l bulk liquid diffusivity (cm^2/s)
D_s surface diffusion coefficient (cm^2/s)
i.d. internal diameter (cm)
k_f film diffusion coefficient (cm/s)
K_F Freundlich isotherm constant
L carbon bed depth (cm)
M carbon dose (mg adsorbent)
n Freundlich isotherm exponent (dimensionless)
N number of target species in solution
P_i empirical competition coefficient for species i, IAST (dimensionless)
q_e equilibrium solid-phase solute concentration (μg/mg)
q_i solid-phase concentration of species *i*, IAST (μmol/mg)
q_i^* solid-phase concentration of species *i* in single-solute system evaluated at the bisolute mixture spreading pressure, IAST (μmol/mg)
$q_{o,i}$ initial solid-phase concentration of species *i* (μmol/mg or μg/mg)
q_T total solid-phase concentration of solutes, IAST (μmol/mg)
V liquid volume in reactor (L)
v_z interstitial flux velocity in carbon bed (cm/s)
z_i solid-phase mole fraction of species *i*, IAST (dimensionless)
ϵ bed void fraction (dimensionless)
η inverse of Freundlich exponent, *n* (dimensionless)
ν kinematic viscosity (cm^2/s)
π spreading pressure, IAST ($kcal/cm^2$)
ρ carbon particle density (g/cm^3)

Acknowledgments

The authors express their appreciation to Brett Farver, Barbara Jacobs, and Daniel Peters for their contributions to the experimental aspects of this project. The work was supported in part by Award No. CEE–8112945 from the National Science Foundation. The contents do not necessarily reflect the views and policies of the NSF, and the mention of trade names or commercial products does not constitute endorsement.

References

1. Weber, W. J., Jr.; Pirbazari, M. *J. Am. Water Works Assoc.* **1982,** *74*, 203.
2. Frick, B. R.; Sontheimer, H. In *Treatment of Water by Granular Activated Carbon*; Suffet, I. H.; McGuire, M. J., Eds.; American Chemical Society: Washington, DC, 1983; p 247.
3. Tien, C. In *Proceedings, First International Conference on the Fundamentals of Adsorption*; Myers, A. L.; Belfort, G., Eds.; Engineering Foundation and American Institute of Chemical Engineers: New York, 1984; p 647.

4. Endicott, D. D.; Weber, W. J., Jr. *Environ. Prog.* **1985,** *4*, 105.
5. Crittenden, J. C.; Luft, P.; Hand, D. W. *Water Res.* **1985,** *19*, 1537.
6. Smith, E. H.; Tseng, S.; Weber, W. J., Jr. *Environ. Prog.* **1987,** *6*, 18.
7. Crittenden, J. C.; Weber, W. J., Jr. *J. Environ. Eng. Div.* (*Am. Soc. Civ. Eng.*) **1978,** *104*, 1175.
8. Weber, W. J., Jr. In *Proceedings, First International Conference on the Fundamentals of Adsorption*; Myers, A. L.; Belfort, G., Eds.; Engineering Foundation and American Institute of Chemical Engineers: New York, 1984; p 647.
9. Weber, W. J., Jr.; Liu, K. T. *Chem. Eng. Commun.* **1980,** *6*, 49.
10. Weber, W. J., Jr.; Wang, C. K. *Environ. Sci. Technol.* **1987,** *21*, 1096.
11. Halfon, E. *Environ. Sci. Technol.* **1985,** *19*, 747.
12. Smith, E. H. Ph.D. Dissertation, University of Michigan, 1987.
13. Yonge, D. R. Ph.D. Dissertation, Clemson University, 1982.
14. Thacker, W. E.; Crittenden, J. C.; Snoeyink, V. L. *J. Water Pollut. Control Fed.* **1984,** *56*, 243.
15. Kataoka, T.; Yoshida, H.; Ueyama, K. *J. Chem. Eng. Jpn.* **1972,** *5*, 132.
16. Weber, W. J., Jr. *Physicochemical Processes for Water Quality Control*; Wiley: New York, 1972.
17. Wershaw, R. L.; Burcar, P. J.; Goldberg, M. C. *Environ. Sci. Technol.* **1969,** *3*, 271.
18. Carter, C. W.; Suffet, I. H. *Environ. Sci. Technol.* **1982,** *16*, 735.
19. Calloway, J. Y.; Gabbita, K. V.; Vilker, V. L. *Environ. Sci. Technol.* **1982,** *18*, 890.
20. Weber, W. J., Jr.; Voice, T. C.; Jodellah, A. J. *J. Am. Water Works Assoc.* **1983,** *75*, 612.
21. Williamson, J. E.; Bazaire, K. E.; Geankoplis, C. J. *Ind. Eng. Chem. Fundam.* **1963,** *2*, 126.
22. Wilson, E. J.; Geankoplis, C. J. *Ind. Eng. Chem. Fundam.* **1966,** *5*, 9.
23. Roberts, P. V.; Cornel, P.; Summers, R. S. *J. Environ. Eng. Div.* (*Am. Soc. Civ. Eng.*) **1985,** *111*, 891.
24. Ohashi, H.; Sugawara, T.; Kikuchi, K. I.; Konno, H. *J. Chem. Eng. Jpn.* **1981,** *14*, 433.
25. Dwivedi, P. N.; Upadhyay, S. N. *Ind. Eng. Chem. Process Des. Dev.* **1977,** *16*, 157.
26. Wilke, C. R.; Chang, P. J. *AIChE J.* **1955,** *1*, 264.
27. Awuwa, A. A. Ph.D. Dissertation, Illinois Institute of Technology, 1984.
28. Fettig, J.; Sontheimer, H. *J. Environ. Eng. Div.* (*Am. Soc. Civ. Eng.*) **1987,** *113*, 780.

RECEIVED for review September 17, 1987. ACCEPTED for publication February 23, 1988.

31

Frontal Chromatographic Concepts To Study Competitive Adsorption

Humic Substances and Halogenated Organic Substances in Drinking Water

Ronald J. Baker and I. H. Suffet

Environmental Studies Institute, Drexel University, Philadelphia, PA 19104

Thomas L. Yohe

Philadelphia Suburban Water Company, Bryn Mawr, PA 19010

This chapter introduces the use of frontal chromatographic theory to describe the breakthrough of solutes in granular activated carbon (GAC). Depletion of adsorption sites during use of GAC contactors can be expressed in terms of changes in moving concentration profiles, or fronts. These fronts can be defined in chromatographic terms. Data from a pilot-scale carbon adsorption study are presented and used as an example of how frontal chromatographic theory can be applied. Current models for describing and predicting solute breakthrough from GAC columns cannot predict breakthrough of a wide variety of compounds under water-treatment conditions. Frontal chromatography theory, as applied in this chapter, is useful for understanding the displacement and breakthrough phenomena in carbon contactors.

GRANULAR ACTIVATED CARBON (GAC) HAS RECENTLY BEEN PROPOSED by the U.S. Environmental Protection Agency's Office of Drinking Water Quality as the best available technology for removal of several categories of synthetic organic substances, including trihalomethanes (THM) and other halogenated organic compounds (*1*). One consequence of this regulatory

0065–2393/89/0219–0533$06.00/0

activity will probably be increased use of GAC for halogenated-organic-substance removal. However, many low-molecular-weight organic materials, including THM and other halogenated disinfection byproducts, are not strongly adsorbed by GAC. Considering the substantial capital and operating costs involved in GAC adsorption, optimization of this process in terms of equipment required and carbon consumption will be a high priority.

Many models for describing and predicting solute breakthrough from GAC columns have been and are being developed (2–5). Although progress is being made, current models cannot predict breakthrough of a wide variety of compounds under water-treatment conditions, where influent concentrations and types of organic mixtures vary and most contaminants are in the nanogram- to microgram-per-liter concentration range. In water- and wastewater-treatment applications the complexity and variability of influent quality make it difficult to develop models to predict breakthrough of individual solutes.

Because competitive effects from humic materials are difficult to predict, they greatly complicate the process of predicting specific solute breakthrough and thus of designing effective and efficient GAC installations. Site-specific water-quality differences add to the complexity. Therefore, the National Academy of Science (NAS) has recommended that each water-treatment design for GAC be piloted at that location and that only general principles be assumed to apply between plants (6). This chapter introduces ideas that may lead to a new approach for evaluation of pilot-plant column data. The approach is based on the concepts of frontal chromatography.

Adsorption-site distribution on a GAC bed and depletion upon use of the GAC are expressed in terms of changes in moving concentration profiles, or fronts, that represent movement of specific solutes down the column. The shapes of these fronts can be defined by using frontal chromatographic theory. Data from a pilot-scale carbon adsorption study are presented and used as an example of how some aspects of frontal chromatographic theory can be applied to observed GAC behavior at specific locations. It is hoped that this concept of using pilot-plant data will enhance general understanding for the control of future full-scale operations.

Analogy: Similarities Between GAC and Frontal Chromatography

Conceptually, a GAC column can be thought of as a slow, inefficient liquid chromatographic (LC) column operated in the frontal elution mode. Table I lists operational similarities and differences between the two systems. The distribution coefficient or capacity factor, k', in chromatographic theory (7) is analogous to points on a carbon isotherm (8), where each point on the isotherm indicates the distribution between the solid and liquid phases. In both systems diffusion is thought to control the mass-transfer rate (7, 8).

Table I. Comparison Between Frontal Elution Liquid Chromatography and Water-Treatment GAC Operation

Attribute	*Frontal Elution Liquid Chromatography*	*Water-Treatment GAC Operations*
Definable distribution coefficients	yes (k')	yes (isotherm points)
Diffusion-controlled mass-transfer kinetics	yes	yes
Theoretical plate concepts applicable	yes	yes
Competition for adsorption sites	yes	yes
Displacement of adsorbed compounds by others	yes	yes
Local equilibrium maintained throughout system	yes	yes
Constant number of theoretical plates	no	no
Time to breakthrough	minutes	days or months
Constant influent composition	yes	no
Irreversible adsorption at some sites	no	yes
Homogeneity of adsorption sites	preferably	no
Homogeneity of adsorption media	yes	no

Displacement of some solutes by others through competitive adsorption also occurs in both systems. The most significant differences between GAC and frontal chromatography (FC) are time to breakthrough (minutes for FC, months for GAC), influent variability (constant for FC, variable for GAC), and homogeneity of the adsorbent (GAC is nonhomogeneous, most chromatographic media are homogeneous). In both FC and GAC systems the solutes compete for adsorption sites, and less strongly adsorbed species are displaced by those more strongly adsorbed.

The humic materials that are irreversibly adsorbed onto GAC (*9*) effectively reduce the number of theoretical plates in the system in a nonuniform manner. It is not known what fraction of humic materials adsorbs irreversibly, and a priori prediction of how the humic substances as represented by nonvolatile total organic carbon (NVTOC) in a feedwater will interact with GAC is not now possible. This type of irreversible fouling would not be experienced in FC, as analysts minimize decreases in column efficiency by sample pretreatment (e.g., cleanup by base extraction (*10*) to remove humic substances before GC or LC analysis).

Each system (FC and GAC) can be described as a column reactor with advection, diffusion, and reaction terms as shown in equation 1. This general equation for liquid-phase concentration as it changes with time results from a mass-balance procedure (*11*) and has been applied to FC (*12*) and GAC (*7*)

$$\frac{\delta C}{\delta t} = -\left(V\frac{\delta C}{\delta x}\right) + \left(D\frac{\delta^2 C}{\delta x^2}\right) + r(C) \qquad (1)$$

total rate of change = advection + diffusion + reaction

where C is solute concentration, t is time, V is flow velocity, D is diffusivity, x is distance down the column, and $r(C)$ is a reaction term.

The reaction term $r(C)$ in both systems refers to rates of adsorption and desorption. Treatment of this term varies somewhat between GAC and FC, although in both cases thin-film and intraparticle mass transfer and distribution equilibrium are considered rate- and capacity-limiting. In order to define and quantify the $r(C)$ term, most GAC models rely on estimation of mass-transfer and competitive equilibrium parameters for each solute; then aqueous and adsorbed concentrations of the solutes are calculated as gradients through the GAC particles at different bed depths through time. Although this approach has also been used in FC, another approach is to describe the front or wave of solute concentration (solid or liquid phase or both) in terms of its statistical moments, and to project how the wave will change with time and distance down the column. Application of this approach to GAC should be possible.

Another chromatographic approach that may be applicable to GAC is the coherence concept developed by Helfferich (*13*). In this approach solutes are considered to be interactive (competitive) in their movement through chromatographic columns, and multicomponent wave movement occurs in a predictable "coherent" manner. Rates of wave and solute movement are found as solutions to eigenvector problems. Application of the coherence theory to GAC in water treatment will be mathematically complex because of the constantly changing influent and because some simplifying assumptions that apply to many chromatographic systems (*13*) may not hold true for GAC with its nonconstant separation factors. This essentially means that relative solute retention times (time spent in the column by the solutes) are significantly affected by relative aqueous-phase concentrations in GAC. The effect of this phenomenon on the use of coherence theory for describing GAC breakthrough has yet to be seen.

Some of the physical factors responsible for front shape and adsorption site depletion in column systems are shown in Figure 1. These factors apply to both frontal chromatography and GAC adsorption. For each column in Figure 1 (I–IV) a profile of bed depth versus surface loading concentration of a compound is shown for a column taken off-line after a period of loading. Each successive column shows additional physical factors that further complicate the system and reduce the column's capacity to adsorb compounds from the mobile phase.

Column I (isothermal) in Figure 1 shows the ideal situation for maximum adsorption of a solute in a column: a single solute (no competition) and instantaneous mass transfer. The advancing front is asymptotic because the initial influent equilibrates with carbon at the column inlet and is reduced in concentration when it contacts carbon farther down the column. This reduction results in lower surface loading at points farther down the column. Because phase equilibrium exists at each point in the front, the isotherm could be constructed from the surface loading–mobile phase concentration

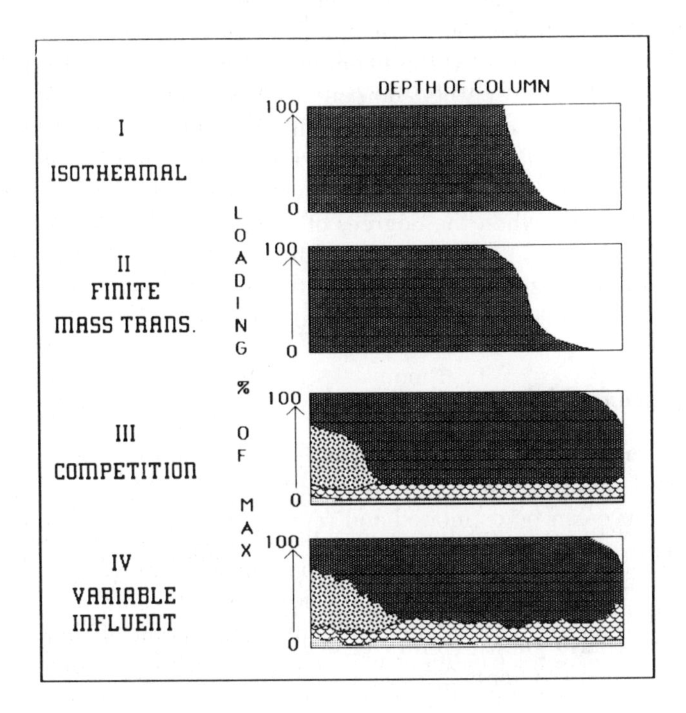

Figure 1. Front shaping mechanisms.

ratio at points in the front. Conversely, the front shape could theoretically be drawn from the isotherm.

The effect of noninstantaneous mass transfer is shown in column II (Finite Mass Transfer). Here the front has a sigmoidal or half-Gaussian appearance characteristic of frontal chromatography and sometimes observed in GAC loading profiles. The slowest (rate-limiting) mass-transfer steps control the front shape in the loading profile and the breakthrough profile. Both GAC modeling theory and chromatographic theory consider diffusion as a rate-limiting step in mass transfer. Chromatographic theory additionally considers adsorption and desorption as potential rate-limiting mass-transfer steps. The current practice of not considering rates of adsorption and desorption as finite in GAC modeling (5) may be an important oversight in breakthrough prediction for low-molecular-weight, rapidly diffusing compounds. In this case, points on the frontal curve of Column B would not represent isotherm points because stationary and mobile phases are not in equilibrium in the frontal region.

Column III (Competition) shows the loss of adsorption capacity resulting from competition from other compounds. A more strongly adsorbing compound is shown concentrated at the column inlet, a weakly adsorbing species has its highest loading farther down the column, and there is some irreversible adsorption throughout the column, typical of some fractions of humic materials (*13*). All three of these generalized types of competition reduce the adsorption capacity for other compounds. The reduction results in earlier breakthrough in GAC and earlier front elution in FC.

Column IV (Variable Influent) shows the effect of varying the influents of the adsorbing species. This variable poses the greatest challenge for GAC modeling, especially when the degrees of variation are not known. Mobile-phase effluent profiles can assume virtually any shape and are difficult to describe mathematically. This is the reason that site specificity for GAC treatment of different water quality is invoked (*6*).

The Statistical Moment Approach

Any curve can be described in terms of its statistical moments, which are mathematical descriptions of area distribution under the curve (*14*). The shape of a curve can be estimated and regenerated if enough moments are known. Figure 2 shows the procedure of approximating curve shape from statistical moments. If only the 0th moment is known (curve area), nothing is known about the curve shape between the boundaries. However, if the first moment is also known (center of mass), the estimated peak begins to take shape. When moments 2–4 are added, the estimated peak approaches

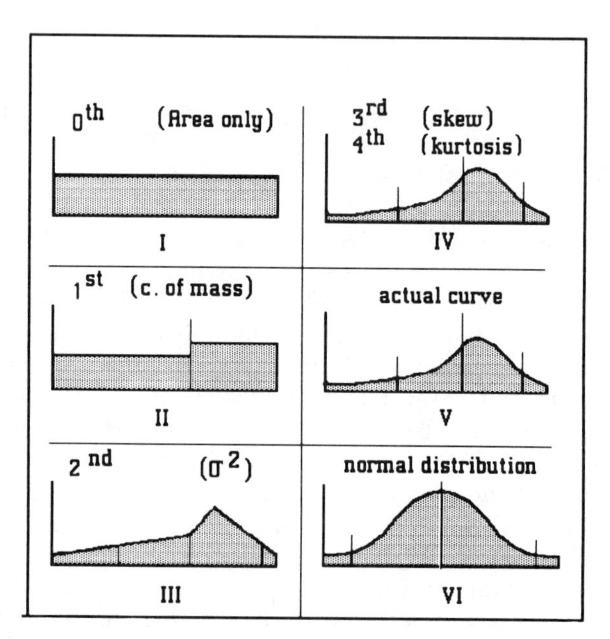

Figure 2. Curve description by statistical moments.

the actual peak in shape. If an infinite number of moments were applied, the actual peak would be reproduced exactly. The Gram–Charlier series (*15*) (equation 2) is one function that can be used to approximate a distribution from its moments. This approach, used extensively in FC, should be applicable to GAC.

$$C = \frac{1}{\sigma\sqrt{2\pi}} e^{\left[\frac{-(t - m_1)^2}{2\sigma^2}\right]} \left[1 + \sum_{i=3}^{\infty} \frac{C_i}{i!} H_i \left(\frac{t - m_i}{\sigma}\right)\right] \quad (2)$$

where C is concentration, which is considered a dependent variable of time; t is time; σ^2 is variance of the peak; H_i is the ith Hermite polynomial; and C_i is functions of statistical moments 3–5:

$$C_3 = \frac{m_3}{m_2^{3/2}}$$

$$C_4 = \left(\frac{m_4}{m_2^2}\right)^{-3}$$

$$C_5 = \left(\frac{m_5}{m_2^{3/2}}\right) - 10\left(\frac{m_3}{m_2^{3/2}}\right)$$

where m_i is the ith statistical moment.

To apply the statistical moment approach to FC, we must convert the elution fronts to distributions that can be described in terms of moments. Kalinichev (*16*) developed one approach, which is shown in Figure 3. The first derivatives of adsorption or desorption fronts are "bell-shaped" curves that have the necessary moments. It is not suggested that GAC breakthrough curves will generate Gaussian curves with this procedure. However, whatever shape they assume can be described by statistical moments.

Theoretical plate concepts are used in frontal chromatography to describe and predict the times of wave breakthrough. From frontal chromatographic plate theory, the effluent profile of a chemical can be related directly to the number of theoretical plates (n). Then n can be calculated from the statistical moments of breakthrough curves by using the method of Reilley et al. (*17*):

$$n = 2\left(\frac{t_p}{w}\right)^2 \quad (3)$$

where t_p is the time the point of inflection reaches the column outlet and w is a constant that is a function of front diffusion.

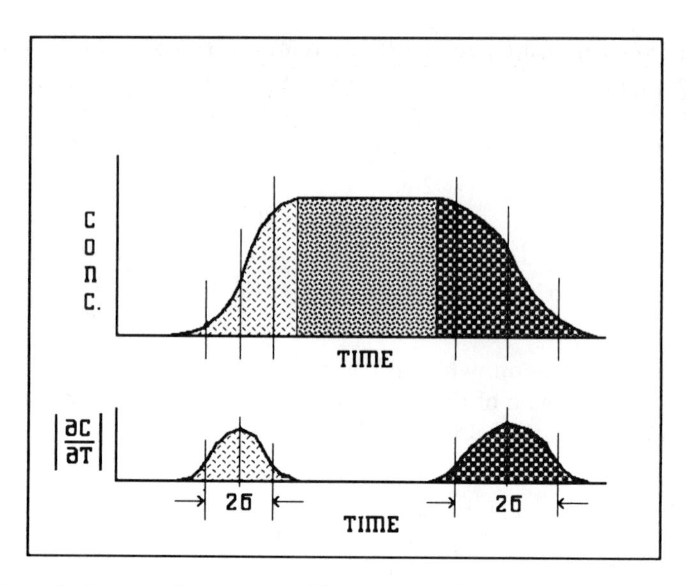

Figure 3. Statistical moments of fronts. (Based on a figure in ref. 12.)

This method of plate calculation relies only on the zeroth and first moments of the breakthrough curve (i.e., peak areas and mass centers, respectively). It was used for calculation of plate numbers for nonvolatile total organic carbon (NVTOC) and specific micropollutants in the study described in the next section, which will show how meaningful information for estimating remaining column capacity can be obtained.

Example: Application of Statistical Frontal Chromatographic Theory to GAC

Some aspects of frontal chromatographic theory were applied to data from a pilot-scale GAC evaluation. System and operating parameters are shown in List 1 and are published in detail elsewhere (*18*). Many parameters and compounds were monitored during the 300-day pilot study. However, for this example only four compounds (Chart 1) and NVTOC will be discussed. Influent concentration variability of the four compounds discussed are shown in Table II.

Relative positions of the Freundlich isotherms (*19*) for four of the compounds are shown in Figure 4. To further the GAC–FC analogy, relative affinity of the solutes for the stationary phase should determine breakthrough (or elution) order; greater affinity should result in later breakthrough. This result was found to be true in the pilot study. Compounds broke through in increasing order of affinity, as defined by their isotherm positions. This effect may not hold true for compounds of widely different concentrations

List 1. System and Operating Parameters in the Pilot-Plant Study

Activated carbon:	Calgon F-400, Calgon Corp, Pittsburgh, PA
Column:	4-in. i.d., 3-ft GAC bed depth, glass construction
Hydraulic parameters:	0.392-gpm flow rate; 5.0-min contact time; 4.5-gpm/ft^2 surface loading
Water:	ground-water-fed reservoir, Upper Merion, PA, operated by Philadelphia Suburban Water Co., Bryn Mawr, PA

1,1-dichloroethane (DCE)	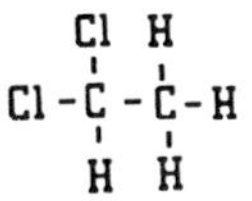 Cl H Cl - C - C - H H H
1,1,1-trichloroethane	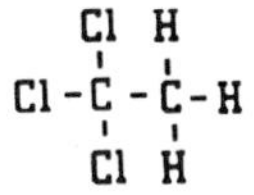 Cl H Cl - C - C - H Cl H
1,2,3-trichloropropane (TCP)	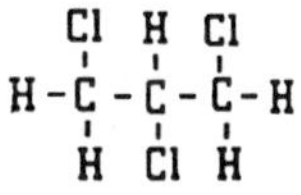 Cl H Cl H - C - C - C - H H Cl H
trichloroethene (TCE)	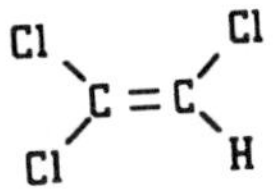 Cl Cl C = C Cl H
tetrachloroethene (PCE)	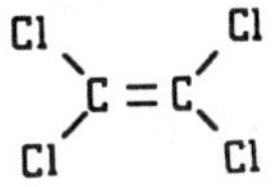 Cl Cl C = C Cl Cl
nonvolatile TOC (NVTOC)	

Chart 1. Compounds monitored.

because position on the isotherm determines affinity to some degree. Figure 5 shows the relative order of breakthrough for the four compounds (*18*). After 180 days 1,1-dichloroethane had reached saturation in the column and was apparently being displaced by competing compounds. Influent concentration exceeded effluent concentration much of the time after 180 days. This result is an example of the so-called chromatographic effect (*20, 21*), which is really a frontal chromatographic effect. The other three compounds approached saturation in reverse order of their affinities for the GAC.

The column was taken off-line and sampled after 300 days. Samples of the core were extracted with hexane, and the hexane extracts were analyzed

Table II. Average Weekly Organic Influent Levels (μg/L)

Week	*$CHCl_2CH_3$* [a]	*CCl_3CH_3* [b]	*TCE* [b]	*PCE* [b]	*TCP* [a]	*NVTOC* [c]
1	1.0	1.0	4.5	0.6	1.5	600
2	1.0	1.1	5.6	0.7	1.4	640
3	1.0	1.0	5.5	0.6	2.1	560
4	1.2	0.8	8.3	0.6	2.1	540
5	1.0	0.6	9.6	0.5	1.3	570
6	1.6	0.6	11.3	0.5	1.8	660
7	1.8	1.2	13.4	0.8	4.3	690
8	2.0	1.3	14.1	0.8	2.9	610
9	2.4	1.4	15.4	0.7	5.3	660
10	3.5	1.7	12.7	0.3	8.9	500
11	2.3	1.6	12.3	0.6	6.5	660
12	2.4	2.0	18.9	0.9	8.2	530
13	2.1	2.0	17.6	0.7	7.8	600
14	1.7	2.5	19.4	1.2	8.2	760
15	3.3	1.9	17.7	1.1	9.1	800
16	2.3	3.4	22.6	1.2	8.6	700
17	2.3	2.5	19.0	1.6	5.0	670
18	1.3	3.6	15.9	2.2	4.7	—
19	1.9	3.0	13.1	1.9	3.9	700
20	1.8	3.9	14.5	2.0	3.6	1240
21	—	3.5	13.7	1.7	—	—
22	2.0	3.3	12.3	1.7	3.1	620
23	1.7	3.8	13.4	1.9	4.0	1040
24	1.9	3.4	11.2	1.7	4.8	—
25	1.5	2.8	13.9	2.1	4.1	640

[a] Weekly average from two to four purge-and-trap analyses of grab samples.
[b] Weekly average from three hexane extractions (Monday, Wednesday, and Friday) of grab samples.
[c] Nonvolatile TOC (samples purged to remove CO_2).

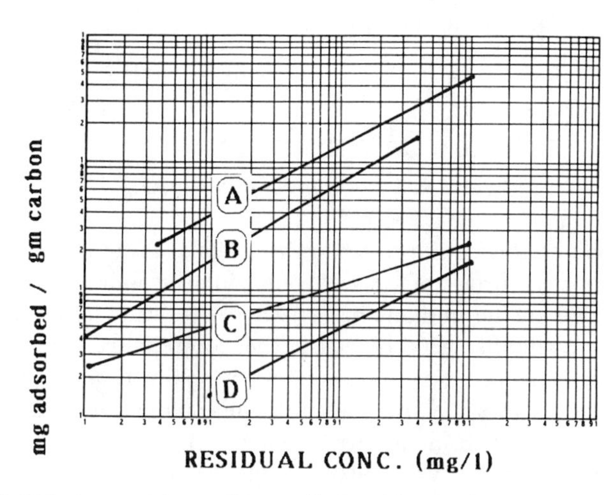

Figure 4. Relative positions of Freundlich isotherms for compounds observed in Upper Merion Reservoir.

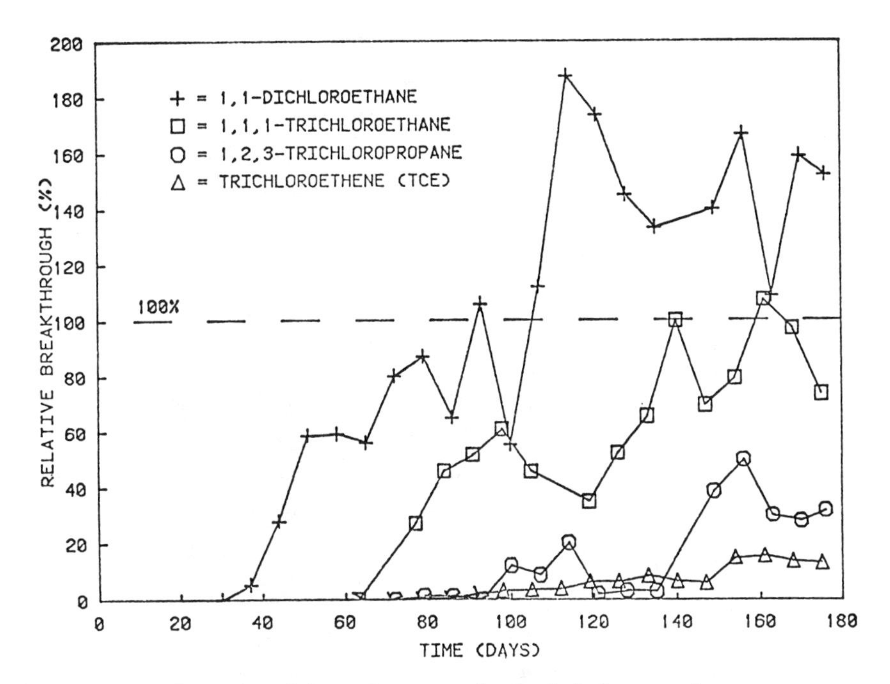

Figure 5. Relative breakthrough curves of volatile halogenated organic compounds in Upper Merion Reservoir. (Reproduced with permission from ref. 21. Copyright 1981 American Water Works Association.)

by gas chromatography (GC). This analysis provided information about the relative distribution of adsorbed solutes in the column. This relative distribution is shown in Figure 6. Compounds with lower affinity have been displaced from the column entrance, and those with greater affinity have not yet saturated lower areas of the column. This pattern of adsorption to saturation followed by competitive displacement is a typical pattern seen in frontal chromatography.

In this example cumulative mass-loading curves (e.g., Figure 7) will be described statistically. The moments zero and one can be used to calculate the number of theoretical plates (n) in the column relative to each compound by applying the relationships of Reilley et al. (*17*) and Yau (*22*) to the mass-loading curves. Figure 8 shows the decrease in n with time for NVTOC and 1,1-dichloroethane. First n decreases drastically between $t = 0$ and 90 days; then it increases slightly up to 110 days, and decreases to nearly zero at 170 days. The increasing period would be made possible by elution of competing species from the column because of a decrease in their influent concentration. The NVTOC showed a relatively stable n after an initial decline at $t = 20$ days. Figure 5 shows the displacement of 1,1-dichloroethane (DCE) by other compounds and an extreme chromatographic effect (effluent exceeding influent concentration) after $t = 110$ days. At $t = 300$ days (Figure 9), virtually

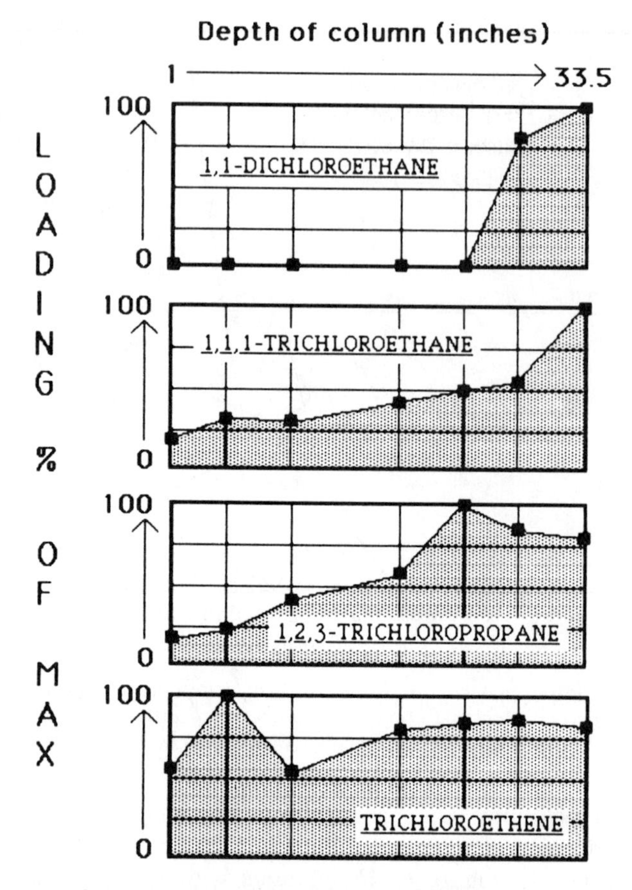

Figure 6. Distribution of compounds in GAC column, Calgon F-400, Upper Merion, 300 days.

all of the DCE had been displaced from the column (i.e., $k' \rightarrow 0$, $n \rightarrow 0$) and the carbon was not capable of retaining DCE. Longer column runs would have been required to observe the same type of displacement in the other micropollutants from effluent evaluation.

NVTOC was adsorbed at a relatively consistent, high rate during the first 180 days; the volatile halogenated compounds were competitively displaced during that time. Also, distributions of adsorbed solutes appear to have moved down the column in a chromatographic manner when the loading profiles at 180 days are compared to those at 300 days (Figure 9). It therefore appears likely that some of the competitive effects were from humic materials. One could also assume from Figures 5 and 6 that the more strongly adsorbed halogenated hydrocarbons are displacing the less strongly adsorbed ones.

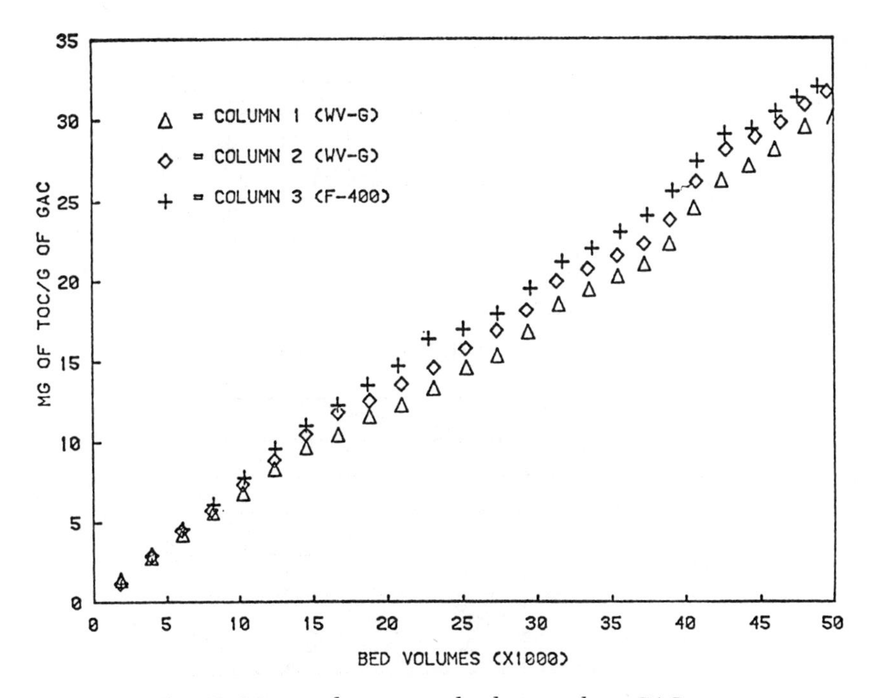

Figure 7. NVTOC cumulative mass loading on three GAC contactors.

Both of these assumptions may be premature, and it can only be concluded that, in general:

1. More strongly adsorbed compounds are competing effectively for sites and displacing less strongly adsorbed compounds.
2. Humic substances may be involved in the competitive displacement process.
3. Irreversible adsorption by humic materials may permanently remove sites from the competitive process.

Proposed Future Work and Summary

A new approach to describing solute movement in GAC contactors has been presented, where concepts of frontal chromatography are applied. The effect of competition between solutes is expressed in terms of decreases in column efficiency (number of theoretical plates).

The micropollutants studied behaved much like solutes in a frontal chromatographic system, although the time frames are much longer (tens

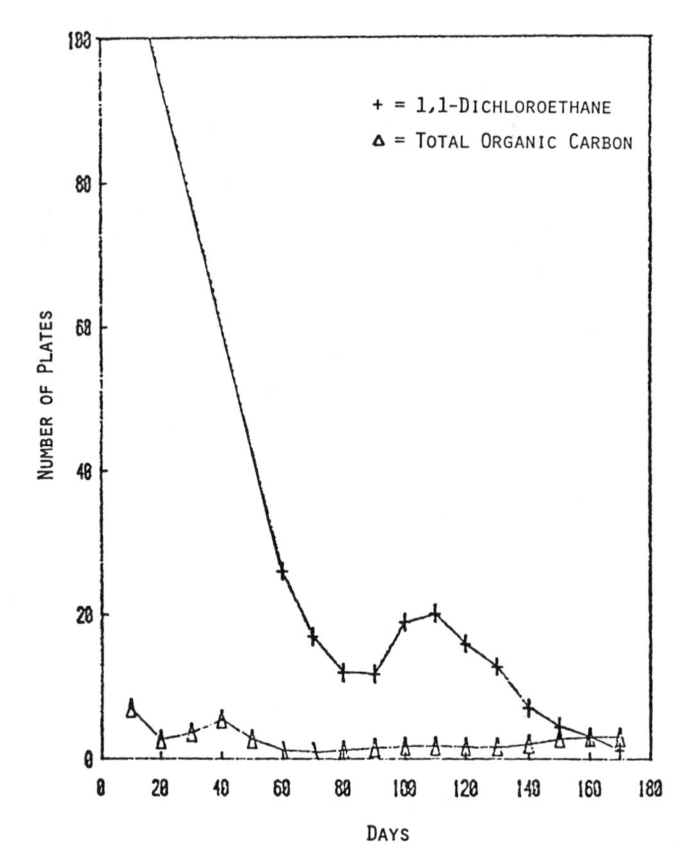

Figure 8. Competition-induced reduction of equilibrium stages over time for 1,1-dichloroethane.

or hundreds of days to breakthrough versus minutes in frontal chromatography). Other frontal chromatographic concepts may be applicable to GAC.

The traditional approach to GAC research is to develop methods to predict breakthrough on the basis of chemical, physical, and engineering parameters. Then the predictive value of the procedure is tested against actual data. The frontal chromatographic concepts developed in this chapter could be applied in a different way. For example, pilot-plant data could be described in chromatographic terms employing a statistical moment or coherence approach; then this chromatographic data base could be used to describe the adsorption and breakthrough characteristics of the source water evaluated on a pilot scale. This information could then be used for scaling up the process to the plant scale. However, complexities from influent variability will make comprehensive modeling in real water systems difficult, a situation already familiar to those working with models based on other adsorption and mass-transfer criteria.

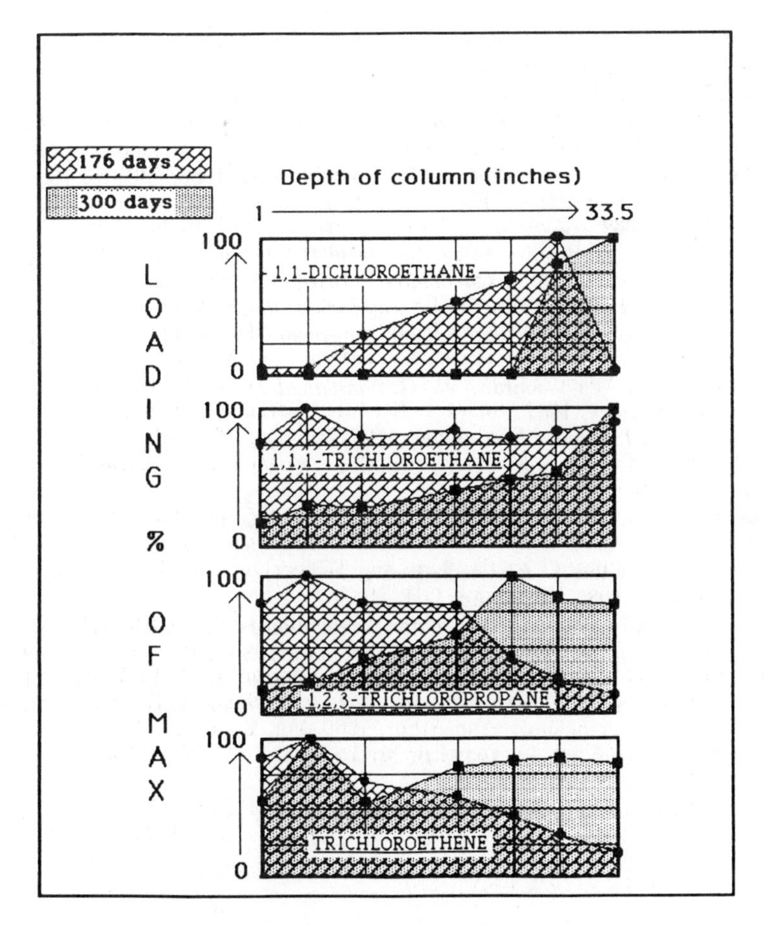

Figure 9. Distribution of compounds in GAC column, Upper Merion, 176 and 300 days.

References

1. American Water Works Association. *Mainstream* **1986,** *September*, pp 5 and 10.
2. Weber, W. J., Jr.; Pirbazari, M. *J. Am. Water Works Assoc.* **1982,** *74*, 203–209.
3. Peel, R. G.; Benedek, A. *J. Environ. Eng. Div.* (*Am. Soc. Civ. Eng.*) **1980,** *106*(EE2), 797–813.
4. Narbaitz, R. M. Ph.D. Thesis, McMaster University, 1985.
5. Crittenden, J. C.; Hand, D. W.; Berrigan, J. K. *J. Water Pollut. Control Fed.* **1986,** *58*, 312–319.
6. *Drinking Water and Health*; National Academy of Science, Safe Drinking Water Committee; National Research Council: Washington, DC, 1980; Vol. 2.
7. Karger, B. L.; Snyder, L. R.; Horvath, C. *An Introduction to Separation Science*; Wiley: New York, 1972; Chapters 1–5 and 13.

8. Weber, W. J., Jr. *Physicochemical Processes for Water Quality Control*; Wiley: New York, 1972; Chapter 5.
9. Keinath, T. M. *Environ. Sci. Technol.* **1985,** *19*, 690–694.
10. Gibs, J.; Suffet, I. H. In *Organic Pollutants in Water: Sampling, Analysis, and Toxicity Testing*; Suffet, I. H.; Malaiyandi, M., Eds.; Advances in Chemistry 214; American Chemical Society: Washington, DC, 1986; Chapter 19.
11. Bird, R. B.; Stewart, W. E.; Lightfoot, E. N. *Transport Phenomena*; Wiley: New York, 1962.
12. Kalinichev, A. I.; Pronin, A. Y.; Zolotarev, P. P.; Goryacheva, N. A.; Chmutov, K. V.; Filimonov, V. Y. *J. Chromatogr.* **1976,** *120*, 249–256.
13. Helfferich, F. In *Adsorption from Aqueous Solution*; Weber, Walter J.; Matijević, E., Eds.; Advances in Chemistry 79; American Chemical Society: Washington, DC, 1968.
14. Snedecor, G. W.; Cochran, W. G. *Statistical Methods*, 6th ed.; Iowa State University: Ames, 1967.
15. Grushka, E. *J. Phys. Chem.* **1972,** *16*, 2586–2593.
16. Kalinichev, A. I.; Pronin. A. Y.; Chmutov, K. V.; Goryacheva, N. A. *J. Chromatogr.* **1978,** *152*, 311–322.
17. Reilley, C. N.; Hildebrand, G. P.; Ashley, J. W., Jr. *Anal. Chem.* **1962,** *34*, 1198–1223.
18. Suffet, I. H.; Gibs, J.; Chrobak, R. S.; Coyle, J. A.; Yohe, T. L. *J. Am. Water Works Assoc.* **1985,** *77*, 65–72.
19. Dobbs, R. A.; Cohen, J. M. *Carbon Adsorption Isotherms for Toxic Organics*; Environmental Protection Agency: Cincinnati, 1980; EPA–600/8–80–023.
20. McGuire, M. J.; Suffet, I. H. *J. Am. Water Works Assoc.* **1977,** *69*, 621–636.
21. Yohe, T. L.; Suffet, I. H.; Cairo, P. P. *J. Am. Water Works Assoc.* **1981,** *73*, 402–410.
22. Yau, W. W. *Anal. Chem.* **1977,** *49*, 395–398.

RECEIVED for review January 13, 1988. ACCEPTED for publication May 27, 1988.

32

Adsorption of Micropollutants on Activated Carbon

Massoud Pirbazari, Varadarajan Ravindran, Sau-Pong Wong, and Mario R. Stevens

Department of Civil Engineering, University of Southern California, Los Angeles, CA 90089–0231

This chapter addresses the effect of humic substances (HS) on the activated-carbon adsorption of several contaminants, including trichloroethylene, chloroform, geosmin, and polychlorinated biphenyls. Complexation potentials of these compounds with HS were investigated to determine the extent of their association. Adsorption experiments performed included equilibrium, minicolumn, and high-pressure minicolumn studies. The ideal adsorbed solution theory (IAST) model was employed to predict adsorption equilibria of mixtures containing each pollutant and HS from their single-solute isotherms. The theoretical predictions obtained from the model were in good agreement with the experimental results. Furthermore, the reduction in adsorption capacity attributed to preloading of carbon with HS was substantial for chloroform, but marginal for polychlorinated biphenyls.

CHARCOAL BEDS WERE REGARDED AS A PANACEA for purification of contaminated ground and surface waters over 2000 years ago. A more sophisticated version of charcoal adsorption that involves the use of activated carbon is widely recognized today as one of the fundamental technologies in water and wastewater treatment. It has proven effective in removing a broad spectrum of potentially hazardous contaminants. The presence of dissolved organic matter (DOM), a major portion of which is dissolved humic substances (DHS), complicates the design and operation of adsorption treatment processes because DOM may interact with activated carbon and organic contaminants. An intensive study of such interactions, their associated mechanisms and transformations, and their effect on adsorption appears to be

0065–2393/89/0219–0549$08.50/0

necessary in order to develop mathematical models for effective appraisal of adsorption systems, from an engineering viewpoint.

The formulation of mathematical models to describe adsorption processes for treating complex organic mixtures is one of the most challenging areas in the research and development of adsorption technology. Predictive mathematical models help in the design of fixed-bed adsorbers by estimating the pattern of adsorbate concentrations as a function of time or volume of influent treated. Models appropriately calibrated for specific influent and operating conditions can be effectively used for optimal design and operation of pilot-scale and eventually full-scale adsorbers. Predictions of such models depend on accurate characterization of the various physical phenomena affecting adsorber performance (*1*). Combined film and particle diffusion models are of interest because they represent the two most common rate-limiting steps in the adsorptive mechanism, namely, film transfer coefficient and intraparticle mass transfer (*2*).

The general approach to adsorber modeling and design involves the solution of differential equations with appropriate boundary conditions representing the mass balance of solid and liquid phases, coupled with the relation that describes the equilibrium distribution of the adsorbate between the two phases, using a suitable numerical technique. Crittenden and Weber (*3, 4*) developed a model of this type that required several input variables, including kinetic parameters that represent film and intraparticle diffusivities, and isotherm constants that describe adsorption-phase equilibria.

Adsorption models have recently attained such high levels of mathematical sophistication that questions often arise about the analytical certainty of experimentally determined model inputs. Nevertheless, such adsorption models show substantial promise for reducing the gap between the design of bench-scale and field-scale adsorbers. They have deservedly become an integral aspect of engineering design.

Adsorption modeling for organic micropollutants in the presence of DHS is complicated by a number of factors. Competition occurs for adsorption sites among the micropollutants and dissolved humic materials (DHM). Furthermore, complexation between the organic pollutants (OP) and DHS adds a new dimension to adsorption modeling, because competition may arise among the OP, DHS, and OP–DHS complexes. All equilibrium models focus on the competitive effect of various adsorbent species, but few models consider the effects of complexation. Study of the theoretical aspects of complexation between OP and DHS is therefore needed to obtain a realistic picture of its contribution to competitive adsorption.

Theoretical Background

Humic Substances and Micropollutants in the Environment. The ubiquitous presence of DHS as the predominant naturally occurring organic species in most surface and ground waters is environmentally significant. The complexation of several organic micropollutants with DHS largely controls their distribution in air, water, biota, suspended solids, and sediments. Complexation can also modify the mechanisms and kinetics of their phase transport and chemical reactions (5). For instance, rates of degradation, volatilization, photolysis, transfer to sediments, and biological uptake may change considerably when a pollutant is bound to DOM. Thus the extent of complexation between the pollutant and DHM would significantly affect the chemical interactions, distribution, and environmental transport of the pollutant. This aspect affects activated-carbon adsorber design, as the degree of complexation controls the equilibrium and kinetics of adsorbate transfer from the liquid phase to the adsorbent phase.

Characteristics of Humic Substances. Aquatic humic substances (HS) constitute about 40–60% of the dissolved organic carbon (DOC). They are nonvolatile and have a molecular weight of 500–5000. Their elemental composition is approximately 50% carbon, 35–40% oxygen, 4–5% hydrogen, 1–2% nitrogen, and less than 1% sulfur and phosphorus. The important functional groups present are carboxylic acids, phenolic hydroxyls, carbonyls, and hydroxyls. Thurman (6) provided a good review of HS characteristics that compared the properties of aquatic and soil HS. Although it has not been proven that both types of HS are the same, they have similar general characteristics (7). They are polyelectrolytic, colored organic acids, with comparable molecular weights and similar elemental compositions and functional groups (6).

HS concentrations vary for different natural waters. Ground-water concentrations are generally low, 0.1–1.0 mg of C/L. In streams, rivers, and lakes the concentrations are much higher, generally 0.6–4.0 mg of C/L. Fulvic acid constitutes about 85–90% and humic acid 10–15% of nearly all surface and ground waters (6). Besides the carboxyl, hydroxyl, and carbonyl groups, humic substances contain lesser amounts of phenolic hydroxyl and trace amounts of carbohydrates and amino acids. In general, humic substances from rivers and streams originate from soil and plant matter; those in lakes are of both terrestrial and aquatic origins (8).

Adsorption of Pollutants in the Presence of Humic Substances. The adsorption system (target organic compound, DOM, and activated carbon) is rather complex because the contaminant can exist in one or more of at least four states (2): free form in aqueous solution, adsorbed state on activated carbon, in a complex with DHM, or associated to a molecule of DHM adsorbed on carbon. Simultaneous existence of all four forms is highly possible, with the distribution of various species governed by mutual equilibria, as well as by the kinetics and degree of reversibility of the reactions.

Hence, in the present context, the extent of complexation and the factors indicating it are quite important.

Factors Indicating the Degree of Complexation. Organic compounds have several properties that determine their degree of complexation with DHS. The most important among these properties are the octanol–water partition coefficient (K_{ow}) and aqueous solubility (S). Both properties, to a certain extent, reflect the compound's polarity, hydrophobic nature, and affinity for association with DHM (*9*). In general, higher K_{ow} or lower S signifies lower polarity, higher hydrophobicity, and therefore a larger potential for forming complexes with DHM. On the basis of this principle, Carter and Suffet (*5*) stated that compounds with $\log_{10} K_{ow}$ values of 4.0 and above could be classified as hydrophobic, and therefore should form a strong complex with DHM. However, those compounds with values less than 4.0 could be considered less hydrophobic and expected to form weak complexes or no complexes at all.

Table I lists the $\log_{10} K_{ow}$ and S values for several hydrophobic organic compounds ($\log_{10} K_{ow}$ in the range of 4.0–8.0) classified as priority pollutants by the U.S. Environmental Protection Agency (*10*). A statistically significant linear relationship exists between the $\log_{10} K_{ow}$ and $\log_{10} S$ values listed in Table I ($\log_{10} K_{ow} = -0.542 \log_{10} S + 8.163$; correlation coefficient = –0.80).

In the present scenario, the relationship between K_{ow} and the extent of complexation represented by the binding or association constant, K_c, is significant. K_c is defined as the ratio of the mass of bound compound per unit mass of total dissolved humic acid (DHA) to the concentration of unbound compound in solution. Table II compares the $\log_{10} K_{ow}$ and $\log_{10} K_c$ values for a number of hydrophobic compounds and also shows the range of degree of complexation for a DHA concentration range of 1–10 mg/L. A statistically significant linear relation can be observed between the two parameters listed in Table II ($\log_{10} K_c = 0.747 \log_{10} K_{ow} + 0.661$; correlation coefficient = 0.86).

This functional dependence can facilitate an a priori estimation of the extent of complexation. Nevertheless, although the correlations between $\log_{10} K_{ow}$ and $\log_{10} S$, and between $\log_{10} K_{ow}$ and $\log_{10} K_c$ may be statistically significant, they are not necessarily practically meaningful. This and other factors that control complexation make such estimation more complicated.

Factors Determining the Degree of Complexation. The extent of binding of organic pollutants to DHM is determined by a number of factors, including the source, characteristics, molecular sizes, and concentrations of the DHM, as well as the nature and concentration of the pollutant (*11*), pH, metallic ion concentration, and ionic strength (*9*). Carter and Suffet (*5*) reported that DHA and sewage-effluent DOM tend to bind DDT most strongly, and that dissolved fulvic acids (DFA) showed considerably less

Table I. Octanol–Water Partition Coefficients, K_{ow}, and Aqueous Solubilities, S, for Several Hydrophobic Organic Compounds

No.	*Compound*	log_{10} K_{ow}	S × 10^{-6} (*ng/L*)	log_{10} S
1	Hexachlorocyclopentadiene	3.99	1.80	6.26
2	4-Chlorophenyl phenyl ether	4.08	3.30	6.52
3	1,2,4-Trichlorobenzene	4.26	30.0	7.48
4	Hexachlorobenzene	6.18	0.006	3.78
5	Pentachlorophenol	5.01	14.0	7.15
6	Di-*n*-butyl phthalate (DBP)	5.20	13.0	7.11
7	Butylbenzyl phthalate	5.80	2.90	6.46
8	Acenaphthene	4.33	3.42	6.53
9	Acenaphthylene	4.07	3.93	6.59
10	Fluorene	4.18	1.98	6.30
11	Anthracene	4.45	0.073	4.86
12	Fluoranthene	5.33	0.26	5.42
13	Phenanthrene	4.46	1.29	6.11
14	Benz[*a*]anthracene	5.61	0.009	3.95
15	Chrysene	5.61	0.002	3.30
16	Pyrene	5.32	0.14	5.15
17	Benzo[*ghi*]perylene	7.23	0.00026	2.42
18	Benzo[*a*]pyrene	7.23	0.0038	3.58
19	Dibenz[*a,h*]anthracene	5.97	0.0005	2.70
20	DDD (*p,p′* isomer)	5.99	0.09	4.95
21	DDD (*o,p′* isomer)	6.08	0.1	5.0
22	DDE (*p,p′* isomer)	5.69	0.12	5.08
23	DDE (*o,p′* isomer)	5.78	0.14	5.15
24	DDT (*p,p′* isomer)	6.19	0.025	4.40
25	Endrin	5.60	0.25	5.40
26	δ-Hexachlorocyclohexane	4.14	21.3	7.33
27	Lindane (γ-hexachlorocyclohexane)	3.72	6.60	6.82
28	PCB (Aroclor 1254)	6.03	0.024	4.38
29	2-Chloronaphthalene	4.12	6.74	6.83

NOTE: DDD, dichlorodiphenyldichloroethane; DDE, dichlorodiphenyldichloroethylene; and DDT, dichlorodiphenyltrichloroethane.
SOURCE: Data are from ref. 10.

Table II. Binding Constants, K_c, Octanol–Water Partition Coefficients, K_{ow}, and Binding Potential of Several Organic Compounds

			Percent Binding[a]	
Compound	log_{10} K_{ow}	log_{10} K_c[b]	*DOC = 1 mg/L*	*DOC = 10 mg/L*
Lindane	3.72[c]	3.04[d]	0.11	1.09
DBP	4.91[e]	3.70[e]	0.50	4.76
Anthracene	4.45[e]	4.46[e]	2.82	22.48
PCB (Aroclor 1254)	6.05[c]	5.07[f]	10.45	53.85
DDT (*p,p′* isomer)	6.19[e]	5.45[e]	21.88	73.68

NOTE: DBP, di-*n*-butyl phthalate; and DDT, dichlorodiphenyltrichloroethane.
[a]Percent binding, $(10^{-6} \times K_c \times DOC)(10^{-6} \times K_c \times DOC + 1)^{-1} \times 100$.
[b]K_c is expressed in milliliters per gram.
[c]Data from ref. 10.
[d]Data from ref. 9.
[e]Data from ref. 11.
[f]Values determined in this study.

potential for complexation. These same investigators made similar observations for anthracene and di(2-ethylhexyl) phthalate (DEHP). There have been a variety of unsuccessful attempts to investigate the characteristics of DHM that determine the differences in the degree of binding, including measurements of carbon, ash, and iron content; spectroscopic techniques; pyrolysis gas chromatography; and molecular size estimates. The degree of complexation is highly system-specific and, with the limited data available, cannot be theoretically predicted for any type of DHM and OP. Actual experimental determinations of binding constants are therefore necessary (5). DHS chemical structures and the nature of binding between them and OP, quite important from the standpoint of understanding the various factors affecting complexation, will be discussed later.

The effects of pH, metallic ion concentration (notably Ca^{2+}), ionic strength, and DHA concentration on the complexation of DDT (1,1′-(2,2,2-trichloroethylidene)bis(4-chlorobenzene)) with DHA were reported by Carter and Suffet (9). Their observations were quite consistent with the previously established theories on the aqueous behavior of DHM. A number of researchers (*12, 13*) have observed that the size of the humic polymer is enhanced by an increase in hydrogen-ion or metallic-ion concentration. Thurman and Malcolm (7) observed that a decrease in pH or an increase in ionic strength causes a coiling effect in humic molecules, making them more prone to sorption on a number of surfaces. The charge on the humic polymer decreases when the hydrogen-ion or metal-ion concentration increases as a result of charge neutralization (*14, 15*). Similarly, as calcium concentration increases, the humic polymer becomes less hydrophilic and tends to bind more easily with hydrophobic organic compounds.

The effect of increased ionic strength on binding may be attributed to the "salting out" of the hydrophobic organic compound. The chemical potential of a nonionic compound dissolved in water is enhanced by increased ionic strength, and the compound tends to reduce its chemical potential by binding itself to the humic polymer (9). The effect of DHA concentration on the binding of organic substances has not been supported by any theory, although the binding constant, K_c, decreases with increasing DHA concentration (9). In an analogous situation, O'Connor and Connolly (*15*) observed that the sorption coefficient of organic compounds to sediments containing DOM decreased as the total amount of sediment used was increased.

Mechanisms of Complexation. A number of mechanisms contribute to the bonding of organic compounds with HS. Such mechanisms include ion exchange and protonation, hydrogen bonding, van der Waals forces, coordination bonding through attached metal ion (ligand exchange), and hydrophobic bonding.

Association of organic molecules to HS by ion exchange and protonation is largely restricted to compounds that either exist in cationic form or become positively charged through protonation (*16*). In the case of certain anionic organic substances, there is a possibility of anion exchange with positively charged areas on the humic molecules. For example, less basic organic substances such as *s*-triazines may become cationic through protonation and bind with HS molecules. The mechanism, as explained by Weber and coworkers (*17*), depends on the acidic or basic strength of the compound (indicated by its pK_a value) and on the proton-supplying power of the HS. Anionic organic substances such as phenoxyalkanoic acids (PAA) may repulse the predominantly negatively charged humic molecules.

Hydrogen bonding is an important mechanism for polar organic compounds such as phenyl carbamates and substituted ureas (*16*), but it is limited to acid conditions where –COOH groups are un-ionized. In the case of PAA such as 2,4-dichlorophenoxyacetic acid (2,4-D), van der Waals forces may contribute significantly to association with HS.

Hydrophobic bonding may be important in the case of highly nonpolar compounds with low aqueous solubilities. The degree of binding depends primarily on the compound's affinity for HS and its hydrophobic nature. Weber and Weed (*18*) observed that humic molecules, by virtue of their aromatic structural framework and nonpolar functional groups, may contain both hydrophilic and hydrophobic parts. Highly nonpolar organic compounds can be squeezed out of aqueous solutions because of their water-repellent characteristics, and they can be adsorbed easily onto the hydrophobic sites of humic molecules (*19*).

Mathematical Modeling

Adsorption Modeling for Micropollutants. There are several approaches for modeling adsorption data of organic compounds in the presence of DHS. In the first approach the adsorption data are modeled in terms of the target organic compounds alone. Compounds such as dissolved humic substances are treated not as components, but as system-specific background materials. The equilibrium parameters of the model for the specific target compounds are experimentally determined in the presence of the background materials (*20, 21*).

A number of researchers have adopted a different theoretical approach that hinges on the mathematical treatment of all background organic substances as a single component. This background is represented by a lumped parameter such as total organic carbon (TOC) or DOC (*1, 22, 23*). Lumped-parameter characterization of background materials (and subsequent application of multicomponent models for prediction of adsorption equilibria) is particularly advantageous from the viewpoint of water and wastewater applications because identification and quantification of all aquatic species present can be a formidable task.

The third approach, which is a refinement of the second, includes the lumped-parameter characterization of the background substances. These substances are broken up into a number of hypothetical pseudospecies, whose individual adsorption equilibrium parameters are determined from lumped-parameter data, suitable mathematical curve-fitting techniques on ideal adsorbed solution theory (IAST) (*22*), or simplified competitive adsorption model (SCAM) (*24*). These hypothesized parameters of pseudospecies can be used in the IAST or SCAM models for the approximate prediction of adsorption equilibria of specific compounds in the presence of background substances. Fettig and Sontheimer (*25*) employed this technique for predicting DFA adsorption isotherms. It can potentially be used for certain compounds in the presence of different types of DHS.

A further extension of this idea has generated a new method in which several pseudocomponents are postulated for the unknown background substance. The isotherm parameters of these components are determined by using a weakly adsorbable tracer compound and observing the displacement of the multisolute isotherm from the single-solute isotherm (*26*, *27*). An offshoot of this procedure employs the concept of species grouping. Its reduced number of pseudospecies in the multicomponent system, characterized by average value of parameters, results in a considerable simplification of multicomponent equilibrium computations (*28*).

Equilibrium Modeling for the Adsorption of Micropollutants. The general approach we employed for the prediction of adsorption equilibria of various organic compounds in the presence of DHS is a straightforward application of the IAST model. Crittenden and coworkers (*26*, *27*) used this theory, developed by Radke and Prausnitz (*29*), to predict the multicomponent competitive interactions between several volatile organic compounds (VOC) from their single-solute isotherm parameters. The model assumes that the adsorbed phase forms an ideal solution and that there is no area change of the various components upon mixing at the spreading pressure of the mixture. The IAST model is based on the concept of a Gibbs dividing surface and an inert adsorbent. A two-dimensional planar surface is postulated at the interface between the adsorbent phase and liquid phase. The adsorption of the solute onto the surface is considered analogous to the accumulation of excess concentration at the dividing surface. The thermodynamic framework for the concept, including the application of the Gibbs–Duhem equation and the excess internal energies and entropies, is discussed in the next section.

Thermodynamic Aspects of IAST Model. The thermodynamic framework for the development of the IAST model is based on Gibbs free energy considerations for the interfacial surface, referred to as the Gibbs dividing surface, between a solid adsorbent phase and a liquid solution phase.

The free energy change of the dividing surface, and its relationship with the molar composition of the components and the surface tension of the solution, are explained in detail by Atkins (*30*).

The excess surface free energy, G_s, at equilibrium is given by

$$G_s = \sigma\gamma + \sum_{j=1}^{N} \mu_j n_{j,s} \tag{1}$$

where σ is the area of the dividing surface; γ is the surface tension of the liquid; μ_j is the chemical potential of the component *j*; $n_{j,s}$ is the number of moles of components *j* in the surface; and *N* is the number of components in solution.

The relationship between the change in surface free energy corresponding to change in surface tension is established by the Gibbs–Duhem equation

$$dG_s = \sigma\, d\gamma + \gamma\, d\sigma + \sum_{j=1}^{N} n_{j,s}\, d\mu_j + \sum_{j=1}^{N} \mu_j\, dn_{j,s} \tag{2}$$

The change in the Gibbs free energy of the surface at constant *T*, *P*, σ, and n_j is as follows:

$$dG_s = -S_s\, dT + \sigma\, d\gamma + \sum_{j=1}^{N} \mu_j\, dn_{j,s} \tag{3}$$

where *T*, *P*, and S_s represent the temperature, pressure, and entropy of the dividing surface, respectively. Comparison of equations 2 and 3 under isothermal conditions ($dT = 0$), yields the result

$$\sigma\, d\gamma + \sum_{j=1}^{N} n_{j,s} d\mu_j = 0 \tag{4}$$

or

$$d\gamma = -\sum_{j=1}^{N} \Gamma_j\, d\mu_j \tag{5}$$

where $\Gamma_j = n_{j,s}/\sigma$ denotes the excess amount of component *j* present per unit surface area. The chemical potential of any component *j* in solution phase is given by the relation

$$\mu_j = \mu_j^o + RT \ln a_j \tag{6}$$

where μ_j^o is the chemical potential under standard conditions of temperature and pressure and a_j represents the activity of component j in solution. For an ideal solution dilute in component j, the activity term can be replaced by the mole fraction in bulk aqueous solution, x_j. Thus equation 5 can be rewritten as follows:

$$d\gamma = -\sum_{j=1}^{N} \Gamma_j RT \, d \ln x_j \tag{7}$$

Because $d \ln x_j$ can be approximated by dC_j/C_j, where C_j denotes the molar concentration of j, the spreading pressure Π_j that corresponds to each component can be computed by integrating the modified form of equation 7, as follows:

$$\Pi_j = \gamma_o - \gamma = -\int_0^{x_j} \Gamma_j RTC_j^{-1} \, dC_j \qquad \text{for } j = 1 \text{ to } N \tag{8}$$

From the basic definition, $\Gamma_j = Q_j/A$, where Q_j is the loading of component j on the adsorbent surface and A is the area per unit mass of the adsorbent. Because the chemical potential of j should be equal in the solution and adsorbent phases, equation 8 can be rewritten as

$$\Pi = \int_0^{C_j} RTQ_jC_j^{-1}A^{-1} \, dC_j \qquad \text{for } j = 1 \text{ to } N \tag{9}$$

where it is assumed that the spreading pressure of each component in the mixture is the same. The formulation of the final IAST equations was explained by Crittenden et al. (*26*) and Radke and Prausnitz (*29*).

The IAST model requires the single-solute isotherm parameters for the various components or hypothetical pseudospecies in solution. A good empirical description of single-solute equilibrium is provided by the Freundlich model, according to which:

$$q_e = K_f C_e^{\frac{1}{n}} \tag{10}$$

where q_e and C_e represent the equilibrium solid-phase and liquid-phase solute concentrations, respectively, and K_f and $1/n$ are characteristic isotherm parameters obtained from a logarithmic linearization of the equilibrium data. When the Freundlich isotherm is used to represent the single-solute behavior, the IAST model yields the following simplified system of equations for the prediction of completely mixed batch reactor (CMBR) isotherms for each adsorbate after eliminating their liquid-phase equilibrium concentrations:

$$C_{m,o} - \frac{M}{V} Q_m - \frac{Q_m}{\sum_{j=1}^{N} Q_j} \left[\frac{\sum_{j=1}^{N} n_j Q_j}{n_m K_m} \right]^{n_m} = 0 \quad \text{for } m = 1 \text{ to } N \tag{11}$$

Here Q_m is the solid-phase concentration (μM/g); $C_{m,o}$ is the initial concentration of adsorbate m; K_m and n_m are single-solute Freundlich isotherm constants for adsorbate m; Q_j denotes the solid-phase concentration for solute j; and M and V represent the mass of carbon and the volume of solution used, respectively. In this system of simultaneous nonlinear equations with N unknowns, Q_m can be solved to estimate the final equilibrium state. Multicomponent equilibria of up to six adsorbates have been successfully predicted by the Newton–Raphson numerical technique (*26*, *27*).

The specific application of the IAST model to predict the adsorption equilibria of OP in the presence of DHM is accomplished by tailoring equation 10 to a bisolute system as shown:

$$C_{1,o} - \frac{M}{V} Q_1 - \frac{Q_1}{Q_1 + Q_2} \left[\frac{n_1 Q_1 + n_2 Q_2}{n_1 K_1} \right]^{n_1} = 0 \tag{12}$$

$$C_{2,o} - \frac{M}{V} Q_2 - \frac{Q_2}{Q_1 + Q_2} \left[\frac{n_1 Q_1 + n_2 Q_2}{n_2 K_2} \right]^{n_2} = 0 \tag{13}$$

where subscripts 1 and 2 represent the OP and DHM, respectively.

The IAST model has a few limitations in its applicability. It cannot be used in the case of hydrophobic pollutants that form strong complexes with DHM, and it may be applied only in situations where the organic substances are either weakly bound or not bound at all. The term "ideal adsorption" implies that the thermodynamics of ideal solutions are obeyed (*29*). Complexation between any of the adsorbates in solution is tantamount to a chemical reaction, with free energy changes not accounted for in the thermodynamic formulation of the model.

The applicability of the model is further restricted to situations where the adsorbent dosage effect is not significant (*31*). When the dosage effect is pronounced, the IAST model cannot estimate adsorption isotherm parameters for adsorption column predictions and design. At high carbon dosages and low surface loadings, the spreading pressures of the mixture components are evaluated by using extrapolated versions of single-solute isotherms, at concentrations well below those corresponding to single-solute data. A similar argument applies to low dosages and high surface loadings at concentrations exceeding the upper limits of single-solute data.

Experimental Materials

Activated Carbon. Commercial Filtrasorb-400 (F-400) granular activated carbon (GAC) (Calgon, Pittsburgh, PA) was used in all experiments. Each batch was sieved; the fraction passing a 100-mesh U.S. standard sieve and retained on a 200-mesh sieve was isolated for this study. The selected fraction was repeatedly washed with organic-substance-free water to remove leachable impurities and fines. The carbon was then dried to constant weight in an oven at 120 °C, immediately desiccated to ambient temperature, and stored in airtight glass bottles.

Solutes. Reagent-grade chloroform and trichloroethylene (TCE) were purchased from J.T. Baker Chemical Co. (Phillipsburg, NJ). The polychlorinated biphenyl (PCB) congener selected for this study was 2,2′,4,5,5′-pentachlorobiphenyl (Supelco, Bellefonte, PA). This congener represents 48% of the polychlorinated compounds composing the commercial product Aroclor 1254 (*32*); according to another group of investigators (*33*), it represents 59.9% of Aroclor 1254. The standards for *trans*-1,10-dimethyl-*trans*-9-decalol (geosmin) were provided by the Water Supply Research Division of the U.S. Environmental Protection Agency (Cincinnati, OH).

Background Solutions. Organic-substance-free water (OFW) was prepared by passing deionized–distilled water through a F-400 GAC column with an empty-bed contact time (EBCT) of about 15 min. The different background solutions used in the study were OFW and solutions of commercial DHA and DFA. In this context it is important to discuss the applicability of commercial HS to simulate DOM in an aquatic environment. Malcolm and MacCarthy (*34*) pointed out that commercial HS may not be representative of those species occurring in water and soil. Their investigations revealed that cross-polarization and magic-angle spinning ^{13}C NMR spectroscopy demonstrate pronounced differences between commercial and naturally occurring HS. Elemental analysis and IR spectroscopy do not show such clear-cut differences.

Nevertheless, commercial humic substances are often well characterized and, despite their limitations in water research, prove useful in explaining some of their fundamental characteristics in aquatic systems. Furthermore, the process of extracting, isolating, and purifying HS from waters, soils, and sediments is so tedious (taking several weeks or a few months) that it would defeat the primary objective of studying their behavior and characteristics in the aquatic environment.

The DHA background solution was prepared to the desired concentration by dissolving commercial humic acid (Aldrich Chemical Co., Milwaukee, WI) in OFW. DFA background solution was prepared by dissolving the material extracted from Prince Edward Island Podzol (Concordia University, Ottawa, Canada). The DHA and DFA stock solutions prepared were refrigerated in dark glass bottles.

Experimental Methodologies

Equilibrium Studies. Adsorption isotherm experiments were performed with the CMBR technique. A series of 100–1000-mL aliquots (depending upon the requirement) of adsorbate solutions of known concentrations were added to CMBR glass bottles containing accurately weighed quantities of carbon. The bottles were pre-equilibrated with adsorbate solution in the case of PCB to minimize glass-adsorption losses, as described by Pirbazari and Weber (*35*). Mass-balance calculations for all compounds subsequently revealed that there were no losses of adsorbate due to glass adsorption. Additional control bottles containing the same adsorbate concentrations were prepared with no carbon.

The bottles were then sealed airtight and agitated in specially designed rotary tumblers for a few days at a constant temperature of 20 °C. At the end of the equilibration period the bottles were removed from the tumbler and allowed to remain quiescent for about 12 h. An aliquot was removed from each bottle and centrifuged to eliminate carbon fines. The sample was then analyzed for the specific adsorbate according to appropriate analytical procedures described later. Upon measurement of the equilibrium concentration, C_e, the corresponding equilibrium solid-phase concentration, q_e, was calculated by the relation:

$$q_e = (C_o - C_e)\frac{V}{M} \tag{14}$$

where C_o denotes the initial concentration of the solution, V is the volume of the solution, and M is the mass of activated carbon in each bottle.

High-Pressure Minicolumn Studies. The high-pressure minicolumn (HPMC) technique, which is based on the principles of high-pressure liquid chromatography, was adopted to study the dynamic behavior of PCB adsorption with differing background solutions. The details of the HPMC apparatus are provided elsewhere (*36*, *37*). A concentrated PCB solution was prepared in acetone carrier–solvent. Appropriate amounts of this concentrate were spiked into a reservoir containing OFW and uniformly mixed with a magnetic stirrer. A high-pressure pump was used to pump the solution through a stainless steel column (80-mm × 2-mm-i.d., packed with 20 mg of 100–200-mesh carbon). A flow rate of 2.5–3 mL/min was maintained throughout the experimental run. Composite effluent samples were periodically collected and analyzed to obtain the breakthrough profile. The influent samples were analyzed daily, monitored, and kept under constant surveillance to maintain constant concentrations.

Complexation Studies. Complexation experiments were performed with the dialysis technique to determine the degree of binding between several organic pollutants and DHA. Cellulose dialysis tubes were cut into 13-cm pieces, washed thoroughly with OFW, and soaked in OFW for 24 h prior to use, to remove the sodium azide preservative. One end of the tube was then tied with polyester thread that had been washed with acetone and rinsed with OFW. Ten milliliters of DHA (as 5 mg of TOC/L) was transferred to the dialysis tube. The other end was tied to form a bag. The bag was introduced into a 250-mL glass bottle containing about 200 mL of OFW and agitated for about 24 h at 20 °C. A number of similar bags containing DHA were prepared in the same manner.

Each bag was then taken out and placed inside a 250-mL bottle containing 200 mL of OFW. The bottles were spiked with different concentrations of organic pollutants and agitated in a rotary tumbler for 5–7 days at 20 °C. Solution samples withdrawn from inside and outside the dialysis bags were analyzed for that compound.

Analytical Techniques. TCE AND CHLOROFORM. Extracts in isooctane (2,2,4-trimethylpentane) were prepared and analyzed for TCE or chloroform with a gas chromatograph (GC) (Hewlett Packard 5790A Series) equipped with an electron-capture detector (ECD), interfaced with a recorder–integrator (Hewlett Packard 3390A). The GC was fitted with a 2.5-mm × 2-mm-i.d. glass column, packed with polyethylene glycol (0.2% Carbowax 1500, 60–70 Carbopack C, Supelco, Bellefonte, PA), and operated with injector, oven, and detector temperatures of 150, 120, and 250 °C, respectively. The carrier gas employed was a mixture of 95% argon and 5% methane, maintained at a 40-mL/min flow rate.

GEOSMIN. The determination was performed with closed-loop stripping analysis (CLSA) followed by gas chromatography (38). The CLSA technique is essentially a purge-and-trap method, where the headspace of the stripping bottle constitutes the purging gas and a 1.5-mg activated-carbon filter serves as the trap. The ionic strength of the water samples containing geosmin was enhanced by the addition of sodium sulfate prior to stripping. The purged gas was circulated for 1.5 h, bubbling through the solution and passing through the filter. The organic species collected in the filter were then extracted with 25 μL of carbon disulfide. Small aliquots (1–5 μL) of the extracts were injected into a GC (Fractovap Series 4160, Carlo Erba, Milan, Italy) equipped with a flame-ionization detector. The GC employed a 25-m × 0.32-mm-i.d. fused-silica column coated with siloxane polymer (OV–1701, Scientific Glass Engineering, Austin, TX). A 2-m retention gap was prepared on the inlet side of the column and deactivated with Carbowax 20–M. The carrier gas used was hydrogen, with a 42-cm/s linear flow velocity. The injector was maintained cold, and the detector temperature was set at 200 °C. The column temperature was held at 40 °C for 1 min and then programmed to increase at 2.1 °C/min to a final temperature of 130 °C. The internal standard employed for quantification was 1-chlorodecane; 1-chlorooctane and 1-chlorododecane were used for obtaining the relative retention times.

PCB CONGENER. Hexane extracts were prepared according to the method described by Pirbazari and Weber (35) and analyzed for the PCB congener with a GC (Perkin-Elmer 3920) equipped with a ^{63}Ni ECD, interfaced with an integrator. The GC was fitted with a 1.8-m × 2-mm-i.d. glass column and packed with 80–100-mesh methyl silicon fluid (SP-2100, Supelco, Bellefonte, PA). The operating temperatures of the injector, column, and detector were 210, 200, and 300 °C, respectively. The carrier gas employed was 95% argon and 5% methane, maintained at a flow rate of 20 mL/min.

DHA AND DFA. The DHA or DFA analysis was performed by UV spectrophotometry, at a 254-nm wavelength.

Results and Discussion

Carbon adsorption studies were conducted with several organic compounds of varying degrees of hydrophobicity to observe the effect of competition with respect to background DHS. The different classes of organic substances studied were volatile organic compounds (VOC) such as TCE and chloroform, which manifest low hydrophobicity and no tendency to form complexes with DHS; semivolatile compounds that do not exhibit any potential for complexation with DHS, represented by geosmin; and nonpolar hydrophobic organic substances such as PCB, which exhibit high hydrophobicity and ability to bind to DHS. The investigations included equilbrium studies to determine the adsorption capacities of GAC for different compounds. Limited HPMC studies were performed to investigate the dynamic behavior of PCB adsorption systems.

Compounds Not Complexing with Dissolved Humic Substances. Complexation studies revealed no binding between HS and TCE, chloroform, or geosmin. Figures 1 and 2 illustrate the effect of background DHA on the adsorption capacities of TCE and chloroform, respectively, and indicate that the presence of background DHA substantially reduces the adsorption capacity for chloroform, while the reduction for TCE is marginal. The Freundlich capacity constant, K_f, for TCE is reduced from 140 to 96.7. The intensity constant, $1/n$, is decreased from 0.36 to 0.31, as can be seen in Figure 1. The corresponding reduction in K_f for chloroform attributable to introduction of DHA is from 22 to 18.9, while $1/n$ undergoes a change from 0.39 to 0.46. As evident from Figures 1 and 2, the predictions of multicomponent equilibria using the IAST model are in good agreement with experimental data (within 5–10%).

Similar adsorption data and IAST predictions for geosmin are presented in Figure 3. The effect of competition is more pronounced in the case of geosmin than in either TCE or chloroform. Systematic deviations of experimental data from IAST model predictions (10–38%) can be observed, perhaps partly because of analytical imprecision inherent with adsorption studies. The K_f value is reduced from 7500 to 6557, and $1/n$ is altered from 0.72 to 0.47. The deviations of experimental data from predicted equilibria are significantly higher in the case of geosmin than for either TCE or chloroform. The ratio of solid-phase loading to liquid-phase concentration at equilibrium is higher for geosmin. This ratio indicates its greater adsorbability, which leads to higher deviations, in accord with the observations of Radke and Prausnitz (29).

Compounds Complexing with Dissolved Humic Substances. Our experimental investigations demonstrated the high potential of PCB mixtures to bind with DHS (39). Figure 4 presents adsorption equilibrium data for PCB congener in the presence and absence of DFA. The solid lines are the statistically best fit representation of data by the Freundlich isotherm

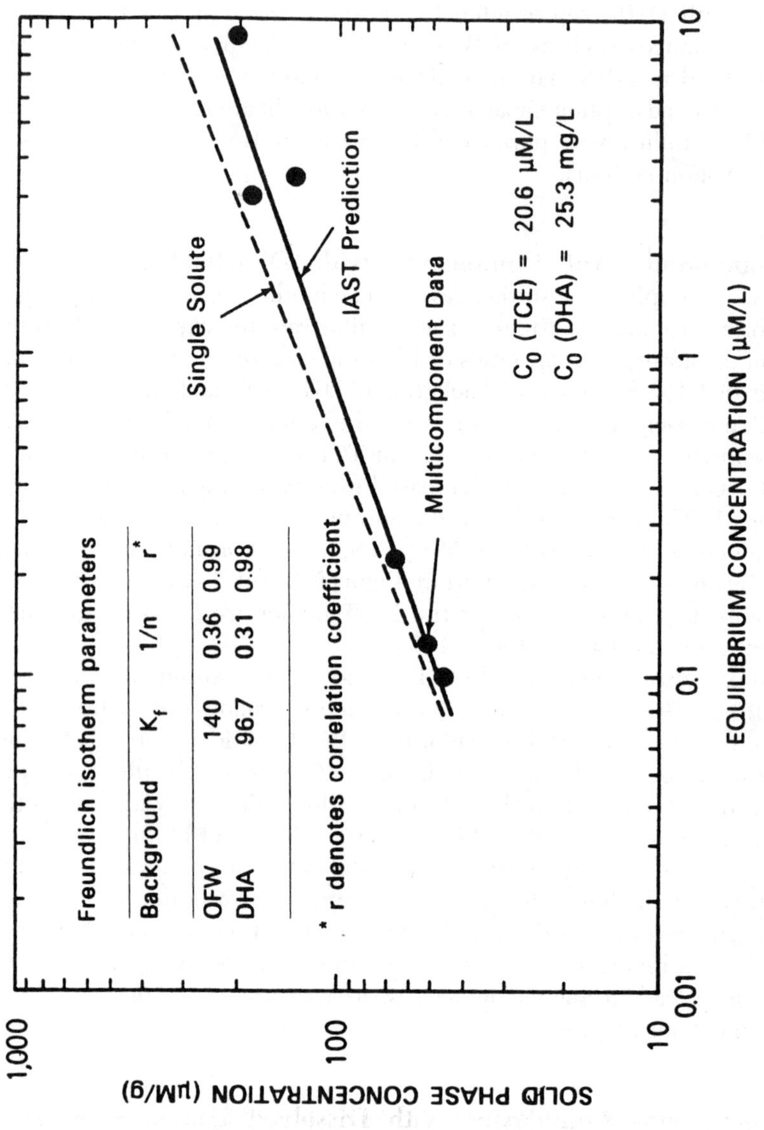

Figure 1. Adsorption isotherm for TCE in DHA background solution; single-solute isotherm and IAST predictions.

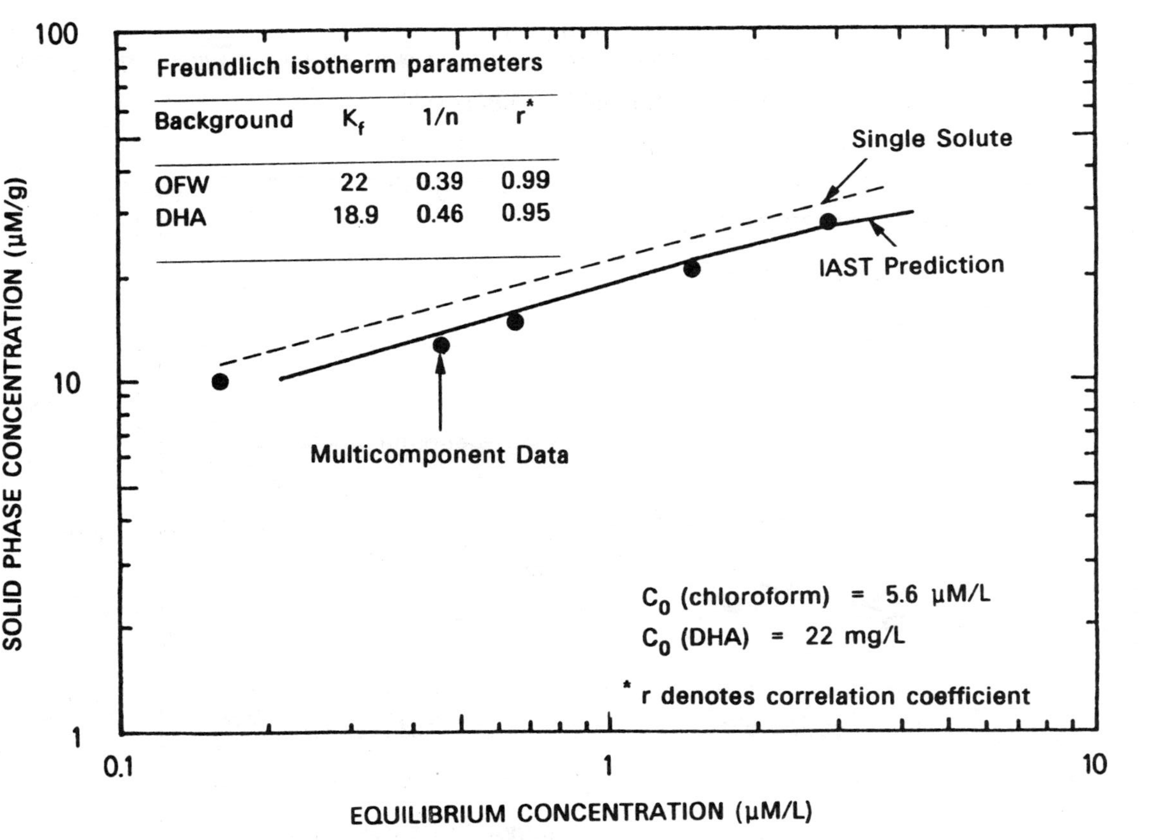

Figure 2. Adsorption isotherm for chloroform in DHA background solution; single-solute isotherm and IAST predictions.

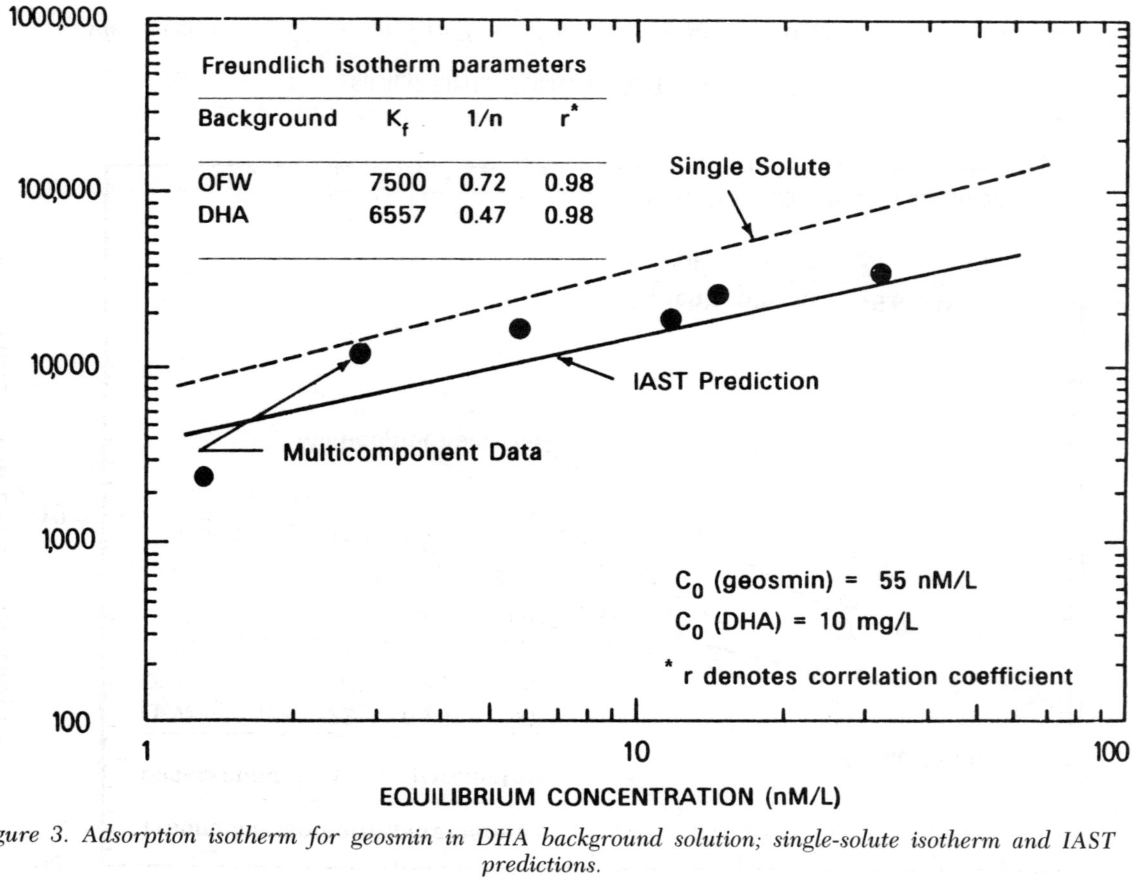

Figure 3. Adsorption isotherm for geosmin in DHA background solution; single-solute isotherm and IAST predictions.

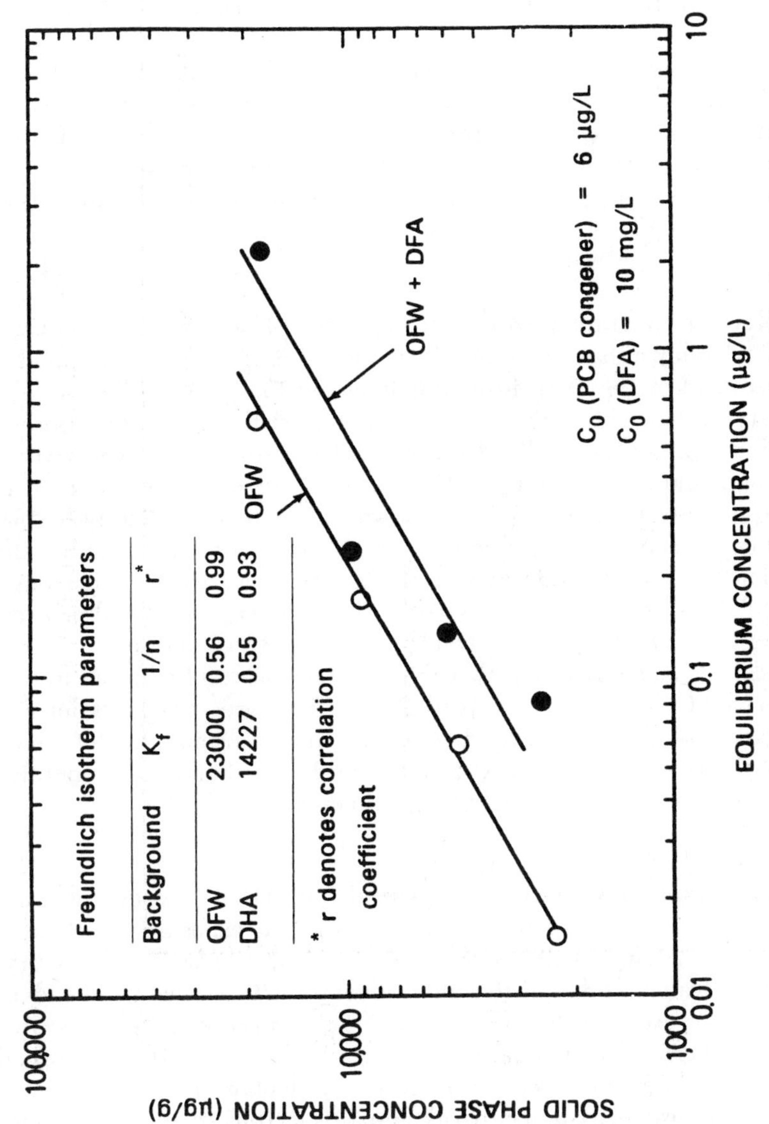

Figure 4. Adsorption isotherms for PCB congener in OFW and DFA background solutions.

equation. PCB adsorption substantially decreases in the presence of DFA. As discussed earlier, the application of the IAST model to predict adsorption equilibria may not be appropriate when the organic compound in question forms complexes with background DHS. The model assumes ideal adsorption and does not distinguish between the effects of complexation and competition.

To investigate the dynamic behavior of PCB adsorption, HPMC studies for the PCB congener were conducted in the presence and absence of DHA background solutions (*40*). The resulting breakthrough curves in Figure 5 indicate that the carbon capacity for PCB is considerably reduced by introduction of DHA. This phenomenon is quite consistent with the findings of equilibrium studies.

Effect of Calcium Ions on Adsorption. The introduction of Ca^{2+} ions considerably enhances DHA adsorption (*41, 42*). This enhancement is well illustrated by the DHA isotherms shown in Figure 6. Earlier work by Pirbazari (*43*) showed that DHA-bound organic substances may be adsorbed better in the presence of Ca^{2+}. However, HPMC studies conducted with a background solution containing DHA and Ca^{2+} revealed a retardation in the adsorption of PCB congener, as shown in Figure 7. A plausible explanation for this anomaly might be that calcium ions not only enhanced the adsorption of both DHA and PCB–DHA complex, but also simultaneously increased the competition for adsorption sites among various adsorbates, including PCB–DHA, PCB–Ca–DHA, and Ca–DHA complexes, as well as unbound PCB and DHA, and that this competition resulted in an overall reduction in PCB removal. Further studies are in progress to make a detailed investigation of the associated mechanisms of interaction to determine whether the reduction in PCB removal could be attributed to competitive phenomena arising from competition, complexation, or both.

Adsorption of Organic Compounds on DHA-Presorbed Carbon. Zimmer and Sontheimer (*44*) reported that exposure of activated carbon to background DHA progressively reduced its adsorption capacity. These investigators observed that the reduction is functionally dependent on the time of exposure to background DHA in adsorption columns. Adsorption isotherms for chloroform on virgin carbon and DHA-presorbed carbon shown in Figure 8 clearly indicate a reduction in carbon capacity attributable to presorption or preloading of DHA, consistent with the findings of Zimmer and Sontheimer.

Figure 9 makes a comparison of isotherms for chloroform on virgin carbon and carbon presorbed with DHA and Ca^{2+}. The adsorption capacity for DHA-presorbed carbon is not significantly altered by introduction of

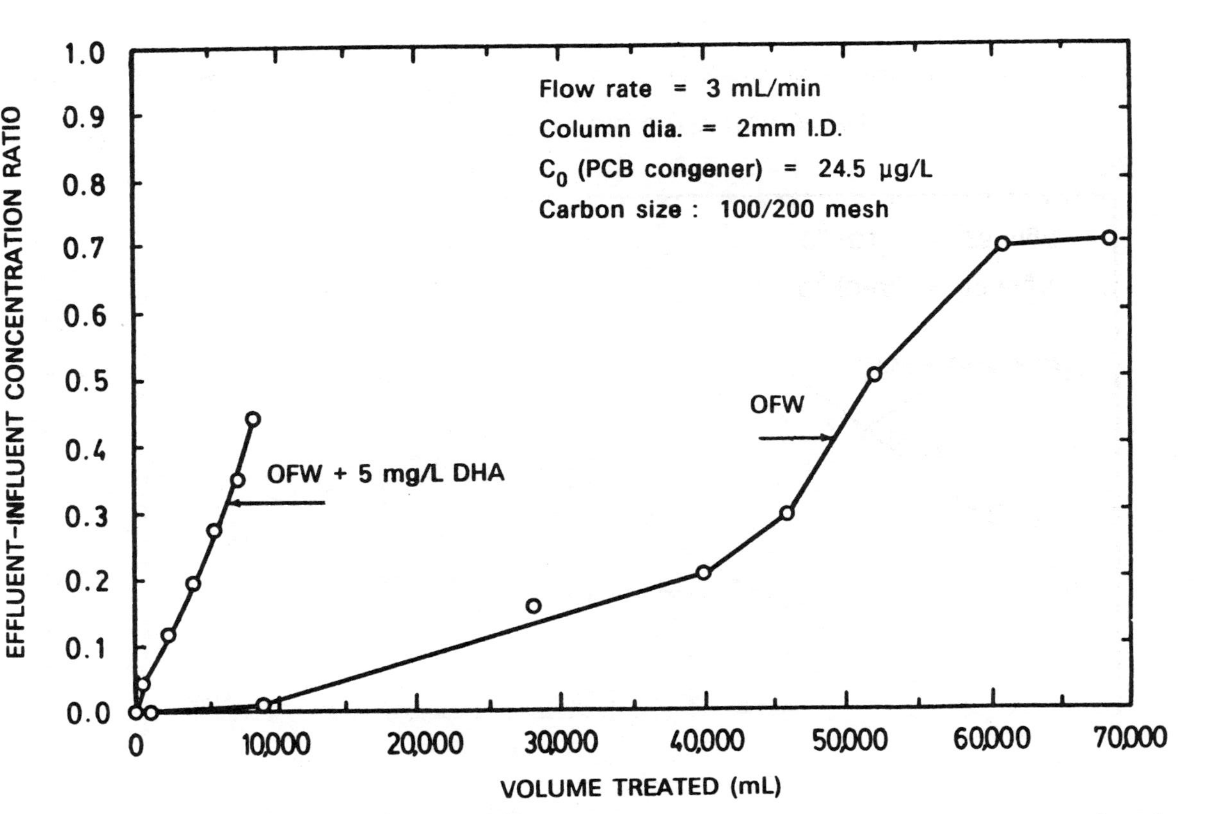

Figure 5. HPMC breakthrough curves for PCB congener in OFW and DHA background solutions. (Adapted from ref. 40.)

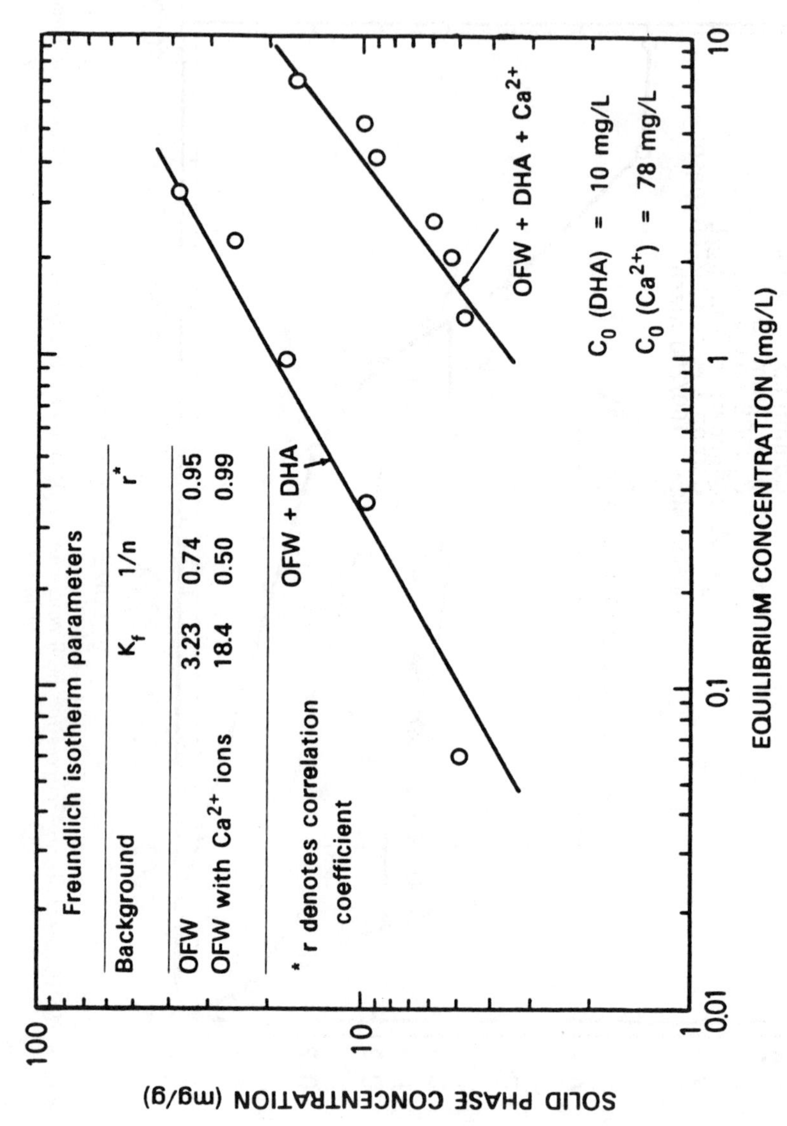

Figure 6. Adsorption isotherms for DHA in the presence and absence of calcium ions. (Adapted from ref. 40.)

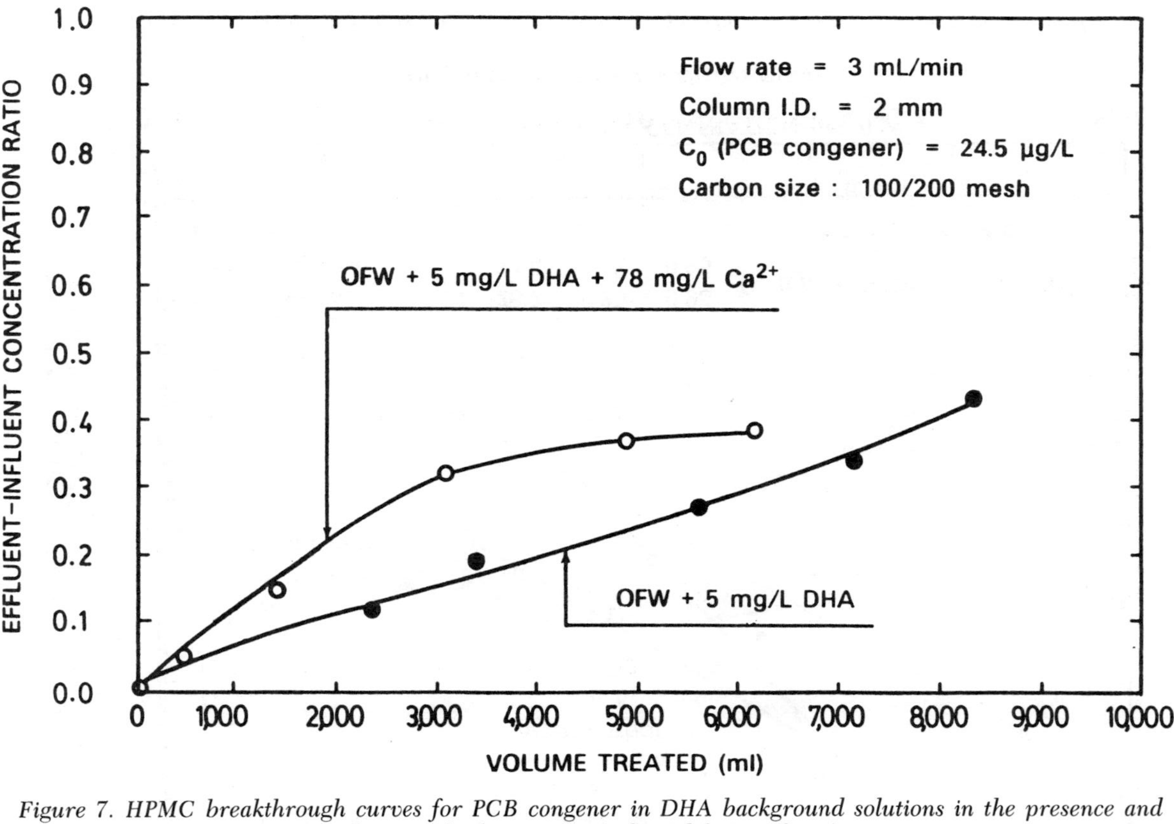

Figure 7. HPMC breakthrough curves for PCB congener in DHA background solutions in the presence and absence of calcium ions. (Adapted from ref. 40.)

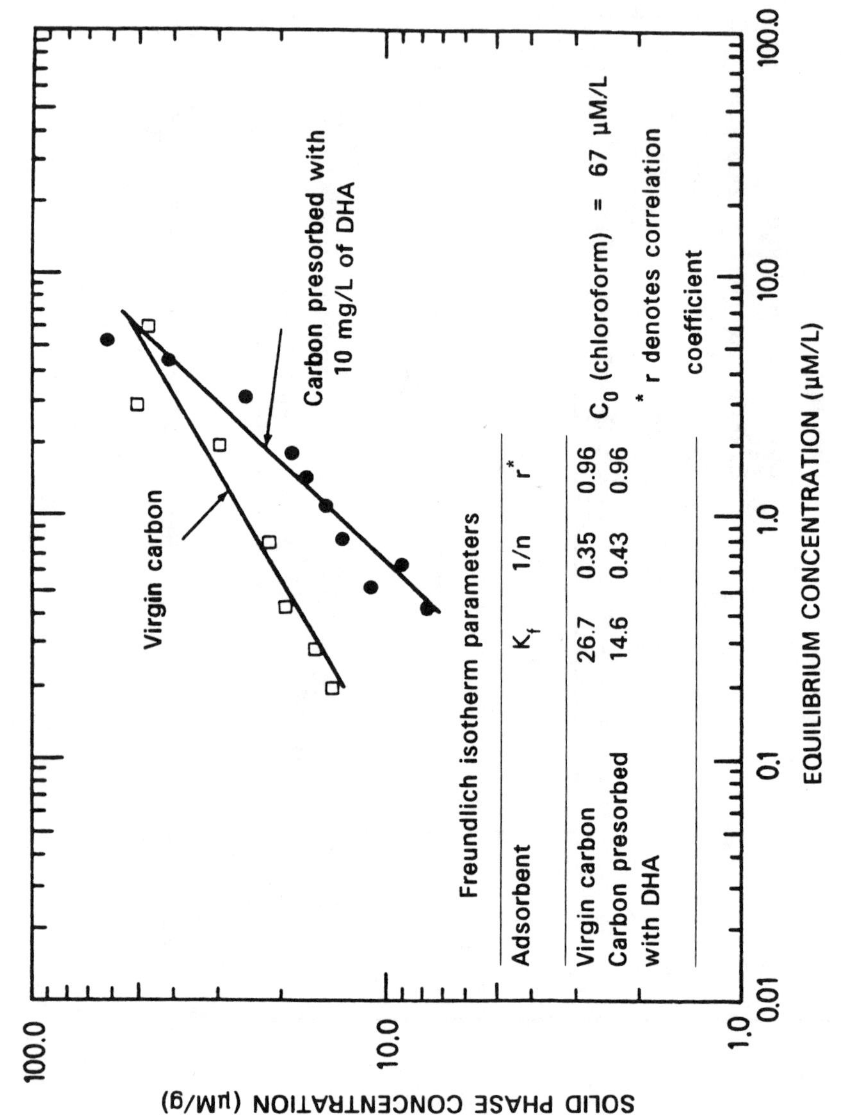

Figure 8. Adsorption isotherms for chloroform on virgin carbon and carbon presorbed with DHA.

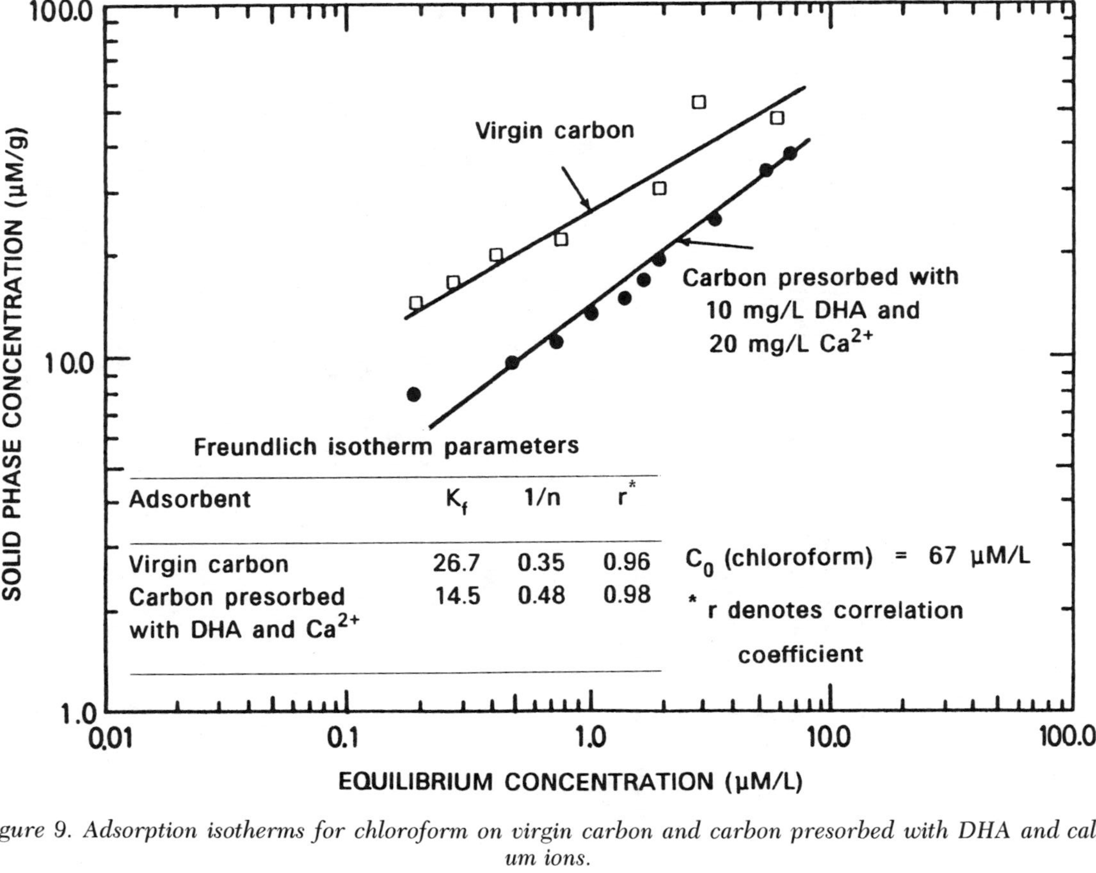

Figure 9. Adsorption isotherms for chloroform on virgin carbon and carbon presorbed with DHA and calcium ions.

Ca^{2+}, as observed from Figures 8 and 9, apparently because the competition for adsorption sites on carbon between DHA and chloroform molecules is not significant. Possibly the adsorption of DHA and chloroform molecules occurs on specific adsorption sites and through pores of different size ranges, so that, even if Ca^{2+} ions enhance adsorption of DHA, the capacity for chloroform remains unaffected.

Earlier minicolumn adsorption studies conducted by Pirbazari and Weber (*20, 35*) compared the performances of virgin carbon and DHA-preloaded carbon for compounds that form complexes with DHA, including dieldrin and PCB (Aroclor 1254), as depicted in Figures 10 and 11, respectively. It is evident from the concentration profiles that the carbon retains its ability to remove traces of PCB and dieldrin even after being preloaded with DHA. A plausible explanation is that, because PCB and dieldrin manifest a high potential for complexation with DHA, they become bound to presorbed humic polymer. Further, because PCB demonstrates a stronger tendency for association with DHA than dieldrin, its adsorption is less affected by carbon preloading.

Summary and Conclusions

Complexation of organic contaminants with HS essentially governs their behavior in processes designed for water purification, such as activated-carbon adsorption. Aqueous solubility and octanol–water partition coefficient are to a large extent indicative of the potential of an organic contaminant to associate with HS.

The IAST model adequately predicted the adsorption equilibria of TCE, chloroform, and geosmin in the presence of background DHA. In general, the model may be employed for approximate prediction of multicomponent equilibria, provided there is no complexation between the adsorbates.

The presence of background DHA reduced the adsorption capacity for PCB, as indicated by equilibrium studies. However, because PCB forms complexes with DHA, it is not possible at this stage of research to differentiate between the effects of complexation and competition on adsorption.

Contrary to expectation, the presence of calcium ions retarded the removal of PCB in DHA background solution, as observed from HPMC studies. This retardation is probably caused by competition for adsorption sites among the various adsorbates, including PCB–DHA, PCB–Ca–DHA, and Ca–DHA complexes, as well as unbound PCB and DHA. More investigations are necessary to arrive at any definite conclusion.

The preloading or presorption of carbon with DHA caused a substantial reduction in the chloroform adsorption capacity. However, the effect varied from insignificant to marginal levels for PCB and dieldrin, respectively, because they formed complexes with presorbed carbon.

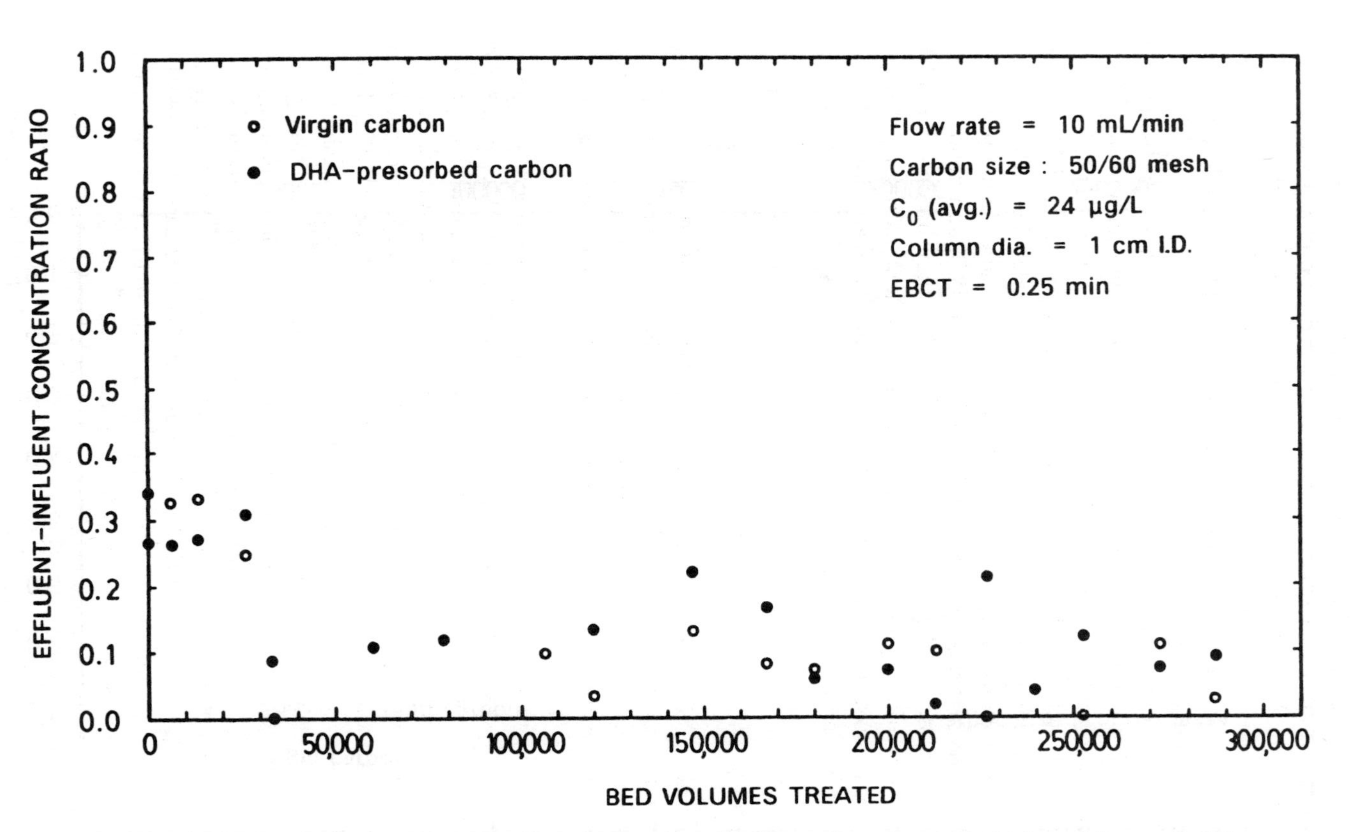

Figure 10. Effect of DHA-presorbed carbon on the removal of PCBs (Aroclor 1254). (Adapted from ref. 35.)

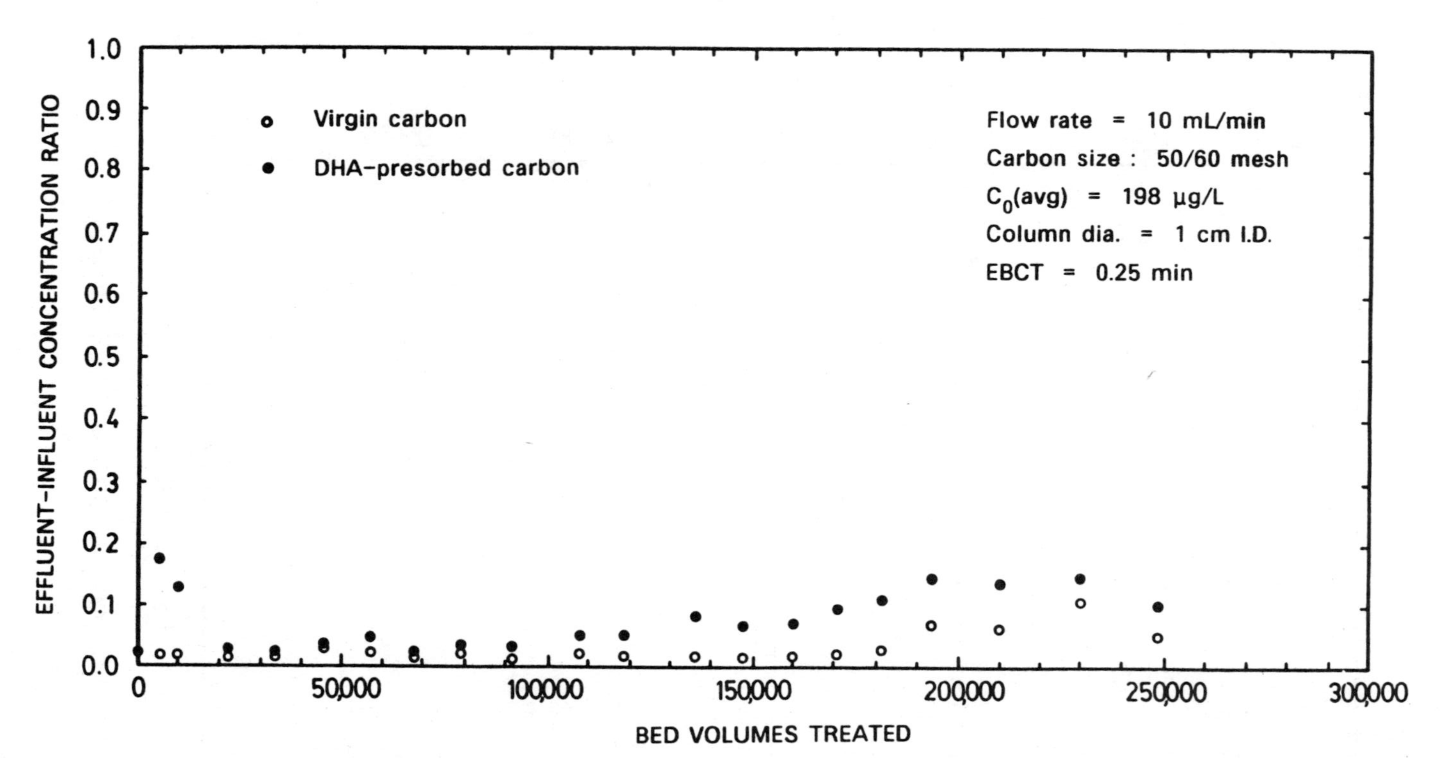

Figure 11. Effect of DHA-presorbed carbon on the removal of dieldrin. (Adapted from ref. 20.)

References

1. Endicott, D. D.; Weber, W. J., Jr. *Environ. Prog.* **1985,** *4*(2), 105–111.
2. Smith, E. H.; Tseng, S. K.; Weber, W. J., Jr. *Envir. Prog.* **1987,** *6*(2), 18–25.
3. Crittenden, J. C.; Weber, W. J., Jr. *J. Environ. Eng. Div. Am. Soc. Civ. Eng.* **1978,** *104*(EE2), 185–197.
4. Crittenden, J. C.; Weber, W. J., Jr. *J. Environ. Eng. Div. Am. Soc. Civ. Eng.* **1978,** *104*(EE3), 433–443.
5. Carter, C. W.; Suffet, I. H. In *Fate of Chemicals in the Environment*; ACS Symposium Series 225; American Chemical Society: Washington, DC, 1983; pp 215–219.
6. Thurman, E. M. *Organic Geochemistry of Natural Waters*; Martin Nijhoff/Dr W. Junk: Dordrecht, Netherlands, 1985.
7. Thurman, E. M.; Malcolm, R. L. *Environ. Sci. Technol.* **1981,** *15*(4), 463–466.
8. Thurman, E. M.; Wershaw, R. L.; Malcolm, R. L.; Pickney, D. J. *Org. Geochem.* **1982** *4*, 27–35.
9. Carter, C. W.; Suffet, I. H. *Environ. Sci. Technol.* **1982,** *16*(11), 735–740.
10. *Water Related Environmental Fate of 129 Priority Pollutants*; U.S. Environmental Protection Agency. Office of Wastewater Management. U.S. Government Printing Office: Washington, DC, 1979; EPA–440/479–0290.
11. Landrum, P. F.; Nihart, S. R.; Eadle, B. J.; Gardner, W. S. *Environ. Sci. Technol.* **1984,** *18*(3), 187–192.
12. Lee, M. C.; Snoeyink, V. L.; Crittenden, J. C. *J. Am. Water Works Assoc.* **1981** *73*(8), 440–446.
13. Ghosh, K.; Schnitzer, M. *Soil Sci.* **1980** *129*(5), 266–276.
14. Schnitzer, M.; Khan, S. V. *Humic Substances in the Environment*; Marcel Dekker: New York, 1972.
15. O'Connor, D. J.; Connolly, J. P. *Water Res.* **1980** *14*(12), 1517–1523.
16. Stevenson, F. J. *Humus Chemistry: Genesis, Composition, Reactions*; Wiley: New York, 1982.
17. Weber, J. B.; Weed, S. B.; Ward, T. M. *Weed Sci.* **1969,** *17*(4), 417–421.
18. Weber, J. B.; Weed, S. B. In *Pesticides in Soil and Water*; Guenzi, W. D., Ed.; American Society of Agronomy: Madison, WI, 1974; pp 223–256.
19. *Physicochemical Processes for Water Quality Control*; Weber, W. J., Jr., Ed.; John Wiley: New York, 1972; pp 199–259.
20. Pirbazari, M. ; Weber, W. J., Jr. *J. Environ. Eng.* **1984,** *110*(3), 656–669.
21. Smith, E. H.; Weber, W. J., Jr. *Proceedings of 1985 Annual Conference*; American Water Works Association: Washington, DC, 1985; pp 553–574.
22. Frick, B.,; Sontheimer, H. In *Treatment of Water by Granular Activated Carbon*; McGuire, M. J.; Suffet, I. H., Eds.; Advances in Chemistry 202; American Chemical Society: Washington, DC, 1983; pp 247–268.
23. Summers, R. S.; Roberts, P. V. In *Treatment of Water by Granular Activated Carbon*; McGuire, M. J.; Suffet, I. H., Eds.; Advances in Chemistry 202; American Chemical Society: Washington, DC, 1983; pp 503–524.
24. DiGiano, F. B.; Baldauf, G.; Frick, B.; Sontheimer, H. *Chem. Eng. Sci.* **1978,** *33*(12), 1667–1673.
25. Fettig, J.; Sontheimer, H. *J. Environ. Eng.* **1987,** *113*(4), 795–810.
26. Crittenden, J. C.; Luft, P.; Hand, D. W.; Oravitz, J. L.; Loper, S. W.; Arl, M. *Environ. Sci. Technol.* **1985** *19*(11), 1037–1043.
27. Crittenden, J. C.; Luft, P.; Hand, D. W. *Water Res.* **1985,** *19*(12), 1537–1548.
28. Calligaris, M. B.; Tien, C. *Can. J. Chem. Eng.* **1982,** *60*, 772–780.
29. Radke, C. J.; Prausnitz, J. M. *AIChE J.* **1972,** *18*, 761–768.
30. Atkins, P. W. *Physical Chemistry*; Oxford University Press: London, 1982.

31. Wong, S. P.; Pirbazari, M. *Proceedings of the 1987 Specialty Conference on Environmental Engineering*, Orlando, FL; Dietz, J. D., Ed.; American Society of Civil Engineers: New York, 1987; pp 567–572.
32. Hutzinger, S.; Safe, S.; Zitko, V. *The Chemistry of PCBs*; Chemical Rubber Company: Akron, OH, 1974.
33. Alford-Stevens, A. L.; Bellar, T. A.; Eichelberger, J. W.; Budde, W. L. *Anal. Chem.* **1986,** *58*(9), 2014–2022.
34. Malcolm, R. L.; McCarthy, P. *Environ. Sci. Technol.* **1986,** *20*(9), 904–911.
35. Pirbazari, M.; Weber, W. J., Jr. In *Chemistry in Water Reuse*; Cooper, W. J., Ed.; Ann Arbor Science: Ann Arbor, 1981; pp 309–339.
36. Rosene, M. R.; Deithorne, R. T.; Lutchko, J. R.; Wagner, N. J. In *Activated Carbon Adsorption of Organics from the Aqueous Phase*; Suffet, I. H.; McGuire, M. J., Eds.; Ann Arbor Science: Ann Arbor, 1980; Vol. 1, pp 309–316.
37. Bilello, L. J.; Beaudet, B. A. In *Treatment of Water by Granular Activated Carbon*; McGuire, M. J.; Suffet, I. H., Eds.; Advances in Chemistry 202; American Chemical Society: Washington, DC, 1983; pp 213–246.
38. Lalezary, S.; Pirbazari, M.; McGuire, M. J. ; Krasner, S. W. *J. Am. Water Works Assoc.* **1984** *76*(3), 83–87.
39. Environmental Engineering Program, University of Southern California—Los Angeles. *Adsorption of Several Toxic and Carcinogenic Compounds from Drinking Water*; submitted to the Drinking Water Division, U.S. Environmental Protection Agency: Cincinnati, OH, 1987.
40. Stevens, M. R. Ph.D. Dissertation, University of Southern California—Los Angeles, 1987.
41. Weber, W. J., Jr.; Pirbazari, M.; Long, J. B.; Barton, D. A. In *Activated Carbon Adsorption of Organics from the Aqueous Phase*; Suffet, I. H.; McGuire, M. J., Eds.; Ann Arbor Science: Ann Arbor, 1980; Vol. 1, pp 317–336.
42. Randkte, S. J.; Thiel, C. E.; Liao, M. Y.; Yamaya, C. N. *J. Am. Water Works Assoc.* **1982** *74*(4), 192–202.
43. Pirbazari, M. Ph.D. Dissertation, University of Michigan, 1980.
44. Zimmer, G.; Sontheimer, H. *Abstracts of Papers*, 193rd National Meeting of the American Chemical Society, Denver, CO; American Chemical Society: Washington, DC, 1987; ENVR 119.

RECEIVED for review September 17, 1987. ACCEPTED for publication February 29, 1988.

33

Activated-Carbon Adsorption of Organic Pollutants

Gerhard Zimmer, Heinz-Jürgen Brauch, and Heinrich Sontheimer

Engler-Bunte-Institut, Universität Karlsruhe, 7500 Karlsruhe, Federal Republic of Germany

Interaction with organic background substances is important in the removal of micropollutants by adsorption. Adsorption analysis characterizes the organic matter in drinking water by adsorbability. The approach is applied for a comparison of varying humic substances and water-treatment processes, along with a description of pH effects. The influence of very low concentrations of natural organic matter on the removal of halogenated pollutants is shown. Organic background significantly affects the range of adsorption capacities for a particular compound between different activated carbons. The main effect could be caused by the slow kinetic properties of humic substances and a long-term preloading of the carbon in a column. Despite different water sources, organic-matter concentrations, and activated-carbon types, a single relationship is found for the maximum column capacities of a halogenated pollutant.

WATER SUPPLY FACILITIES USING GROUND WATER in nearly all countries have encountered chlorinated hydrocarbons in their raw water within the last few years. This situation has often prompted the construction of treatment plants that use activated-carbon filters.

In Germany about 30 waterworks operate plants for the removal of chlorinated hydrocarbons. Although air stripping is used as a pretreatment for very high initial concentrations, all such plants use activated-carbon filters as the final or only treatment (*1–4*). In addition to the chlorinated hydrocarbons, humic substances are present in these ground waters. Optimal design and operation of a carbon filter plant must allow for adsorption com-

0065–2393/89/0219–0579$06.00/0

petition between a small amount of micropollutants and humic substances, which are present in a much higher concentration (5).

Design and operation of activated-carbon treatment plants requires a detailed quantification of the equilibrium state between dissolved and adsorbed substances. Isotherm measurements are necessary to determine this relationship (6). An example for three volatile chlorinated hydrocarbons is given in Figure 1.

An adsorption isotherm describes how the solid-phase concentration, q, depends on the concentration, c, in solution after a long enough contact time. Figure 1 shows that the isotherms for these three important chlorinated hydrocarbons are fit by a linear relationship in a log–log diagram. The well-known Freundlich isotherm equation ($q = K \cdot c^n$) is valid for these hydrocarbons over a large concentration range. K and n in this equation are empirically determined constants.

Figure 1 also indicates further a large difference in the adsorbability of the three substances, although they have similar structures. The nonpolar tetrachloroethene is the most adsorbable ($K = 12.02$) and 1,1,1-trichloroethane is the least adsorbable ($K = 0.367$).

The single-solute isotherms given in Figure 1 were measured in deionized–distilled water without any additional substances present. Isotherm tests with ground water containing micropollutants and humic substances

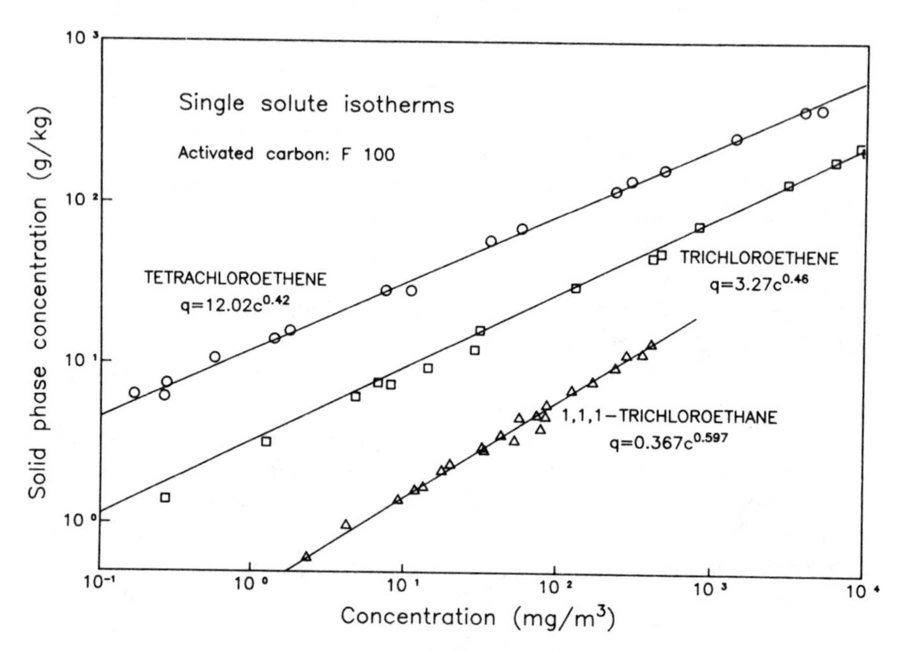

Figure 1. Activated-carbon adsorption isotherms for volatile chlorinated hydrocarbons. Data are from ref. 7.

result in a very small reduction in the solid-phase micropollutant concentration at the same aqueous-phase concentration. This reduction is ascribed to adsorption competition by humic substances. Because of the low natural-organic-matter content (0.4–1.5 mg/L of dissolved organic carbon) and the nearly neglectable impact on isotherm results, we did not expect a large influence on filter behavior. This assumption has been proven wrong, as demonstrated in Figure 2, where the predicted breakthrough of a micropollutant is compared with observations.

Figure 2 indicates that only about 30,000 bed volumes could be reached for a 50% breakthrough, instead of the 130,000 bed volumes predicted from the isotherm measurements. In order to understand the large influence of natural organic matter on filter efficiency, the adsorption behavior of these ubiquitous substances must be investigated in more detail. However, it is important to realize that natural organic matter consists of various substances, which cannot be isolated or separately analyzed (*8*, *9*).

Adsorption Analysis for Humic Substances

Adsorption analysis is used to describe the adsorption equilibrium for multicomponent mixtures of unknown substances in water (*10–12*). In this

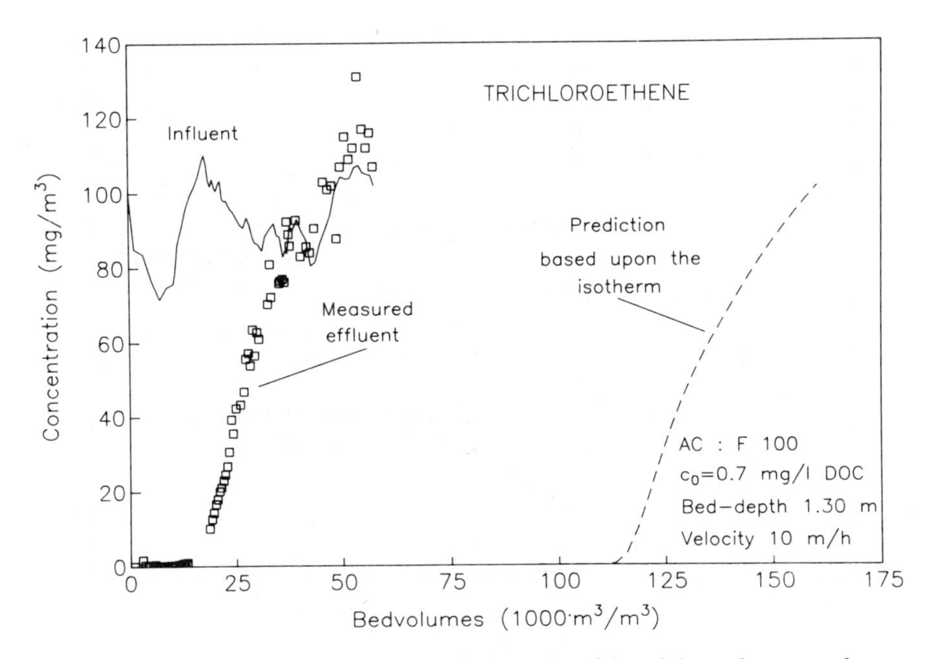

Figure 2. Comparison of measured and predicted breakthrough curves for trichloroethene.

method, described in detail by Sontheimer et al. (*13*), the mixture of unknown substances is replaced with three to six selected components of different adsorbabilities. If we assume that the adsorption equilibria can be described by the Freundlich equation,

$$q_i = K_i c_i^{n_i} \tag{1}$$

then the overall isotherm concentration ($C_{T,\mathrm{DOC}}$), measured by dissolved organic carbon (DOC), represents the summation of single concentrations (c_i).

$$C_{T,\mathrm{DOC}} = \sum_{i=1}^{N} c_i \tag{2}$$

where N is the number of components. Extrapolation at the lowest solid-phase concentrations results in a nonadsorbable portion that corresponds to the Freundlich parameter $K = 0$. The competition of other DOC compounds at different isotherm points can be calculated by using known models for multisolute systems like the IAS (ideal adsorbed solution) theory and the SCA (simplified competitive adsorption) model (*13*).

For a given number of selected components, the individual parameter K_i and concentrations $c_{o,i}$ are adjusted to fit the measured DOC isotherm. The Freundlich parameter $n = 0.20$ is the best result for many adsorption analyses of similar systems (*14*). With this method it is possible to calculate isotherms with variable initial concentrations. In Figure 3, the three isotherms show the characteristic adsorption behavior of a mixture. They are not describable with equilibrium equations for single-solute systems.

Figure 3 shows the analyzed and predicted isotherm data for three initial concentrations of a humic substance that was isolated from Lake Constance (Bodensee) with macroporous resins. The K_i values and concentrations represent the result for an adsorption analysis, assuming there are four adsorbable components and one nonadsorbable component. These parameters allow the simultaneous description of the three isotherms (*15*).

Such adsorption behavior is expected for a mixture of different organic compounds at different initial concentrations. Figure 4 illustrates this behavior for a mixture of *p*-nitrophenol and benzoic acid (*14*).

Figure 4 also shows the DOC isotherms for the two single substances, as well as the mixture isotherms for constant composition but different initial concentrations. The mixture curves connect the isotherms for the two single substances, and competitive adsorption yields the different isotherm shape. The influence of initial concentration on the isotherm of mixtures of organic substances can be described by other methods. An interesting example was presented by Summers (*16*), who showed that the isotherms for different initial concentrations can result in a unique isotherm if the ratio between

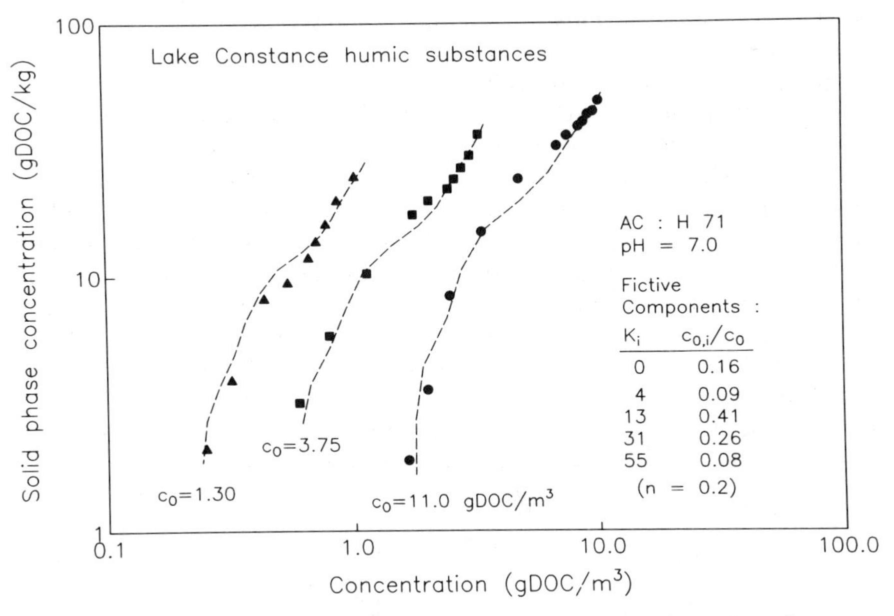

Figure 3. Influence of initial concentration on the DOC isotherm for a humic acid from Lake Constance (Bodensee). Data are from ref. 15.

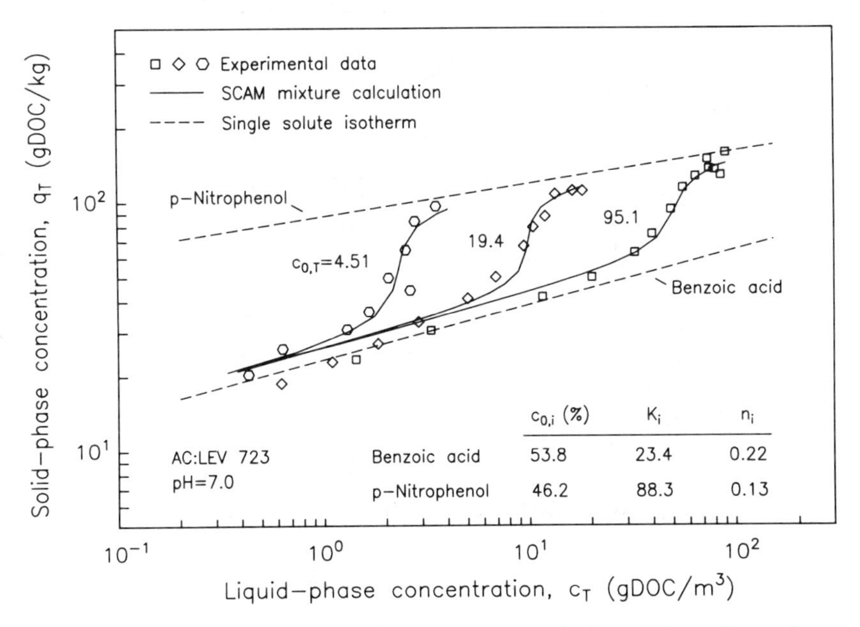

Figure 4. Influence of the initial concentration on the DOC isotherm for a mixture of p-*nitrophenol and benzoic acid. Data are from ref. 14.*

equilibrium solution and mass of adsorbent used is related to the solid-phase concentration. This method is very difficult to use for filter calculations and for prediction of the adsorbability of the effluent of a carbon filter after different operation times. However, filter calculations are possible with the selected-component approach (*17*).

Adsorption analysis also allows comparison of the adsorption behavior of different humic substances, as shown in Figure 5. Hohloh Lake (Hohlohsee) is a muddy lake in the Black Forest with a humic concentration of more than 50 mg/L. Lake Constance (Bodensee) has a very low DOC concentration of 1.5–1.7 mg/L from the Alpine Rhine River. The organic substances in these lakes undergo changes along the 800-km distance that the Rhine River flows from Lake Constance to Düsseldorf–Flehe; DOC increases by a factor of 2. These changes are characterized by the results of the adsorption analyses of three samples, given in Figure 5 (*15*). Organic matter from Hohloh Lake is much better adsorbed than that from the Rhine River. In addition, the

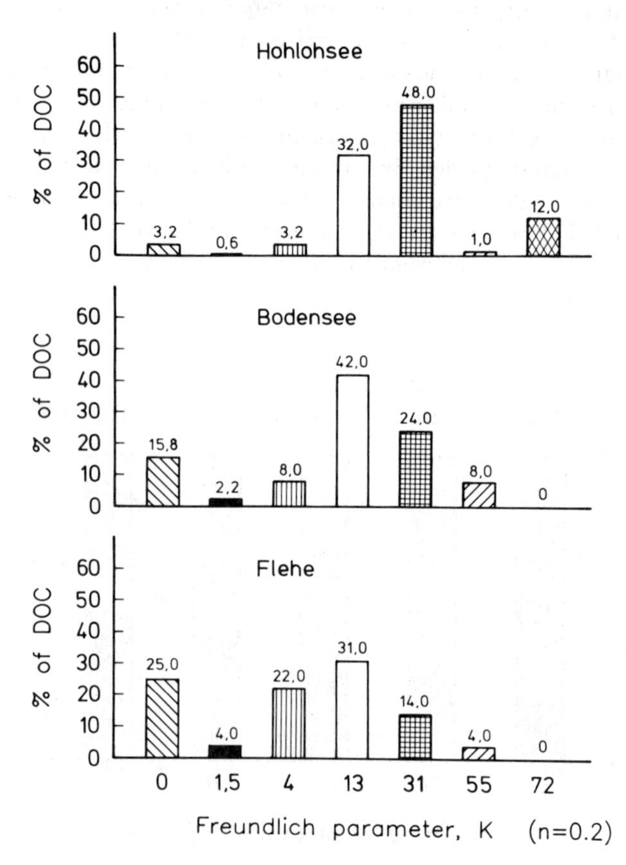

Figure 5. Concentration distribution of selected-component K *values for different humic substances. Data are from ref. 15.*

nonadsorbable portion is lowest in Hohloh Lake and increases along the Rhine River.

Applications of Adsorption Analysis for Treatment Control

Adsorption analysis can be very helpful for treatment considerations. For example, Figure 6 indicates the influence of ozone on the adsorption behavior of a humic substance. An increase in nonadsorbable substances ($K = 0$) and a decrease in the better-adsorbable substances ($K = 72$) occurs with additional amounts of ozone (*15*). The oxidation reaction leads to more polar substances, which are not as adsorbable by activated carbon.

For an actual treatment problem it is significant to know that ozone increases biodegradability. Adsorbability increases again after biological degradation, as has been shown by Hubele (*17*), who proposed a method for the optimization of such a combined treatment.

The same adsorption analysis can be used with a few additional assumptions to describe the effect of pH on the adsorbability of humic substances. Figure 7 shows adsorption isotherms for a humic substance at different pH values. It indicates that adsorbability increases with decreasing pH; nondissociated forms of weak organic acids have a higher adsorbability than the dissociated species. But humic acids are considered as a mixture of organic acids, with more than one pK value. Therefore, an accurate description of pH influence on adsorbability implies that the components of humic substances have different pK values and compete with each other for adsorption sites.

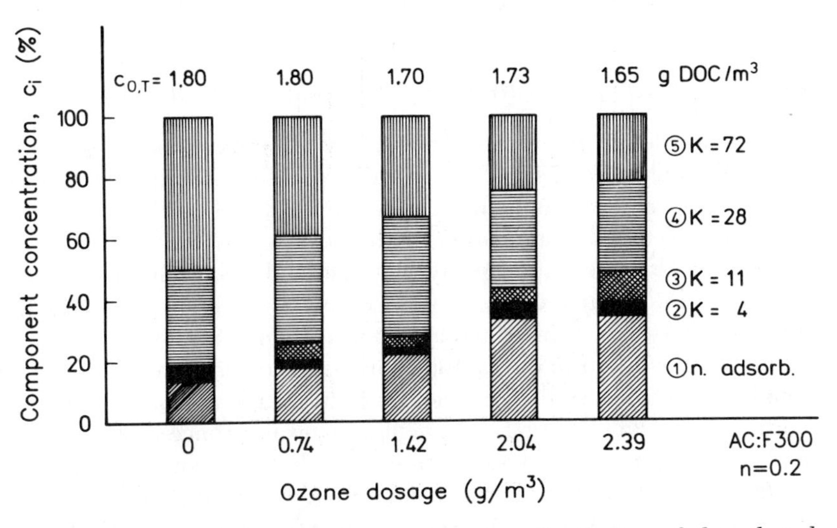

Figure 6. Effect of ozone on the concentration distribution of the selected components. Data are from ref. 15.

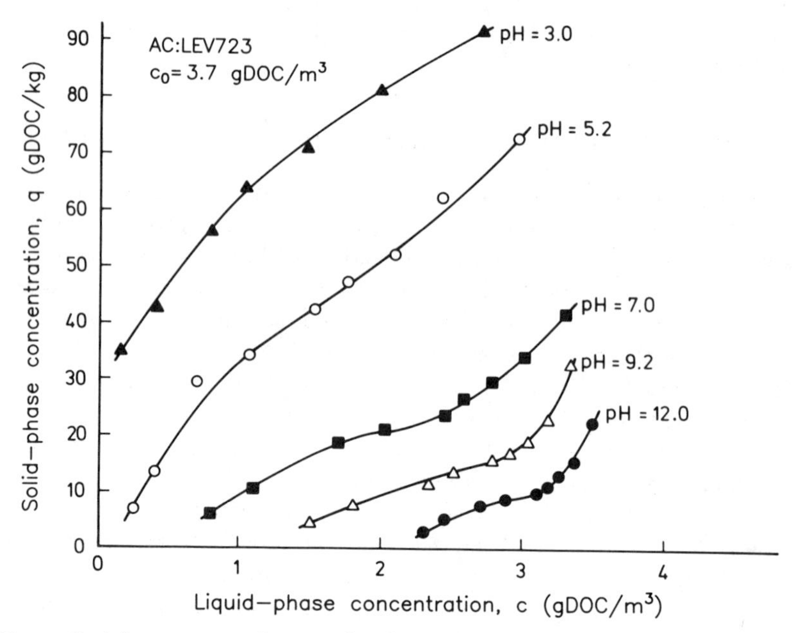

Figure 7. Adsorption isotherms of Lake Constance (Bodensee) humic acid at different pH values. Data are from ref. 15.

Using this implication about humic substances, we can predict the effect of pH on adsorbability by summation of the amount of each selected single component adsorbed (dashed lines, Figure 8). This procedure also allows the prediction of adsorption behavior for other initial concentrations (*15*).

Removal of Micropollutants by Activated-Carbon Filters

Given this knowledge about the adsorbability of humic substances, we investigated the influence of small amounts of ground-water humic substances on the removal of micropollutants by an activated-carbon filter. The best approach is a detailed examination of the results observed in treatment plants. In a study based upon many full-scale and pilot columns, Baldauf and Zimmer (*18*) compared the results of several water utilities that had different ground-water sources and activated carbons. As shown in Table I, the operating conditions of the carbon columns were very similar, especially the depths and the empty-bed contact time. In nearly all cases the influent concentration was 25–250 μg/L. For higher raw-water concentrations, using an air-stripping operation as a pretreatment step has been found worthwhile.

Total amount adsorbed at exhaustion provides a useful way to compare different column runs. We calculated solid-phase concentration in the various filters by integrating the breakthrough curve at the total saturation of the

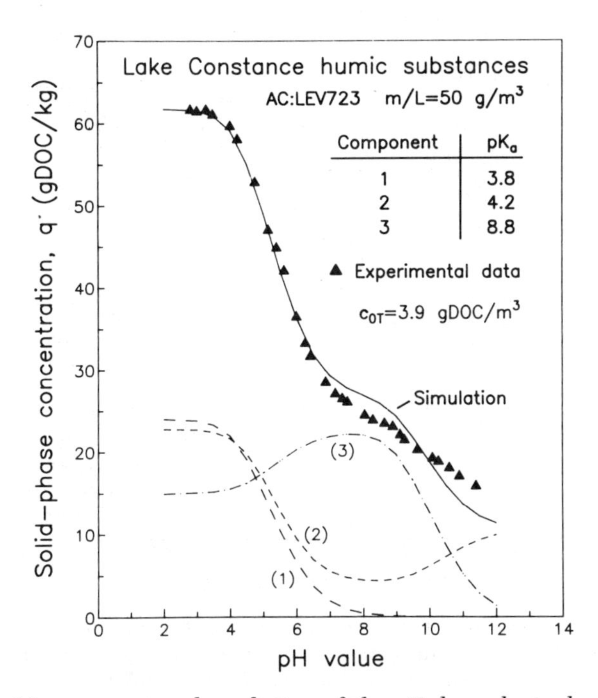

Figure 8. Measurement and prediction of the pH-dependent adsorption of a Lake Constance (Bodensee) humic acid using three selected organic acids. Data are from ref. 15.

Table I. Range of Operating Conditions for Carbon Filters

Operating Conditions	*Range*
Carbon bed depth	2.5–3 m
Filter velocity	10–15 m/h
Empty bed contact time	12–15 min
DOC before treatment	0.4–1.5 mg/L
Chlorinated hydrocarbons	25–250 μg/L
Maximum effluent concentration	5 μg/L

carbon. For raw water that contained several substances, each micropollutant was evaluated separately, with the assumption that the main effect was caused by the organic background.

The combined results for the important volatile hydrocarbon trichloroethene show a very unexpected result (Figure 9). Despite the varying capacities of the diverse activated-carbon types in the single-solute systems and the different adsorbability of aquatic organic matter, a single adsorption relationship appeared for trichloroethene with all the activated-carbon types and raw water sources examined. For the adsorber sizes we used, this filter

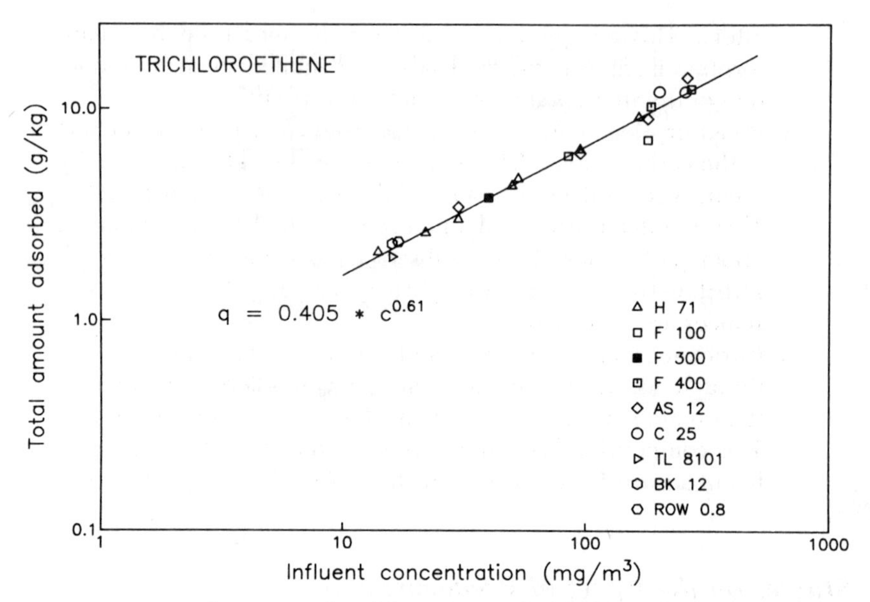

Figure 9. Total amount of trichloroethene adsorbed by full-scale columns at saturation for different raw waters, activated-carbon types, and influent concentrations. Data are from ref. 18.

correlation provides a connection between the influent concentration and the maximum solid-phase concentration of the carbon at total breakthrough.

Activated-carbon column adsorption capacities (i.e., the total amount adsorbed at saturation) for a particular halogenated organic compound are significantly reduced in the presence of humic substances. This reduction can be seen by comparison of the high K value (3.27) for humic-free solution shown in Figure 1 with the lower value shown in Figure 9. Earlier observations had suggested a lower carbon loading. However, we were surprised to find this filter correlation independent of the type of carbon used, the concentration and source of the humic substances, and the presence of other micropollutants.

Similar results are found for the strongly adsorbing tetrachloroethene and the weakly adsorbing 1,1,1-trichloroethane. Despite diverse adsorbabilities in humic-free water (Figure 1), the filter correlations are very similar (*18*).

Discussion

It is difficult to understand and explain these unexpected results. However, the following points seem to be important:

- Adsorption competition during simultaneous micropollutant and humic adsorption occurs in the upper part of the large filters. This competition, which can be measured by co-adsorption isotherm studies, leads to an immediate small reduction of the micropollutant adsorption capacity.

- Preadsorption of humic substances takes place in the lower part of the carbon filters. The organic-matter breakthrough usually occurs very quickly because of the slow adsorption kinetics. This rate difference results in a separation of the filter into an upper part, where humic substances and micropollutants adsorb together, and a lower part, where only the humic substances are adsorbed.
- Enrichment of the better-adsorbable humic substances occurs through adsorption competition between different fractions of the humic material and through slow displacement processes. This competition leads to a further reduction of the micropollutant adsorption capacity, which continues over a long period.

Studies on the Effect of Preadsorption

The proposed explanations can be illustrated by the influence of the preadsorption process on the micropollutant adsorption equilibrium. For this purpose a commonly used activated carbon was preloaded over a range of adsorption exposure times with humic matter from Karlsruhe tap water that contained a DOC concentration of nearly 1 mg/L. We found no significant biological degradation of organic substances from this anaerobic ground water. At given time intervals, this carbon was used for isotherms with chlorinated substances. The trichloroethene and tetrachloroethene results for a 25-week preloading time are demonstrated in Figure 10. Over a wide concentration range, the isotherm is parallel to the isotherm of organic-free water. The impact of the humic uptake is the same for high and low concentrations of the micropollutant. This behavior is not modeled by theories of multicomponent adsorption.

Figure 11 shows the isotherms of different preloading times, which display this time-related parallel shift. Thus, when the Freundlich equation ($q = K \cdot c^n$) is used to describe the adsorption data, the decrease in adsorbability can be adequately described by a decrease in the Freundlich parameter K. The n value describes the slope, which is constant. The lower part of Figure 11 shows a great reduction in the beginning. Then the slow diffusivity of the humic matter causes only a small reduction in the K value, which is still active after 1 year.

The filter correlation presented in Figure 9 is inserted as the lower dashed line in Figure 11. This curve is steeper than the other isotherms, because for a given adsorber size the operation time depends on the influent concentration. The lower the concentration of pollutants, the longer the time to saturation will be. Thus, there is an equivalent diminishing of adsorption

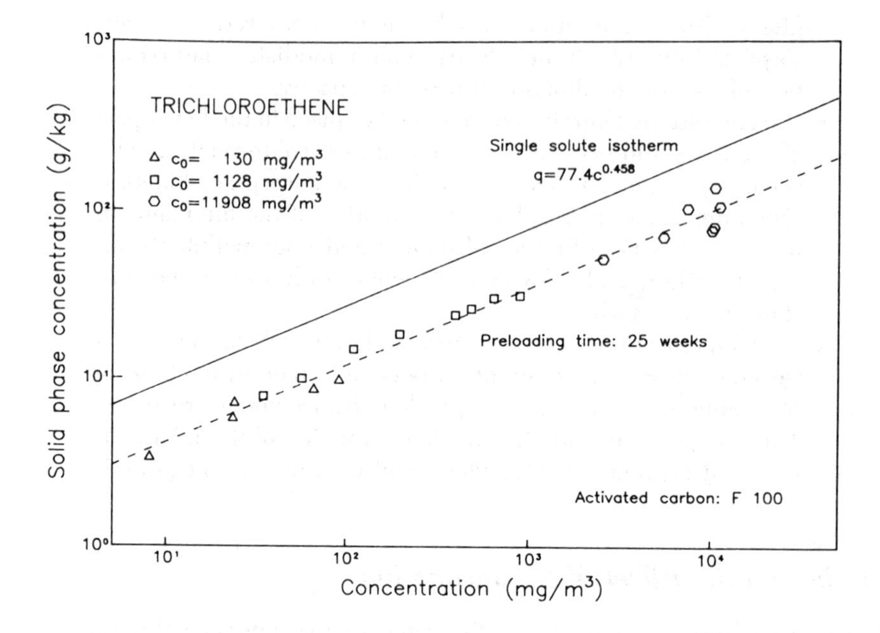

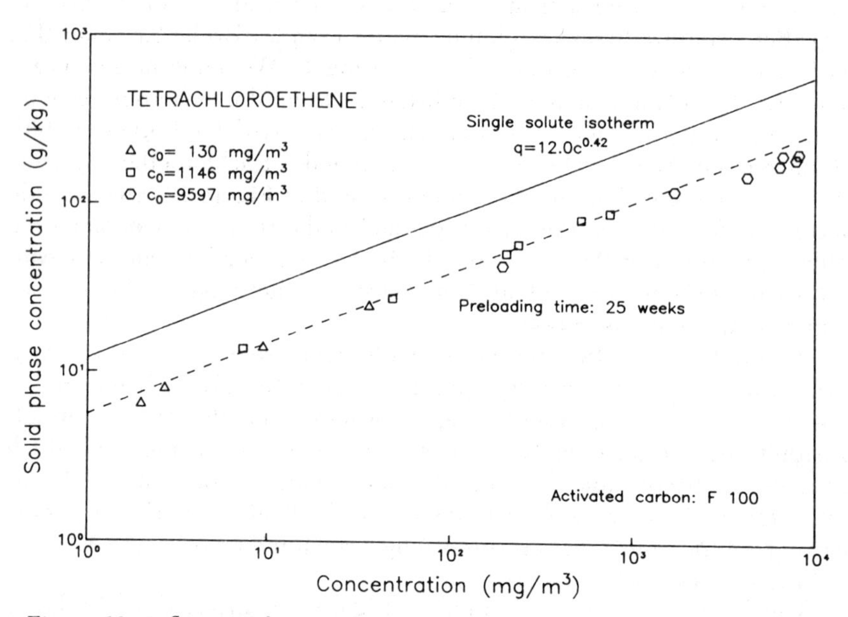

Figure 10. Influence of 25 weeks of preloading with organic matter on the isotherm of trichloroethene and tetrachloroethene. Data are from ref. 7.

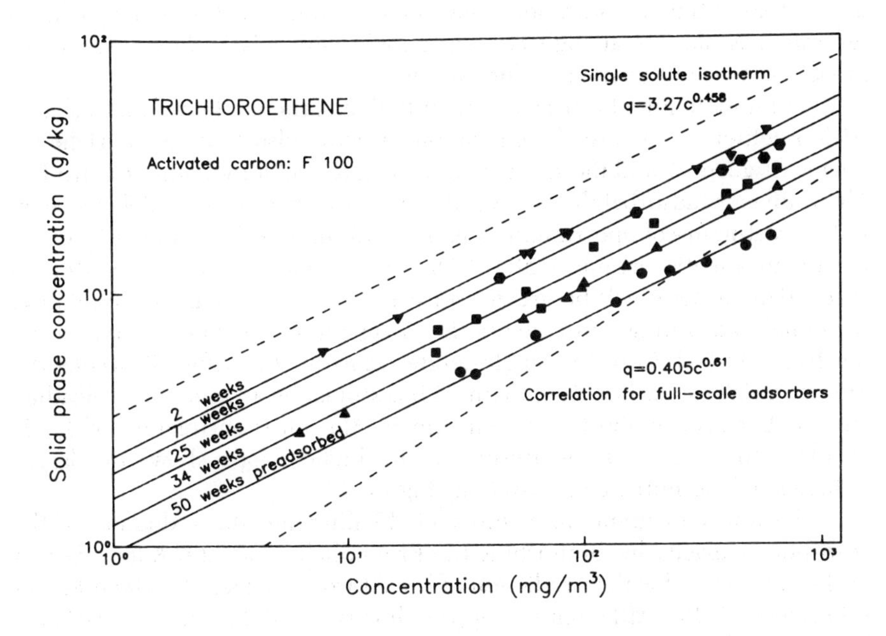

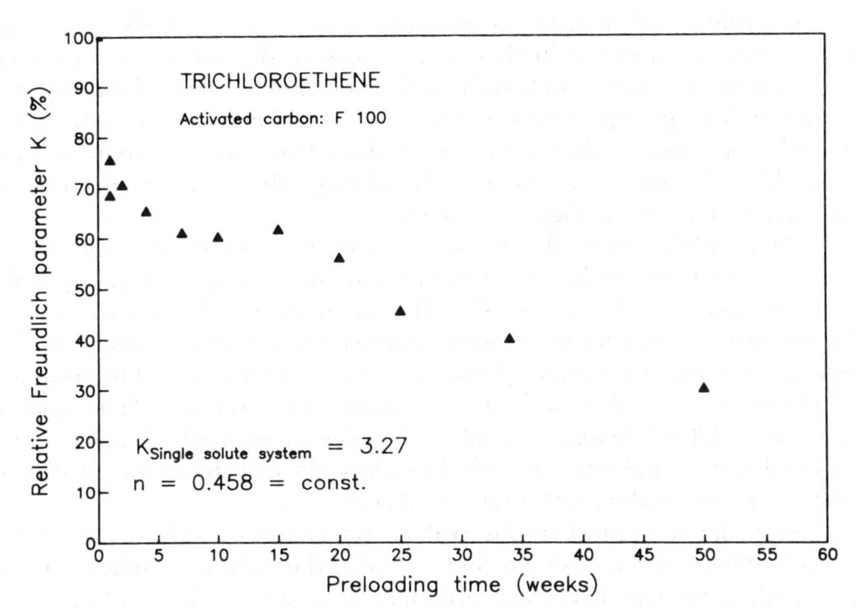

Figure 11. Reduction in the adsorption isotherms and Freundlich parameter K for trichloroethene by preloading with organic matter. Data are from ref. 7.

and increased time to saturation at lower concentrations. The time to achieve saturation is shorter at high concentrations, with adsorption capacity approaching that of the single-solute system.

A time-dependent capacity, similar to that of the moderately adsorbable trichloroethene, was also found for the weakly adsorbing 1,1,1-trichloroethane (Figure 12) and the strongly adsorbing tetrachloroethene (Figure 13). Their diverse adsorbabilities give these substances a great difference in water-treatment-column running times, resulting in different lengths of exposure to aquatic organic matter. Thus the impact of organic matter on adsorption capacity will be greater for compounds with high adsorbability. A strongly adsorbing substance with a long retention time (over 2 years) reaches only 5–10% of the single-solute capacity (*18*). After 50 weeks the filter correlation (dashed line, Figure 13) is still far away from the preloading curves. Because of shorter running times, the filter correlation of 1,1,1-trichloroethene crosses the isotherms. Its higher slope shows the larger influence of the initial concentration (Figure 12).

The lower diagrams in Figures 11–13 illustrate the reduction of the adsorption capacity by diminishing the Freundlich parameter K for different preloading times for the prediction of breakthrough curves for the different substances (*7*, *19*). Although this approach is successful in the calculation of breakthrough curves, the activated-carbon types and different humic substances and concentrations lack influence on these filter correlations.

Regarding carbon type, carbons with a larger original adsorption capacity showed greater reduction in that capacity through the presence of natural organic matter. Additional preadsorption studies showed that organic substances occupy especially high energetic sites. Therefore, the advantage of further activation of the carbon is neutralized with increased running time of the filter (*7*). After 1 year of operation all the carbons have a fairly similar adsorption capacity, as shown in Figure 9.

Figure 14 shows the effect of preadsorption on different micropollutants. In this diagram the reduction of the adsorption capacity through a preadsorption step is much larger for the chlorinated aliphatic hydrocarbons than for the two aromatic micropollutants studied. Only a 30% reduction of the adsorption capacity for *p*-nitrophenol is observed after 1 year of preloading. However, an 85% reduction for 1,1,1-trichloroethane is found. This explains why it is so difficult to remove weakly adsorbing chlorinated substances from polluted waters and why an early breakthrough and displacement effects occur in plants dealing with these substances (*10*).

Although many practical observations can be explained by the approach described here, it is not yet possible to predict the behavior of other organic micropollutants. One important result may be a better understanding of the adsorption kinetic properties of humic substances, but not enough experimental studies exist for further conclusions.

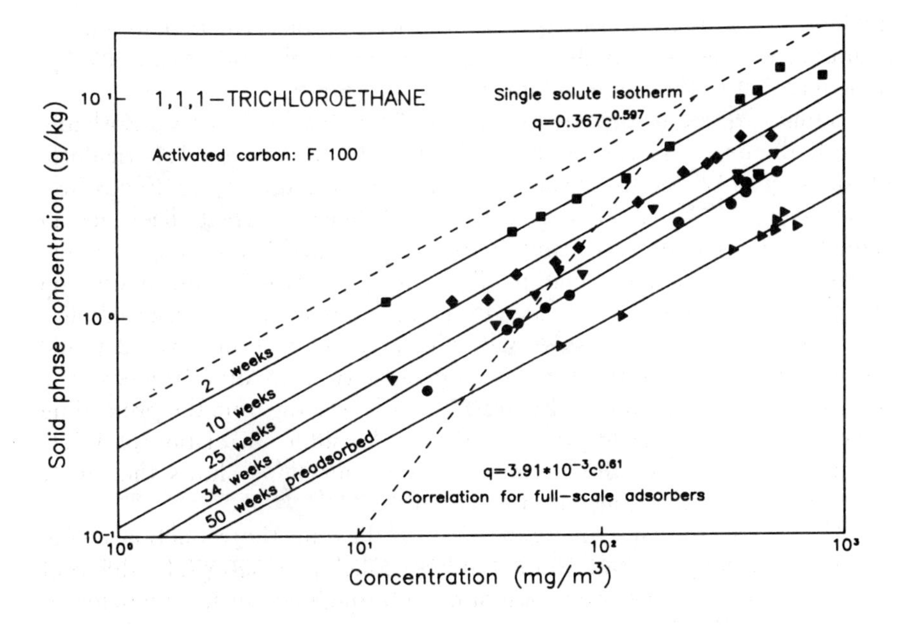

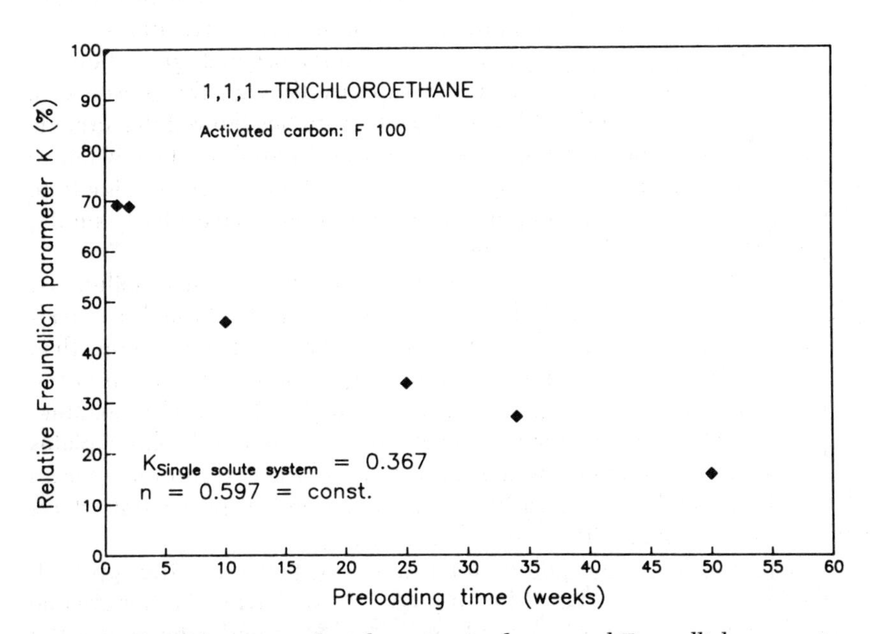

Figure 12. Reduction in the adsorption isotherms and Freundlich parameter K for 1,1,1-trichloroethane by preloading with organic matter. Data are from ref. 7.

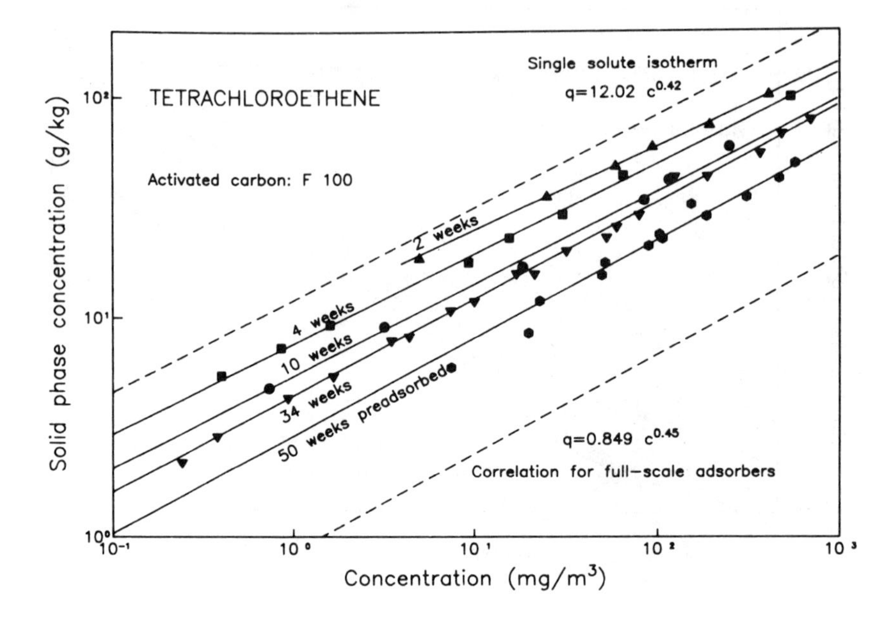

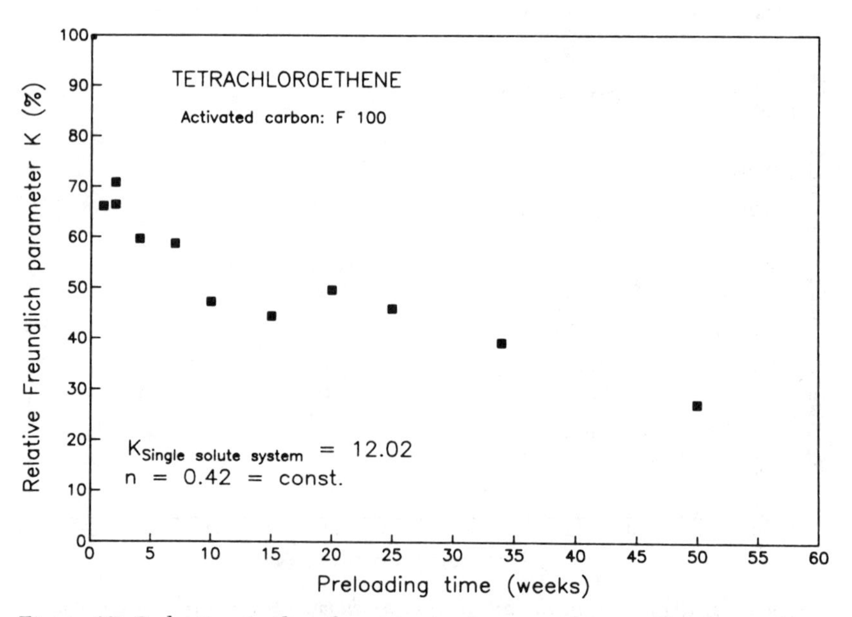

Figure 13. Reduction in the adsorption isotherms and Freundlich parameter K *for tetrachloroethene by preloading with organic matter. Data are from ref. 7.*

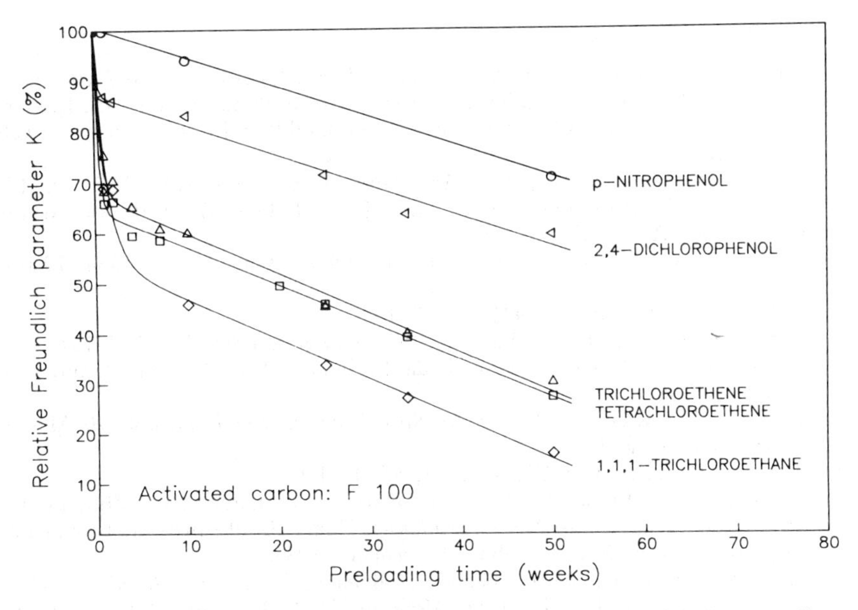

Figure 14. Reduction in the Freundlich parameter K *for several micropollutants as a function of preloading time with organic matter. Data are from ref. 7.*

Summary

The results presented can be summarized as follows. Adsorption analysis is a good method to characterize the activated-carbon adsorption of humic substances and can be used as an aid for drinking-water-treatment design. Removal of micropollutants with an activated-carbon filter depends on adsorption competition with the humic substances that are always present in the waters to be treated.

The large effect of humic substances on micropollutant adsorbability is due to the long time available for preadsorption. This preadsorption may lead to an enrichment of the organic substances on the activated carbon. This process is very slow, but leads to a large reduction of the adsorption capacity for the chlorinated hydrocarbons.

The effect of preadsorption depends on the chemical structure of the micropollutant, but is nearly independent of the humic concentration and the activated-carbon type.

The impact of humic substances is a general reduction of the carbon capacity, which can be adequately described by diminishing of the Freundlich parameter K. With this time dependency, the breakthrough of micropollutants in treatment plants can be calculated.

References

1. Sontheimer, H. *Verfahrenstechnische Grundlagen fur Anlagen zur Entfernung von Halogenkohlenwasserstoffen aus Grundwassern;* Publication Series, Engler-Bunte-Institut, Universität Karlsruhe (TH): Federal Republic of Germany, 1983; Volume 21.
2. McKinnon, R. J.; Dyksen, J. E. *J. Am. Water Works Assoc.* **1984,** *76*(5), 42–47.
3. Hand, D. W.; Crittenden, J. C.; Gehin, J. L.; Lykins, B. W. *J. Am. Water Works Assoc.* **1986,** *78*(9), 87–98.
4. Kavanaugh, M. C.; Trussell, R. R. *J. Am. Water Works Assoc.* **1980,** *72*(12), 684.
5. Baldauf, G. *Water Supply* **1985,** *3,* 187–196.
6. Crittenden, J. C.; Luft, P.; Hand, D. W. *Water Res.* **1985,** *19,* 1537–1548.
7. Zimmer, G. Ph.D. Thesis, Universität Karlsruhe (TH), Federal Republic of Germany, 1988.
8. Schnitzer, M.; Khan, S. U. *Humic Substances in the Environment;* Marcel Dekker: New York, 1972.
9. Fuchs, F.; Raue, B. *Vom Wasser* **1981,** *57,* 95–106.
10. Sontheimer, H.; Frick, B.; Fettig, J.; Hörner, G.; Hubele, C.; Zimmer, G. *Adsorptionsverfahren zur Wasserreinigung;* Engler-Bunte-Institut, Universität Karlsruhe (TH): Federal Republic of Germany, 1985.
11. Frick, B.; Bartz, R.; Sontheimer, H.; DiGiano, F. A. In *Activated Carbon Adsorption,* Volume 1; Suffet, J. H.; McGuire, M. J., Eds.; Ann Arbor Sciences: Ann Arbor, 1980; pp 229–242.
12. Fettig, J.; Sontheimer, H. *J. Environ. Eng. (N.Y.)* **1987,** *113*(4), 795–810.
13. Sontheimer, H.; Crittenden, J. C.; Summers, R. S. *Activated Carbon for Water Treatment;* Engler-Bunte Institut, Universität Karlsruhe: Federal Republic of Germany; distributed in the United States by the American Water Works Association, 1988.
14. Sontheimer, H.; Volker, E. *Charakterisierung von Abwassereinleitungen aus der Sicht der Trinkwasserversorgung;* Engler-Bunte-Institut, Universität Karlsruhe (TH): Federal Republic of Germany, 1987; Volume 31.
15. Brauch, H. J. Ph.D. Thesis, Universität Karlsruhe, Federal Republic of Germany, 1984.
16. Summers, R. S. Ph.D. Thesis, Stanford University, 1986.
17. Hubele, C. Ph.D. Thesis, Universität Karlsruhe (TH), Federal Republic of Germany, 1985.
18. Baldauf, G.; Zimmer, G. *Vom Wasser* **1986,** *66,* 21–31.
19. Zimmer, G.; Haist, B.; Sontheimer, H. *Proc. AWWA Annu. Conf.* **1987,** Kansas City, 815–826.

RECEIVED for review July 24, 1987. ACCEPTED for publication February 26, 1988.

34

Trihalomethane Precursor and Total Organic Carbon Removal by Conventional Treatment and Carbon

Benjamin W. Lykins, Jr., and Robert M. Clark

Drinking Water Research Division, Risk Reduction Engineering Laboratory, U.S. Environmental Protection Agency, Cincinnati, OH 45268

Data from four water-treatment plants were used to describe the performance of conventional treatment and granular activated carbon for removing trihalomethane precursors to meet various treatment goals. Also presented are data for total organic carbon removal, which has been suggested as an organic surrogate for measuring the effectiveness of water treatment. Conventional treatment, as used in the four water-treatment plants evaluated, substantially reduced total organic carbon and trihalomethane precursor concentrations. Granular activated carbon may be a treatment alternative to consider for meeting trihalomethane standards as low as 50 μg/L.

DISINFECTION BYPRODUCTS ARE BEING CONSIDERED FOR REGULATION under the Safe Drinking Water Act Amendments of 1986 (*1*). One of the most significant disinfection byproducts for utilities that use chlorine is total trihalomethanes (TTHMs). Pressure is growing to reconsider the existing TTHM standard of 0.1 mg/L (100 μg/L) and to lower it to some as yet unspecified level. Trihalomethane levels as low as 10–50 μg/L may be considered. Utilities may be forced to investigate disinfectants other than chlorine and to evaluate treatment modifications. New options might range from improved conventional treatment to granular-activated-carbon (GAC) adsorption.

0065–2393/89/0219–0597$07.25/0

Most water utilities are able to meet a TTHM level of 0.10 mg/L (100 μg/L) by using properly operated conventional treatment. However, if the standard is reduced substantially, adding GAC to conventional treatment may be an acceptable option. The length of time during which GAC can remove THMs to meet a 10-, 25-, 50-, or 100-μg/L standard will determine its efficacy as a viable treatment option.

The U.S. Environmental Protection Agency's Drinking Water Research Division has collected extensive treatment data for removal of organic substances, including TTHM, their precursors, total organic carbon (TOC), and total organic halide (TOX) at several water utilities under actual operating conditions. In these studies GAC was used at some sites—including Cincinnati, Ohio; Jefferson Parish, Louisiana; Manchester, New Hampshire; and Evansville, Indiana—to determine its ability to remove those organic compounds present after conventional treatment.

Literature Survey

Conventional Treatment. TTHM precursors can be reduced by proper conventional treatment (coagulation, flocculation, sedimentation, and filtration). The extent of reduction can depend on several factors, such as type of coagulant, pH, and temperature. The effects of pretreatment processes for removal of humic substances are site-specific because of raw water quality variables, treatment-plant operating conditions, and treatment-plant design (*2*, *3*). The literature shows some diversity of findings that make it difficult to understand the THM-precursor removal process during coagulation.

Reckhow and Singer (*4*) reported that alum coagulation of aquatic fulvic acid removed TOC and THM formation potential proportionately. Jodellah and Weber (*5*) observed that high levels of TOC removal may yield no selective removal of THM precursors. Just as there were differences in the findings of investigators during bench studies, water-treatment plants also showed varying removals for TOC and THM precursors (*6*). Under slightly acidic pH conditions, Edzwald and co-workers (*2*) reported that similar TOC and THM-precursor removals were achieved despite differences in raw water quality.

GAC Treatment. The specific coagulation process influences both the amount and the THM reactivity of the residual organic matter remaining after treatment prior to chlorination (*7*). Higher-molecular-weight organic compounds were most effectively removed during pretreatment, and lower-molecular-weight organic materials were effectively reduced by GAC (*7*, *8*). Jodellah and Weber (*5*) indicated that increased TOC removal by activated-carbon treatment resulted in decreased THM formation in treated water.

Proper pretreatment appears to benefit activated-carbon adsorption. Randtke and Jepsen (*9*) reported significant increases in the adsorption capacity of organic substances after alum coagulation. Lee and co-workers (*10*) showed that alum coagulation enhanced both carbon adsorption capacity and the rate of uptake. Semmens and co-workers (*11*) observed improved GAC performance with greater levels of pretreatment. Weber and Jodellah (*3*) noted that alum coagulation improved overall adsorbability of TOC.

Treatment at Research Locations

Various conventional treatment methods were used at the research sites to remove or reduce the mix of compounds present in the source water. The type of treatment (conventional and GAC) used at these utilities is as follows.

Cincinnati, Ohio. The primary source water for the Cincinnati Water Works is the Ohio River. To aid settling, 17 mg/L of alum was added to the raw water. Prior to flocculation and clarification, 17 mg/L of lime, ferric sulfate (8.6 mg/L for high turbidity and 3.4 mg/L for low turbidity), and chlorine (plant effluent concentration 1.8 mg/L of free chlorine) were added. Postfiltration adsorption was evaluated by deep-bed GAC contactors with an ultimate empty-bed contact time (EBCT) of 15.2 min.

Jefferson Parish, Louisiana. The Mississippi River provides source water to the Jefferson Parish treatment plant. Potassium permanganate (0.5–1.0 mg/L) was added for taste and odor control. A cationic polyelectrolyte (diallyldimethyl diammonium chloride; 0.5–8.0 mg/L) was added as the primary coagulant, with lime (7–10 mg/L) fed for pH adjustment to 8.0–8.3. Chlorine and ammonia (3:1 ratio) were added for chloramine disinfection (1.4–1.7 mg/L residual after filtration). A sand filter was converted to a postfilter GAC adsorber with about 20 min EBCT. In addition, four GAC pilot columns were operated in series, providing 11.6, 23.2, 34.7, and 46.3 min EBCT.

Manchester, New Hampshire. The principal water source for the Manchester Water Works is Lake Massabesic. Alum and sodium aluminate were added for coagulation, pH adjustment, and alkalinity control at dosage levels averaging about 12 and 8 mg/L, respectively. Chlorine was added prior to sand filtration at an average dose of 1 mg/L. At the clearwell, chlorine was again added in the range of 2–3 mg/L to produce an average-distribution free chlorine residual of 0.5 mg/L. A GAC filter normally used for taste and odor control was used for postfiltration adsorption with 23 min EBCT.

Evansville, Indiana. The Evansville Water Works uses Ohio River water as its source. Chlorine and alum were added before primary settling, with average concentrations of 6 and 28 mg/L, respectively. A free chlorine residual of 1.5–2.0 mg/L was maintained after sand filtration. Approximately

12 mg/L of lime was added after primary settling for pH control to 8.0. A pilot plant operating parallel with the full-scale plant used chlorine dioxide for disinfection. Average alum and polymer (anionic high molecular weight) dosages of 12 and 0.8 mg/L, respectively, were added to the raw water of the pilot plant. An average lime dose of about 6 mg/L was used for pH control to 8.0. Post-pilot-plant GAC contactors had an EBCT of 9.6 min.

Results

TOC removal has been suggested as a means of measuring treatment performance. Although TOC is relatively easy to analyze and incorporates all organic compounds, it does not relate to any specific regulatory requirements. In the following evaluation, however, TOC was used as a general surrogate parameter to determine the performance of conventional treatment and GAC adsorption.

Removal of instantaneous trihalomethanes and their precursors to meet a TTHM standard was also evaluated by using the terminal trihalomethane (terminal THM) parameter. Because the utilities studied used various disinfectants that affected the trihalomethane concentrations, terminal THM (instantaneous THM plus THM formation potential) allows a comparison among utilities by indicating the maximum trihalomethane in the distribution system at a given time. In this evaluation, ambient pH and temperature were maintained. Chlorine dosages were chosen to ensure a chlorine residual after a storage time that simulated the time from the treatment plant to the farthest point in the distribution system.

Conventional Treatment. The TOC raw water concentration at Evansville, Indiana, varied from 2.8 to 3.6 mg/L during one 85-day operational phase. Average raw water TOC concentration was 3.0 mg/L. Average TOC removal was 37% with full-scale conventional treatment and 40% for the pilot plant. Average sand filter TOC concentration was 1.9 and 1.8 mg/L for the full-scale and pilot plant, respectively. Evansville's 3-day raw water terminal THM concentrations ranged from 95 to 178 μg/L, for an average of 140 μg/L. After conventional treatment the average concentration was 82 μg/L for the full-scale plant (an average reduction of 41%) and 34 μg/L for the pilot plant (a 76% reduction). More efficient THM precusor removal in the pilot plant for this operational phase was attributed to the addition of a polymer (anionic high molecular weight) for effective turbidity removal.

The initial raw water TOC concentration at Manchester was 4.6 mg/L. It varied from 3.8 to 4.8 mg/L, with an average concentration of 4.5 mg/L for 130 days of operation. The raw water TOC concentration was reduced about 47% to an average of 2.4 mg/L. Three-day terminal THM concentrations for Manchester's raw water at ambient temperature ranged from 104 to 191 μg/L (an average of 151 μg/L). Precursor removal through conventional treatment reduced the 3-day terminal THM to an average of 70 μg/L, a 54% reduction.

In Cincinnati, where ferric sulfate was used as the primary coagulant, a 41% reduction in the TOC concentration was seen through conventional treatment. Raw water TOC concentrations ranged from 1.9 to 5.9 mg/L, for an average of 3.4 mg/L. After conventional treatment the TOC concentrations ranged from 1.1 to 3.4 mg/L, for an average of 2.0 mg/L. The average reduction through conventional treatment was 41%. Three-day terminal THM concentrations for the raw water ranged from 64 to 211 μg/L, for an average of 146 μg/L. After conventional treatment, the terminal THM concentrations ranged from 39 to 181 μg/L, for an average of 89 μg/L, producing an average terminal THM reduction of 39%.

At Jefferson Parish, polymers were used as the primary coagulant. The raw water (Mississippi River) TOC concentration ranged from 2.9 to 5.9 mg/L, with an average of 4.0 mg/L. After conventional treatment the TOC concentrations ranged from 2.3 to 3.8 mg/L, with an average of 2.9 mg/L, for a reduction of 27.5%. Five-day terminal THM concentrations for the raw water ranged from 133 to 511 μg/L, with an average of 281 μg/L. After conventional treatment the range was 82 to 364 μg/L, for an average of 175 μg/L. Average 5-day terminal THM reduction through conventional treatment was 37.7%.

Table I summarizes the removal of the TOC through conventional treatment. Table II shows removal of terminal trihalomethanes through various steps in the treatment process. In this case, terminal trihalomethanes are used because they represent the formation potential of TTHM in the dis-

Table I. Average Total Organic Carbon Removal During Conventional Treatment

Water Utility	*Raw Water (mg/L)*	*Sand Filter Effluent (mg/L)*	*Percent Removal*
Cincinnati, OH	3.4	2.0	41
Jefferson Parish, LA	4.0	2.9	28
Manchester, NH	4.5	2.4	47
Evansville, IN	3.0	1.9	37

Table II. Average Terminal Trihalomethane Removal During Conventional Treatment

Water Utility	*Terminal Day*	*Raw Water (μg/L)*	*Sand Filter Effluent (μg/L)*	*Percent Removal*
Cincinnati, OH	3	146	89	39
Jefferson Parish, LA	5	281	175	38
Manchester, NH	3	151	70	54
Evansville, IN	3	140	82	41

tribution system itself. The time to the most distant customer in the distribution system is represented by the terminal day.

As can be seen from Tables I and II, the utilities examined experienced variable performance in average percent removal of both TOC and terminal THM. This variability may be due in part to source water quality. For example, Cincinnati, Jefferson Parish, and Evansville (with a river water source) had lower percent removal efficiency for terminal THM than did Manchester (with a lake source).

Granular Activated Carbon Treatment. GAC performance for removing both TOC and terminal THM also varied for the different utilities evaluated. For instance, at Evansville, Indiana, after conventional treatment, GAC further reduced the TOC concentration during about 30 days of operation, after which the GAC effluent tracked just below the filter effluent (Figure 1). The 3-day terminal THM concentration of the GAC effluent was essentially the same as the filter effluent after about 30 days of operation (Figure 2).

At Manchester, New Hampshire, the TOC concentration of the GAC effluent was about 0.5 mg/L at the start of one evaluation. It increased in concentration until about 35 days of operation, before tracking just below the sand-filter effluent (Figure 3). The 3-day terminal THM concentration was initially about 10 μg/L, increasing to about 45 μg/L after 40 days of operation, and then tracking below the sand-filter effluent (Figure 4).

The TOC effluent GAC concentration at Cincinnati, Ohio, was about 0.2 mg/L at the start of one of the runs and increased to about 1.1 mg/L after approximately 100 days of operation. As with Manchester and Evansville, the TOC then tracked just below the sand-filter effluent (Figure 5). The 3-day terminal THM concentration of the GAC effluent was about 3 μg/L at the start of an adsorption study, and "breakthrough" occurred after about 50 days of operation. From about day 110, the 3-day terminal THM effluent was approximately the same increment below the sand-filter effluent throughout the 320-day study (Figure 6).

The full-scale GAC adsorber at Jefferson Parish, Louisiana, seemed to remove the TOC concentration steadily for about 160 days. Initial concentration was 0.2 mg/L, increasing to about 2.0 mg/L (Figure 7). The 5-day terminal THM GAC effluent concentration for the full-scale system at Jefferson Parish was about 15 μg/L at the start of one run and, like the TOC, steadily increased for 140 days (Figure 8).

Effect of Empty-Bed Contact Time

The length of GAC operation before replacement or reactivation depends on several factors, one of which is empty-bed contact time (EBCT). If a drinking-water utility is required to use existing filters, very little flexibility is available for selection of EBCT. In designing a new system, however, the

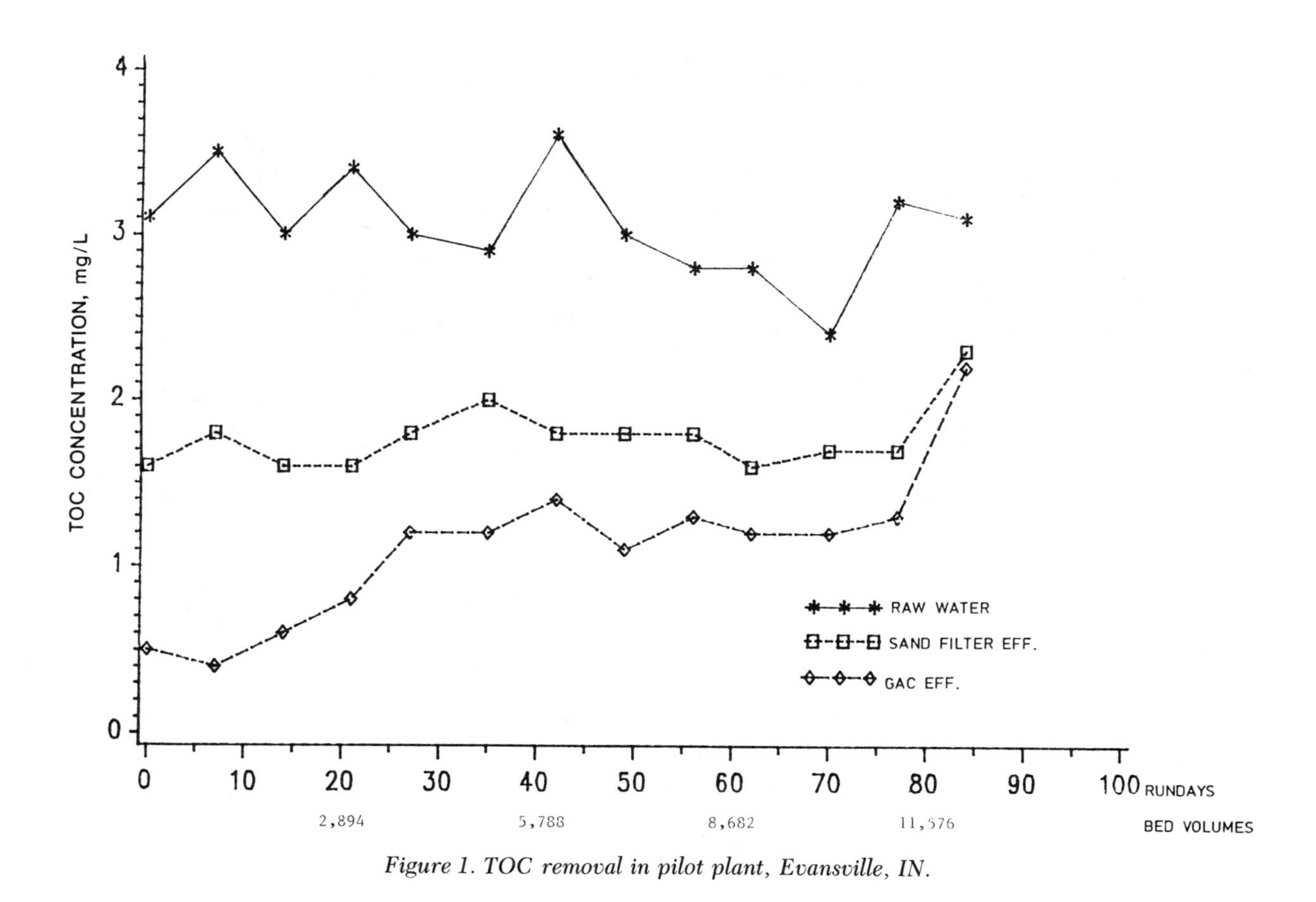

Figure 1. TOC removal in pilot plant, Evansville, IN.

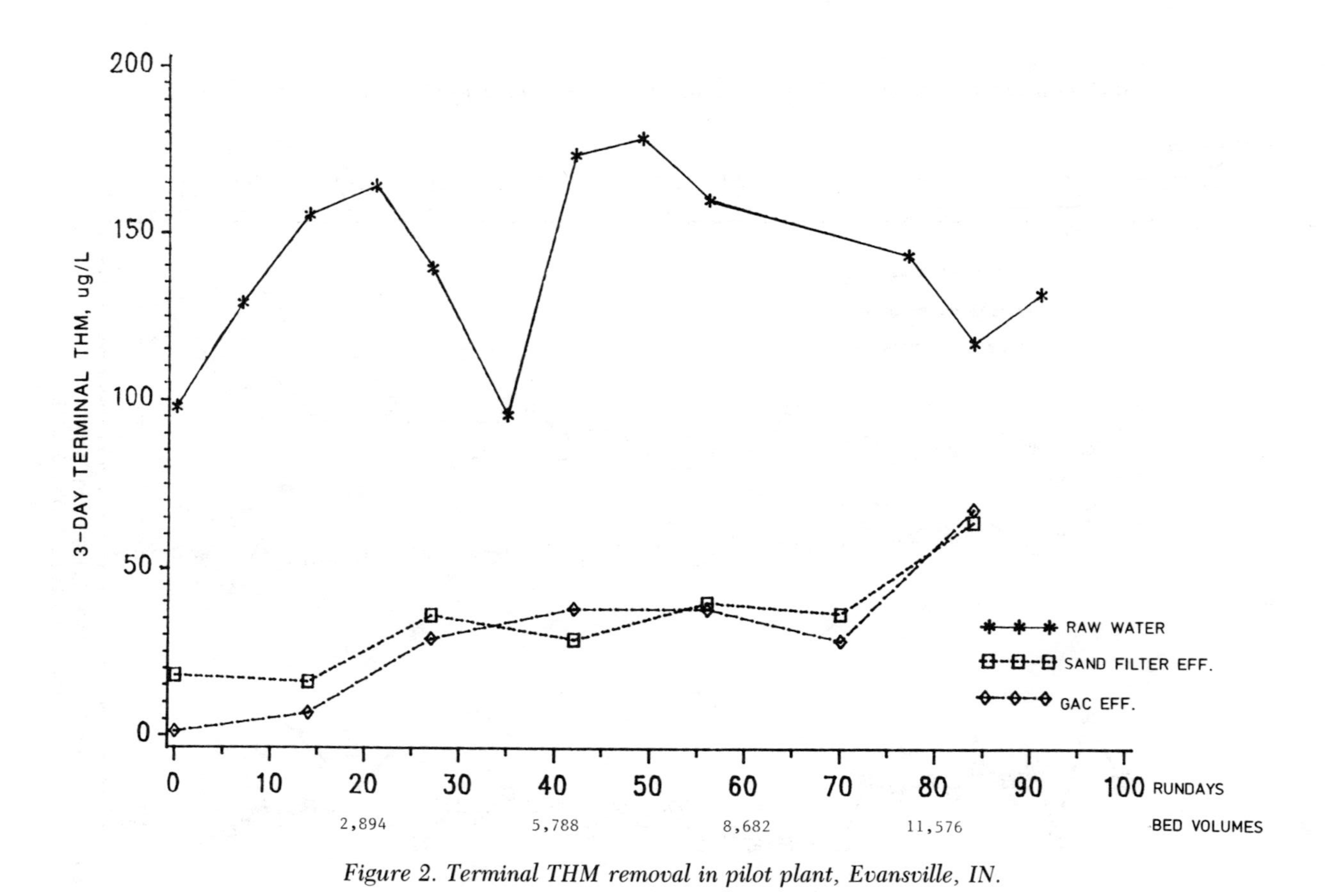

Figure 2. Terminal THM removal in pilot plant, Evansville, IN.

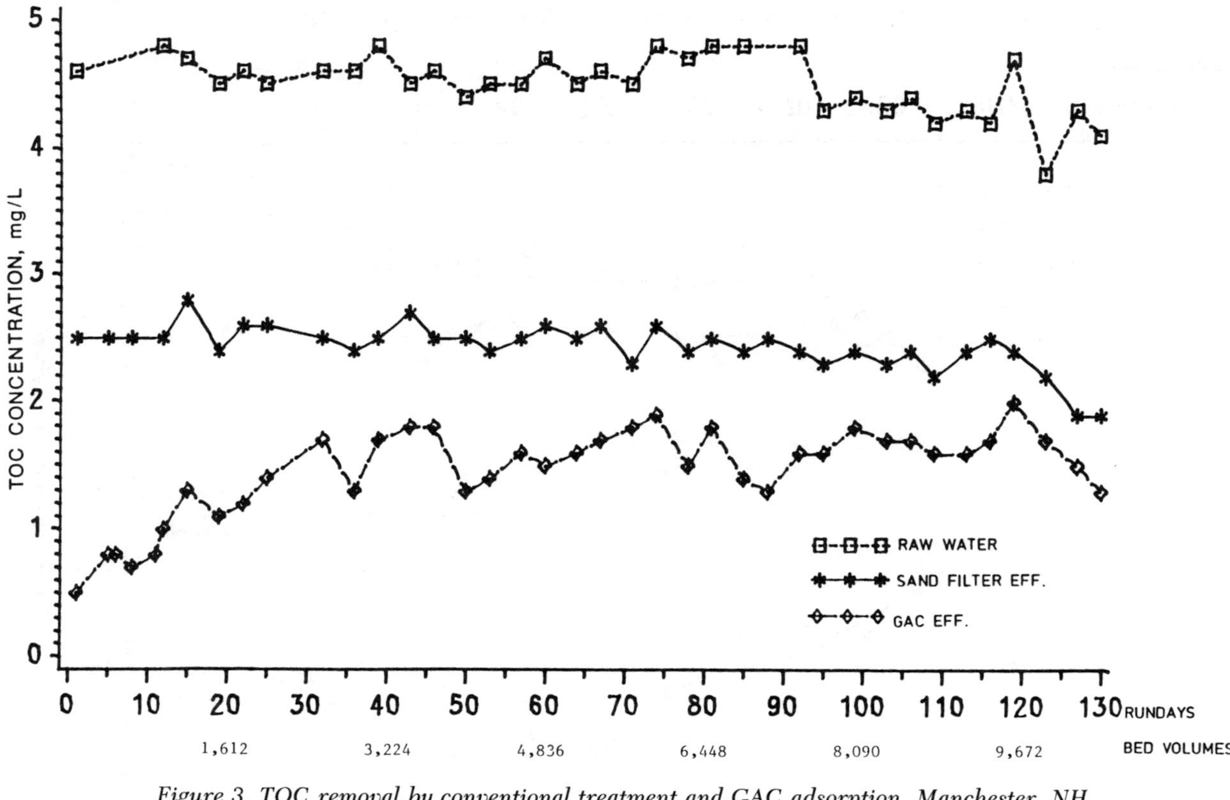

Figure 3. TOC removal by conventional treatment and GAC adsorption, Manchester, NH.

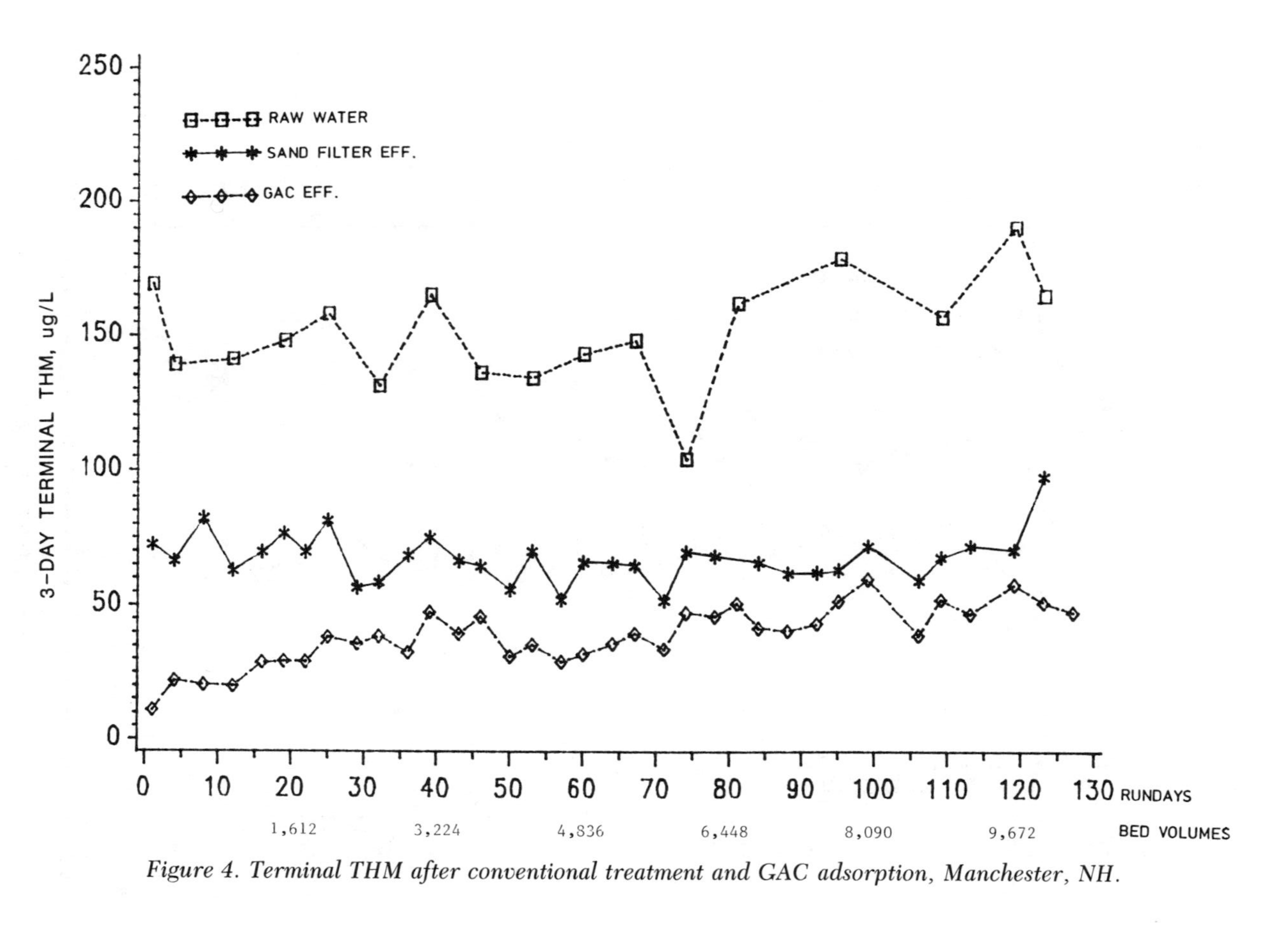

Figure 4. Terminal THM after conventional treatment and GAC adsorption, Manchester, NH.

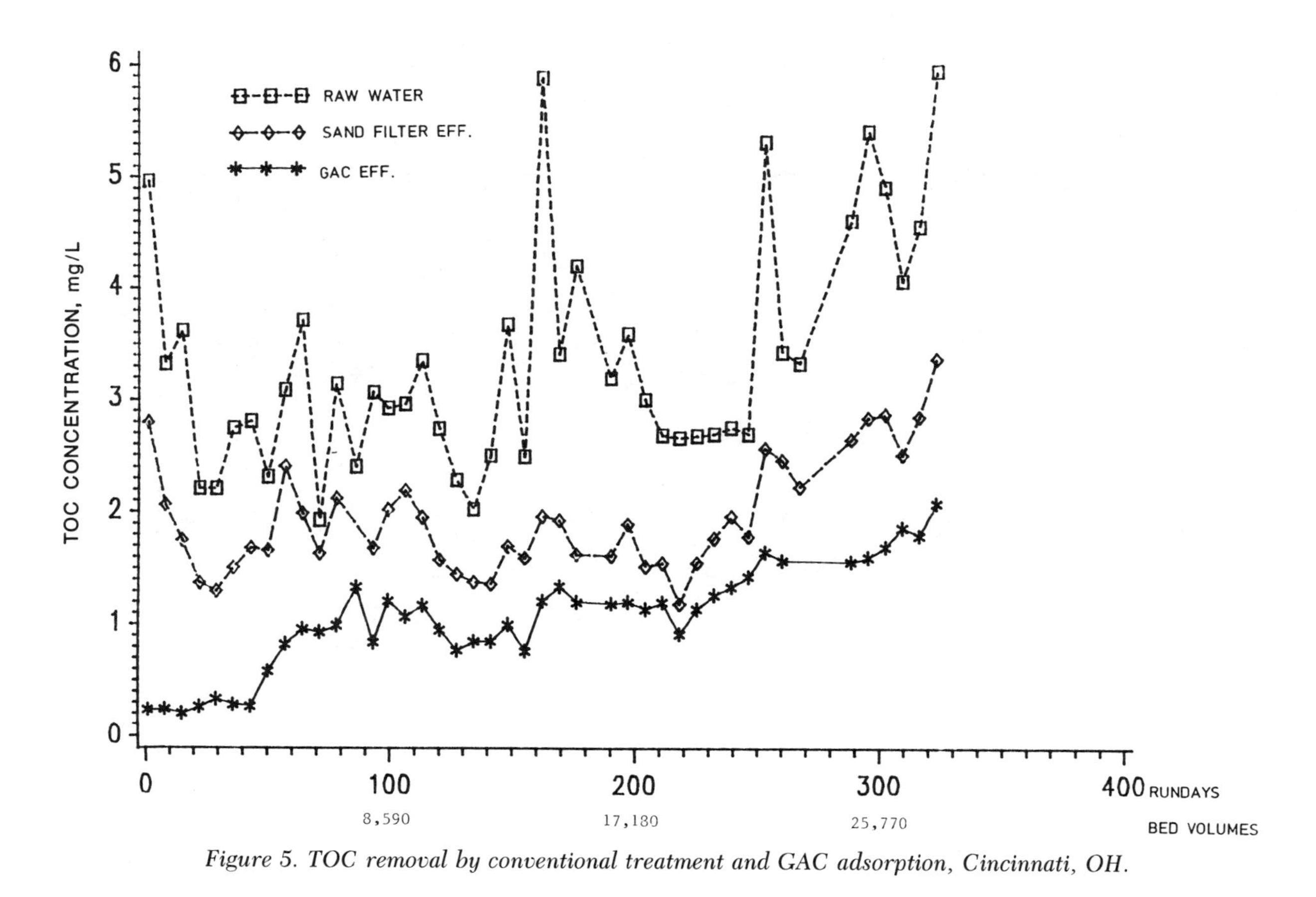

Figure 5. TOC removal by conventional treatment and GAC adsorption, Cincinnati, OH.

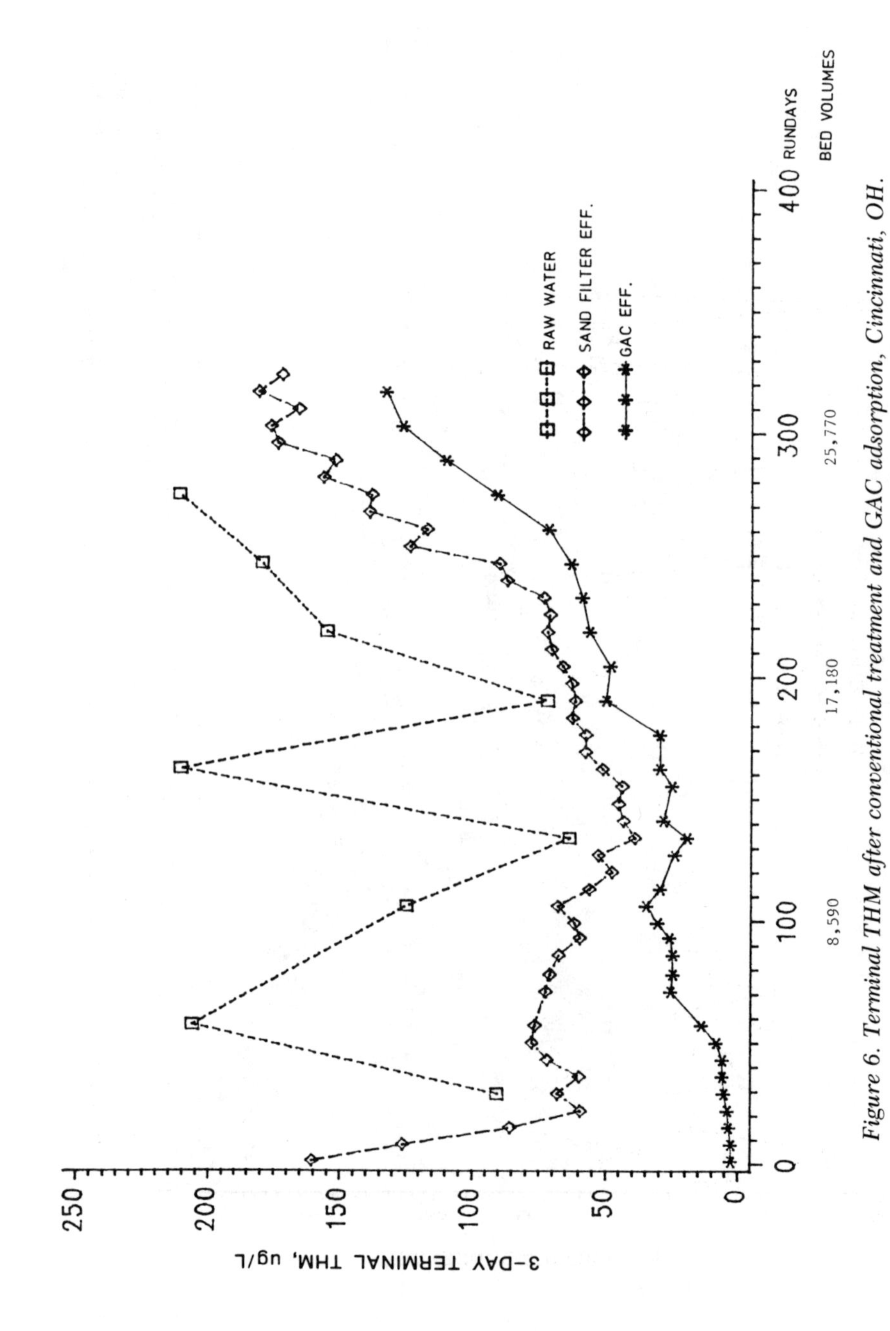

Figure 6. Terminal THM after conventional treatment and GAC adsorption, Cincinnati, OH.

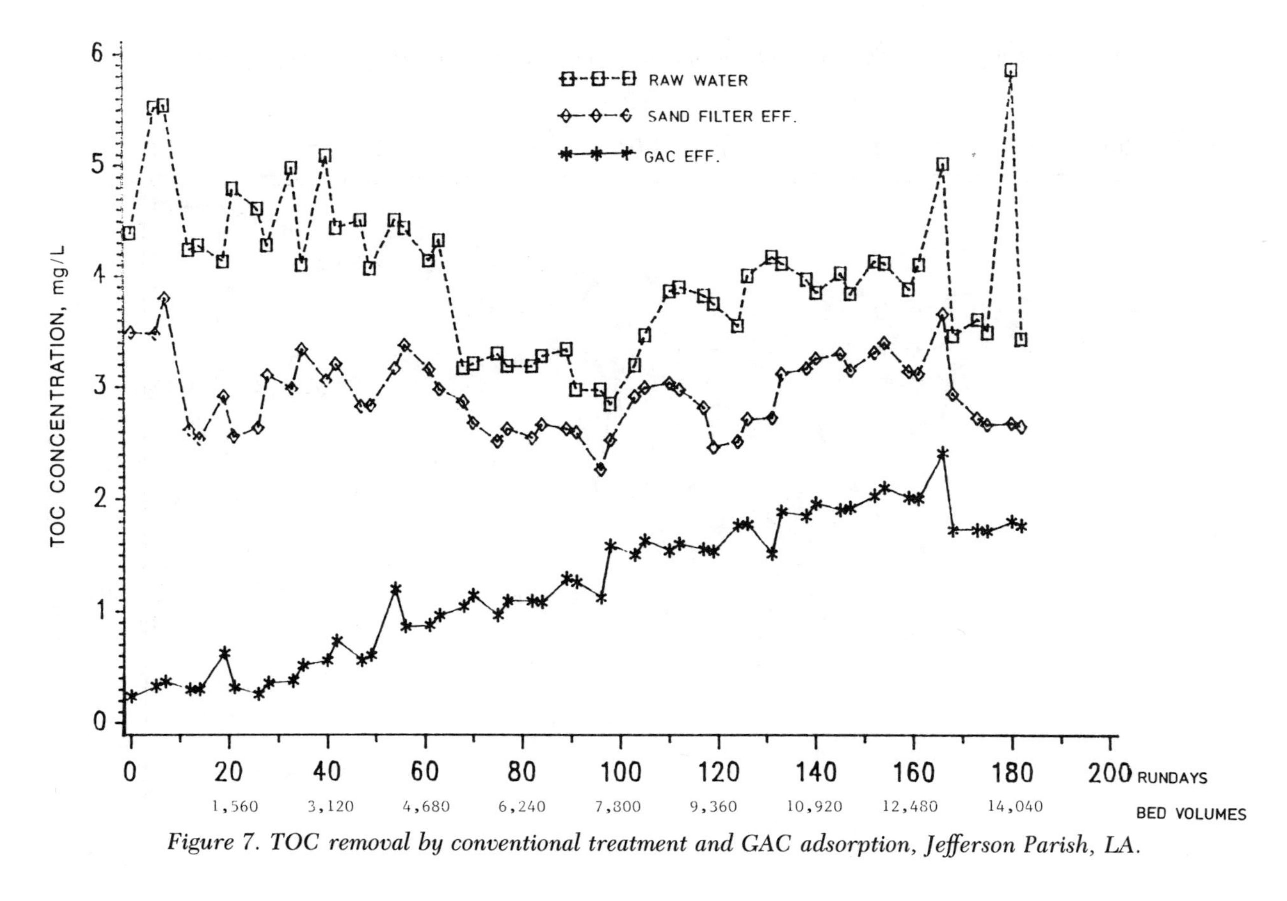

Figure 7. TOC removal by conventional treatment and GAC adsorption, Jefferson Parish, LA.

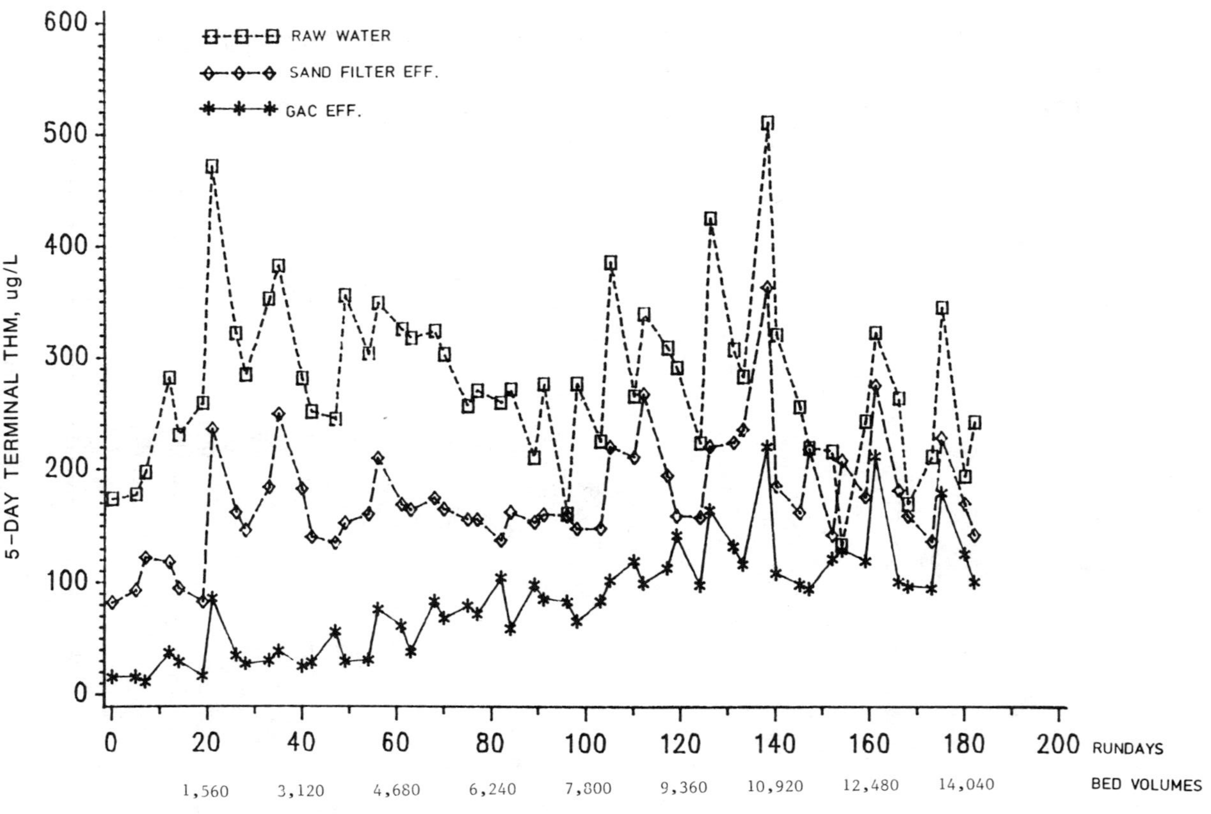

Figure 8. Five-day terminal THM after conventional treatment and GAC adsorption, Jefferson Parish, LA.

utility has an opportunity to determine the best EBCT, relative to performance and cost.

A GAC exhaustion criteria of 1.0 mg/L of TOC has been suggested as a reasonable performance standard (*12*). By applying this criterion to TOC breakthrough curves for virgin GAC, one can see the effects of EBCT. The GAC contactor at Cincinnati, Ohio, had sampling ports located at 4.3 ft (1.3 m), 7.0 ft (2.1 m), and 15.0 ft (4.6 m), yielding EBCTs of 4.4, 7.2, and 15.2 min, respectively. Length of GAC operation to the TOC exhaustion criterion and the carbon-use rate as shown in Table III indicate that, for the Cincinnati evaluation, longer EBCTs provided more efficient use of the GAC.

At Jefferson Parish, Louisiana, pilot columns in series were used to produce EBCTs of 11.6, 23.2, 34.7, and 46.3 min. By applying the same TOC exhaustion criterion (1.0 mg/L), one can see a longer operational time with increased EBCT. After 23.2 min, however, the incremental advantage of longer EBCT is questionable, as shown in Table IV.

The terminal THM removal through various EBCTs showed the same general trend as noted with TOC. For Cincinnati, Ohio, increased EBCT up to 15.2 min produced additional removals (Figure 9). For Jefferson Parish, Louisiana, the closeness of the terminal THM concentrations for 23.2 min EBCT and higher indicates that relatively little, if any, advantage is gained at higher EBCTs (Figure 10).

Granular Activated Carbon for Trihalomethane Control

Some water utilities are able to maintain their THM concentrations below the existing promulgated standard of 0.10 mg/L (100 μg/L) by proper con-

Table III. Summary Data at TOC Exhaustion of 1.0 mg/L

GAC Depth	EBCT (min)	Hydraulic Loading (m/hr)	Hydraulic Loading (gpm/ft²)	TOC Exhaustion Time (days)	Carbon Use Rate (kg/mL)	Organic Loading (g/kg)
1.3 m (4.3 ft)	4.4	17.8	7.4	22	46	15
2.1 m (7.0 ft)	7.2	17.8	7.4	71	34	25
4.6 m (15.0 ft)	15.2	17.8	7.4	204	26	51

NOTE: Virgin bituminous coal 12 × 40 GAC.

Table IV. Summary Data at TOC Exhaustion of 1.0 mg/L

GAC Depth	EBCT (min)	Hydraulic Loading (m/hr)	Hydraulic Loading (gpm/ft²)	TOC Exhaustion Time (days)	Carbon Use Rate (kg/mL)	Organic Loading (g/kg)
0.9 m (3.0 ft)	11.6	4.9	2.02	42	73	32
1.8 m (6.0 ft)	23.2	4.9	2.02	105	30	78
2.7 m (9.0 ft)	34.7	4.9	2.02	140	22	107
3.7 m (12.0 ft)	46.3	4.9	2.02	159	20	126

NOTE: Virgin bituminous coal GAC.

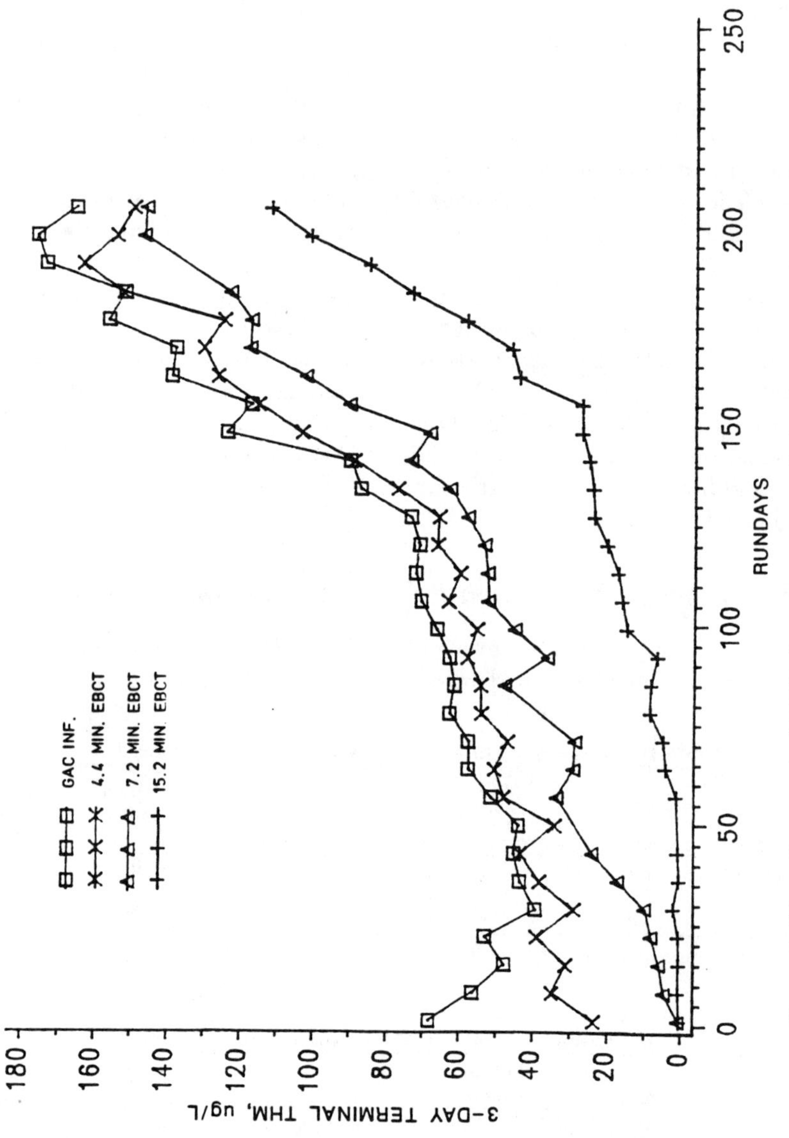

Figure 9. Three-day terminal THM for various empty-bed contact times, Cincinnati, OH.

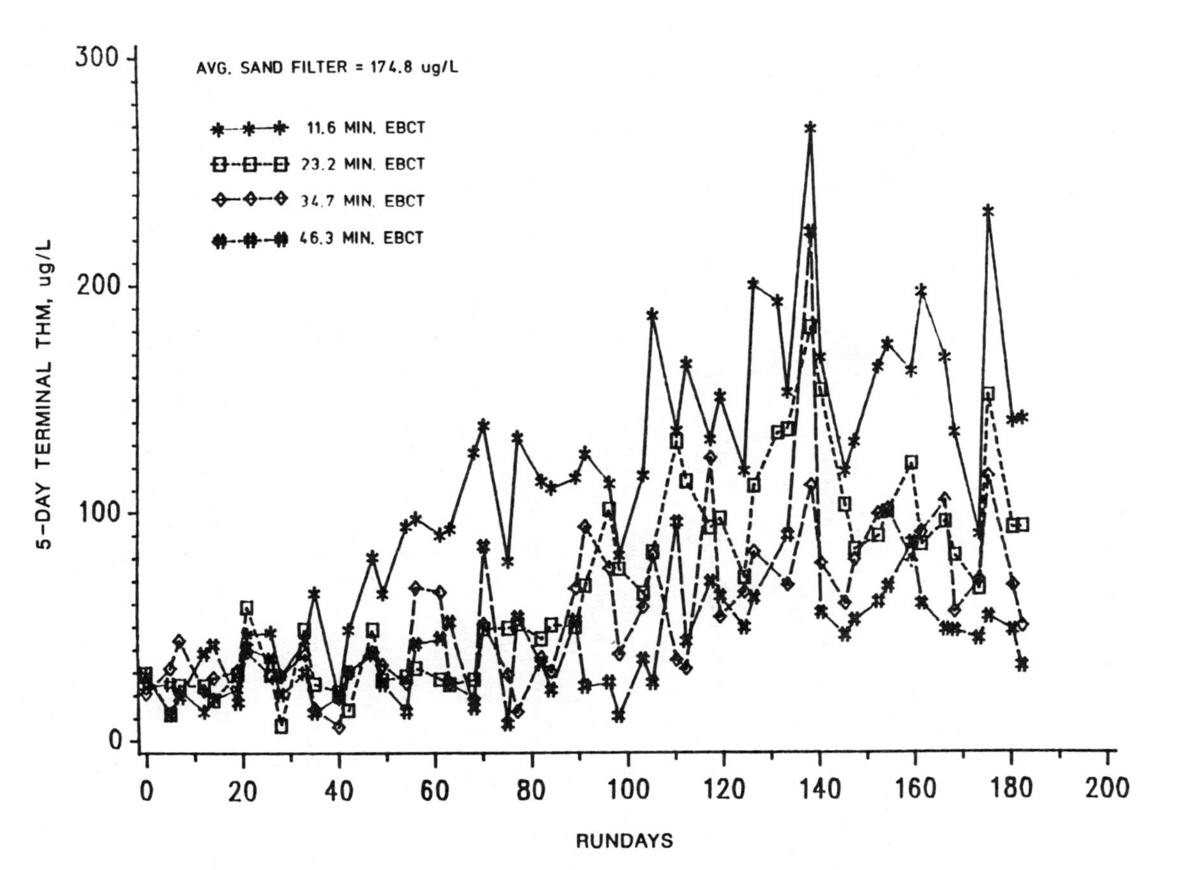

Figure 10. Five-day terminal THM for GAC series contactors, Jefferson Parish, LA.

ventional treatment. If, however, the standard is reduced substantially, other treatment alternatives will be required. GAC may be an alternative worth evaluating. The length of time that GAC can remove trihalomethanes to meet a standard of 10, 25, 50, or 100 μg/L will determine its efficacy as a viable treatment option.

Because terminal THM values can simulate concentrations in the distribution system, one can estimate the length of GAC operation for meeting THM goals. Table V gives an indication of how long GAC can remove various concentrations of THMs.

As can be seen from Table V, establishment of a 10-μg/L trihalomethane standard will probably negate the use of GAC. In addition, use of GAC to meet a 25-μg/L standard may not be feasible. However, GAC may be more attractive at the 50-μg/L trihalomethane concentration. Carbon-use rates for the operational days at the 50-μg/L trihalomethane concentration were 0.216, 0.716, 0.690, and 0.459 lb/1000 gal for Cincinnati, Jefferson Parish, Manchester, and Evansville, respectively.

Data Normalization and Prediction of THM Concentrations

The data reported here have shown the performance of GAC in removal of terminal THM over various days of operation and bed volumes through GAC adsorbers at different locations. Normalization of the data by using percent removal shows that the GAC adsorbers used at Cincinnati produced the overall highest removal rate for terminal THM (Figure 11). Evansville had the lowest percent removal.

TOC has been suggested as a surrogate for prediction of THM concentrations. If TOC is removed through GAC adsorption, will THM precursors be selectively removed? A definite pattern of TOC with 3-day THM formation potential (THMFP) and TOC with 7-day THMFP was noted in Cincinnati for the GAC effluent. This pattern indicates that TOC may be used as a predictive tool at that location (Figures 12 and 13). With Jefferson Parish, Louisiana (another plant using river water as its source), TOC and 5-day

Table V. Length of GAC Operation Before Exceeding Terminal THM Levels

	10 μg/L[a]		*25 μg/L*[a]		*50 μg/L*[a]		*100 μg/L*[a]	
Location	*Day*	*Inflow (μg/L)*	*Day*	*Inflow (μg/L)*	*Day*	*Inflow (μg/L)*	*Day*	*Inflow (μg/L)*
Cincinnati, OH (3-day term, 15.2-min EBCT)	50	75	155	45	208	70	280	150
Jefferson Parish, LA (5-day term, 18.8-min EBCT)	—	—	20	80	63	170	103	220
Manchester, NH (3-day term, 23-min EBCT)	2	73	16	70	98	65	—	—
Evansville, IN (3-day term, 9.6-min EBCT)	—	—	6	96	56	53	—	—

[a]THM goals.

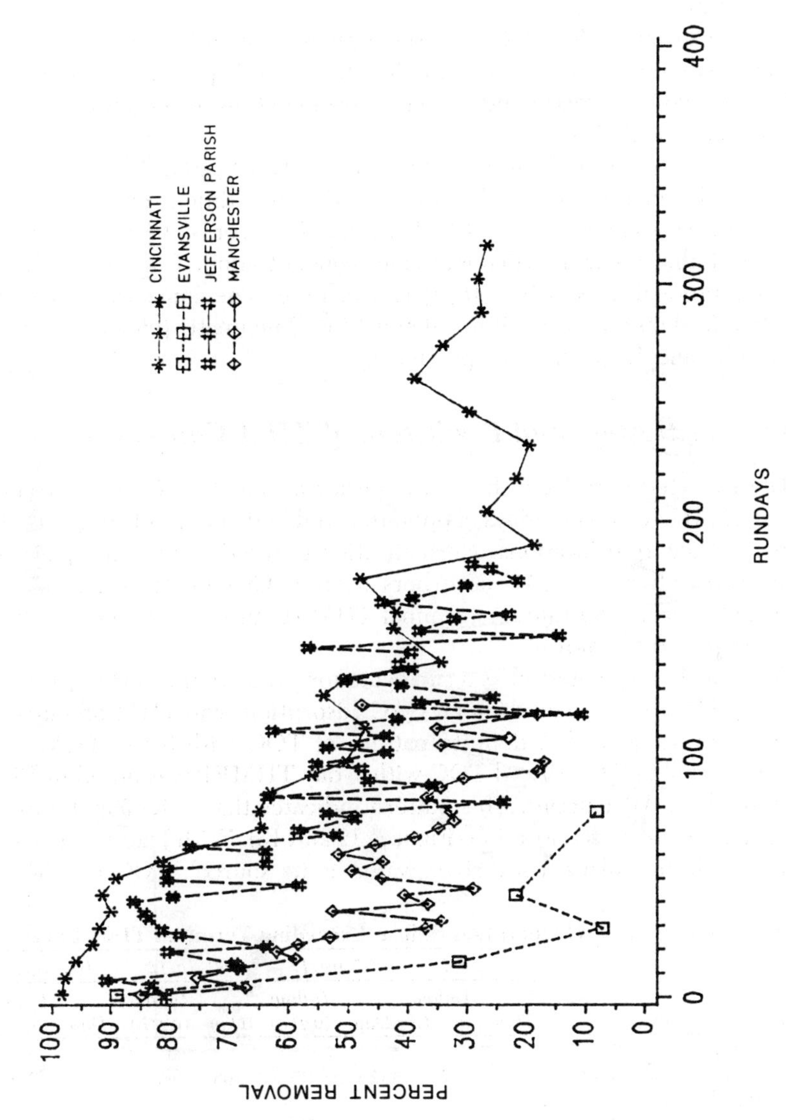

Figure 11. Terminal THM percent removal for GAC effluent.

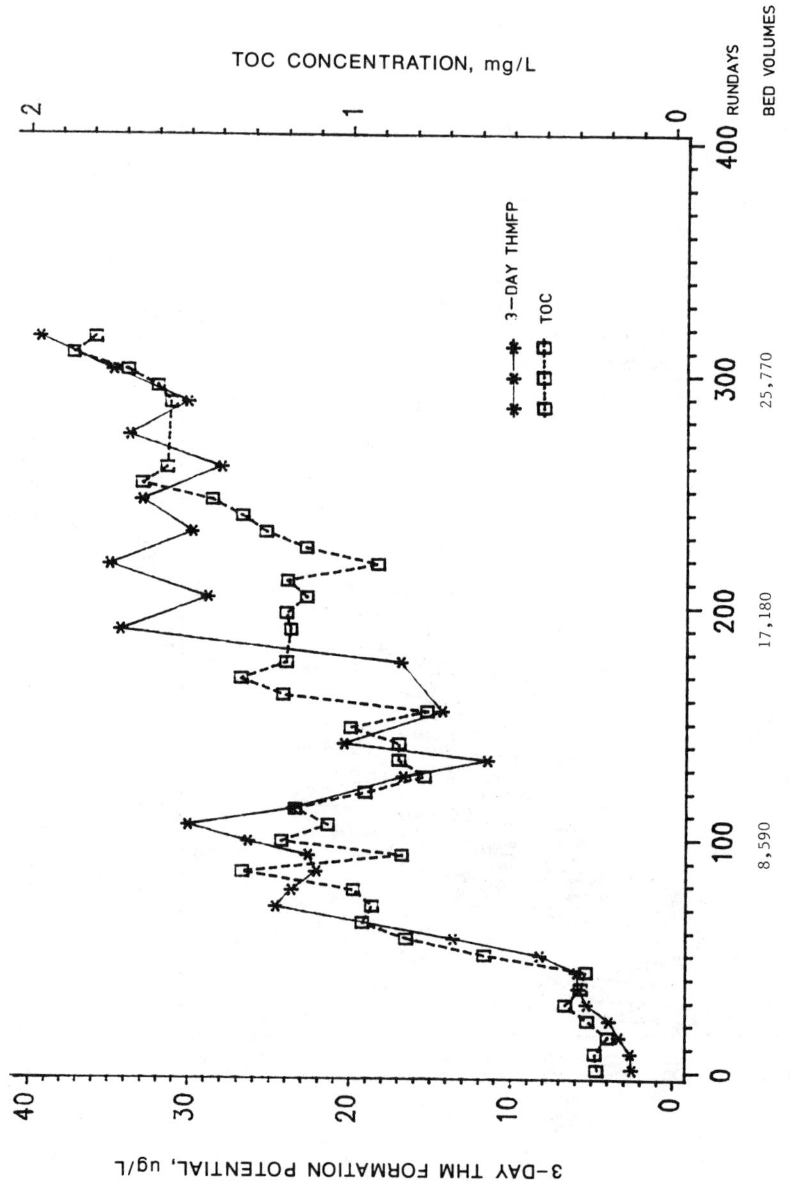

Figure 12. Comparison of GAC effluent 3-day THM formation potential and TOC, Cincinnati, OH.

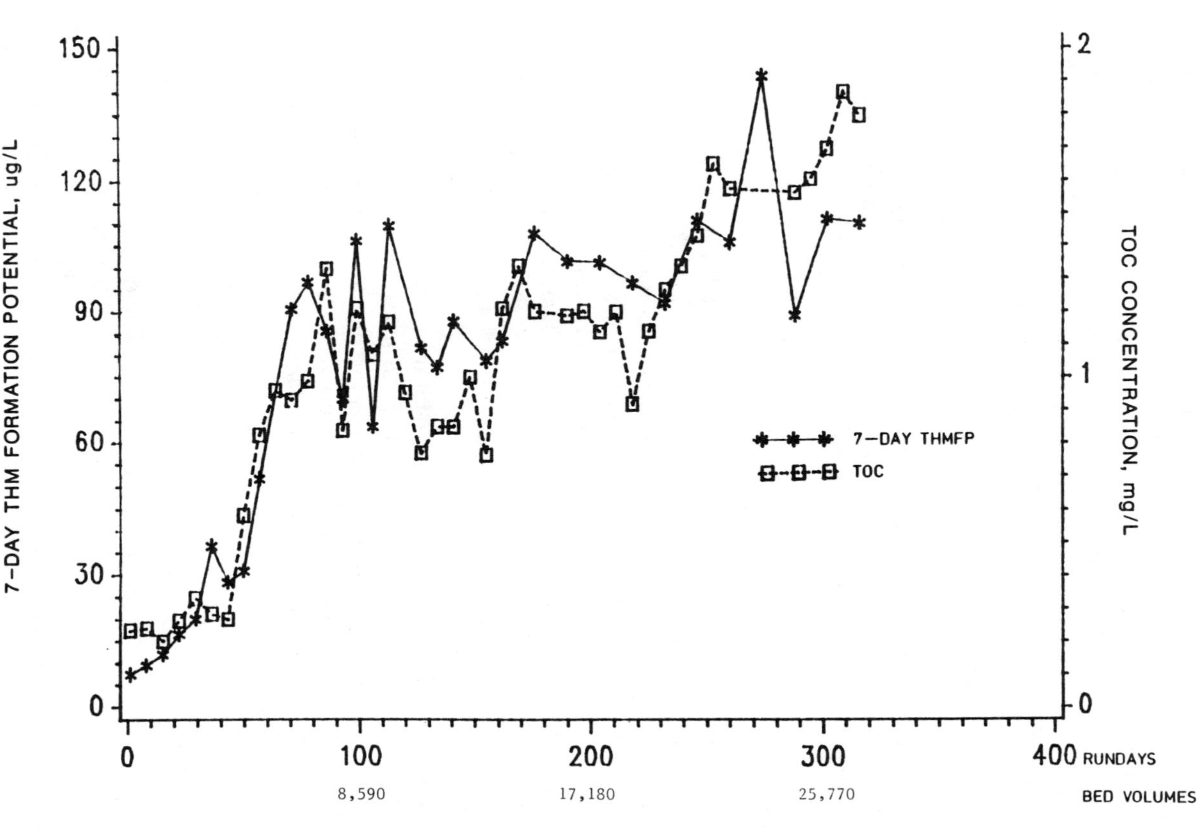

Figure 13. Comparison of GAC effluent 7-day THM formation potential and TOC, Cincinnati, OH.

THMFP also seemed to follow a pattern. A close correlation was also seen for Manchester, New Hampshire, GAC effluent (lake water source) with TOC and 3-day THMFP.

Removal of TOC by GAC may give an indication of THMFP removal. In a comparison of percent THMFP removal to percent TOC removal for the four utilities evaluated, Figure 14 shows a 45° line of equal percent removal. Regression of this data, however, indicates that removing TOC does not necessarily mean removing an equal percentage of THMFP. The following equation describes the data from the four utilities:

$$\text{THMFP} = 7.83 + 0.87\ \text{TOC}\ (R^2 = 0.73) \tag{1}$$

The instantaneous organic halide did not seem to be as good a predictor of THMFP as TOC. However, the instantaneous organic halide might be used to predict GAC instantaneous THM breakthrough (Figure 15).

Summary

Proper conventional treatment can reduce TOC and THM precursors substantially. For the full-scale systems, average percent TOC removal was variable, possibly because of the type of coagulants used. Average percent terminal THM removal, however, seemed to follow a pattern. The river water sources were about the same, with lower average terminal THM percent removal through conventional treatment than for the lake water at Manchester. This result may be attributed to better TOC removal during conventional treatment and different THM precursors in the lake water than in the river source water.

Although THM precursors are reduced during conventional treatment, this reduction will probably not be enough to meet THM concentrations much below 100 μg/L if chlorine is used as the primary disinfectant. With GAC adsorption, however, additional precursors are removed. For the utilities evaluated, meeting a 50-μg/L THM standard appears to be possible after GAC treatment.

Acknowledgments

The authors thank Sue Campbell for her direction and coordination in producing the graphics. The authors also thank Sandi Dryer for typing the manuscript. This paper has been reviewed in accordance with the U.S. Environmental Protection Agency's peer and administrative review policies and approved for presentation and publication. Mention of trade names or commercial products does not constitute endorsement or recommendation for use by the U.S. Environmental Protection Agency.

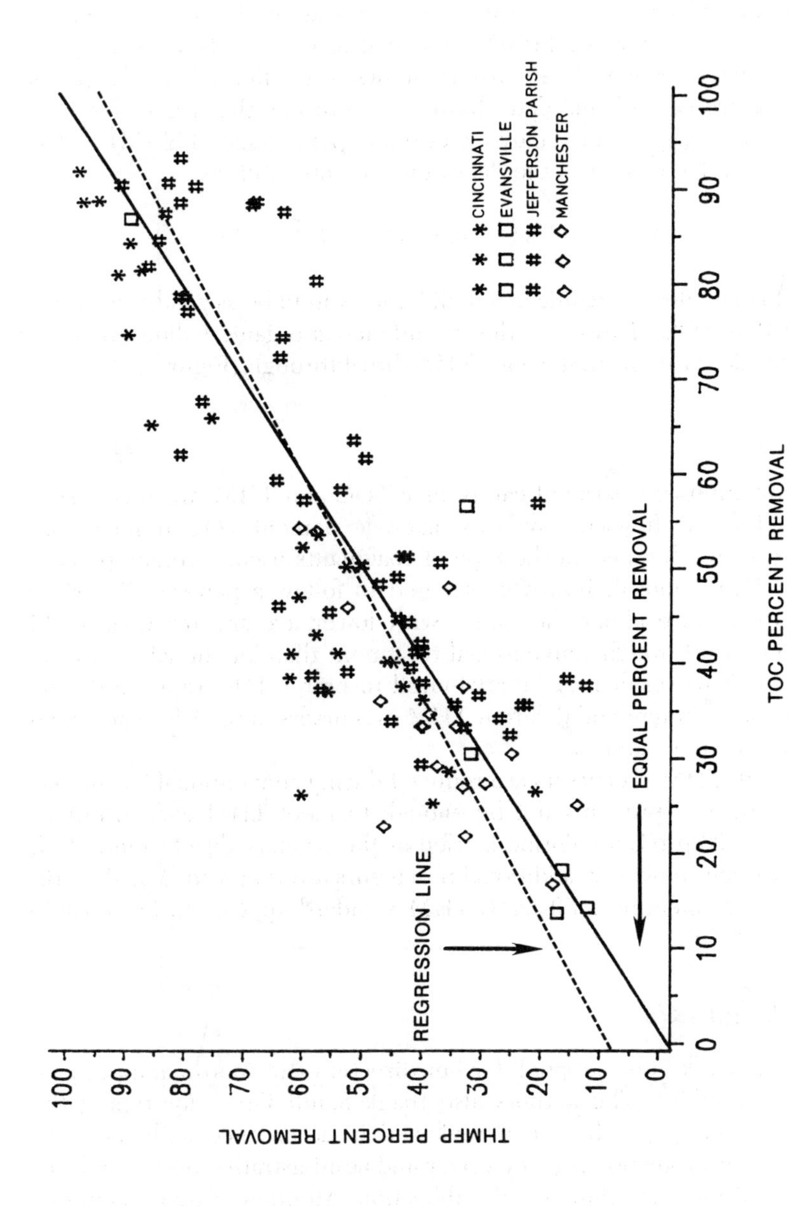

Figure 14. THMFP versus TOC percent removal for GAC effluent.

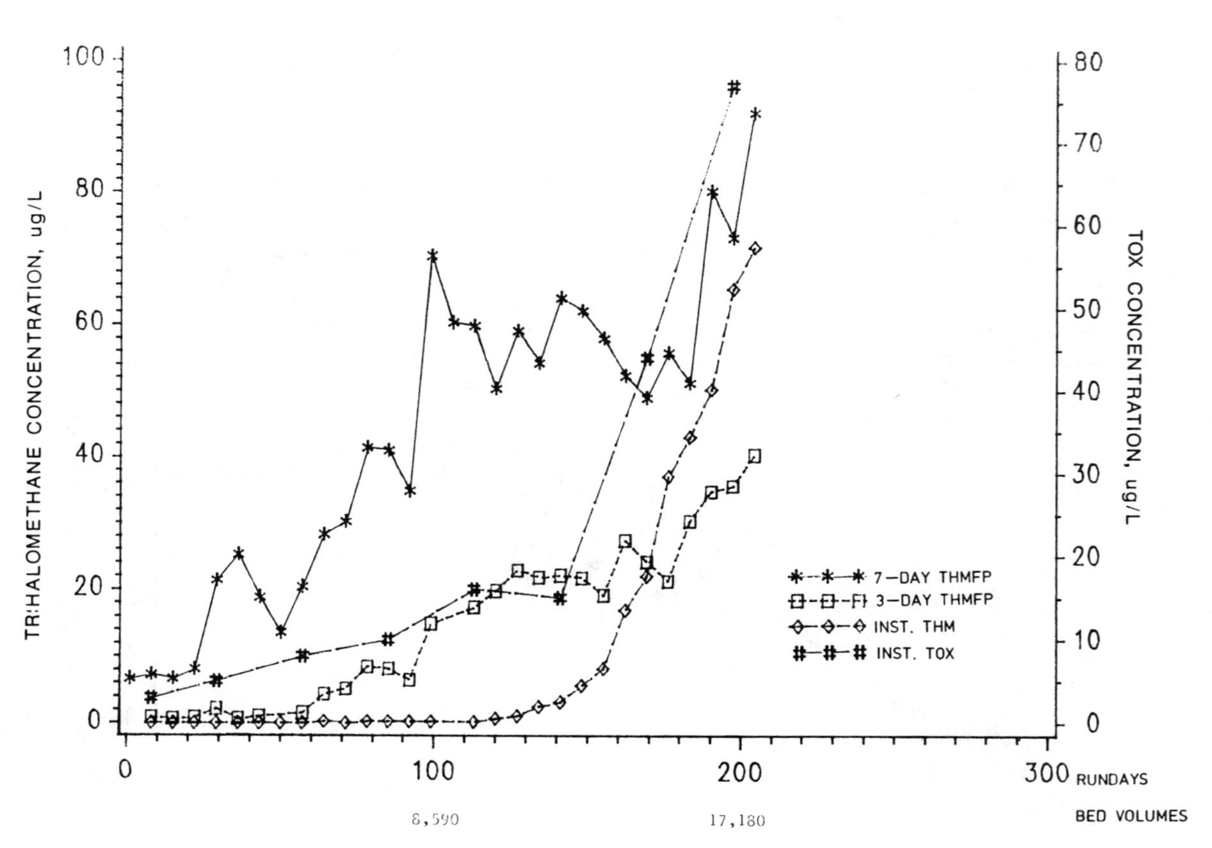

Figure 15. Comparison of GAC effluent trihalomethanes and TOX, Cincinnati, OH.

References

1. "Safe Drinking Water Act" as amended by the "Safe Drinking Water Act Amendments" of 1986; Public Law 99–339, 1986.
2. Edzwald, J. K.; Becker, W. C.; Wattier, K. L. *J. Am. Water Works Assoc.* **1985,** 77(4), 122–132.
3. Weber, W. J.; Jodellah, A. M. *J. Am. Water Works Assoc.* **1985,** 77(4), 132–137.
4. Reckhow, D. A.; Singer, P. C. *Proc. Am. Water Works Assoc. Annu. Conf.* Las Vegas, NV; American Water Works Association: Denver, CO, 1983.
5. Jodellah, A. M.; Weber, W. J. *J. Am. Water Works Assoc.* **1985,** 77(10), 95–100.
6. Ohio River Valley Water Sanitation Commission. *Water Treatment Process Modification for Trihalomethane Control and Organic Substances in the Ohio River*; U.S. Environmental Protection Agency; National Technical Information Service: Springfield, VA, 1980; EPA–600/2–80–028.
7. Collins, M. R.; Amy, G. L.; King, P. H. *J. Environ. Eng.* (*N.Y.*) **1985,** *111*(6), 850–864.
8. Semmens, M. J.; Staples, A. B. *J. Am. Water Works Assoc.* **1986,** *78*(2), 76–81.
9. Randtke, S. J.; Jepsen, C. P. *J. Am. Water Works Assoc.* **1981,** 73(8), 411–419.
10. Lee, M. C.; Snoeyink, V. L.; Crittenden, J. C. *J. Am. Water Works Assoc.* **1981,** 73(8), 440–446.
11. Semmens, M. J.; Staples, A. B.; Hohenstein, G.; Norgaard, G. E. *J. Am. Water Works Assoc.* **1986,** *78*(8), 80–84.
12. Lykins, B. W., Jr.; Geldreich, E. E.; Adams, J. Q.; Ireland, J. C.; Clark, R. M. *Granular Activated Carbon For Removing Nontrihalomethane Organics From Drinking Water*; U.S. Environmental Protection Agency; National Technical Information Service: Springfield, VA, 1984; EPA–600/2–84–165.

RECEIVED for review July 24, 1987. ACCEPTED for publication March 10, 1988.

35

The Fate and Removal of Radioactive Iodine in the Aquatic Environment

R. Scott Summers[1], Friedrich Fuchs, and Heinrich Sontheimer

Engler-Bunte-Institut, Universität Karlsruhe, 7500 Karlsruhe, Federal Republic of Germany

The reaction of iodine with aquatic humic substances (HS) and the subsequent removal of the products by typical drinking-water-treatment processes was investigated. Both iodine and iodide react completely with isolated HS in the concentration range below 0.03 mg of I per mg of HS and behave similarly with Rhine River water. The reaction is independent of pH, initial HS concentration, and HS molecular size. However, at higher I–HS ratios iodine reacts slightly more than iodide. Kinetic studies indicate that the reaction is complete within 10 min. No interaction was found between methyl iodide and HS. Flocculation and activated-carbon (AC) adsorption were effective for the removal of the I–HS complex, and the dissolved organic carbon measurement served as a good surrogate parameter. Volatilization and AC adsorption were effective for methyl iodide removal.

AFTER A NUCLEAR REACTOR ACCIDENT the release of radionuclides poses a problem for drinking-water-treatment facilities using surface waters as their raw water source. A recent nuclear power plant accident resulted in high levels of radioactivity in the environment throughout Europe, as shown in Table I for the Federal Republic of Germany. The highest activity levels occurred in the southern part of the country where, fortunately, 95% of the potable water originates from ground-water aquifers that were not directly contaminated. However, for communities that use surface waters, an un-

[1]**Current address: Civil and Environmental Engineering Department, University of Cincinnati, Cincinnati, OH 45221–0071**

0065–2393/89/0219–0623$06.00/0

Table I. Radioactivity Levels in the Federal Republic of Germany after the Chernobyl Accident

Source	*Maximum Reported Activity*	*Main Component*	*Ref.*
Air (Bq/m^3)	150	I-131	1
Ground (Bq/m^2)	280,000	I-131	1
Aqueous (Bq/L)			
Rain	35,000	I-132 or I-131	2
River	370	I-131	2
Reservoir	570	I-131	2
Ground	<1	—	2
Lake	360	I-131	2
Sewage	1,600	I-131	2
Sediments (Bq/kg)			
River	18,900	Cs-137	2
Lake	6,000	Cs-137	3
Drinking-water treatment sludge	104,000	Ru-103	4
Sewage sludge	780,000	Ru-103	2
Biota (Bq/kg)			
Fish	1,640	Cs-137	3
Plankton	38,000	Cs-137	3

derstanding of the reduction in radioactivity during drinking-water treatment (DWT) is of great importance.

The high activity levels in both DWT and sewage sludges, shown in Table I, qualitatively indicate that the processes commonly employed are removing some of the radioactivity. A survey of Swiss water-treatment plants showed that flocculation and filtration were able to reduce the radioactivity by 50–90% (*5*). Table I also shows that the main atmospheric and aqueous component was I-131, while the activity accumulated in sediments and biota are mainly from Cs-137 and Ru-103. This situation is partially due to the relatively short half-life, 8 days, of I-131 and the time periods before the activity of sediments and biota are measured. However, residence time in a water-treatment plant and distribution system is only a few days; this condition makes I-131 a problem for communities using surface waters. The proposed U.S. Environmental Protection Agency maximum recommended level for I-131 in drinking water is 3 pCi/L, that is, 0.11 Bq/L (*6*).

The objective of this study was to determine the effectiveness of commonly employed treatment processes for the removal of radioactive iodine from drinking-water sources. In the first phase of the study, the reaction between iodine and aquatic humic-substances was investigated. In the second phase, the removal of radioactive iodine in the concentration range below 1000 Bq/L (1.0 mBq/m^3) by bench-scale processes was examined. The dissolved organic carbon (DOC), UV absorption ($\lambda = 254$), and turbidity were also measured in order to find an indicator parameter that could be measured more easily than radioactivity.

One problem encountered was the determination of the appropriate species of iodine to use in such an investigation. Metal iodide is the chemical form that escapes from the reactor core (7); however, the form of iodine predominating after exposure to the atmosphere is not completely understood. The form is thought to be dependent on the conditions in the containment building; the volatile elemental iodine and methyl iodide are important with respect to atmospheric release.

Experimental Details

The humic substances used in this study were isolated by a strong basic anionic resin (Lewatit MP 500 A, Bayer Chemical Co.) used in the treatment of ground water with a high humic content (7 g of DOC/m^3) in Fuhrberg, Federal Republic of Germany. The resins were regenerated with a solution containing 10% NaCl and 2% NaOH (8). The regenerate has a molecular size (MS) range of 200–4000, with an average of 1500 as estimated by gel-permeation chromatography (GPC). The experimental solution properties for the Fuhrberg humic-substances (FHS) are shown in Table II.

The GPC of the Rhine River sample indicates a MS range of 160–5000, with an average of about 1500. Samples of the Rhine River were taken at Karlsruhe, Federal Republic of Germany, approximately 360 km downstream of Lake Constance. Their properties are also shown in Table II. The radionuclide I-131 was supplied in a carrier solution of NaI at a ratio of 4.72×10^4 Bq/μg of I, and the solution activity was analyzed by γ-ray spectrometry. The concentrations of I^- and I_2 were analyzed by the leuco crystal violet method (6) with a detection limit of 5 mg/m^3 in the Rhine River and FHS solutions. The DPD (*N*,*N*-diethyl-*p*-phenylenediamine) photometric method was used to measure chlorine (6). The concentration of CH_3I was measured with an electron-capture detector–gas chromatograph with a detection limit of less than 0.1 mg/m^3.

The reaction experiments between I^-, I_2, or CH_3I and the FHS or Rhine River water were conducted in 0.1- or 0.25-L closed volumetric flasks. Flocculation was conducted in 1-L glass beakers with the addition of iron sulfate at high mixing intensities (250 rpm) for 5 min, followed by 10 min of flocculation at 50 rpm and 30 min of settling. The cationic polyelectrolyte, polyacrylamide (PAA), when used, was added 2.5 min after the addition of iron sulfate. Filtration utilized glass-fiber filters or 0.45-μm membrane filters. Adsorption experiments were conducted with closed 0.25-L bottles on a shaker-table (250 rpm) at a contact time of 2 days with pulverized activated carbon (F300, Chemviron Corp.). In the combined floccula-

Table II. Properties of Fuhrberg Humic Substances and Rhine River Sample

Sample	*DOC* (*g/m³*)	*UV-254* (*m⁻¹*)	*pH*	*Redox* (*mV*)	*Temperature* (*°C*)	*Turbidity* (*FTU*[a])
Rhine River	2.62	6.52	7.65	240	19.6	2.9
Fuhrberg humic substances	4.20	16.3	6.5	515	20	—

[a]Formazine turbidity units.

tion–activated-carbon adsorption experiments, the pulverized carbon was added to the beaker at high mixing intensities 10 min prior to the addition of iron sulfate; it was filtered out with the sludge; the result was a 1-h contact time.

Results and Discussion

Reaction with Humic Substances. The reaction kinetics of elemental iodine, I_2, and iodide, I^-, with FHS over a 2-week period are shown in Table III. All initial solutions of I_2 used throughout this study contained 27% iodide. For both iodine and iodide the reaction with FHS is very fast, with no additional reduction in solution concentration occurring after 10 min. The reaction may even be faster, but this possibility could not be assessed, as 10 min was required for analytical sample preparation. The reaction kinetics of iodine and chlorine with the FHS can be seen in Figure 1. The iodine reaction is much faster than that of chlorine, which displays continued formation of organic-bound chlorine over a 17-h period.

All nonorganic-bound iodine is reduced to iodide in the reaction between I_2 and FHS, as shown in Table III. This result can also be seen in Figure 2, where the initial concentrations of iodine were five times higher than in Table III. In the system of I_2 and FHS, the total amount of iodine in solution decreases from 4.38 to 1.02 g/m^3 after 0.5 h; the I_2 component decreases from 3.20 to 0.23 g/m^3. After 3.5 h the total amount of iodine does not change, but the I_2 component is completely reduced to iodide. In the system with both I^- and Cl_2, all iodide is initially oxidized to I_2. After reaction with FHS (0.5 h), most of the iodine left in solution has been reduced to iodide. After 3.5 h all solution iodine is in the form of iodide, but no additional organic-bound iodine was formed.

The relationship between organic-bound iodine and the added or total system iodine is shown in Figure 3 for both elemental iodine and iodide. Below a total iodine concentration, normalized for the DOC concentration of FHS, of 0.05 g of I per g of DOC, both forms of iodine react completely with the FHS. This reaction represents iodine concentrations as high as

Table III. Reaction of Iodine and Iodide with Fuhrberg Humic Substances

	Iodine (I_2)			*Iodide* (I^-)		
Time	I_2	I^-	ΣI	I_2	I^-	ΣI
0	650	240	890	nd[a]	650	650
10 min	nd	370	370	nd	280	280
8 h	nd	400	400	nd	270	270
1 day	nd	380	380	nd	285	285
4 day	nd	400	400	nd	290	290
14 day	nd	400	400	nd	290	290

NOTE: All values are concentration in milligrams per cubic meter. Solution conditions: 3.2 g of DOC/m^3; pH 6.5.
[a]nd, not detected.

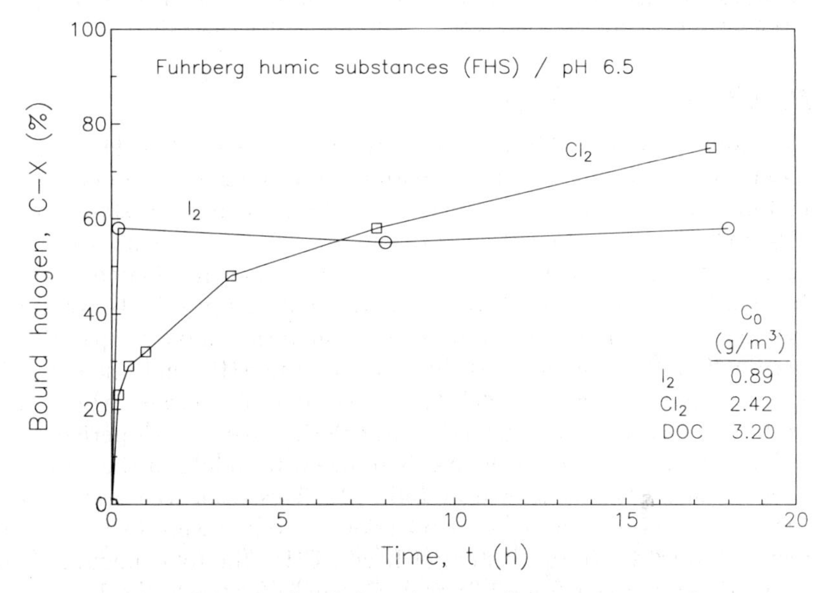

Figure 1. Halogen and Fuhrberg humic substances reaction kinetics.

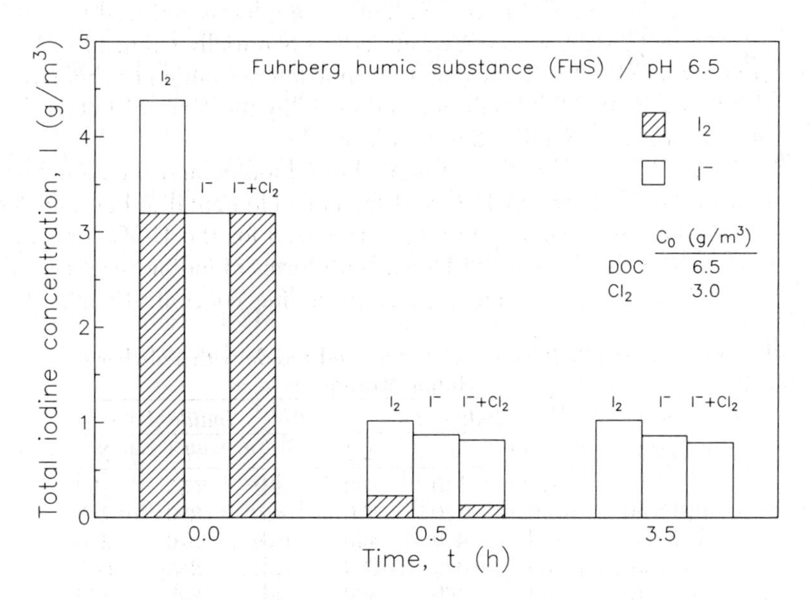

Figure 2. Reaction kinetics and iodine species distribution for the reaction of I_2, I^-, or $I^- + Cl_2$ with Fuhrberg humic substances.

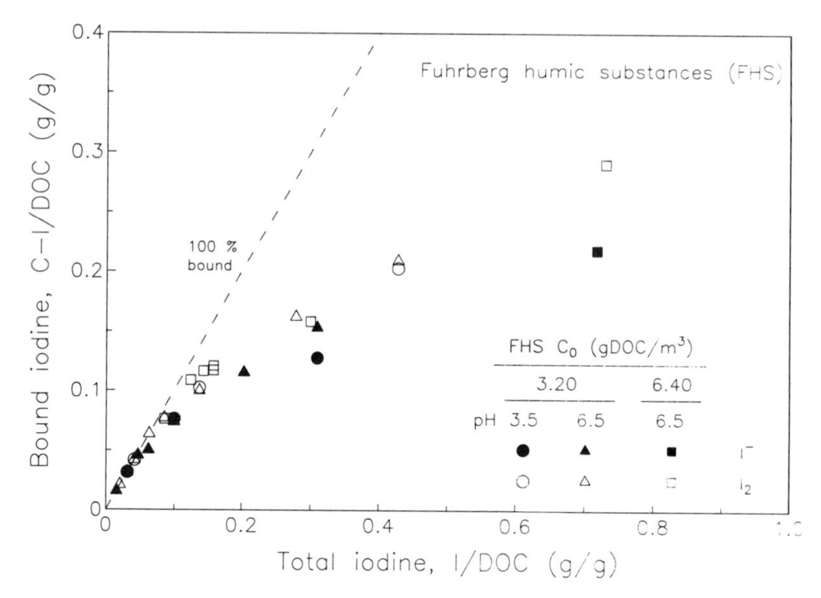

Figure 3. Organic-bound iodine as a function of the total iodine in the system with Fuhrberg humic substances.

160 mg/m^3 for a FHS concentration of 3.2 g of DOC/m^3. At higher total iodine concentrations the percent of total iodine bound to the FHS decreases, but in the concentration region below 0.8 g of I per g of DOC, a saturation of the FHS by iodine is not found. Elemental iodine is slightly more reactive than iodide, but the reaction is independent of pH and FHS concentration, as shown in Figure 3. However, Skogerboe and Wilson (9) found that low pH values reduce the extent of reaction between a soil fulvic acid and both I_2 and I_3^-.

FHS was separated by ultrafiltration into high (>1000) and low (<1000) molecular size (MS) fractions. The reaction of both I_2 and I^- with these MS fractions was examined. For both high and low MS fractions at two levels of iodine addition, as shown in Table IV, both I_2 and I^- reacted to an extent similar to that of the unfractionated FHS.

Two doses, 50 and 100 mg/m^3 each, of I_2 and I^- were added to samples of the Rhine River water. One sample of the Rhine water had been glass-fiber filtered prior to the experiments. At the lower dose both I_2 and I^- reacted completely with both the filtered and unfiltered Rhine water. At the higher dose both I_2 and I^- reacted completely with the unfiltered Rhine water, but reacted slightly less, 93%, with the filtered sample. This decrease indicates that part of the reaction is with particulate organic matter found in the Rhine River, and that iodine reacts to a similar extent with both FHS and the organic matter in the Rhine River. Several mechanisms have been proposed for the reaction between iodine and organic compounds, including electrophilic substitution, charge-transfer, and biochemical oxidation (*10–12*). However, the specific mechanism involved with humic substances is not clear because of the complex nature of this heterogeneous macromolecular material.

The interaction of methyl iodide and FHS is shown in Table V. Although both sample concentrations, with and without FHS, decrease with time, no significant difference exists between the samples at a given time. This similarity indicates little interaction between CH_3I and FHS. The decrease in solution concentration for both samples is an indication of the volatility of CH_3I. It also appears from this data that FHS has no effect on the Henry's coefficient, as the liquid-to-gas volume ratio of both samples was the same.

Removal by DWT Processes. RHINE RIVER AND I-131. The radionuclide I-131 as NaI was added to a sample of the Rhine River water to yield an activity level of 1.17 mBq/m^3 (1170 Bq/L), which is a factor of 3 larger than any river water value reported in the Federal Republic of Germany after Chernobyl (Table I). The sample was mixed for 0.5 h before chlorine was added at a concentration of 1.0 g of Cl/m^3, which raised the redox potential from 240 to 350 mV. Preoxidation is commonly practiced in DWT

Table IV. Influence of Fuhrberg Humic Substance Molecular Size on Organic-Bound Iodine

I	C_0[a] (mg/m^3)	*Bound Iodine, C/I (%)*		
		FHS	*High MS*[b]	*Low MS*[c]
I_2	274	88	81	85
	890	58	62	58
I^-	200	90	90	85
	650	57	46	52

NOTE: Solution conditions: 3.2 g of DOC/m^3; pH 6.5.
[a]Initial C concentration.
[b]Molecular size >1000.
[c]Molecular size <1000.

Table V. Interaction Between Methyl Iodide and Fuhrberg Humic Substances

Time (h)	*Without FHS*	*With FHS*
0.5	42.0	41.1
1.0	—	39.0
5.5	39.2	38.4
22.0	34.9	34.6

NOTE: All values are CH_3I concentration in milligrams per cubic meter; solution conditions: 3.2 g of DOC/m^3; 1.0 mM Na_2HPO_4; pH 6.5.

plants that directly utilize surface waters to control problems associated with biological growth.

The results of the flocculation experiments with iron sulfate are shown in Figure 4. The results with no addition of iron sulfate indicate the removal by volatilization and sedimentation. The maximum removal effectiveness by flocculation for I-131 and DOC is about 30%. Effectiveness nearly doubles for UV_{254} and triples for turbidity. The addition of PAA at 0.1 g/m^3 with an iron sulfate dosage of 15 g/m^3 did not improve the removal effectiveness for any of the measured parameters. Similar removal values for I-131 and DOC and higher removal values for turbidity indicate that most of the I-131 has reacted with dissolved organic matter and not particulate organic matter. This finding is also supported by the filtration results shown in Table VI. Both glass-fiber and 0.45-μm membrane filters are effective in the removal of turbidity, but significantly less so for I-131, DOC, and UV_{254}. With all

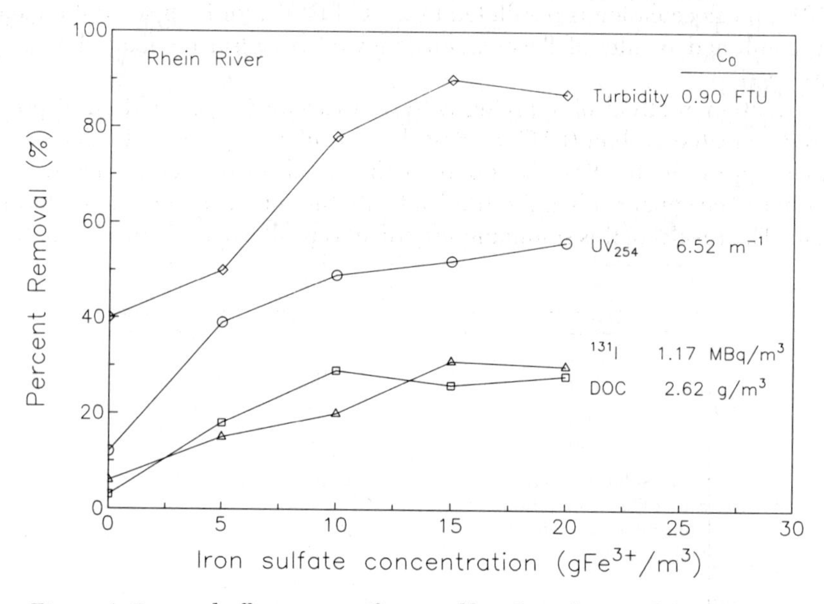

Figure 4. Removal effectiveness of iron sulfate flocculation of the Rhine River sample.

Table VI. Removal by Filtration of the Rhine River Sample

Parameter	*Glass Fiber*	*Membrane*
I-131	7.6	11
DOC	3.0	4.6
UV-254	8.0	11
Turbidity	69	80

NOTE: All values are percents.

parameters, membrane filtration is slightly more effective. The removal as measured by UV_{254} most closely matches that of I-131, while DOC is slightly less removed.

Adsorption by activated carbon (AC) in the dosage range of 5–1000 g of AC per m^3 is shown in Figure 5 for the Rhine River sample prior to pretreatment. At nearly all dosages, the adsorption as measured by UV_{254} and DOC is greater than that of I-131, although for DOC the difference is normally less than 15%. The maximum removal at 1000 g of AC per m^3 is 70%, 83%, and 97% for I-131, DOC, and UV_{254}, respectively. However, when AC is applied after flocculation, as shown in Figure 6, removal by adsorption is increased. This result can be seen by comparing the removal at a dosage of 100 g of AC per m^3 (Table VII). Prior to flocculation this dosage of activated carbon results in a removal of 57%, 71%, and 87% for I-131, DOC, and UV_{254}, respectively. After flocculation the respective removals by adsorption increased to 79%, 76%, and 91%. The removal of I-131 after flocculation is paralleled by that of DOC, for both adsorption and the combined results of flocculation followed by adsorption, as shown in Table VII.

Another method of applying activated carbon is the addition of powdered activated carbon (PAC) during the flocculation process. In this alternative approach the PAC is added to the rapid-mixing tank prior to the addition of coagulant aids and settled out with the sludge in the sedimentation basin. The results of this simultaneous process are shown in Figure 7. Contact

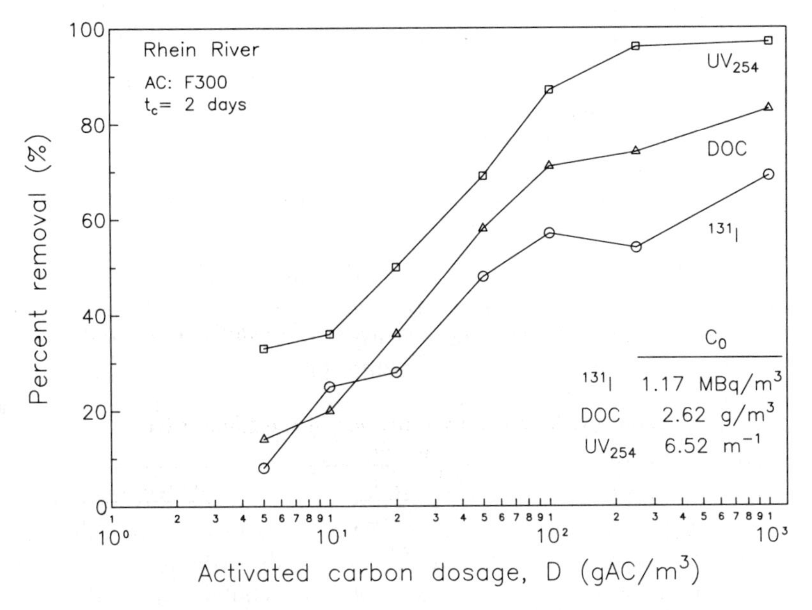

Figure 5. Activated-carbon adsorption with the Rhine River sample.

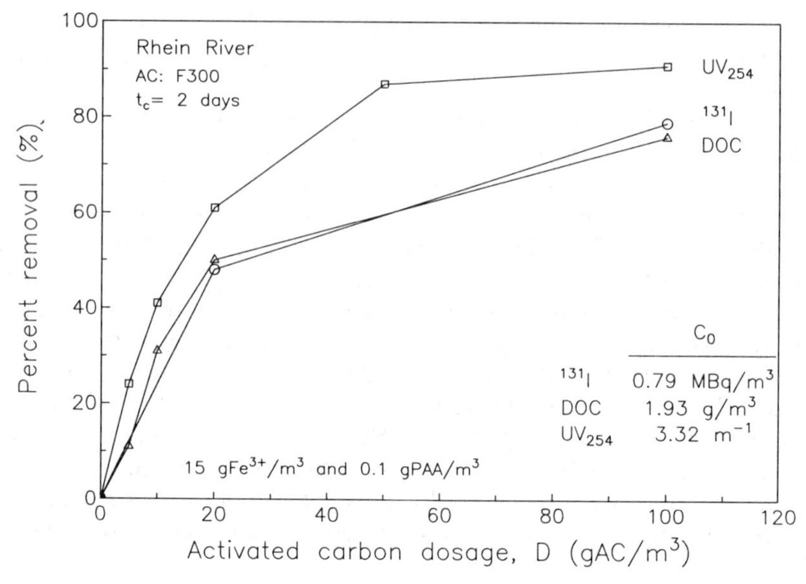

Figure 6. Activated-carbon adsorption after flocculation of the Rhine River sample.

Table VII. Removal by Activated Carbon with the Rhine River Sample

Pretreatment	*I-131*	*DOC*	*UV_{254}*
None	57	71	87
Flocculation			
Adsorption	79	76	91
Combined	85	83	96
Simultaneous adsorption and flocculation	82	75	91

NOTE: All values are percents; activated carbon dosage: 100 g/m^3.

time for the PAC process is short, 1 h in the case of Figure 7, compared to the adsorption residence time in a typical granular-activated-carbon column, modeled by the 2-day contact time in Figures 5 and 6. Even at this short contact time, the PAC–flocculation results in comparable removal, as can be seen in Table VII. The results for all three parameters after a 1-h contact time are nearly the same as those of the combined removal of flocculation followed by adsorption at a 2-day contact time.

The good removal achieved with the simultaneous PAC–flocculation process has significant implications for DWT plants that do not utilize granular-activated-carbon adsorbers, but do include sedimentation. In emergency situations of high levels of radioactivity or other contaminants, PAC could be added to the flocculation system and settled out in the sedimentation

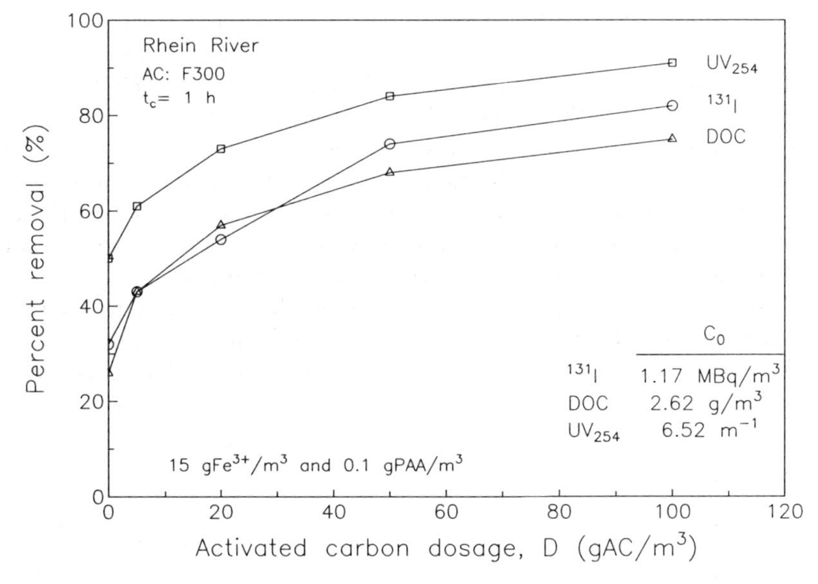

Figure 7. Simultaneous activated-carbon adsorption and flocculation of the Rhine River sample.

basin with the sludge. This procedure would require the ability to add the PAC and to handle increased amounts of a now-hazardous sludge. This approach could also be applied to wastewater-treatment plants.

FUHRBERG HUMIC SUBSTANCES AND I-131. I-131 was also added to a solution of FHS with the properties shown in Table II, to yield an activity level of 1.17 mBq/m^3. The bulk solution was not chlorinated. Membrane filtration of the solution after mixing yielded no removal of the three parameters measured: I-131, DOC, and UV_{254}. Aeration in a 0.5-L bubble column for 20 min yielded an insignificant removal of 5% and 4% for I-131 and DOC, respectively, a result that indicates that the I–HS complex is not any more volatile than the original FHS.

Removal effectiveness by activated-carbon adsorption in the dosage range 5–1000 g of AC per m^3 is shown in Figure 8. Adsorption was found to be less effective for the FHS in the low dosage, <100 g of AC per m^3, than for the Rhine River sample.

However, activated carbon was more effective in the high-dosage range, >250 g of AC per m^3, with removals of 94%, 86%, and 99% for I-131, DOC, and UV_{254}, respectively, at an adsorbent dosage of 1000 g of AC per m^3. Again DOC seems to be an adequate indicator of the filtration, volatilization, and adsorption behavior of I-131.

FUHRBERG HUMIC SUBSTANCES AND METHYL IODIDE. As shown in Table V, methyl iodide does not react with the FHS. However, the data in this table indicate that methyl iodide is volatile and has an estimated Henry's constant, H_c, of 0.2. To estimate the volatilization of methyl iodide during DWT, a solution of 95.4 mg/m^3 in a background of 3.2 g of DOC per m^3 of FHS was tested with the flocculation apparatus without coagulant aid addition. Membrane filtration of this solution yielded no removal of CH_3I. The mixing conditions were the same as for the other flocculation tests. A control solution was placed into the mixing vessel and exposed to the atmosphere for the same time but without mixing. The results, shown in Table VIII, indicate a 23% removal by volatilization during mixing; standing in an open vessel yielded an 8.2% removal. No significant change in DOC or UV_{254} was found. The removal of CH_3I could probably be improved if the process was optimized for CH_3I removal. Significant additional removals

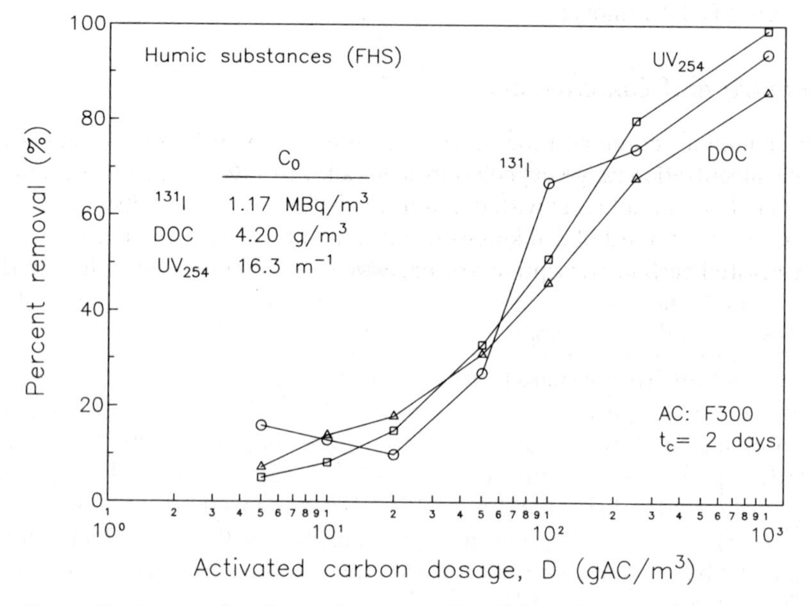

Figure 8. Activated-carbon adsorption of Fuhrberg humic substances and I-131.

Table VIII. Volatilization of Methyl Iodide in the Flocculation Apparatus

Condition	*Time (min)*	*Concentration C (mg/m^3)*	*CH_3I Percent Removal*
Initial	0	95.4	—
After mixing	45	72.1	23
Open vessel	45	87.2	8.2

NOTE: Fuhrberg humic substances in solution with 3.2 g of DOC/m^3 and 1.0 mM Na_2HPO_4 at pH 6.5.

would be expected if a packed column or other processes designed for the removal of volatile substances were used.

The activated-carbon adsorption CH_3I–FHS solution in the dosage range 9–150 g of AC per m^3 is shown in Figure 9. The adsorption process is much more effective for the removal of CH_3I than for DOC and UV_{254}, and for I-131, as shown in Figure 8. At a carbon dosage of 100 g of AC per m^3, 92% of the CH_3I is removed. Only 67% of the I-131 is removed in a similar FHS solution. This result seems to be due to the solution interaction between I-131 and the FHS. Because this interaction is missing, CH_3I can independently adsorb.

These results indicate that both volatilization and activated-carbon adsorption are effective removal pathways if the radioactive iodine occurs as CH_3I. However, DOC and UV_{254} are not good surrogate parameters for monitoring CH_3I removal.

Summary and Conclusions

Both elemental iodine and iodide react completely with humic substances in the concentration range expected to occur after a nuclear reactor accident. Methyl iodide did not react with the humic substances. The results of bench-scale processes typical of drinking-water treatment indicate that flocculation and activated-carbon adsorption are effective in the removal of radioactive

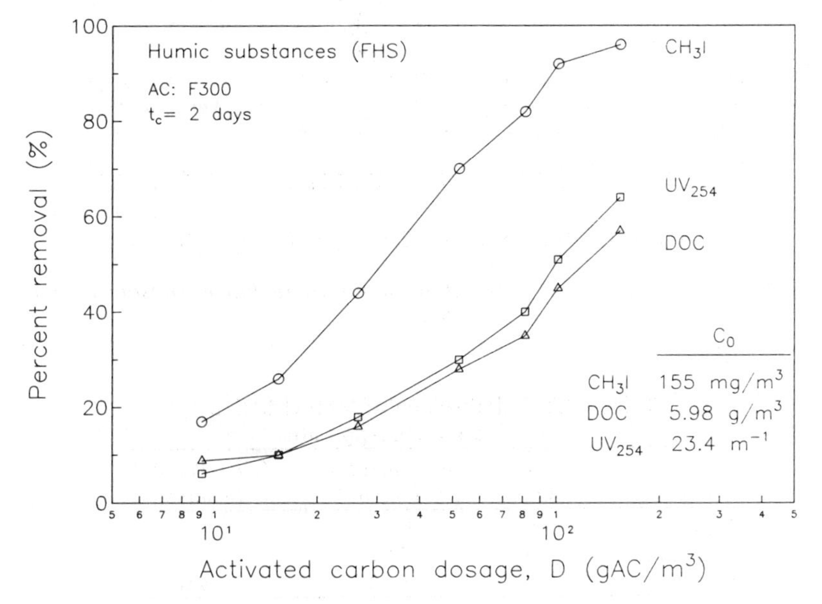

Figure 9. Activated-carbon adsorption of Fuhrberg humic substances and methyl iodide.

iodine when it is complexed with organic matter. DOC was found to be a good surrogate parameter to more easily assess the removal of complexed radioactive iodine in both humic substances and the Rhine River. Volatilization and activated-carbon adsorption were found to be effective for the removal of methyl iodide in the presence of humic substances. However, DOC and UV_{254} are not good indicators of CH_3I removal.

Additional characterization of the removal effectiveness of treatment processes is needed at a scale larger than the bench scale investigated herein for I-131 and other radionuclides likely to contaminate water supply sources, such as Cs-137 and Ru-103. Pilot-plant investigations should include the flocculation process with the addition of powdered activated carbon and should include fixed-bed columns of granular activated carbon, as this is the mode in which carbon is most commonly used in European drinking-water treatment. The effectiveness of aeration by packed columns and the impact of processes involving biological degradation should also be examined.

References

1. Haberer, K. *GWF Gas Wasserfach: Wasser/Abwasser* **1986,** *127,* 597–603.
2. Friedman, L.; Amann, W.; Lux, D. *GWF Gas Wasserfach: Wasser/Abwasser* **1986,** *127,* 604–613.
3. Laschka, D.; Herrmann, H.; Hübel, K.; Lunsmann, W. *GWF Gas Wasserfach: Wasser/Abwasser* **1987,** *128,* 128–135.
4. Eberle, S. H.; Fuchs, F.; Haberer, K.; Summers, R. S.; Sontheimer, H. *Radioaktiv kontaminierte Rohwässer bei der Trinkwassergewinnung.* Agrar- und Umweltforschung in Baden-Württemberg, Band 17, Ministerium für Ernährung, Landwirtschaft, Umwelt und Forsten, 1986.
5. Massarotti, A. *Gas Wasser Abwasser* **1986,** *66,* 827–832.
6. *Standard Methods for the Examination of Water and Waste Water*; American Public Health Association: Washington, DC, 1985.
7. Campbell, D. O.; Malinauskas, A. P.; Stratton, W. R. *Nucl. Technol.* **1981,** *53,* 111–119.
8. Kölle, W. In *Adsorption Techniques in Drinking Water Treatment*; Roberts, P. V.; Summers, R. S.; Regli, S., Eds.; EPA 570/9–84–005; U.S. Environmental Protection Agency, Office of Drinking Water: Washington, DC, 1984; pp 805–811.
9. Skogerboe, R. K.; Wilson, S. A. *Anal. Chem.* **1981,** *53,* 228–232.
10. Dore, M.; Merlet, N.; De Laat, J.; Goichon, J. *J. Am. Water Works Assoc.* **1982,** *74,* 103–107.
11. Foster, R. *Organic Charge-Transfer Complexes*; Academic Press: London, 1969.
12. Behrens, H. In *Environmental Migration of Long-Lived Radionuclides*; IAEA–SM–257/36, International Atomic Energy Agency: Vienna, Austria, 1981.

RECEIVED for review July 24, 1987. ACCEPTED for publication December 21, 1987.

INFLUENCES OF OZONATION AND CHLORINATION PROCESSES ON WATER TREATMENT

Catalytic–Competition Effects of Humic Substances on Photolytic Ozonation of Organic Compounds

Gary R. Peyton and Chai S. Gee

Illinois State Water Survey, Champaign, IL 61820

John Bandy and Stephen W. Maloney

U.S. Army Construction Engineering Research Laboratory, Champaign, IL 61820

During the treatment of organic compounds in water with free-radical processes such as photolytic ozonation, humic substances are expected to compete with the target compound for hydroxyl radicals. Experimental studies were performed on the competitive effect of macromolecular humic substances during photolytic ozonation and H_2O_2–UV treatment of a model pollutant compound. None of the humic substances interfered as effectively with the photolytic ozonation of diethyl malonate as was expected from an estimate of the hydroxyl radical reaction rate constant based on the molecular weight of the humic substances and from the competitive behavior of polyethylene glycols in a similar molecular weight range. The apparent noncompetitive behavior of the humic substances implies the production of secondary species that catalyze the generation of additional hydroxyl radical from ozone and thus counteract the competitive effect.

INCIDENTS OF GROUND WATER AND SURFACE WATER CONTAMINATION due to the improper storage and disposal of organic chemicals are widespread. Many cleanup technologies in use today, such as activated-carbon adsorption or air stripping, merely transfer the pollutant to another phase, rather than

0065–2393/89/0219–0639$06.75/0

eliminate it. We need environmentally sound water-treatment methods that will destroy the pollutant instead of merely relocating it.

Oxidation to carbon dioxide is the only way to eliminate organic pollutants completely. Biological oxidation, although frequently cost-effective, is not applicable to all organic compounds and waste streams. Incineration of wastewater can be costly if the fuel value of the waste stream is low. Wet-air oxidation is a promising technique, but it requires either a deep hole or pressurized equipment. A definite need exists for an oxidative treatment method that can be taken to completion, is universally applicable to all organic compounds, and can be implemented with a minimum of land use and construction. Photolytic ozonation is one such process.

Photolytic ozonation is the simultaneous ozonation and UV irradiation of water to be treated. Hydroxyl radical is generated by the process (*1–3*) and, if the reaction is taken to completion, is capable of converting most organic compounds completely to carbon dioxide and water. Photolytic ozonation thus has the potential for being an environmentally "clean" treatment process for organic contaminants in water.

However, free-radical scavengers such as humic material and bicarbonate alkalinity are naturally present in many waters. These scavengers may interfere significantly with treatment efficiency by competing with target solutes for hydroxyl radical. This study addresses the extent to which humic material interferes with hydroxyl-radical treatment processes such as photolytic ozonation.

Experimental Details

Reactor System. The reactor was a continuously sparged stirred-tank photochemical reactor (CSTPR), with standard relative dimensions (*4, 5*) and four quartz lamp wells mounted in the quadrants created by the baffles (*5*). The reactor body (10.65-L total volume, 8.5-L liquid volume) was made from a 12-in. piece of borosilicate process pipe (Corning Pyrex), with a nominal 9-in. i.d. The reactor heads were machined from 0.5-in.-thick sheet poly(tetrafluoroethylene) (PTFE), as were the baffles, sparger, six-blade impeller, and o-ring glands to secure the lamp wells. The stirring gland was glass (Cole-Parmer, Chicago) with a PTFE-coated viton o-ring seal and glass shaft. The impeller was secured to the shaft with a PTFE pin, inserted through a hole bored in the glass shaft. The stirring motor was a 1/8-hp, variable-speed DC motor with SCR controller (W.W. Grainger, Decatur, IL), the speed of which was set at 750 rpm by using a phototachometer. Gas fittings and liquid sample valve were PTFE, as was all connecting tubing. All wetted surfaces were either PTFE or glass.

Ozone was generated from dry oxygen with an ozone generator (Grace, model LG–2–L2). Inlet and off gas flows were regulated to within 0.1% of full scale (usually 1% of the measured value) by two UFC–1000 mass-flow controllers attached to a URS–100 power supply and digital readout (Unit Instruments, Orange, CA). This system can respond quickly to a reactor pressure change (such as that caused by switching the ozone monitor from feed gas to off gas), and restore the flow rate to within ±2% of the set point in a period of 2–4 s. Ozone concentration was followed by a high-concentration ozone monitor (PCI, model HC), which gave digital readout and provided an analog signal to a strip-chart recorder for later calculation of ozone doses. Factory calibration of this monitor was checked by wet-chemical methods for ozone analysis by bubbling ozone into reagents in the CSTPR and withdrawing samples as a function of time.

The reactor and manifold system used are shown in Figure 1. The ozone stream from the ozone generator (OG) is split and sent to two mass-flow controllers (MFC). The stream through MFC_2 may either be sent as feed to the reactor (R) or bypassed to the vent (V) with PTFE solenoid valve V_c, as was done during generator warm-up and initial concentration adjustment. The slip stream through MFC_1 to V_a is diverted through V_a and V_b to the ozone monitor (OM) for feed gas concentration measurement or sent to vent. Total gas flow through the ozone generator is kept

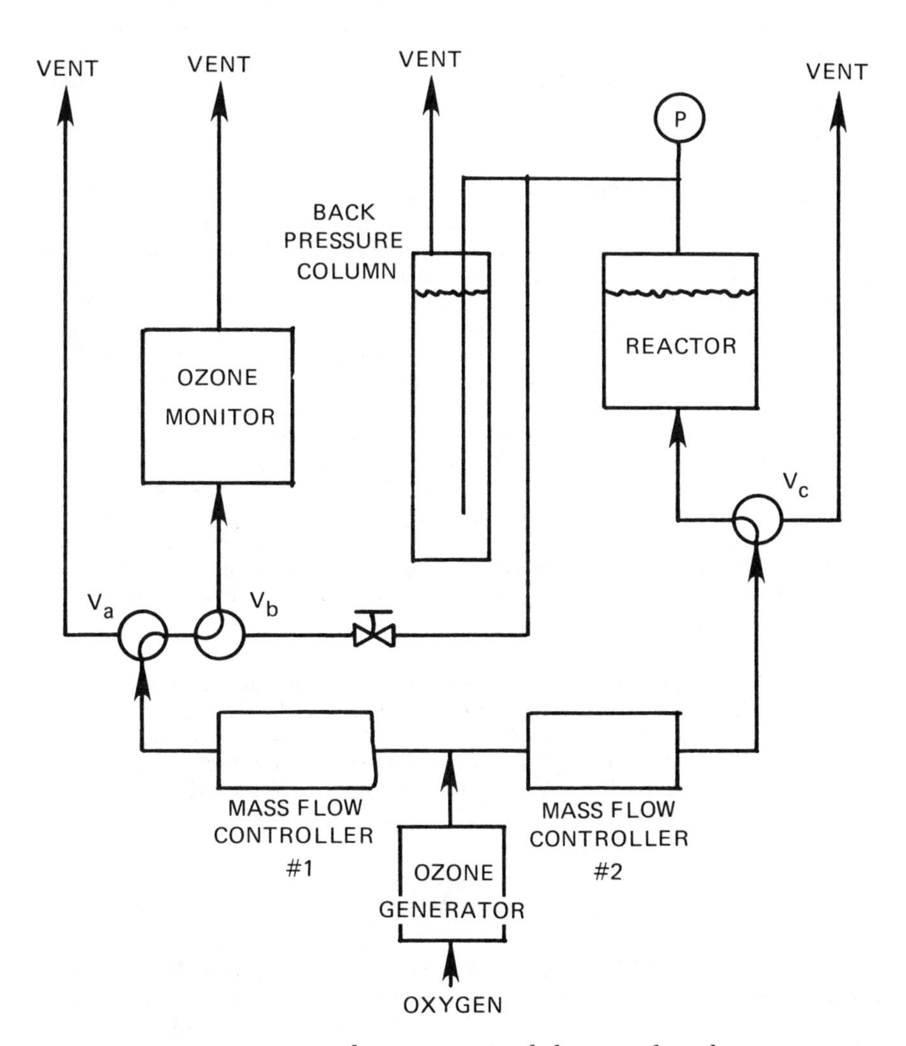

Figure 1. Experimental apparatus. Symbols are explained in text.

constant because it is the sum of the flows through the two mass-flow controllers, MFC_1 and MFC_2. Off gas from the reactor is kept at constant pressure by using the back pressure column (BPC), which doubles as a crude ozone kill unit. The back pressure is sufficient to force off gas through the ozone monitor when V_b is appropriately positioned. The mass-flow controllers (which see only dry gas), the back pressure column (downstream of the system), and the spectrophotometric cell in the ozone monitor are the only components made of materials other than PTFE or glass, since severe decomposition of ozone by stainless steel tubing was noted in previous work (*3*).

The UV lamps (American Ultraviolet, Chatham, N.J.; model G10T5–1/2) were rated at 5.5 W of UV power at 100 h of life. Their output is primarily at 254 nm. Lamp intensities, measured both radiometrically and actinometrically, differed considerably from those specifications. In the course of this work, 0.25–3.50 lamps were used in an experiment. Fractional lamp values were obtained by using a foil shroud on the lamps.

No attempt was made to optimize reaction conditions or mass transfer during this study. Conditions were chosen to favor precise and accurate data collection for mechanistic and mass balance determination.

Oxidant Analysis. Ozone in the aqueous phase was analyzed by the indigo method of Bader and Hoigné (*6, 7*), with the disulfonate rather than the trisulfonate as originally described by those authors. This method was calibrated in purified water against the iodimetric method of Flamm (*8*) and checked by UV absorbance with the extinction coefficient of Hart et al. (*9*). The iodimetric method was, itself, calibrated by quantitative iodine liberation with excess iodide and standard iodate solution, prepared by using dried potassium iodate as a primary standard. Ozone in the gas phase was measured by UV absorbance, with the factory calibration checked against the wet methods by adsorbing the gas in reagent solution contained in the reactor.

Hydrogen peroxide was measured colorimetrically by complexation with Ti(IV) (*10*) or by the method of Masschelein et al. (*11*). As ozone appears to interfere with hydrogen peroxide measurement in the course of the titanium method, ozone was quickly and vigorously sparged from solution with oxygen before peroxide measurements were made. The method of Masschelein was not used on ozone-containing solutions. Total oxidants were measured iodimetrically by the method of Flamm (*8*), but with the addition of a small quantity of ammonium molybdate to catalyze the reaction with peroxides.

Diethyl Malonate. Diethyl malonate (DEM) was measured by microextraction into ethyl acetate, which contained diethyl oxalate as an internal standard. The extract was analyzed by gas chromatography on a 5-ft, 1/8-in. o.d. stainless steel column packed with 10% biscyanopropylphenyl polysiloxane (SP–2340) on a diatomite support (100–120 Chromosorb W AW, Supelco, Bellefonte, PA). The temperature program used was 100 °C for 1 min, then 20 °C/min to 140 °C.

Reagents. All chemicals were reagent grade and were used without further purification. Deionized carbon-filtered water was used in all experiments.

Procedures. Aldrich humic acid (Aldrich Chemical Company, Milwaukee, WI) and Suwannee River humic and fulvic acids (International Humic Substance Society, Arvada, CO) were used as purchased. Concentrates were prepared by dissolving the humic substances in the minimum amount of freshly prepared dilute sodium hydroxide, adding the concentrate to the reactor, and adjusting the pH with dilute sulfuric acid. In some experiments in which DEM was present in the reactor before the humic concentrate was added, the amount of acid necessary to neutralize the basic humic solution was added first, to avoid basic hydrolysis of DEM. Reactions were run unbuffered at pH 5.5, because that was the pH toward which the mixture tended as the reaction proceeded. Ambient laboratory temperature was about 23 °C. DEM was analyzed before and after humic addition to determine that neither appreciable DEM hydrolysis nor irreversible adsorption of DEM to the humic substances was occurring.

In experiments in which the water was pretreated ("burned") by ozone–UV, the UV lamps were turned off to allow the O_3–H_2O_2 reaction to destroy H_2O_2. Then the power to the ozone generator was shut off to sparge any remaining ozone from solution with oxygen flow. The resulting purified water was analyzed to confirm the absence of ozone and hydrogen peroxide before DEM or humic substances were added.

Results

Figure 2 illustrates data taken during a typical photolytic ozonation experiment, with diethyl malonate (DEM) as the model pollutant. The applied ozone dose rate of 1.3×10^{-5} mol/L·min and one UV lamp were kept constant throughout the study, except as noted here. As diethyl malonate disappeared, hydrogen peroxide accumulated and a residual ozone concentration of about 1×10^{-5} M was maintained. The hydrogen peroxide concentration decreased later in the experiment to near zero. Figure 3 shows the diethyl malonate disappearance curves for three different experiments under identical conditions and is thus representative of the reproducibility of the data. In the third experiment, the water was "burned" for 1 h at 10^{-4} mol of O_3/L·min with 3.5 UV lamps, before oxidants were destroyed (*see* **Experimental Details** section), and diethyl malonate was added. The results indicate that no significant interferences arise from organic compounds in the laboratory water source.

Figure 4 shows the DEM disappearance curves that resulted from experiments performed as described, but with the addition of 1, 5, and 10 ppm of Aldrich humic acid (AHA) to the 5-ppm DEM solution. Although slight differences can be seen, the results lie within experimental scatter of the DEM measurement. From the point of view of water treatment to remove the target compound, virtually no difference exists, and removal of DEM is complete within 35–40 min. No attempt was made to optimize treatment conditions, but rather to perform successive experiments under comparable conditions.

Because the use of AHA to simulate aquatic humic material is in some respects questionable, the experiments were repeated with reference Suwannee River humic and fulvic acids (SRHA and SRFA, respectively), which were obtained from the International Humic Substance Society. These results are shown in Figure 5. Although the SRHA gave results very similar to those for AHA, the DEM disappearance curve obtained using SRFA

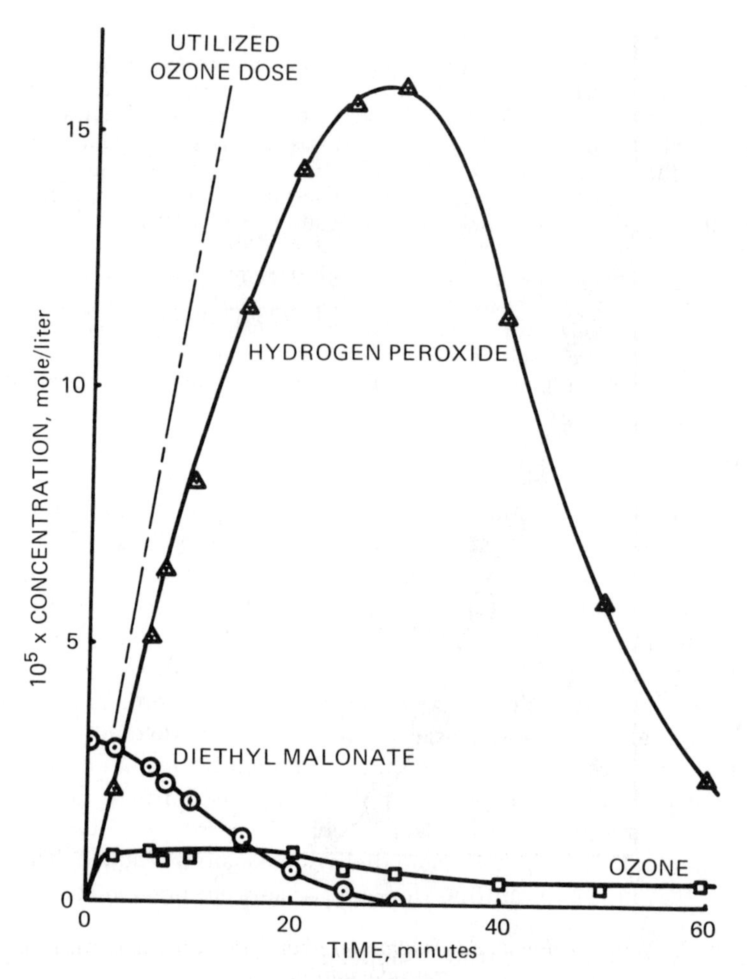

Figure 2. Diethyl malonate destruction by photolytic ozonation; typical experiment.

(number-averaged molecular weight = 825 daltons) was discernibly different, and slightly slower than for the humic substances.

For comparison with a system of known structure, similar experiments were run with 10-ppm solutions of polyethylene glycol (PEG) 400 and 8000. The numbers 400 and 8000 represent approximate number-averaged molecular weights of the polymers. The results, shown in Figure 6, display a striking difference in the effect of PEG compared to humic material, even though the molecular weights of the PEG samples span that of at least the SRFA. Furthermore, the results from the two PEG experiments are identical

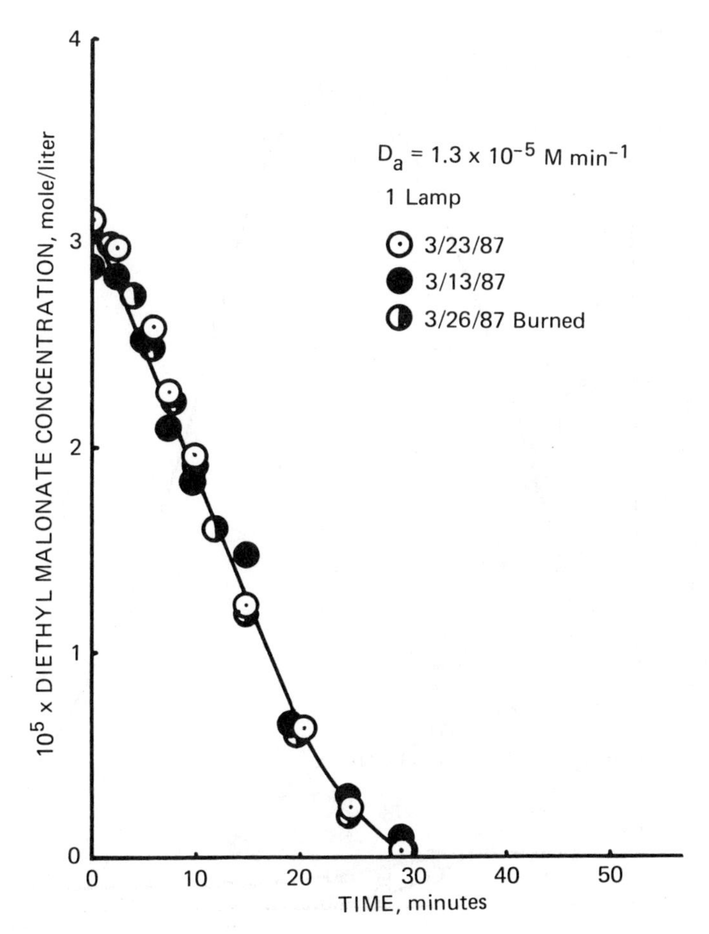

Figure 3. Diethyl malonate destruction by photolytic ozonation; experimental reproducibility.

within experimental error, even though there is a factor of 20 difference in their average molecular weights.

In order to separate different possible effects, the reaction system was simplified. Hydrogen peroxide photolysis experiments were performed in the presence of DEM, with and without the addition of macromolecules. The UV photolysis of hydrogen peroxide yields hydroxyl radical directly. The results are shown in Figures 7 and 8. Hydrogen peroxide disappearance (Figure 7) was fastest in the absence of organic material, slower with diethyl malonate added, still slower with PEG added, and slowest in the presence of AHA plus DEM. Unlike the photolytic ozonation system, diethyl malonate disappearance upon H_2O_2–UV treatment (Figure 8) was fastest in the ab-

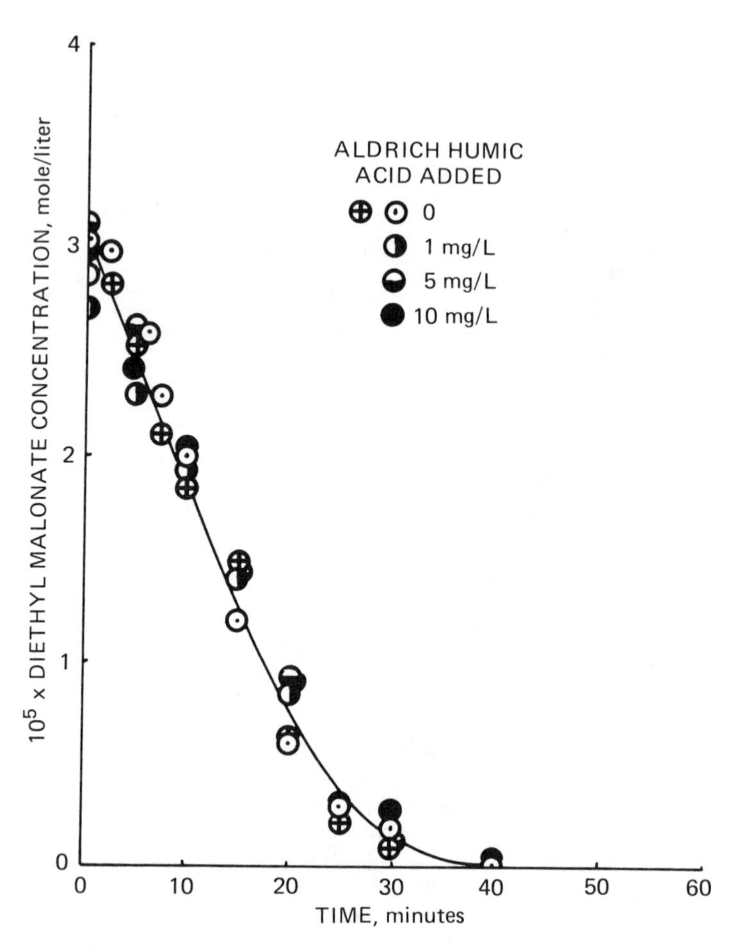

Figure 4. Effect of Aldrich humic acid addition on diethyl malonate destruction by photolytic ozonation.

sence of macromolecule, slower in the presence of PEG, and much slower in the presence of AHA.

Discussion

Reaction Mechanism. The differences in apparent competitive effects of macromolecules in the O_3–UV system compared to the H_2O_2–UV system can be understood within the mechanistic framework of the total O_3–H_2O_2–UV system. Building on the work of Staehelin and Hoigné (*12–15*) on the base- and peroxide-catalyzed decomposition of ozone, and that of Peyton and Glaze (*1–3, 16*) on aqueous ozone photolysis, Peyton et al. (*17*) have proposed a mechanistic scheme that aids in the understanding of the

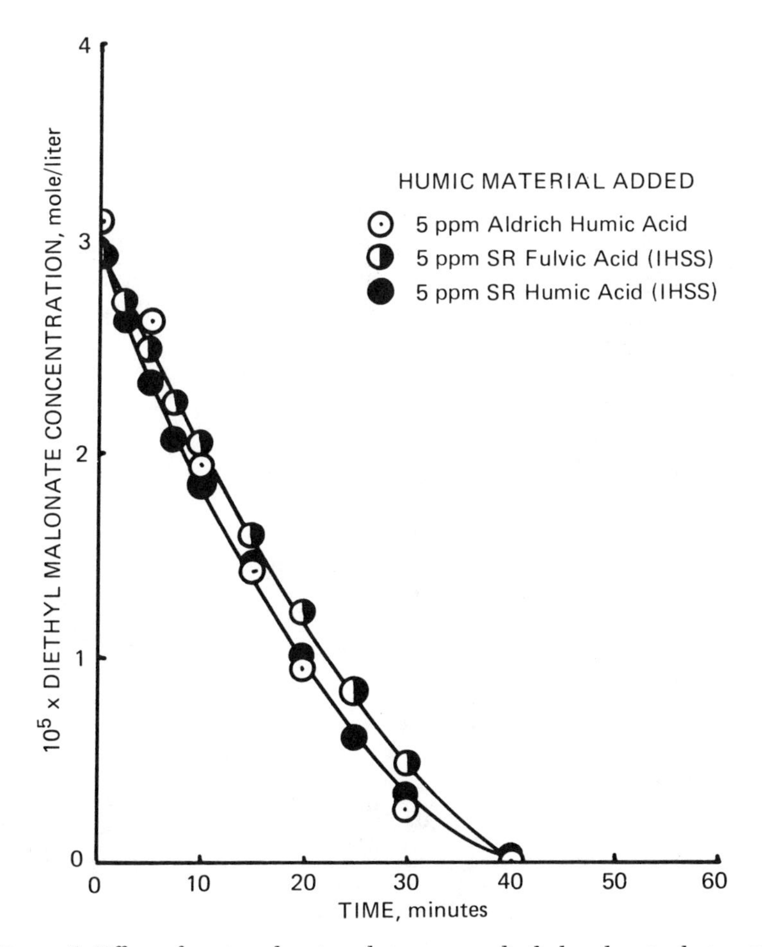

Figure 5. Effect of various humic substances on diethyl malonate destruction by photolytic ozonation.

behavior of all or any part of the O_3–H_2O_2–UV–organic system. The scheme, which is illustrated in Figure 9, has so far been shown (*17*) to quantitatively describe the relatively simple system with methanol–formaldehyde–formic acid as the organic degradation chain, with rate constants from the literature. Work is currently under way to verify the usefulness of the model for larger organic molecules.

In this scheme (Figure 9), ozone is photolyzed to produce hydrogen peroxide, which then reacts with ozone through its conjugate base, HO_2^-, as described by Staehelin and Hoigné (*12*), to produce O_3^-, HO_3, and finally hydroxyl radical. At higher hydrogen peroxide concentrations, peroxide photolysis can also be a significant contributor of hydroxyl radical. In the case of a saturated aliphatic compound as shown in Figure 9, hydroxyl radical abstracts a hydrogen atom from the organic compound, after which the

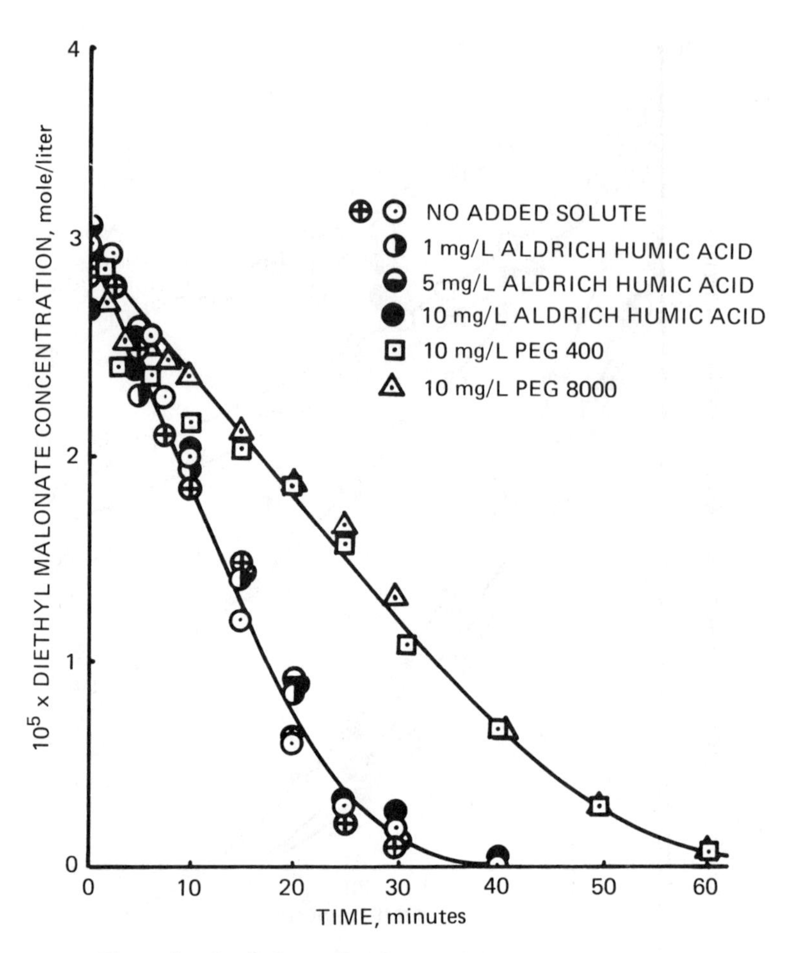

Figure 6. Effect of polyethylene glycols on photolytic ozonation of diethyl malonate.

organic radical that is formed quickly reacts with dioxygen to yield an organic peroxy radical (*18*).

The peroxy radical can suffer several fates, and there is still not agreement in the literature concerning which pathways are followed. A general observation in our laboratory, however, is that compounds that produce a peroxy radical structurally incapable of eliminating superoxide will produce some hydrogen peroxide on their way to stable organic product molecules. Conversely, lack of appreciable H_2O_2 production may be taken as an indication that superoxide is being generated. This effect is illustrated in the peroxide accumulation curves of Figure 10. The curves resulted from experiments that were run identically, except that methanol (a superoxide producer) was used in one case and diethyl malonate in the other. The

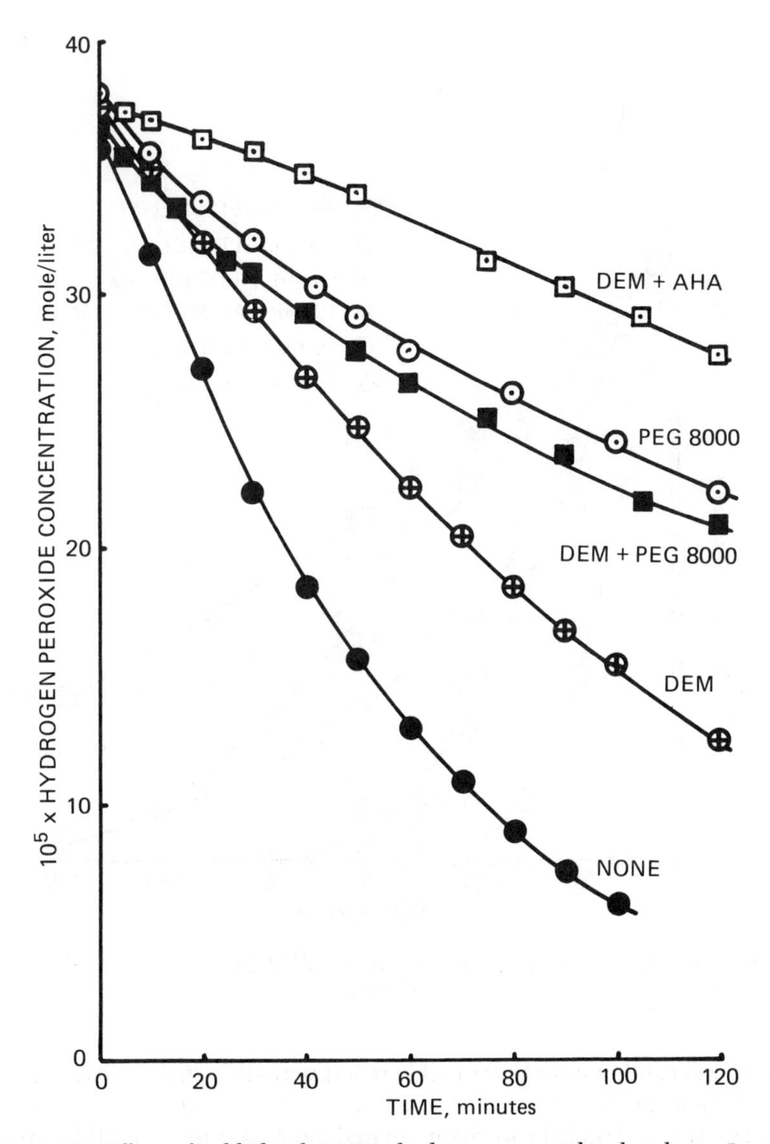

Figure 7. Effect of added solutes on hydrogen peroxide photolysis. In all experiments the concentration of DEM is 5 mg/L and that of other scavengers is 10 mg/L.

dramatic difference in the curves suggests that DEM is primarily a peroxide producer, rather than a superoxide producer.

For simplicity, direct ozone reaction with organic compounds has been left out of the scheme. As Hoigné has noted (*19*), many of the reaction products are identical with those obtained from the radical pathways. The base-catalyzed decomposition of ozone has been omitted because it is insig-

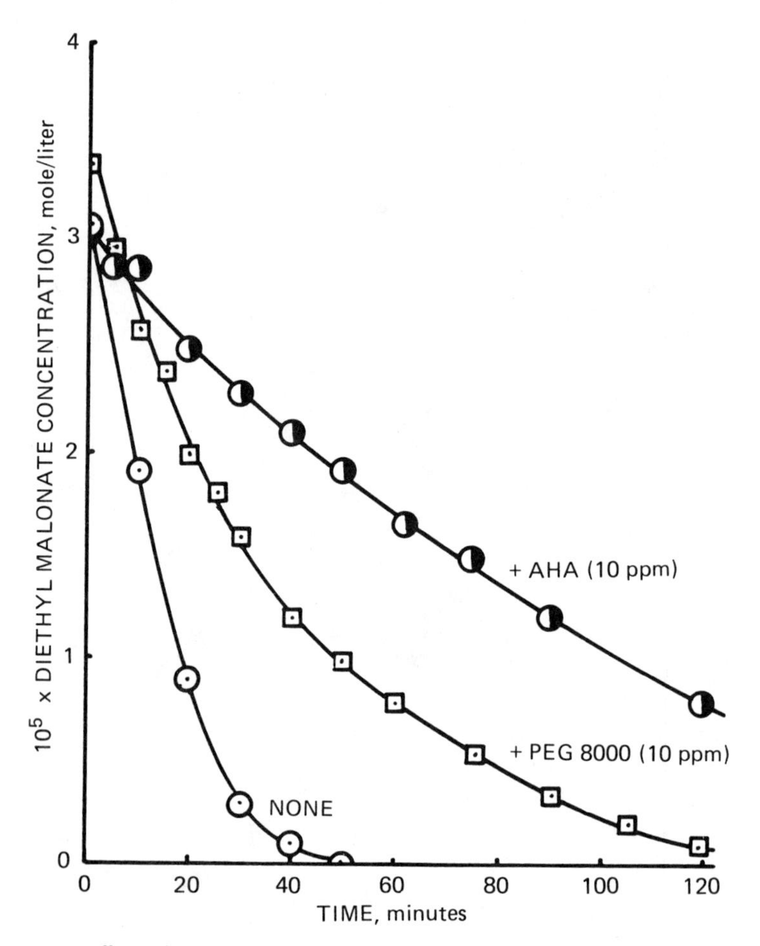

Figure 8. Effect of added macromolecular solutes on H_2O_2–UV destruction of diethyl malonate.

nificant (*12*) compared to the reaction with peroxide, which is always present in photolytic ozonation systems. Similarly, disproportionation of superoxide–hydroperoxyl radical to H_2O_2 is insignificant when ozone is present, because of the very fast reaction between ozone and superoxide.

To apply the scheme shown in Figure 9 to the H_2O_2–UV system, all reactions that involve O_3, O_3^-, or HO_3 are left out. Disproportion of superoxide to hydrogen peroxide (not shown in Figure 9) must then be included.

Competition Kinetics. Consumption of hydroxyl radical, which is produced in systems such as these, can be assumed to be essentially complete because of the very high reactivity of the hydroxyl radical. Hydroxyl radical

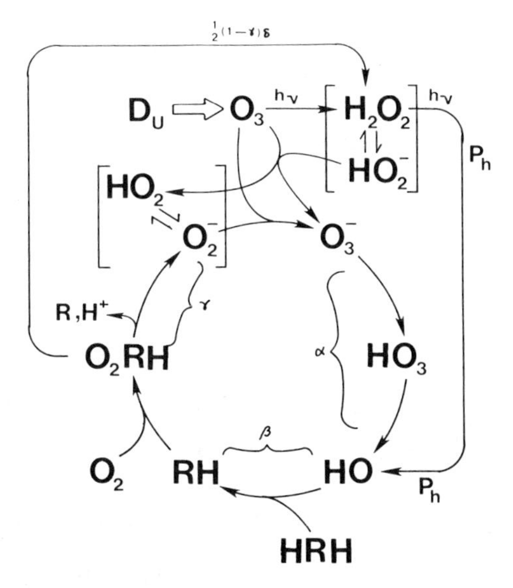

Figure 9. Mechanism of photolytic ozonation. HRH represents a neutral closed-shell organic molecule.

is partitioned among the various species, i, in solution, according to their concentrations (S_i) and reactivities:

$$R_p = \text{OH production rate} = \sum_i k_i[S_i][\text{OH}] = \sum_i k_i'[\text{OH}] \tag{1}$$

where $k_i' = k_i[S_i]$ is the pseudo-first-order rate constant and k_i is the second-order rate constant for reaction of species i with hydroxyl. The instantaneous rate of disappearance, R_D, of target solute in the presence of competing scavengers is given by

$$R_D = R_p \frac{k_t[S_t][\text{OH}]}{\sum_i k_i[S_i][\text{OH}]} = \frac{R_p k_t'}{\sum_i k_i'} \tag{2}$$

where the subscript t denotes target solute and the primes carry the same meaning as before. The competition factor, F_c, is defined as the relative target-compound disappearance rate in a scavenged (i.e., added macromolecule) versus an unscavenged experiment, and is given by equation 3:

$$F_c = \frac{R_{D,\text{scavenged}}}{R_{D,\text{unscavenged}}} \tag{3}$$

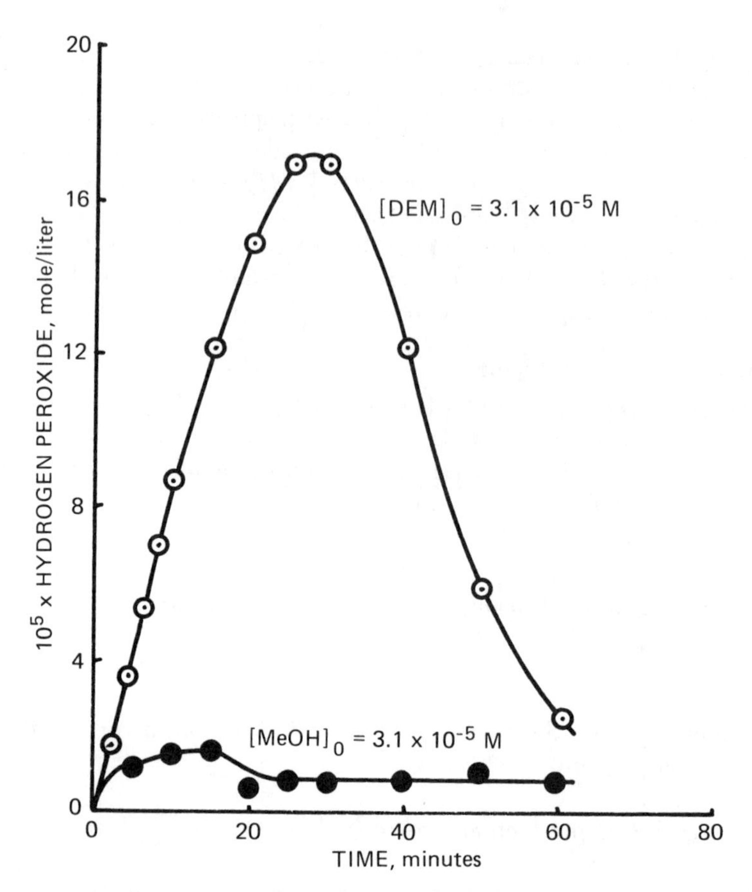

Figure 10. Hydrogen peroxide production, diethyl malonate versus methanol.

Substitution of equation 2 into equation 3, using appropriate literature values of the rate constants and experimentally determined concentrations, can be used to determine an unknown rate constant if all of the others are known.

Hydroxyl Radical Reaction Rate Constants of Macromolecules. A second-order rate constant value of slightly over 10^{10} M^{-1} s^{-1} is frequently referred to in the literature as the "diffusion-controlled limit", meaning that reaction occurs upon virtually every collision and that the reaction rate is limited by the rate at which the two reacting species can reach an encounter by molecular diffusion. However, it should be easier to encounter a macromolecule by diffusion than a small molecule, considering that the macromolecule commands a significantly larger hydrodynamic volume. The limit of 10^{10} referred to is, in fact, the calculated limit for two molecules of "normal" (i.e., small) size.

Braams and Ebert (*20*) used a modified version of the Smoluchowski equation (*21*) to calculate the collisional frequency rate constant of hydroxyl radicals with macromolecules and compared those calculated rate constants with observed reaction rate constants. Figure 11 shows a log plot of their tabulated data, which lie in a straight line with approximately unit slope and approximately equal abscissa and ordinate values. These features merely reflect the fact that hydroxyl radical reacts with these molecules on virtually every collision. A more striking effect is that the rate constants reach 3 orders of magnitude above what is frequently quoted as the diffusion-controlled limit. When the log of these observed rate constants and those for selected small molecules (22) are plotted versus the log of the molecular weight, a strong linear correlation is seen (open and half-open symbols in Figure 12) in the region $10^2 \leq M \leq 10^6$.

The line shown in the plot is given by $k_{obs} = AM_s^{\ m}$, where k_{obs} is the rate constant, M_s is the molecular weight of the macromolecular scavenger, $A = 10^{6.8}$, and $m = 0.87$. Under this constraint, the pseudo-first-order rate constant k_s' would be given by the equation

$$k_s' = k_s[S] = \frac{AM_s^{0.87}C_{ws}}{M_s} = AC_{ws}M_s^{-0.13} \tag{4}$$

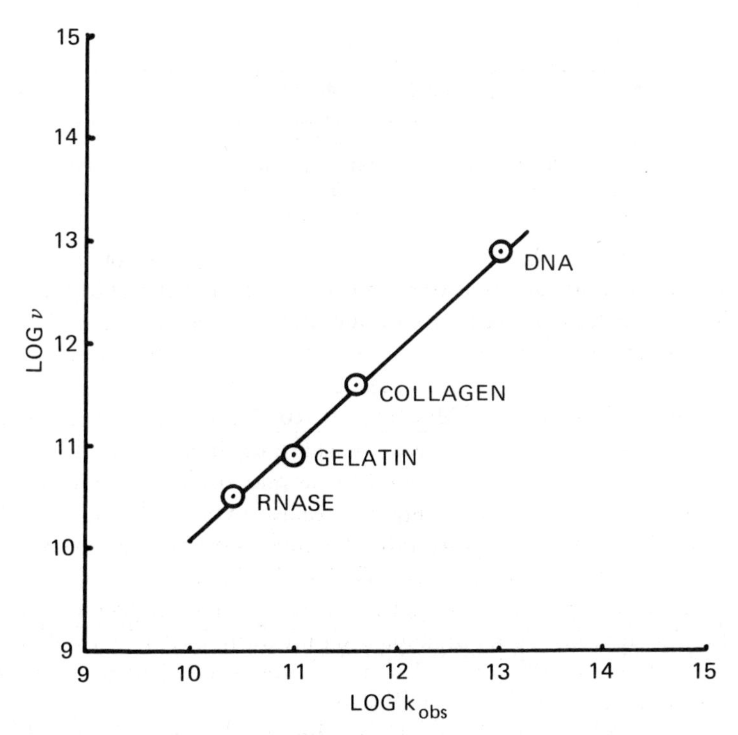

Figure 11. Correspondence between collisional frequency and hydroxyl radical rate constant for macromolecules. (Data are from ref. 20.)

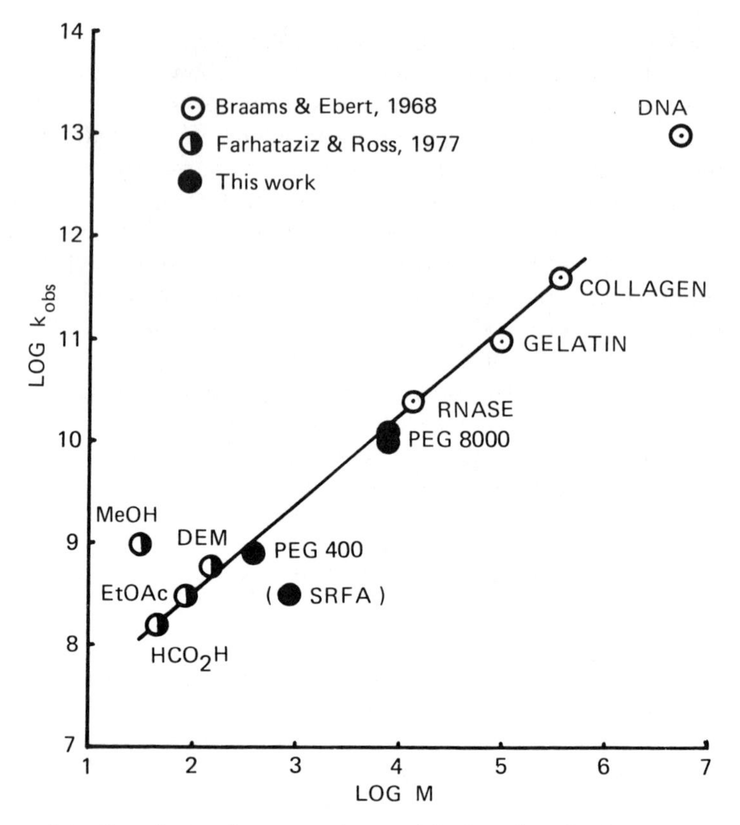

Figure 12. Correlation between observed hydroxyl radical rate constant (M^{-1} s^{-1}) and molecular weight of substrate (daltons).

because $[S] = C_{ws}/M$. C_{ws} is the weight concentration of the scavenger. Thus, k_s' is proportional to the weight concentration of scavenger, although the dependence on M_s is relatively weak, as shown by the following calculated values:

M_s	$M_s^{-0.13}$
100	0.55
1,000	0.41
10,000	0.30
100,000	0.22

Because the equation $k = AM^m$ also holds for the target compound, the competition factor (equation 3) reduces to

$$F_c = \frac{k_t'}{k_t' + k_s'} = \frac{AC_{wt}M_t^{-0.13}}{AC_{wt}M_t^{-0.13} + AC_{ws}M_s^{-0.13}} = \left[1 + \frac{M_s^{-0.13}C_{ws}}{M_t^{-0.13}C_{wt}}\right]^{-1} \quad (5)$$

where the subscript t refers to the target compound. The ratio r of $M_s^{-0.13}/M_t^{-0.13}$ is in the range $\frac{1}{3} < r < 1$ for $M_t > 24$ and $M_s \leq 10^5$, so that equal weight concentrations of target compound and scavenger should result in $\frac{1}{2} < F_c < \frac{3}{4}$. This result is in agreement with observation for the PEG experiments, but not for the AHA–SRHA–SRFA experiments.

Although some scatter around the line in Figure 12 is expected because of variations in molecular shape, the approximate linearity of the plot suggests that macromolecules should be able to compete for hydroxyl radicals with equal weight concentrations of smaller compounds. The lower molar concentration of the macromolecule should be compensated for somewhat by the higher molar rate constant. Thus, the apparent behavior of the humic and fulvic acids suggests that some additional effect may be operative.

Interpretation of Experimental Results. In view of information presented in the previous sections, the slower DEM disappearance rate (Figure 6) in the presence of PEG (compared to in its absence) appears to be a result of competition by PEG for hydroxyl radical. PEG does not absorb UV appreciably at 254 nm. A 10-ppm solution of AHA, however, had an absorbance of 0.281 with a 1-cm path length. A preliminary calculation indicated that the decrease in UV intensity due to absorption by AHA may account fully for the decreased DEM destruction rate seen in Figure 8.

The slower rate of peroxide disappearance seen in the curve labeled DEM (5) in Figure 7 actually represents peroxide regeneration, as was seen in the H_2O_2–UV–acetic acid system by Baxendale and Wilson (*23*). Because DEM does not absorb 254-nm radiation appreciably at the concentration used in these experiments, the hydrogen peroxide photolysis rate should be the same in the presence and absence of diethyl malonate. Even greater peroxide regeneration is seen in the PEG experiments because of its higher concentration (by weight) and a higher efficiency of regeneration from intermediate organic peroxy species. The fact that regeneration in the DEM–PEG system is slightly slower than with PEG alone suggests that the latter reason is at least partly responsible for the effect. Again, the rate decrease seen upon addition of AHA can be explained entirely on the basis of its UV absorption.

These interpretations are supported by the behavior of the ozone and hydrogen peroxide concentrations during ozonation and photolytic ozonation experiments. Shown in Figure 13 are the ozone accumulation profiles for ozonation of water (labeled "O_3 only") and photolytic ozonation of water, using one UV lamp (labeled "O_3/UV"). In the ozonation of water, ozone accumulated until the autodecomposition rate was equal to the mass-transfer rate, at which point the aqueous ozone concentration leveled out at about 3×10^{-5} M. By contrast, the ozone concentration in the ozone–UV experiment stabilized at about $0.2–0.3 \times 10^{-5}$ M because photolysis of aqueous ozone produces hydrogen peroxide (*1, 2, 16, 24*) and subsequent free-radical chain reactions were consuming ozone (*1, 2, 12, 16, 25*). When the ozonation

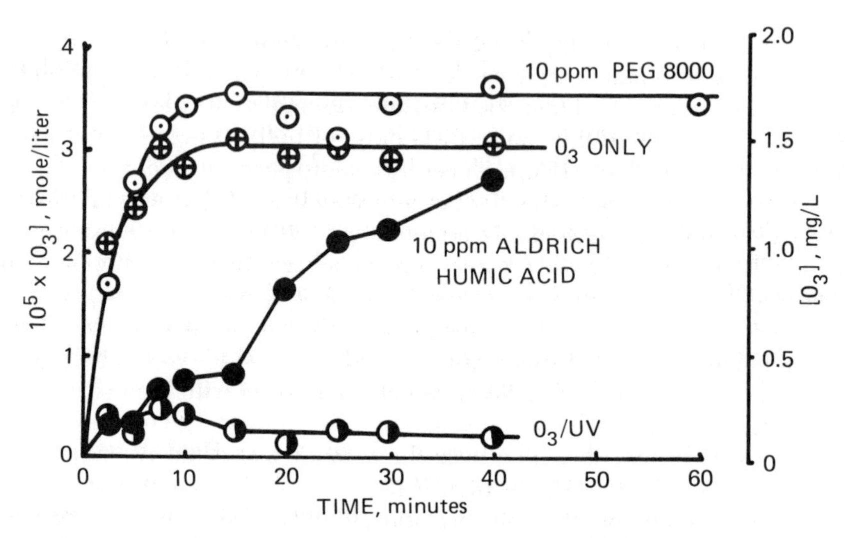

Figure 13. Effect of added macromolecular solutes on ozone accumulation.

experiment was rerun with the addition of 10 mg/L of PEG 8000 (Figure 13), the aqueous ozone concentration reached an even higher level, a result indicating that ozone did not react appreciably with PEG and that PEG was effectively scavenging free radicals that would otherwise attack ozone (autodecomposition).

When this experiment was rerun, after PEG was replaced with 10 mg/L of AHA, ozone accumulated only slowly during the first 10–20 min; thus ozone consumption by reaction occurred. The amount of ozone consumed in the reactor during the first 20 min was 2×10^{-4} mol/L. Assuming the AHA to be 50% carbon, 10 mg/L of AHA represents 4×10^{-4} mol/L of carbon atoms. If ozone reacted only with carbon–carbon double bonds, the amount of ozone consumed would be enough to involve every carbon atom in the molecule. This unreasonable result implies that reactions with secondary species were also consuming ozone. The ozone concentration later approached that attained in the absence of AHA ("O_3 only" in Figure 13), a result indicating that the source of the secondary species was eventually used up. The implication is that ozone reacted with humic material, consuming ozone and producing secondary species that further consumed ozone. Most likely, the secondary species could be an electron-transfer reagent such as superoxide, which is capable of converting ozone to hydroxyl radical. Once hydroxyl radicals begin to be formed, however, they can react with phenolic groups in the humic material. The OH adduct of phenol itself is known to react quickly with oxygen ($k = 1.5 \times 10^9\ M^{-1}\ s^{-1}$) to form a peroxy radical with a half-life of less than a millisecond, before yielding $HO_2(O_2^-)$ and a hydroquinone (*26*).

Further evidence supporting this interpretation is seen in Figure 14, which shows accumulation curves for hydrogen peroxide during photolytic ozonation experiments. Dramatic peroxide accumulation is seen in the experiments containing DEM only, PEG only, or both. However, photolytic ozonation of 10 mg/L of AHA produces little more peroxide than does similar treatment of pure water. This finding indicates that AHA is not a peroxide producer in $\cdot OH{-}O_2$ systems. In aqueous systems hydrogen peroxide production is the fate of most organic peroxy radicals that are not primarily superoxide producers; thus, these results indicate that AHA is a superoxide producer. That conclusion is consistent with the argument from the ozone accumulation data, in that superoxide, if produced, would react quickly with ozone to produce hydroxyl radicals, which could react with both ozone and AHA. These reactions of superoxide and hydroxyl radical with ozone account for the excess ozone consumed over the amount that could reasonably be expected to react with AHA in Figure 13.

To further support this interpretation, diethyl malonate was ozonated without UV light, both in the absence and in the presence of AHA. In the absence of AHA, DEM disappears slowly (data not shown), probably because

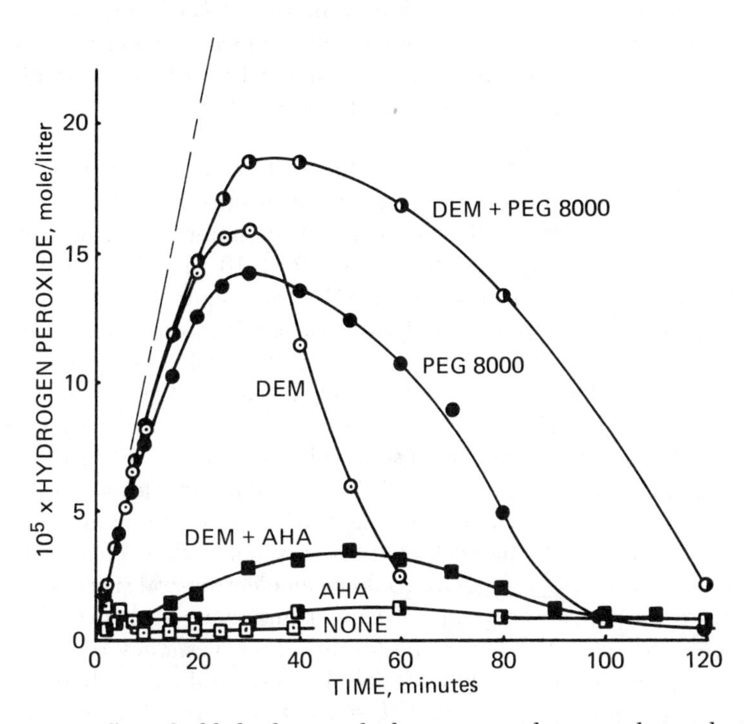

Figure 14. Effect of added solutes on hydrogen peroxide accumulation during photolytic ozonation. In all experiments the concentration of DEM is 5 mg/L and that of other scavengers is 10 mg/L.

of the free radical component of ozonation, because DEM reacts only slowly with ozone ($k = 0.06 \pm 0.02\ M^{-1}\ s^{-1}$) (*27*). In the presence of AHA, DEM disappears rapidly during the first 10–20 min of treatment (Figure 15), after which the disappearance rate slows to that seen upon ozonation in the absence of AHA. At the same time, ozone accumulation more nearly resembles that seen in the ozonation of 10-ppm PEG solution (Figure 6) than of 10-ppm AHA, a result indicating that the free radicals produced were being scavenged by DEM rather than by ozone. All of the ozone-reactive sites in AHA are consumed within the first 10–20 min, and production of the catalytic species ceases, whereupon the DEM disappearance rate returns to that observed during ozonation.

Several investigators (*28–30*) have suggested that irradiation of humic material in aqueous solution produces oxyradicals. We do not feel that those reactions are adequate to explain the magnitude of the effect observed in the present experiments, for the following reasons. In our experiments, the process of oxidation of humic materials should use up hydroxyl radicals at a rate similar to that of diethyl malonate disappearance (i.e., about 10^{-6} M min^{-1} (Figure 4) $\sim 2 \times 10^{-8}\ M\ s^{-1}$). Only OH radicals are effective in destroying DEM, because the reaction of HO_2, O_2^-, or $RO_2\cdot$ radicals with organic substances is too slow to be significant in these systems. $RO_2\cdot$ reactions are dominated by unimolecular superoxide elimination and bimolecular reaction to form the tetroxide (*31*), and $HO_2 \leftrightarrows O_2^-$ reacts almost

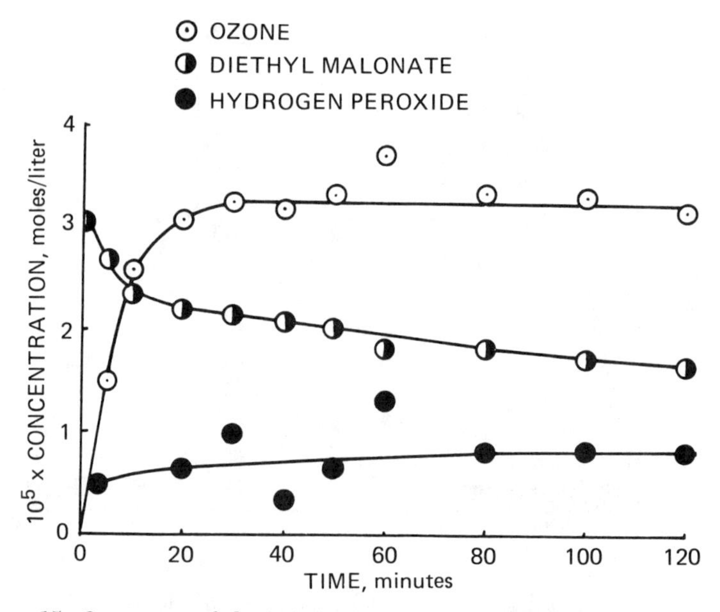

Figure 15. Ozonation of diethyl malonate and Aldrich humic acid mixture. Both components are present at 5 mg/L.

entirely with ozone, with $k = 1.6 \times 10^9$. The fastest total radical generation rate (including $RO_2\cdot$) measured by Korte (*28*) was 6×10^{-10} M s^{-1}, and even slower rates of $\sim 10^{-11}$ M s^{-1} were measured by Cooper and Zika (*29*) and Mill et al. (*30*). Therefore, radical generation by humic acid irradiation is probably insignificant compared to that generated by the ozone–UV system. Furthermore, the control run, in which a solution of DEM and humic acid was irradiated with UV light while oxygen was bubbled through the reactor, showed less than 4.4×10^{-7} M decrease in DEM concentration over a period of 1 h (i.e., less than 1.2×10^{-10} M s^{-1}).

Thus, AHA very well may scavenge hydroxyl radical with a rate constant consistent with the correlation shown in Figure 12. However, rather than consuming the hydroxyl radical, AHA merely "borrows" it and produces a species that results in conversion of another ozone to hydroxyl radical. By the time the ozone-reactive sites on AHA are used up, DEM has broken down to the point that superoxide producers are accumulating in solution to continue the cycle. This catalytic effect may of course be considerably less significant when very dilute solutions of substrate are treated, because of the decreased ability to replace the destroyed superoxide producers. However, the competitive effect of the humic material tested here appears to have been offset by the increase in hydroxyl radical production rate caused by byproducts of the competitive reaction. Because of this effect, the rate constant calculated for SRFA (shown in parentheses, Figure 12) is undoubtedly too low. The actual hydroxyl radical production rates were not known and were, according to this interpretation, probably significantly higher in the scavenged (by SRFA) compared to the unscavenged experiments.

Conclusions

- Macromolecular solutes may have k_{OH} in excess of 10^{10} M^{-1} s^{-1} and may effectively compete with smaller molecules for hydroxyl radical.
- Some substances, upon reaction with ozone or hydroxyl radical, produce secondary species that catalyze the conversion of ozone to hydroxyl radical. This catalytic effect opposes their competitive effect.
- All humic substances tested appeared to exhibit this catalytic effect, and thus did not interfere to the extent predicted by radical scavenging alone.

Acknowledgments

This work was funded in part by the U.S. Army Construction Engineering Research Laboratory, Champaign, Illinois, and by the Hazardous Waste

Research and Information Center, Savoy, Illinois. The laboratory assistance of Michelle Smith and Brent Peyton is gratefully acknowledged. We particularly thank J. Keith Carswell and USEPA–WERL–DWRD for their assistance through the loan of ozonation equipment used in these studies.

References

1. Peyton, G. R.; Glaze, W. H. Presented at the Sixth World Ozone Congress, Washington, DC; International Ozone Association: Washington, DC, 1983.
2. Glaze, W. H.; Peyton, G. R.; Sohm, B.; Meldrum, D. A. *Pilot Scale Evaluation of Photolytic Ozonation for Trihalomethane Precursor Removal*; Final Report to U.S. Environmental Protection Agency/Drinking Water Research Division/Municipal Environmental Research Laboratory: Cincinnati, OH, 1984; Cooperative Agreement CR–808825.
3. Peyton, G. R.; Glaze, W. H. In *Photochemistry of Environmental Aquatic Systems*; Zika, R.; Cooper, W. J., Eds.; ACS Symposium Series 327; American Chemical Society: Washington, DC, 1986; pp 76–88.
4. Horak, J.; Pasek, J. *Design of Industrial Chemical Reactors from Laboratory Data*; Heyden: Philadelphia, 1978; p 325.
5. Prengle, H. W., Jr.; Hewes, C. G., III; Mauk, C. E. *Proceedings of 2nd International Symposium on Ozone Technology*; Rice, R. G.; Richet, P.; Vincent, M. A., Eds.; International Ozone Institute: Syracuse, NY, 1975; p 211.
6. Bader, H.; Hoigné, J. *Water Res.* **1981,** *15,* 449.
7. Bader, H.; Hoigné, J. *Ozone: Sci. Eng.* **1982,** *4,* 169.
8. Flamm, D. L. *Environ. Sci. Technol.* **1977,** *11,* 978.
9. Hart, E.; Sehested, K.; Holcman, J. *Anal. Chem.* **1983,** *55,* 46.
10. Parker, G. A. In *Colorimetric Determination of Nonmetals*; Boltz, D. R.; Howell, J. A., Eds.; Wiley: New York, 1928; p 301.
11. Masschelein, W. J.; Davis, M.; Ledent, R. *Water Sewage Works* **1977,** *August,* 69.
12. Staehelin, J.; Hoigné, J. *Environ. Sci. Technol.* **1982,** *16,* 676.
13. Staehelin, J.; Hoigné, J. *Vom Wasser* **1983,** *61,* 337–348.
14. Bühler, R. F.; Staehelin, J.; Hoigné, J. *J. Phys. Chem.* **1984,** *88,* 2560.
15. Staehelin, J.; Hoigné, J. *Environ. Sci. Technol.* **1985,** *19,* 1206–1213.
16. Peyton, G. R.; Glaze, W. H. *Environ. Sci. Technol.* **1988,** *22,* 761–767.
17. Peyton, G. R.; Smith, M. A.; Peyton, B. M. *Photolytic Ozonation for Protection and Rehabilitation of Ground-Water Resources: A Mechanistic Study,* Research Report No. 206; University of Illinois Water Resources Center: Urbana–Champaign, IL, 1987.
18. Swallow, A. J. *Prog. React. Kinet.* **1978,** *9,* 195–366 and references therein.
19. Hoigné, J.; Bader, H. *Ozone Sci. Eng.* **1979,** *1,* 73–85.
20. Braams, R.; Ebert, M. In *Radiation Chemistry—I*; Hart, E. J., Ed.; Advances in Chemistry 81; American Chemical Society: Washington, DC, 1968; pp 464–471.
21. Smoluchowski, M. S. *Z. Physik. Chem.* **1918,** *92,* 129.
22. Farhataziz; Ross, A. B. *Selected Specific Rates of Reactions of Transients From Water in Aqueous Solution. III. Hydroxyl Radical and Perhydroxyl Radical and Their Radical Ions*; National Standard Reference Data System–National Bureau of Standards, 59, January 1977 (available from National Technical Information Service).
23. Baxendale, J. H.; Wilson, J. A. *Trans. Faraday Soc.* **1957,** *53,* 344.

24. Taube, H. *Trans. Faraday Soc.* **1957,** *53,* 656.
25. Taube, H.; Bray, W. C. *J. Am. Chem. Soc.* **1940,** *62,* 3357.
26. Micic, O. I.; Nenadovic, M. T. *J. Phys. Chem.* **1976,** *80,* 940.
27. Hoigné, J.; Bader, H. *Water Res.* **1983** *17,* 173–183.
28. Kotzias, D.; Herrman, M.; Zsolnay, A.; Russi, H.; Korte, F. *Naturwissenschaften* **1986,** *73,* 35–36.
29. Cooper, W. J.; Zika, R. G. *Science (Washington, D.C.)* **1983,** *220,* 711–712.
30. Mill, T.; Hendry, D. G.; Richardson, H. *Science (Washington, D.C.)* **1980,** *207,* 886–887.
31. Schuchmann, M. N.; Zegota, H.; von Sonntag, C. Z. *Naturforschung* **1985,** *406,* 215–221.

RECEIVED for review July 24, 1987. ACCEPTED for publication February 12, 1988.

37

Effect of Ozonation and Chlorination on Humic Substances in Water

Joop C. Kruithof, Marten A. van der Gaag, and Dick van der Kooy

Kiwa Ltd., P.O. Box 1072, 3430 BB Nieuwegein, Netherlands

Ozonation converts humic substances, as can be seen by a small decrease in dissolved organic carbon (DOC) and a substantial decrease in UV extinction. This conversion is related to the formation of low-molecular-weight biodegradable compounds, which enhance regrowth of organic substances in water during distribution. Post-treatment by coagulation and filtration processes can remove these types of compounds. Chlorination of humic substances causes the production of trihalomethanes (THMs), high-molecular-weight organohalides, and mutagenicity. A partial THM-precursor removal by pretreatment does not reduce the THM content under practical conditions and causes a shift to production of more highly brominated THMs. Extensive pretreatment with ozonation and granular activated carbon (GAC) filtration lowers the adsorbable organohalogen (AOX) content and the mutagenic activity in the Ames test.

WATER SOURCES USED FOR DRINKING-WATER PREPARATION contain humic substances, which account for roughly 75% of the total dissolved organic carbon (DOC). These humic substances strongly interfere with water-treatment processes, especially in surface-water treatment. Therefore, the Netherlands Waterworks Testing and Research Institute Kiwa Ltd. and the Netherlands Waterworks have been carrying out investigations into the removal and conversion of bulk organic materials (mainly humic substances) by water-treatment processes.

0065–2393/89/0219–0663$06.00/0

Many analytical methods are available for a general characterization of humic substances in water: determination of DOC content as a measure of the concentration of humic substances; spectrophotometric analysis (measurement of UV extinction and color) as a measure of the concentration and, in combination with the DOC content, as a first indication of the character of the humic substances; and gel permeation chromatography and X-ray scattering to determine the molecular-size distribution. Other methods are used to assess the interference of humic substances with treatment processes. Examples of these methods are determination of easily assimilable organic carbon (AOC) content as a measure of the regrowth potential of the organic materials, especially after oxidative treatment, and determination of trihalomethane (THM) precursors as a measure of the potential THM formation upon chlorination.

All of these parameters have been used to characterize the concentration and nature of humic substances during Dutch drinking-water treatment. For this chapter we will concentrate on

- DOC content as a measure of the content of humic substances
- UV_{254} extinction at a wavelength of 254 nm as an indication of the changing character of the humic substances during oxidative treatment
- AOC content as a measure of the regrowth potential of the water
- THM-precursor content as a measure of potential THM formation upon chlorination.

Treatment processes influencing the concentration of humic substances are coagulation and granular activated carbon (GAC) filtration. Standardized jar test equipment was developed (*1*) to study the effect of coagulation on the content of natural organic compounds. GAC filtration is applied at 12 full-scale plants, primarily to remove toxic compounds and to improve taste (*2*, *3*). Removal of humic substances (removal of color, UV extinction, and dissolved organic carbon) is an important secondary aim.

This chapter will consider the interaction of oxidative processes with humic substances in water. The processes studied most extensively are ozonation and chlorination.

Data on oxidation of humic substances by ozonation are available from 12 pilot-scale and eight full-scale experiments (*4*). Ozonation of humic substances leads to the production of low-molecular-weight compounds, which increase the availability for bacteria of organic compounds in water. This phenomenon is responsible for the regrowth of bacteria during distribution (*5*). Removal of these biodegradable compounds by posttreatment of the ozonated water is necessary to restrict bacterial regrowth during distribution

(6). The effects of ozonation will be illustrated through DOC, UV extinction, and AOC measurements gathered from pilot-scale and full-scale experiments carried out at the treatment plants at Driemond (Municipal Waterworks of Amsterdam) and at Kralingen (Waterworks of Rotterdam).

In the years since the discovery of THM production during drinking-water chlorination (7, 8), measures have been taken to reduce this side effect of chlorination as much as possible (9). One option was the removal of humic substances by GAC filtration before chlorination. Under Dutch chlorination conditions, partial removal of humic substances by GAC filtration did not give a substantial reduction of the THM content for long GAC-filter running times. Moreover, chlorination of GAC-filter effluents caused a shift to the production of more highly brominated THMs, especially at short GAC-filter running times (*10–12*).

In addition to a relatively high adsorbable organic halogen (AOX) content, mutagenic activity was found after postchlorination with a relatively low chlorine dosage of about 0.5 mg/L (*13, 14*). Both side effects of chlorination will be illustrated on the basis of THM and THM-precursor measurements gathered from full-scale experiments carried out at the treatment plant at Zevenbergen (Waterworks of North West Brabant) and on the basis of AOX and mutagenic activity measurements carried out at the pilot plant at Nieuwegein (Kiwa Ltd).

Effects and Side Effects of Ozonation

Experimental Parameters. An impression of the character and the content of humic substances in the water was obtained by determination of the UV extinction and the DOC content. In samples of nonozonated and ozonated water, the UV extinction was determined by spectrophotometer (Perkin Elmer, Type 500S) at a wavelength of 254 nm. The DOC content was determined subsequently, after acidification to pH 2 with concentrated hydrochloric acid and membrane filtration, with an ultra-low-level organic analyzer system (Dohrmann DC–54).

To determine the concentration of AOC, 600 mL of water was heated in thoroughly cleaned Pyrex Erlenmeyer flasks at 60 °C for 30 min to inactivate the bacteria orginally present in the water. After cooling, pure cultures of selected bacteria were inoculated into the samples, which were incubated at 15 °C. Growth of these bacteria in the water samples was measured by periodic determinations of the number of viable organisms. The maximum colony count is considered a measure of the amount of AOC available for the organism in the water used in the growth experiment.

Two bacterial strains were used: *Pseudomonas fluorescens* strain P17, able to metabolize a great variety of organic compounds such as amino acids, carbohydrates, and aromatic acids: AOC (P17); and *Spirillum* species strain NOX, specialized in the use of carboxylic acids such as formic acid, glyoxylic acid, and oxalic acid: Δ AOC (NOX). Yield values of P17 for acetate and of NOX for oxalate have been determined for use in calculation of the AOC concentration. The total AOC (AOC_T) is AOC_T = AOC (P17) + Δ AOC (NOX). In this equation, AOC (P17) is expressed in micrograms of acetic acid carbon (Ac C) per liter and Δ AOC (NOX) is expressed in micrograms of oxalic acid carbon (Ox C) per liter. Detailed information about the determination has been published elsewhere (*6, 15, 16*).

Application of Ozonation. In the Netherlands ozone is applied at eight full-scale treatment plants. The two largest plants are the Driemond plant of the Municipal Waterworks of Amsterdam with a production of 22×10^6 kL/year and the Kralingen plant of the Waterworks of Rotterdam with a production of 33×10^6 kL/year. The treatment systems of these plants are shown in Figures 1 and 2. The average ozone dosage at both plants is about 2.5–2.7 mg/L.

Pilot-scale ozonation experiments have been carried out with rapid filtrate from lake water at the Driemond plant and surface water after coagulation at the Kralingen plant. The effect of the ozonation on humic substances is expressed as the reduction of the UV extinction and DOC as a function of the ozone dosage (mg of O_3/mg of DOC). The side effect of the ozonation on humic substances is expressed as the AOC_T = AOC (P17) + Δ AOC (NOX) fraction of the DOC content as a function of the ozone dosage.

Full-scale experiments were carried out to investigate the variation in AOC (P17) from raw to finished water at both treatment plants. In addition, full-scale data were gathered about the formation of AOC_T during the ozonation step.

Effect of Ozonation on Humic Substances. The reduction of the UV extinction as a function of the ozone dosage is presented in Figure 3. At a maximum ozone dosage of about 2.0 mg of O_3/mg of DOC, the UV

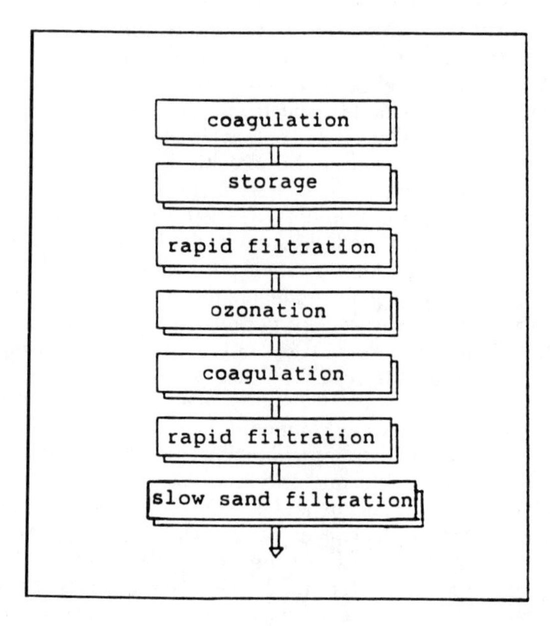

Figure 1. Treatment system at Driemond, Municipal Waterworks of Amsterdam.

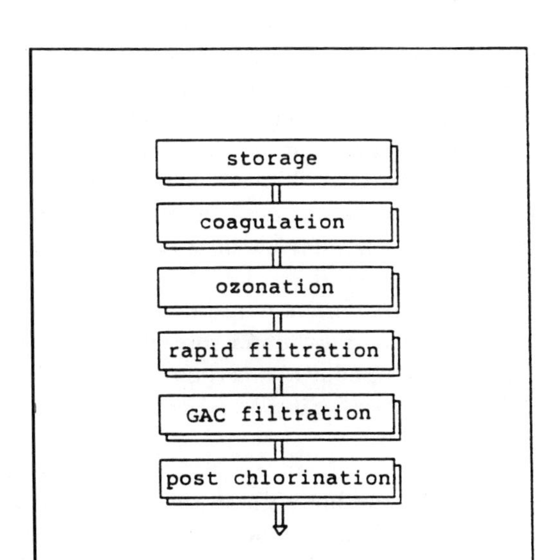

Figure 2. Treatment system at Kralingen, Waterworks of Rotterdam.

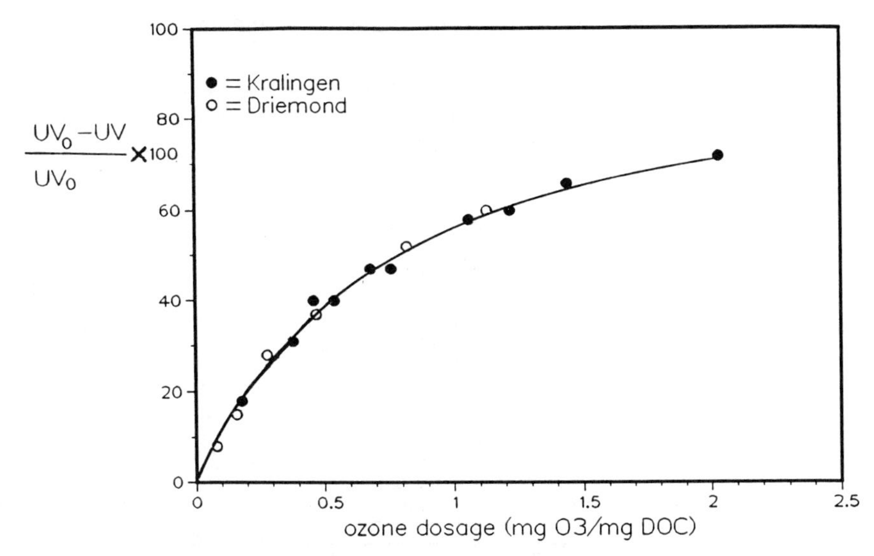

Figure 3. Reduction of UV extinction by ozonation; pilot-scale experiments at Driemond and Kralingen.

extinction was reduced by about 60–70% for both water types. About 40% of the initially present UV-absorbing compounds showed the greatest susceptibility for ozonation at ozone dosages up to 0.5 mg of O_3/mg of DOC. This conversion of UV-absorbing compounds is reached under practical conditions. The reduction of the DOC content is given in Figure 4. At maximum ozone dosage, the DOC at Kralingen was reduced by about 8%. At Driemond a maximum conversion of 18% could be reached. Under practical ozonation conditions, these percentages proved to be about 4% and 8%, respectively.

These experiments indicate that for the maximum ozone dosage the UV extinction is reduced 60–70% and the DOC removal 8–18%. This preferential decrease of the 254-nm chromophore indicates that ozonation causes a change in the character of the humic substances, without a complete oxidation of the organic compounds to carbon dioxide. This conversion of humic substances is the cause of the increased concentration of biodegradable compounds.

AOC Formation. For both water types, pilot-scale experiments showed an increase of AOC_T as a function of the ozone dosage. At the maximum ozone dosage AOC_T amounted to 550 μg/L for ozonated rapid filtrate at Driemond and 800 μg/L for ozonated coagulated water at Kralingen. The AOC_T/DOC ratio as a function of the ozone dosage is presented in Figure 5. For both water types the AOC_T/DOC ratio showed a gradual rise. The values for ozonated water at Kralingen were much higher than those at Driemond (0.3 and 0.09, respectively). These data indicate that the

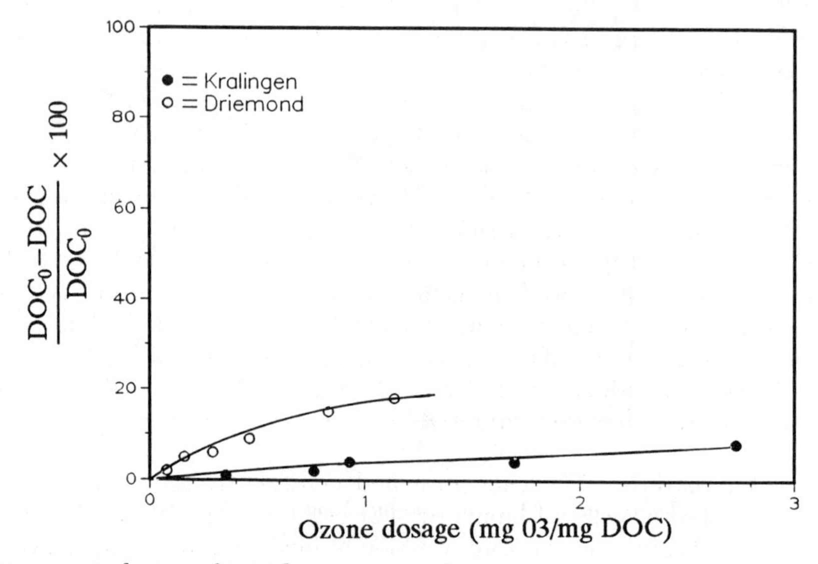

Figure 4. Reduction of DOC by ozonation; pilot-scale experiments at Driemond and Kralingen.

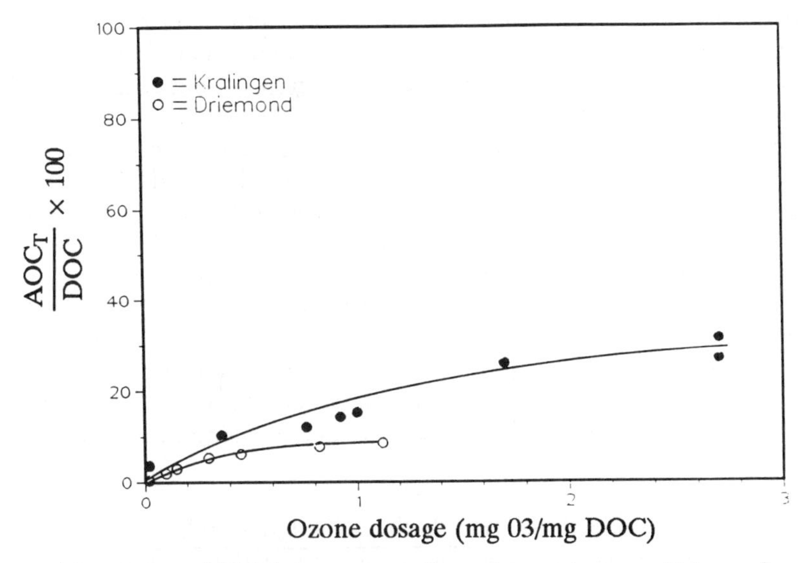

Figure 5. Formation of AOC by ozonation; pilot-scale experiments at Driemond and Kralingen.

extent of AOC production by ozonation is strongly affected by the nature of the humic substances present in the water.

For both full-scale plants, the variation in AOC (P17) has been determined from raw water to finished water. At Driemond (Figure 6) the AOC (P17) content after storage of lake water is 46 μg/L. Rapid filtration gives a reduction to 14 μg/L. Ozonation causes a large increase to 120 μg/L. Finally, coagulation, rapid filtration, and slow sand filtration lower the AOC (P17) content to 18 μg/L. Roughly the same pattern is obtained for the treatment of surface water after storage at Kralingen (Figure 7). The AOC (P17) content after storage is lowered by coagulation from 27 to 12 μg/L. Ozonation once again causes a large rise to about 100 μg/L. Rapid filtration and especially GAC filtration remove assimilable compounds to a concentration of 12 μg/L. A final application of chlorine gives a small rise to 22 μg/L. Full-scale data for AOC_T have been gathered for the effect of ozonation (*see* Table I). At Driemond, with an ozone dose of 2.7 mg/L, 4.3% of the total organic carbon (TOC) content of 6.5 mg/L is converted into assimilable compounds. At Kralingen, with an ozone dose of 2.5 mg/L, 12.6% of the TOC content of 2.6 mg/L is converted into AOC.

Discussion. The data presented here clearly show that ozonation changes the character of humic substances (change in UV/DOC ratio) and causes an increase in the concentration of biodegradable compounds. The UV extinction was reduced up to 60–70% for both surface (reservoir) water and lake water. The effect of ozone decreased as ozone dosage increased.

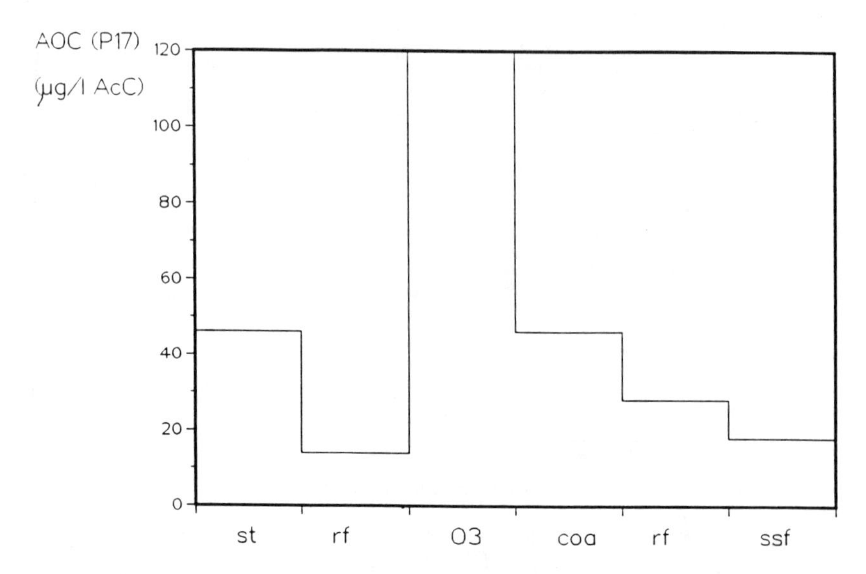

Figure 6. Variations in AOC (P17) content from raw to finished water; full-scale data at Driemond.

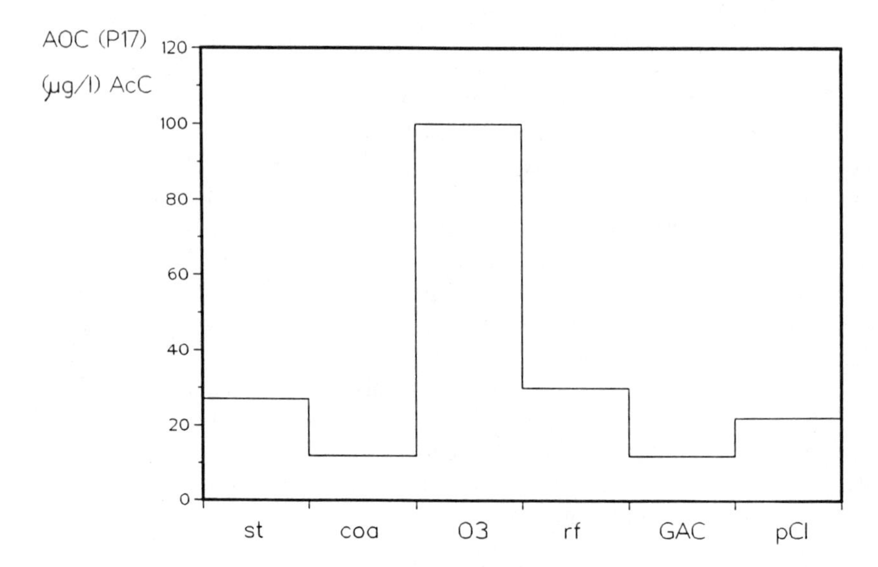

Figure 7. Variations in AOC (P17) content from raw to finished water; full-scale data at Kralingen.

Table I. Formation of Biodegradable Compounds

Plant	*AOC(P17)* (*μg/L Ac C*)	*ΔAOC(NOX)* (*μg/L Ox C*)	AOC_T (*μg/L*)	$\frac{AOC_T}{DOC} \times 100$
Driemond[a]	97	175	272	4.3
Kralingen[a]	95	220	315	12.6

[a]Full-scale data.

This correlation indicates a growing resistance of humic substances toward ozone. DOC was reduced by 18% for lake water at Driemond and by 8% for surface water at Kralingen. Hydrogen carbonate concentration and pH of both water types are in the same order of magnitude, so different oxidation mechanisms may be excluded. Therefore the percentage of complete oxidation to CO_2 is dependent on the type of humic substances. Both reduction percentages show clearly that the effect of ozonation is dependent on the type of humic substances present in the water and that ozonation causes a change in the character of the humic substances.

This change in the composition of the humic substances is once more established by an increase in the concentration of biodegradable compounds expressed by an increase in AOC_T. This rise in AOC_T concentration explains the pronounced bacterial regrowth in the distribution systems of waterworks supplying ozonated water without any further posttreatment. Oxidation of humic substances as present in lake water at Driemond produced a large DOC reduction (18%) and a relatively small AOC formation (AOC_T/DOC = 9%). On the other hand, oxidation of humic substances as present in surface water at Kralingen showed a small DOC reduction (8%) and a relatively extensive AOC formation (AOC_T/DOC = 31%). Excluding other reaction mechanisms, this result again demonstrates dependence on the type of humic substances present in the water. The biodegradable organic compounds can be removed by coagulation and filtration processes such as rapid filtration, slow sand filtration, and GAC filtration. Up to 70% of the AOC can be removed by coagulation and rapid filtration. This posttreatment of ozonated water is needed to prevent an extensive regrowth in the distribution system. However, final AOC concentrations in the water can be higher than they were before ozonation. Extensive treatment with GAC filtration, slow sand filtration, or even postdisinfection with chlorine may be necessary to prevent regrowth.

Effects and Side Effects of Chlorination

Experimental Parameters. The total trihalomethane (TTHM) concentration in aqueous samples was measured with a headspace technique that uses gas chromatography with a capillary column and an electron-capture detection system.

Samples for the THM formation potential (THMFP) were prepared by using a chlorine dose of 35 mg/L and a reaction time of 48 h.

The AOX content of water samples was determined after they are purged with nitrogen to remove volatile compounds. Then the water was mixed with carbon, sodium nitrate, and nitric acid. The sample was filtered over a polycarbonate filter. Carbon and filter were washed with a sodium nitrate solution and used for AOX measurements. This isolation procedure enabled the determination of the nonvolatile, high-molecular-weight organohalide fraction only. AOX was measured with a microcoulometric titration system after pyrolysis of the sample at 850–1000 °C in an oxygen atmosphere. The AOX samples were manually introduced into the oven tube through a boat inlet system.

The determination of mutagenicity in the Ames test with strains TA98 and TA100 was carried out in styrene–divinylbenzene resin (XAD–4) isolates according to Maron and Ames (*17*), with slight modifications (*18*). After removal of the volatile organic material by a nitrogen purge, the nonvolatile organic material was adsorbed sequentially on two XAD–4 columns at pH 7 and 2. Each fraction was eluted and concentrated in ethanol (*19*). The XAD samples were tested at six dose levels (10–140 μL of ethanol per plate), depending on the type of water tested. In all samples, organic substances from 1 L of water were concentrated in 25 μL of ethanol. The liver homogenate (S9) was prepared from Aroclor-induced Sprague-Dawley rats. (Arochlor is a series of polychlorinated polyphenyls.) In the assays with S9 mix, 0.5 mL of S9 mix containing 0.075 mL of S9 was added to the top agar.

Application of Chlorination. In the Netherlands chlorination is applied at 12 full-scale treatment plants. One of these plants, the surface-water plant at Zevenbergen of the Waterworks of North West Brabant, has an annual production of 2.5×10^6 kL. The treatment system of this plant, shown in Figure 8, includes breakpoint chlorination and postchlorination. The average dosage for breakpoint chlorination amounts to 1.7 mg/L; postchlorination is applied with a chlorine dosage of 0.5 mg/L. The application of chlorine and the side effects of chlorination are investigated extensively in pilot-plant studies. One of the pilot plants used for this type of investigation is the Kiwa pilot plant at Nieuwegein, with a capacity of 2.5 kL/h. The treatment system of this pilot plant is represented in Figure 9. The rapid filtrate, the ozonated water, and both GAC filtrates are postchlorinated according to a criterion of 0.2 mg of free chlorine per liter after a 20-min contact time.

With dual-media filtrate and GAC filtrate at the Zevenbergen plant, pilot-scale chlorination experiments have been carried out with chlorine dosage corresponding to the dose used in practice for postchlorination (0.5 mg/L). With these experiments the side effects of the chlorination on humic substances are expressed as the THM formation, measured as a function of the GAC-filter running time. Extensive attention has been paid to THM production in relation to THMFP reduction.

With dual-media filtrate and both GAC filtrates, chlorination experiments have been carried out at the Nieuwegein pilot plant with a chlorine dosage of 0.2 mg of free chlorine per liter after a 20-min contact time. These experiments led to the characterization of the side effects of chlorination,

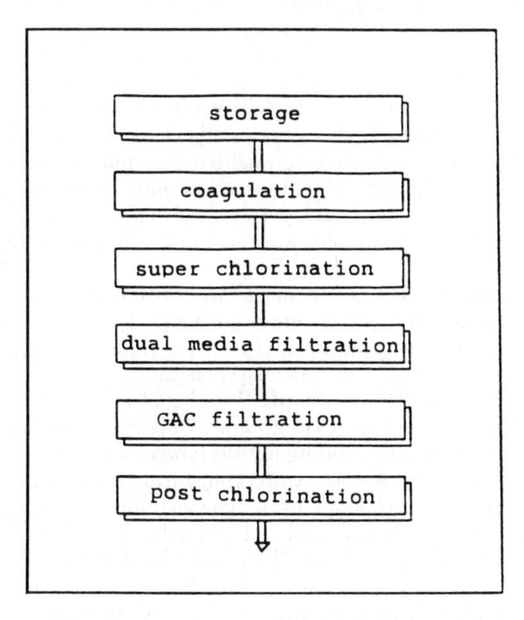

Figure 8. Treatment system at Zevenbergen, Waterworks of North West Brabant.

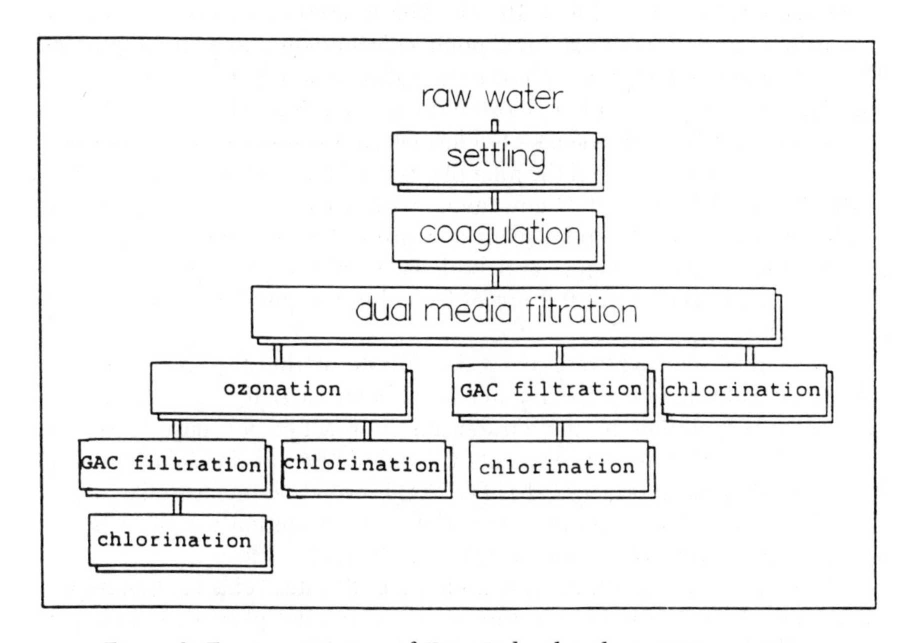

Figure 9. Treatment system of Kiwa Ltd. pilot plant at Nieuwegein.

with AOX formation and the formation of mutagenicity in the Ames test as a function of the GAC-filter running time.

Formation of THMs. For dual-media filtrate (the GAC-filter influent) and GAC filtrate at Zevenbergen, the THMFP was determined as a function of the GAC-filter running time (Figure 10). The THMFP of dual-media filtrate varied from 150 to 240 μg/L. THMFP in GAC filtrate increased with increasing carbon life from 20 to 160 μg/L. Marked THMFP reduction occurred only during very short filter runs. A removal of about 80% could be achieved for no more than 3 weeks. Precursor removal rapidly decreased and amounted to only 35% after a 26-week filter run. The THM content in chlorinated dual-media filtrate amounted to 13–25 μg/L, with $CHCl_3$ in the highest concentrations, followed by $CHBrCl_2$, $CHBr_2Cl$, and $CHBr_3$ (Figure 11a). The THM content in chlorinated GAC filtrate increased from 5 to 30 μg/L. By far the highest THM levels were $CHBr_3$ (up to 20 μg/L), followed by $CHBr_2Cl$, $CHBrCl_2$, and $CHCl_3$ (Figure 11b). Thus, in comparison with the chlorinated dual-media filtrate, no significant reduction of the THM content was achieved. Moreover, more highly brominated THMs are formed in GAC filtrates, especially for short carbon filter runs of about 10 weeks.

Thus, although carbon filtration leads to a significant THMFP reduction, this is not accompanied by a reduction of the THM content under practical conditions. It leads to formation of more highly brominated THMs, especially at relatively short filter runs. This phenomenon is illustrated in Figure 12. The THMFP increased as a function of DOC. The THM content rose to a

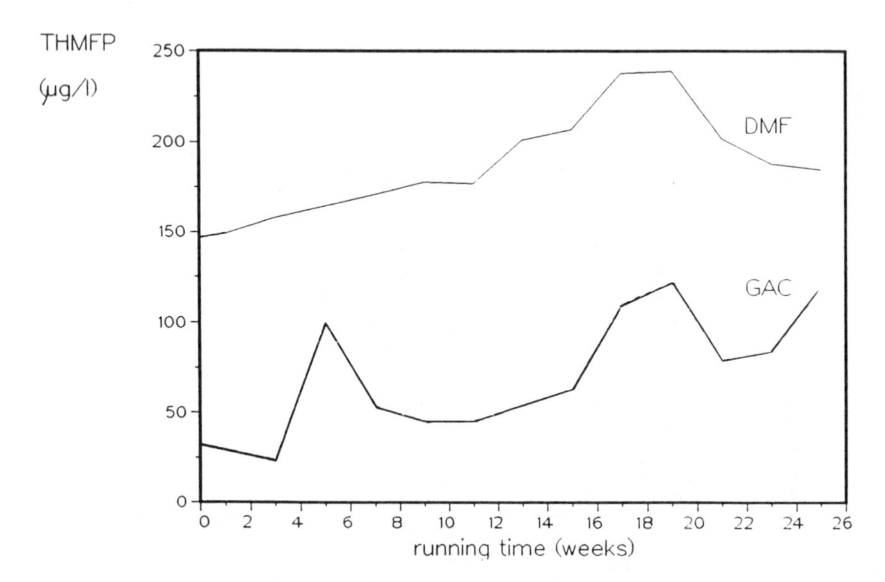

Figure 10. THMFP in chlorinated rapid filtrate and GAC filtrate as a function of the GAC-filter running time; full-scale data of Zevenbergen.

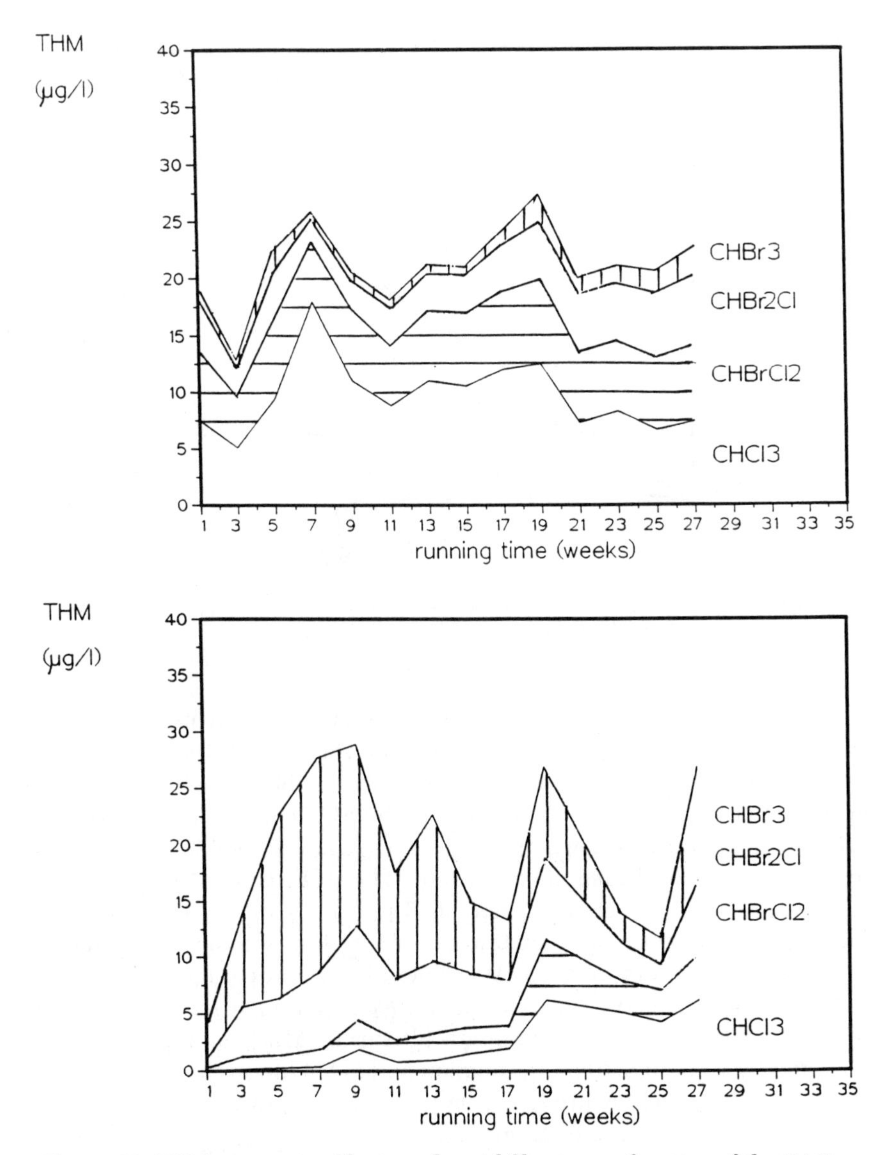

Figure 11. THM content in chlorinated rapid filtrate as a function of the GAC-filter running time; full-scale data at Zevenbergen.

DOC level of about 1.2 mg/L, then remained constant. The $CHBr_3$ content rose to a somewhat lower DOC level of 1.0 mg/L, then decreased. Only THM-precursor removal to a THMFP lower than 0.5 μmol/L (60 μg/L) led to a reduction of the THM content under practical conditions. This coincided with a DOC reduction to 1.2 mg/L. Maximum formation of $CHBr_3$ took place at about the same DOC concentration.

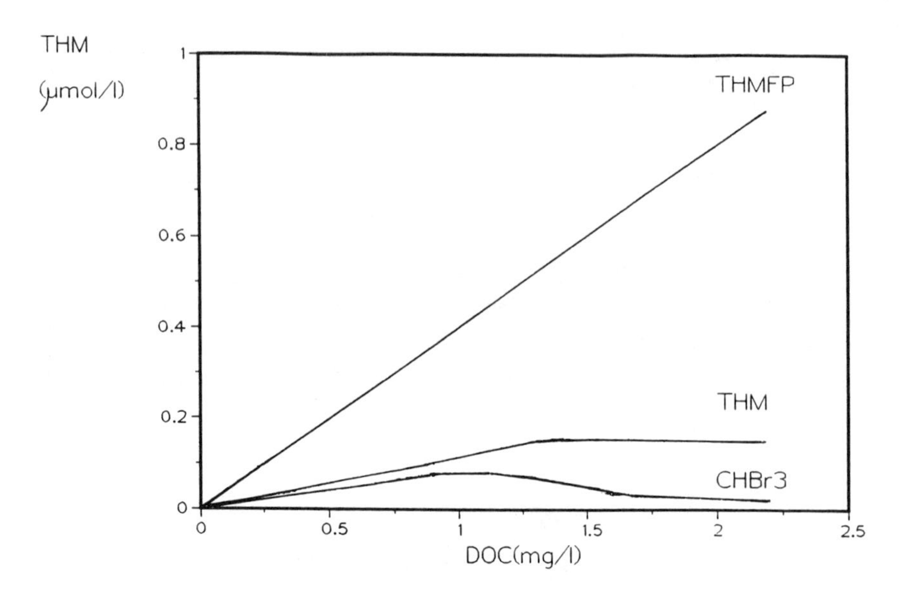

Figure 12. THMFP, THM, and $CHBr_3$ content as a function of the DOC content; full-scale data at Zevenbergen.

Formation of AOX and Mutagenicity in the Ames Test. For chlorinated dual-media filtrate and both chlorinated GAC-filter effluents at Nieuwegein, the AOX and the mutagenicity in the Ames test have been expressed as a function of GAC filter run. The AOX content for all chlorinated effluents is shown in Figure 13. High concentrations of AOX were formed upon chlorination of dual-media filtrate. The AOX content varied from 22 to 104 µg/L, with an average value of 65 µg/L. Formation of AOX in GAC filtrates started after more than 6 weeks of filtration and increased gradually for both carbon filters. In the GAC effluent fed with dual-media filtrate, the AOX content reached values of about 20 µg/L after 6 months and about 65 µg/L at the end of the 18-month filter run. Much better results were obtained by a combination of ozonation and GAC filtration. Chlorination of this GAC-filter effluent produced only 15 µg/L of AOX at the end of the GAC-filter running time of 18 months.

The mutagenicity in the Ames test for chlorinated dual-media filtrate and ozonate is shown in Table II for strain TA100–S9 mix. Mutagenicity was about 10 times higher than before chlorination. Part of the mutagenicity was inactivated by liver homogenate. Ozonation produced substantial removal of the mutagenic effect. Chlorination of both GAC effluents caused a gradual increase of the mutagenicity as a function of the filter run (Figure 14). Mutagenicity of the chlorinated GAC effluent (influent dual-media filtrate) for TA100–S9 mix was first observed after 5 weeks. The breakthrough was initially higher in the pH 2 fraction, but after about 40 weeks the mutagenic

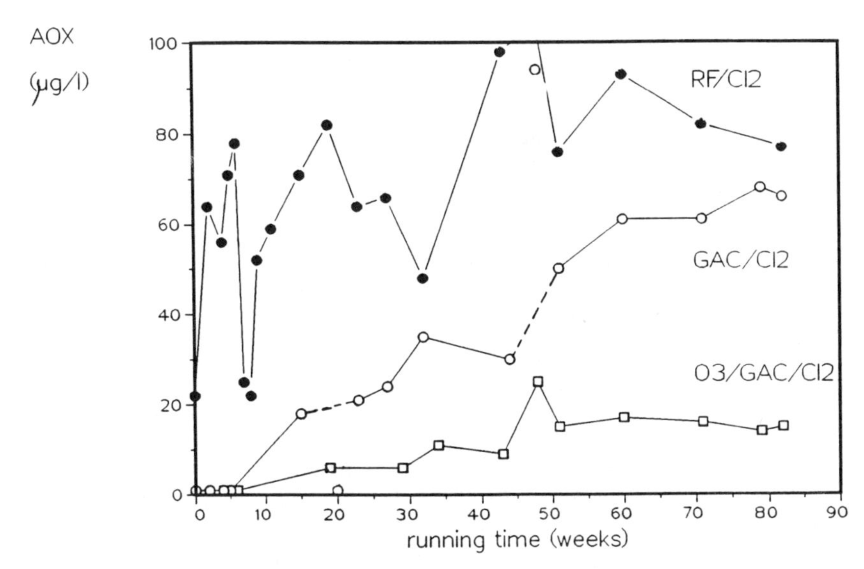

Figure 13. AOX content in chlorinated effluents; pilot-plant data at Kiwa.

Table II. Mean Mutagenicity of Chlorinated Rapid Filtrate and Ozonate

Effluent	*TA98–S9*	*TA98 + S9*	*TA100–S9*	*TA100 + S9*
Chlorinated rapid filtrate				
pH 7	550	790	1240	460
pH 2	190	100	880	210
Chlorinated ozonate				
pH 7	160	60	190	110
pH 2	35	10	170	70

activity of the pH 7 fraction reached the same level. Mutagenic effects following chlorination of the GAC effluent (influent ozonate) started appearing only after 10 weeks in the pH 2 fraction and after 40 weeks in the pH 7 fraction. In both fractions, this mutagenic activity leveled off at 150–200 induced revertants per 1.6 L equivalents, where it remained for the rest of the filter run. This level was the same as in the chlorinated ozonate, which means that there is no additional effect of the GAC filtration after that filter running time.

Discussion. From the data presented in the second section of this chapter, it can be concluded that

- partial THMFP reduction did not lead to a decrease in THM concentration under practical chlorination conditions. In ad-

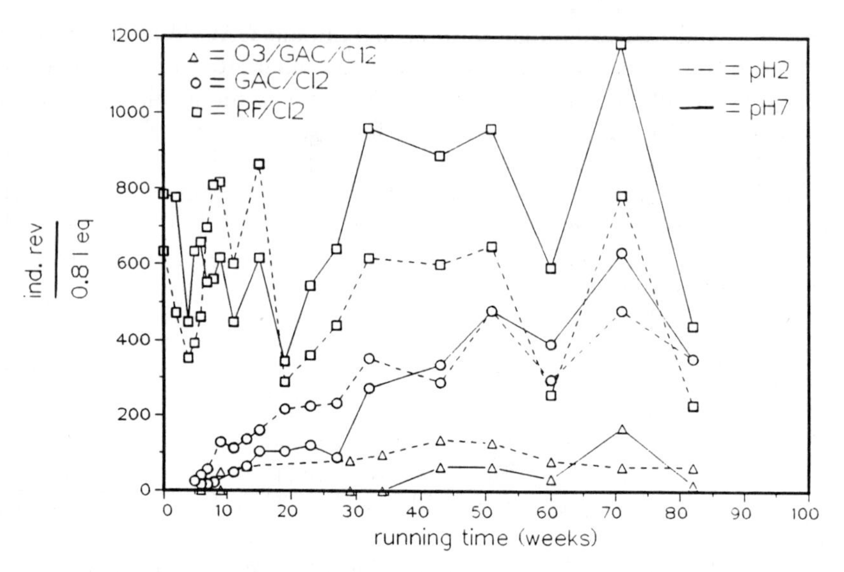

Figure 14. Mutagenicity in chlorinated effluents; pilot-plant data at Kiwa.

dition, a shift took place toward the formation of more highly brominated THMs.

- besides conversion into THMs, humic substances are converted by chlorination into adsorbable organohalides and thereby cause a strong rise in the mutagenic activity in the Ames test. In view of the applied isolation technique, both organohalide content and mutagenic activity represent nonvolatile, high-molecular-weight organic substances.

The THMFP is reduced by GAC filtration as a function of the filter run. Initially 80% of all THM precursors are adsorbed; after a filter run of 5 weeks the removal is about 70%. By the end of the running time, removal of THM precursors has dropped to 35%. The THM concentration under customary chlorination conditions in the Netherlands (a chlorine dosage of about 0.5 mg/L) is reduced only for a THMFP reduction of at least 70%. Therefore, at GAC filter runs longer than 5 weeks, THMFP removal does not lead to a decrease in the THM concentration under practical conditions, in combination with a shift toward the formation of more highly brominated THM.

In chlorinated dual-media filtrate the average AOX content amounted to 65 μg/L. GAC filtration initially caused a complete reduction of the AOC content upon chlorination. Formation of AOX after chlorination started after short filter runs. After a complete 18-month filter run, there was only a small difference compared to chlorinated dual-media filtrate. Much better results were achieved by a combination of ozonation and GAC filtration. Even at the end of the filter run the AOX production amounted to only 15 μg/L, an improvement of about 80% over chlorinated dual-media filtrate. Roughly the same conclusions are valid for the formation of mutagenicity in dual-media filtrate and both GAC-filter effluents. From the results it can be concluded that the performance of the O_3/GAC system in controlling AOX formation and formation of mutagenic activity was superior to that of GAC filtration alone. The total difference between the two systems could be attributed to the effect of ozonation.

Evaluation

The results of the studies presented in this chapter show that most humic substances present in the water are not converted in CO_2 by oxidative processes such as ozonation and chlorination. Therefore, both processes suffer from side effects.

Ozonation changes the character of humic substances strongly. UV absorbance of humic substances is greatly lowered, but there is hardly any decrease of the DOC content. Therefore complete oxidation to CO_2 is unlikely. Instead, easily assimilable low-molecular-weight organic compounds are formed, enhancing regrowth of bacteria in the distribution system. Posttreatment by coagulation and filtration processes is needed to remove these biodegradable compounds in order to prevent regrowth. Even further treatment of the filtrate (e.g., by disinfection) may be necessary to prevent regrowth. These negative side effects of ozonation must be considered in making decisions about the use of ozonation in a drinking-water treatment plant.

Chlorination of humic substances causes the formation of THMs, nonvolatile high-molecular-weight organohalides, and mutagenicity. Partial removal of THM precursors (humic substances) by pretreatment does not produce THM reduction under practical postchlorination conditions, and it causes a shift to the formation of more highly brominated THMs. Extensive pretreatment, especially with a combination of ozonation and GAC filtration, lowers the AOX content and the mutagenic activity. As long as the evaluation of the health effects of these halogenated compounds is not completed, the concentration of these compounds in water should be kept as low as possible. Therefore, postchlorination should be omitted when the biological quality (hygienic aspects, aftergrowth) is sufficient.

Based on these principles, the following disinfection philosophy is followed:

- No chemical disinfection is applied when sufficient physical, mechanical, and biological barriers are present.

- When chemical disinfection is needed, alternatives to chlorine can be considered if their side effects have proven to be less important than those of chlorine.
- In the meantime, when chemical disinfection is needed, a limited dose of chlorine may be applied.

References

1. Meijers, A. P.; de Moel, P. J.; van Paassen, J. A. M. Kiwa communication no. 70; Kruithof, J. C.; Kostense, A., Eds.; KIWA Ltd.: Nieuwegein, Netherlands, July 1984.
2. Kruithof, J. C.; Hess, A. F.; Manwaring, J. F.; Beville, P. B. *Aqua* **1983,** *2,* 89–99.
3. Kruithof, J. C.; van der Leer, R. Chr. In *Activated Carbon in Drinking Water Technology*; Kruithof, J. C.; van der Leer, R. Chr., Eds.; AWWA RF: Denver, 1983; pp 57–118.
4. Meijers, A. P. *Water Res.* **1977,** *11,* 647–652.
5. Stalder, K.; Klosterkötter, W. *Zentralbl. Bakteriol. Mikrobiol. Hyg. Abt. 1, Orig. B* **1976,** *161,* 474–481.
6. van der Kooij, D.; Visser, D.; Hijnen, W. A. M. *J. Am. Water Works Assoc.* **1982,** *74,* 540–544.
7. Rook, J. J. *Water Treat. Exam.* **1974,** *23,* 234–245.
8. Bellar, T. A.; Lichtenberg, J. J.; Kroner, R. C. *J. Am. Water Works Assoc.* **1974,** *66,* 703–706.
9. Sybrandi, J. C.; Meijers, A. P.; Graveland, A.; Poels, C. L. M.; Rook, J. J.; Piet, G. J. Kiwa communication no. 57; KIWA Ltd.: Rijswijk, Netherlands, 1978.
10. Kruithof, J. C.; Nuhn, P. A. N. M.; van Paassen, J. A. M. *H_2O* **1982,** *15,* 277–284.
11. Graveland, A.; Kruithof, J. C.; Nuhn, P. A. N. M. *Abstracts of Papers,* 181st National Meeting of the American Chemical Society, Atlanta, GA; American Chemical Society: Washington, DC, 1981; ENVR 63.
12. Bassie, W.; Kruithof, J. C.; van Puffelen, J.; Smeenk, J. G. M. M. In *Chlorination By-products; Production and Control*; Kruithof, J. C., Ed.; AWWA RF: Denver, 1986; pp 129–156.
13. Kruithof, J. C.; Noordsij, A.; Puijker, L. M.; van der Gaag, M. A. In *Water Chlorination: Chemistry, Environmental Impact and Health Effects,* Vol. 5; Jolley, R. L.; Bull, R. J.; Davis, W. P.; Katz, S.; Roberts, M. H., Jr.; Jacobs, V. A., Eds.; Lewis Publishers: Chelsea, MI, 1985; pp 1137–1163.
14. van der Gaag, M. A.; Kruithof, J. C.; Puijker, L. M. In *Organic Micropollutants in Drinking Water and Health*; de Kruijf, H. A. M.; Kool, H. J., Eds.; Elsevier Science Publishers: Amsterdam, 1985; pp 137–154.
15. van der Kooij, D.; Hijnen, W. A. M. *Appl. Environ. Microbiol.* **1984,** *47,* 551–559.
16. van der Kooij, D.; Visser, A.; Oranje, J. P. *Antonie van Leeuwenhoek* **1982,** *48,* 229–243.
17. Maron, D. M.; Ames, B. N. *Mutat. Res.* **1983,** *113,* 173–215.
18. van der Gaag, M. A.; Oranje, J. P. *H_2O* **1984,** *17,* 257–261.
19. Noordsij, A.; van Beveren, J.; Brandt, A. *Int. J. Environ. Anal. Chem.* **1983,** *13,* 205–217.

RECEIVED for review July 24, 1987. ACCEPTED for publication February 26, 1988.

38

Chlorinated Humic Acid Mixtures

Criteria for Detection of Disinfection Byproducts in Drinking Water

Alan A. Stevens, Leown A. Moore, Clois J. Slocum, Bradford L. Smith, Dennis R. Seeger[1], and John C. Ireland

Drinking Water Research Division, Water Engineering Research Laboratory, U.S. Environmental Protection Agency, Cincinnati, OH 45268

This chapter reports on the feasibility of using a chlorinated humic acid byproduct data base, developed in-house, as a drinking water quality screening tool. Specifically, a gas chromatographic–mass spectral (GC–MS) data base of more than 780 compounds identified during experiments involving the reaction of humic acids with chlorine has been compiled and systematically compared to GC–MS profiles from extracts of finished drinking water sampled from 10 preselected operating utilities. A major goal of the research was to narrow this library down to a smaller, more significant target compound list that would be representative of the chlorination byproducts most frequently encountered in the finished drinking water of utilities practicing chlorine disinfection. In addition, the study demonstrates the practicality of using the described methodology for concentrating and identifying specific compounds from water samples at low concentrations.

DRINKING WATER CHLORINATION FOR DISINFECTION PURPOSES produces numerous organic byproducts other than trihalomethanes (THMs) (*1–11*).

[1]Current address: University Hygienic Laboratory, Oakdale Campus, University of Iowa, Iowa City, IA 52242

0065–2393/89/0219–0681$06.00/0

Naturally occurring humic substances in water serve as precursor material for reaction with chlorine to produce a variety of non-THM compound classes, a large percentage of which are halogenated (*1–13*). The results of most available studies on this topic suggest that the number and identity of all possible chlorination byproducts have not yet been determined. Because a need exists to estimate the true extent of the chlorination byproduct problem in full-scale drinking-water-treatment systems, we set out to develop an experimental screening approach that would reveal the maximum possible number of such byproducts.

A relatively large data base of compounds formed by the reaction of chlorine with humic acids had already been compiled in-house in previous studies. We attempted to use this data base (via computerized gas chromatographic–mass spectral (GC–MS) searching techniques) to screen for water chlorination byproducts in finished drinking water sample extracts obtained from 10 representative treatment facilities. Modifications to a previously reported sample concentration technique (*11*) were also tested for improved separation and recovery of trace organic materials from aqueous solution.

Experimental Procedures

Materials. Commercial humic acid was obtained from Fluka Chemicals (Ronkonkoma, NY). Household bleach (Clorox) was used as the chlorinating agent. Chlorine content was determined by diluting 5 mL of Clorox to 1 L with demand-free water (Milli-Q) and titrating 50 mL of this solution to a KI–starch–iodine endpoint. Diazomethane gas was generated fresh from *p*-tolylsulfonylmethylnitrosamide (Diazald, Aldrich Chemical Co., Milwaukee, WI) and was stripped by nitrogen gas from the generation tube into sample vials. For the resin–granular-activated-carbon (XAD–GAC) extractions, dried, unpreserved, peroxide-free ether was prepared by treatment with acidified ferrous sulfate and sodium sulfate crystals. Buffer solutions were prepared from reagent-grade potassium dihydrogen phosphate and adjusted to the required pH with HCl or NaOH. All glassware (except volumetric flasks) was heated for 1 h at 400 °C in a muffle furnace to remove trace organic substances.

Methods. Humic acid solutions at organic carbon concentrations representative of drinking water sources were chlorinated in the laboratory to produce chlorination–oxidation byproducts. Confidence in the use of the readily available commercial humic acid as a reaction–product model was based on the comparative studies reported previously by Seeger et al. (*11*). For typical experiments, concentrated humic acid (HA) solutions were made by mixing 800 mg of HA in 1 L of 0.02 N sodium hydroxide solution and stirring for 30 min. This mixture was neutralized with sulfuric acid and filtered through glass fiber filters (Whatman 934 AH).

The filtered HA concentrate was then split equally, and the 500-mL portions were each diluted to 20 L with demand-free water in separate glass containers. The solution in one container was used to produce a chlorinated sample, and the solution in the other container served as an unchlorinated control. The resulting total organic carbon (TOC) of these samples (both chlorinated and unchlorinated) was 5 to

7 mg/L, in the range of that found in drinking water sources. Experiments were performed in duplicate, starting with a fresh concentrate of HA each time. The chlorination of HA solutions was studied at three different pH levels, 5, 7, and 11. In addition, a separate experiment with bromide was performed at pH 7. A phosphate buffer was used to control pH. The chlorine was added in sufficient quantity to produce a free chlorine residual at the end of a 3-day reaction period. After the 3-day period, the concentration of free (and total) chlorine residual was determined (*14*), and an excess of sodium sulfite was added to destroy that chlorine residual. The pH was then lowered to 2.0 with HCl and the XAD–GAC extraction was initiated.

XAD–GAC Adsorption Analysis. This analysis was a variation of the procedure reported by Seeger et al. (*11*) to concentrate products of reactions of chlorine with natural humic materials from large volumes (10–20 L) of dilute solution. Figure 1 presents a schematic diagram of the experimental apparatus. Briefly, the XAD–GAC adsorption analysis was carried out as follows: Three 1-gal finished water samples were combined in a 20-L glass carboy. Chlorine residual was then determined by *N*,*N*-diethyl-*p*-phenylenediamine (DPD) titrimetric analyses (*14*) for free and total Cl. Anhydrous Na_2SO_3 (2 mg) was then added for each 1 mg of total Cl residual and allowed to react for 30 min. An influent sample was taken for THM and total organic halogen analysis, and a check was made to be sure no chlorine remained. Phosphate buffer (100 mL) and 35 mL of 6 N HCl were added to reduce the pH to the range of 2.0–2.2.

After approximately 10 L had been passed through the columns in the adsorption mode (Figure 1A)—the XAD resin above the GAC—the columns, inverted so the GAC was above the XAD, were drained of water. Special fittings were then connected to the top and bottom of the connected columns so that ether could be distilled through a side arm leading to a condenser at the top. The condensed ether drained directly onto the GAC and eventually flooded the entire column length. In this way, ether was continuously refluxed (Figure 1B) through the connected GAC and XAD columns, backwashing the previously adsorbed organic compounds. The extracted organic compounds were concentrated in the receiving flask at the bottom.

After 3 h of continuous extraction, the ether was cooled; 1 mL of 1 N HCl was added (to ensure a pH <1), and the aqueous layer (about 10 mL) was separated from the ether layer. The ether layer was dried over Na_2SO_4. The aqueous layer was re-extracted with an additional 25 mL of ether and was dried with the same Na_2SO_4. The ether extracts were combined and concentrated to approximately 3 mL in a concentrating apparatus (Kuderna–Danish) followed by the addition of 0.25 mL of methanol. The concentrate was then washed into a 7-mL vial (marked at 5 mL) and was diluted to 5 mL with additional previously dried ether. Finally, samples were derivatized with diazomethane (*21*).

An internal standard, 1-chloro-*n*-dodecane, was added to a 0.1-mL aliquot of the 5-mL sample just prior to analysis by splitless injection onto a 50-m by 0.20-mm i.d. (SP–2100, Hewlett Packard) or a 60-m by 0.256-mm i.d. (DB–5+, J & W Scientific) fused silica capillary column. A gas chromatograph (Finnigan, model 9500) was used with the column exiting directly into the ion source of a quadrupole mass spectrometer (Finnigan, model 3300) operated at 70 eV in the electron-impact mode. The data were collected and processed with the data system (Finnigan, INCOS, model 3200) previously described (*11*).

All peaks were normalized on the area of the internal standard. A library was developed containing all compounds detected in the extracts of the chlorinated HA samples. This library contained 782 total entries. Ten finished drinking water samples

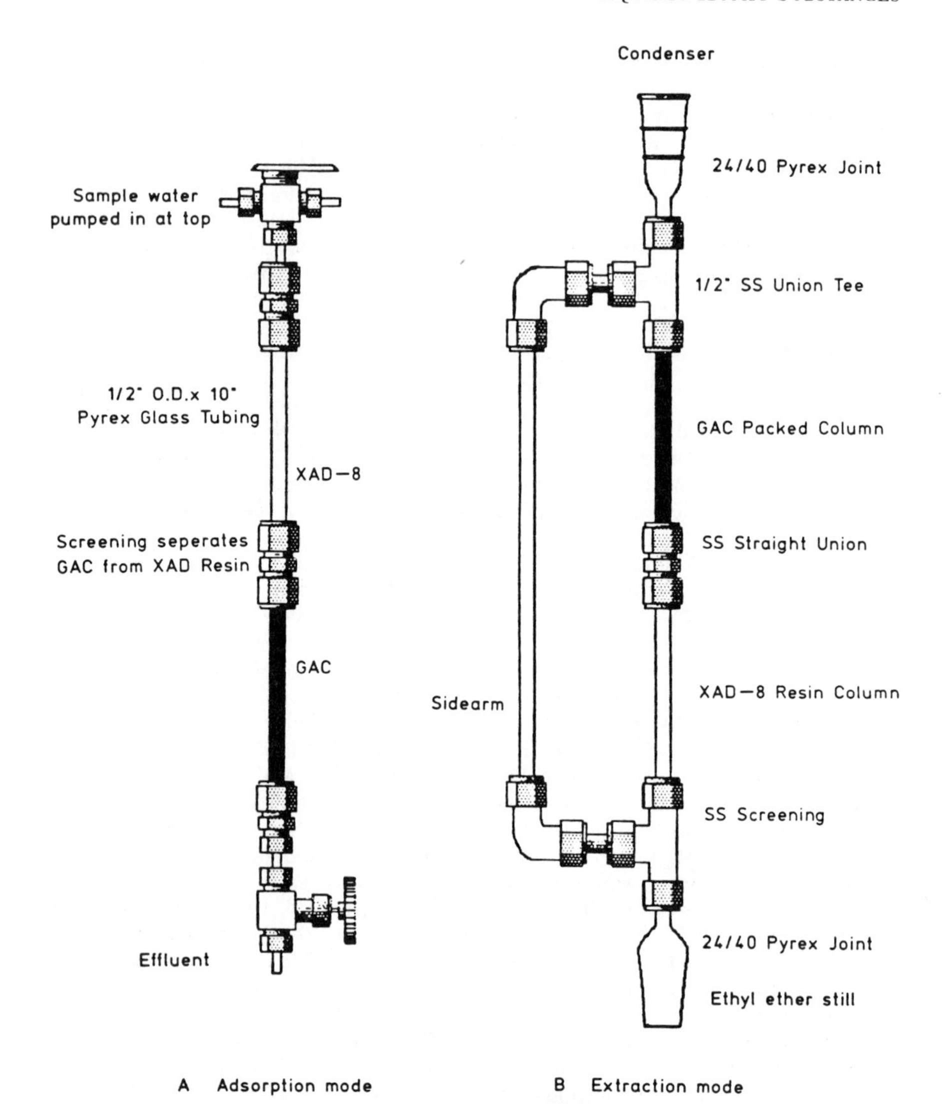

Figure 1. Schematic diagram of apparatus used to concentrate trace organic compounds from aqueous solution. A, apparatus configured in adsorption mode; and B, apparatus configured in desorption–extraction mode.

were analyzed with the same techniques as for the HA studies. The compounds found in these finished water samples were then compared to those previously entered into the library from the HA studies.

Selection of Drinking Water Utilities. The 10 utilities intentionally selected for this study represent a wide geographic area and several source types. Represented are ground and surface water, large and small populations served, and low and high organic carbon contents. Utilities A–C were selected because of known disinfection byproduct problems. All of the utilities use free chlorine as a disinfectant at some point in their treatment schemes. Sampling was performed between July and December 1985.

Results and Discussion

Before we could perform a cross comparison between GC–MS data derived from the 10 utilities with the chlorinated HA data base, it was necessary to establish a meaningful criterion that would define which of the 782 compounds in our library could be attributed solely to the chlorination process. After extensive evaluation of the data, we defined a chlorination byproduct as any single compound entry whose total ion current area count was at least 3 times the area count of the control for both of two duplicate runs and at each of three pH values investigated (pH 5, 7, and 11). Approximately 500 of the 782 compounds in our chlorinated humic substances GC–MS library fit our criterion for a chlorination byproduct. This 3× criterion served as the basis for screening the GC–MS extract data from the 10 utilities; the same 3× criterion was applied to each sample extract and associated blanks.

Computerized reverse searching was applied (*11*) to compare the chlorinated humic substances subset library of approximately 500 chlorination byproduct compounds to all data collected at the 10 locations. Table I displays the search results and lists 196 compounds found (cumulatively) that may be attributed to the chlorination process.

Eight compounds were manually excluded from the automated search procedure because their ubiquitous presence at multiple sites was anticipated from earlier studies (*15–19*): chloroform, bromodichloromethane, dibromochloromethane, bromoform, dichloroacetonitrile, bromochloroacetonitrile, dibromoacetonitrile, and chloropicrin were all detected in significant quantities. As expected, locations A–C had the highest THM concentrations because a conscious effort had been made to locate and sample utilities with acknowledged byproduct problems. These three utilities also exhibited some of the higher dichloroacetonitrile and chloropicrin concentrations relative to the others in the group. The presence of these compounds is by no means surprising, and their formation during water chlorination is well known (*15–19*).

In contrast to the eight compounds mentioned, much less is known about the 196 remaining compounds listed in Table I and found at one or more of the 10 utilities. For example, 128 of the 196 compounds have unknown structures. That is, the mass spectra of these compounds have no available reference spectrum match in a full U.S. Environmental Protection Agency–National Bureau of Standards data base search (*20*) and are awaiting manual interpretation. Of these 128 entries, 63 are known to contain at least chlorine. Tentative structures or functional groups other than chlorine have

Table I. Methylated Byproducts of Chlorination of Humic Acid Found at 10 Locations

Compound[a]	*RRT*[b]	*Utility/Log Area Count*[c]										*Count*[d]
		A	*B*	*C*	*D*	*E*	*F*	*G*	*H*	*I*	*J*	
1 1-Ethoxy-1-methoxymethane	0.180	5.8	4.6	4.6	5.0	5.4	5.0	5.3	4.2	4.2	5.8	10
2 Acetic acid, dichloro-, methyl ester	0.363	6.0	5.8	5.6	5.2	5.2	4.7	5.3	5.9	5.0	4.1	10
3 Oxirane, trichloromethyl-	0.432	4.6	4.0	4.2	4.0	4.3	3.7	4.2	4.4	3.3	2.7	10
4 Butanedioic acid, 2,2-dimethyl-, dimethyl ester	0.632	4.4	4.3	4.0	3.6	4.1	3.3	3.2	3.7	3.4	3.1	10
5 Benzoic acid, methyl ester	0.640	4.4	4.3	4.2	4.2	4.1	4.1	3.8	4.6	3.9	2.8	10
6 Dodecanoic acid, methyl ester	1.040	3.2	3.5	3.4	3.7	3.9	3.5	3.5	3.7	3.8	2.9	10
7 Propanoic acid, 2,2-dichloro-, methyl ester	0.381	4.7	4.6	4.6	3.8	4.0	3.0	3.9	4.5	3.5	—	9
8 (*X*)Cl compound? (GAC #31)	0.582	4.1	3.8	3.8	3.6	3.7	3.2	3.3	4.4	3.2	—	9
9 Pentanedioic acid, dimethyl ester isomer	0.588	4.4	4.7	4.4	3.9	4.5	3.4	3.6	3.9	3.6	—	9
10 Methylfurancarboxylic acid, methyl ester?	0.613	4.7	3.4	3.6	3.7	3.6	3.2	3.4	3.8	3.4	—	9
11 Chlorobutanedioic acid, dimethyl ester isomer	0.681	4.4	4.5	4.0	4.0	4.2	3.6	3.8	4.4	3.4	—	9
12 Hexanedioic acid, dimethyl ester isomer (GAC #39)	0.717	3.2	3.7	3.0	2.5	3.1	1.9	2.3	2.8	2.5	—	9
13 Chlorine compound? (LM #71)	0.723	4.8	4.4	3.6	3.9	3.5	3.2	3.4	4.7	3.1	—	9
14 Unknown dioic acid, dimethyl ester (GAC #43)	0.943	3.5	3.6	3.5	2.9	3.1	2.5	2.9	2.9	3.1	—	9
15 Benzenedicarboxylic acid, dimethyl ester isomer	0.993	3.9	4.1	3.9	3.9	4.0	3.9	4.2	4.3	4.0	—	9
16 Benzenetricarboxylic acid, trimethyl ester isomer	1.273	3.6	3.7	3.3	2.8	—	2.8	2.7	3.4	3.4	2.9	9
17 Ethene, trichloro-	0.205	3.5	3.3	3.2	3.0	3.2	3.0	3.2	—	—	3.9	8
18 Propanoic acid, 2-chloro-, methyl ester	0.292	3.8	4.0	3.7	3.6	3.6	3.3	3.3	—	—	3.1	8
19 Unknown (#139)	0.406	—	3.7	—	3.7	4.0	3.4	3.4	3.0	3.5	3.3	8
20 Hexanoic acid, methyl ester	0.453	3.7	3.9	—	3.5	3.8	3.4	3.6	3.7	3.4	—	8
21 (*X*)Cl compound? (LM #210)	0.654	4.0	—	4.5	4.1	4.3	3.8	4.4	5.1	3.5	—	8
22 (*X*)Cl compound? (ML #10)	0.735	4.2	3.5	3.5	3.6	3.7	2.8	4.1	4.3	—	—	8
23 Nonanoic acid, methyl ester	0.773	3.0	3.6	3.0	3.4	3.6	3.3	3.4	—	3.3	—	8
24 Methylfurandicarboxylic acid, dimethyl ester isomer	0.959	4.4	—	4.0	3.4	3.3	3.2	4.0	4.4	3.3	—	8
25 Unknown (#585)	1.059	3.7	3.4	—	3.2	3.7	3.0	3.1	3.9	3.1	—	8
26 Unknown (#576)	0.227	3.6	3.5	4.2	4.7	4.6	4.3	—	—	—	3.9	7
27 Pentanoic acid, methyl ester	0.334	3.4	3.6	3.3	3.1	3.5	3.1	3.4	—	—	—	7
28 2-Hexene, 1-methoxy-3-methyl-	0.381	—	—	3.9	4.0	—	3.9	2.9	2.9	4.0	4.0	7
29 Propanedioic acid, dimethyl ester	0.459	—	—	3.7	3.8	3.6	3.4	3.4	3.5	3.4	—	7
30 (*X*)Cl compound? (LM #209)	0.506	3.1	2.7	—	3.7	4.1	3.3	4.1	4.6	—	—	7
31 Unknown (GAC #30)	0.518	—	3.9	—	3.4	3.9	2.8	2.7	—	2.2	2.4	7

32 C3-Benzene isomer (#473)	0.534	—	—	3.2	3.2	3.2	3.1	3.2	3.5	3.5	—	7
33 (*X*)Cl compound? (LM #53)	0.597	4.1	3.0	3.4	—	2.3	—	2.5	3.3	3.0	—	7
34 (*X*)Cl compound? (ML #13)	0.767	4.3	4.2	4.2	—	3.6	—	3.4	4.4	3.2	—	7
35 Decanoic acid, methyl ester	0.867	—	3.1	2.9	3.3	—	3.3	3.1	3.3	3.2	—	7
36 Octanedioic acid, dimethyl ester	0.966	—	—	2.1	2.5	2.6	2.1	2.3	3.0	2.5	—	7
37 Benzenedicarboxylic acid, dimethyl ester isomer	1.027	3.4	3.1	—	3.3	2.9	3.2	—	3.1	3.5	—	7
38 Methylbenzenedicarboxylic acid, dimethyl ester isomer	1.073	—	—	3.2	2.7	3.1	2.7	2.5	3.4	3.0	—	7
39 Unknown (LM #178)	1.315	—	—	—	3.2	3.1	2.9	2.9	3.0	3.0	3.0	7
40 1,4-Dioxane	0.207	—	—	3.8	4.7	4.6	4.4	4.3	—	—	3.0	6
41 Acetic acid, chloro-, methyl ester	0.268	—	4.2	—	3.8	3.7	3.5	3.6	—	—	3.3	6
42 Methylbutanoic acid, methyl ester	0.282	4.4	4.1	3.9	—	3.4	3.7	3.4	—	—	—	6
43 Unknown (#138)	0.367	—	—	—	3.5	3.4	2.9	—	3.1	3.6	2.9	6
44 Chlorinated methyl ester (GAC #29)	0.414	3.8	3.7	3.6	—	—	—	2.4	2.8	2.4	—	6
45 Acetic acid, trichloro-, methyl ester	0.448	5.3	4.9	4.6	—	—	—	3.8	5.2	3.6	—	6
46 Heptanoic acid, methyl ester	0.566	—	—	—	2.8	3.2	2.9	3.1	3.1	2.8	—	6
47 Unknown (#175)	0.754	3.1	3.3	2.7	2.0	2.5	—	—	2.3	—	—	6
48 Unknown (#412)	0.876	—	—	—	2.9	3.0	2.6	3.2	4.0	2.6	—	6
49 Benzenedicarboxylic acid, dimethyl ester isomer	1.017	3.5	3.1	2.9	3.3	—	3.2	—	3.0	—	—	6
50 Nonanedioic acid, dimethyl ester	1.049	—	—	—	2.5	2.9	2.4	2.9	3.4	2.8	—	6
51 Aliphatic acid, methyl ester (FL #14)	1.395	—	—	—	3.3	3.1	2.8	2.8	3.2	3.0	—	6
52 2-Methyl-3-pentanone	0.248	—	—	—	3.3	3.4	3.2	3.3	—	—	3.2	5
53 Chlorinated methyl ester (GAC #27)	0.318	—	—	—	3.2	3.1	2.9	3.2	—	4.0	—	5
54 Aliphatic cpd (#435)	0.400	—	—	—	3.3	3.5	3.3	3.3	—	—	3.0	5
55 3,3-Dichloropropenoic acid, methyl ester	0.467	4.7	4.0	—	2.9	2.9	—	—	—	—	2.6	5
56 Benzaldehyde	0.490	—	—	—	3.0	—	3.0	3.8	3.5	3.5	—	5
57 Benzonitrile	0.502	4.6	—	4.4	4.2	3.4	3.9	—	—	—	—	5
58 Unknown (#596)	0.601	—	3.3	—	2.3	2.6	1.8	—	3.1	—	—	5
59 (*X*)Cl compound (ML #8)	0.603	3.0	—	—	2.8	2.9	—	2.9	3.7	—	—	5
60 (*X*)Cl compound (#512)	0.690	—	—	—	3.6	3.6	2.5	2.7	3.5	—	—	5
61 Benzeneacetonitrile	0.694	—	—	—	2.8	2.7	2.2	—	3.2	3.1	—	5
62 (*X*)Cl compound? (#168)	0.698	—	—	—	2.2	1.7	1.7	—	2.9	2.0	—	5

Continued on next page.

Table I. Methylated Byproducts of Chlorination of Humic Acid Found at 10 Locations (Continued)

		Utility/Log Area Count[c]										
Compound[a]	*RRT*[b]	*A*	*B*	*C*	*D*	*E*	*F*	*G*	*H*	*I*	*J*	*Count*[d]
63 Dichloroacetamide isomer (#513)	0.705	5.1	4.9	4.5	—	—	—	—	4.0	2.3	—	5
64 (*X*)Cl compound? (LM #315)	0.731	—	—	—	2.8	2.5	2.0	2.8	3.7	—	—	5
65 Unknown (LM #260)	0.739	3.5	—	—	3.1	2.7	2.3	2.6	—	—	—	5
66 2,2-Dimethylpentanedioic acid, dimethyl ester	0.757	—	3.3	2.7	2.7	3.1	—	—	—	—	3.3	5
67 (*X*)Cl compound? (#176)	0.763	3.8	3.5	3.4	3.0	—	—	—	3.9	—	—	5
68 Hexanedioic acid, dimethyl ester	0.783	—	—	—	2.9	3.2	—	2.6	3.0	2.5	—	5
69 (*X*)Cl compound? (#195)	0.889	3.5	2.7	2.7	2.6	—	—	—	3.2	—	—	5
70 Unknown (#209)	1.034	—	—	—	2.4	—	2.0	2.3	2.6	2.4	—	5
71 Unknown (#390)	1.208	—	2.5	—	2.9	—	1.7	2.8	3.0	—	—	5
72 Unknown #646	0.383	—	—	—	3.4	—	—	—	—	—	—	4
73 Unknown #651	0.416	—	—	—	3.6	—	—	—	—	—	—	4
74 Unknown #656	0.444	—	—	—	4.1	—	—	—	—	—	—	4
75 Bromochloroacetic acid, methyl ester (ML #7)	0.453	—	5.2	4.5	—	—	—	—	4.0	4.4	—	4
76 Cycloalkane or olefinic cpd (GAC #20)	0.541	—	—	—	2.1	3.6	2.0	—	—	2.4	—	4
77 Unknown #669	0.559	—	—	—	3.4	—	—	—	—	—	—	4
78 Hexanedioic acid, dimethyl ester (#329)	0.694	—	3.2	2.8	—	2.8	2.1	—	—	—	—	4
79 (*X*)Cl compound? (ML #11)	0.739	—	—	—	3.4	3.3	—	3.2	3.9	—	—	4
80 Unknown (#331)	0.744	—	—	—	2.3	—	2.2	3.4	3.4	—	—	4
81 Unknown (#181)	0.785	—	3.2	3.5	—	—	—	—	3.6	2.8	—	4
82 Unknown (#186)	0.800	—	—	—	3.2	3.6	—	3.1	2.9	—	—	4
83 Unknown (#LM #263)	0.907	—	—	—	3.3	3.0	—	—	3.7	2.9	—	4
84 Aliphatic acid, methyl ester (#202)	0.956	—	—	—	2.6	2.6	—	2.4	2.5	—	—	4
85 Aliphatic acid, methyl ester (#450)	1.218	3.5	3.9	4.5	—	4.3	—	—	—	—	—	4
86 Aliphatic cpd (#488)	1.235	—	—	—	—	2.9	2.4	—	2.8	2.7	—	4
87 Methane, tetrachloro-	0.178	2.9	3.4	2.7	—	—	—	—	—	—	—	3
88 Unknown (#594)	0.259	—	3.7	—	4.0	—	3.9	—	—	—	—	3
89 Unknown (#314)	0.294	—	—	—	4.0	—	—	3.8	—	—	4.1	3
90 Tetrachloroethene	0.319	—	—	2.6	—	3.0	—	—	—	—	3.4	3
91 C_8H_{18} isomer	0.322	—	1.0	—	2.5	—	—	—	—	3.7	2.5	3
92 Unknown #635	0.326	—	—	—	—	—	—	—	—	—	—	3
93 Hexamethylcyclotrisiloxane	0.353	—	3.8	—	—	—	—	—	3.9	4.0	—	3

94 C3-benzene isomer (#463)	0.506	—	—	—	3.2	2.9	—	—	—	—	2.8	3
95 Aliphatic cpd (#403)	0.625	—	—	—	—	2.8	2.4	—	2.3	—	—	3
96 Unknown (#158)	0.626	—	—	—	2.5	2.9	—	—	3.2	—	—	3
97 Aliphatic compound (#404)	0.628	—	—	—	2.7	2.9	2.4	—	—	—	—	3
98 Unknown (#162)	0.650	—	—	—	2.8	2.3	—	1.8	—	—	—	3
99 (*X*)Cl compound? (LM #65)	0.695	3.0	—	—	—	—	—	1.8	3.2	—	—	3
100 Unknown (#590)	0.730	—	3.6	3.5	2.0	—	—	—	—	—	—	3
101 Unknown (#431)	0.840	—	—	—	2.5	—	2.2	2.0	—	—	—	3
102 Unknown (#194)	0.871	—	—	—	2.9	—	2.4	—	3.2	—	—	3
103 (*X*)Cl compound (#567)	0.871	—	—	—	—	2.4	—	2.6	4.1	—	—	3
104 (*X*)Cl compound? (#196)	0.901	—	—	—	—	2.6	—	2.2	3.1	—	—	3
105 Dioic acid, dimethyl ester isomer (FL #20)	0.993	—	—	—	2.3	—	—	—	2.6	2.3	—	3
106 (*X*)Cl compound? (ML #18)	0.993	3.3	—	—	—	—	—	2.4	3.4	—	—	3
107 Unknown (#212)	1.064	—	—	—	3.0	3.4	—	—	3.5	—	—	3
108 Unknown (#214)	1.086	—	—	—	—	2.5	—	2.3	—	2.5	—	3
109 Aliphatic acid, methyl ester	1.118	—	—	—	2.9	—	2.6	—	—	2.7	—	3
110 (*X*)Cl compound? (LM #143)	1.230	—	—	—	2.6	—	—	2.9	—	2.7	—	3
111 Dibromomethane	0.198	3.9	—	—	—	3.2	—	—	—	—	—	2
112 Acetaldehyde, trichloro-	0.221	3.5	3.4	—	—	—	—	—	—	—	—	2
113 2,2-Dichlorobutanoic acid, methyl ester	0.500	3.3	—	3.0	—	—	—	—	—	—	—	2
114 (*X*)Cl compound? (#154)	0.553	3.3	2.7	—	—	—	—	—	—	—	—	2
115 Unknown (#325)	0.563	—	—	—	—	2.8	—	—	3.1	—	—	2
116 (*X*)Cl compound? (LM #314)	0.595	—	—	—	—	—	—	2.0	2.8	—	—	2
117 (*X*)Cl compound? (#156)	0.596	3.2	—	—	—	—	—	—	2.6	—	—	2
118 (*X*)Cl compound? (#161)	0.643	—	—	—	—	—	—	2.2	3.6	—	—	2
119 (*X*)Cl compound? (ML #9)	0.677	3.5	—	3.2	—	—	—	—	—	—	—	2
120 Pentanedioic acid, dimethyl ester isomer (GAC #34)	0.679	—	—	—	—	—	2.6	—	—	2.6	—	2
121 (*X*)Cl compound? (LM #307)	0.701	—	—	—	—	—	—	2.4	3.2	—	—	2
122 Aliphatic cpd (#514)	0.735	—	—	—	2.6	3.0	—	—	—	—	—	2
123 (*X*)Cl compound? (#173)	0.741	—	—	2.5	—	—	—	—	3.3	—	—	2
124 Benzoic acid, 3-methyl-, methyl ester	0.752	—	—	2.7	—	2.5	—	—	—	—	—	2

Continued on next page.

Table I. Methylated Byproducts of Chlorination of Humic Acid Found at 10 Locations (Continued)

Compound[a]	RRT[b]	Utility/Log Area Count[c]										Count[d]
		A	B	C	D	E	F	G	H	I	J	
125 Benzoic acid, 4-methyl-, methyl ester	0.760	—	—	—	—	2.5	—	—	2.3	—	—	2
126 Cl compound? (LM #77)	0.768	3.7	—	—	2.6	—	—	—	—	—	—	2
127 (X)Cl compound? (#184)	0.793	3.8	—	—	—	—	—	—	3.1	—	—	2
128 (X)Cl compound? (LM #85)	0.818	2.8	—	—	—	—	—	—	2.8	—	—	2
129 (X)Cl compound (#517)	0.819	—	—	—	2.3	—	—	—	3.0	—	—	2
130 Unknown (LM #262)	0.822	—	—	—	3.5	—	2.6	—	—	—	—	2
131 (X)Cl compound (#519)	0.830	3.8	—	—	—	—	—	—	2.7	—	—	2
132 Unknown (#338)	0.839	—	—	—	—	—	—	—	2.4	2.4	—	2
133 Heptanedioic acid, dimethyl ester	0.876	—	—	—	—	3.4	—	1.9	—	—	—	2
134 2-Methylpropanoic acid, 2,2-dimethyl-1-(2-hydroxy)-	0.903	—	—	—	2.3	—	2.4	—	—	—	—	2
135 Benzoic acid, 3-methoxy-, methyl ester	0.911	—	—	—	—	3.0	—	2.3	—	—	—	2
136 Unknown (#343)	0.938	—	—	—	—	—	—	2.9	3.1	—	—	2
137 (X)Cl compound (GAC #42)	0.939	—	—	—	2.1	—	—	2.4	—	—	—	2
138 (X)Cl compound (#418)	0.952	—	—	—	2.8	3.2	—	—	—	—	—	2
139 3,4-Dimethoxybenzoic acid, methyl ester	0.082	—	—	—	—	—	—	—	3.1	3.0	—	2
140 Acid methyl ester ?	1.127	—	—	—	2.2	2.5	—	—	—	—	—	2
141 Unknown (LM #129)	1.144	—	—	—	—	—	—	2.8	3.1	—	—	2
142 Benzenetricarboxylic acid, trimethyl ester isomer	1.260	—	—	—	2.8	—	2.3	—	—	—	—	2
143 Aliphatic acid, methyl ester (LM #204)	1.264	—	—	—	—	3.7	2.4	—	—	—	—	2
144 Benzenetricarboxylic acid, trimethyl ester isomer	1.316	—	—	—	2.7	—	2.2	—	—	—	—	2
145 Unknown (#362)	1.366	—	—	—	—	2.5	—	—	2.5	—	—	2
146 Unknown (#477)	0.230	—	—	—	—	3.7	—	—	—	—	—	1
147 Unknown (#398)	0.298	3.7	—	—	—	—	—	—	—	—	—	1
148 2-Propanone, 1,1,1-trichloro-	0.350	—	—	—	—	4.0	—	—	—	—	—	1
149 (X)Cl compound (#326)	0.621	—	—	—	—	—	—	—	3.0	—	—	1
150 (X)Cl compound? (#436)	0.623	4.0	—	—	—	—	—	—	—	—	—	1
151 Butenedioic acid isomer (#437)	0.638	—	3.7	—	—	—	—	—	—	—	—	1
152 Unknown (#328)	0.659	—	—	—	—	—	2.4	—	—	—	—	1
153 Unknown (#164)	0.669	2.7	—	—	—	—	—	—	—	—	—	1
154 (X)Cl compound? (#482)	0.673	4.5	—	—	—	—	—	—	—	—	—	1
155 Pentanedioic acid, 2-methylene-, dimethyl ester	0.690	—	—	3.0	—	—	—	—	—	—	—	1

156 Unknown (#169)	0.702	—	—	—	—	2.5	—	—	—	—	—	1
157 Unknown (#407)	0.708	—	—	—	—	—	—	2.6	—	—	—	1
158 Unknown (#170)	0.711	—	—	—	—	—	—	—	2.6	—	—	1
159 Unknown (#330)	0.727	—	—	—	—	—	—	—	2.9	—	—	1
160 (*X*)Cl compound? (#172)	0.737	—	—	—	—	—	—	—	3.7	—	—	1
161 (*X*)Cl compound? (#438)	0.739	2.8	—	—	—	—	—	—	—	—	—	1
162 Unknown (#178)	0.770	—	—	—	2.4	—	—	—	—	—	—	1
163 Acetamide, 2,2-dichloro-	0.774	—	3.9	—	—	—	—	—	—	—	—	1
164 (*X*)Cl compound? (#440)	0.777	—	—	—	—	—	—	—	—	2.1	—	1
165 (*X*)Cl compound? (#180)	0.780	—	—	—	—	—	—	—	2.9	—	—	1
166 3-Methyl-1,2,4-cyclopentanetrione	0.781	—	—	—	—	2.7	—	—	—	—	—	1
167 (*X*)Cl compound? (#183)	0.798	—	—	—	—	—	—	—	3.6	—	—	1
168 (*X*)Cl compound? (#185)	0.801	—	—	—	—	—	—	—	2.9	—	—	1
169 (*X*)Cl compound (#516)	0.818	—	—	—	—	—	—	—	3.6	—	—	1
170 (*X*)Cl compound? (#189)	0.827	—	—	—	—	—	—	—	2.6	—	—	1
171 (*X*)Cl compound (#336)	0.833	—	—	—	—	—	—	—	2.6	—	—	1
172 Unknown (#527)	0.848	—	—	—	2.9	—	—	—	—	—	—	1
173 (*X*)Cl compound? (#190)	0.848	—	—	—	—	2.3	—	—	—	—	—	1
174 1,4-Benzodioxin	0.860	—	—	—	2.9	—	—	—	—	—	—	1
175 (*X*)Cl compound (LM #212)	0.901	—	—	—	—	—	—	2.9	—	—	—	1
176 Unknown (#522)	0.904	—	—	—	—	—	—	—	2.9	—	—	1
177 Unknown (#199)	0.918	—	—	—	—	—	—	—	2.6	—	—	1
178 (*X*)Cl compound (FL #16)	1.018	—	—	—	—	—	—	—	3.1	—	—	1
179 Unknown (#210)	1.047	—	—	—	—	2.5	—	—	—	—	—	1
180 Dichloromethoxybenzoic acid, methyl ester isomer	1.057	—	—	—	—	—	—	—	3.5	—	—	1
181 (*X*)Cl compound (#532)	1.073	—	—	—	—	—	—	—	3.4	—	—	1
182 Unknown (#213)	1.077	—	—	—	—	—	1.7	—	—	—	—	1
183 Unknown (#421)	1.131	—	—	—	2.5	—	—	—	—	—	—	1
184 Unknown (#217)	1.132	—	—	—	—	—	—	—	2.3	—	—	1
185 Cl–($C_{10}H_6O$)–$COOCH_3$ isomer (ML #20)	1.138	—	—	—	—	—	—	—	2.7	—	—	1
186 (*X*)Cl compound? (LM #127)	1.141	—	—	—	—	—	—	—	3.0	—	—	1

Continued on next page.

Table I. Methylated Byproducts of Chlorination of Humic Acid Found at 10 Locations (Continued)

Compound[a]	*RRT*[b]	*Utility/Log Area Count*[c]										*Count*[d]
		A	*B*	*C*	*D*	*E*	*F*	*G*	*H*	*I*	*J*	
187 Unknown (#221)	1.151	—	—	—	—	2.1	—	—	—	—	—	1
188 Unknown (#223)	1.172	—	—	—	—	—	—	—	2.6	—	—	1
189 (*X*)Cl compound (LM #136)	1.186	—	—	—	—	—	—	—	3.2	—	—	1
190 Aliphatic acid, methyl ester	1.201	—	—	—	—	—	—	—	2.3	—	—	1
191 Unknown (LM #313)	1.239	—	—	—	—	—	—	—	3.3	—	—	1
192 $C_{14}H_{10}$ aromatic (#475)	1.240	—	—	—	—	—	2.5	—	—	—	—	1
193 Chloromethoxydicarboxylic acid, dimethyl ester	1.247	—	—	—	—	—	—	—	3.2	—	—	1
194 Aromatic methyl ester (#238)	1.267	—	—	—	—	—	—	—	2.9	—	—	1
195 Tridecanedioic acid, dimethyl ester	1.339	—	4.0	—	—	—	—	—	—	—	—	1
196 Unknown (#247)	1.404	—	—	—	1.5	—	—	—	—	—	—	1

[a]Either the known chemical compound name or some descriptive information about the entry as determined by its mass spectrum.
[b]Compound retention time (on the DB–5 column) relative to the internal standard.
[c]The columns labeled *A* through *J* are the water utility designations.
[d]Number of locations with an occurrence.

been assigned to 68 byproducts. Of these 68 byproducts, 44 are aliphatic or aromatic acids, some of which are halogenated. Five of the remaining 22 compounds are non-THM volatiles, four are ethers, two are aldehydes, two are ketones, two are nonhalogenated aromatic substances, four are organonitrogen compounds, and four do not represent any single class. Work is currently in progress to confirm the identities of these compounds by comparison of retention time and full-scan mass spectra with authentic standards.

The logarithm of the individual MS area counts for each location are listed under the respective utility code headings across the top of the table. These values represent relative order-of-magnitude concentrations for a given compound across the utilities in the study. The transformed area counts are not relative to each other across compounds. A minus sign in the log area count column indicates that the compound was not detected. Of the 196 entries, 122 (62%) have average areas of 1000 or more. The area counts were reproducible from one run to the next within a factor of 2 approximately 85% of the time during analysis of the duplicate data for each of the three varied-pH studies.

Table II lists the frequency of occurrence of the entries with counts by number of locations with that entry and cumulative totals of locations. This format provides easy reference to the number of locations with counts above or below a given level. For example, 110 entries can be found at three or more locations and at three or fewer locations. Half the locations sampled (5) had 71 of the entries present.

Conclusions

A substantial amount of work remains to be done in identifying the unknowns in Table I. We do, however, consider the reduction of the chlorination byproduct candidate list from over 700 entries down to around 500 (via the

Table II. Frequency of Occurrence of the 196 Chlorination Byproduct Data Base Entries Found at 10 Locations

Parameter	*10*	*9*	*8*	*7*	*6*	*5*	*4*	*3*	*2*	*1*
Number of entries occurring at this number of locations	6	10	9	14	12	20	15	24	35	51
Cumulative number of entries at this (or greater) number of locations	6	16	25	39	51	71	86	110	145	196
Cumulative number of entries at this (or fewer) number of locations	196	190	180	171	157	145	125	110	86	51

NOTE: The numbers at the top of the column are locations 1–10.

3× criterion) to be a significant first step toward the overall goal of identifying major disinfection byproducts in drinking water. Almost 200 compounds out of the reduced collection of 500 were found in one or more field test locations. This fact indicates that the chlorinated HA library matching scheme we have devised offers a meaningful approach to organic contaminant screening of drinking water. Further, the approach seems to focus on those humic acid chlorination products that may be most important in drinking water-treatment work, health studies, and possibly drinking water regulatory considerations.

In spite of previous criticism (*22*), it has recently been demonstrated that chlorination of both commercially available humic acids and humic material derived from various natural soil sources produces essentially the same major chlorination byproducts (*23*). This equivalence is true in particular for the known mutagenic chlorination byproducts, including the highly potent Ames mutagen MX (*24–25*), which was not included in our 782-compound data base at the time of the study.

This study augments (and greatly expands) previous investigations (*1–10*) into the characterization of aqueous chlorination products of humic substances. In particular, the combination XAD–GAC technique employed for isolation and concentration of sample components has revealed nearly an order of magnitude more compounds than previously reported. Future work will focus on identification of unknowns and on treatment techniques for effective removal of the major contaminants found with the greatest frequency at multiple sites.

References

1. Miller, J. W.; Uden, P. C. *Environ. Sci. Technol.* **1983,** *17*, 150–156.
2. Glaze, W. H.; Peyton, G. R.; Saleh, F. Y.; Huang, F. Y. *Int. J. Environ. Anal. Chem.* **1979,** *7*, 143–160.
3. Rook, J. J. *Water Chlorination: Environ. Impact Health Eff.* **1980,** *3*, 85–98.
4. Peters, C. J.; Young, R. J.; Perry, R. *Environ. Sci. Technol.* **1980,** *14*, 1391–1395.
5. Quimby, B. D.; Delaney, M. F.; Uden, P. C.; Barnes, R. M. *Anal. Chem.* **1980,** *52*, 259–263.
6. Christman, R. F.; Johnson, J. D.; Norwood, D. L.; Liao, W. T.; Haas, J. R.; Pfaender, F. K.; Webb, M. R.; Bobenrieth, M. J. *Chlorination of Aquatic Humic Substances*; U.S. Environmental Protection Agency Project Summary, EPA-600/2-81-016; U.S. Environmental Protection Agency: Cincinnati, OH, 1981; 178 pp.
7. McCreary, J. J.; Snoeyink, V. L. *Environ. Sci. Technol.* **1981,** *15*, 193–197.
8. McCreary, J. J.; Snoeyink, V. L. *Water. Res.* **1980,** *14*, 151.
9. Christman, R. F.; Johnson, J. D.; Pfaender, F. K.; Norwood, D.; Webb, M.; Hass, J. R.; Bobenrieth, M. J. *Water Chlorination: Environ. Impact Health Eff.* **1980** *3*, 75–82.
10. Norwood, D. L.; Thompson, G. P.; St. Aubin, J. J.; Millington, D. S.; Christman, R. F.; Johnson, J. D. In *Safe Drinking Water: The Impact of Chemicals on a Limited Resource*; Rice, R. G., Ed.; Lewis: Chelsea, MI, 1985; pp 109–121.

11. Seeger, D. R.; Moore, L. A.; Stevens, A. A. *Water Chlorination: Chem. Environ. Impact Health Eff.* **1985,** *5,* 859–873.
12. Christman, R. F.; Norwood, D. L.; Millington, D. S.; Johnson, J. D.; Stevens, A. A. *Environ. Sci. Technol.* **1983,** *17,* 625–628.
13. Norwood, D. L.; Johnson, J. D.; Christman, R. F. *Water Chlorination: Environ. Impact Health Eff.* **1983,** *4,* 191–200.
14. *Standard Methods for the Examination of Water and Wastewater*, 16th ed.; Greenberg, A. E.; Trussell, R. R.; Clesceri, L. S., Eds.; American Pharmaceutical Association, American Water Works Association, and Water Pollution Control Federation: Washington, DC, 1985; pp 306–309.
15. Bellar, T. A.; Lichtenberg, J. J.; Kroner, R. C. *J. Am. Water Works Assoc.* **1974,** *66,* 703–706.
16. Rook, J. J. *Water Treat. Exam.* **1974,** *23,* 234–243.
17. Oliver, B. G. *Environ. Sci. Technol.* **1983,** *17,* 80–83.
18. Duguet, J. P.; Tsutsumi, Y.; Mallevialle, J.; Fiessinger, F. *Water Chlorination: Chem. Environ. Impact Health Eff.* **1985,** *5,* 1201–1213.
19. Trehy, M. L.; Yost, R. A.; Miles, C. J. *Environ. Sci. Technol.* **1986,** *20,* 1117–1122.
20. EPA/NBS Mass Spectral Database Magnetic Tape; Office of Standard Reference Data, National Bureau of Standards: Gaithersburg, MD.
21. Schlenk, H.; Gellerman, J. L. *Anal. Chem.* **1960,** *32,* 1412–1414.
22. Malcolm, R. L.; MacCarthy, P. *Environ. Sci. Technol.* **1986,** *20,* 904–911.
23. Kopfler, F. C.; Ringhand, H. P.; Meier, J.; Kaylor, W. *Water Chlorination: Environ. Impact Health Eff.* **1988,** *6,* in press.
24. Holmbom, B.; Kronberg, L.; Hemming, J.; Reunanen, M.; Backlund, P.; Tikkanen, L. *Water Chlorination: Environ. Impact Health Eff.* **1988,** *6,* in press.
25. Meier, J. R.; Knohl, R. B.; Merrick, B. A.; Smallwood, C. L. *Water Chlorination: Environ. Impact Health Eff.* **1988,** *6,* in press.

RECEIVED for review July 24, 1987. ACCEPTED for publication February 8, 1988.

Removal of Aquatic Humus by Ozonation and Activated-Carbon Adsorption

Ellen Kaastrup[1] **and Terje M. Halmo**[2]

Department of Civil Engineering, Norwegian Institute of Technology, N–7034 Trondheim–NTH, Norway

The separate and combined effects of treatment with ozone and activated carbon were studied for three different humus sources: Norwegian brook water, Norwegian bog water, and commercial humic acid. The effect of ozonation on solution properties was determined by ultrafiltration, color, UV, and dissolved organic carbon (DOC) analysis. Adsorption prior to and after ozonation was studied in laboratory isotherm studies and a pilot-scale column experiment. Ozonation caused significant reductions in the content of high-molecular-weight material, UV extinction, and color; DOC reductions were insignificant for ≤1 mg of ozone per mg of DOC. Both isotherm and pilot-scale studies showed significant increases in adsorption capacities resulting from preozonation.

SURFACE WATERS HIGH IN COLOR-IMPARTING HUMIC SUBSTANCES are commonly used for drinking water in Norway. Such water used to be considered harmless, and the brownish color was treated as an aesthetic nuisance. The discovery of possible health threats caused by formation of trihalomethanes (THM) and heavy metal complexes (*1*) has led to more restrictive treatment requirements and extensive research on treatment alternatives.

[1]Current address: Ebasco Services, Inc., 143 Union Boulevard, Lakewood, CO 80228
[2]Current address: Elf-Aquitaine Norway A/S, P.O. Box 168, Dusavik, N–4001 Stavanger, Norway

0065–2393/89/0219–0697$08.50/0

It is difficult to find suitable treatment methods because of the wide variety of compounds in such water. The treatment combination investigated in this study was ozonation and activated-carbon adsorption. Neither of these methods has been successful in efficient removal of humic materials when used alone. The study objectives were to

- study the adsorption of organic matter from different humic-water sources onto activated carbon,
- determine how ozonation affects solution properties such as molecular-size distribution, color, UV extinction, and dissolved organic carbon (DOC),
- determine if the adsorbability of organic matter is increased (or changed) as a result of preozonation, and
- relate the changes in adsorptive properties to the changes in solution properties.

Background

Humic Substances in Water. Humic substances in water are generally divided into humic acids and fulvic acids. The humic acid fraction, containing the larger base-soluble molecules, precipitates on acidification to pH <2. Fulvic acids generally have lower molecular weights and higher polarity and hydrophilicity, and are soluble in both acid and base.

The relative distribution of the two fractions varies from one water source to another, depending on factors such as climatic and hydrogeological conditions. The concentration of aquatic humus is higher in cold and wet climates than in more arid areas; this results in higher DOC and color values. The higher concentration of organic matter in wet areas is caused by the high water extractability of organic matter from soil (*1*, *2*).

Humic acids have traditionally been thought of as extremely large molecules, with molecular weights ranging from thousands to hundreds of thousands and structures dominated by aromatic compounds. Recent studies suggest significantly lower molecular weights and higher content of aliphatic structures. Thurman et al. (*3*) used X-ray scattering to determine the molecular size of different humic substances. This study concluded that humic acids range from 1000 to more than 10,000 amu; in contrast, fulvic acids have molecular weights in the range 500–2000 amu.

Commercial humic substances have often been used as model substances in reported studies of humus treatment. Malcolm and MacCarthy (*4*) showed that the structures of such products are completely different from natural aquatic humus. Various commercial products were compared to humic and fulvic acids isolated from water and soil. The commercial humic substances were found to be almost identical to each other but significantly different from natural humus. The commercial materials had a lower content of aromatic molecules, longer chains of carbon–carbon double bonds, and lower carboxyl-group content.

Activated-Carbon Adsorption of Humic Substances. Most organic compounds are removed from water by activated-carbon adsorption. The adsorbability of individual compounds depends on a number of factors, such as polarity and hydrophilicity, solubility, molecular size and structure, and pH.

Generally, relatively insoluble and nonpolar compounds are most easily adsorbed. Size and structure may be limiting factors for molecules that are too large to have access to the smaller carbon pores that make up the major part of the surface area. The carbon surface can be divided by size into different groups of pores. According to the IUPAC (International Union of Pure and Applied Chemistry) definition, pores with diameter <2 Å are micropores, those in the range 2–20 Å are mesopores, and the larger pores are macropores. The total surface area of most carbons used in water treatment is 700–1000 m^2/g, most of which comprises smaller pores. A low pH is favorable for adsorption if the ionized form of the adsorbate is anionic.

Pore-size distribution is the most important factor in the adsorbability of humic compounds onto activated carbon. Carbons with high percentages of macropores have the highest capacities for adsorption of humic substances, and capacities for the smaller fulvic acid molecules are higher than for humic acids. Poor adsorbability is reflected in unfavorable adsorption isotherms, slow adsorption kinetics, and immediate breakthrough of color-imparting substances in adsorption columns.

McCreary and Snoeyink (*5*) found in studies with humic and fulvic acids that lower-molecular-weight species are more readily adsorbed. Lee and Snoeyink (*6*) studied the adsorption of humic and fulvic acid onto different types of carbon. They concluded that carbon pore-size distribution is the most important factor in determining adsorption capacity. Weber et al. (*7*) found that the adsorption rate for humic acid is inversely related to carbon-particle size. In a later study (*8*) higher adsorption capacities were observed for carbons with higher fractions of large pores, and the isotherms were divided into segments of different adsorbability.

Because natural waters are heterogeneous mixtures of unknown organic and inorganic compounds, collective parameters have to be used for characterizing the solutions. This approach makes it possible to model the adsorption equilibria by "pseudo-single-solute isotherms". The parameters most commonly used for characterization of humic substances are DOC, UV extinction (UV absorption measured at 254 nm), and color (visible transmission at 450 nm, defined relative to a cobalt–platinum standard).

For modeling of adsorption of organic substances from unknown mixtures of natural organic matter, the Freundlich equation (or modifications of it) is most often used. Previous investigators have proposed that the collective organic matter in humic water can be considered as consisting of fractions of varying adsorbability, often including a nonadsorbable fraction (*8–10*). The Freundlich model can be modified to account for a nonadsorbable fraction of organic matter, and different parameters can be used for segments of different adsorbability. By use of a substitution term for nonadsorbable matter, the equation takes the form

$$q_e = K_F(C_e - C_n)^{1/n} \tag{1}$$

where q_e is the amount adsorbed per unit weight of carbon at equilibrium (or most often "pseudoequilibrium"), C_e is the equilibrium concentration of organic substances, C_n is the concentration of nonadsorbable matter, K_F is the Freundlich coefficient, and $1/n$ is the Freundlich exponent.

Ozonation. Ozone is one of the strongest chemical oxidants known. It is used in water and wastewater treatment for disinfection, for decolorization, and as pretreatment for filtration and adsorption processes. The ozonation products are generally smaller, more polar, and hydrophilic than their precursors. Therefore, they are considered less adsorbable and more easily biodegradable.

The decolorization of humic water occurs because ozone is a typical double-bond reagent. It reacts with double bonds in the conjugated chains of large, color-imparting molecules, and thereby reduces the color and size of the molecules. Efficient and satisfactory decolorization of humic water has been reported in several studies (*11–14*).

Ozonation breaks up molecules into smaller units, but does not remove organic matter to any significant extent. Complete oxidation to carbon dioxide and water is too slow to be of significance in water treatment, and only minor reductions in the content of organic matter are therefore observed. Ozone treatment is consequently not sufficient as a removal method for organic matter.

Ozonation and GAC Adsorption. Ozonation is commonly used as a pretreatment for granular activated carbon (GAC) adsorption. Because the formation of more polar molecules generally causes reduced adsorbability, the intention of such pretreatment is usually to enhance biological activity in the filter and thereby to increase the overall removal efficiency.

Several studies have reported increased organic-substance removal in adsorption filters as a result of preozonation. However, in most cases ozonation had an adverse effect on adsorption, and the improved removal resulted from increased biological growth that led to increased biodegradability. Such results have been reported in several studies (*15–18*). However, the character of the organic matter is important. Most of the reported studies have dealt with organic matter that was relatively easily adsorbable prior to ozonation and in which intermediate-to-small molecules, rather than extremely large molecules, dominated. The situation is different for large humic molecules, for which adsorption is restricted by carbon pore size. Such molecules are adsorbed to a very low extent; they use large pores, which are only a small fraction of the carbon surface. These large molecules are easily oxidized by ozone. The reaction between ozone and the larger molecules will dominate as long as these are present because of ozone's high double-bond reactivity. On the basis of these observations, the following hypotheses were proposed for this study:

- Ozone will react with large color-imparting molecules and thereby reduce the color and molecular size.
- The adsorption capacity of organic matter from humic solutions will be increased as a result of reduced molecular size, as long as the reduced size dominates the increase in polarity.
- The adsorption rate will increase as a result of reduced diffusional restrictions.
- Increased biological activity is expected, in addition to increased adsorbability, in long-term studies.

Approach

Because two treatment processes were involved in this study, it was important to study them both separately and in combination. To study the influence of ozonation on solution properties, a thorough characterization was required. It was therefore decided to use three different concentration parameters: color, UV extinction, and DOC. Another goal was to determine the molecular-weight distributions for the different doses of ozone applied.

The color reduction resulting from ozonation demonstrates the bleaching efficiency of ozone for different ozone doses and different humic substances. UV extinction is commonly used as a measure for the content of natural organic matter in water. Most small, easily biodegradable molecules are not detected by this parameter. The DOC value represents a collective measure of all the organic substances in solution.

Because of the confusion and disagreements among studies that deal with humic substances and adsorption of organic matter in general, it was decided to use humic substances of different origins. These substances included a commercial product that has frequently been used in adsorption studies but is considered very different from aquatic humus (*4*), in addition to two different natural humus sources.

Two characteristic properties of the solution to be treated, adsorption rate and adsorption capacity, are important for the design of adsorber systems. The emphasis of this work was to study the adsorption capacities of the different humic substances and changes in them caused by preozonation. Most of the reported studies of ozonation–adsorption systems have been carried out in full-scale plants. Because biological degradation is likely to interfere when long contact times are used, the effect of preozonation on adsorptive properties was studied in laboratory experiments, with precautions against biodegradation.

Adsorption capacities before and after ozonation were determined from isotherm experiments. The Freundlich model is known for its wide applicability to natural water and mixtures of differing adsorbability. This model was therefore selected to describe the adsorption isotherms and changes in adsorbability. The adsorption was modeled by "pseudo-single-solute" isotherms, with DOC and UV extinction, respectively, as concentration parameters.

Absolute equilibrium for large humic molecules is not reached within the reaction times chosen in this study. Summers (*19*) found that, for commercial humic acid, equilibrium was not reached even after 50 days of reaction time. He did, however, choose 5 days of reaction time in his study because a pseudoequilibrium state was reached within this period. This reaction time also agreed with the theoretical equilibration time calculated by the method of Suzuki and Kawazoe (*20*).

Finally, it was important to see how well the laboratory results agree with larger-scale results. A pilot study for comparison of organic-substance removal from nonozonated and ozonated water was therefore designed. The experimental conditions were chosen to allow for biological growth, and thus to determine the importance of biological removal compared to adsorptive removal.

Experimental Procedures

Humus Sources. Different humus sources were used in the experiments in order to be able to make general conclusions. Three kinds of humic materials were studied: a commercial humic acid and two different Norwegian waters from Hellerudmyra and Heimdalsmyra.

The commercial humic acid was prerinsed according to the procedure described by Altmann (*21*). Stock solutions of humic acid were made by dissolving 1 g of the solid in 1 L of distilled deionized water (Milli-Q) at pH 11, obtained by adding 1.0 N NaOH. This solution was placed in an ultrasonic bath for 30 min to achieve complete dissolution and then stored at 4 °C. Dilutions of the stock solution were made up prior to each experiment. The pH was adjusted by addition of 1.0 N HCl and a phosphate buffer.

Brook water was collected from Hellerudmyra, a marsh area outside Oslo, Norway. The water was immediately filtered through glass fiber filters (Whatman GF/C) and stored at 4 °C until use. Prior to each adsorption experiment, the water was filtered through 0.45-μm microfilters (Millipore).

Heimdalsmyra water was collected from a bog outside Trondheim, Norway. This water has a color of approximately 700 mg of Pt per liter, and was diluted with tap water for experimental use. The filtration procedure was the same as for Hellerudmyra water.

Carbon. The carbon selected for this study (Calgon F-400) was ground and sieved to the actual particle sizes, washed in distilled deionized water (Milli-Q), and dried at 110 °C for 48 h. It was stored in an airtight bottle placed in a desiccator until use.

Ozonation. Ozonation was carried out with an ozone generator (BBC LN 103). The apparatus was calibrated, and unreacted ozone was quantified by the potassium iodide absorption method, as described by *Standard Methods for the Examination of Waters and Wastewater* (22).

The contact column used for batch ozonation was a 7-L glass column. Ozone was applied through a glass sinter of porosity 0 in the bottom of the column. The ozone dose was adjusted by keeping the gas flow constant and varying the contact time.

Continuous ozonation for the pilot-scale study was carried out in a 1.5-L glass column with countercurrent flow of ozone through a sinter in the bottom. The applied dose was 1 mg of ozone per milligram of DOC.

Analytical Methods. Various total organic carbon (TOC) analyzers were used for DOC analysis (Dohrmann DC-80, Astro Model 1850, Sybron–Barnstead Photochem Analyzer). The instruments were calibrated with a potassium hydrogen phthalate (KHP) standard and tested with humus standards. Excellent agreement between the different instruments was obtained.

UV–visible absorption spectra were determined by transmission measurements on spectrophotometers (Bausch and Lomb Spectronic 600, Perkin Elmer Model 554, Hitachi 110–20). The instruments were tested with humus standards, and again excellent agreement was obtained.

Spectral absorption coefficients, reflecting the UV/DOC ratios, were determined according to Zepp and Schlotzhauer (23):

$$k = 2.303\ A/C \tag{2}$$

where k is the spectral absorption coefficient ($mg^{-1}\ m^{-1}$), A is absorption at some wavelength (m^{-1}), and C is the concentration of organic matter (mg of DOC/L).

Color was determined from transmission measurements at 450 nm (Bausch and Lomb Spectronic 88). The apparatus was calibrated with a cobalt–platinum standard according to the definition of color measured as milligrams of Pt per liter. Procedure and definitions are given in ref. 22.

Preservation of Samples. Most samples were analyzed immediately after the experiments were completed, and such samples did not need any preservation. Samples that had to be stored for DOC analysis were preserved by the addition of 2 mg of $HgCl_2$ per liter.

Ultrafiltration. The molecular-size distributions of DOC in the various solutions were determined by ultrafiltration. A 400-mL ultrafiltration cell (Amicon) was used and operated at a pressure of 4.1 atm. The membranes used (XM 50, YM 10, and YM 2) had molecular-weight cutoff values of 50,000, 10,000, and 1,000, respectively.

Ultrafiltration was used to determine changes occurring in the individual solutions as a result of preozonation, but not for comparisons of different humus solutions, because the method is strongly dependent on solution properties. The calculation procedure used for determining the molecular-weight distribution was given by Reinhard (*24*).

Adsorption Isotherms. The isotherm experiments were carried out by the bottle point method, with constant initial organic-substance concentration and varying carbon dose. The carbon-particle size used was 200–400 mesh (28 μm). Different carbon doses were weighed by the difference in 160-mL hypovials (Wheaton Scientific), and 150 or 100 mL (SD = ±0.5 mL) of humic acid solution was added to each vial. The vials were sealed with poly(tetrafluoroethylene) (Teflon) face rubber and aluminum crimp caps, and the samples were placed on a 270-rpm shaker table or in a shaker bath (Julabo SW1). The reaction time was 5 days, as determined by reaction-time studies, and the different experimental conditions gave identical results. The isotherms were plotted as amount adsorbed per unit weight of carbon (Q_e) versus the corresponding equilibrium concentration (C_e).

For the experiments carried out in connection with the ozonation study, 1–10 mg of sodium azide per liter was added as an inhibitor for biological growth. These concentrations of inhibitor did not affect adsorption in studies with nonozonated samples.

Microcolumn Adsorption. Short-time microcolumn studies were performed with columns of 1-cm diameter that contained 1.00 and 2.50 g of carbon. The columns were operated in the upflow mode with gravity flow, at flow rates of 10 and 7 mL/min, respectively. Carbon dose and flow rate were varied to account for differences in solution character. The carbon particle size in these experiments was 48–65 Tyler mesh (i.e., 0.210–0.297 mm). This size was selected to avoid wall effects and optimize the homogeneity of the particles.

Pilot-Scale Column Experiments. Parallel column experiments were carried out with preozonated and nonozonated water from Heimdalsmyra. The water concentration prior to ozonation was approximately 5 mg of DOC per liter, corresponding to a UV extinction of 20 m^{-1} and a color of 52 mg of Pt per liter. Figure 1 shows a schematic illustration of the pilot plant.

The water was mixed in a 2.3-m^3 tank, and then pumped to a rapid sand filter for removal of particulate matter. The effluent from the filter was separated into two flow streams; one was pumped directly to one of the GAC filters, and a second was pumped to the ozone contact column. The ozonated water passed through an equalization column (retention time 82 min) and a rapid sand filter before it was pumped to the other GAC filter. A minor fraction was pumped through a slow sand filter to determine the importance of biological growth in the GAC filter. The rapid sand filters were backwashed after 2-day intervals.

Both carbon columns were operated in the downflow mode, with sampling points at 30-, 60-, and 150-cm (effluent) bed depth. The carbon (Chemviron F-400, 12–48 mesh) was used as received, and each column contained 1 kg of carbon. Prior to the experiment the carbon was washed with tap water to get rid of carbon fines. The experimental conditions for the sand filters and the ozonation columns are given in Table I, while those for the carbon columns are listed in Table II.

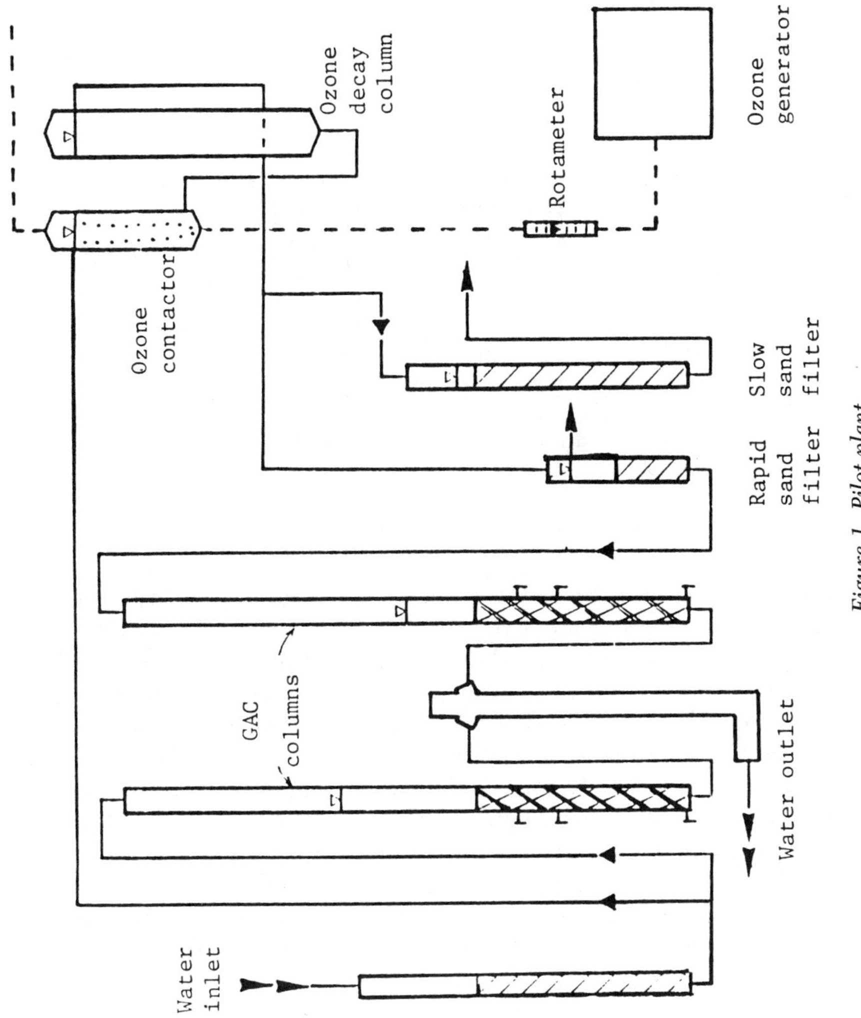

Figure 1. Pilot plant.

Table I. Experimental Conditions in Ozonation System and Sand Filters

Parameter[a]	*Ozonation Reactor*	*Retention Column*	*Slow Sand Filter*	*Rapid Sand Filter*
A (cm^2)	20.3	81.1	15.2	15.2
L (cm)	72.0	86.3	150.0	50.0
V (L)	1.46	7.0	2.20	0.76
Q (L/h)	5.1	5.1	0.0005	5.1
v_L (m/h)	—	—	0.0033	5.0
EBCT (min)	17.2	82.4	273.6	8.9

[a]Abbreviations are A, cross-sectional area; L, length of column; V, volume; Q, flow rate; v_L, linear velocity; and EBCT, empty-bed contact time.

Table II. Experimental Conditions for Carbon Columns at Bed Depth L_i (i = 1, 2, or 3)

Parameter	L_1 = *30 cm*	L_2 = *60 cm*	L_3 = *150 cm*
A (cm^2)	15.2	15.2	15.2
Q (L/h)	4.56	4.56	4.56
v_L (m/h)	3.00	3.00	3.00
V_i (L)	0.46	0.91	2.28
$EBCT_i$ (min)	6.0	12.0	30.0

[a]Abbreviations are A, cross-sectional area; Q, flow rate; v_L, linear velocity; V_i, bed volume; and $EBCT_i$, empty-bed contact time at bed depth.

Results

The influence of ozonation on the humus solutions was evaluated through molecular-size determinations by ultrafiltration, color, UV, and DOC analysis. The results of applying a dose of 1.0 mg of ozone per mg of DOC to the three humus solutions are represented in Table III by initial DOC values and spectral absorption coefficients, initial and final values, and percent reduction for UV extinction and color. The spectral absorption coefficient, representing the UV/DOC ratio, is significantly higher for commercial humic acid than for the other two solutions. The UV reduction measured per unit weight of ozone added was also slightly higher, 2.3 m^{-1} per milligram of ozone for commercial humic acid and 2.0 m^{-1} per milligram of ozone for the other two solutions. A color reduction of 72% was observed for Hellerudmyra water; the reduction for Heimdalsmyra water was close to 63%. These values correspond to color removals of 4.3 and 7.9 mg of Pt per milligram of ozone, respectively.

Table IV shows molecular-size distributions for commercial humic acid, Heimdalsmyra water, and Hellerudmyra water prior to and after ozonation. The highest-molecular-weight fraction of commerical humic acid was reduced from 91.3 to 58.4% when an ozone dose of 1.0 mg per milligram of DOC was applied. The ozonated water still had a dark brown color. Signif-

Table III. Ozonation Results

Humus Source	*Initial Spectral Absorption Coefficients, k (mg^{-1} m^{-1})*	*Initial DOC (mg/L)*	*Initial–Final (% Reduction)* UV Extinction (m^{-1})	Color (mg of Pt/L)
Commercial humic acid	20.0	5.4	44.4–31.4 (29.3)	—
	19.1	14.2	125.4–91.0 (27.4)	—
Hellerudmyra humus water	10.0	13.8	63.7–35.1 (44.9)	84.0–23.5 (72.0)
Heimdalsmyra humus water	9.6	5.1	21.2–12.4 (41.3)	56.5–21.0 (62.8)
	11.3	14.9	75.5–42.2 (44.1)	—

NOTE: Results are for an ozone dose of 1 mg of DOC per milligram.

icant reductions in the highest-molecular-weight fraction, with corresponding increases in the lower-molecular-weight fractions, result from ozonation of Hellerudmyra water. Application of an ozone dose of 1.0 mg per milligram of DOC resulted in a reduction of the highest-molecular-weight fraction from approximately 60 to 5%, corresponding to almost complete decolorization. The two lowest-molecular-weight fractions were significantly increased as a result of ozonation. A significant reduction of the highest-molecular-weight fraction is also observed for Heimdalsmyra water as a result of ozonation with 1.0 mg per milligram of DOC. Corresponding increases are again observed in the two lowest-molecular-weight fractions. Ozonated water has a brownish color.

Figure 2 shows DOC adsorption isotherms for Hellerudmyra water with initial concentrations of approximately 13 mg of DOC per liter and ozonated with 0, 0.5, 1.0, and 2.0 mg of ozone per milligram of DOC. The isotherm is presented as amount of DOC adsorbed per unit weight of carbon versus the corresponding DOC concentration. The plotted isotherms were determined by the Freundlich equation. The isotherm for nonozonated water is almost linear. Ozonation with 0.5 and 1.0 mg of ozone per milligram of DOC resulted in increased adsorption capacities and favorable adsorption. Further increase of the ozone dose resulted in lower adsorption capacities and unfavorable adsorption behavior. Almost all high-molecular-weight material was destroyed at an ozone dose of 1.0, and an ozone dose of 2.0 resulted in a 30% DOC loss that indicated significant loss of CO_2.

Figure 3 shows UV isotherms corresponding to the DOC isotherms in Figure 2 for ozone doses of 0, 0.5, and 1.0 mg per milligram of DOC. The initial UV extinctions are represented by A_0. Significant increases in the adsorption of UV-active compounds with increasing ozone doses are observed. All isotherms exhibit favorable adsorption behavior.

Figure 4 shows the adsorption isotherms for Heimdalsmyra water corresponding to ozone doses of 0, 1.0, and 2.0 mg per milligram of DOC. In this case, all isotherms show favorable adsorption behavior. Ozonation again resulted in significantly increased adsorption, but the isotherms for ozone doses of 1.0 and 2.0 are only insignificantly different. UV isotherms for

Table IV. Molecular-Size Distributions for the Different Humic Water Sources

Humus	*Initial DOC (mg/L)*	*Applied Ozone (mg/mg of DOC)*	*% of Total DOC in MW Range*				
			>50,000	*10,000–50,000*	*1,000–10,000*	*<1,000*	*Loss*
Commercial	13.09	0	91.3	4.4	1.4	5.6	−2.7
humic acid	13.28	1.0	58.4	2.5	6.8	15.4	16.9
Hellerudmyra	12.74	0		54.0	1.5	23.5	12.5
humus water	12.74	0	61.0	7.0	17.5	9.5	5.0
	12.57	0.5	28.0	12.0	32.5	13.0	15.5
	12.57	0.5	30.0	9.0	—	—	—
	11.49	1.0	5.0	8.0	46.5	29.0	11.5
Heimdalsmyra	14.94	0	84.1	−1.0	6.7	10.4	−0.2
humus water	14.53	1.0	34.6	6.8	18.9	25.5	14.2
	13.10	1.0	24.6	9.9	20.8	25.0	19.7

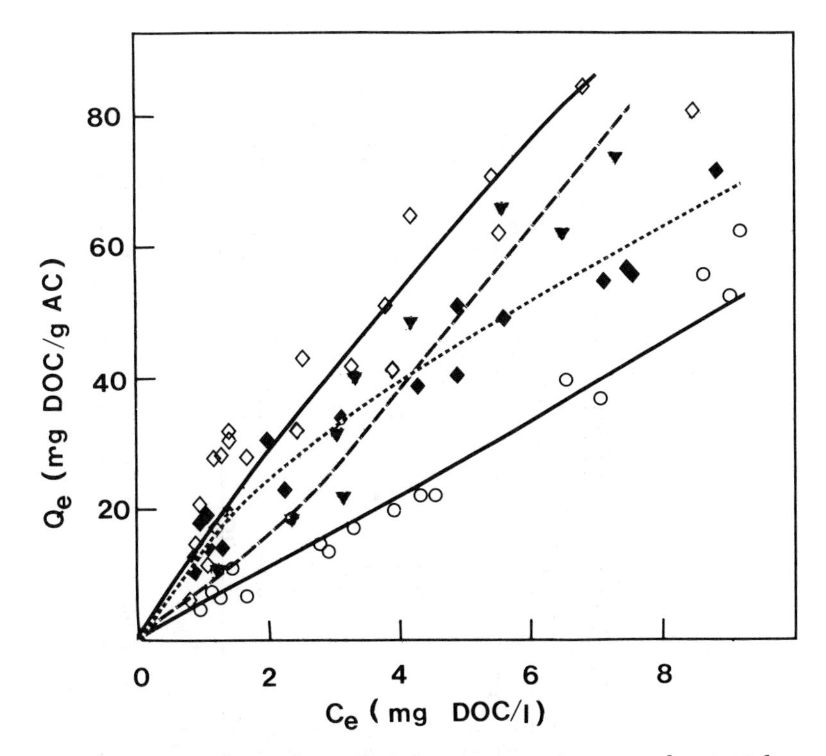

Figure 2. DOC isotherms for Hellerudmyra water. Corresponding initial concentrations and Freundlich parameters are given in Table V. Key: ○, solid line, nonozonated; ◆, dotted line, 0.5 mg of O_3 per milligram of DOC; ◇, solid line, 1.0 mg of O_3 per milligram of DOC; ▼, dashed line, 2.0 mg of O_3 per milligram of DOC.

Heimdalsmyra water corresponding to ozone doses of 0 and 1.0 mg per milligram of DOC are shown in Figure 5. Increased adsorption resulting from preozonation and favorable isotherms are observed.

Figure 6 shows isotherms for commercial humic acid ozonated with 0, 0.5, and 1.0 mg of ozone per milligram of DOC. All isotherms show unfavorable adsorption behavior characteristic of large molecules for which pore-size restrictions exist, and ozonation caused significantly enhanced adsorption. The UV isotherms for commercial humic acid corresponding to ozone doses of 0 and 1.0 mg per milligram of DOC are given in Figure 7. In this case the UV isotherms, like the DOC isotherms, are unfavorable. A significant increase in adsorption of UV-active organic matter is observed as a result of preozonation.

Parameters corresponding to the Freundlich model with and without a substitution term for a nonadsorbable fraction for the isotherms described are listed in Table V. The modified model gave a better fit only for an ozone

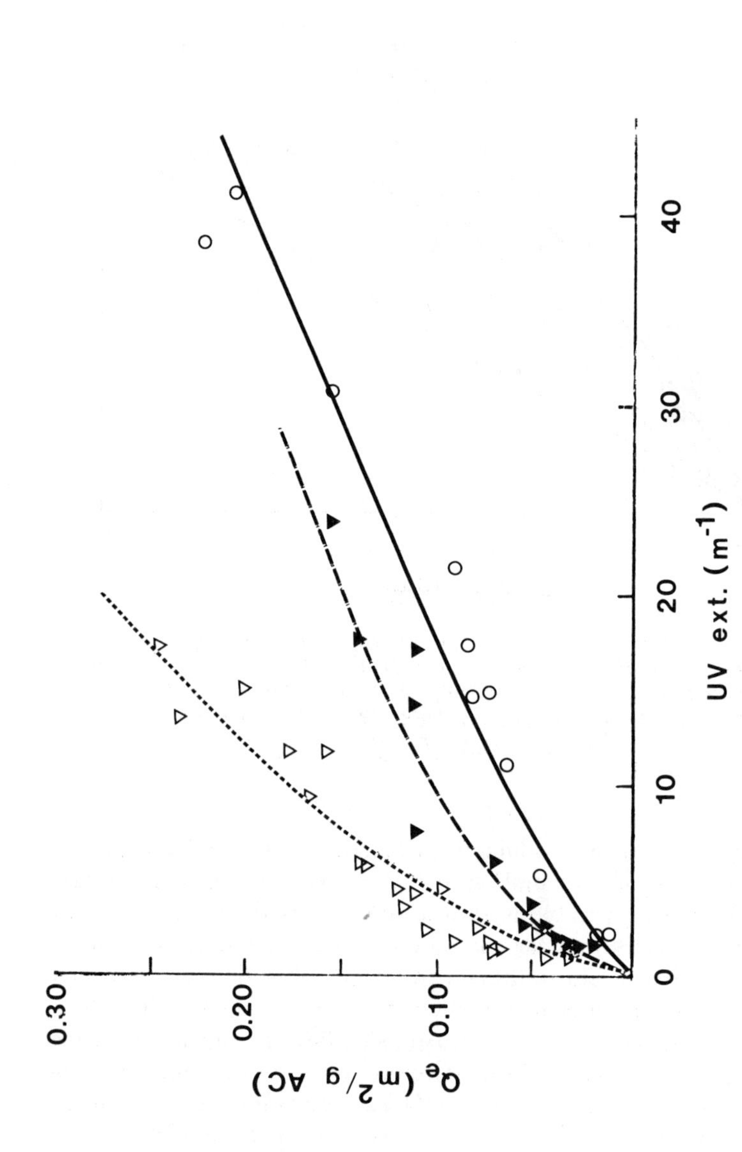

Figure 3. UV isotherms for Hellerudmyra water. Key: ○, *solid line, nonozonated* (A_0 = 59.6 m^{-1}); ▼, *dashed line, 0.5 mg of* O_3 *per milligram of DOC* (A_0 = 39.9 m^{-1}); ▽, *dotted line, 1.0 mg of* O_3 *per milligram of DOC* (A_0 = 29.3 m^{-1}).

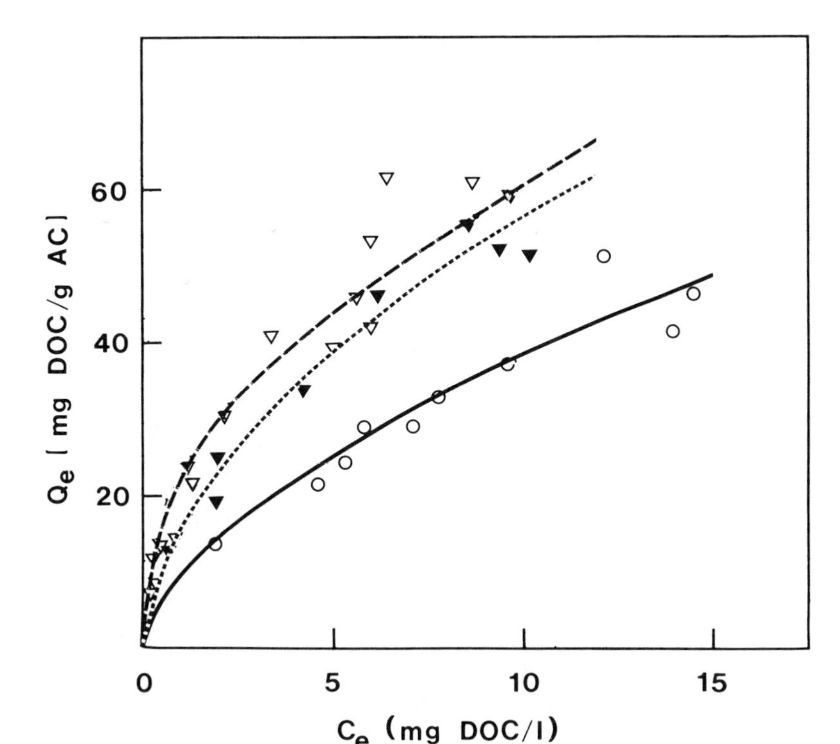

Figure 4. DOC isotherms for Heimdalsmyra water with high initial DOC. Initial concentrations and Freundlich parameters are given in Table V. Key: ○, solid line, nonozonated; ▽, dashed line, 1.0 mg of O_3 per milligram of DOC; ▼, dotted line, 2.0 mg of O_3 per milligram of DOC.

dose of 1.0 mg of ozone per milligram of DOC applied to Hellerudmyra water, resulting in a significantly higher K_F value and a lower $1/n$ value.

Figure 8 shows DOC isotherms for a lower initial concentration of commercial humic acid, approximately 5 mg/L, ozonated with 0, 0.5, and 1.0 mg of ozone per milligram of DOC. The trend of increased adsorbability is the same as for the higher initial concentration.

Results from microcolumn studies for commercial humic acid, 2.5 g of carbon, and ozone doses of 0 and 1.0 mg per milligram of DOC are shown in Figures 9 and 10, represented by DOC and UV extinction breakthrough profiles, respectively. The initial concentrations for the microcolumn experiments, indicated in the figures, were constant throughout the experiment. The breakthrough curves for nonozonated water show immediate breakthrough of DOC as well as UV-absorbing compounds. This breakthrough indicates low adsorption capacity and slow adsorption kinetics. A significant increase in removal efficiency was obtained as a result of preozonation.

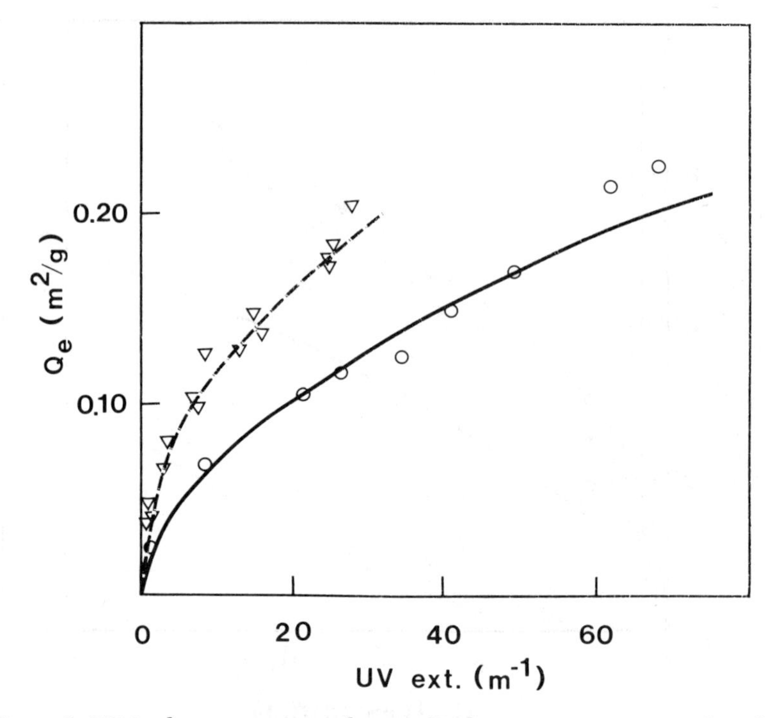

Figure 5. UV isotherms corresponding to the data in Figure 4. Key: ○, *solid line, nonozonated* (A_0 = *77.5* m^{-1}); ▽, *dashed line, 1.0 mg of* O_3 *per milligram of DOC* (A_0 = *41.4* m^{-1}).

Breakthrough curves for Hellerudmyra water for columns with 1 g of carbon are plotted in Figures 11–13 for DOC, UV extinction, and color. The curves for ozonated and nonozonated water exhibit steep concentration increases from the beginning of the experiments, whereafter both curves reach plateau concentrations. The removal efficiency was, however, significantly higher for ozonated than for nonozonated water.

Results from the pilot-scale column study carried out with Heimdalsmyra water are presented in Figures 14–22. Figures 14–16 show influent concentration profiles and breakthrough curves for the upper sampling point with respect to color, UV extinction, and DOC. Corresponding data for the middle and effluent sampling points are shown in Figures 17–19 and 20–22, respectively.

The DOC influent curves in Figure 14 show that the influent value for ozonated water is only slightly lower than that for nonozonated water during the first 3 weeks of the experiment; the difference corresponds to that observed in the isotherm experiments. After this time the influent concentration for ozonated water drops significantly. This drop was assumed to be a

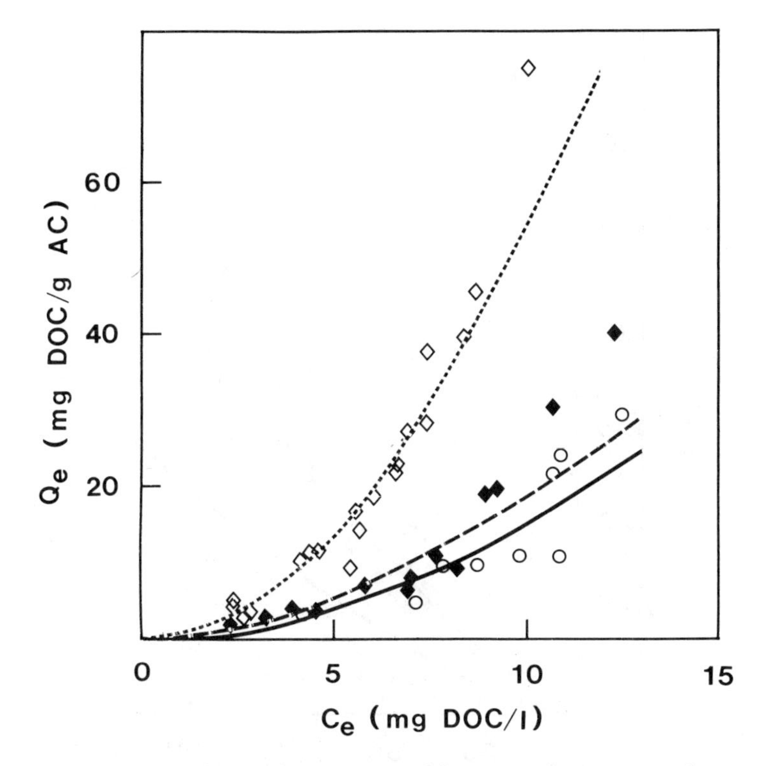

Figure 6. DOC isotherms for commercial humic acid. Corresponding initial concentrations and Freundlich parameters are given in Table V. Key: ○, solid line, nonozonated; ♦, dashed line, 0.5 mg of O_3 per milligram of DOC; ◇, dotted line, 1.0 mg of O_3 per milligram of DOC.

result of biological growth that was observed in the tubing system at this time. The influent concentration decrease is not observed for the UV and color parameters. This distinction suggests that only compounds that are not detected by these analyses are degraded biologically.

The effluent profile for nonozonated water shows a significant, immediate concentration increase in all three parameters. This increase demonstrates that the UV-absorbing, color-imparting compounds are responsible for the early DOC breakthrough. A significantly better DOC removal is observed for ozonated water with a much lower content of large color-imparting molecules. Ozonation reduced the UV extinction and color by approximately 40 and 60%, respectively.

The curves in Figures 17–19 show that higher adsorption capacities are obtained for the 60-cm sampling point as a result of longer contact times, but again a more rapid and sharper breakthrough is observed for nonozonated water.

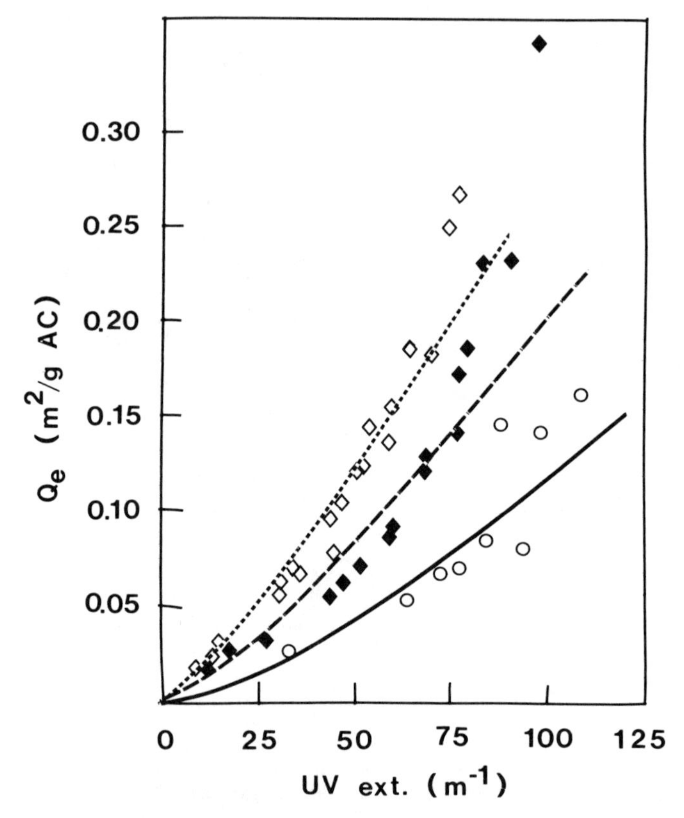

Figure 7. UV isotherms for commercial humic acid. Key: ○, *solid line, non-ozonated* (A_0 = *117.9* m^{-1}); ◆, *dashed line, 0.5 mg of DOC;* ◇, *dotted line, 1.0 mg of* O_3 *per milligram of DOC* (A_0 = *87.6/90.1* m^{-1}).

Table V. Freundlich Parameters

Humus Source	*Applied Ozone (mg/mg of DOC)*	*Initial DOC (mg/L)*	*Nonadsorbent DOC (mg/L)*	K_F	*1/n*	r^2
Commercial humic acid	0	14.20	0	0.19	1.93	0.85
	0.5	13.69–13.09	0	0.48	1.71	0.84
	1.0	13.29–13.77	0	0.57	1.97	0.92
Hellerudmyra humus water	0	13.80	0	5.43	1.02	0.94
	0.5	12.89–13.04	0	14.72	0.69	0.96
			0.70	24.66	0.42	0.96
	1.0	12.37	0	15.70	0.88	0.85
			0.75	81.33	0.43	0.90
	2.0	9.09–9.60	0	7.29	1.20	0.85
			0.50	8.48	1.00	0.81
Heimdalsmyra humus water	0	15.81	0	9.35	0.61	0.95
	1.0	14.01	0	18.65	0.52	0.96
	2.0	13.05	0	15.19	0.57	0.96

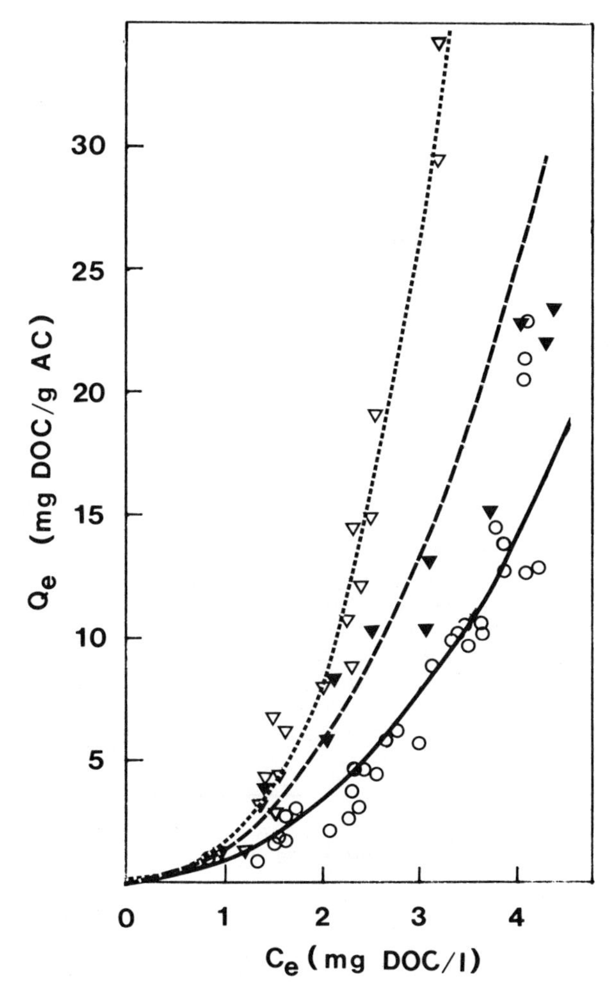

Figure 8. DOC isotherms for commercial humic acid. Key: ○, solid line, nonozonated; ▼, dashed line, 0.5 mg of O_3 per milligram of DOC; ▽, dotted line, 1.0 mg of O_3 per milligram of DOC.

The effluent curves in Figures 20–22 show less pronounced differences between the DOC curves for ozonated and nonozonated water because the carbon is farther from saturation. The UV and color curves do, however, demonstrate why adsorption alone does not give satisfactory treatment results. Even though the DOC values are low, a color breakthrough to unacceptable levels is observed in nonozonated water after a few days.

No significant difference was observed between DOC removal in the slow sand filter and the rapid sand filter. Such a difference would indicate biological removal in the slow sand filter.

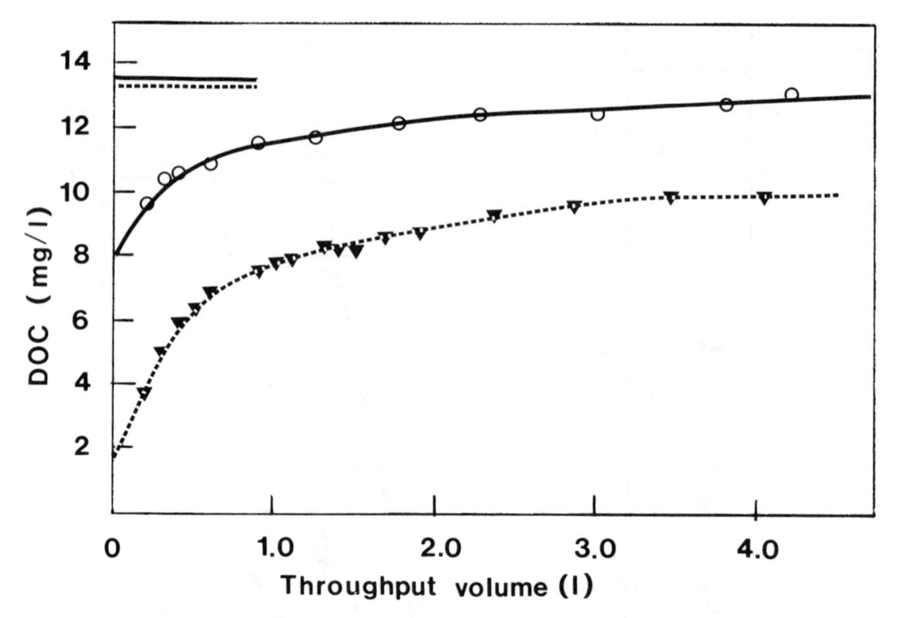

Figure 9. DOC breakthrough curves for nonozonated and ozonated solutions of commercial humic acid. Key: ○, solid line, nonozonated effluent; solid line, nonozonated influent; ▼, dotted line, ozonated effluent (1.0 mg of O_3 per milligram of DOC); dotted line, ozonated influent. Carbon: 2.50 g (48–65 Tyler mesh), column diameter: 1.0 cm, empty-bed contact time: 0.80 min.

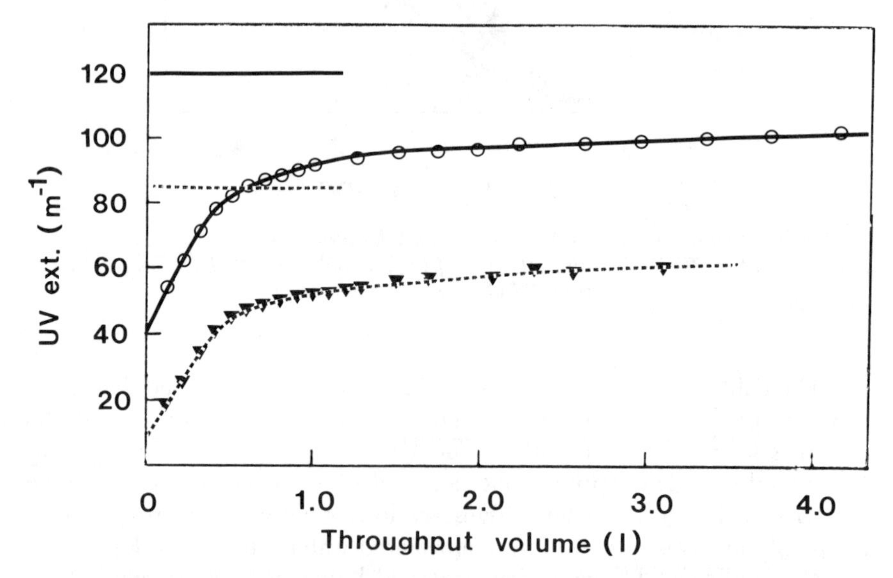

Figure 10. UV breakthrough curves for experiments in Figure 9. Key: ○, solid line, nonozonated effluent; solid line, nonozonated influent; ▼, dotted line, ozonated effluent (1.0 mg of O_3 per milligram of DOC); dotted line, ozonated influent.

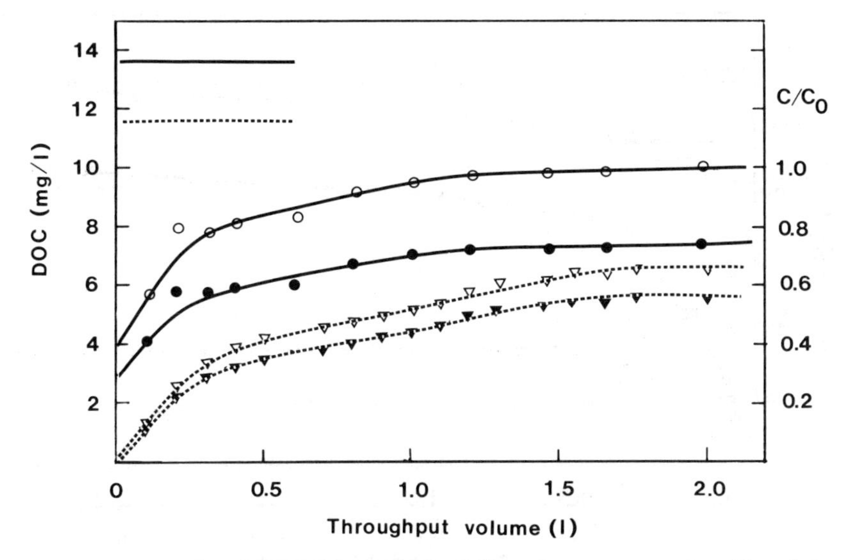

Figure 11. DOC breakthrough curves for Hellerudmyra water. Key: ○, *solid line, nonozonated effluent;* ●, *C/C₀; solid line, nonozonated influent;* ▽, *dotted line, ozonated effluent (1.0 mg of O_3 per milligram of DOC);* ▼, *C/C_0; dotted line, ozonated influent. Carbon: 1.00 g (48–65 Tyler mesh), column diameter: 10 mL/min, empty-bed contact time: 0.22 min.*

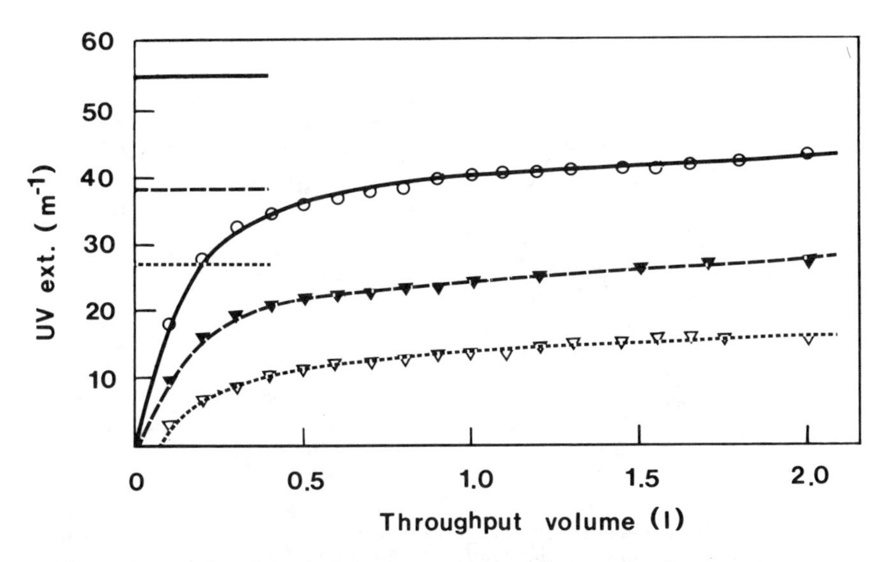

Figure 12. UV breakthrough curves corresponding to results in Figure 11. Key: ○, *solid line, nonozonated effluent; solid line, nonozonated influent;* ▼, *dashed line, ozonated effluent (0.5 mg of O_3 per milligram of DOC); dashed line, ozonated influent;* ▽, *dotted line, ozonated effluent (1.0 mg of O_3 per milligram of DOC); dotted line, ozonated influent.*

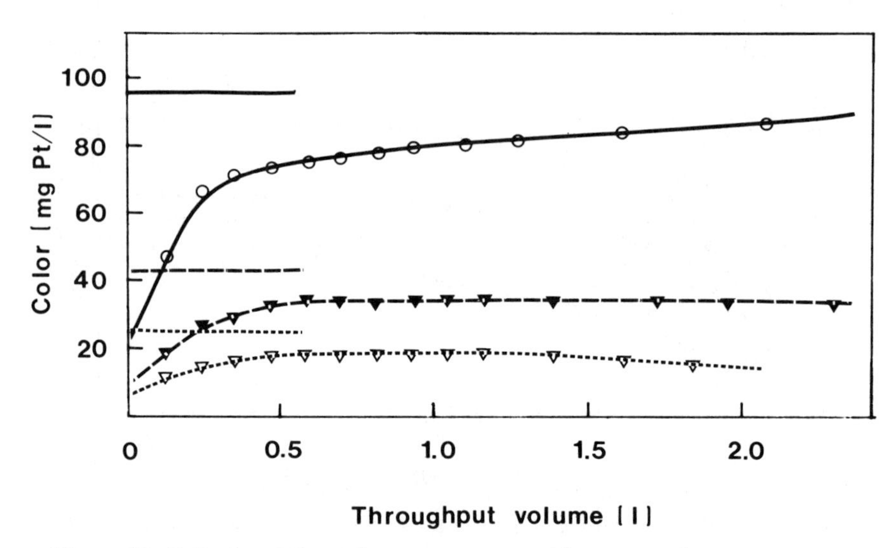

Figure 13. Color breakthrough curves corresponding to results in Figures 11 and 12. Key: ○, solid line, nonozonated effluent; solid line, nonozonated influent; ▼, dashed line, ozonated effluent (0.5 mg of O_3 per milligram of DOC); dashed line, ozonated influent; ▽, dotted line, ozonated effluent (1.0 mg of O_3 per milligram of DOC); dotted line, ozonated influent.

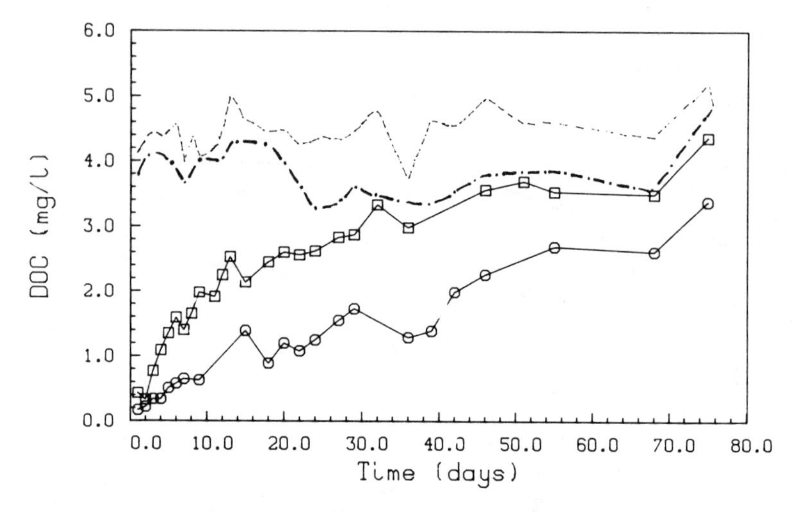

Figure 14. DOC concentration profiles for influent and 30-cm sampling point. Key: Dashed line, nonozonated influent; dot–dashed line, ozonated influent; □, nonozonated breakthrough profile; ○, ozonated breakthrough profile.

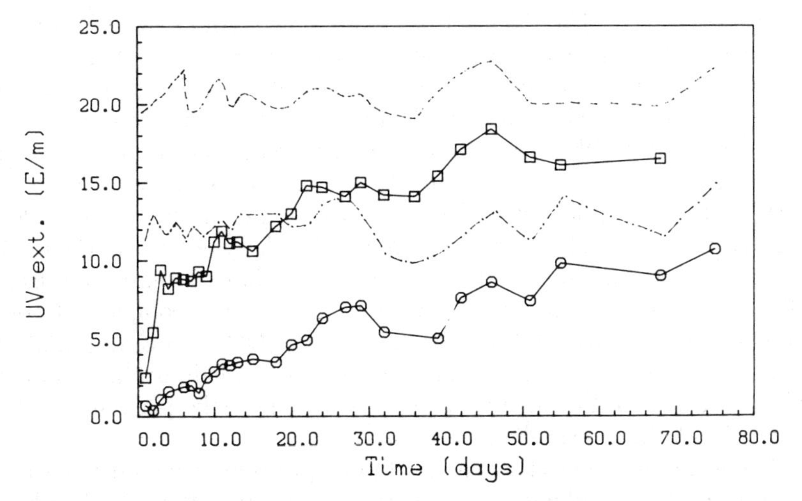

Figure 15. UV profiles for influent and 30-cm sampling point. Key: Dashed line, nonozonated influent; dot–dashed line, ozonated influent; □, nonozonated breakthrough profile; ○, ozonated breakthrough profile.

Discussion

Of the three humus sources investigated, only the two natural sources can be considered representative of the humic-substance problems confronted in drinking-water treatment. The commercial humic acid did, however, serve as an interesting comparison because of its large fraction of high-molecular-

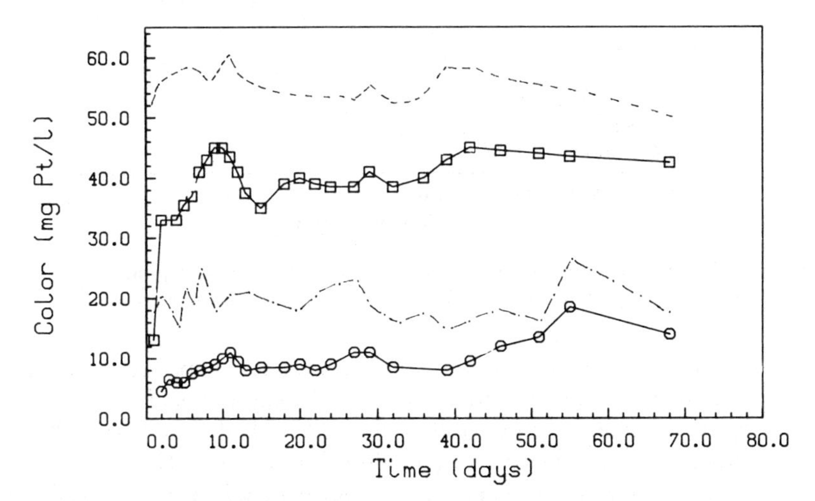

Figure 16. Color profiles for influent and 30-cm sampling point. Key: dashed line, nonozonated influent; dot–dashed line, ozonated influent; □, nonozonated breakthrough profile; ○, ozonated breakthrough profile.

weight compounds. The different character of the commercial humic acid was reflected in a significantly higher spectral absorption coefficient.

Table III shows significantly lower percentage reductions in UV extinction for commercial humic acid than for the other humus sources. These results, combined with the molecular-size distributions in Table IV, indicate that significant amounts of easily oxidized high-molecular-weight material are still left after application of 1 mg of ozone per milligram of DOC to commercial humic acid. Most of the high-molecular-weight material in Hellerudmyra water has been oxidized at this ozone dose. A significant drop in DOC value was observed for Hellerudmyra water after complete decolorization at an ozone dose of 2 mg per milligram of DOC. This drop indicates that the large color-imparting molecules are most easily oxidized and that further oxidation of the smaller molecules does not take place until all the large molecules are degraded.

The isotherm studies leave no doubt that preozonation with ozone doses normally used in drinking-water treatment enhances the adsorbability of organic matter from highly colored humic water. The results indicate that the optimal ozone dose depends on the initial spectral absorption coefficient

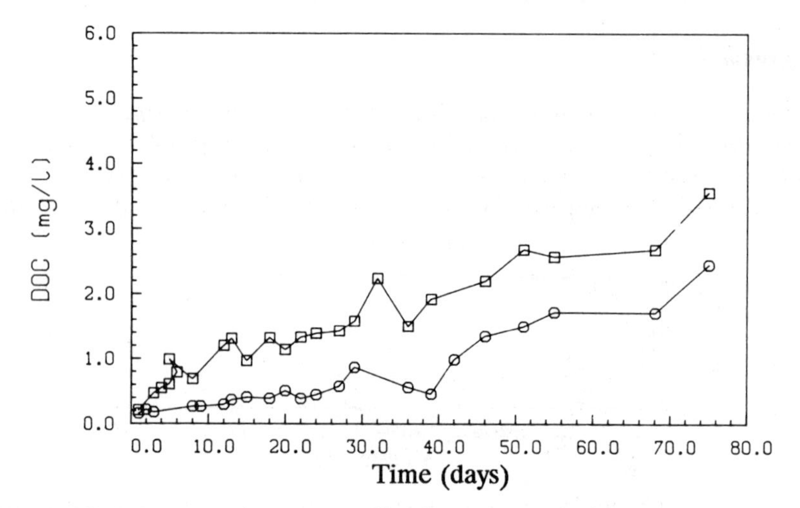

Figure 17. DOC concentration profiles for 60-cm sampling point. Key: □, nonozonated breakthrough profile; ○, ozonated breakthrough profile.

and molecular-weight distribution; higher doses can be applied to solutions with higher coefficients and higher fractions of large molecules.

An ozone dose of 2 mg per milligram of DOC had an adverse effect on the adsorption of organic matter from Hellerudmyra water. This effect, which indicates that an optimal dose of ozone is approximately 1 mg per milligram of DOC, can be explained by complete removal of large molecules and further ozonation of the products to polar, less-adsorbable compounds. For

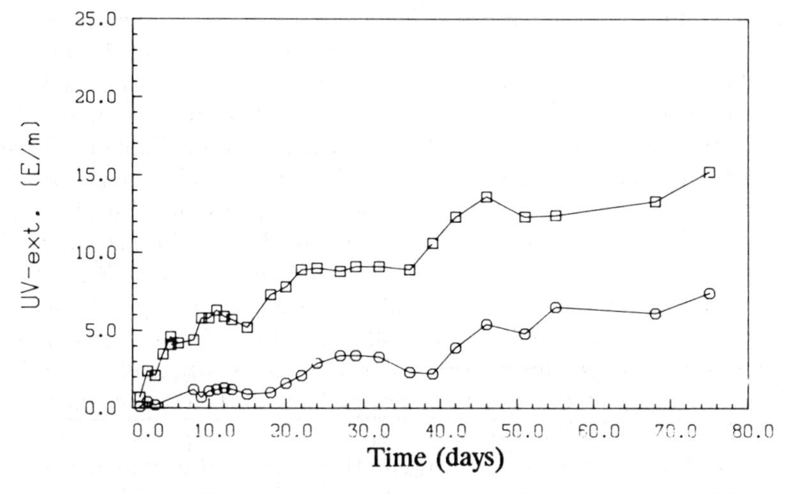

Figure 18. UV profiles for 60-cm sampling point. Key: □, nonozonated breakthrough profile; ○, ozonated breakthrough profile.

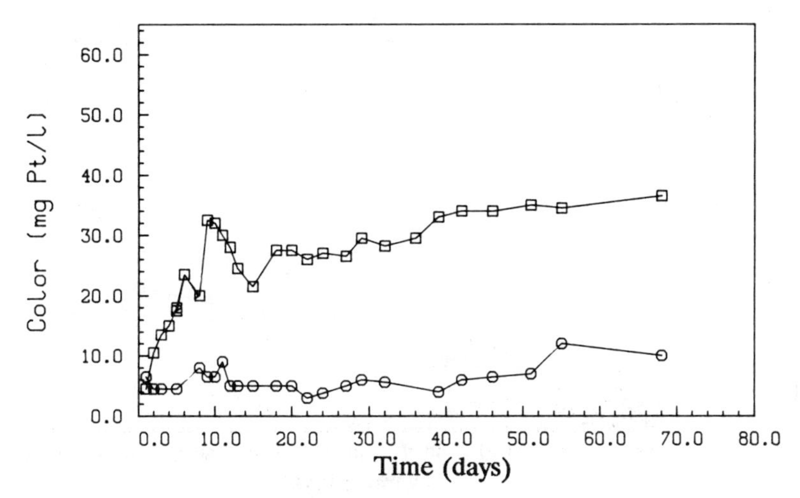

Figure 19. Color profiles for 60-cm sampling point. Key: □, nonozonated breakthrough profile; ○, ozonated breakthrough profile.

Heimdalsmyra water, the difference between isotherms for ozone doses of 1 and 2 mg per milligram of DOC were insignificant. In this case there was no significant decrease in DOC value for the higher dose.

The large improvement in adsorbability for commercial humic acid that results from an increase in the ozone dose from 0.5 to 1.0 mg per milligram of DOC indicates that even higher doses would benefit the adsorption. A relatively strong color indicated that a significant amount of color-imparting

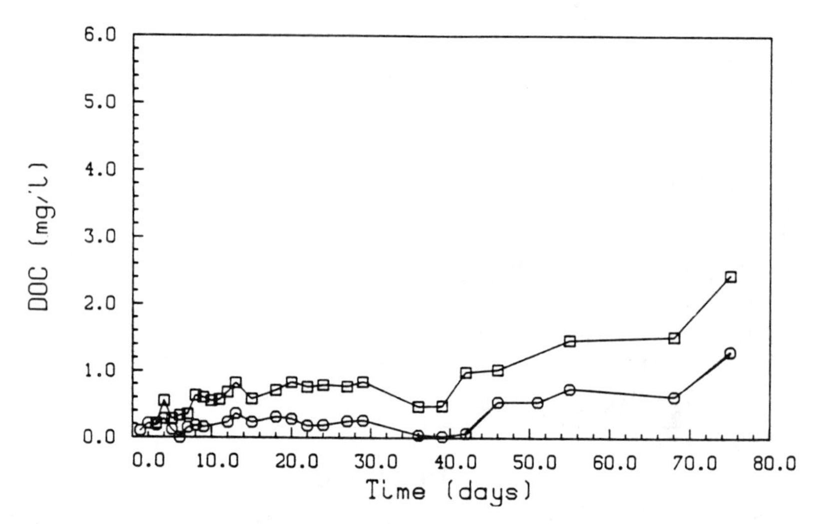

Figure 20. DOC concentration profiles for effluent. Key: □, nonozonated effluent; ○, ozonated effluent.

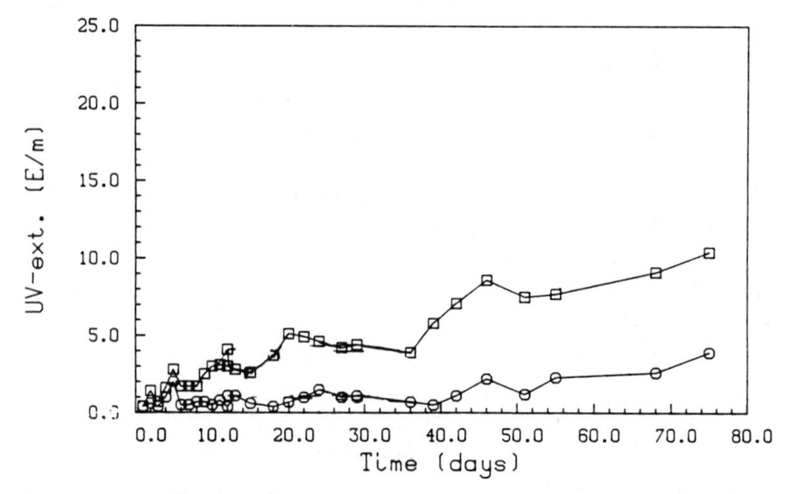

Figure 21. UV profiles for effluent. Key: □, nonozonated effluent; ○, ozonated effluent.

substances was still left in the solution after application of the ozone dose of 1.0 mg per milligram of DOC. An adverse effect of increased polarity would not be expected until the color-imparting substances had been oxidized.

The agreement between experimental data and the Freundlich parameters in Table IV for Hellerudmyra water was good, even though the data corresponding to ozone doses of 0.5 and 1.0 indicate the presence of non-

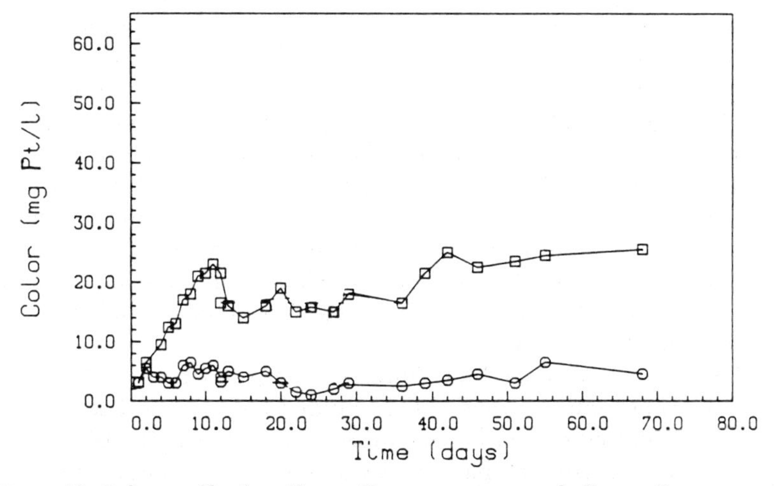

Figure 22. Color profiles for effluent. Key: □, *nonozonated effluent;* ○, *ozonated effluent.*

adsorbable organic matter. The modified model gave a better fit in the case of an ozone dose of 1, with a significantly higher K_F value and a lower $1/n$ value. Good agreement between experimental data and the Freundlich model was obtained for Heimdalsmyra water. For commercial humic acid, the model gives a good fit to experimental data at lower concentrations, but the estimated adsorption capacities are too low at higher concentrations.

Pilot-Scale Study. The results from the pilot-scale study confirm the trend of increased adsorbability of organic substances as a result of preozonation observed in the laboratory studies. The importance of the contact time is clearly illustrated by the differences among color retentions at the different sampling points. The color-imparting substances that cause the immediate breakthrough at the top of the column are adsorbed when the contact time becomes sufficiently long. The effluent color values were, however, too high after a few days to meet drinking-water standards when activated carbon was used as the only treatment.

Biological Activity in Pilot Plant. The pilot-scale study leaves no doubt that the major improvement in removal efficiency resulting from ozonation is caused by enhancement of adsorptive properties. Some biological growth in the system handling ozonated water was observed, whereas none was observed in the other system. The improvement of adsorption capacity and rate is illustrated by the initial concentration profiles. No biological activity was detected until 3 weeks after startup. The drop in DOC concentration after this time, with no corresponding drop in UV extinction or color, shows that the microorganisms preferentially take care of the oxidation products that are not detected by UV or color measurements. The color-imparting substances must therefore be eliminated by adsorptive removal.

Filter systems where biological activity occur are often characterized as biological filters. In this study, the biological activity did not seem to take place in the carbon filter, but rather in the system ahead of the filter. Because sand filters have been shown to be comparable to carbon filters in biological removal efficiency (25), a slow sand filter was used in this study to simulate biological removal in the carbon filter. Comparison with the rapid sand filter indicated that the sand filter did not give significant additional removal of organic matter and that the slight removal observed was caused by filtration rather than biodegradation.

Flowing water is a good medium for biological activity if the nutrient requirements are satisfied, and it is likely that the biological activity often occurs in the tubing system between different process units. Biological removal in the carbon beds may, however, be more important in systems treating water with a higher fraction of easily biodegradable matter. The reason why biological activity seemed to be limited to the tubing system in this study is probably that only a minor fraction of the organic matter was biodegradable. By assuming that the nutrient requirements were satisfied, the biodegradable fraction was estimated to be approximately 10%, on the basis of the DOC data for the last part of the study and the results discussed.

The pilot study showed that ozonation of humic substances prior to activated-carbon adsorption is advantageous not only because of improved adsorption capacity and adsorption rate, but also because of formation of biodegradable organic matter.

Comparison of Results. The results for all humic substances studied agree with the hypothesis of increased adsorbability resulting from preozonation. The ozone dose that can be applied without adverse effects on adsorption seems to be strongly dependent on the color/DOC ratio and spectral absorption coefficients; the higher these values, the more ozone can be applied. These observations confirm the composition-dependence of ozonation that formed the basis for this study.

High color values correspond to large fractions of molecules with very high molecular weights. High doses of ozone have to be applied before these molecules are broken up into smaller units for which more of the activated-carbon surface is accessible. The results indicate that the water has to be completely decolorized before the lower fractions are degraded to small, polar, poorly adsorbable molecules. This finding is in agreement with a very high reactivity toward ozone for the color-imparting conjugated chains in the very large molecules. The high reactivity makes this the dominating reaction as long as such material is present.

In most of the ozonation adsorption studies reported earlier, the adsorption of organic substances onto activated carbon was adversely affected by preozonation. The positive effect of preozonation observed in this study does not contradict the results from previous studies. The explanation for this difference lies in the character of the solution. Several of the previous studies have been performed with water containing relatively small and easily adsorbable organic molecules. Ozonation of such water is likely to reduce the adsorbability because of increased hydrophilicity and polarity of the organic substances. The compounds studied here were large, poorly adsorbable humic compounds that utilized only a small fraction of the adsorption area prior to ozonation. For such compounds the size and structural limitations hindering adsorption are reduced by ozonation; consequently, their adsorbability is increased. The results illustrate the importance of determining the character of the organic substances present in the water before deciding on treatment schemes.

Conclusions

Ozonation of water with a high content of color-imparting humic substances resulted in significant reductions in the content of high-molecular-weight material, color, and UV absorption. However, the organic-matter content (DOC) was almost unaffected by ozone doses normally used in drinking-water treatment.

Preozonation increased the adsorption of organic matter from the three humic solutions studied here. An ozone dose of 1 mg per milligram of DOC seemed optimal for adsorption of organic matter from the two natural solutions. In both cases, the adsorption capacities were reduced by application of 2 mg of ozone per milligram of DOC. The adsorption results for the commercial humic acid combined with a higher fraction of the very large molecules and higher spectral absorption coefficients indicate that the optimal ozone dose for commercial humic acid is higher than 1 mg per milligram of DOC.

Pilot-scale studies of Heimdalsmyra water confirmed the increased adsorbability that was observed in the laboratory studies. In addition, biological growth was observed in the system treating ozonated water. The biodegradable fraction of organic matter was estimated to be approximately 10% at the applied ozone dose of 1 mg per milligram of DOC. The major biological activity seemed to take place in the tubing system ahead of the carbon filters.

Acknowledgments

Several people contributed help and stimulation during the course of this study. The authors thank Paul V. Roberts, whose advice has been of great importance for this work. Special thanks also go to Egil T. Gjessing for many helpful discussions on humic substances and to R. Scott Summers and Helge Brattebo for inspiring discussions on adsorption matters.

The funds for this study were provided in part by NTNF (The Royal Norwegian Council for Scientific and Industrial Research), SINTEF (the

Foundation for Scientific and Industrial Research), and NTH (the Norwegian Institute of Technology).

References

1. Christman, R. F.; Gjessing, E. T. In *Aquatic and Terrestrial Humic Materials*; Christman, R. F.; Gjessing, E. T., Eds.; Ann Arbor Science: Ann Arbor, 1983; pp 517–528.
2. Gjessing, E. T. In *Proc. Conjunc. 10th IAWPR Conf.*; Edmonton, Alberta, June 1980; Smith, D. W.; Hrudey, E. W., Eds.; Pergamon: Oxford, England, 1981.
3. Thurman, E. M.; Wershaw, R. L.; Malcolm, R. L. *Org. Geochem.* **1982,** *4*, 27–35.
4. Malcolm, R. L.; MacCarthy, P. *Environ. Sci. Technol.* **1986,** *20*, 904–911.
5. McCreary, J. J.; Snoeyink, V. L. *Water Res.* **1980,** *14*, 151–160.
6. Lee, M. C.; Snoeyink, V. L. *Humic Substances Removal by Activated Carbon*; Water Resources Center, University of Illinois at Urbana–Champaign: Urbana, 1980.
7. Weber, W. J., Jr.; Pirbazari, M.; Herbert, M. D. *Proc. Am. Water Works Assoc. Annu. Conf.* **1978,** *98*, paper 15–1.
8. Weber, W. J.; Voice, T. C.; Jodellah, A. *J. Am. Water Works Assoc.* **1983,** *75*, 612.
9. Frick, B. R.; Bartz, R.; Sontheimer, H.; DiGiano, F. A *Carbon Adsorption of Organics from the Aqueous Phase*; Suffet, I. H.; McGuire, M. V., Eds.; Ann Arbor Science: Ann Arbor, 1982; Vol. 1A, p 229.
10. Frick, B. R.; Sontheimer, H. *Treatment of Water by Granular Activated Carbon*; McGuire, M. V.; Suffet, I. H., Eds.; Advances in Chemistry 202; American Chemical Society: Washington, DC, 1983; p 247.
11. Samdal, J. E. *Vattenhygien* **1966,** *4*, 161–165.
12. Meijers, A. P. *Water Res.* **1976,** *11*, 647–652.
13. Nebel, C. *Public Works* **1981,** *112*(6), 86–90.
14. Flogstad, H. *Behandling av humus med ozon*; SINTEF report STF21 A84037; Foundation for Scientific and Industrial Research: Trondheim, Norway, 1984.
15. Benedek, A. International Ozone Institute Symposium on Advanced Ozone Technology, Toronto, Ontario; International Ozone Institute: Ontario, 1977.
16. Kuhn, W.; Sontheimer, H.; Steiglitz, L.; Maier, D.; Kurz, R. *J. Am. Water Works Assoc.* **1978,** *70*, 326–331.
17. Lienhard, H.; Sontheimer, H. *Ozone: Sci. Eng.* **1979,** *1*, 61–72.
18. Hubele, C.; Sontheimer, H. *Environmental Engineering, Proceedings of the 1984 Specialty Conference*; Pirbazari, M.; Devinney, S., Eds.; American Society of Civil Engineers: New York, 1984; University of Southern California, Los Angeles, June 1984.
19. Summers, R. S. Ph.D. Dissertation, Stanford University, 1985.
20. Suzuki, M.; Kawazoe, K. *Seisan Kenkyu* **1974,** *26*, 383.
21. Altmann, R. S. Ph.D. Dissertation, Stanford University, 1984.
22. *Standard Methods for the Examination of Water and Wastewater*, 14th ed.; American Public Health Association: Washington, DC, 1975; p 455.
23. Zepp, R. G.; Schlotzhauer, P. F. *Chemosphere* **1981,** *10*, 479–481.
24. Reinhard, M. *Environ. Sci. Technol.* **1984,** *18*, 410–415.
25. Maloney, S. W.; Bancroft, K.; Pipes, W. O.; Suffet, I. H. *J. Environ. Eng. Div. (Am. Soc. Civ. Eng.)* **1984,** *110*, 519–533.

RECEIVED for review July 24, 1987. ACCEPTED for publication April 20, 1988.

40

Effect of Coagulation, Ozonation, and Biodegradation on Activated-Carbon Adsorption

Gregory W. Harrington[1], Francis A. DiGiano, and Joachim Fettig[2]

Department of Environmental Sciences and Engineering, University of North Carolina, Chapel Hill, NC 27514

The ability to describe humic substance adsorption is important for the design of activated-carbon filters in water treatment. Humic solutions are composed of a multitude of unknown molecular species, and competitive adsorption among these species, as well as with trace anthropogenic organic chemicals, needs to be understood better. In this research, ideal adsorbed solution theory was used to describe an aquatic humic solution as a set of several pseudocomponents and to evaluate the effects of two treatment processes (alum coagulation and a combination of coagulation, ozonation, and biodegradation) on the solution's equilibrium and kinetic adsorption behavior.

OUR LIMITED UNDERSTANDING OF THE STRUCTURE of aquatic humic substances (HS) is manifested in our poor understanding of how these materials are adsorbed on activated carbon. When measured collectively by a surrogate parameter such as total organic carbon (TOC), HS solutions are not considered well adsorbed by activated carbon. Nevertheless, their adsorption is important because the removal of synthetic organic chemicals may be decreased by the presence of HS. Moreover, the removal of HS by adsorption

[1]Current address: Malcolm Pirnie, Inc., Newport News, VA 23606
[2]Current address: Division of Hydraulics and Sanitary Engineering, Norwegian Institute of Technology, N–7034 Trondheim–NTH, Norway

0065–2393/89/0219–0727$06.00/0

may become important if more stringent maximum contaminant levels (MCLs) for trihalomethanes and other chlorination byproducts are set. Water treatment facilities will be forced to rely more on adsorption if future MCLs cannot be met by moving the point of chlorination or increasing the efficiency of coagulation. Therefore, the ability to describe HS adsorption and its effect on the adsorption of pollutants is of utmost importance in the design of activated-carbon filters.

Because of the heterogeneous nature of HS solutions, this research was aimed at testing the applicability of a competitive adsorption model to describe equilibrium and kinetic adsorption behavior. The approach is similar to that used by Frick and Sontheimer (*1*) and Crittenden et al. (*2*), wherein an unknown mixture is described as a set of several pseudocomponents by using the ideal adsorbed solution theory (IAST) (*3*). This approach was used to evaluate the effects of coagulation, ozonation, and biodegradation on the adsorption behavior of HS solutions.

Preparation of Samples

Raw water was obtained from Lake Drummond in southeastern Virginia. The water is highly colored, low in alkalinity (~50 mg/L as $CaCO_3$), and pH 4. Upon return to the laboratory, the water was prefiltered with 1.0-μm honeycomb filters to remove leaves and sediment. The water was then stored in a cool, dark storage area prior to treatment.

The first stage of treatment used was alum coagulation. The pH of the raw water was adjusted to 6.5 by the addition of 2.4×10^{-2} M sodium carbonate. Alum was added at high mixing intensity and constant pH to a concentration of 205–210 mg/L as $Al_2(SO_4)_3 \cdot 18\ H_2O$ to achieve destabilization. This step was followed by 40 min of flocculation, overnight settling, and filtration with 0.45-μm membrane filters to achieve 76% removal of UV absorbance at 254 nm and 50% removal of TOC. The coagulated water was then stored in a refrigerator for prevention of biodegradation.

The coagulation stage was followed by ozonation and biodegradation. Distilled, deionized water was ozonated to a concentration of 25 mg of O_3/L and then combined with an equal volume of the coagulated sample to yield an ozone dosage of 1.15 mg of O_3/mg of TOC. The ozonated sample was placed in the reservoir of a recycle batch reactor, the column of which was previously seeded with return activated sludge from the local wastewater-treatment plant. The reactor was allowed to run until no further reduction in TOC was observed (10 days of run time) and the sample was then stored in a refrigerator. TOC values of each stage are given in Table I.

Adsorption Isotherms

Equilibrium studies were performed by using the bottle point technique for determining adsorption isotherms. Granular activated carbon (GAC, F–400)

Table I. Total Organic Carbon Levels in the Humic Mixtures Tested

Humic Mixture	*TOC (mg/L)*
Prefiltered	43.8
Alum coagulated	21.7
Ozonated and biostabilized	7.0

was washed, dried, stored, and ground to a 200–325 mesh size to ensure that a representative sample of carbon was used (*4*). In order to adequately describe the isotherms, bottles were filled with powdered carbon dosages ranging from 5 to 4000 mg/L. Samples were buffered with a 5 mM phosphate buffer to yield a pH of 6.5. Each bottle was then filled with 100 mL of sample and tumbled for 7–10 days at 23 °C. After the period of tumbling, the powdered carbon was removed by using 0.45-μm membrane filters and the TOC of each sample was measured. Each component in a given mixture must satisfy the following mass balance equation for each bottle:

$$q_i = \frac{C_{0,i} - C_i}{M/V} \quad (1)$$

where q_i is the surface loading of component i at equilibrium, C_i is the bulk liquid-phase concentration of component i at equilibrium, $C_{0,i}$ is the initial liquid-phase concentration of component i, and M/V is the carbon dosage.

Prediction of Multicomponent Equilibrium

IAST was assumed capable of describing the multicomponent nature of the resulting HS isotherms through the use of the following five equations:

$$q_T = \sum_{i=1}^{N} q_i \quad (2)$$

$$z_i = \frac{q_i}{q_T} \quad \text{for } i = 1 \text{ to } N \quad (3)$$

$$\frac{1}{q_T} = \sum_{i=1}^{N} \frac{z_i}{q_i^0} \quad (4)$$

$$C_i = z_i C_i^0 \quad \text{for } i = 1 \text{ to } N \quad (5)$$

$$\frac{\pi_T A_{\text{ads}}}{RT} = \frac{\pi_i^0 A_{\text{ads}}}{RT} = \int_0^{q_i^0} \frac{d[\ln (C_i^0)]}{d[\ln (q_i^0)]} \, dq_i^0 \quad \text{for } i = 1 \text{ to } N \quad (6)$$

Total surface loading, q_T, is defined as the sum of individual solute surface loadings by equation 2; the surface mole fraction, z_i, of an individual solute is defined by equation 3. Equation 4 defines the surface loading of the mixture as a function of single-solute surface loadings, q_i^0, achieved when the single-solute systems adsorb at the same temperature and spreading pressure as the mixture. By equating the chemical potentials of a solute in the adsorbed and liquid phases, one arrives at equation 5, where C_i^0 would be in equilibrium with q_i^0 in a single-solute system. Equations 4 and 5 are key equations in IAST, because both assume that the adsorbed phase forms an ideal solution. Finally, equation 6 equates the spreading pressure of the mixture, π_T, with the spreading pressures of the single-solute systems, π_i^0. A_{ads} is the adsorbent surface area per unit mass of adsorbent. The spreading pressure of a single-solute system is evaluated by the integral shown in equation 6.

Each HS solution was assumed to contain a set of several individual pseudocomponents, each having a single-solute adsorption behavior that could by described by the Freundlich isotherm equation. The linearized form of the Freundlich equation is given by

$$\ln (q_i^0) = \ln (K_i) + \left[\frac{1}{n_i}\right] \ln (C_i^0) \qquad (7)$$

where K_i and n_i are the Freundlich isotherm constant and exponent, respectively, for component i. When equation 7 is substituted into equation 6, the following expression results:

$$n_1 q_1^0 = n_j q_j^0 \qquad \text{for } j = 2 \text{ to } N \qquad (8)$$

Combining equations 1–5, 7 and 8 yields the following objective function, which was derived by Crittenden et al. (2):

$$F_i = 0 = C_{0,i} - \left[\frac{M}{V}\right] q_i - \left[\frac{q_i}{\sum_{j=1}^{N} q_j}\right]\left[\frac{\sum_{j=1}^{N} (n_j q_j)}{(n_i K_i)}\right]^{n_i} \qquad \text{for } i = 1 \text{ to } N \qquad (9)$$

Thus, the equilibrium state of each isotherm bottle may be described by a set of N equations when the Freundlich parameters, n_i and K_i, and the initial concentration, $C_{0,i}$, of each component are known and when the carbon dosage, M/V, is also known. The set of equations may be solved by a Newton–Raphson algorithm, as shown by Crittenden et al. (2).

In this case, the Freundlich constants and initial concentrations of each component were unknown, although the initial concentration of the total mixture was known. Therefore, a search routine was combined with the Newton–Raphson algorithm to find the Freundlich constants and initial concentrations of the pseudocomponents that yielded the best fit to the isotherm of the total mixture. The objective of the search routine was to minimize

$$\mathrm{SSR} = \sum_{i=1}^{N_{\mathrm{obs}}} \left[\frac{C_{Ti,\mathrm{obs}} - C_{Ti,\mathrm{calc}}}{\sigma_i} \right]^2 \tag{10}$$

where SSR is the residual sum of squares, N_{obs} is the number of observations, $C_{Ti,\mathrm{obs}}$ and $C_{Ti,\mathrm{calc}}$ are the observed and calculated concentrations of the mixture in bottle i, and σ_i is the standard deviation of replicate measurements of $C_{Ti,\mathrm{obs}}$.

Because of the unknown nature of HS mixtures, their concentrations must be measured through the use of surrogate parameters such as TOC. The IAST equations (equations 2–6), however, are based on a thermodynamic derivation in which concentrations are expressed in molar rather than mass units. If the Freundlich model (equation 7) is used to describe single-solute adsorption on a TOC basis, equation 8 is still valid on a molar basis. However, equation 9 is arrived at only by assuming that each component contains the same number of carbon atoms per molecule. The need for this assumption produces a dilemma in using TOC data in the IAST model to search for Freundlich adsorption constants of individual components if, in fact, these components do not all contain the same number of carbon atoms per molecule. This dilemma has not been adequately addressed by others who have used the IAST model (*5, 6*). On-going work by this research group is addressing further manipulation of the IAST equations to include the number of carbon atoms per molecule for each unknown fraction. However, for the purposes of this chapter, we assume that all fractions have the same number of carbon atoms per molecule.

Prior to determining pseudocomponent properties, the number of statistically valid pseudocomponents was determined. This procedure was begun by determining the root mean square error (RMSE) of the fit obtained using one adsorbing pseudocomponent. The value of RMSE was calculated from

$$\mathrm{RMSE}_{N_{\mathrm{ads}}} = \left[\frac{\mathrm{SSR}_{N_{\mathrm{ads}}}}{N_{\mathrm{obs}} - p_{N_{\mathrm{ads}}}} \right]^{0.5} \tag{11}$$

where N_{ads} is the number of adsorbing pseudocomponents, $p_{N_{\mathrm{ads}}}$ is the number of adjustable parameters for N_{ads} pseudocomponents, $\mathrm{SSR}_{N_{\mathrm{ads}}}$ is the residual sum of squares for N_{ads} pseudocomponents, and N_{obs} is the number of observations. For the IAST–Freundlich model used in this work,

$$p_{N_{\mathrm{ads}}} = 3N_{\mathrm{ads}} - 1 \tag{12}$$

Equation 11 shows that the value of RMSE can increase with increasing values of N_{ads}. The remainder of the procedure involved calculating $RMSE_{N_{ads}}$ for succeeding values of N_{ads} and comparing the values obtained. For the isotherms presented here, $RMSE_{N_{ads}}$ for $N_{ads} = 3$ was always greater than or not sufficiently smaller than the value of $RMSE_{N_{ads}}$ for $N_{ads} = 2$. Thus, the largest number of statistically valid adsorbing pseudocomponents was limited to two.

Another modeling technique, reported by Jayaraj and Tien (7), reduces the number of parameters to

$$p_{N_{ads}} = N_{ads} - 1 \tag{13}$$

and, as a result, reduces computational limits to the maximum number of pseudocomponents. However, this reduction in parameters is obtained by arbitrarily assigning Freundlich parameters to pseudocomponents and searching only for pseudocomponent concentrations, whereas the technique employed in this work searches for all of these parameters. Statistical limitations to the Jayaraj and Tien technique are unknown, but the fact that they exist is demonstrated by the inability of this technique to produce a unique description of one of the wastewaters tested when using five adsorbing pseudocomponents (7). The use of three or four adsorbing pseudocomponents may have produced a more valid result. In any event, further research is required to determine the statistical limitations to this approach.

Rate Tests and Kinetic Models

External diffusion rates were studied through the use of the minicolumn rate test (8). This test uses a high flow rate and a short GAC bed length to allow for the domination of external mass transfer. External mass transfer coefficients were calculated from

$$K_l = \frac{Q}{A_{eff}M} \ln \frac{C_{T,t}}{C_{T,t=0}} \tag{14}$$

where Q is the volumetric flow rate through the column, M is the mass of GAC in the column, A_{eff} is the effective external specific surface area of the adsorbent (as determined by the use of a solute with a known diffusion coefficient), and $C_{T,t}$ and $C_{T,t=0}$ are the effluent plateau and influent concentrations of the mixture, respectively. Free liquid diffusivities (D_l) were calculated from Gnielinski's correlation; these measures of D_l were independent of the equilibrium parameters. Gnielinski's correlation was used because experiments by Roberts et al. (8) determined that the particle shape factor and, therefore, A_{eff} were not affected by the Reynolds number. The D_l

obtained in this manner describes an average free liquid diffusivity for the mixture. No attempt was made to describe rates of diffusion for each pseudocomponent.

Data from batch rate tests were used in a heterogeneous diffusion model to determine the internal diffusion characteristics of a given mixture. The kinetic model incorporates pore and surface diffusion mechanisms and an IAST description of multicomponent equilibrium. Thus, the model not only requires batch rate data as input; it also requires the input of pseudocomponent equilibrium parameters. In addition, an external mass-transfer coefficient is required, and this requirement implies that the free liquid diffusivity must also be known. When these inputs are combined with GAC characteristics (particle radius, apparent density, pore void fraction, and dosage), average diffusion coefficients can be determined by guessing values of pore diffusivity (D_P) and surface diffusivity (D_S) until an adequate description of the rate data is achieved. This step is best accomplished by fixing one of the coefficients and then adjusting the other coefficient until the smallest residual sum of squares is obtained.

Unfortunately, the search for D_P and D_S in the heterogeneous diffusion model will not yield a unique set of D_P and D_S values that describe the test data. At the two extremes, the heterogeneous model simplifies to homogeneous diffusion; that is, a pore diffusion model (PDM) is easily obtained by setting D_S equal to zero and a surface diffusion model (SDM) is obtained by setting D_P equal to zero. Fettig and Sontheimer (6) presented a methodology by which another experimental measurement can yield a unique combination of D_S and D_P. The procedure entails the collection of equilibrium data for the HS mixture remaining after the rate test. The objective of the kinetic modeling is to find the combination of D_S and D_P that allows the proper amount of adsorption of each pseudocomponent to occur during the rate test such that the remaining concentrations in solution, which become the initial concentrations for subsequent equilibrium modeling, yield the best equilibrium description of the residual mixture. The resulting values of D_S and D_P will be referred to as those of the heterogeneous diffusion model (HDM).

Both types of rate tests employed 18–20 mesh GAC (F–400), prepared and stored as described previously. The minicolumn experiments were performed in a lucite minicolumn with the GAC packed between two layers of glass beads. This bed was fixed between two stainless steel screens that were 5 cm apart. Solutions were fed from a constant-head reservoir at 25 mL/min for 10 min. A particle shape factor of 1.54 was obtained with *p*-nitrophenol (PNP) in calibration runs.

For the batch rate tests, 3 L of a given mixture was placed in a 4-L glass beaker and allowed to contact the GAC for 190–225 h. Mixing was provided by a poly(vinyl chloride) impeller with an attached chamber to allow flow-through contact with the GAC. GAC dosages were designed to achieve approximately 90% removal of the total initial concentration at the equilibrium state. The HS mixture remaining at the end of the rate test was immediately used in a bottle point isotherm experiment to describe the equilibrium behavior of the residual HS solution.

Equilibrium Results

Figure 1 shows that the equilibrium adsorption behavior of the coagulated HS solution changes when the initial total concentration is changed. This observation, not seen in single-solute systems, confirms the multicomponent nature of HS solutions. With the technique described earlier, a best fit was obtained for the data having an initial TOC of 10.85 mg/L. This fit yielded a set of one nonadsorbing pseudocomponent and two adsorbing pseudocomponents, each having the Freundlich constants and percent initial concentrations given in Table II for the coagulated HS mixture. The pseudocomponent properties listed in Table II should apply to both of the mixtures studied in Figure 1, because the only difference between the two mixtures was their initial total concentration.

Initial pseudocomponent concentrations can be calculated from Table II by knowing the initial total concentration. Thus, the IAST model should be able to predict the isotherm of the more concentrated solution (TOC = 21.70 mg/L) from the Freundlich constants and the percent initial concentrations of the pseudocomponents as obtained by the best fit of the dilute solution (TOC = 10.85 mg/L). As Figure 1 shows, the IAST model makes a very reasonable prediction (dashed line) in the upper and lower regions of the more concentrated isotherm, but overpredicts adsorbability in the middle region. Additional data points in the middle region may improve the performance of the IAST model.

By assuming that the IAST model makes adequate predictions of the effect of initial concentration on HS isotherms, the model can be used to compare how various treatments change HS adsorbability. For instance, Figure 2 shows isotherm data for uncoagulated and coagulated HS solutions that reveal definite improvement of adsorbability upon coagulation. However, the data points do not reveal whether the improvement was due to decreased initial concentration, changes in solution composition, or both.

The model was used to test whether the observed shift in isotherm position upon coagulation could be accounted for solely by the decrease of initial concentration. First, best fits were obtained for both sets of isotherm data to yield the pseudocomponent parameters shown in Table II. As noted in the previous paragraph, the parameters obtained for the coagulated HS mixture can be used by the IAST model to predict the position of the coagulated HS isotherm at various initial concentrations. A comparison between the best fit of the prefiltered HS isotherm at an initial concentration of 14.60 mg of TOC/L and the prediction of the coagulated HS isotherm at

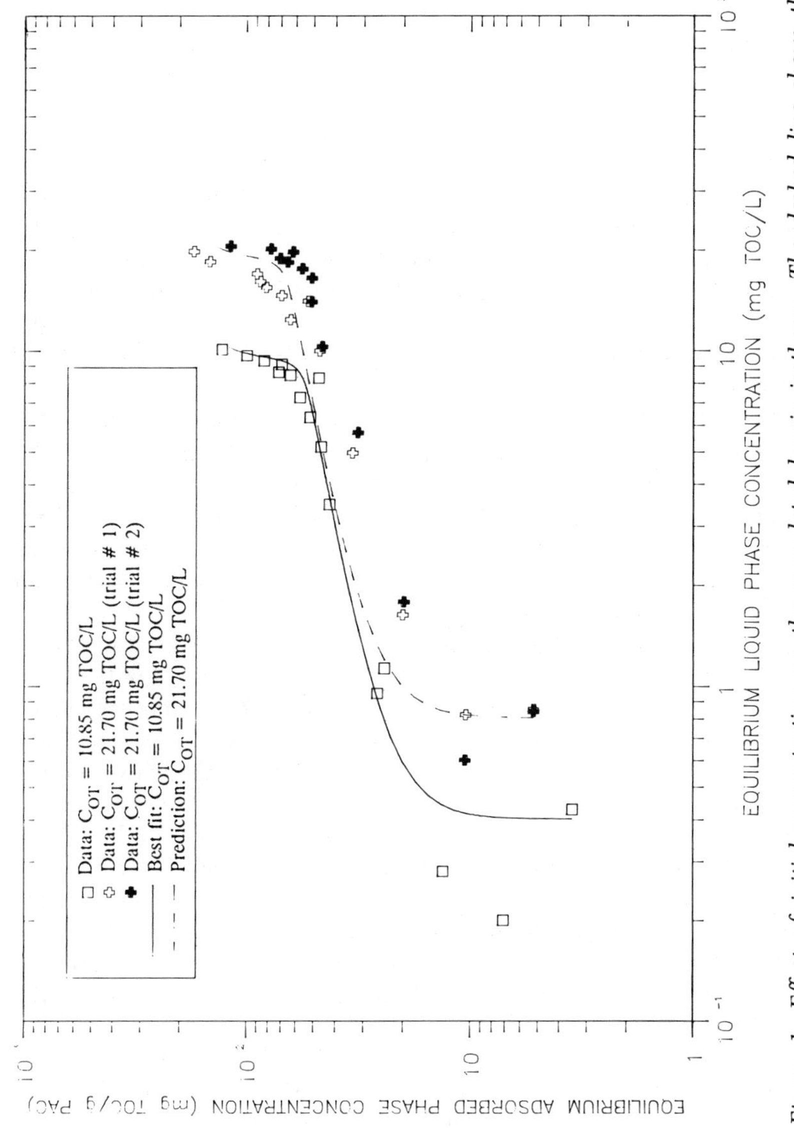

Figure 1. Effect of initial concentration on the coagulated humic isotherm. The dashed line shows the IAST prediction of the more concentrated isotherm, based on parameters obtained from the best fit of the more dilute isotherm.

Table II. Ideal Adsorbed Solution Theory Equilibrium Model Results

	Pseudocomponent 1			*Pseudocomponent 2*			*Nonadsorbing Pseudocomponent*
Humic Mixture	K	*1/n*	% $C_{T,0}$	K	*1/n*	% $C_{T,0}$	% $C_{T,0}$
Prefiltered[a]	25	0.36	92	—	—	—	8
Coagulated	31	0.27	84	127	0.20	12	4
Ozonated and biostabilized	23	0.17	71	87	0.14	25	4

[a]A second pseudocomponent was found to be statistically insignificant for the prefiltered mixture.

an initial concentration of 14.60 mg of TOC/L should indicate whether the improvement in adsorbability upon coagulation was merely due to the decrease of initial concentration. For instance, if the prediction of the coagulated HS isotherm were to match or closely agree with the best fit of the prefiltered HS isotherm at the indicated initial concentrations, then the improvement in adsorbability could be attributed to the decrease of initial concentration.

The results of the actual comparison, shown in Figure 2, reveal that the prediction (dashed line) is in poor agreement with the best fit (solid line), thereby implying that the decrease of initial concentration upon coagulation is not enough to explain the change in isotherm position. Therefore, coagulation must also create compositional changes, such as the preferential removal of larger and more poorly adsorbed organic molecules, that result in improved adsorbability.

The combined effect of ozonation and biodegradation was also studied through the use of the same comparison technique. The results of this comparison, depicted in Figure 3, show that the prediction of the coagulated HS isotherm at an initial concentration of 7.00 mg of TOC/L (dashed line) is quite similar to the best fit of the ozonated and biodegraded HS isotherm at an initial concentration of 7.00 mg of TOC/L (solid line). A slight decrease in adsorbability upon ozonation and biodegradation is observed in the middle region of the comparison, although the upper and lower regions are virtually the same. Ozonation is known to decrease the adsorption capacity for HS by creating solutions that are highly polar in nature. Therefore, these results indicate that such polar components are preferentially removed by the biodegradation process to produce a solution similar in adsorbability to the coagulated solution.

Table II may be used to compare the equilibrium properties of each solution in a quantitative manner. The fact that coagulation caused the IAST modeling procedure to call for the addition of a second and more strongly adsorbing pseudocomponent (i.e., in the Freundlich model, a larger K and a smaller $1/n$) suggests improved adsorbability of the HS mixture. The percentage of the nonadsorbing pseudocomponent also decreased. Subsequent ozonation and biodegradation, however, caused the IAST model to

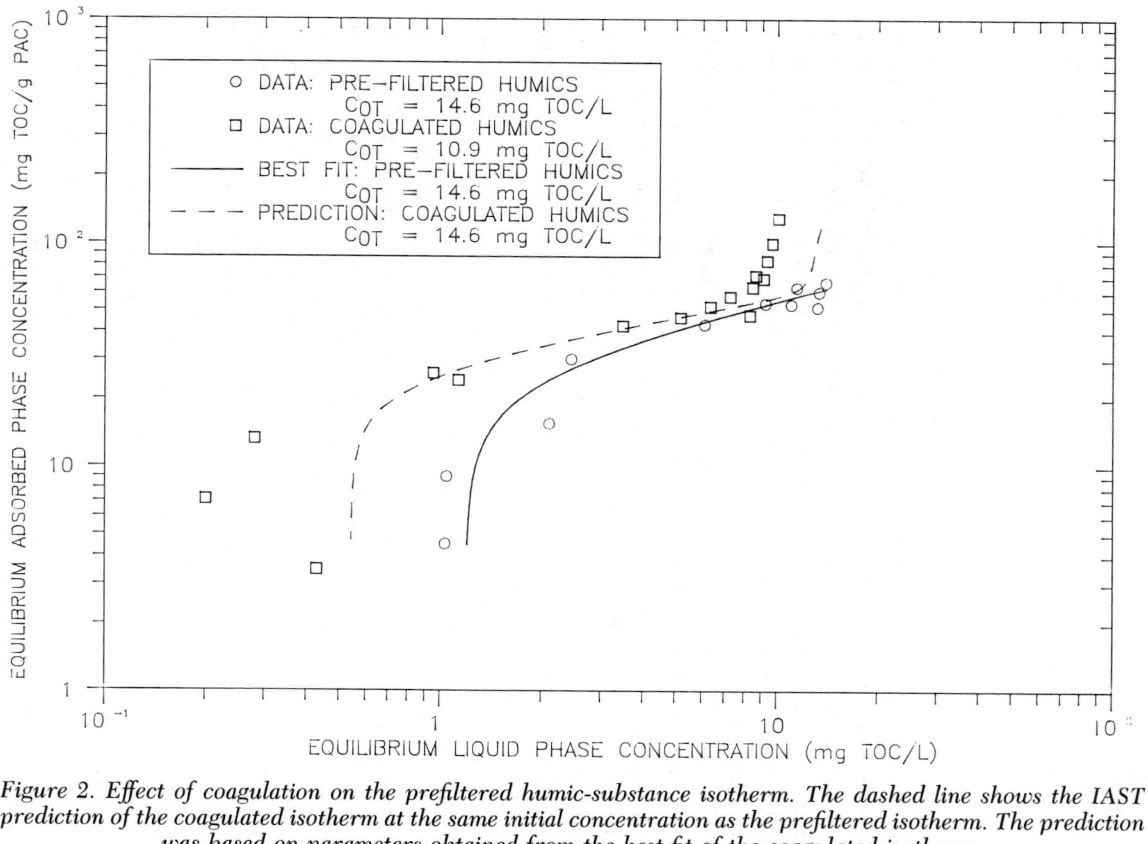

Figure 2. Effect of coagulation on the prefiltered humic-substance isotherm. The dashed line shows the IAST prediction of the coagulated isotherm at the same initial concentration as the prefiltered isotherm. The prediction was based on parameters obtained from the best fit of the coagulated isotherm.

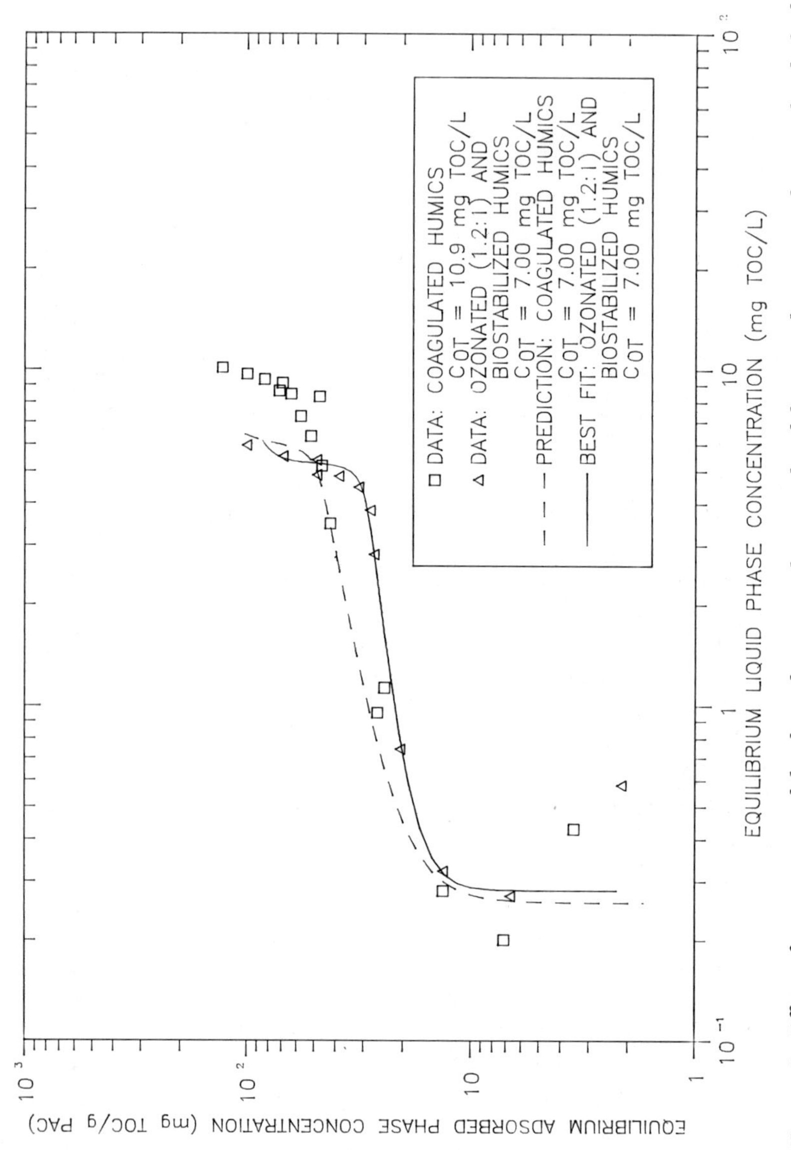

Figure 3. Effect of ozonation and biodegradation on the coagulated humic-substance isotherm. The dashed line shows the IAST prediction of the coagulated isotherm at the same initial concentration as the ozonated and biostabilized isotherm. The prediction was based on parameters obtained from the best fit of the coagulated isotherm.

predict further changes in Freundlich parameters and percent compositions that are less easily explained. The decrease in the Freundlich $1/n$ value for both pseudocomponents and the increase in percent composition of Pseudocomponent 2, the more strongly adsorbing pseudocomponent, from 12% to 25% can be taken as evidence of increased adsorbability. However, the Freundlich K value for both pseudocomponents decreases significantly upon ozonation and biodegradation, a result leading to the conclusion that adsorbability has decreased. Therefore, the Freundlich constants and the percent compositions do not necessarily yield conclusive results on their own.

Kinetic Results

The results of the minicolumn and batch rate tests (see Tables III and IV) revealed small changes in average external and homogeneous internal diffusion rates upon treatment. Coagulation was found to increase the average rate of external diffusion, as was expected because of a decrease of average molecular size. Although ozonation and biodegradation would be expected to produce even smaller molecules, this result was not manifested by any further increase in D_l. However, the batch rate tests showed an increase in average internal diffusion rates with each treatment stage.

As noted earlier, results from the IAST equilibrium and HDM kinetic models should predict the equilibrium behavior of an HS solution remaining at the end of a batch rate test. Data from adsorption rate tests of the coagulated HS mixture and the coagulated, ozonated, and biostabilized HS mixture are presented in Figure 4, with the accompanying HDM predictions. Figure 5 presents isotherm data obtained for the coagulated HS mixture before and after 190 h of batch rate testing. This figure also shows the IAST model prediction (solid line) of the isotherm for the HS mixture remaining after the kinetic testing. A good agreement between the prediction and the data indicates that the proper internal diffusion coefficients were

Table III. Minicolumn Rate Test Results

Humic Mixture	D_l (cm^2/s)
Prefiltered	1.1×10^{-6}
Coagulated	2.0×10^{-6}
Ozonated and biostabilized	2.0×10^{-6}

Table IV. Kinetic Model Results

	PDM	*SDM*	*HDM*	
Humic Mixture	D_P (cm^2/s)	D_S (cm^2/s)	D_P (cm^2/s)	D_S (cm^2/s)
Prefiltered[a]	3.2×10^{-7}	6.1×10^{-11}	—	—
Coagulated	4.0×10^{-7}	6.8×10^{-11}	4.0×10^{-7}	2.2×10^{-12}
Ozonated and biostabilized	6.3×10^{-7}	2.0×10^{-10}	6.3×10^{-7}	1.6×10^{-17}

[a]An HDM test was not performed for the prefiltered humic mixture.

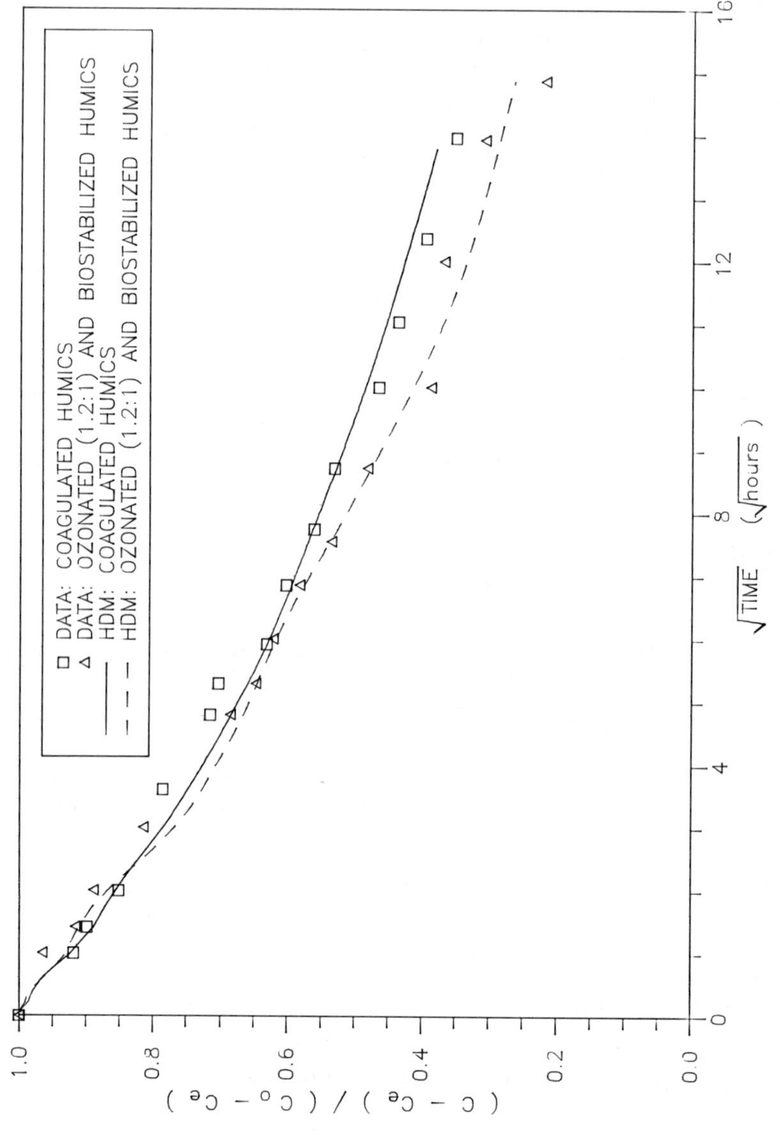

Figure 4. Effect of ozonation and biodegradation on the internal diffusion rate of coagulated humic substances.

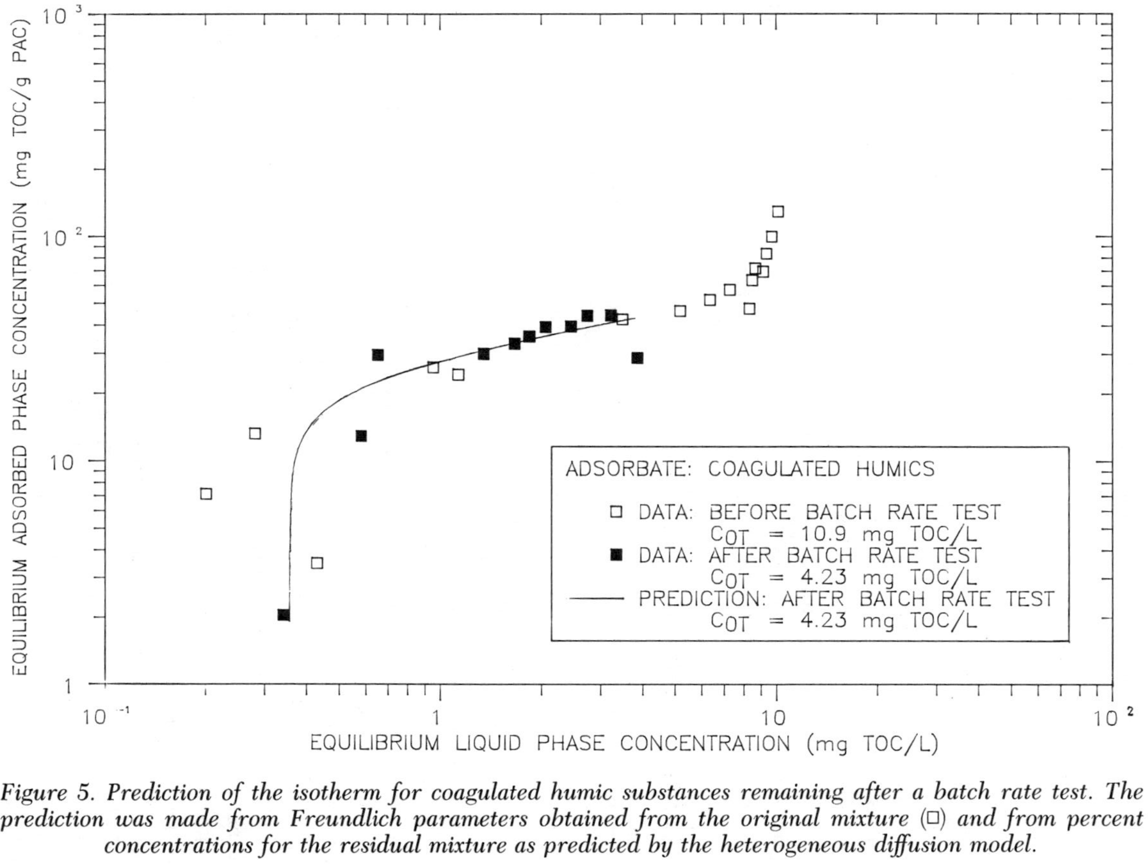

Figure 5. Prediction of the isotherm for coagulated humic substances remaining after a batch rate test. The prediction was made from Freundlich parameters obtained from the original mixture (□) and from percent concentrations for the residual mixture as predicted by the heterogeneous diffusion model.

used in the HDM to yield pseudocomponent concentrations that provided the proper input to the IAST equilibrium model. Similar results were obtained for the ozonated and biodegraded mixture, as shown in Figure 6. Table V shows that the HDM predicted the preferential adsorption of the more strongly adsorbed pseudocomponent (Pseudocomponent 2) during both batch rate tests.

The values of D_S and D_P that yielded these predictions are posted in Table IV and show that, according to this modeling approach, pore diffusion was the dominant mechanism for internal HS transport. However, all three kinetic models (PDM, SDM, and HDM) provided an adequate description of the rate data. Moreover, a simplifying assumption was made that each pseudocomponent had the same diffusion coefficient, even though this situation is unlikely. Thus, caution is needed in stating that pore diffusion is actually the dominant mechanism. Nevertheless, this result agrees with recent discussion (*8*) that the macromolecular nature of HS mixtures may provide a tremendous obstacle to mass transport in the adsorbed phase and thereby force pore diffusion to account for transport.

Modeling Limitations

When modeling a complicated process such as multicomponent adsorption, the procedure is likely to have limitations. First, several of the assumptions used to derive the IAST model are violated by the HS–activated-carbon system. For instance, an adsorbed-phase solution of HS is not likely to be in conformance with the assumption of an ideal adsorbed solution, and activated carbon is certainly not the inert solid-phase material it is assumed to be. In addition, the IAST model assumes that adsorption takes place by physical means and ignores the phenomenon of chemisorption. The polyelectrolytic nature of HS molecules, combined with the ionic surfaces of activated carbon, implies that chemisorption of humic substances is highly likely. Thus, the model will not work if some mechanism other than physical adsorption is primarily responsible for the removal of HS from solution. Finally, the intraparticle diffusion coefficients listed here are dependent upon equilibrium characterizations and may not represent the true kinetic properties of an HS mixture.

The ultimate goal of this approach is to provide a means to predict the effects of HS mixtures on the adsorption of pollutants in water-treatment systems. Thus, the severity of such limitations will be tested when this approach is eventually extended to predicting breakthrough curves of pollutants in HS solutions.

Summary

The equilibrium adsorption behavior of aquatic HS solutions was described by the use of the multicomponent IAST model. The IAST model is simplistic

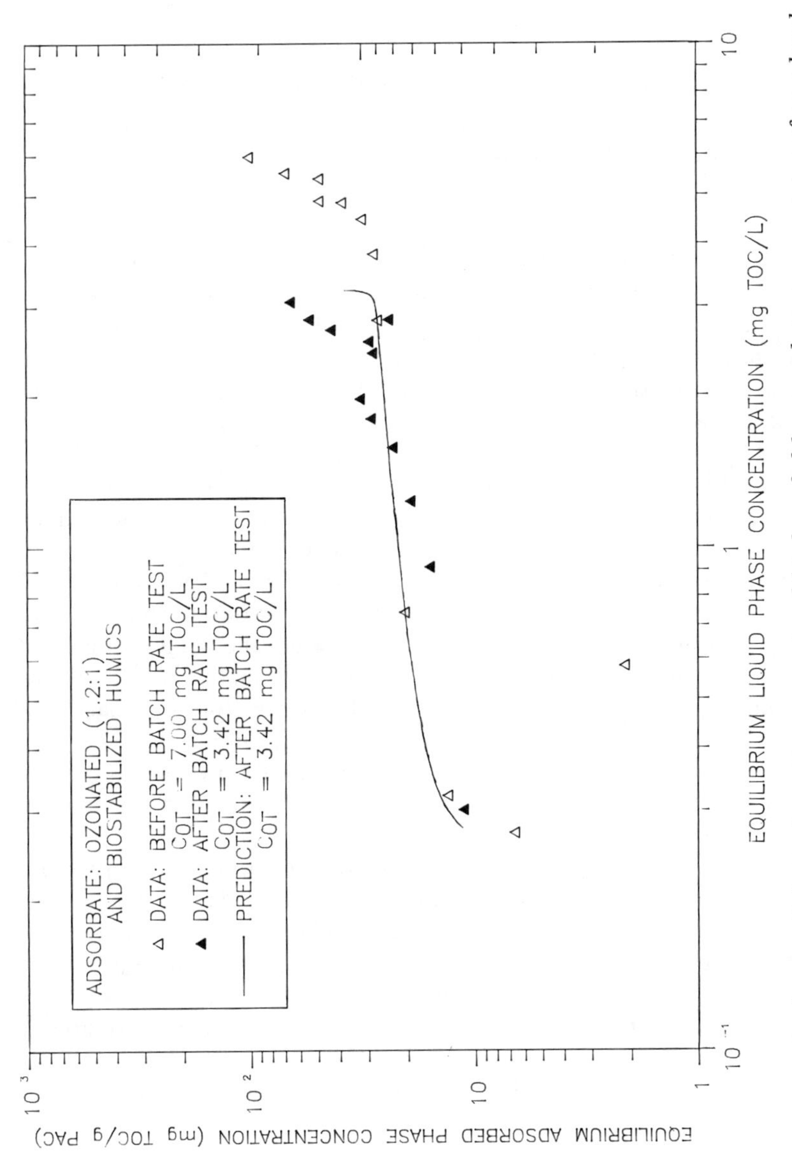

Figure 6. Prediction of the isotherm for ozonated and biodegraded humic substances remaining after a batch rate test. The prediction was made from Freundlich parameters obtained from the original mixture (**Δ**) *and from percent concentrations for the residual mixture as predicted by the heterogeneous diffusion model.*

Table V. Mixture Compositions Before and After Batch Rate Testing

	Pseudocomponent 1[a]		*Pseudocomponent 2*[a]		*Nonadsorbing Pseudocomponent*	
Humic Mixture	*Before*[b]	*After*[c]	*Before*[b]	*After*[c]	*Before*[b]	*After*[c]
Coagulated	84	92	12	0	4	8
Ozonated and biostabilized	71	87	25	5	4	8

NOTE: All values are % $C_{T,0}$.

[a]Freundlich parameters for each adsorbing pseudocomponent are listed in Table II and were not changed during the course of kinetic testing.

[b]Before batch rate test; determined by the IAST equilibrium model.

[c]After batch rate test; predicted by the HDM.

in that the pseudocomponents comprising an HS mixture are assumed to have the same number of carbon atoms per molecule. This deficiency in modeling needs more attention. Nevertheless, the model was used to evaluate the adsorption equilibrium and kinetic effects of two different treatment processes—coagulation and a combination of ozonation and biodegradation following coagulation.

The IAST model provided a reasonable prediction of changes in the equilibrium adsorption behavior of the coagulated HS solution upon dilution or concentration. This result allows the study of treatment effects without the interference of initial concentration effects. Coagulation improved the adsorbability of the HS mixture by changing its composition, as well as by decreasing its initial concentration. The improvement due to compositional changes is probably a result of the preferential removal of larger-molecular-weight species by the coagulation process. After ozonation and biodegradation, the HS mixture showed no significant changes from the coagulated mixture in terms of adsorption equilibrium behavior. Therefore, since ozonation is thought to reduce adsorbability because of the production of polar organic species, biodegradation must remove these compounds to yield the observed results.

Kinetic tests showed a small increase in free liquid diffusivity upon coagulation and no change upon ozonation and biodegradation. In addition, the competitive diffusion models were found to be useful in describing the internal kinetics of an HS mixture, as well as in predicting the equilibrium behavior of an HS mixture remaining at the end of a batch rate test. The results show modest increases in internal diffusion rates with each treatment stage and the domination of pore diffusion as the mechanism of internal mass transfer. However, the actual diffusion mechanism is unknown without further testing.

Nomenclature

A_{ads} Adsorbent surface area per unit mass of adsorbent (length^2/mass)

A_{eff} Effective surface area of the adsorbent per unit mass of adsorbent (length^2/mass)

C_i	Equilibrium liquid-phase concentration of component i (mass/length^3)
C_i^0	Equilibrium liquid-phase concentration of component i in a single-solute system at the same temperature and spreading pressure as the mixture (mass/length^3)
$C_{0,i}$	Initial liquid-phase concentration of component i (mass/length^3)
$C_{T,0}$	Initial liquid-phase TOC concentration of the humic mixture (mass/length^3)
$C_{Ti,calc}$	IAST-calculated equilibrium liquid-phase concentration of the mixture in isotherm bottle i (mass/length^3)
$C_{Ti,obs}$	Observed equilibrium liquid-phase concentration of the mixture in isotherm bottle i (mass/length^3)
$C_{T,t}$	TOC concentration of the minicolumn effluent mixture at a given time, t, of a rate test (mass/length^3)
$C_{T,t=0}$	TOC concentration of the minicolumn influent mixture during a rate test (mass/length^3)
D_l	Free liquid diffusion coefficient (length^2/time)
D_P	Pore diffusion coefficient (length^2/time)
D_S	Surface diffusion coefficient (length^2/time)
K_i	Freundlich isotherm constant for component i [(mass/mass) (length^3/mass)$^{1/n_i}$]
K_l	External mass-transfer coefficient (length/time)
M	Mass of carbon in an isotherm bottle or in the minicolumn reactor (mass)
N_{ads}	Number of adsorbing pseudocomponents in a mixture
n_i	Freundlich isotherm exponent for component i
N_{obs}	Number of isotherm observations
$p_{N_{ads}}$	Number of adjustable parameters for N_{ads} components in the IAST equilibrium model
q_i	Equilibrium solid-phase concentration of component i (mass/mass)
q_i^0	Equilibrium solid-phase concentration of component i in a single-solute system at the same temperature and spreading pressure as the mixture (mass/mass)
q_T	Sum of the solid-phase concentrations of each solute (mass/mass)
Q	Volumetric flow rate (length^3/time)
R	Ideal gas constant (mass·length^2/time^2·temperature·mole)
RMSE	Root mean square error
SSR	Residual sum of squares
T	Absolute temperature (temperature)
V	Liquid volume in an isotherm bottle (length^3)
z_i	Surface mole fraction of component i
π_i^0	Spreading pressure of component i in a single-solute system (mass/time^2)
π_T	Spreading pressure of the mixture (mass/time^2·length)
σ_i	Standard deviation of the TOC observations for isotherm bottle i (mass/length^3)

Acknowledgments

The authors thank Emery Kong for his help in setting up and performing the ozonation and biostabilization experiments. The heterogeneous diffusion model was provided by John Crittenden's group at Michigan Technological University. This research was supported by the U.S. Environmental Protection Agency's Exploratory Research Program, Grant No. R811824.

References

1. Frick, B. R.; Sontheimer, H. In *Treatment of Water by Granular Activated Carbon*; American Chemical Society: Washington, DC, 1983; p 247.
2. Crittenden, J. C.; Luft, P.; Hand, D. W. *Water Res.* **1985,** *19*, 1537.
3. Radke, C. J.; Prausnitz, J. M. *AIChE J.* **1972,** *18*, 761.
4. Randtke, S. J.; Snoeyink, V. L. *J. Am. Water Works Assoc.* **1983,** *75*, 406.
5. Calligaris, M. B.; Tien, C. *Can. J. Chem. Eng.* **1982,** *60*, 772.
6. Fettig, J.; Sontheimer, H. *J. Environ. Eng. Div.* (*Am. Soc. Civ. Eng.*) **1987,** *113*, 795.
7. Jayaraj, K.; Tien, C. *Ind. Eng. Chem. Process Des. Dev.* **1984,** *24*, 1230.
8. Roberts, P. V.; Cornel, P.; Summers, R. S. *J. Environ. Eng. Div.* (*Am. Soc. Civ. Eng.*) **1985,** *111*, 891.
9. Summers, R. S. Ph.D. Thesis, Stanford University, 1986.

RECEIVED for review October 15, 1987. ACCEPTED for publication February 9, 1988.

INFLUENCES OF ION-EXCHANGE AND MEMBRANE PROCESSES ON WATER TREATMENT

Effects of Humic Substances on Membrane Processes

J. Mallevialle, C. Anselme, and O. Marsigny

Centre de Recherche Lyonnaise des Eaux-Degrémont, 38 rue du Président Wilson, 78230 Le Pecq, France

Analytical methods are being developed to help characterize the mechanism of membrane fouling. For example, pyrolysis–gas chromatography–mass spectrometry provides global information on the nature of macromolecules present in the water and in the cake that forms at the membrane surface. Scanning electron microscopy, transmission electron microscopy, elemental analysis, X-ray microanalysis, secondary ion mass spectroscopy, and Fourier transform IR spectroscopy also can be used to define the encrustation.

SUSPENDED MATTER AND MACROMOLECULES are removed from water primarily by coagulation and sand filtration. Many water-treatment companies are involved in research into ways to optimize this clarification step, but so far the complexity of the problem has led to very involved solutions that create difficulties with regard to any automation. One alternative is to develop new liquid–solid separation processes such as membrane filtration, which seems to be promising from the viewpoints of both economics and automation. Some limitations inherent in this process must be resolved; permanent fouling, adsorption, and concentration polarization all lead to a quick flux decline (75–90%) during the first period of use (*1*). These phenomena must be understood if we are to overcome their disadvantages and develop an economically viable process.

Figure 1 illustrates our approach to the goal of optimization of filtered-water production by cleaning cycles (backflushing, pulse flux, etc.). The main problem in achieving this goal is permanent fouling, which appears to

0065–2393/89/0219–0749$06.00/0

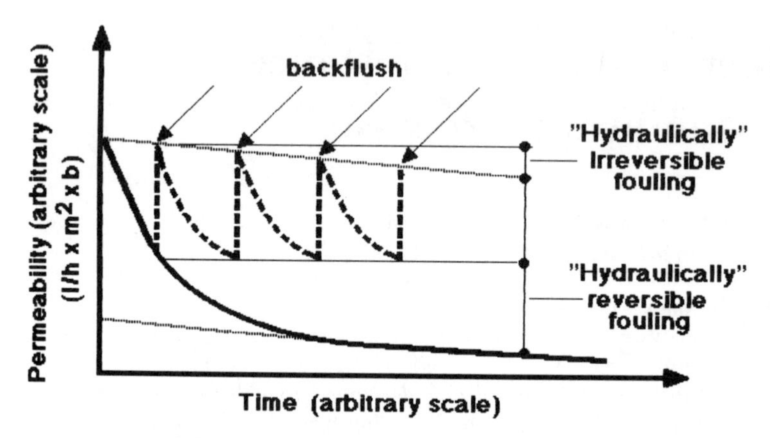

Figure 1. Permeability of a membrane as a function of time and reversibility of fouling.

be tied to the aqueous organic matrix. Inorganic compounds such as clay do not induce any definitive fouling, but filtration of water containing complex mixtures of organic and inorganic matter can lead to irreversible fouling (*2*).

Three types of studies were conducted to characterize this fouling. The first study included measurement of general water-quality parameters (e.g., turbidity, total organic carbon (TOC), and elemental analysis). In the second study, a separation was performed by gel permeation chromatography (GPC) to determine the apparent molecular weight of the organic content. These samples were studied by pyrolysis–gas chromatography–mass spectrometry (Py–GC–MS) (*3*, *4*). This process reveals the distribution of the different families of organic macromolecules in the feedwater, filtrate, and retentate and enables us to determine the organic mass balance of the effluents. This Py–GC–MS method can also be applied directly to any matter deposited in the membrane. Finally, methods such as electron microscopy coupled with analytical detectors (X-ray microanalyzer) and secondary ion mass spectrometry (SIMS) were applied to obtain a structural view of the membrane and the deposit on it.

The goals of this work were to characterize the composition and structure of filtration "cakes" and to point out the importance of the adsorption mechanism of some macromolecular compounds in the phenomenon of permanent fouling. From this complete characterization, further research works could be implemented in the area of fouling reduction. Such results could help to improve the fouling-reduction methods that are used in membrane-filtration processes:

- pretreatment of raw water (coagulation or addition of powdered activated carbon before membrane filtration)
- modification of the membrane surface to avoid adsorption of specific foulant macromolecules
- dynamic cleaning cycles (backflush, pulse flux, reverse circulation)
- chemical cleaning of membrane (oxidants, surfactants)

Experimental Materials and Methods

Water Samples. Four different water samples were used in this work. One was a ground water from Normandy (France) at Bernay with low TOC and high mineral content. This water was rapidly recharged by rain, and that led to an occasional sharp increase in turbidity. A second water sample, from the Allier River at Bellerive, had a mountain origin. Turbidity in this case was normally between 1.5 and 10 nephelometric turbidity units (NTU), with some peaks surging to 70 NTU. This water had a low mineral content and medium TOC. The third sample was reservoir water from Nebias, France, with strong turbidity and high TOC. The last sample was river water from Dunkirk, in the North of France, with high TOC, medium turbidity and very high algae content. These water samples will be referred to in this chapter as Bernay, Bellerive, Nebias, and Dunkirk.

Membrane Samples. Various microfiltration and ultrafiltration hollow fibers were used, with porosities ranging from 0.2 μm to 20 Å. The constitutive polymers were polypropylene, polyacrylonitrile, polysulfone, poly(ether sulfone), and a cellulosic polymer.

Water Concentration Methods. Water samples were first concentrated by rotary evaporation to obtain a TOC between 100 and 200 mg/L. These concentrated samples (in 10-mL aliquots) were injected into a gel permeation chromatograph packed with a fine gel (Sephadex G25, Sephadex Pharmacia, Uppsala, Sweden). The conditions were as follows: column size, 0.25 × 90 cm; eluant, ultrapure water; flow rate, 150 mL/h. Elution was followed by a continuous UV absorbance detector (254 nm) and TOC measurement on a TOC meter (Dohrman DC80, Envirotech Corp., Santa Clara, CA). The samples were then dried for pyrolysis.

Flash pyrolysis was performed with a temperature-control system (Pyroprobe 100, Chemical Data System, Oxford, PA). Each sample (300–500-μg aliquots in a quartz tube) was pyrolyzed at 200–750 °C, with a temperature program of 20 °C/ms and a final hold for 20 s. After pyrolysis, the fragments were separated on a 30-m fused-silica column (DBWAX) that was temperature-programmed from 30 to 220 °C at 3 °C/min, with a final period of 10 min at 220 °C. This separation was followed by mass spectrometry (R 10–10 C, Nermag, Rueil Malmaison, France) operating in electron impact at 70 eV and scanning from 20 to 400 *m/z*. The chromatograms were analyzed with a computer (PDP 11/73, DEC), using a 70,000-spectra library (WHILEY) on a hard disk.

Pyrolysis IR is a step-by-step temperature-programmed pyrolysis technique that transforms organic carbon into CO_2, which is subsequently measured by IR spectroscopy. The various forms of carbon (aliphatic, aromatic, and inorganic) are detected at discrete temperatures; this method allows for the determination of the various sources of carbon in the sample from different fractions.

Deposit. The deposits were obtained from the fouled membranes by submitting them to sonication. The solution was centrifuged and the fouling-matter fraction was freeze-dried. The dry residue was then pyrolyzed with flash Py–GC–MS, pyrolysis IR analysis, or other analytical techniques.

Overall Parameters. The parameters studied are shown in Table I. UV absorbance at 254 nm (Perkin Elmer, Lambda 3) and TOC (Dohrman, DC 80) were measured for each water sample. Some determinations were also made by standard methods for parameters such as NO_3^-, Fe, alkalinity, and number of algae cells (5).

Microscopic Techniques. This study was performed with transmission (TEM) and scanning electron microscopy (SEM) (Figure 2). The scanning electron microscope (JEOL JSM2) was coupled with an energy-dispersing system (6) that provided information on inorganic elements contained in the samples, for $Z > 20$ amu.

The samples for SEM were frozen with liquid nitrogen. Cryofracture of the hollow fiber was performed to allow observation within the fiber, and all samples were then lyophilized. Some samples were compared with air-dried membranes. The preparation of samples for TEM (Philips 300) was more difficult than for SEM because we used resin inclusion and ultramicrotom sampling equipment.

The SIMS apparatus (CAMECA SMI 300) gives an exact spatial distribution (7) for selected elements in cross sections of samples (Figure 3). Some fouled-membrane data were also obtained by means of Fourier transform IR spectroscopy with an attenuated total reflection technique. The general scheme of analysis given in Figure 4 represents all of the analytical tools used in this study.

Results and Discussion

Organic Matrix of the Water. Table I shows the data for the overall parameters for each water, with few differences among the waters. Bernay water has a very low TOC, which indicates a low organic content, and this agrees with the elemental analysis (Table II). The UV measurement showed that a few aromatic structures are present in the organic matrix. For Bellerive water, the TOC is low-to-medium (3 ppm); UV absorbance at 254 nm is medium; turbidity is also medium (<10); and the algae content is medium, with values up to 50,000 cells/L. For Nebias water, the TOC and the UV absorbance were high. For Dunkirk water, TOC and UV are very high; this level indicates that the organic matrix is very aromatic. Algae counts for Dunkirk water are also very high, with values up to 1,000,000 cells/L. All these characteristics also indicate a high alkalinity for Bernay and Dunkirk waters, and an iron concentration of 100–300 μg/L in all these raw waters. Elemental analysis was performed for Bernay, Bellerive, and Dunkirk raw waters and for Bernay, Bellerive, and Nebias deposit samples.

Table I. Characteristic Values of Overall Parameters for Feedwaters

Parameter	*Bernay (Normandy)*	*Nebias (Aude)*	*Bellerive sur Allier*	*Dunkirk*
Type of water	ground water	reservoir water	river water	river water
UV absorbance	2.7	12.7	7.0	20–25
TOC (mg/L)	1–1.5	7–8	2–3	8–10
Turbidity (NTU)	2–20	10–70, max 200	8–10, max 70	8–20
Fe (μg/L)	130	330	100–300	200–300
NO_3^- (mg/L)	22	0.5	3–6	29
Alkalinity (mg/L of $CaCO_3$)	200	—	80	310
Algae (cells/L)	0	3,400	50,000	2,000,000 1,000,000

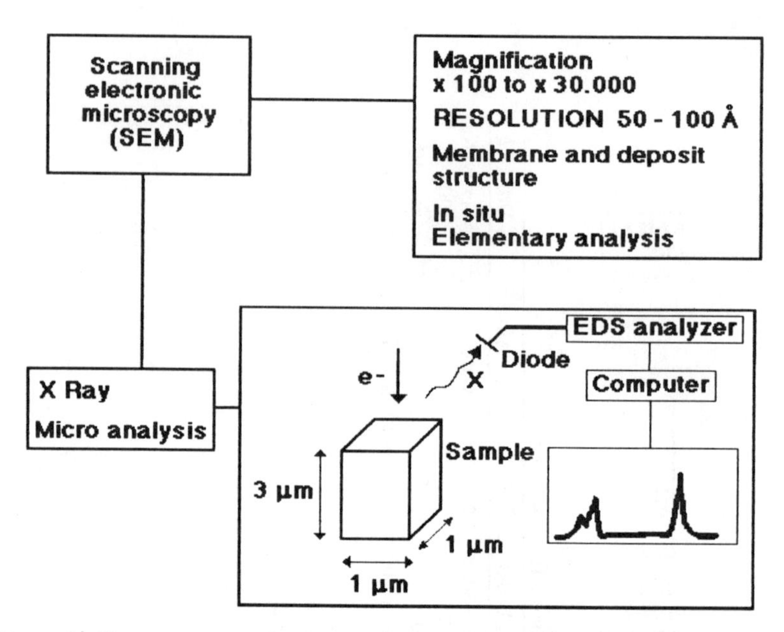

Figure 2. Characteristics of scanning electron microscopy–X-ray spectrometric (SEM–EDS) system.

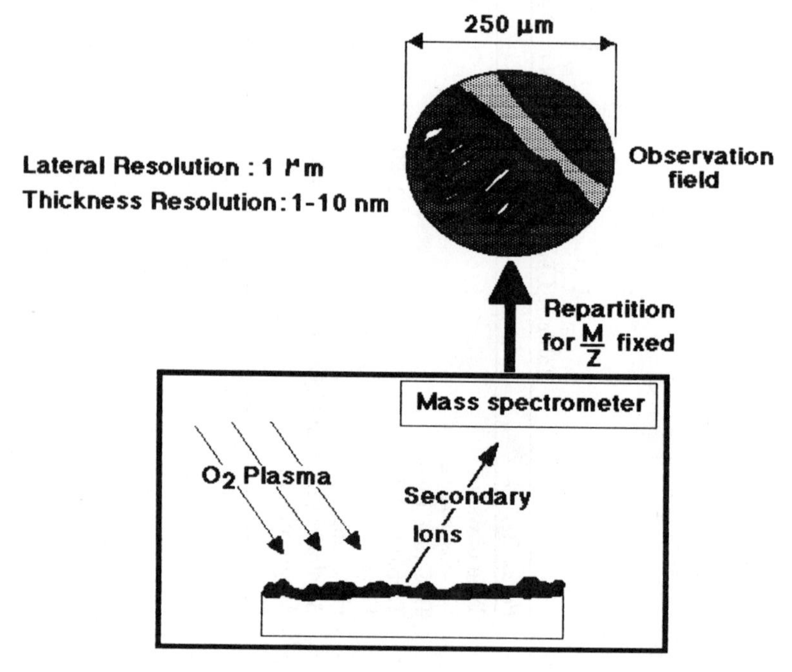

Figure 3. Principles and characteristics of secondary ion mass spectrometry (SIMS).

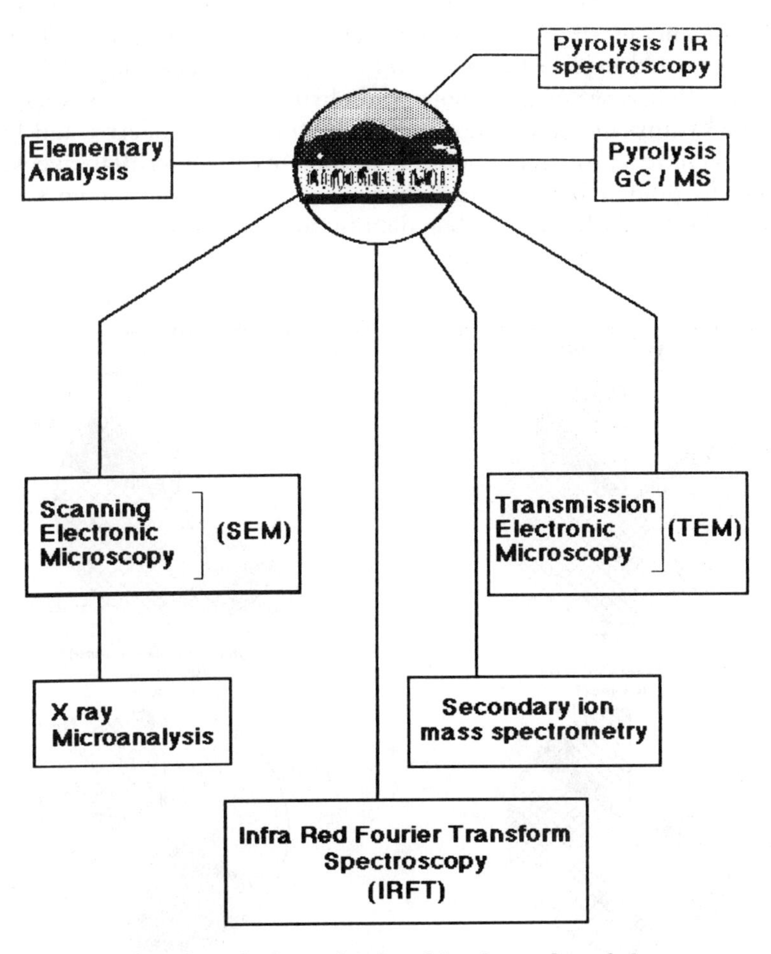

Figure 4. Total analytical scheme developed for the study and characterization of deposits.

Table II. Elemental Analyses of Deposits and Raw Waters

Sample	*% C*	*% H*	*% N*	*% O*	*% S*	*% Ash*
Bernay raw water	8.5	0.95	0.19	34.1	0.74	47.0
Bernay deposit	4.6	1.7	0.4	16.2	0.1	72.0
Nebias deposit	10.8	2.0	7.0	16.3	1.8	63.0
Bellerive raw water	8.4	1.7	0.9	—	—	5.1
Bellerive deposit	13.3	2.9	1.6	—	—	58.3
Dunkirk raw water	9.0	1.7	0.57	—	—	40.0

Figure 5 shows the distribution of the organic content in these raw waters by Py–GC–MS. The Bernay water (in the case of high turbidity) clearly contains mostly carbohydrates and aromatic compounds. The carbohydrates could explain the low level of UV absorbance and TOC measurements. Nebias water also shows carbohydrates and amino sugars, but has a polyhydroxyaromatic component. In addition, the Bellerive and Dunkirk waters have a protein fraction. Repartition of carbon for Bellerive water is more distributed between carbohydrates, proteins, and amino sugars, although the Dunkirk water carbon distribution is characterized by aromatics, proteins, and carbohydrates.

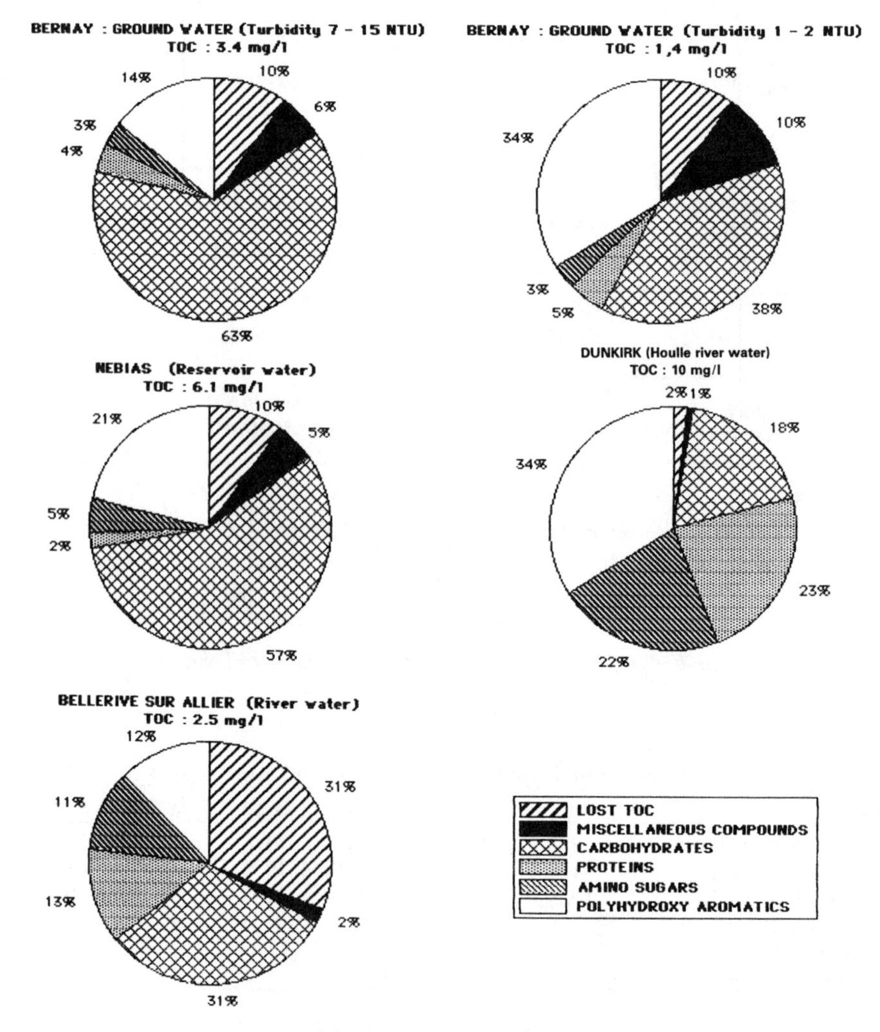

Figure 5. Repartition of feedwater TOC into macromolecular species.

Finally, a comparison of high and low turbidity in Bernay water indicates the importance of carbohydrates at low turbidity levels. Many authors have assumed that the organic content of water is essentially composed of polyhydroxyaromatic compounds (*8*). This hypothesis is based on UV absorbance results, trihalomethane formation potential, and other overall parameter values obtained after oxidation pretreatment of the sample. The situation is probably not that simple; water can contain sizable quantities of carbohydrates, proteins, and other natural macromolecular derivatives (*9*). This approach differs from a former (*10*) attempt to isolate compounds of natural origins. In this study we used a separation method and chose not to do selective acid–base or XAD resin extraction, or ion-exchange technique, and Py–GC–MS allowed a more precise final analysis of the GPC fractions. Three types of macromolecular compounds seem to be predominant in the TOC of the waters under study: carbohydrates, aromatic compounds, and proteins.

Organic Balance for Feedwater, Retentate, and Filtrate. Py–GC–MS was applied to different effluents after membrane filtration to determine the organic families involved in the membrane fouling. Figure 6 shows the distribution of organic carbon in the filtrates. This distribution must be compared with the Py–GC–MS results for fouling matter (Figure 7) and raw water (Figure 5). In both cases, significant differences exist between the distribution of TOC in feedwater and filtrate if the amount of TOC is taken into account.

TOC is generally 30–40% removed from feedwater as the stream progresses to filtrate. Sephadex gel permeation involves mainly compounds larger than 5000 daltons. Molecular-weight distributions in feedwaters are generally 40% TOC for MW >5000 daltons, 30% TOC for the first fraction <1500 daltons, and 30% TOC for the second fraction <1500 daltons. An exception is Nebias raw water, in which TOC is represented at 97% by a molecular-weight fraction >5000 daltons. In this last case, TOC removal during membrane ultrafiltration is generally more important and can reach values around 50%. Specific macromolecules, determined by Py–GC–MS, can be removed during the filtration. Three main conclusions can be made about these macromolecules. Carbohydrates seem to stay in the deposit or in the retentate; proteins and polyhydroxyaromatic compounds seem to be removed from the feedwaters during the filtration; amino sugars and low-molecular-weight compounds seem to cross the membrane preferentially to reach the filtrate.

Characterization of the Deposit. PYROANALYSIS, ELEMENTAL ANALYSIS. The purpose of pyroanalysis is to discriminate between the organic and inorganic carbon involved in fouling. Two facts appear to be important in this discrimination. First, the percentage of residue after analysis is significant. It indicates that the deposit on the filter is mostly inorganic (Table

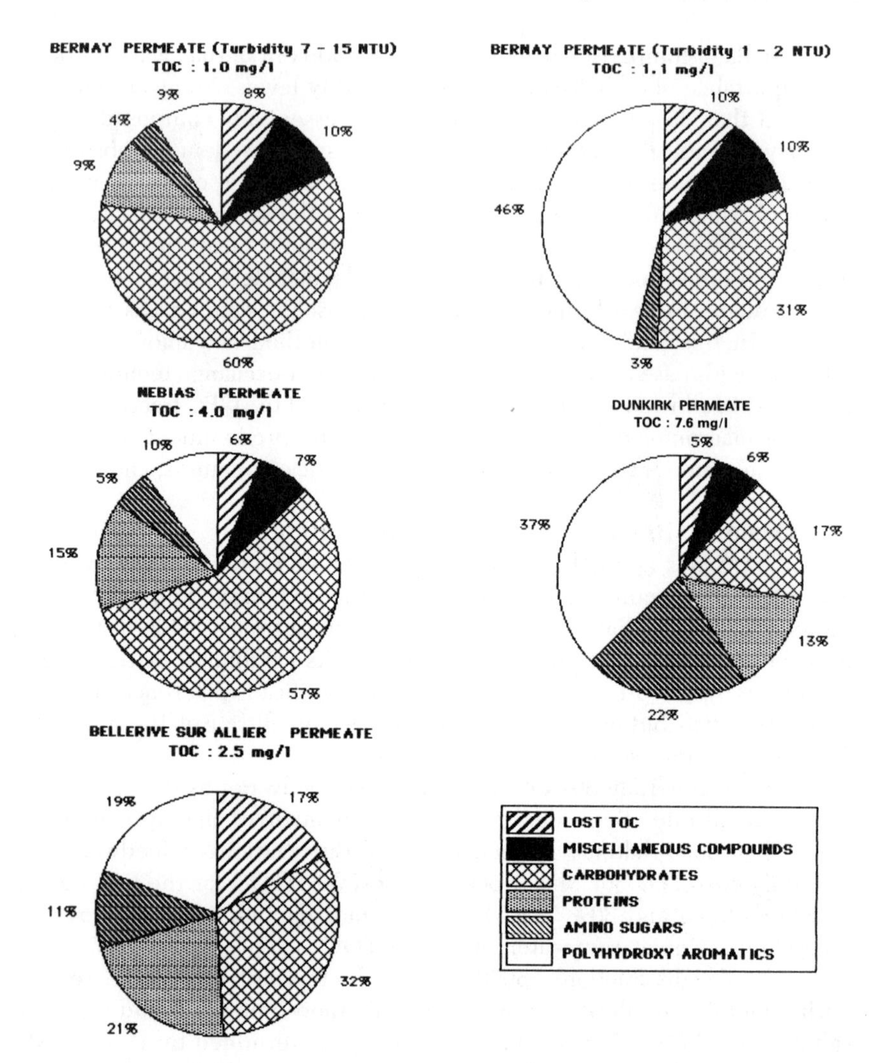

Figure 6. Repartition of filtrate TOC into macromolecular species.

II). The amount of Si, Al, Fe, and Mg in the composition of the deposit confirms this fact (Table III) and makes it possible to assume that clays represent the principal component of the inorganic substances. The Si/Al ratio, expected to be between 1 and 2, is 1.4 in this case. Table IV shows that more carbonates exist in the Bernay feedwater, whereas the deposit is composed mostly of organic carbon. These facts, confirmed by the distribution of organic and inorganic carbon in all the deposits, leads to the unambiguous conclusion that organic carbon contributes greatly to fouling.

As a confirmation, ash analysis indicates that calcium and $CaCO_3$ are minor components of the fouling matter for each type of membrane, even

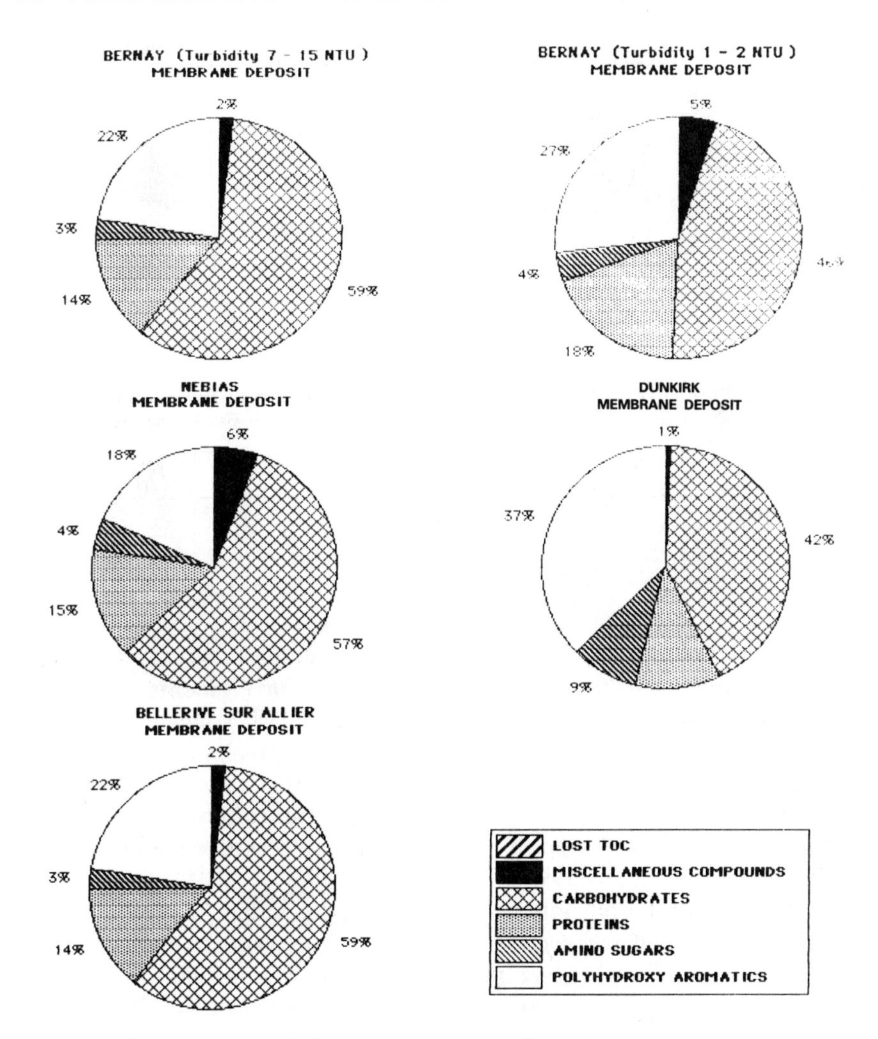

Figure 7. Repartition of the organic carbon of membrane deposits into macromolecular species with Py–GC–MS analysis.

if calcium is present at high concentration in the feedwater (as in the case of Bernay and Dunkirk).

PY–GC–MS. Results of Py–GC–MS indicate that the organic components of the deposit are mostly carbohydrates, proteins, and polyhydroxyaromatic substances (Figure 7). Therefore, it was possible to assume that this deposit was composed mostly of clays, with organic matter acting as a cement. The formation of this gel could be explained as an effect of the concentration polarization (*1*) (increase of the concentration close to the membrane). To verify this hypothesis, some other methods had to be used.

Table III. Analysis of Various Elements in the Bernay Deposit

Element	*Quantity*[a]
Si	280
Al	200
Mg	7
Fe	93
Mn	0.3
Ca	20

[a]Milligrams of element per gram of deposit.

Table IV. Repartition of Organic and Inorganic Carbon in Bernay Deposit, Nebias Deposit, and Bernay Raw Water Using Pyrolysis–IR Analysis

Sample	*Inorganic Carbon %* (*$CaCO_3$*)	*Organic Carbon %* (*220–700 °C*)
Bernay deposit	6.4	>90
Bernay raw water	7.3	25
Nebias deposit	2.5	>95

The filtration "cakes" are generally more than 80% nonorganic matter; iron, aluminium, silicon, and calcium are the major elements. The Al/Si ratio is characteristic of kaolinite clays. Iron, particularly concentrated at the internal "skin" of the membrane, could come from precipitation of hydroxides, complexation with organic compounds, or the natural residual iron contained by clays. In contrast, few inorganic elements (other than calcium, sodium, and potassium) seem to enter the structure of the membrane and foul its pores.

Organic matter from raw waters could be included in such a fouling mechanism. Calcium carbonate is not heavily involved in the fouling of ultrafiltration membranes, even in highly alkaline waters (200–300 mg/L of alkalinity as $CaCO_3$). The carbon content in the deposit is generally between 10 and 20%, and more than 90% of this carbon has an organic origin. Organic carbon is concentrated on the internal skin of the membrane during filtration. The nature of these organic macromolecules does not depend on the types of waters treated. The organic composition of these "cakes" is about 50% carbohydrates (MW >5000 daltons) and 25% proteins and polyhydroxyaromatic substances. The average organic composition of the waters suggested a great enrichment of proteins and polyphenols in the cake, probably caused by adsorption phenomena in the internal skin of the membrane and in the cake. Carbohydrates generally have low adsorptive affinity with the membrane polymer. However, their high concentration in raw water leads to an important contribution to the permanent fouling of the membrane.

MORPHOLOGICAL STUDY. This study, conducted with SEM and TEM, enabled us to determine the structure of the unused membrane and its initial porosity. The fouled samples, suitably prepared, were observed with SEM at magnifications between ×100 and ×30,000. The membranes, except for the microfiltration membranes, had an internal skin whose porosity seemed to be the first limiting factor of filtration. On this internal skin, a deposit (between 2 and 30 μm thick) showed holes corresponding to the placement of water before the preparation (Figure 8). Structures such as diatoms and other microorganisms were also present, but in very small quantities. We observed some types of small plates with typical inorganic structures. Figure 9 shows kaolinite (clay) plates encased in an organic cement.

With appropriate preparation conditions for TEM, the deposit also showed this plate structure on its external part (Figure 10); near the internal skin, we found some globular structures that could correspond to organic matter or macromolecules.

X-ray microanalysis, which uses an energy-dispersing system with a window before the diode, detected elements by their α- and β-rays (*6*). It provided a general spectrum specific to each sample. For the Bellerive river water, this analysis showed significant quantities of Fe, Al, and Si. It confirmed the presence of clay in the membrane deposit and the absence of carbonates. However, this instrument was not able to give any further quantitative information on organic compounds.

Secondary ion mass spectrometry was applied to fouled membranes. The first example was filtration of the Bellerive water with a poly(ether sulfone) membrane. The plate showing C gives the structure of the mem-

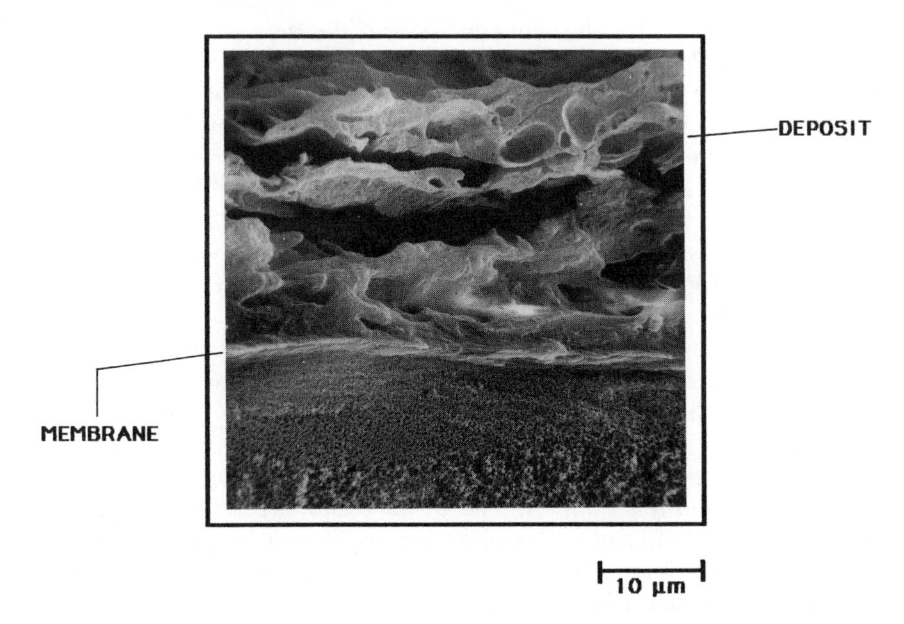

Figure 8. SEM micrograph of a freeze-dried fouled cellulosic membrane.

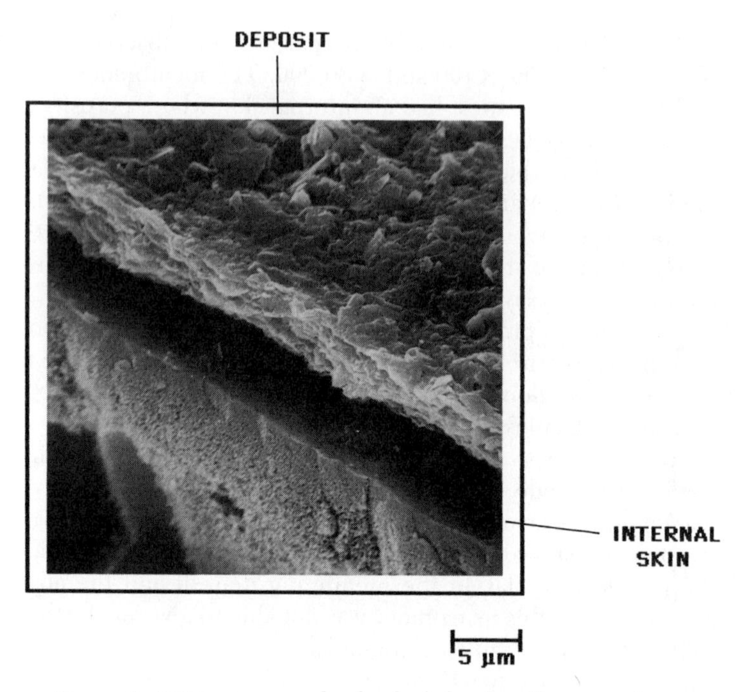

Figure 9. SEM micrograph of a fouled polysulfone membrane.

brane and of the main organic part of the deposit, in conjunction with the N picture (Figure 11). The inorganic part of the deposit is shown in this case by the distribution of Al and Si (Figure 12), indicating clays. A difference of distribution is evident between two aspects of this deposit. The organic substances seem to be packed under the inorganic layer, an observation that could confirm the hypothesis of organic gel formation.

The second example reports on a membrane washed with HNO_3 (0.3%). The plates were taken for C, Fe, Ca (Figure 13), and K. The Fe remains in the deposit, close to the membrane skin. This fact may be explained by precipitation of Fe or complexation with the organic matter. The localization of Ca and K, deep in the thickness of the membrane, indicates that chemical cleaning with HNO_3 can make these ions soluble, and therefore change the physical and chemical properties of the membrane. This solubilization could explain why, in this case, HNO_3 does not lead to any flux recovery.

Finally, the measurements on different plates enable us to estimate the thickness of the deposit between 30 and 50 µm, and to confirm the SEM results on lyophilized samples. The SIMS technique appears to be very efficient in observing and analyzing fouled membrane samples, even in the case of fouling elements in the thickness of the membrane.

Fourier transform IR–attenuated total reflection was conducted with a Bruker IFS 88 and applied on virgin membrane samples as well as on fouled

Figure 10. TEM micrograph showing detail of clay in the deposit.

membrane. Characteristic alkyl-group absorption may be observed for virgin membrane (Figure 14). In the case of 10-day fouled membrane, these peaks are attenuated and characteristic peaks of clays appear in the 1000-cm^{-1} area. The inorganic part of the deposit shows more clearly than the organic part.

Conclusion

The first purpose of this study was to set up an analytical scheme to characterize the foulant matter. We also attempted to determine which polymer is most appropriate to use for filtration with each water we study. Such

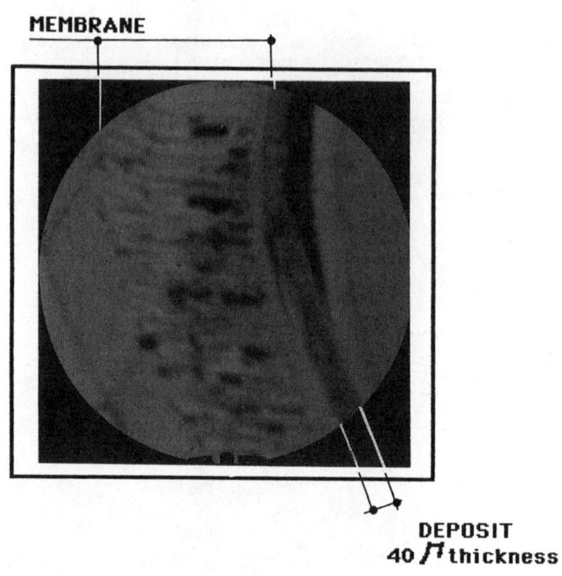

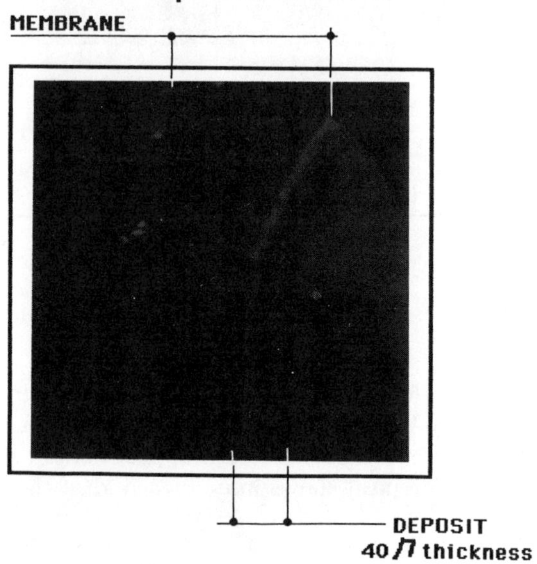

Figure 11. SIMS repartition of carbon and nitrogen in a poly(ether sulfone) membrane and its foulant deposit.

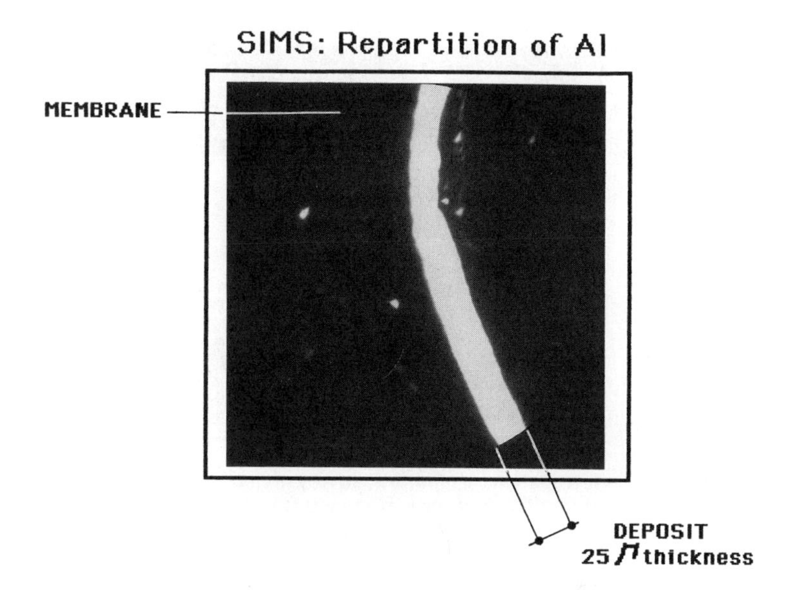

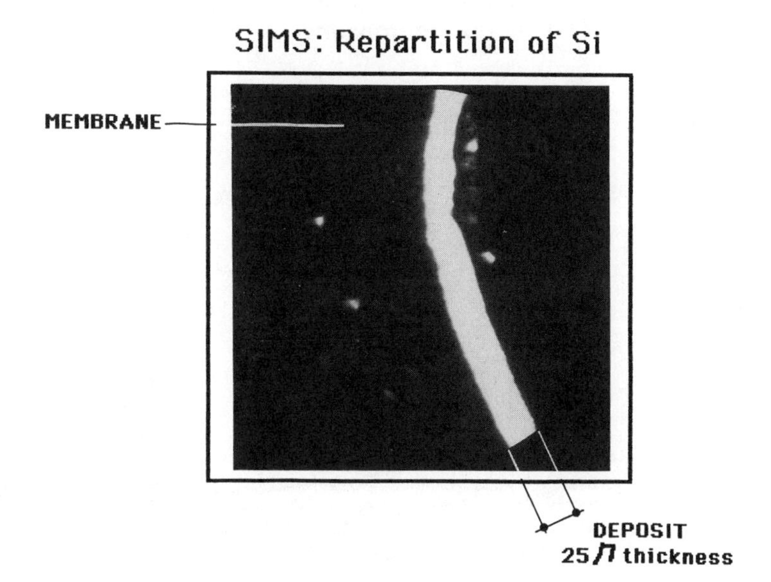

Figure 12. SIMS repartition of aluminum and silicon in a poly(ether sulfone) membrane and its foulant deposit.

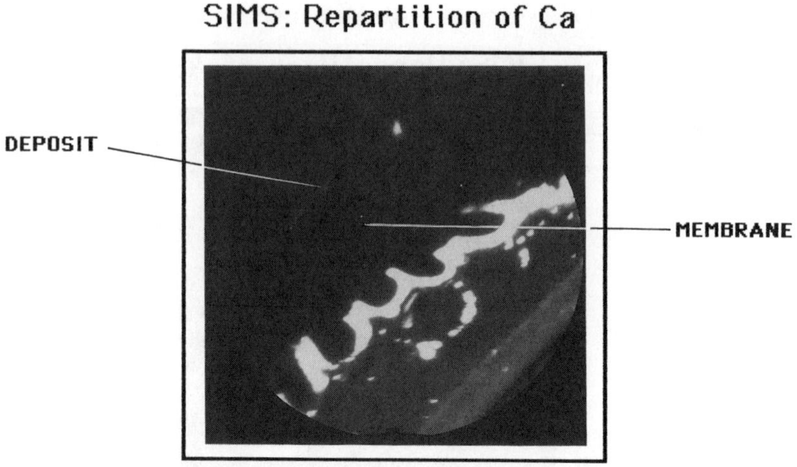

SIMS: Repartition of Ca

DEPOSIT

MEMBRANE

Figure 13. SIMS repartition of calcium in a poly(ether sulfone) membrane washed with HNO_3 (0.3%).

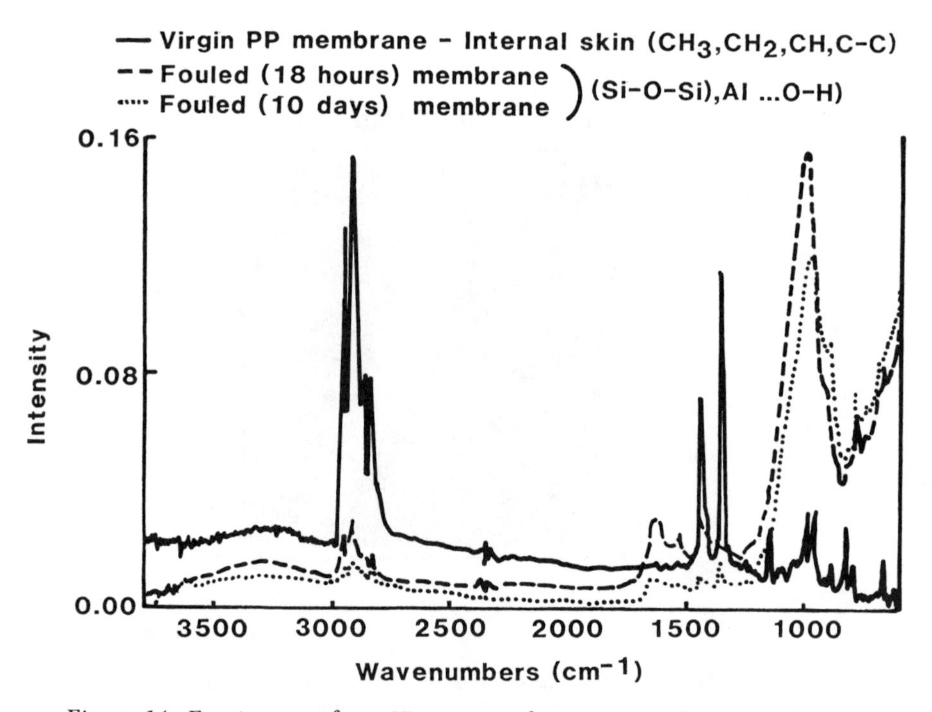

Figure 14. Fourier transform IR spectra of various membranes with total attenuated reflection.

information could suggest combinations of types of water and membrane polymers that would help us to optimize the process. Our results show that fouling could be linked to the organic matrix, and especially to carbohydrates, proteins, and polyhydroxyaromatic compounds. This link has been shown by the analytical procedure used in this study, which effectively couples the balances of organic matter for different effluants to results obtained by observation and microanalysis.

A future goal is to study the fouling mechanism by using the same testing setup and analytical procedure with synthetic solutions. We hope to confirm the contribution to the fouling of each family of each organic compound. The final step of this study would be to implement pretreatment methods for feedwater and dynamic cleaning cycles in order to limit the influence of the fouling in the ultrafiltration process. This work could probably help to minimize the influence of fouling in membrane-filtration processes, or help to select the right chemical cleaning agent to remove the organic substances responsible for permanent fouling.

References

1. Cheryan, M. *Handbook of Ultrafiltration*; Technomic Publishing: Lancaster, Basel, Switzerland, 1986; pp 171–174.
2. Matthiason, E. *J. Membrane Sci.* **1983,** *16,* 23–36.
3. Bruchet, A. Thése de 3éme Cycle, Poitiers University, France, 1985.
4. Bruchet, A.; Anselme, C.; Marsigny, O.; Mallevialle, J. *THM Formation Potential and Organic Content: A New Analytical Approach*; Presented at the Grutte Association conference on humic substances, Rennes, France, October 1986.
5. *Recueil des Normes Francaises. Eaux méthodes d'essais,* 2nd ed.; AFNOR: Paris, France, 1985; ISBN 2–12–179021–7.
6. *Seminaire Microscopie Analytique.* Laboratoire de spectrochimie infrarouge et Raman, Centre National de Recherche Scientifique: Thiais, France, 1986.
7. Truchet, M. *J. Microscopie* **1975,** *24*(1), 1–22.
8. Schnitzer, M.; Khan, S. *Humic Substances in the Environment*; Marcel Dekker, New York, 1972.
9. Thurman, E. M. *Organic Geochemistry of Natural Waters*; Martinus Nijhoff/ Dr. W. Junk: Dorchester, Boston, Lancaster, 1985.
10. Malcolm et al. *Reconnaissance Samplings and Characterization of Aquatic Humic Substances at the Yuma Desalting Test Facility, Arizona*; U.S. Geological Survey: Denver, 1981; p 42; Water Resources Investigations.

RECEIVED for review October 15, 1987. ACCEPTED for publication May 18, 1988.

42

Removal of Humic Substances by Membrane Processes

Hallvard Ødegaard and Thor Thorsen

Norwegian Institute of Technology, Division of Hydraulic and Sanitary Engineering, N–7034 Trondheim–NTH, Norway

Both laboratory and pilot-plant experiments have been carried out to evaluate the use of membrane processes for the removal of humic substances. These processes are competitive for small waterworks with high raw-water color. Cellulose acetate membranes with a molecular weight (MW) cutoff of 800–1000 may be used favorably at a pressure of 7–10 bars. The capacity of the membrane will be reduced, even when optimal membrane washing is performed. The washing solution should be citric acid, sodium citrate, and sodium alkylaryl sulfonate in the proportions given in this chapter. The long-term capacity of the spiral-wound cellulose acetate membrane was found to be 25 L/m²·h at optimal membrane washing. The lifetime of a membrane at this capacity is estimated as 4 years.

MEMBRANE FILTRATION HAS NOT BEEN USED yet in full-scale waterworks for the prime objective of removing humic substances. Membrane filtration is well-known, however, from the analytical practice of fractionating humic substances. In existing drinking-water plants that use reverse osmosis, humic substances in raw water are often considered a nuisance because of their tendency to clog the membranes. These plants are not specifically designed for removing humic substances.

This chapter summarizes our findings with respect to the use of membrane processes in a research program on the removal of humic substances at small Norwegian waterworks (*1*). (Chapter 45 summarizes our findings with the use of macroporous anionic resins.) The philosophy behind our research was that, because humic molecules are so big, the use of open

0065–2393/89/0219–0769$06.00/0

membranes and low pressures might make this traditionally expensive water-treatment method economically competitive with traditional humic-substance removal techniques. Our aim was therefore to evaluate this process, to recommend operational guidelines, and to give design criteria for typical Norwegian surface waters, which are high in color but low in turbidity.

Several experiments have been performed, both short-term laboratory-scale and long-term pilot-scale (*2–5*). Only the major experiences will be given here. Because the results of the laboratory-scale experiments are presented in more detail elsewhere (*3*), this chapter will concentrate primarily on the long-term experiments.

Laboratory-Scale Experiments

Experimental Methods. The raw water used in the laboratory-scale experiments typically had a raw-water color of 60–70 mg of Pt/L, permanganate number of 6–8 mg of O_2/L, conductivity of 45–55 μS/cm, and iron concentration of 0.08–0.12 mg/L. It was soft humic water of a type commonly found in Norway.

The experiments were performed in laboratory reverse osmosis units as shown in Table I. Several cellulose acetate membranes were tested with respect to treatment efficiency in terms of color, permanganate number, conductivity, and specific flux (capacity per bar of operating pressure).

Discussion. Results from the laboratory-scale experiments are summarized in Table II. A great deal of color could be removed (>80%) even with very open membranes (MW cutoff = 3000). In order to have the same kind of removal of organic matter in terms of permanganate number, membranes that were less open had to be used.

Table II shows that treatment efficiency increased and specific flux decreased when the MW cutoff was lowered. However, systematic relationships between the parameters could not be derived from the data. For a given membrane, the pressure did not seem to have any impact on treatment efficiency. Moreover, the membrane flux was not significantly influenced by the raw-water humic-substance concentration.

When both treatment efficiency and flux were taken into consideration, it was concluded that low humic-substance concentration (5 mg of Pt/L) in treated water could be achieved with cellulose acetate membranes with a MW cutoff in the range of 500–2000 operated at a pressure of 7–15 bars.

Table I. Laboratory Units Used in Laboratory Experiments

Manufacturer	*Type*	*Membrane Surface Area (m^2)*	*Pressure (bar)*	*Module Type*
DDS	20–laboratory	0.36	0–80	plate and frame
Osmonics	519–SB	0.48	0–15	spiral
PCI	BRD MK 2	0.10	0–80	tubular

Table II. Treatment Efficiency and Specific Capacity of 13 Membranes Tested in Laboratory-Scale Experiments

Membrane	*Molecular Weight Cutoff*	*Treatment Efficiency (%)*			*Capacity*	*Operating Pressure (bar)*	
		Color	*Permanganate*	*Conductivity*	*(L/m²·h·bar)*	*Experimental*	*Recommended Value*
SEPA–20KCA	20,000	80	—	—	—	3.6	—
SEPA–OPS	2,000	83	—	—	14	3.6	15
SEPA–OCA	1,000	85	60	17	7	7–15	15
SEPA–50CA	600	98	80	65	2.2	11	21
SEPA–89CA	400	100	90	82	1.6	11–15	43
SEPA–97CA	200	97	98	85	1.0	7–15	60
DDS–600	20,000	70	58	13	21	5–10	10
DDS–800	6,000	75	66	18	10	10–20	20
DDS–865	500	95	90	60	3.1	10–50	40
DDS–870	500	100	96	78	2.1	10–50	50
PCI–T4A	3,000	80	67	23	6.3	5–10	10
PCI–T2A	800	100	95	44	4.3	10–20	25
PCI–T2/15N	300	100	97	92	0.9	10–40	80

SOURCE: Reproduced with permission from ref. 1. Copyright 1986 Pergamon.

Long-Term Pilot-Scale Experiments

The laboratory-scale investigations gave indications of a considerable flux drop after some time of operation. The flux could, to a certain extent, be regained by washing the membranes, but a residual flux drop of about 20% during 170 h of operation was experienced even when the membranes were washed according to manufacturers' specifications.

Because of the promising treatment results obtained during the laboratory-scale investigation, we built a comprehensive pilot plant to study the process further, especially the long-term effects of membrane washing and flux reduction.

Experimental Methods. The pilot plant shown in Figure 1 actually consisted of three separate plants: A, B, and C. Plant A was equipped for recirculation of the concentrate and allowed additions in the recirculation tank. This plant was primarily used to evaluate the efficiency of bacteria and virus removal. Membranes were frequently changed. Plant B was planned for long-term operation and was not changed significantly during the experiments. Experiments for the evaluation of treatment efficiency and flux of various membranes (MW cutoff in the range of 1,000–20,000) were performed in Plant C. The water was pretreated in automatically backflushed bag filters with nominal light-openings in the range of 1–20 μm.

This chapter concentrates on the results of the long-term experiments in Plant B, where membranes with MW cutoffs of 800 and 1000 were installed (OSMONICS, 25 CA and OCA).

Discussion. Soon after start-up the capacity of the continuously operated plant sank more rapidly than in the laboratory experiments, and the pressure loss through the plant increased. Both of these changes are indications of membrane fouling (film formation). The thickness of the film was calculated on the basis of flow channel geometry and measured pressure loss.

The basis for calculation, data on pressure drop versus bulk flow through one module, was supplied by the membrane manufacturer. The channel height between the membranes and the dimensions of the turbulence promoter wire were known. According to the laws of fluid flow, pressure drop will increase when the channel height decreases because of fouling, and the linear velocity will thus increase. The film thickness model based on these facts included total pressure drop through the whole plant, feed flow, concentrate flow, and permeate flow from each module in a series of 10 modules. By assuming constant layer thickness in each module, a mean layer thickness could be calculated because the mean flow in each module and total pressure drop were known.

In spite of washing procedures as recommended by the membrane manufacturer, the calculated thickness of the film increased almost linearly with time, up to about 100 μm after about 1900 h of operation (*see* Figure 2).

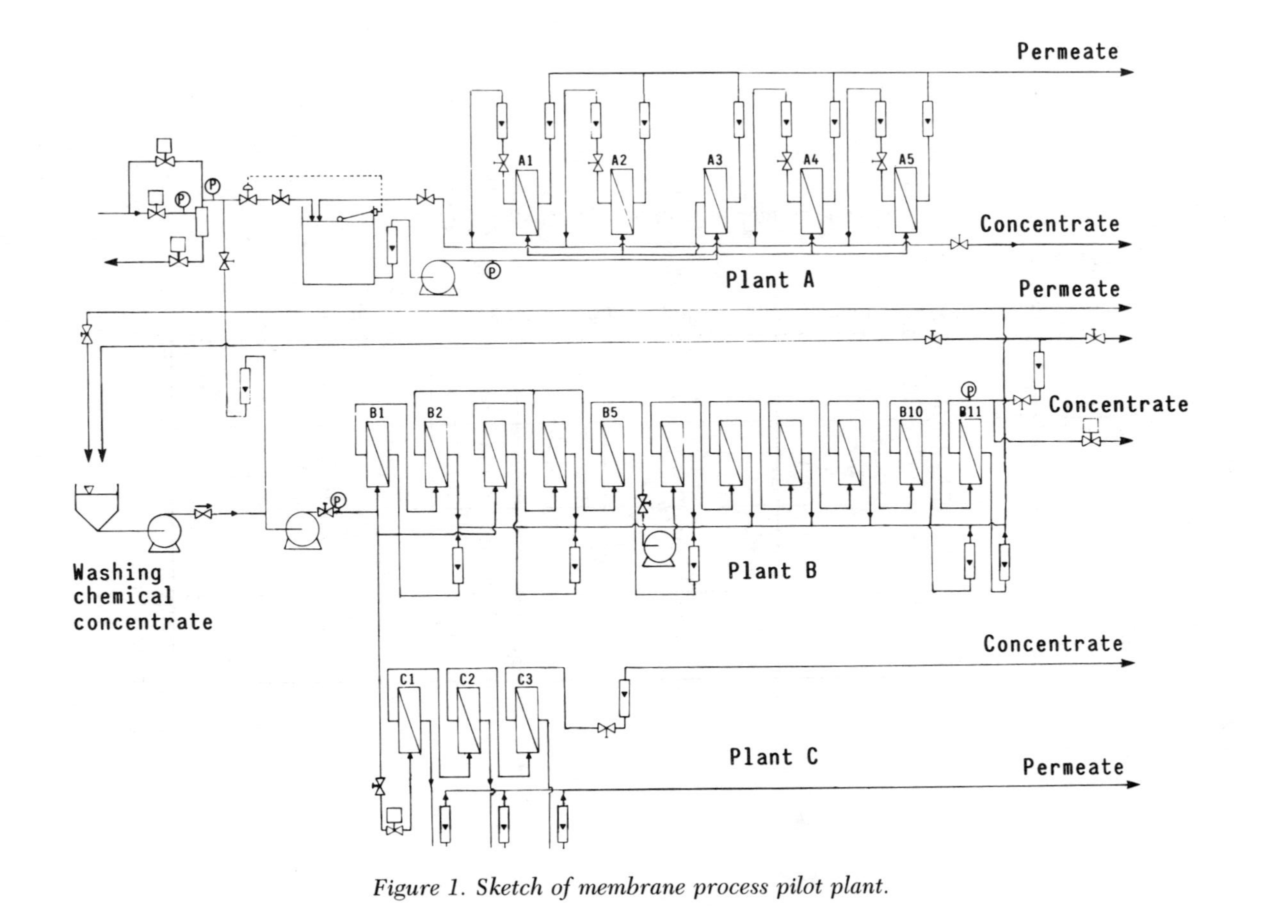

Figure 1. Sketch of membrane process pilot plant.

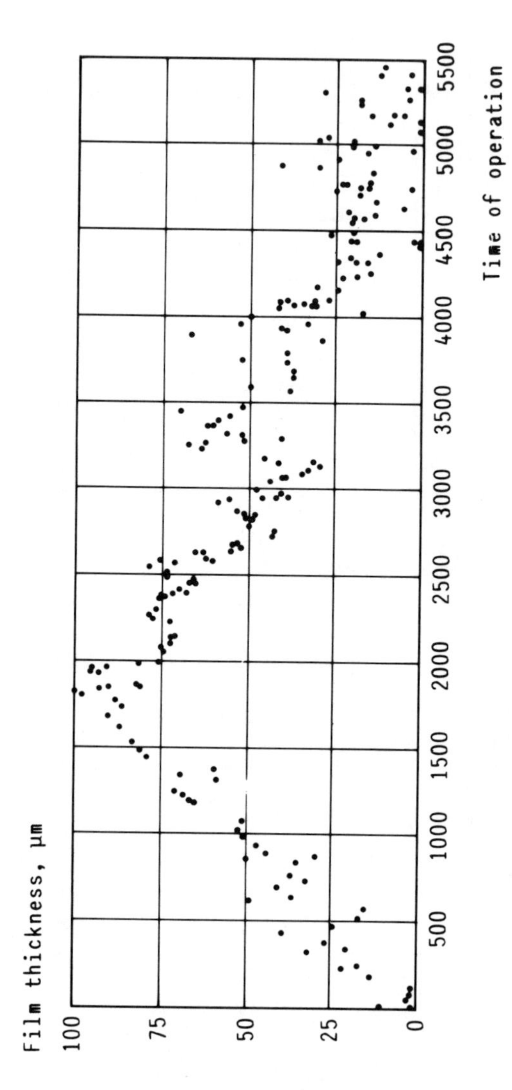

Figure 2. Development of film thickness on membrane.

At this point some membranes were replaced, and the film formed on the membrane was investigated. The thickness of the film was indeed of the same magnitude as calculated. The film was found to be soft, dark brown, and loosely connected to the surface of the membrane. Several experiments with a broad range of washing solutions were performed with sheets of fouled membranes.

Citrate was effective in terms of regaining capacity, but citrate alone could not prevent film formation over a longer period. An anionic detergent and an optimal pH were needed in addition to citrate. The most effective detergent was found to be sodium lauryl sulfate. This detergent is, however, not chemically stable at the optimal pH for routine wash (pH 3.4–3.7). An alternative detergent, stable at this pH, is sodium alkylaryl sulfonate.

Routine wash performed with only one washing solution leaves a residual capacity reduction. Therefore, two different washing solutions are recommended. The primary solution for routine wash should maintain stable short-term capacity, and the secondary solution should be used only now and then to maintain acceptable long-term capacity. We recommend the washing routine shown in Table III.

When we implemented the washing procedure (at 2000 h), the membrane film formation problem improved considerably. After 4500 h of operation, the film thickness was only about 10 μm. A certain capacity reduction was still experienced, though much smaller.

It was concluded that the flux reduction experienced in membrane filtration of humic substances could be divided into two categories. Temporary flux reduction is recoverable by use of a proper washing routine. Permanent flux reduction, primarily as a result of a different sort of scaling, is not removable by washing. However, it might also be due to some membrane compaction.

Figure 3, based on the total experience from the experiments, shows the mean capacity for properly washed cellulose acetate membranes at 10 bars of operating pressure.

The greatest flux reduction is experienced during the first 2 years of the membrane life, and very limited capacity reduction occurs during the next 2 years.

Reduction in long-term flux is heavily dependent upon the absolute flux ($L/m^2 \cdot h$). Therefore, membranes with different initial fluxes gradually approach each other in terms of flux.

Figure 3 also shows that temperature influences the capacity and indicates that capacity may be highest during summer. Likewise, capacity will be high after membrane replacement. In order to obtain smooth operation, 25% of the membranes should be replaced every year. The replacement of membranes should be carried out in a period of falling raw-water temperature (during autumn). When the variation in temperature from 2 to 14 °C is taken into account, the design capacity for a plant treating water from the source

Table III. Recommended Washing Routine

Wash	*Frequency*	*Composition, Washing Solution*[a]	*Solution Concentration (%)*	*Duration of Wash (h)*
Primary wash	2–7 times per week	2–3 parts citric acid 1 part sodium citrate 2 parts sodium alkylaryl sulfonate	0.3–1	1–2
Secondary wash	5–50 times per year (seasonal variation)	2 parts sodium citrate 1 part sodium lauryl sulfate	0.5–1	1–10

[a]Temperature of washing solution: 25–30 °C.
SOURCE: Reproduced with permission from ref. 1. Copyright 1986 Pergamon.

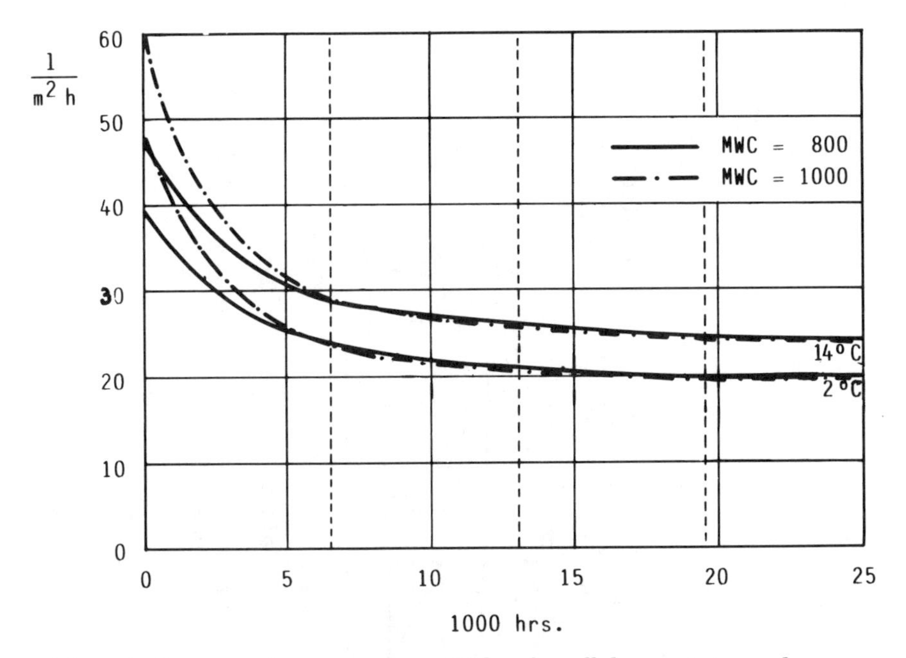

Figure 3. Mean membrane capacity at 10 bar for cellulose acetate membranes with a MW cutoff of 800–1000 separating humic substances at 2 and 14 °C. (Reproduced with permission from ref. 1. Copyright 1986 Pergamon.)

used in our experiments would be 24–26 $L/m^2 \cdot h$, with 3–4 years of membrane life.

On the basis of our experiences, we recommend that a reverse osmosis plant for the removal of humic substances be designed according to Figure 4. Specifications for the plant would be according to Table IV, and for the washing routine according to Tables III and V.

Costs

Estimating the cost of reverse osmosis for the removal of humic substances is very difficult. Because the cost will vary from one country to the other, we shall primarily estimate the cost of this process for Norwegian conditions, relative to other alternative processes.

In Figure 5 the investment costs are shown in Norwegian kroner [$1 (U.S.) = 6.65 Nkr] for three levels of raw water color and three levels of plant size. Figure 6 illustrates the corresponding unit cost (Nkr/m^3) (*6*).

Reverse osmosis may be economically competitive at very small plants and high raw-water colors. Membrane processes are particularly attractive at very high colors, because neither the investment cost nor the operating

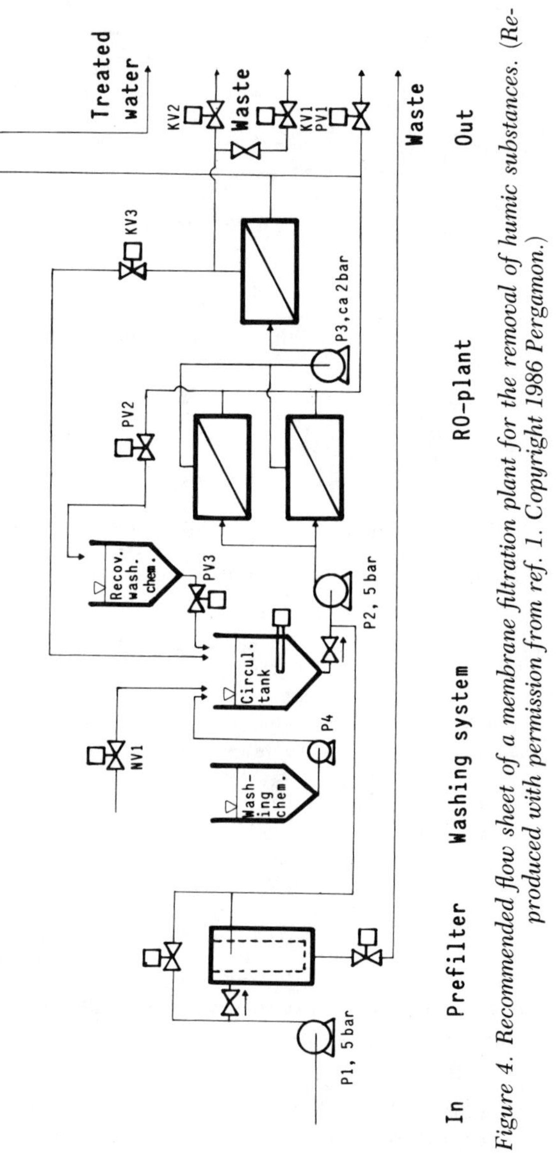

Figure 4. Recommended flow sheet of a membrane filtration plant for the removal of humic substances. (Reproduced with permission from ref. 1. Copyright 1986 Pergamon.)

Table IV. Specifications of a Membrane Filtration Plant for the Removal of Humic Substances

Component	*Measure*
Prefilter	Sieve opening: 20–50 μm
Module	Spiral wound
Membrane	Cellulose acetate
Molecular weight cutoff	800–1000
Pressure	7–10 bar
Membrane replacement	25% annually
Washing solution	According to Table III
Temperature of washing solution	25–30 °C

SOURCE: Reproduced with permission from ref. 1. Copyright 1986 Pergamon.

Table V. Washing Routine for the Primary Wash

Sequence	*Duration (min)*	*Pump/Valve in Operation*
Pumping in	4	P2, P3, KV2, PV1
Pause	6 × 10[a]	PV2, KV3
Circulation	6 × 5[a]	P2, P3, KV3, PV2
Pumping out	4	P2, P3, KV2, PV1
Flushing	10	P1, P3, KV2, PV1
Afterfilling	1	P4, NV1, PV3

NOTE: Data are according to the flow sheet in Figure 4.
[a]Repeated 4–8 times.
SOURCE: Reproduced with permission from ref. 1. Copyright 1986 Pergamon.

cost is very dependent upon raw-water concentration, which is the case for the other processes.

Summary

Membrane filtration is an interesting alternative for the removal of humic substances in small waterworks when the raw-water concentration is very high. Cellulose acetate membranes with a MW cutoff of 800–1000 may be used effectively at an operating pressure of 7–10 bars. The film forming on the membrane consists of a loosely connected layer readily removable by proper washing and a nonremovable compressible layer that will cause capacity reduction even with optimal membrane washing.

At the design capacity and washing routine recommended here, 25% of the membranes in a plant should be replaced every year. The washing solution should be made up of citric acid, sodium citrate, and sodium alkylaryl sulfonate in the proportions recommended in this chapter. When the washing routine and replacement follow this schedule, the long-term design capacity of the membranes can be set at 25 $L/m^2{\cdot}h$.

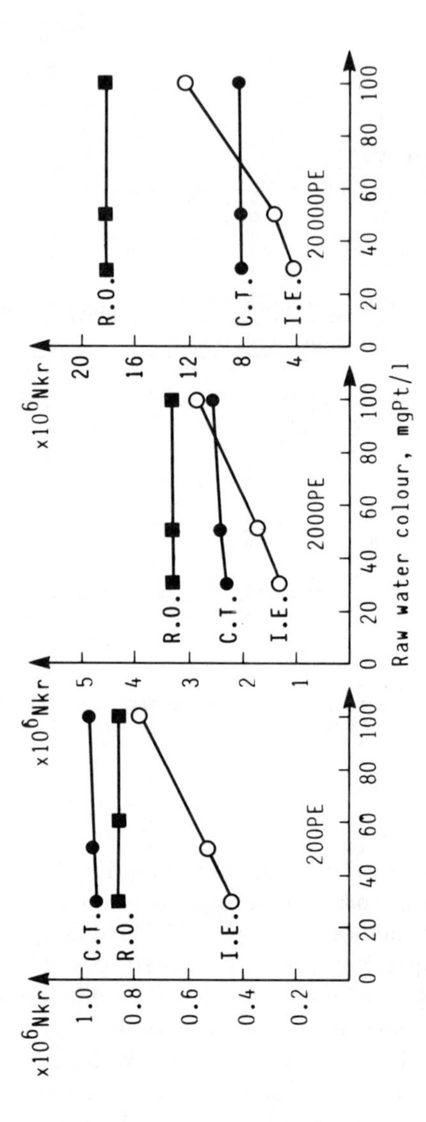

Figure 5. Investment cost of reverse osmosis (RO) as compared to conventional treatment (CT) and ion exchange (IE) for the removal of humic substances.

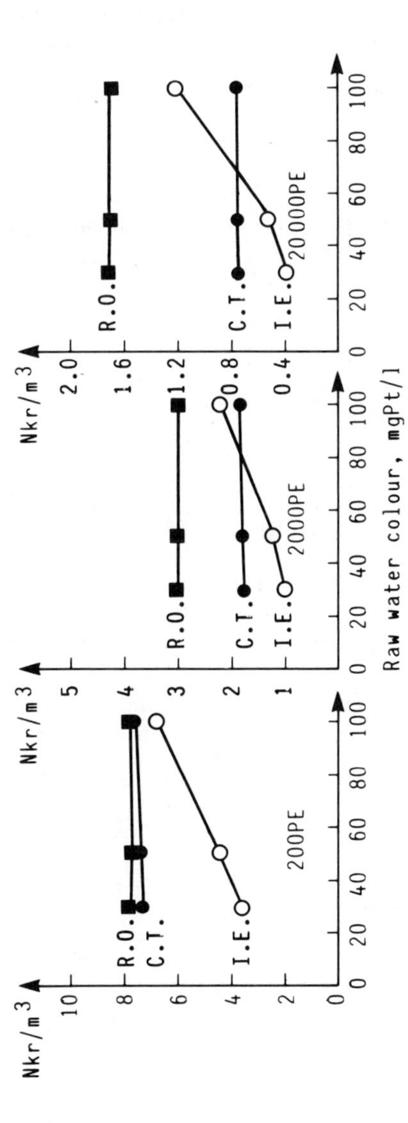

Figure 6. Unit cost of reverse osmosis (RO) as compared to conventional treatment (CT) and ion exchange (IE) for the removal of humic substances.

References

1. Ødegaard, H.; Brattebø, H.; Eikebrokk, B.; Thorsen, T. *Water Supply* **1986,** *4,* 129–158.
2. Koottatep, S. Dr. ing. Dissertation, Norwegian Institute of Technology, 1979.
3. Ødegaard, H.; Koottatep, S. *Water Res.* **1982,** *16,* 613–620.
4. Thorsen, T. SINTEF report STF 21 A 84071. Trondheim, Norway, 1984 (in Norwegian).
5. Thorsen, T. SINTEF report STF 21 A 84094. Trondheim, Norway, 1984 (in Norwegian).
6. Hem, L. J. SINTEF report STF 60 A 86161. Norwegian Institute of Technology, 1986 (in Norwegian).

RECEIVED for review July 24, 1987. ACCEPTED for publication February 11, 1988.

43

Dissolved Organic Components in Process Water at the Los Banos Desalting Facility

Julius Glater, Lawrence C. Wilson, and Johannes B. Neethling

School of Engineering and Applied Science, University of California, Los Angeles, CA 90024

Agricultural wastewater reclamation is under serious consideration in the California Central San Joaquin Valley. A reverse osmosis (RO) pilot plant at Los Banos was designed with an elaborate pretreatment system to protect membranes from fouling and degradation. This chapter reports a surveillance of dissolved organic materials in process water throughout this plant. Samples were analyzed by gas chromatography–mass spectrometry (GC–MS) with standard extraction techniques. Untreated water contains a highly complex mixture of essentially nonvolatile organic substances. Following chlorination, the mixture increases in complexity and includes three trihalomethane (THM) fractions, $CHCl_2Br$, $CHClBr_2$, and $CHBr_3$. Bromoform was present at the highest concentration and chloroform, if present, was below the GC detection limit. Dissolved organic concentrations were observed to decline progressively following clarification and filtration. RO membranes tested at Los Banos did not significantly reject THMs reported in this study.

CONVENTIONAL PRETREATMENT for reverse osmosis membrane plants involves unit processes for removal of suspended solids, pH adjustment, control of scale, and suppression of biological fouling. These strategies are generally adequate for most brackish and sea-water applications, but other factors must be considered with feedwaters of greater chemical complexity.

0065–2393/89/0219–0783$06.00/0

Such a situation arises in the treatment of wastewaters containing high levels of dissolved organic compounds. The present study is concerned specifically with reclamation of agricultural drainage water in the San Joaquin Valley of California. Desalination of this type of water by reverse osmosis (RO) technology has been under serious study since 1971. Pilot plants operated at Firebaugh, California (*1*); Yuma, Arizona (*2*); and Los Banos, California (*3*) have provided useful data, but a variety of technical problems remain to be solved. One of the principal issues involves dissolved organic substances in feedwater and products resulting from the reaction of these compounds with the chlorine used to control biological fouling.

Studies at the Yuma Desalting Facility (*2*) suggest that cellulose acetate membrane performance may be adversely affected by the presence of various dissolved organic compounds in process water. Unusually high rates of product flux decline were observed during the 1980 proof testing, amounting to approximately 2–4% per 1000 h of operation. Continuation of this process was estimated to shorten membrane life by more than 50%.

According to Moody et al. (*2*), organic compounds causing flux decline were classified as purgeable and nonpurgeable. The first category designates low-molecular-weight halocarbons, and the second includes all other dissolved organic materials. Following those observations at Yuma, a laboratory study reported by Milstead and Riley (*4*) showed that both bromoform and chloroform cause flux decline in cellulose acetate membranes. These experiments were conducted in a synthetic Yuma feedwater at an organic concentration approximately 1000 times the ambient level.

This chapter will summarize our efforts in sampling and characterization of dissolved organic components at the Los Banos Desalting Facility. The fate of feedwater organic substances through the process flow train will be reported. Figure 1 shows the schematic process flowsheet of the Los Banos Desalting Facility. Pretreatment included chlorination, clarification, dual-media filtration, and ion exchange. Dechlorination, threshold treatment, and acid injection were also applied just prior to RO processing. The desalting unit consisted of three RO installations, representing a wide variety of membrane types and configurations.

Feedwater for the Los Banos Desalting Facility results primarily from subsurface tile drainage of the Westlands Irrigation District. Composite drainage is collected in the San Luis drain. Routine analyses of this water over several years show total dissolved solids (TDS) values averaging approximately 9600 mg/L (*5*). Dissolved organic carbon (DOC) levels have also been measured and show average values of approximately 7.75 mg/L. These levels are more than twice those reported at the Yuma Desalting Facility (*6*). A wide variety of dissolved organic compounds have also been identified at Yuma. The principal compounds are humic substances, which make up more than 25% of the DOC. The high DOC values reported at Los Banos suggest that this water may contain even greater concentrations

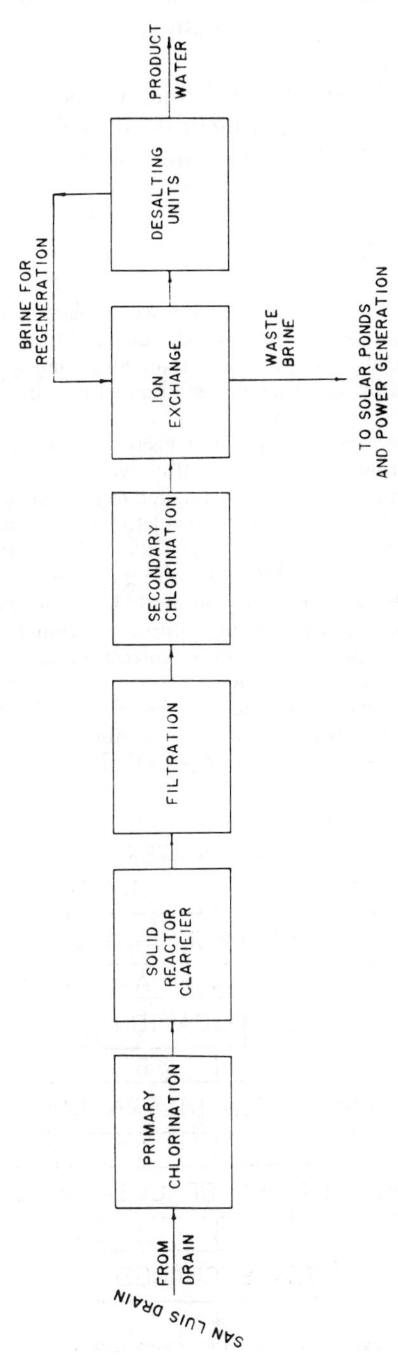

Figure 1. Process flow sheet of the Los Banos Desalting Facility.

of similar substances. The work presented here represents the first effort to characterize organic compounds and their chlorination products in San Joaquin Valley agricultural drainage water.

Analytical data derived from this study will be useful for planning RO product water management and brine disposal. It should also help to establish guidelines for future studies of membrane–chemical interaction.

Experimental Details

All wastewater samples reported in this chapter were collected at Los Banos between May and August 1986. Samples were taken directly from the San Luis drain and at four additional points in the pretreatment train identified as A, B, C, and D on Figure 2. Limited sampling was also conducted on various fractions of RO permeate and brine.

Two types of samples were obtained at each location. The first was collected in 120-mL bottles at ambient pH. The bottles were fitted with poly(tetrafluoroethylene) (Teflon) airtight caps, and care was taken to exclude all air from the system. These samples were scheduled for isolation of volatile organic compounds via pentane extraction. The second type of sample was collected in 1-gal bottles containing 4.0 mL of concentrated sulfuric acid. The resulting diluted sample pH was less than 2.0. These samples would be extracted with methylene chloride and used to detect acidic and neutral organic species over the entire gas chromatographable range. The two types of samples will be referred to as volatile organic substances and acid-extractable organic substances, respectively.

Organic separations were conducted by gas chromatography (GC) with a gas chromatograph (Varian Vista 6000) using a fused silica column (30 m × 0.32 mm i.d.) packed with a polar stationary phase of polyethylene glycol. Organic substances

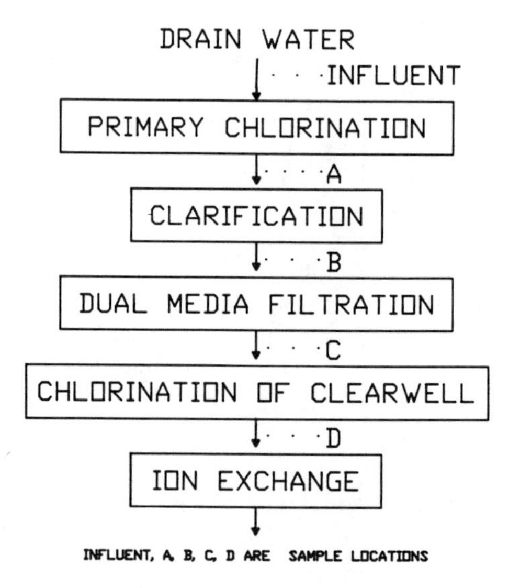

Figure 2. Sampling locations in pretreatment train.

eluted from the column were split into parallel streams and detected with both flame ionization detectors and electron-capture detectors (FID and ECD, respectively). An integrator (Hewlett Packard 3392) recorded the detector signals. The FID is generally responsive to organic compounds, regardless of functionality. The ECD is selective to halogenated species over a wide molecular weight range. ECD sensitivity to halogenated organic substances is considerably greater than that of the FID detector.

Samples containing volatile organic materials were concentrated by pentane extraction in an airtight system as described by Glaze (*7*). Extract aliquots were injected into the GC, which was programmed as follows: held at 50 °C for 4 min; temperature increased at 10 °C per min up to 140 °C; held for 2 min.

Acid-extractable organic compounds were concentrated by methylene chloride extraction as described by Fam (*8*). Aliquots of the final extract were injected into the GC, which was programmed as follows: held at 65 °C for 4 min; temperature increased at 4 °C per min up to 250 °C; held for 30–35 min.

The acid extraction technique was used for the Los Banos sample because prior work (*8, 9*) indicated that agricultural and domestic wastewater contain relatively large concentrations of acidic organic compounds, such as phenols and carboxylic acids. These species and most of their chlorination products are completely protonated at pH 2.0 and readily extractable in methylene chloride.

The methylene chloride extraction technique is capable of isolating compounds over a wide molecular weight range. This range includes trihalomethanes (THMs) and other volatile organic substances. Certain disadvantages, however, must be considered when attempting to quantify volatile organic materials. First, some volatile compounds may be lost between sampling and extraction. Second, solvent interference is usually a problem in the volatile GC region. This situation results from high sensitivity of the ECD to chlorinated organic species. Impurities in the methylene chloride solvent are often detected by the ECD and thus interfere with volatile organic peaks.

As a result of these complications, we chose pentane extraction for analysis of volatile organic compounds. Our confidence in this technique is greater because minimal sample losses occur in airtight containers. In addition, the ECD is unresponsive to the pentane extraction solvent.

Volatile sample fractions were examined by gas chromatography–mass spectrometry (GC–MS) in an attempt to characterize specific compounds present in these samples. The most predominent compounds isolated corresponded to mixed chloro and bromo trihalomethanes (THMs). MS determinations were conducted with a GC–MS (Finnigan 4000) and data system (INCOS 2300).

THM concentrations were quantified by comparing ECD-generated areas of samples with areas obtained from known THM standards. ECD areas are essentially linear with respect to concentration for the range reported in this chapter. Standards were prepared in water to give fixed concentration levels according to procedures described by the U.S. Environmental Protection Agency (*10*). These standards were then pentane-extracted and analyzed in the same manner as our wastewater samples. This method was reproducible for replicate samples to approximately 5%.

Results and Discussion

San Luis drain water sampled at the Los Banos Desalting Facility shows average pH and TDS values of 8.2 and 9600 mg/L, respectively. This water is a sodium chloride type, but contains unusually high levels of calcium and sulfate ions. As a result, an elaborate ion-exchange system was used in the pretreatment train to prevent calcium sulfate scale deposition on RO membrane surfaces. The schematic process flowsheet is given in Figure 2. Points labeled A, B, C, and D are sample locations throughout the pretreatment train prior to ion exchange.

Figure 3 shows gas chromatograms of water samples collected at the labeled sample locations. Influent samples consist of untreated San Luis

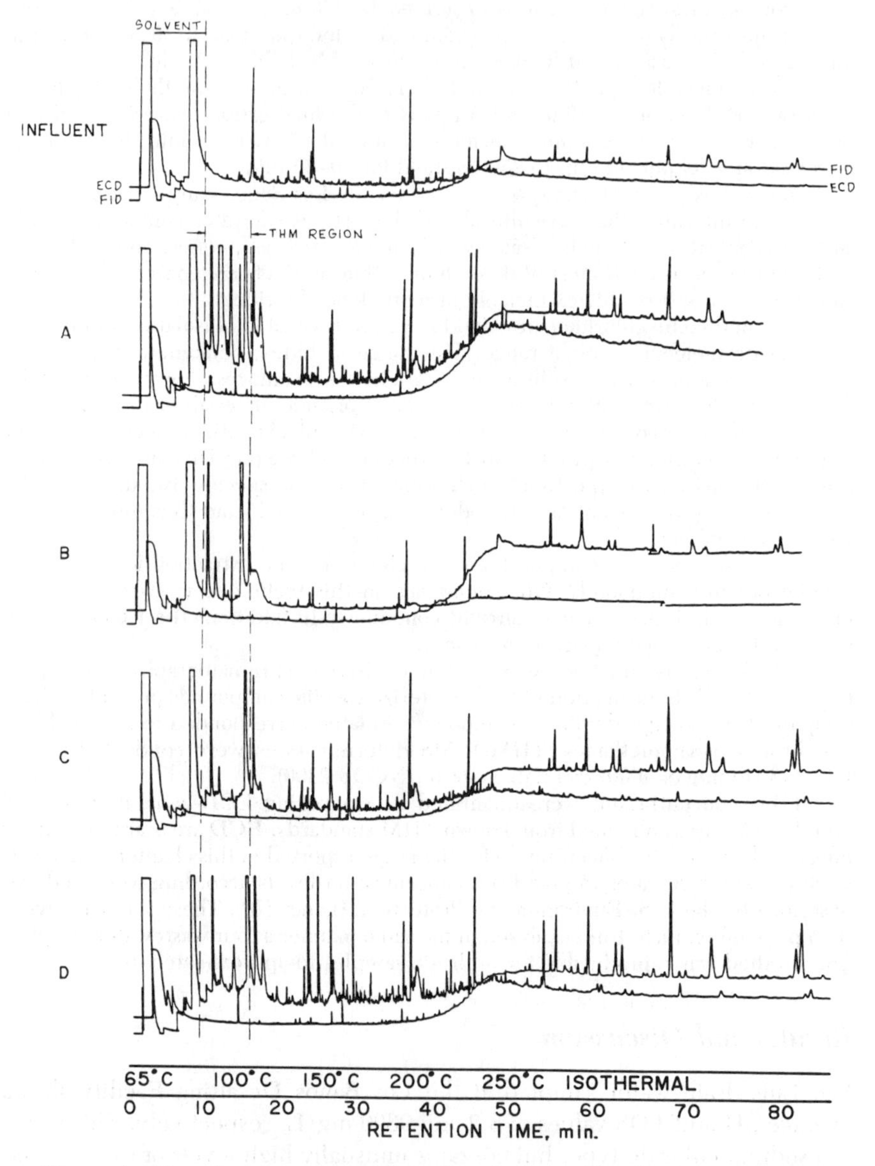

Figure 3. Gas chromatograms of acid methylene chloride extracted water samples collected at locations shown in Figure 2.

Drain water. The horizontal axis on each trace represents a temperature profile increasing at the rate of 4 °C/min. This axis also indicates species retention time within the column. Peak heights and, more specifically, areas under peaks are related to relative concentrations of sample species. The two lines on each trace correspond to compounds detected by FID and ECD detectors. Considerably more detail is shown in the ECD trace because of its greater sensitivity to the halogenated organic compounds that appear in large concentrations following chlorination. Chromatograms shown in Figure 3 were generated from acidified methylene chloride extracted samples and represent both volatile and nonvolatile organic compounds. A wide variety of organic species with molecular weights up to approximately 400 AMU are detected.

Figure 3 indicates that untreated influent drainwater samples contain few GC-detectable organic substances. This condition is especially true of THMs, which show up clearly on traces A–D. Most GC-detectable organic compounds found in these samples are produced from primary chlorination of canal water. The greatest increase occurs in the THM fraction. In addition, a wide variety of both chlorinated and unchlorinated organic products show up on these chromatograms. To date, we have not attempted to positively identify any of these species except for THMs.

Most of the low-molecular-weight organic materials result from cleavage of humic substances present in agricultural drainage water. The humic substances, consisting primarily of fulvic acids, are composed of macromolecules containing a variety of aromatic ring structures. Most prominent among these are carboxylic acids, ketones, ethers, and phenols. The mechanism of THM production has been studied by various investigators (*11–13*) and is believed to involve cleavage of phenol and resorcinol moieties in the presence of chlorine.

Morris and Baum showed (*14*) that methyl ketones are degraded to THMs via the classical haloform reaction. This reaction is not the only mechanism; a variety of reaction schemes may describe oxidative degradation of humic substances following chlorination. In addition to THMs, Christman (*15*) has characterized approximately 50 chlorinated and unchlorinated low- and intermediate-molecular-weight organic compounds derived from fulvic acid chlorination. According to Christman, less than 15% of the organic products are detectable by gas chromatography.

When bromide ion is present in wastewaters, the resulting THMs may contain bromine alone or consist of mixed chloro and bromo compounds. The following reaction sequence is involved.

$$Cl_2 + H_2O \rightarrow HCl + HOCl \quad (1)$$

$$HOCl + Br^- \rightarrow HOBr + Cl^- \quad (2)$$

$$HOCl + HOBr + \text{fulvic acids} \rightarrow \text{chloro} + \text{bromo THMs} \quad (3)$$

The shift toward more highly brominated THMs in Los Banos process water can be explained by the relatively high bromide ion concentration in the San Luis drain (approximately 0.3 mg/L).

In addition, reactions 1–3 are all known to be pH-dependent. For Los Banos water (average pH 8.2), however, reaction 2 is known to be considerably faster than reaction 3. This situation results in a higher concentration of HOBr available for reaction with humic substances.

Figure 4 represents a portion of a typical gas chromatogram for pentane-extracted volatile organic components. The sample was collected following primary chlorination at point A on Figure 2. The three major peaks (II, III, and IV) were scanned by MS and identified as mixed chloro–bromo THMs. Sample identity and relative concentrations are given in Table I. A peak for chloroform ($CHCl_3$) does not show in Figure 4 because of extremely low concentration.

The total THM concentration (Table I) is approximately 122 μg/L, which represents only a very small fraction of the DOC value of San Luis drain water. Evidently a large portion of the dissolved organic products do not react with chlorine or are converted to other types of compounds. According to the literature (*15*), these compounds may be both halogenated and un-

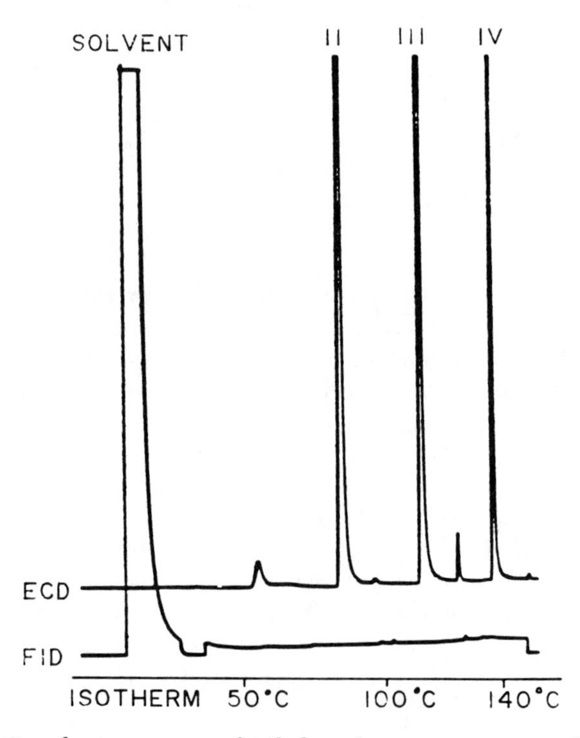

Figure 4. Gas chromatograms of trihalomethanes in water sample following primary chlorination.

halogenated products of oxidative cleavage. We are now attempting to determine what portion of the organic components are halogenated and to identify other prominent compounds in the gas chromatographable region.

We have also determined the total acid-extractable organic concentrations at various locations in the Los Banos pretreatment train. Values of total ECD-integrated areas at sample locations A–D relative to the influent sample area are shown in Figure 5. A significant increase, greater than 400%, occurs

Table I. Comparison of Reverse Osmosis Influent THM Concentrations at Los Banos and Yuma Desalting Facilities

	Los Banos		*Yuma*[a]	
THM	*Concentration (μg/L)*	*Percent of Total*	*Concentration (μg/L)*	*Percent of Total*
$CHCl_3$	—	0	4	4
$CHCl_2Br$	25	21	14	14
$CHClBr_2$	43	35	40	41
$CHBr_3$	54	44	40	41
Total	122	100	98	100

[a]Data are from ref. 6.

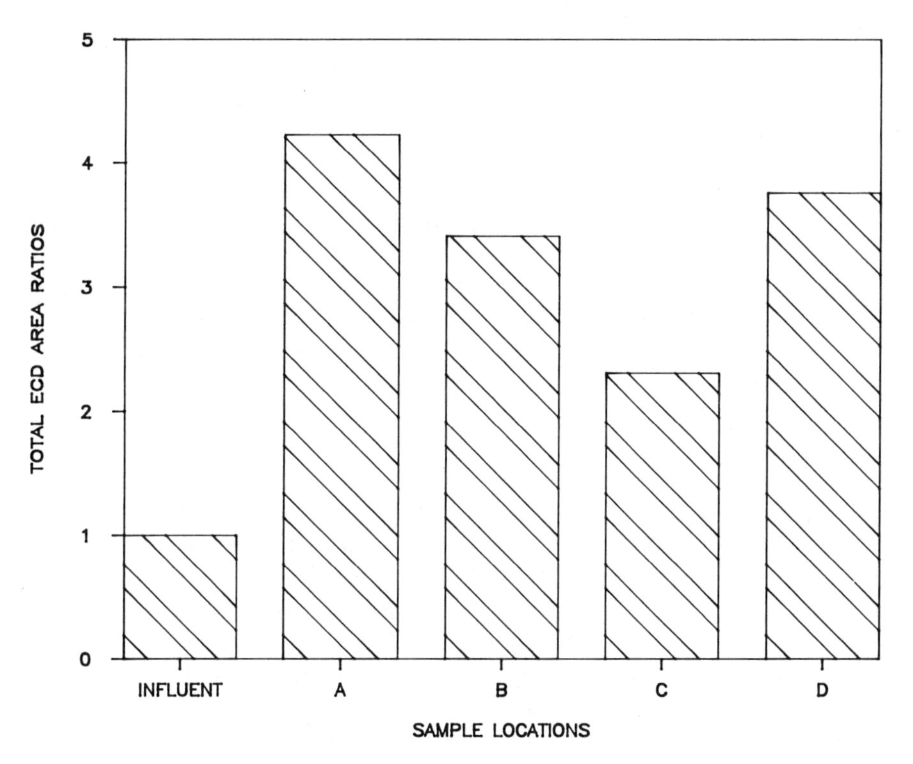

Figure 5. Integrated ECD area ratios of chromatographs shown in Figure 3.

in GC-detectable species following primary chlorination (point A). A significant loss in organic materials follows clarification (point B) and filtration (point C). We are unable to suggest a firm mechanism for this observed decline in organic substances, but it may involve volatility losses or some adsorptive process. This phenomenon needs to be further investigated. ECD areas in Figure 5 are not directly comparable to mass concentrations. This condition is primarily a result of the variable detector response to different types of organic compounds. These data may therefore be considered semiquantitative, but should reflect general changes in overall organic concentrations.

Increased concentrations of chlorinated organic substances in the clear well (point D) occur as a result of rechlorination here. At point A, feedwater was chlorinated at levels between 10 and 20 mg/L. Secondary chlorination (point D) was at 3–4 mg/L. All samples were collected at steady-state conditions, and no effort has been made to study the kinetics of chlorine–organic interaction. This increased concentration seems to indicate, however, that some of the precursors remain unreacted in process water.

Prior to shutdown of the Los Banos facility in late 1986, a single series of samples was collected from the RO units in August of that year. These units were in continuous operation for about a week prior to sampling. The RO systems shown in Figure 6 are brine staged and equipped with the membrane types indicated. Sampling locations for influent, brine, and product water are marked on this diagram as points 1, 2, and 3. In addition, some samples were collected following specific membrane modules to compare the selectivity of various membrane types.

Figure 7 illustrates GC traces for acid-extractable organic materials at the three locations indicated on Figure 6. These traces show that higher molecular weight organic substances are strongly rejected by the RO membranes in place. Volatile samples confirmed, however, that THMs and other small organic substances easily penetrate these RO membranes. Such data is in good agreement with material balances for cellulose acetate membranes reported at Yuma (5).

Conclusions

Agricultural drainage water from the San Luis drain in the Central San Joaquin Valley contains a highly complex mixture of gas chromatographable organic compounds. These chemical species are essentially nonvolatile. Following chlorination, the mixture increases in complexity, with a shift toward lower-molecular-weight organic compounds and the appearance of volatile species, including three THM fractions. These fractions have been identified as $CHCl_2Br$, $CHClBr_2$, and $CHBr_3$. Bromoform was present at the highest concentration, although the chloroform concentration was below the GC

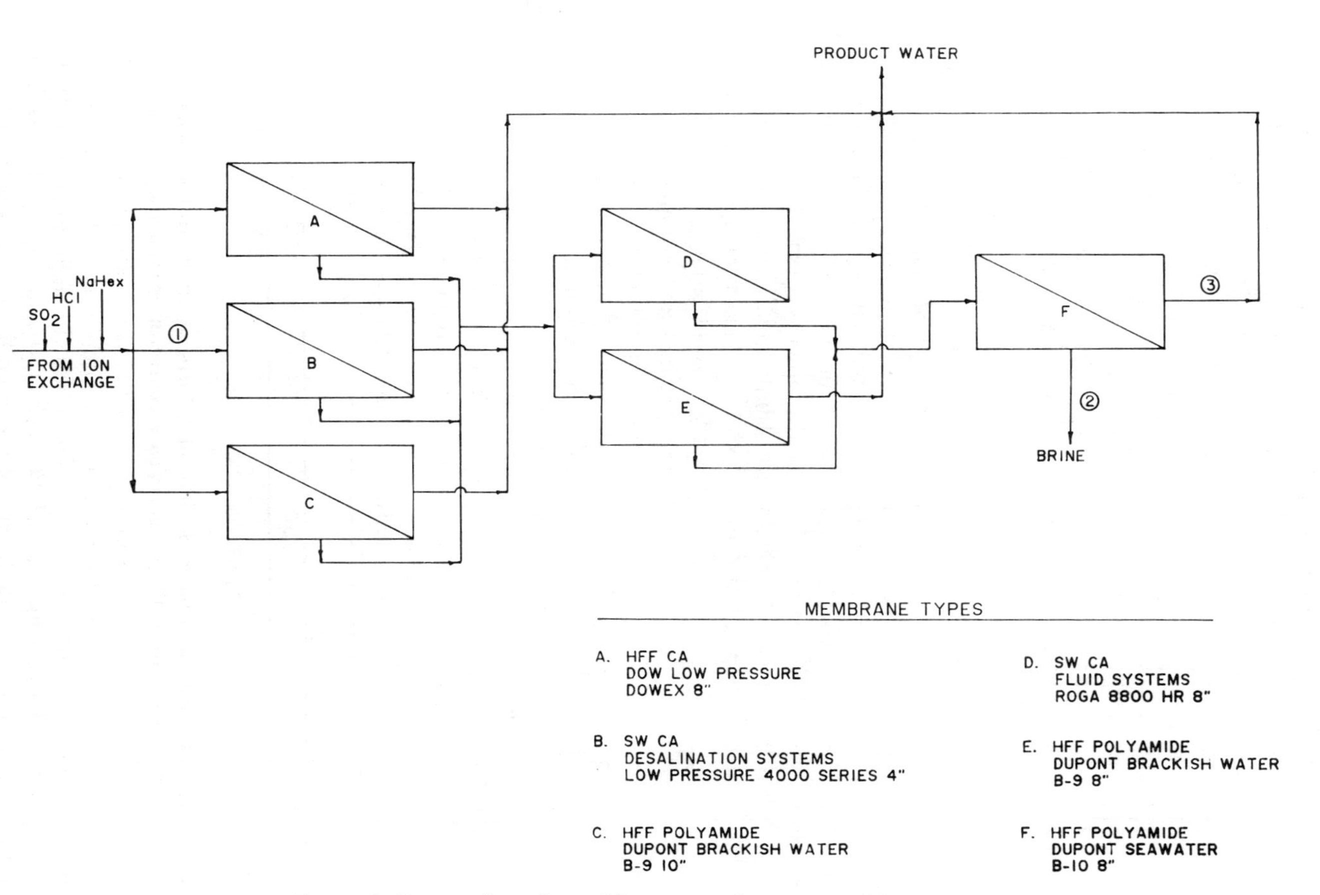

Figure 6. Process flow sheet of brine staged composite RO units.

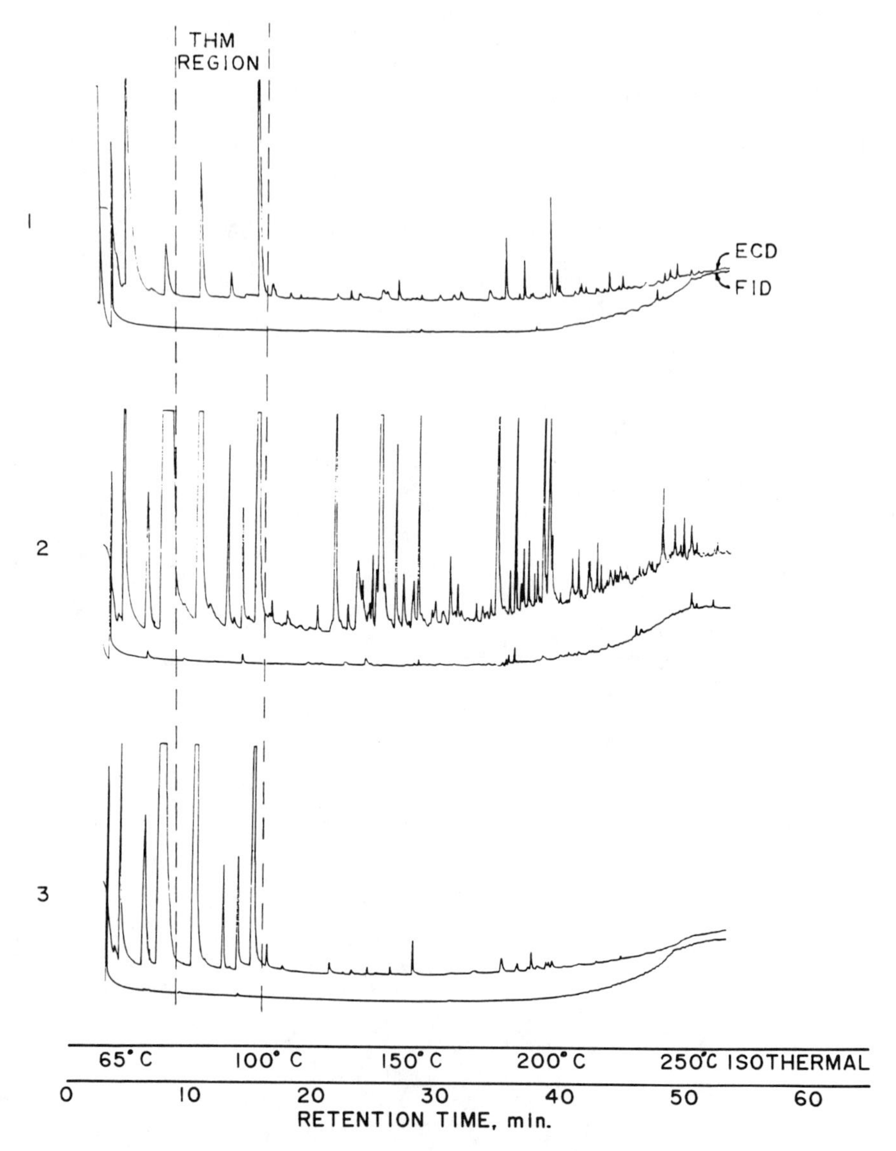

Figure 7. Gas chromatograms of 1, influent; 2, brine; and 3, product samples from RO units (acid methylene chloride extracted samples).

detection limit. Concentrations of THM species were in similar ratios to those reported at the Yuma Desalting Facility. Production of THMs is a relatively small, but important, part of the overall chlorine–organic chemistry of agricultural drainage water. Identification of other prominent organic products is presently underway.

Surveillance of organic compounds throughout the Los Banos flow train has revealed some interesting information. First, the concentration of acid-extractable organic compounds is diminished progressively following clarification and filtration. The exact mechanism of this removal is unclear. Second, the membranes tested at Los Banos did not significantly reject THMs that carried over into product water. Third, RO membranes strongly reject most other gas chromatographable acid-extractable organic substances. Polyamide membranes appear to be more selective than cellulose acetate membranes.

We are concerned not only with the ability of RO membranes to pass or reject dissolved organic compounds, but also with potential interaction between membrane polymers and these organic substances. Such interactions may seriously affect the performance and life expectancy of RO membranes. We are currently engaged in such studies with model compounds and will report our results in a forthcoming paper.

Acknowledgment

The authors acknowledge the support provided for this research by the California Department of Water Resources. Special thanks are due to Brian Smith of DWR for his cooperation in obtaining water samples and for the many useful discussions in his office. We also thank Michael Stenstrom and Kasi Gabbita, who have provided numerous stimuli during the preparation of this manuscript.

References

1. Antoniuk, D.; McCutchan, J. W. *Desalting Irrigation Field Drainage Water by Reverse Osmosis, Firebaugh, California*; University of California—Los Angeles: Los Angeles, 1973; Engineering Report 7368.
2. Moody, C. D.; Kaakinen, J. W.; Lozier, J. C.; Laverty, P. E. *Desalination* **1983**, *47*, 239.
3. Smith, B. E.; Brice, D. B.; Kasper, D. R.; Everest, W. R. "Agricultural Wastewater Desalting in California DWR: Test Facility Description", *Proc. 10th Annu. Conf. Water Supply Improvement Assoc.* Honolulu, HI; Vol. II, July 1982.
4. Milstead, C. E.; Riley, R. L., UOP, Fluid Systems: San Diego, unpublished report.
5. California Department of Water Resources, San Luis Drain Water Quality Analyses, San Joaquin District: Sacramento, 1985.
6. Malcolm, R. L.; Wershaw, R. L.; Thurman, E. M.; Aiken, G. R.; Pickney, D. J.; Kaakinen, J. *Reconnaissance Samplings and Characterization of Aquatic Humic Substances at the Yuma Desalting Test Facility, Arizona,* USGS Water Resources Investigations, 1981; Report USGS/WRI 81–42.
7. Glaze, W. H.; Rawley, R.; Burleson, J. L.; Mapel, D.; Scott, D. R. In *Advances in the Identification and Analysis of Organic Pollutants in Water,* Vol. 1; Keith, L. H., Ed.; Ann Arbor: Ann Arbor, 1981; pp 267–280.
8. Fam, S. Ph.D. Thesis, University of California—Los Angeles, 1986.

9. Miller, J. W.; Uden, P. C. *Environ. Sci. Technol.* **1983,** *17*, 150.
10. U.S. Environmental Protection Agency. *Fed. Regist.* **1984,** *49*, 209.
11. Rook, J. J. *Environ. Sci. Technol.* **1977,** *11*, 478.
12. Peters, C. J.; Young, R. J.; Perry, R. *Environ. Sci. Technol.* **1980,** *14*, 1391.
13. Hirose, Y.; Okitsu, T.; Kanno S. *Chemosphere* **1982,** *11*, 81.
14. Morris, J. C.; Baum, B. In *Water Chlorination—Environmental Impact and Health Effects*, Vol. 2; Jolley, R. L., Ed.; Ann Arbor Science: Ann Arbor, 1978; pp 29–48.
15. Christman, R. F.; Johnson, J. D.; Norwood, D. L.; Liao, W. T.; Hass, J. R.; Pfaender, F. K.; Webb, M. R.; Bobenrieth, M. J. *Chlorination of Aquatic Humic Substances*, U.S. Environmental Protection Agency R&D Report EPA—600/32–81–016, 1981.

RECEIVED for review September 17, 1987. ACCEPTED for publication February 12, 1988.

44

Mechanistic Interactions of Aquatic Organic Substances with Anion-Exchange Resins

Paul L. K. Fu

Camp Dresser and McKee, Inc., P.O. Box 9626, Fort Lauderdale, FL 33310

James M. Symons

Department of Civil Engineering, University of Houston, Houston, TX 77004

Specific interactions of aquatic organic substances isolated from a natural water source on commercial anion-exchange resins were investigated. Focuses were verifications of the mechanisms of organic removal and the influences of resin properties on the removal. The organic compounds used were concentrated by reverse osmosis and then separated by ultrafiltration into four molecular size fractions. Each organic fraction was characterized for organic carbon and carboxyl contents. Specific interactions of the organic fractions on various types of anion-exchange resins were studied by batch equilibrium experiments. Results indicate that about 90% of the dissolved organic carbon recovered from the natural water was removable by anion-exchange resins via an ion-exchange mechanism, and that both skeletal and pore structures were important to the removal, depending on the characteristics of the organic substances.

REMOVAL OF ORGANIC SUBSTANCES FROM DRINKING WATER SUPPLIES has become a major concern since trihalomethanes were found in chlorinated drinking water by Rook (*1*) and Bellar et al. (*2*) in 1974. The U.S. Environmental Protection Agency recommended that water utilities should lower

0065–2393/89/0219–0797$06.00/0

the organic content of their water to the lowest extent possible prior to the use of any oxidant as a disinfectant (*3, 4*).

The anion-exchange process is an attractive method for the removal of aquatic organic substances because the majority of them are acidic (ionic). Several previous studies focused on the removal of organic substances from waters. They were more empirical than fundamental, however, because the complex nature of the natural waters made the verification of mechanistic interactions of organic fractions on anion-exchange resins difficult. As a result, these interactions still remained speculative. The purpose of this research was, therefore, to understand the resin–organic compound interactions via a systematic study using four different molecular size fractions of aquatic organic substances concentrated and isolated from a single natural water source and six anion-exchange resins of different skeletal and pore structures. This chapter focuses on the portion of the study (*5*) concerned with mechanisms of aquatic organic compound removal by anion-exchange resins and influences of resin properties on the removal.

Removal Mechanisms

Abrams (*6*) suggested that ion exchange is the most significant mechanism for the removal of organic acids from water supplies by strong-base anion-exchange resins. He assumed that some adsorption (other than ion exchange) also occurred on the aromatic structure of the styrenic resin polymer.

In their article concerning organic substance removal by anion-exchange resins, Anderson and Maier (*7*) stated, "The mechanism by which these organic materials are retained by anion-exchange resins has not been unequivocally established. Nevertheless, it appears that anion exchange plays a prominent part."

Kunin and Suffet (*8*) indicated that some humic substances are removed by anion-exchange resins via a simple anion-exchange process, and that some portions of the humic material are removed by "true" (surface) adsorption involving covalent bonding via van der Waals forces, particularly in acid media when the humic substances are not protonated. None of these authors, however, presented substantial evidence or data to verify the removal mechanisms.

According to Thurman's book (*9*), the acid dissociation constant (pK_a) of carboxyl groups on aquatic humic substances ranges approximately from 1.5 to 6.0 and that of phenolic hydroxyl groups falls between 8.0 and 12.0. Because natural water is generally neutral (pH 7–8), carboxyl groups on aquatic organic acids are essentially all ionized and phenolic hydroxyl groups are almost completely nonionized. Therefore, the term "carboxyl group" is used in this research to represent the organic functional groups that would contribute to ion exchange.

Two possible mechanisms responsible for the removal of aquatic organic acids are proposed: (1) A mechanism that is responsible for the attachment of the carboxyl groups of an organic acid to the resin ionogenic groups involving the replacement of counterions on the resin is described as "ion exchange." This term does not define whether the exchange is reversible. (2) A mechanism that is responsible for the attachment of the nonionic portion of an organic acid to the internal surface of the resin without interacting with any ionogenic groups is termed "surface adsorption".

An organic acid could be removed by either mechanism 1 alone or mechanism 2 alone, or by a combination of both mechanisms. These three conditions are demonstrated in Figure 1. When only mechanism 1 is involved (shown in Figure 1A), the removal is by ion exchange. When only mechanism

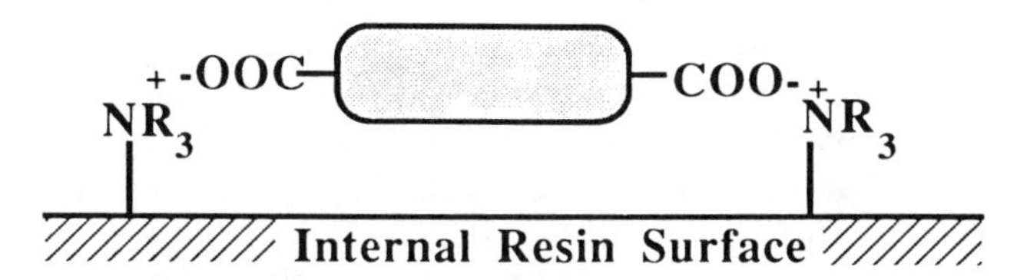

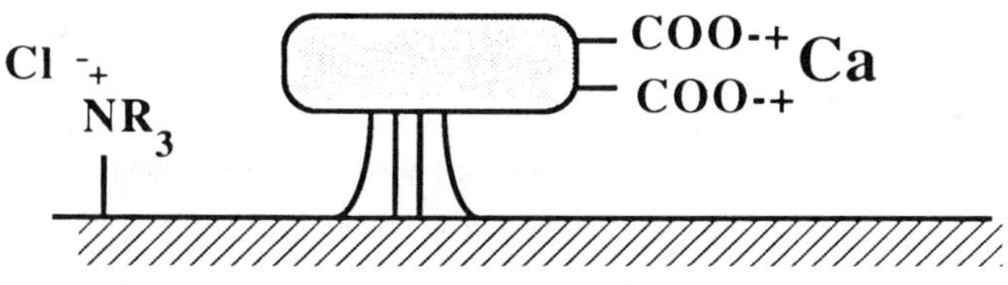

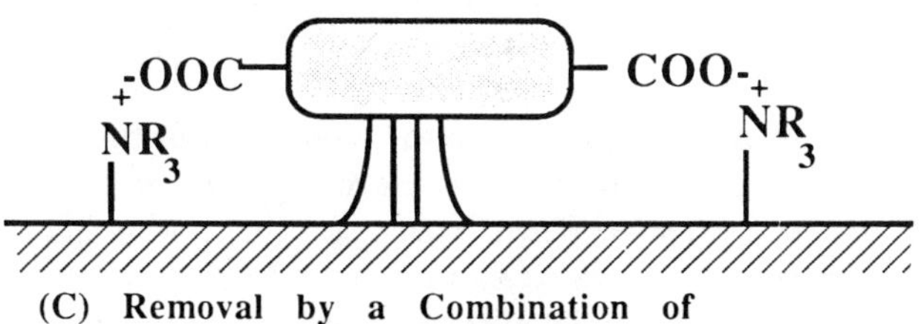

Figure 1. Removal of an organic acid by an anion-exchange resin through various mechanisms.

2 is involved (shown in Figure 1B), the removal mechanism is surface adsorption, and the removal is called "electrolyte adsorption".

A combination of mechanisms 1 and 2 (shown in Figure 1C) occurs when the carboxyl groups of an organic molecule bind with resin amine functional groups (ion exchange) and the nonionic portion of the molecule attaches to the resin inner surface through surface adsorption.

After equilibrating an organic fraction with different quantities of various anion-exchange resins at pH 7.5, a value typical of natural surface waters, the variations of total organic carbon (ΔTOC) and of counterion concentration (ΔCl for chloride-form resins) were measured. The plot of ΔTOC vs. ΔCl would yield an approximately straight line only if the removal of the organic fraction involved the ion-exchange mechanism, because the aqueous-phase chloride concentration can be increased only by the exchange of the organic anions for the resin-phase chloride ions. The straight line, however, cannot allow judging whether electrolyte adsorption of some organic compounds or a combination of ion exchange and surface adsorption also occurred. To clarify whether or not electrolyte adsorption of the organic matter was occurring, the following procedure was used.

The variation of the ionic concentration in milliequivalents per liter of an organic fraction, ΔA, was calculated from ΔTOC by the following equation:

$$\Delta A = \Delta \mathrm{TOC}(\Omega)/1000 \quad (1)$$

where Ω, the concentration (meq/g of TOC) of carboxyl groups of the organic fraction at a given pH, was obtained from titrations of acidic functional groups, and ΔTOC is in milligrams per liter. After ΔA values were calculated, ΔA vs. ΔCl was plotted. A straight line with a slope higher than 1, demonstrating $\Delta A > \Delta \mathrm{Cl}$, indicated the existence of some electrolyte adsorption of the organic acid molecules. If, however, the plot resulted in a straight line with a slope equal to 1, which represents the diagonal where $\Delta A = \Delta \mathrm{Cl}$, the stoichiometry of ion exchange and the absence of electrolyte adsorption was demonstrated.

If $\Delta A = \Delta \mathrm{Cl}$, the following steps were carried out to clarify whether surface adsorption, in addition to ion exchange, existed on the same molecule, as was the case shown in Figure 1C. Selected polymeric adsorbents, which possess skeletal and pore structures similar to the ion-exchange resins but contain no ionogenic functional groups, were used in this investigation. Because of the lack of ionogenic functional groups in polymeric adsorbents, ion exchange was not present. Thus, surface adsorption could be studied without the interference of ion exchange.

To obtain the degree of removal through surface adsorption, an organic fraction was equilibrated with the polymeric adsorbents. If little organic matter is adsorbed, surface adsorption would be insignificant to the uptake of the organic fraction by the ion-exchange resins, and ion exchange would therefore be the dominant removal mechanism. If, however, a significant quantity of the organic matter is adsorbed, the surface adsorption mechanism would, in addition to the ion-exchange mechanism, also be significantly responsible for the uptake of some of the organic matter by the ion-exchange resins. This verification can only be qualitative, however, because no polymeric adsorbent possesses exactly the same skeletal and pore structures as an anion-exchange resin. Nevertheless, these data are useful in understanding the mechanisms involved.

Influences of Resin Properties

Synthetic organic anion-exchange resins and adsorbents have two major types of skeletons: polystyrene and acrylate. Both of these skeletons are commonly cross-linked with divinylbenzene. The main difference between these two skeletons is their hydrophobicity. Because of its higher content of aromatic rings, the styrenic resin sorbs less water (swells less) and is more hydrophobic than the acrylic-type resin.

Using high-performance liquid chromatography (HPLC) columns, Sinsabaugh et al. (*10*) separated aquatic organic matter in the Harwood's Mill Reservoir (in Virginia) water into hydrophobic, mesic, and hydrophilic fractions. The skeleton of anion-exchange resins, which affects resin hydrophobicity, may therefore be an important property for the uptake of aquatic organic materials by anion-exchange resins.

Because the molecular size of aquatic organic substances is polydisperse, the pore structure of anion-exchange resins is important to the removal of aquatic organic fractions. Anion-exchange resins can be made gelular or porous. Gelular resins possess no permanent pores, but they do have "micropores" of atomic dimension that depend markedly on their swelling properties, described by Kunin and Hetherington (*11*). Macroporous resins, however, have a "true" pore phase in addition to the gel phase and have an internal surface area. Because of their pore phase, macroporous resins should be more accessible to organic compounds of large molecular sizes than gelular resins. Some important properties of the various anion-exchange resins and polymeric adsorbents used in this research are summarized in Table I.

In the literature, Macko (*12*) was the only researcher who studied the removal by anion-exchange resins of aquatic organic substances fractionated by ultrafiltration (UF). She, however, focused on only one anion-exchange resin (Amberlite IRA904) in her work and did not use a single batch of source organic matter, but collected water samples and UF-fractionated them at different periods during her investigation. Her results, therefore, contributed little to the understanding of the influence of resin skeleton and pore structure on the removal of aquatic organic fractions. In this research, however, six strong-base anion-exchange (SBA) resins with different skeletal and

Table I. Properties of Anion-Exchange Resins and Polymeric Adsorbents Used in This Research

		NaCl Exchange Capacity[b]		*Surface Area*	*Mean Pore Radius*
Resin	*Type*[a]	*(meq/g)*	*(meq/mL)*	*(m^2/g)*	*(nm)*
IRA458	ACR–GEL–SBA	4.37	1.29	0.1	—[c]
IRA904	STY–MAC–SBA	2.67	0.75	60	35
IRA938	STY–MAC–SBA	3.91	0.58	7	3500
IRA958	ACR–MAC–SBA	4.14	0.87	<5[d]	100–150[d]
XE510	STY–MAC–SBA	1.64	0.70	409	10[d]
IRA402	STY–GEL–SBA	4.10	1.18	0.1	—
XAD–8	ACR–MAC–ADS	~0	~0	160	22.5
XAD–16	STY–MAC–ADS	~0	~0	800	10

NOTE: Resins from the Rohm and Haas Company. Data from the Rohm and Haas Technical Bulletins unless otherwise specified.
[a]ACR, acrylic; STY, styrenic; MAC, macroporous; GEL, gelular; SBA, strong-base anion-exchange resin; and ADS, polymeric adsorbent.
[b]Data determined by the investigator.
[c]No true pores for gelular resins.
[d]Data suggested by Robert Albright of the Rohm and Haas Company (personal communication, August 1985).

pore structures were used in batch equilibrium experiments to investigate the influences of resin properties on the uptake of various organic fractions from a single source.

Experimental Details

Concentration and Fractionation of Aquatic Organic Substances. A reverse osmosis (RO) system (Model AK–300, equipped with FilmTec–30 synthetic membrane, Water Equipment Technologies, Inc.) was used to concentrate organic fractions from 800 gal of Lake Houston water, the source of 50% of the City of Houston's drinking water. A single-cartridge, hollow-fiber ultrafiltration system (Romicon Model HF–LAB–5) was then used to fractionate and dialize the concentrated organic materials into four molecular size fractions. Over 88% of the TOC passing through a 0.4-μm prefilter was recovered in the concentrate.

Titration of Acidic Groups. Each of the organic fractions was converted to acid form by passing it through a strong-acid cation-exchange column packed with hydrogen-form resins (IR120). Fifty milliliters of an organic fraction of known weight concentration (milligrams of TOC/L) were placed in a 140-mL beaker (Pyrex) that was set inside an air-sealed glass container. The solution was stirred at a constant speed and purged with nitrogen gas to prevent carbon dioxide dissolution. Increments of less than 10 μL of 1.00 N standard KOH solution were added to the solution by a microliter syringe, and the pH of the solution was monitored by a pH meter (Fisher Accumet Model 805) equipped with a combinated, gel-filled electrode.

Resin Conditioning. Six anion-exchange resin samples (the important characteristics of which are given in Table I) were screened prior to use to obtain beads of uniform size. The conditioning procedure involved extensive backwashing to remove fines and beads of low densities, followed by a rinse with methanol to remove residue monomers and two cyclic exhaustions using 2.0 N sodium hydroxide and 2.0 N hydrochloric acid with intermediate and final deionized organic-free (activated-carbon adsorption-treated) water rinses. Conditioned acid-form resins were treated with 5 bed volumes of 1.0 N NaCl followed by a 12-h slow rinse with 0.005 N NaCl to convert the resins into chloride form. The conditioned resins were air-dried for convenience in weighing. The moisture content was determined as the loss of weight after oven-drying the resins at 103 °C for 24 h.

Batch Equilibrium Experiments. Two sizes of amber glass bottles were used in equilibrium experiments: 500-mL and 130-mL capacity. The larger bottle was filled with 400 mL of the test solution and the smaller one with 100 mL, and the solution pH was raised to 7.5 by adding 1.0 N or 0.1 N NaOH. After accurately weighed amounts of resins were added, the bottles were capped head-space-free by screw caps (with poly(tetrafluoroethylene)-faced (Teflon-faced) septa). They were then placed in a tumbler rotated at 16 rpm. Preliminary experiments demonstrated that 1 week of equilibration at an ambient temperature of 25 ± 2 °C was sufficient to reach a stable solution TOC concentration for all the organic fractions studied. Therefore, 1 week of tumbling was used in all batch equilibrium experiments. In a typical equilibrium experiment, two replicate bottle samples for each experimental condition were used. A control and a blank were also prepared. The control was identical to the test samples except for the lack of any organic matter, and the blank was identical to the test samples except that it contained no resin.

Results and Discussion

The results from RO-concentration, UF-fractionation, and titrations of acidic groups are summarized in Table II.

Removal Mechanisms. The relationship between ΔTOC and ΔCl for the exchange of the organic fraction of 1000–5000 MW (hereafter called the 1K–5K organic fraction) with chloride ions in the resin phase at pH 7.5

Table II. Summary of the Results from RO-Concentration, UF-Fractionation, and Titrations of Acidic Groups

Description	*TOC (mg/L)*	*Volume (L)*	*DOC[a] (g)*	*% DOC*
RO concentrates	78.6	350	27.5	100
UF fractions				
>10K[b]	178	16.8	3.0	10.9
5K–10K	282	27.3	7.7	28.0
1K–5K	112	23.0	2.6	9.5
<1K	23	620[c]	14.2	51.6
Carboxyl content				
(meq/g of TOC)	11.03	13.13	15.05	—[d]

[a]DOC represents the dissolved organic carbon that passed through a 0.4-μm cartridge prefilter.
[b]Greater than 10,000 apparent molecular weight units.
[c]Volume was higher because deionized water was added for dialyzing.
[d]The <1K fraction was not desalted; therefore, its carboxyl content could not be titrated.

using three different types of strong-base anion-exchange resins—namely, IRA458, IRA938, and IRA958—is shown in Figure 2. A strong linear relationship between ΔTOC and ΔCl, with a squared correlation coefficient (R^2) of 0.980 resulted and showed that ion exchange was at least somewhat responsible for the sorption of the 1K–5K organic fraction and that the fraction contained fairly homogeneous organic matter in terms of milliequivalents of acidic groups per unit weight of TOC. Data points at the upper right portion of the straight line indicate more than 90% of the initial 98 mg/L TOC was removed by the resins at a 1-g/L dose. This finding confirmed that this fraction contained predominantly removable organic acids.

The ΔTOC values were converted to ΔA by equation 1, and the ΔA vs. ΔCl relationship for these three resins is shown in Figure 3. All data points were fitted by linear regression, yielding a straight line with a slope nearly equal to 1 (1.014) and an R^2 of 0.980. The stoichiometry was thus verified for the ion exchange of the 1K–5K organic fraction for chloride ions, and no electrolyte adsorption of the fraction was involved.

According to Oliver and Visser (*13*), on the basis of their information obtained from Amicon Corporation, the average pore sizes of 1K and 10K ultrafiltration membranes are 1.2 and 1.5 nm, respectively. Therefore, the 1K–5K organic fraction studied in this research would probably have an average molecular size between 1.2 and 1.5 nm. The mean pore diameters of polymeric adsorbents methyl methacrylate resin (XAD–8) and polystyrene–divinylbenzene resin (XAD–16) are 45 and 20 nm, respectively. These resin pore diameters are more than a dozen times larger than the average molecular size of the 1K–5K organic fraction. Therefore, there should not

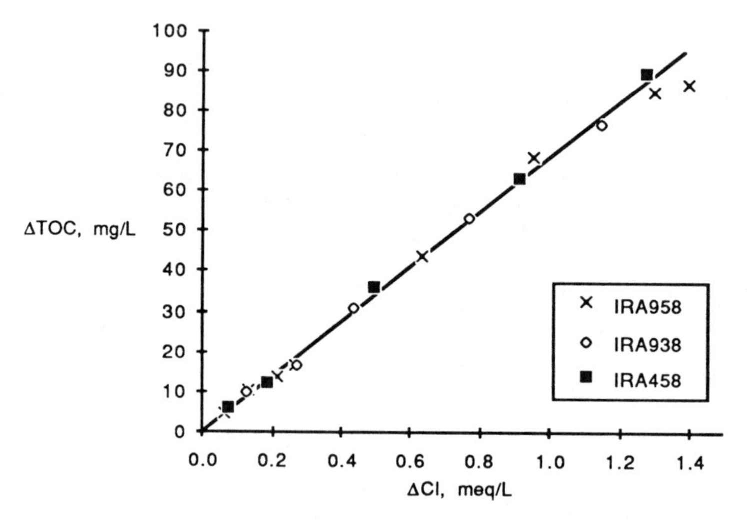

Figure 2. ΔTOC vs. ΔCl for the exchange of the 1K–5K organic substances with chloride ions at pH 7.5 with three types of anion-exchange resins.

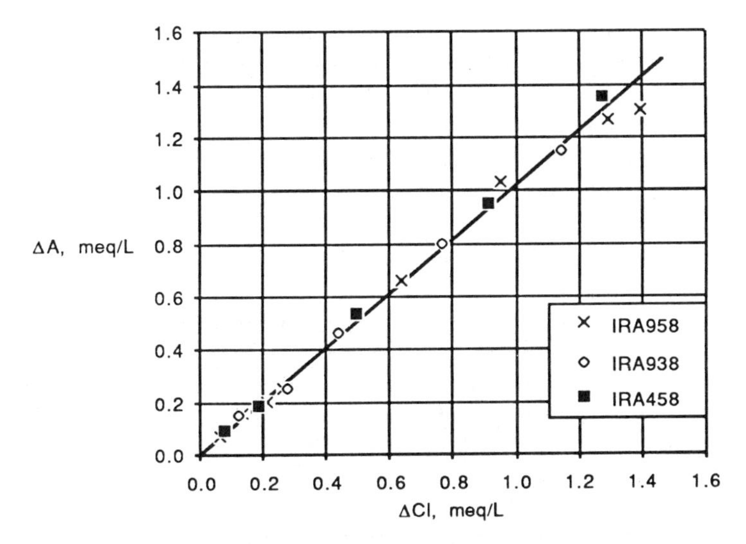

Figure 3. ΔA vs. ΔCl for the exchange of the 1K–5K organic substances with chloride ions at pH 7.5 with three types of strong-base anion-exchange resins.

be significant pore-structure limitations of the polymeric adsorbents with respect to their uptake of the 1K–5K organic fraction. The degree of removal of the 1K–5K fraction by two polymeric adsorbents, XAD–8 and XAD–16, of different skeletal structures are shown in Table III.

Both the acrylic XAD–8 (with a internal surface area of 160 m^2/g) and the styrenic XAD–16 (with a very high internal surface area of 800 m^2/g) adsorbed 6.7% or less of the initial 98 mg/L TOC for the 1K–5K organic fraction, qualitatively verifying that a combination of mechanisms (*see* Figure 1C) did not occur significantly for the removal of this organic fraction at a pH of 7.5 by the anion-exchange resins. Therefore, ion exchange was the dominant mechanism for the uptake of the 1K–5K organic fraction by the anion-exchange resins.

Other organic fractions were studied in a similar manner. Ion exchange was still the dominant removal mechanism, although the presence of some nonionic hydrophobic organic compounds (which were surface-adsorbed by the styrenic resins) in the fraction of molecular weight less than 1000 and some organic–inorganic complex colloids (which were too large to be removed by any of the resins) in the fraction of molecular weight greater than 10,000 complicated the investigations.

Influences of Resin Properties. The influences of resin skeletal and pore structures on the removal of each of the organic fractions are described as follows.

FRACTION >10,000 MW. The resin-removal isotherms for the fraction of molecular weight >10,000 shown in Figure 4 are fairly steep, a result in-

Table III. Removal of the 1K–5K Organic Fraction by XAD–8 and XAD–16 Polymeric Adsorbents at pH 7.5

Adsorbent	*Type*	*Equilibrium TOC (mg/L)*	*% TOC Removal*
XAD–8	acrylic	94.4	3.7
XAD–16	styrenic	91.4	6.7

NOTE: The dose was 1 g/L and the initial TOC was 98.0 mg/L in both cases.

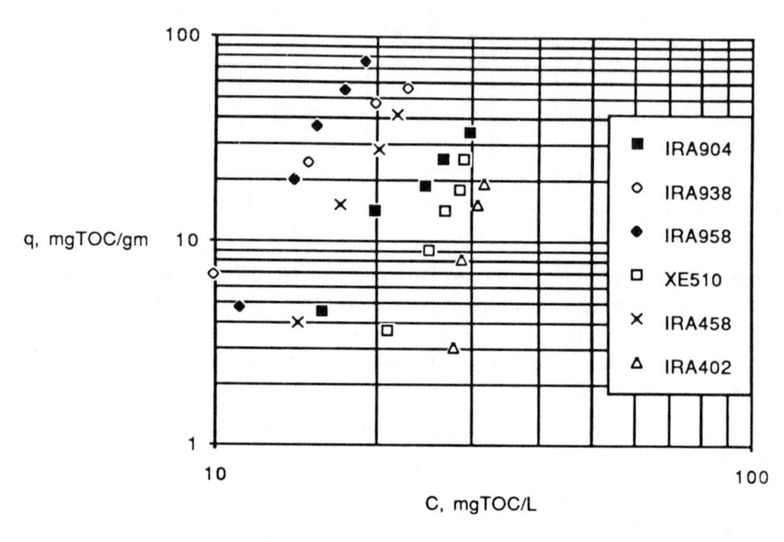

Figure 4. Removal isotherms of the >10K organic-substance fraction at pH 7.5 with an initial TOC of 34 mg/L.

dicating that some of these organic materials were not removable by the resins. The isotherms plotted arithmetically (not shown) were extrapolated to the horizontal axis, and approximately 12–85% of the organic fractions were estimated to be nonremovable by the various resins. The percentage of nonremovable organic fractions and the pertinant resin properties for each of the resins are shown in ascending order in Table IV.

The sequence of increasing nonremoval shown in Table IV is related to the pore structure and the hydrophobicity of the resins. Although the IRA938 resin is hydrophobic (a rigid gel phase), it has the lowest percentage of nonremoval because of its extremely large pores. IRA958, macroporous and hydrophilic (a flexible gel phase), has the second lowest percentage because of its relatively large pores. Although the IRA458 hydrophilic gel resin has no true pores, it swells in water so that its gel phase is flexible enough to be accessible to some very large organic molecules, and it is the third in the sequence.

Table IV. Percentages of TOC (of the >10K Fraction) Not Removed by the Various SBA Resins

% TOC Not Removed	*Resin Type*	*Skeleton*	*Hydrophobicity*	*Pore Structure*	*Mean Pore Radius (nm)*	*Surface Area* (m^2/g)
12	IRA938	styrene	hydrophobic	porous	3500	7
20	IRA958	acrylate	hydrophilic	porous	100	<5
35	IRA458	acrylate	hydrophilic	gel	—[a]	0.1
40	IRA904	styrene	hydrophobic	porous	35	60
50	XE510	styrene	hydrophobic	porous	10	409
85	IRA402	styrene	hydrophobic	gel	—[a]	0.1

[a]No true pores in gel resins.

IRA904, although a macroporous resin, is below IRA458 in the sequence because it has a rigid (hydrophobic) gel phase and a moderately large mean pore radius. The hydrophobic XE510, even with a high internal surface area, is the fifth in the sequence because its mean pore radius is only 10 nm. Lastly, IRA402, a styrenic gel resin (contains no true pores), has too rigid a gel phase to adsorb the very large organic matter (>10,000 MW). This sequence indicates that pore radius and rigidity (hydrophobicity) of the resin gel phase are interrelated and both critical to the adsorption of the >10,000 MW organic matter by anion-exchange resins.

5K–10K AND 1K–5K FRACTIONS. The 5K–10K and 1K–5K fractions (5000–10,000 and 1000–5000 molecular weight) are discussed together because their removal isotherms were very similar. These isotherms of six different types of anion-exchange resins are shown in Figures 5 and 6 for the 5K–10K and the 1K–5K fractions, respectively. Compared to the isotherms of the >10K fraction shown in Figure 4, the isotherms shown in Figures 5 and 6 for a specific resin for the 5K–10K and 1K–5K fractions are flatter and show a higher sorption density at the same aqueous TOC concentration, a result indicating that these two smaller fractions were removed by the resin much better than the >10K fraction.

Unlike the isotherms for the >10K fraction, the two acrylic anion-exchange resins, gel (IRA458) and macroporous (IRA958), have very similar isotherms because the two organic fractions are smaller in molecular size than the >10K fraction, such that the hydrophilicity of resins becomes more dominant than the pore size in terms of influencing the removal of the organic matter. This situation is also the reason that IRA938, with a styrenic hydrophobic skeleton, removed less of the organic matter than the acrylic resins, even though it had a large mean pore radius of 3500 nm. The other resins—IRA904, XE510, and IRA402—are all styrenic and show low adsorbing capacities, especially the gel type IRA402, which has no true pores.

<1K FRACTION. The resin-removal isotherms of the <1K (less than 1000 molecular weight) fraction shown in Figure 7 are quite different from those

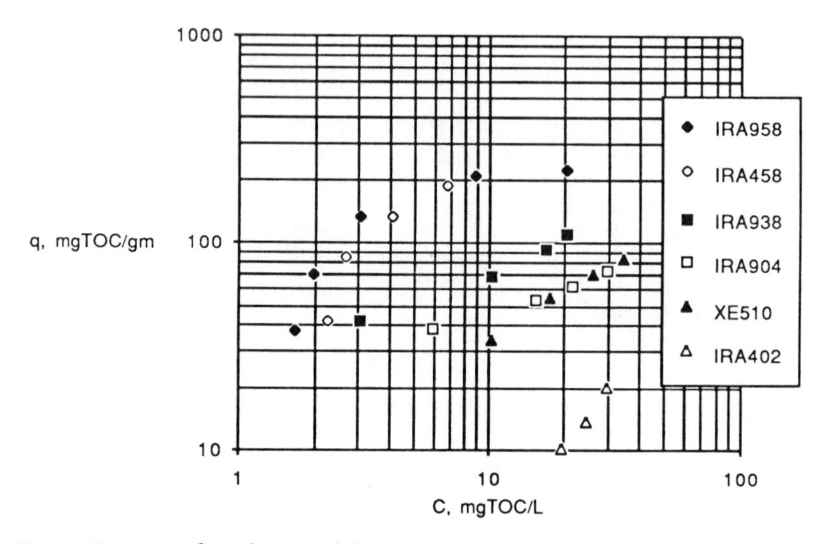

Figure 5. Removal isotherms of the 5K–10K organic-substance fraction at pH 7.5 with an initial TOC of 44 mg/L.

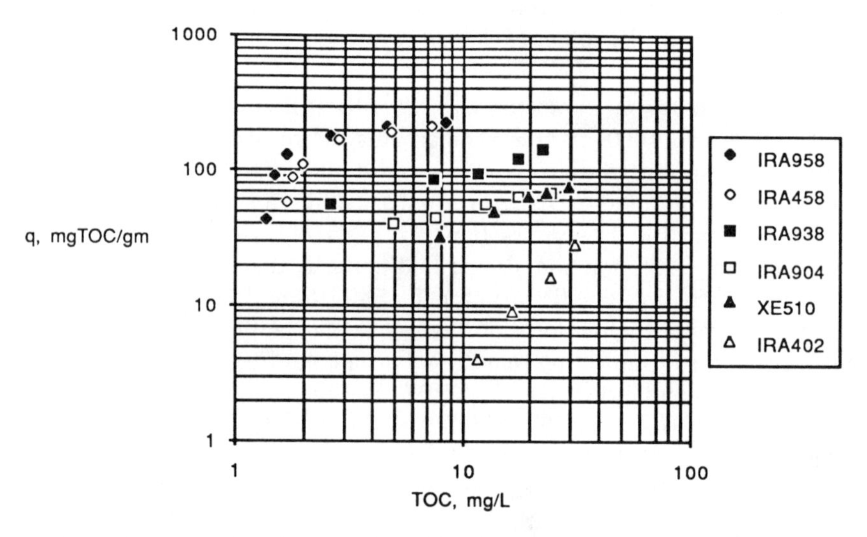

Figure 6. Removal isotherms of the 1K–5K organic-substance fraction at pH 7.5 with an initial TOC of 32 mg/L.

of the 5K–10K and 1K–5K fractions (Figures 5 and 6). The isotherms in Figure 7 are steep, a result indicating that approximately 2.5–3.0 mg/L (15–18%) of the 17-mg/L TOC in the <1K fraction is not removable by either of the resins, regardless of the resin skeleton or pore structure. Because the <1K fraction contributed 51.6% of the total DOC in the RO-

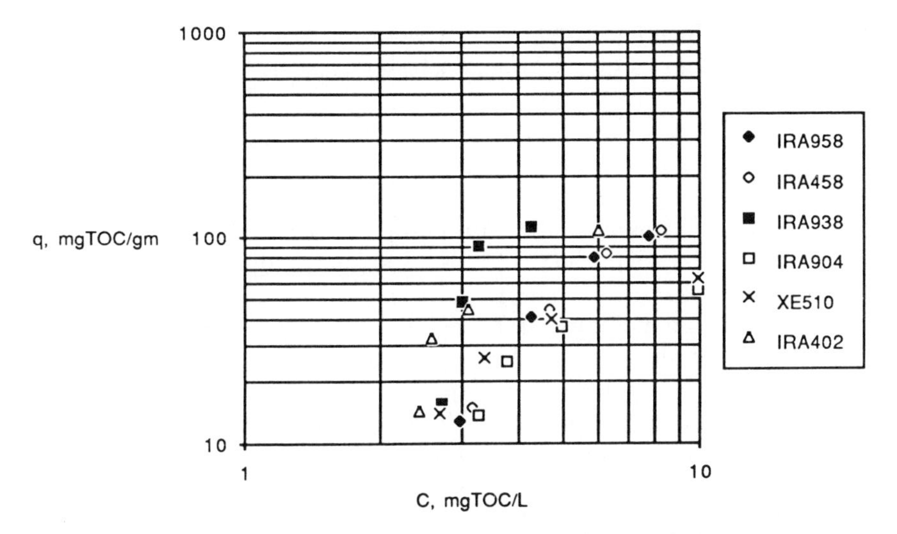

Figure 7. Removal isotherms of the <1K organic-substance fraction at pH 7.5 with an initial TOC of 17 mg/L.

concentrate (*see* Table II), this nonremovable organic matter would contribute approximately 9% (17% × 0.516) of the total DOC in the original prefiltered natural water.

Because the <1K fraction was smaller in molecular size than other three fractions, the screening of the <1K organic molecules by resin pores was less significant. Consequently, the resin-exchange capacity became an important factor influencing the removal capacity of the <1K fraction. This influence is introduced in the following description.

The styrenic IRA938 and IRA402 removed this fraction better (as measured by the sorption density) than the acrylic IRA458 and IRA958, although all four resins have comparable exchange capacities (3.8–4.4 meq/g); thus, variation in exchange capacity was not the cause of sorption differences. The differences in sorption density probably resulted because some hydrophobic nonionic organic compounds existed in the <1K fraction; these compounds were surface-adsorbed (not ion-exchanged) by the styrenic but not by the acrylic anion-exchange resins. For the styrenic gel IRA402, a dramatic increase of removal resulted for the <1K fraction, compared to its ability to remove the other three organic fractions. This result indicates that most of the <1K organic molecules were small enough to permeate into the gel phase of IRA402.

On the basis of sorption density, the XE510 and IRA904, although styrenic and macroporous, removed the <1K organic matter no better than the acrylic resins. This situation probably exists because the two resins have lower exchange capacities—1.6 meq/g for XE510 and 2.6 meq/g for

IRA904—than the other resins. Thus, for this fraction a styrenic (to promote surface adsorption) resin with high exchange capacity (to promote ion exchange) is best.

Conclusions

Ion exchange is the dominant resin-removal mechanism for all organic fractions. Surface adsorption, however, also contributes to the removal of some <1K organic matter.

Resin skeleton and pore structure are interrelated; their impact on the removal of each organic fraction depends on the characteristics of the fraction. Overall, however, resin skeleton, which affects degree of swelling of the resins, is more important than pore structure in influencing the removal of organic substances by resins. Those highly swelled resins with acrylic skeleton performed better than resins with rigid polystyrenic (even macroporous) skeleton in removing the aquatic organic substances. The influences of resin skeleton and pore structure on each organic fraction are listed separately.

>10K Fraction. Both resin skeleton and pore structure were important to the removal. The hydrophilic acrylic resins (as opposed to the styrenic resins that are relatively hydrophobic) removed the >10K organic fraction better than the hydrophobic styrenic resins except for IRA938 (although relatively hydrophobic), whose extremely large pores made it effective for removing the very large >10K organic fraction. Some organic substances in this fraction, however, were too large to be removed by all the resins used.

5K–10K and 1K–5K Fractions. Hydrophilicity of the resins was more dominant than resin pore size in terms of controlling the removal because the acrylic resins removed these organic fractions much better than the styrenic resins. The resin pore size, however, was still important in removal by styrenic resins. The larger the resin pores, the better the resin performed.

<1K Fraction. The removal of this fraction was not influenced by the pore structure of the resins. The hydrophobic styrenic resins removed the organic substances more than the acrylic resins with equivalent exchange capacities. This result occurred because, in addition to the removal of ionic organic compounds, the styrenic resins further adsorbed some hydrophobic nonionic organic substances, which were not adsorbed by the acrylic resins presented in this fraction.

This work has verified that the majority of aquatic organic substances can be removed by selected anion-exchange resins, and that the removal involves predominantly ion-exchange mechanism but little surface adsorption. Because of the dominance of the ion-exchange mechanism, the exchange process was potentially reversible by salt regeneration. In practical applications where organic substances are to be removed from natural waters, the anion-exchange process can be successfully applied by knowing the molecular weight distribution of the organic substances and properly selecting the type of resin for application.

Nomenclature

ΔCl change of the chloride concentration
ΔA change of the ionic concentration of an organic fraction
Ω carboxyl content of an organic fraction

Acknowledgment

This research has been supported by the U.S. Environmental Protection Agency, Office of Research and Development under Grant No. R–812155–01–0.

References

1. Rook, J. J. *Water Treat. Exam.* **1974,** *23*, 234–243.
2. Bellar, T. A.; Lichenberg, J. J.; Kroner, R. C. *J. Am. Water Works Assoc.* **1974,** *66*, 703–706.
3. *Treatment Techniques for Controlling Trihalomethanes in Drinking Water*; U.S. Environmental Protection Agency: Cincinnati, OH, 1981.
4. *Fed. Regist.* **1979,** *44*(231), 68624–68707.
5. Fu, P. L. K. Ph.D. Thesis, University of Houston, 1987.
6. Abrams, I. M. *Chem. Eng. Prog. Symp. Ser.* **1969,** *65*, 106–112.
7. Anderson, C. T.; Maier, W. J. *J. Am. Water Works Assoc.* **1979,** *71*, 278–283.
8. Kunin, R. F.; Suffet, I. H. In *Activated Carbon Adsorption of Organics from the Aqueous Phase*; McGuire, M.; Suffet, I. H., Eds.; Ann Arbor Science: Ann Arbor, 1980; Vol. 2, p 425.
9. Thurman, E. M. *Organic Geochemistry of Natural Waters*; Martinus Nijhoff/Dr. W. Junk: Dordrecht, Netherlands, 1985.
10. Sinsabaugh, R. L.; Hoehn, R. C.; Knocke, W. R.; Linkins, A. E. *J. Am. Water Works Assoc.* **1986,** *78*, 74–82.
11. Kunin, R. F.; Hetherington, R. Presented at the International Water Conference, Pittsburgh, 1969.
12. Macko, C. A. Ph.D. Thesis, University of Minnesota, 1979.
13. Oliver, B. G.; Visser, S. A. *Water Res.* **1980,** *14*, 1137–1141.

Received for review July 24, 1987. Accepted for publication December 8, 1987.

45

Removal of Humic Substances by Ion Exchange

Hallvard Ødegaard, Helge Brattebø, and Ola Halle

Norwegian Institute of Technology, Division of Hydraulic and Sanitary Engineering, N–7034 Trondheim–NTH, Norway

This chapter presents the major results from a research program in which ion exchange has been evaluated as an alternative process for removal of humic substances, especially at small waterworks. It has been demonstrated that ion exchange in a strong-base, anionic, macroporous resin is economically competitive with other alternatives when the raw-water humic-substance concentration is relatively low (<50 mg of Pt/L or 5 mg of total organic carbon per liter). Such resins have a good capacity, but the sorption kinetics are relatively slow because of the slow diffusion of the humic molecules. The process should be carefully designed. Empty-bed contact time is the most important design parameter, and a countercurrent beds-in-series design should be used. Design criteria for the process and for regeneration are given.

THE MAJOR PROBLEM WITH THE QUALITY OF DRINKING WATER in Norway is the humic-substance content of the water. About half of all Norwegian waterworks deliver drinking water with a color that from time to time exceeds 15 mg of Pt/L, which is the present standard.

This problem is the background for a research program on the removal of humic substances at small waterworks. The research has been performed at The Norwegian Institute of Technology since 1979 (*1*). This chapter summarizes our findings with respect to the use of macroporous anionic resins. Chapter 42 summarizes our findings with the use of membrane processes.

Ion-exchange technology is widely used in industrial and potable water treatment. Its application in the removal of humic substances, however, is

0065–2393/89/0219–0813$06.50/0

rather new. Very few municipal waterworks are equipped with such a process. One of the few is the Fuhrberg waterworks in Hannover. However, this facility removes humic substances from ground water (*2*). Humic substances in Fuhrberg ground water can be expected to be quite different from those that we find in Scandinavian surface water.

Several researchers have presented laboratory-scale results from the use of ion exchange. Rüffer and Schilling (*3*) found that strong-base resins were superior to weak-base resins for removing humic materials from surface water. Their results showed adequate rate of uptake, capacity, and regeneration efficiency. Desorption was nearly complete, without any significant decrease in capacity over a 3-year testing period.

Kölle (*2*) did resin (Lewatit MP 500 A) experiments at the Hannover water plant and found that a 10% NaCl and 2% NaOH solution was a proper regenerant. This solution could be reused seven or eight times, with an adjustment of the salt and base content after each regeneration. Kölle (*4*) reported that the treatment efficiency during 2 years of full-scale plant operation decreased about 10%. The reason given was fouling effects and a 3% resin loss per year due to mechanical erosion during regeneration. Currently, the same resin has been successfully used for more than 5 years (Kölle, personal communication).

Jørgensen (*5*) performed experiments with a cellulose-based macroporous resin. The resin had a low capacity, but gave a high treatment efficiency because of the good sorption kinetic characteristics.

Boening et al. (*6*) compared activated carbon to different types of resin for the removal of fulvic acids and a commercial humic acid. They found that activated carbon was the less suitable adsorbent, because the large-molecular-weight humic compounds did not penetrate the micropores. Anderson and Maier (*7*) concluded that a strong-base resin was able to remove most organic compounds from Mississippi River water. Macko (*8*) found that the resin also removed sulfate and several heavy metals complexed to the humic substances.

Baldauf et al. (*9*) arrived at the conclusion that a sorption system consisting of anionic resin filters (Lewatit MP 500 A) followed by granular activated-carbon filters was favorable in the treatment of humic ground water. The benefit of this process combination was not a selective adsorption of humic substances, but rather the rapid sorption kinetics on the resin as compared to the carbon.

Rüffer and Slomka (*10*) and Slomka (*11*) showed that humic-substance compounds were efficiently removed when a macroporous resin was applied to biologically treated wastewater.

Our own experiments with ion exchange started in 1979. The first part of the work dealt with evaluations of the sorption capacity of different types of resin and with studies of column behavior at different loading conditions (*12*). Later, the work was concentrated toward developing the process in terms of optimum design and regeneration (*13*). Finally the process was tested in full scale with commercial ion-exchange equipment (*14*). The most relevant results from all these experiments will be summarized here.

Isotherm Experiments

Experimental Methods. The isotherm experiments were carried out with the bottlepoint technique. For each isotherm experiment 12 Erlenmeyer flasks were filled with 250 mL of the humic-substance solution. A given amount of wet resin was added before the flasks were placed in a shaking water bath (JULABO SW 1) for 24 h. The bath was equipped with a precise shaking frequency and a temperature-control device (JULABO VM). After shaking, the suspension was filtered through a 1.0-μm membrane filter (Whatman) prior to photometric determination of the remaining humic-substance concentration in the filtrate.

Discussion. The raw water used in all the experiments except the ones in the full-scale plant was obtained from the same source. Typical characteristics of the water are given in Table I. The concentration was analyzed in terms of UV extinction (i.e., transmission of UV light at 254 nm), by applying the following formula:

$$\text{UV extinction} \left(\frac{E}{\text{m}}\right) = -\frac{1}{L} \log \frac{T}{100} \tag{1}$$

where L is the light distance (m) and T is the transmission (%).

In the other experiments the humic-substance concentrations have been measured in terms of color and dissolved organic carbon divided by total

Table I. Characteristics of Water Used in Laboratory Experiments

Characteristic		*Measure*
pH		6.9–7.1
Alkalinity		0.2–0.3 meq/L
Turbidity		0.1–0.5 NTU
DOC		3–6 mg C/L
UV extinction		15–35 E/m
Color		30–90 mg Pt/L
Molecular Weight Distribution	*Color (%)*	*DOC (%)*
>50,000	65	84
10,000–50,000	6	1
1,000–10,000	23	12
<1,000	6	10
Loss		−7

organic carbon (DOC/TOC). The relationship between these parameters varies with respect to both season and sampling point in the experimental setup. However, the following typical correlations may facilitate the reading:

$$\text{UV extinction} \left(\frac{E}{\text{m}}\right) = 0.3 \cdot \text{color} \left(\frac{\text{mg Pt}}{\text{L}}\right) + 2 \qquad (2)$$

$$\text{DOC} \left(\frac{\text{mg C}}{\text{L}}\right) = 0.16 \cdot \text{UV extinction} \left(\frac{E}{\text{m}}\right) + 1 \qquad (3)$$

The six different types of resin studied are characterized in Table II. The adsorption capacity could very well be expressed by the Freundlich equation.

$$\frac{x}{M} = K_F \cdot C_e^n \qquad (4)$$

where M is the weight of wet resin multiplied by the dry solid content of resin (%) over 100, x is mass adsorbed, C_e is liquid concentration at equilibrium, K_F is the Freundlich constant, and n is the Freundlich exponent.

As already mentioned, the humic-substance concentration was interpreted in terms of UV extinction (E/m). Even if the humic-substance concentration is no direct measure of the concentration of organic matter, it is indicative of concentration. The correlation between DOC and UV extinction is, for instance, very good. However, in order to use UV extinction as a measure for humic-substance concentration, we had to introduce a new dimension for the x parameter in the Freundlich equation. Because E/m represents a concentration, the corresponding mass may be obtained by multiplication by the volume dimension. Thus the dimension of the x parameter must be $(E/\text{m}) \cdot (\text{m}^3) = (E\text{m}^2)$, which now expresses the mass of matter adsorbed to the resin that before adsorption contributed to the UV extinction in the water. Equivalently, one may transform the concentration term (E/m) to $(E\text{m}^2/\text{m}^3) = (0.001\ E\text{m}^2/\text{L})$. The dimension of x/m will be $(E\text{m}^2/\text{g})$ and the dimension of C_e will be $E\text{m}^2/\text{L}$.

The results from the isotherm experiments are given in Table III. The table gives the K_F and n values of the Freundlich isotherm obtained by linear regression. The square of the correlation coefficient (R^2) is also indicated. The values are representative for a temperature of 10 °C and pH 7.0.

The isotherm data show a considerable difference in the capacity of the six resins. The best one for this water was obviously the macroporous strong-base resin Lewatit MP 500 A.

Table II. Different Types of Resin Tested

Resin Name	*Type*	*Matrix*	*Specific Area* (m^2/g)	*Capacity* (*mekv/mL*)	*Manufacturer*
Lewatit MP 500 A	Strong base type I	Macroporous styrene–DVB	45–50	1.0	Bayer AG
Lewatit OC 1035	Medium base	Macroporous styrene–DVB		1.4	Bayer AG
WRL 200 A		Cellulose		0.35	Water Research Laboratory A/S Copenhagen
Dowex MSA–1	Strong base type I	Macroporous styrene–DVB	23	1.0	Dow
Dowex MSA–2	Strong base type II	Macroporous styrene–DVB			Dow
Dowex MWA–1	Weak base	Macroporous styrene–DVB		1.1	Dow

Table III. Parameter Values in the Freundlich Isotherm for Six Types of Resin

Resin Type	K_F	n	R^2
Lewatit MP 500 A	15.55	1.152	0.950
Lewatit OC 1035	12.07	1.132	0.820
WRL	1.50	0.913	0.941
Dowex MSA–1	4.36	1.019	0.958
Dowex MSA–2	2.54	0.919	0.965
Dowex MWA–1	0.77	0.798	0.967

NOTE: The values are valid for the corresponding dimensions: x/m in (Em^2/g) and C_e in (Em^2/L) C_0 = 15–16 E/m.

SOURCE: Reproduced with permission from ref. 1. Copyright 1986 Pergamon.

Single-Bed Column Experiment

Experimental Methods. The single fixed-bed column experiments were performed with two 4-m-high and 4.4-cm-i.d. down-flow loaded plexiglas columns in parallel. Each column was filled with resin to the 1.5-m level. Each run was stopped when the head loss equalized the available head of 4 m. After each run, the bed was backwashed with tap water and regenerated with 3 bed volumes of a 10% NaCl + 2% NaOH solution. During each run, samples were manually collected at several sampling points along the bed. Variables in these experiments were the influent concentration, the hydraulic load, and the type of resin.

Results. Three types of resin were tested. The breakthrough curves for the sampling point at 0.5-m bed depth are given in Figure 1. The results show that the resin (Lewatit MP 500 A) was best with respect to both dynamic behavior and capacity. Consequently, this resin was selected for further studies.

The resin was tested through 26 runs in the single fixed-bed column. Altogether 1419 observations were analyzed statistically with respect to breakthrough behavior. The two most important factors determining the effluent concentration were the raw-water concentration (C_0) and the empty-bed contact time in the column (t_k). The filter velocity (v), as such, was of minor importance. At a given filter velocity, the empty-bed contact time will be proportional to the length of the column (l). These three factors (t_k, v, and l) are therefore mutually dependent upon each other.

Figure 2 shows results from three different experiments where the contact time was constant, but the column length and filter velocity varied. The figure demonstrates clearly that the empty-bed contact time is the important design parameter, and that a suitable combination of column length and filter velocity should be chosen on the basis of the necessary contact time. The good agreement of the curves in Figure 2 indicates that the sorption kinetics are intraparticle diffusion controlled.

The treatment efficiency, $(1 - C/C_0) \cdot 100\%$, where C/C_0 is the relative concentration at the sampling point, will consequently be primarily depen-

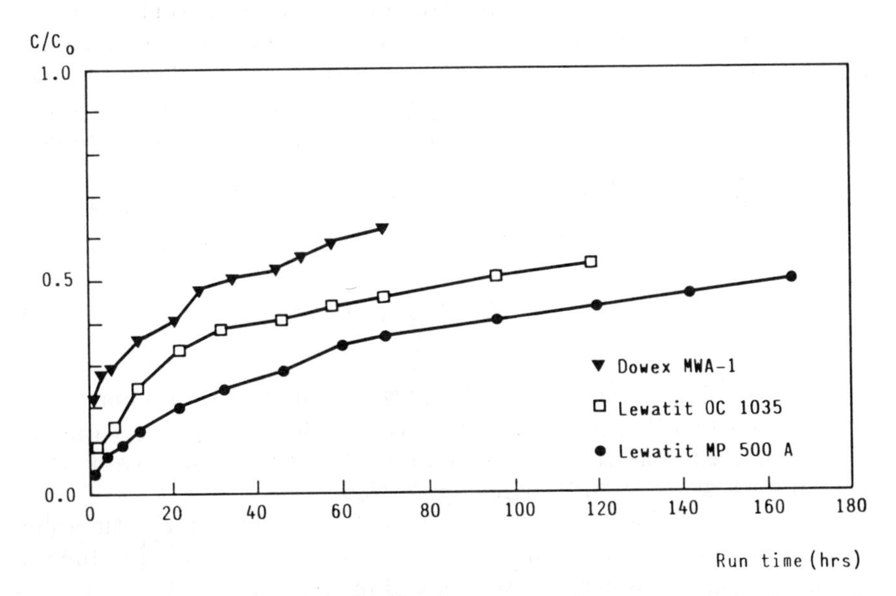

Figure 1. Breakthrough curves for three types of resin at a constant contact time (C_0 = 15 E/m, v = 10 m/h, l = 0.5 m). *(Reproduced with permission from ref. 1. Copyright 1986 Pergamon.)*

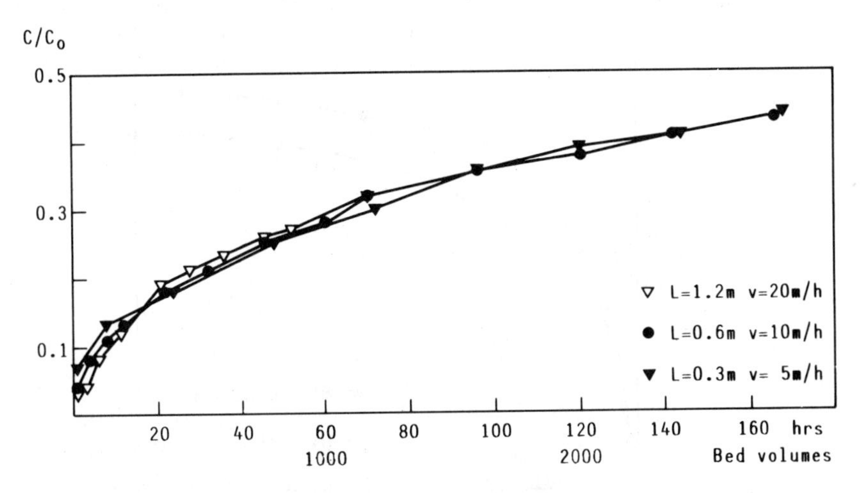

Figure 2. Breakthrough curves for Lewatit MP 500 A at a fixed influent concentration but at varying column length and filter velocity (C_0 = 16 E/m, t_k = 3.6 *min*).

dent upon empty-bed contact time, as shown in Figure 3. However, treatment efficiency will decrease with increasing raw-water concentration. The effluent standard is normally given at a certain level; therefore, the treatment efficiency must be increased when raw-water concentration increases. As a consequence, necessary empty-bed contact time, and therefore investment costs, will increase considerably with increasing raw-water concentration.

Discussion. An empirical breakthrough model was developed on the basis of the single-bed data:

$$\frac{C}{C_0} = k_0 \cdot C_0^{k_1} \cdot t_k^{k_2} \cdot t^{k_3} \tag{5}$$

where C_0 is influent concentration (E/m), t_k is empty-bed contact time (min), and t is run time between regenerations (h). The model gave a squared correlation coefficient, $R^2 = 0.9281$, for the following k values: $k_0 = 0.0396$, $k_1 = 0.4561$, $k_2 = -0.6721$, and $k_3 = 0.3739$.

In Figure 4 this model has been used to illustrate the relationship between the influent concentration, the contact time, and the run time, if one has to satisfy a water quality standard of 5.0 E/m, which corresponds to about 15 mg of Pt/L and 1.3 mg of TOC/L in this water.

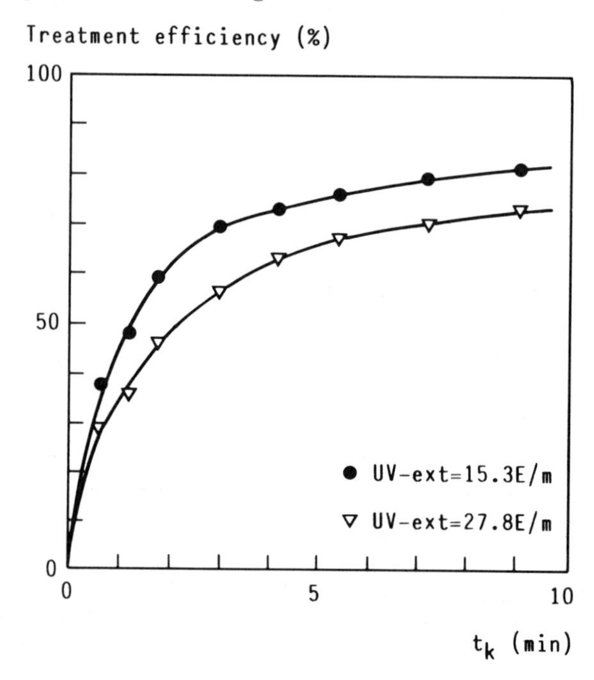

Figure 3. Treatment efficiency versus empty-bed contact time at two influent concentrations. Time of operation was equivalent to 1000 filter-bed volumes.

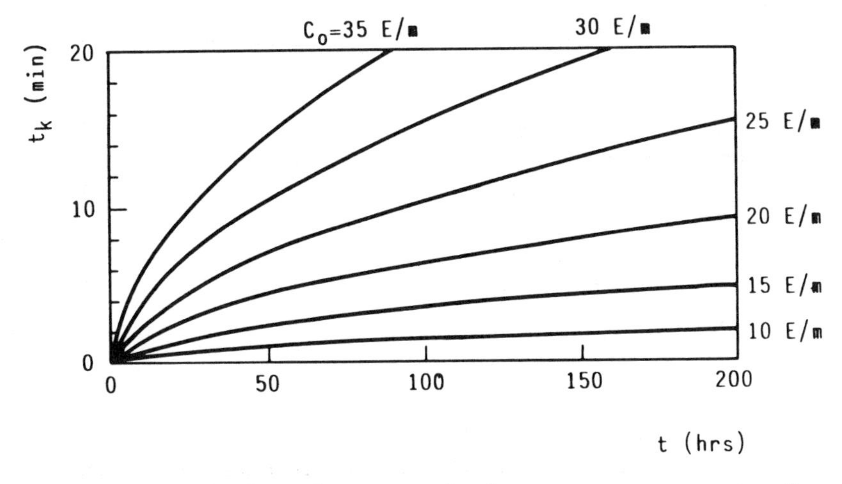

Figure 4. Modeled relationship among the three important parameters for process design when a water quality standard of 5.0 E/m *has to be satisfied. (Reproduced with permission from ref. 1. Copyright 1986 Pergamon.)*

Figure 4 clearly illustrates that the run time between regenerations is heavily influenced by the raw-water concentration. A high raw-water concentration requires a long contact time (high investment cost) or frequent regeneration (high operation cost). We estimate a raw-water color of about 50 mg of Pt/L (about 15 *E*/m) to be an upper economical limit for the use of ion exchange for humic-substance removal.

Regeneration Experiments

Experimental Methods. The regeneration experiments were performed in the columns previously described. One experiment was carried out to study desorption during regeneration with 3.0 bed volumes. This experiment was done in a 1.5-m-high bed of Lewatit MP 500 A. Treated water samples were collected each third minute and analyzed in a TOC analyzer (Barnstead).

In addition, a 0.45-m-high bed of Lewatit MP 500 A was loaded and regenerated in a total of 11 cycles. In these runs we reused 3.0 bed volumes of regenerant. The pore volume of the bed necessarily contained water before regeneration, so the regenerant was diluted somewhat during each regeneration. Thus, we corrected the volume and the strength of the regenerant after each cycle. The experiments were designed to study the extent to which the sorption cycle was affected by the increasing number of previous regenerations, in order to evaluate how many times the regenerant could be reused.

Results. Figure 5 shows the desorption during one regeneration with 3.0 bed volumes of solution over 51 min. Figure 6 illustrates the accumulated removal of color from the water during each cycle, as a function of the number of previous regenerations. In these experiments the volume ad-

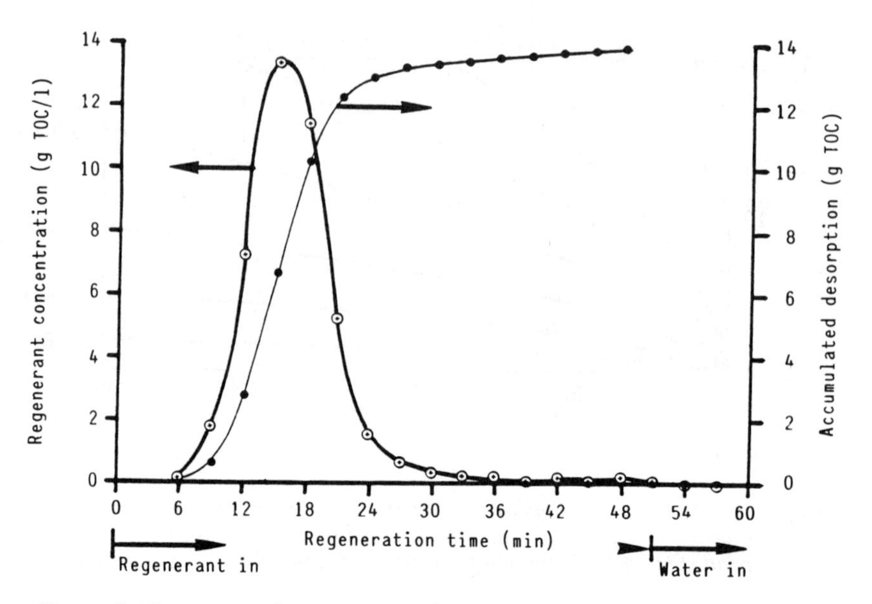

Figure 5. Desorption during a normal regeneration. (Reproduced with permission from ref. 15. Copyright 1987 Pergamon.)

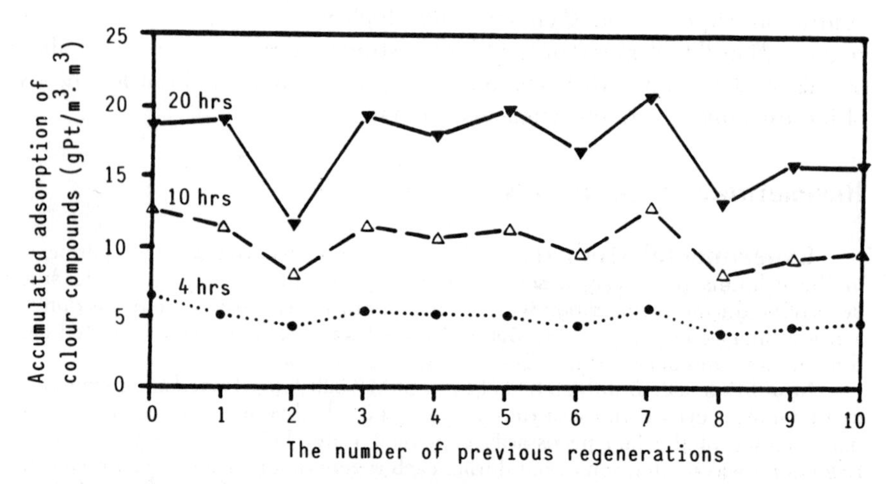

Figure 6. Accumulated color adsorption during each cycle as a function of the number of previous regenerations (C_0 = 50 mg of Pt/L, l = 0.45 m, v = 30 m/h).

justment of the regenerant resulted in 88% as the highest possible recirculation ratio.

Discussion. The results in Figure 5 demonstrate that about 95% of the desorption happened in the first 9–24 min of regeneration. This period corresponds to a regenerant volume of about 0.9 bed volumes, which might be led to a separate waste tank; therefore, about 70% of the total regenerant solution of 3 bed volumes could be recirculated without any significant enrichment of humic substances. We accomplished a thorough regeneration of the bed when 3.0 bed volumes of the regenerant was eluted through the bed for 50–60 min.

Figure 6 indicates that the maximum reuse of regenerant did not affect the uptake of humic substances significantly during the first seven or eight regenerations. The poor result obtained at two early regenerations could be attributed to a poor regeneration procedure. More runs would probably have given a more marked decrease in humic-substance uptake than the graphed results indicate.

The maximum number of reuses with efficient removal of humic substances appears to be 8–10. The use of a fresh regenerant after 10 regenerations seems to be the best way to cut the consumption of regenerant chemicals.

Countercurrent Beds-in-Series Experiments

The reasoning behind the study of a countercurrent beds-in-series system was

1. The relatively poor sorption kinetics require a rather long bed in order to operate continuously for a longer period.
2. The capacity and the kinetics of the process indicate that the use of a countercurrent system would decrease the regeneration frequency significantly.
3. A design should be developed to be operated with a high degree of automation.

Experimental Methods. The beds-in-series system is illustrated in Figure 7. The system consisted of four columns, each 2.0 m high with 4.2-cm i.d. Each column was half filled with Lewatit MP 500 A. Prior to the resin column, a sand filter column was used for particle removal. During the 51-day experiments, only three of the columns were operated at the same time, giving a total bed length of 3.0 m. A pipe system containing 12 magnetic valves connected the four bed units. The operation was automated by the use of a microcomputer, which controlled both the valves and the effluent-analyzing equipment. Regeneration and backwashing of each column was, however, carried out manually.

The device for analyzing the effluent quality is illustrated in Figure 8. Three magnetic valves were connected to the computer. The apparatus included a peristaltic pump that, on a signal from the computer, started pumping water through a UV monitor (ISCO). Distilled water was pumped for a given time, followed by a standard humic-substance solution with a known concentration, in order for the computer to calculate a standard curve. Then the treated water was pumped from the sampling tank and the concentration was determined in relation to the standard curve.

The valves were controlled by computer so that the four columns were operated in sequence. An exhausted column was regenerated while the other three were operating.

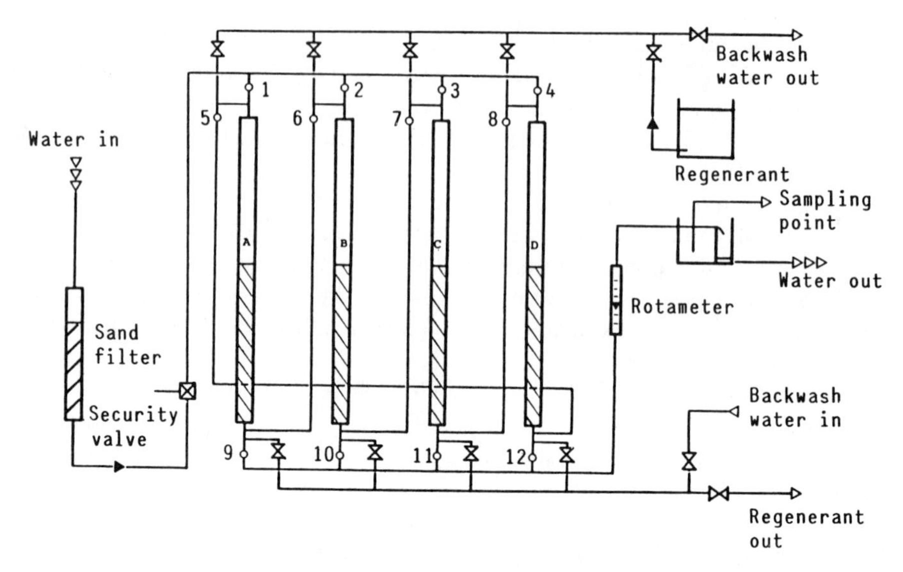

Figure 7. The countercurrent beds-in-series system (1–12 are magnetic valves).

A pump (Seepex) fed the system with water. This particular type of pump could provide a nearly constant amount of water, independent of the head loss in the system. The hydraulic load was constantly 20.8 L/h, which gave a superficial velocity of 15 m/h and an empty-bed contact time of 12 min. The influent water had a nearly constant color level of 52 mg of Pt/L, with a corresponding UV extinction of 19.5 E/m. When required, the columns were regenerated with 3.0 bed volumes of fresh regenerant.

Results. Results from the countercurrent beds-in-series system are given in Figure 9. The operation time of 51 days required six cycles.

Despite some variations in each cycle, the breakthrough results from the countercurrent beds-in-series system follow a consistent pattern. In cycles II and IV, however, the cycles were terminated too early because of a wrong determination of the effluent concentration and to higher hydraulic load than expected, respectively.

The mean run time is a measure of the required regeneration frequency at the given load. The average run time was 8.5 days for the 51-day experiment. Considering that two cycles terminated too early, the correct run time could have been higher. Cycles V and VI were run without any kind of operational problems. For these cycles, the mean run time was 11.8 days. These last cycles were clearly the most successful ones, possibly due to a more skilled regeneration as time passed. In conclusion, the experiments showed that it is possible to obtain a cycle length of at least 11 days at the actual loadings.

In order to evaluate what may be gained by using a countercurrent beds-in-series system rather than a single-bed system, we can calculate the

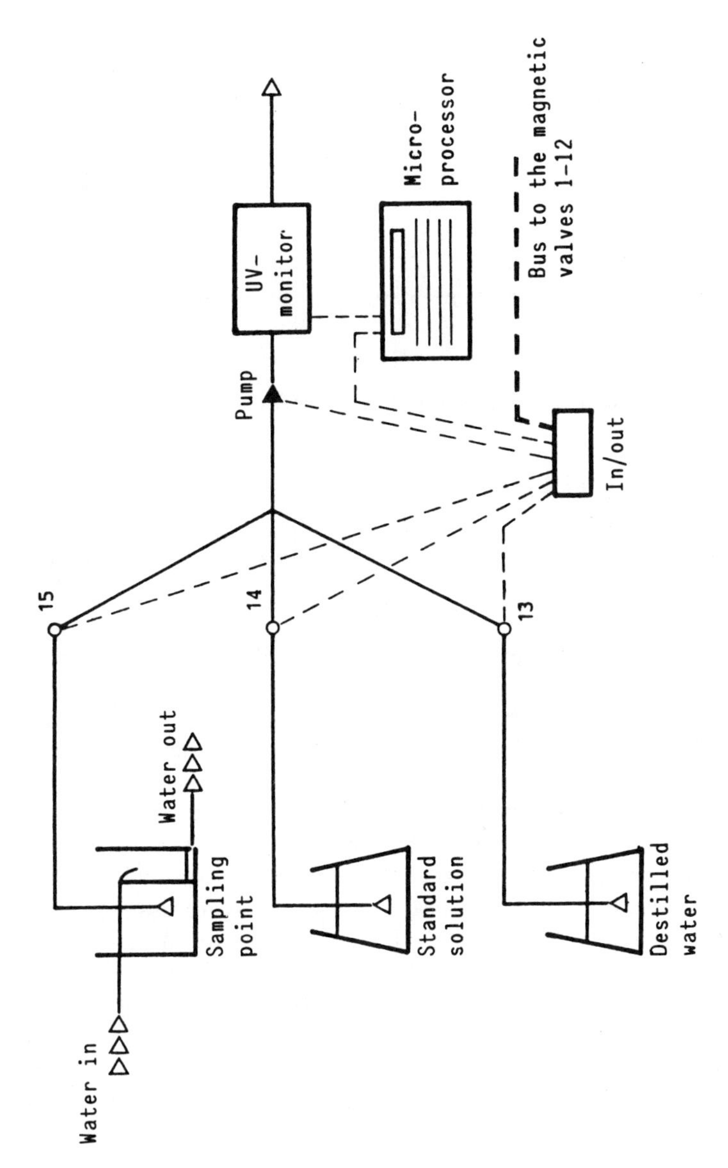

Figure 8. The device for effluent quality control (13–15 are magnetic valves).

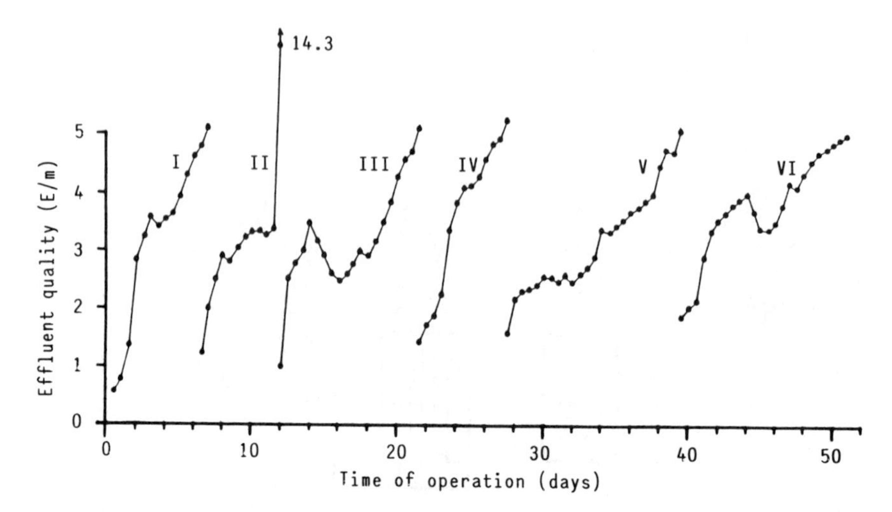

Figure 9. Effluent quality from the countercurrent beds-in-series system during six cycles.

run time in a single-bed system from the model given earlier. With the actual raw-water quality (C_0 = 19.5 E/m) and contact time (t_k = 12 min), the run time would be 342 h (14.3 days) if the treated water standard was 5.0 E/m.

The possible run time for the beds-in-series system was found to be 11 days. In this system, however, only one of the three operating beds was regenerated at a time. The corresponding run time for the whole 3-m bed length would be 33 days. By the use of a beds-in-series system, one may therefore obtain a total run time 2.3 times greater than that of a comparable (3-m) single bed.

An increase in run time by a factor of 2.3 gives a corresponding reduction in the consumption of regeneration chemicals and volume of waste solution. The beds-in-series design will thus improve the total exchange process significantly, because the two major disadvantages of the process are the chemical costs of regeneration and the production of a waste solution.

When using both a countercurrent beds-in-series system and regenerant reuse, one may treat 5910 m^3 of water to an effluent quality of 5 E/m with the production of 1 m^3 of waste solution. These numbers are valid for the loadings that were used in this study (C_0 = 19.5, E/m = 52 mg of Pt/L, v = 15 m/h, t_k = 12 min). The water–waste production ratio will, of course, increase as the loading decreases.

Full-Scale Experiments

This research program has been aimed primarily at the development of reliable treatment methods for small waterworks. A small, but full-scale, treatment plant based on the ion-exchange filters of Eurowater AB was therefore operated during the autumn of 1985.

Experimental Methods. The plant consisted of three columns in series, with alteration of the sequence of operation for each regeneration. Each column was filled with 300 L of resin (Lewatit MP 500 A), giving a column length of 93 cm. Only one column was regenerated at a time, and the water was allowed to pass through the remaining two columns. Each regeneration lasted for 85 min. The valve scheme of the plant is shown in Figure 10.

During October–December 1985, the color of the raw water was in the range of 52–58 mg of Pt/L, the UV extinction 24–27 E/m, and the TOC 6.0–6.5 mg TOC/L.

In the full-scale experiments we decided to evaluate the use of a pure NaCl solution for regeneration instead of the alkaline NaOH–NaCl solution used in the previous experiments. We reasoned that a pure NaCl solution would be much easier for the operators of small waterworks to handle.

The plant was operated at a flow of 3 m^3/h, corresponding to an empty-bed contact time of 6 min per column (18 min total). Filter velocity was 9.3 m/h. The treated water was analyzed with respect to color, UV extinction, TOC, and pH.

Results. Figure 11 shows the effluent concentration with respect to color. Each operation cycle was very much like the others. It is, however, demonstrated that regeneration every 3–4 days was needed to meet the treated water standard of 15 mg of Pt/L. The pH of the treated water was consistently in the 4.9–5.9 range.

Discussion. A comparison of the basic data from the laboratory and the full-scale experiments (as in Table IV) shows that although the loading in the full-scale experiments was lower, the run time was also lower. This result may be attributed primarily to the much lower temperature of the water and the poorer regeneration obtained, but may also be due to differences between the humic materials used in each of the tests.

Practical Considerations

Regeneration. The regeneration was not optimal in the full-scale experiments. Only 78% of the humic substances adsorbed were actually washed out during regeneration. This result indicated that regeneration by a pure NaCl solution is not good enough, and that an alkaline salt solution should be preferred (2–3 bed volumes of a 10% NaCl and 2% NaOH solution).

Influence of Temperature. As demonstrated, the temperature effect was probably quite significant in the full-scale experiments. The influence of temperature was examined in a separate investigation (*16*). The results based on raw water from three different sources and on temperatures varying from 2 to 20 °C were statistically treated versus the single-bed model

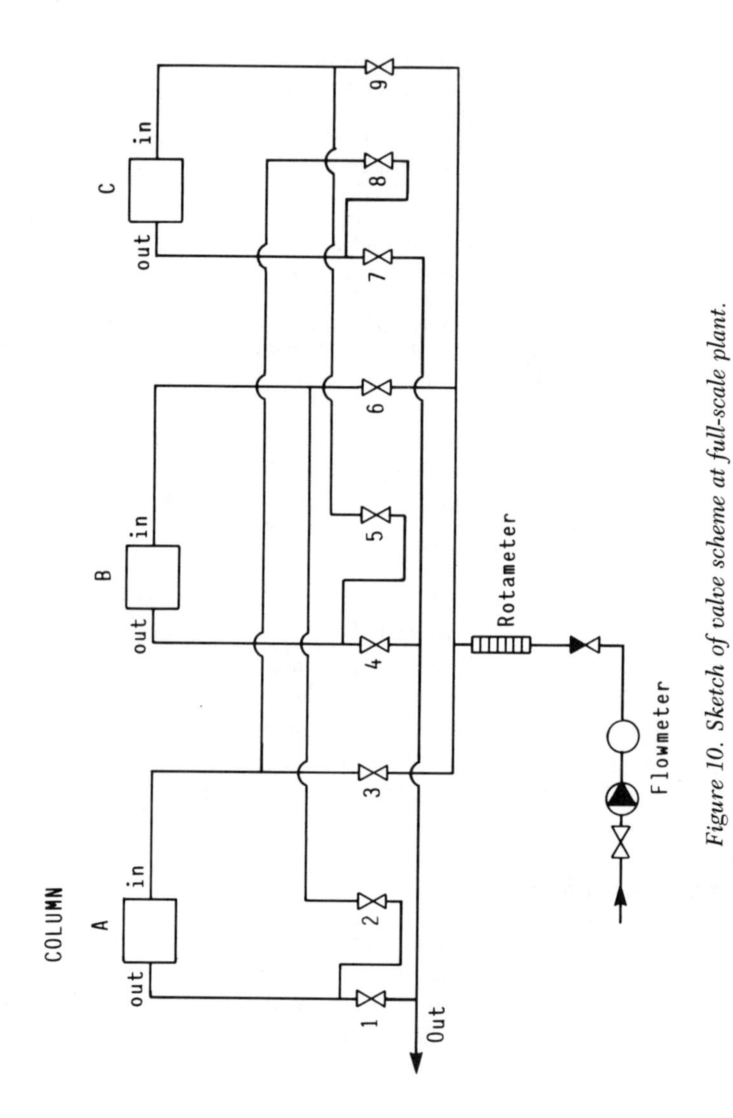

Figure 10. Sketch of valve scheme at full-scale plant.

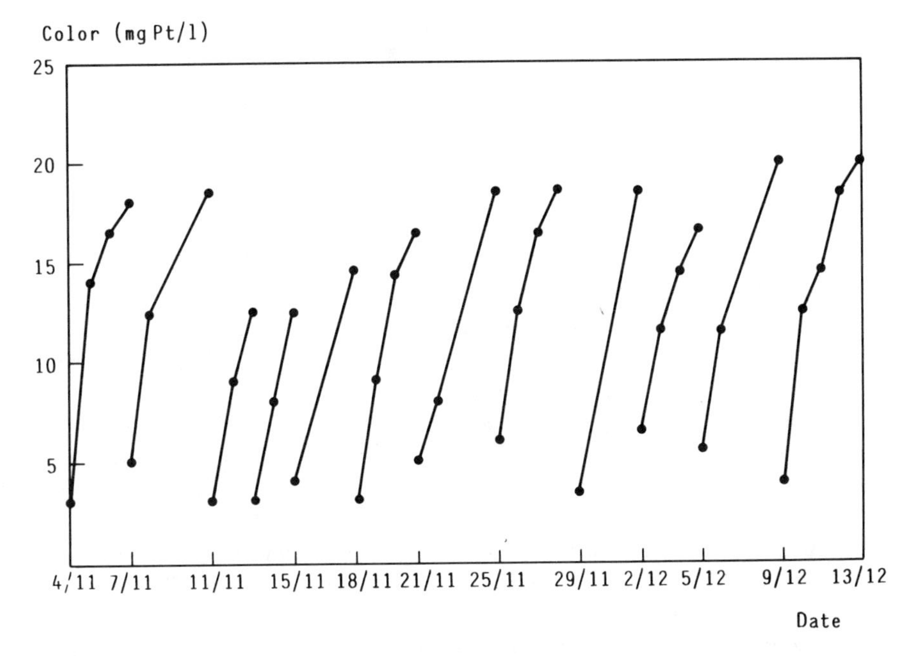

Figure 11. Treated-water color at full-scale plant.

Table IV. Comparison Between Laboratory-Scale and Full-Scale Experiments

Condition	*Laboratory-Scale Experiment*	*Full-Scale Experiment*
Filter velocity (m/h)	15.0	9.3
Contact time (min)	12.0	18.0
Influent concentration (mg Pt/L)	50–55	52–58
Temperature (°C)	16–18	2–5
Run time (days)[a]	11	4.5
Regeneration	10% NaCl + 2% NaOH	6.8% NaCl

[a]At a treated water color of 20 mg Pt/L.

with the addition of a temperature factor. A squared correlation coefficient of $R^2 = 0.914$ was obtained with the following model:

$$\frac{C}{C_0} = 0.0840\ C_0^{0.493} t_k^{-0.7417} t^{0.2993} T^{-0.1268} \tag{6}$$

where T is temperature in degrees Celsius, and the other factors are as explained before.

Equation 7 may be used to calculate with what factor one has to multiply the necessary empty-bed contact time, if the treatment efficiencies are to be the same when the temperature is changed from T_1 to T_2.

$$t_{k_2} = t_{k_1} \left(\frac{T_1}{T_2}\right)^{0.171} \quad (7)$$

Table V shows examples of this factor in the actual temperature range.

The pH in the treated water of the full-scale plant was as low as 4.9–5.9. In the laboratory-scale experiments, where an alkaline salt solution was used as a regenerant, the pH was very high (9–10) in the beginning of each run, but stabilized rather quickly at pH 6–7.

In practice, one has to take care of these pH variations by the use of an equalizing tank or by wasting the first portion of the treated water after regeneration.

Automation. As demonstrated in the countercurrent beds-in-series experiments, the process may be highly automated. A precise effluent water quality measurement is, however, of great importance to optimum operation and utilization of the capacity of the system. The apparatus should be precise in the lower concentration range. One alternative is a photometric determination of color at 450 nm with a 20-cm flow cell. Calibration should be carried out before each measurement by pumping distilled water and a standard solution through the cell. The procedure can easily be automated. The flow cell should be cleaned regularly with a weak sodium hydroxide solution, a procedure that may be incorporated into an automatic analytical sequence.

Economy. The cost of ion exchange for humic-substance removal has been compared to the cost of alternative methods (*17*). In Figure 12 investment costs in Norwegian kroner [$1 (U.S.) = 6.65 Nkr] are shown for three levels of raw-water color and three levels of plant size, and in Figure 13 the corresponding unit cost (Nkr/m^3) is shown.

The primary interest here is the cost of one process relative to the other. It is demonstrated that the cost of ion exchange is far more sensitive to raw-water concentration than the other processes. At raw-water colors less than

Table V. Temperature Correction Factor

T_1/T_2 (°C)	2	5	10	15	20
2	1.00	0.85	0.76	0.71	0.67
5	1.17	1.00	0.89	0.83	0.79
10	1.32	1.13	1.00	0.93	0.89
15	1.41	1.21	1.07	1.00	0.95
20	1.48	1.27	1.13	1.05	1.00

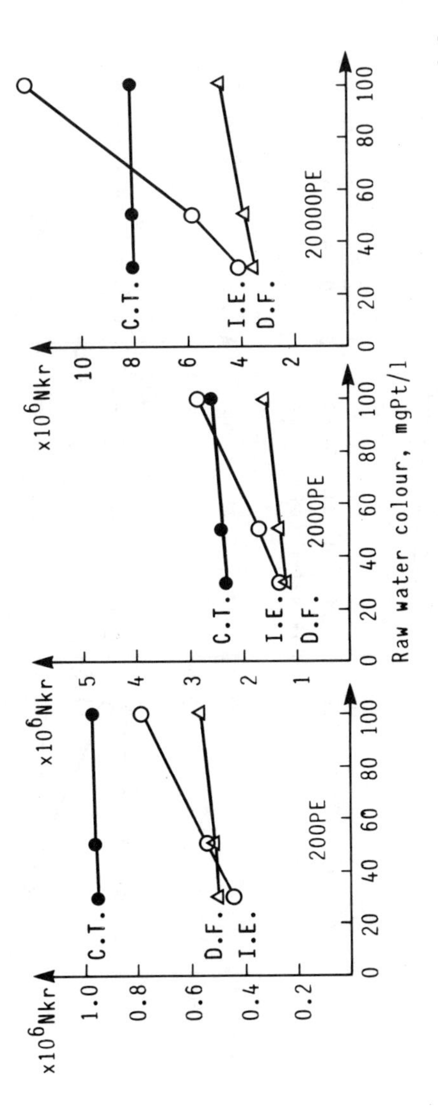

Figure 12. Investment cost of ion exchange (I.E.) as compared to conventional treatment (C.T.) and direct filtration (D.F.).

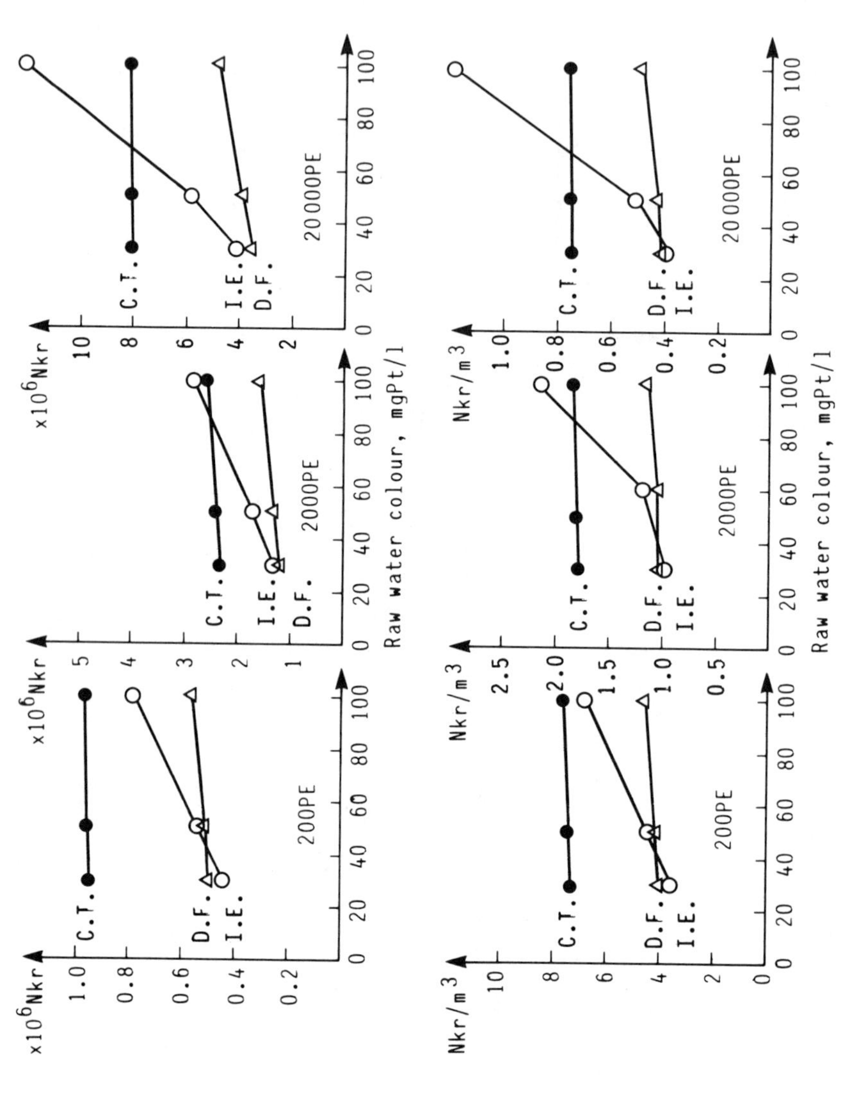

Figure 13. Unit cost of ion exchange (I.E.) as compared to conventional treatment (C.T.) and direct filtration (D.F.).

50 mg of Pt/L, ion exchange seems to be cheaper than the conventional combination of coagulation, flocculation, sedimentation, and filtration at all the plant sizes examined here. For the smallest plants ion exchange even seems to be economically competitive with direct filtration, especially with respect to the unit cost of treatment.

Summary

Ion exchange in a strong-base, anionic, macroporous resin is a viable alternative for the removal of humic substances from surface water, especially for smaller waterworks. In order for the process to be economically competitive with other alternatives, the raw-water color ought to be lower than 50 mg of Pt/L.

The empty-bed contact time is the most important design parameter. It should be >8–10 min. The optimum design is that of a countercurrent beds-in-series system, which will cut the consumption of regenerant chemicals and production of waste solution.

The regenerant may consist of an alkaline salt solution (10% NaCl and 2% NaOH), and 2–3 bed volumes may be used for each regeneration. An average regenerant reuse of about 75% will not reduce the sorption behavior of the resin. Water temperature has a significant influence on necessary contact time, which has to be increased about 20% if the temperature is reduced from 15 to 5 °C.

The process is economically competitive with alternative, traditional humic-substance removal processes, especially at smaller waterworks, when the raw-water color is relatively low.

References

1. Ødegaard, H.; Brattebø, H.; Eikebrokk, B.; Thorsen, T. *Water Supply* **1986,** *4,* 129–158.
2. Kölle, W. *Humic acid removal with macroreticular ion-exchange resins at Hannover*; Nato–CCMS–Congress, Reston, VA, 1979.
3. Rüffer, H.; Schilling, J. *Die Aufbereitung huminsäurehaltigen Oberflächenwassers durch Flockung und Adsorption an makroporösen Ionenaustauschharzen und Übertragung der Ergebnisse auf ähnliche Wässer*; report, Institut für Siedlungswasserwirtschaft der Universität Hannover, 1977.
4. Kölle, W. *Erfahrungen bei der Aufbereitung eines reduzierten huminstoffhaltigen Grundwassers im Wasserwerk Fuhrberg der Stadtwerke Hannover AG*; Abschlussbericht zu dem BMFT-Forschungs-vorhaben 02 WT 606, Hannover, 1981.
5. Jørgensen, S. E. *Water Res.* **1979,** *13,* 1239–1247.
6. Boening, P. H.; Beckmann, D. D.; Snoeyink, V. L. *J. Am. Water Works Assoc.* **1980,** *72,* 54–59.
7. Anderson, C. T.; Maier, W. J. *J. Am. Water Works Assoc.* **1971,** *71,* 278–283.
8. Macko, C. A. Ph.D. Dissertation, University of Minnesota, 1980.
9. Baldauf, G.; Woldmann, H.; Klette, J.; Sontheimer, H. *GWF Gas Wasserfach: Wasser/Abwasser* **1985,** *126*(H3), 107–114.

10. Rüffer, H.; Slomka, T. *Wasser Abwasser Forsch. Prax.* **1981,** *14*(516), 176–180.
11. Slomka, T. *Entwicklung weitergehender Abwasserreinigung durch Einsatz von Flockung, Mehrschichtfiltern und makroporösen Adsorberharzen*; report, Institut für Siedlungswasserwirtschaft der Universität Hannover, 1982.
12. Halle, O. Dr. ing. Dissertation, Norwegian Institute of Technology, 1983 (in Norwegian).
13. Brattebø, H. *Proc. NATO Adv. Study Inst. Conf.* "Ion Exchange: Science and Technology", Troia, Portugal, 1985.
14. Halle, O.; Brattebø, H. SINTEF report, STF 60 A 86051. The Norwegian Institute of Technology, 1986 (in Norwegian).
15. Brattebø, H.; Ødegaard, H.; Halle, O. *Water Res.* **1987,** *21*(9), 1045–1052.
16. Johansen, H. Diploma Thesis, Norwegian Institute of Technology, 1986 (in Norwegian).
17. Hem, L. J. SINTEF report STF 60 A 86161. Norwegian Institute of Technology, 1986 (in Norwegian).

RECEIVED for review July 24, 1987. ACCEPTED for publication February 11, 1988.

INDEXES

Author Index

Affiliation Index

Subject Index

A

D

E

G

I

K

L

Q

T

Copy editing and indexing by Colleen P. Stamm
Production by Paula M. Bérard
Managing Editor: Janet S. Dodd

Typesetting of text by Techna Type, Inc., York, PA
Typesetting of front matter and index by Hot Type Ltd., Washington, DC
Printing and binding by Maple Press Company, York, PA

Titles of Related Interest

Treatment of Water by Granular Activated Carbon
Edited by Michael J. McGuire and I. H. Suffet
Advances in Chemistry Series 202; 599 pages; ISBN 0-8412-0665-1

Organic Marine Geochemistry
Edited by Mary L. Sohn
ACS Symposium Series 305; 427 pages; ISBN 0-8412-0965-0

Evaluation of Pesticides in Ground Water
Edited by Willa Y. Garner, Richard C. Honeycutt, and Herbert N. Nigg
ACS Symposium Series 315; 573 pages; ISBN 0-8412-0979-0

Materials Degradation Caused by Acid Rain
Edited by Robert Baboian
ACS Symposium Series 318; 447 pages; ISBN 0-8412-0988-X

Organic Pollutants in Water: Sampling, Analysis, and Toxicity Testing
Edited by I. H. Suffet and Murugan Malaiyandi
Advances in Chemistry Series 214; 797 pages; ISBN 0-8412-0951-0

Sources and Fates of Aquatic Pollutants
Edited by Ronald A. Hites and S. J. Eisenreich
Advances in Chemistry Series 216; 558 pages; ISBN 0-8412-0983-9

Principles of Environmental Sampling
Edited by Lawrence H. Keith
ACS Professional Reference Book; 458 pages; ISBN 0-8412-1173-6

The Kinetics of Environmental Aquatic Photochemistry
By Asa Leifer
ACS Professional Reference Book; 304 pages; ISBN 0-8412-1464-6